MERCURY

MERCURY

Faith Vilas
Clark R. Chapman
Mildred Shapley Matthews

Editors

With 47 collaborating authors

THE UNIVERSITY OF ARIZONA PRESS
TUCSON

About the cover:

The cover painting shows the surface of Mercury. A large boulder ejected from the background crater hides the solar disk, permitting a view of the spectacular corona and the inner zodiacal light. The brightest "stars" in the upper parts of the zodiacal light are Venus and Earth. Painting by William K. Hartmann. The back cover is a computer mosaic of Mercury's Discovery scarp, one of the largest lobate scarps imaged by Mariner 10. This scarp is probably a thrust fault due to compressive stresses in Mercury's crust as a result of the planet's contraction.

THE UNIVERSITY OF ARIZONA PRESS

Copyright © 1988
The Arizona Board of Regents
All Rights Reserved

This book was set in 10/12 Times Roman.
♾ This book is printed on acid-free, archival-quality paper.
Manufactured in the United States of America.

92 91 90 89 88   5 4 3 2 1

Library of Congress Cataloging-in-Publication Data

Mercury.

    Papers presented at the Mercury Conference held
6–9 Aug. 1986 in Tucson, Ariz., and sponsored by the
Division for Planetary Sciences of the American
Astronomical Society and International Astronomical
Union Commission 16.
    Includes bibliographies and index.
    1. Mercury (Planet) — Congresses. I. Vilas, Faith.
II. Chapman, Clark R. III. Matthews, Mildred S.
IV. American Astronomical Society. Division for
Planetary Sciences. V. International Astronomical
Union. Commission 16. VI. Mercury Conference (1986 :
Tucson, Ariz.)
QB611.M47 1988   523.4'1   88-29515

ISBN 0-8165-1085-7 (alk. paper)

British Library Cataloguing in Publication Data are available.

CONTENTS

[v]

COLLABORATING AUTHORS

D. N. Baker, *513*

W. Benz, *691*

A. G. W. Cameron, *691*

D. B. Campbell, *101*

C. R. Chapman, *ix, 1*

P. E. Clark, *77*

J. E. P. Connerney, *493*

B. Fegley, Jr., *691*

L. Fleitout, *400*

K. A. Goettel, *612*

J. D. Goguen, *37*

J. E. Guest, *118*

B. Hapke, *37*

J. K. Harmon, *101*

P. Helfenstein, *37*

D. M. Hunten, *561*

R. F. Jurgens, *77*

M. A. Leake, *77*

J. S. Lewis, *650*

P. Masson, *400*

W. B. McKinnon, *373*

T. H. Morgan, *561*

H. J. Melosh, *373*

N. F. Ness, *493*

G. Neukum, *335*

S. J. Peale, *460*

G. V. Pechernikova, *666*

R. J. Pike, *165*

M. N Ross, *428*

C. T. Russell, *513*

V. S. Safronov, *666*

G. Schubert, *428*

P. H. Schultz, *274*

D. E. Shemansky, *561*

W. L. Slattery, *691*

J. A. Slavin, *513*

T. Spohn, *428*

P. D. Spudis, *118*

S. A. Stern, *24*

D. J. Stevenson, *428*

R. G. Strom, *335*

P. G. Thomas, *400*

J. Veverka, *37*

F. Vilas, *ix, 24, 59*

A. V. Vityazev, *666*

J. T. Wasson, *621*

G. W. Wetherill, *669*

PREFACE

Perhaps the most commonly asked question prior to the organization and publication of this book was "Why Mercury"? Following the 1974 and 1975 flybys of Mariner 10, two conferences on the science of Mercury were held (Post-Mariner 10 Conference, June 1975; Conference on Comparison of Mercury and the Moon, November 1976). For almost a decade, these meetings and the subsequent volumes of papers remained the core of published research on Mercury. By the late 1970s, Mercury had become the stepchild of the solar system. No mention of Mercury missions was made in a discussion of proposed space probes for the 1980s and 1990s by the Solar System Exploration Committee (Planetary Exploration through the Year 2000—A Core Program, 1983).

Two events have focused new interest on Mercury. Chen-Wan Yen of the Jet Propulsion Laboratory developed the successive Venus-Mercury encounter gravity-assist methods for delivering payloads to Mercurian orbit without any greater expenditure of energy than required for a probe to Mars or Venus. Drew Potter and Tom Morgan of the Johnson Space Center discovered the sodium and potassium components of Mercury's tenuous atmosphere while investigating the effects of infilling of Fraunhofer lines on Mercury and the Moon. Meanwhile, groundbased optical and radar telescopic research quietly continued. Separate efforts using the radar facilities of Goldstone and Arecibo have given us a first, rough glance at the structure of the unimaged Mercurian hemisphere in the 12° latitudinal band around the planet's equator. In-depth geomorphologic analyses and stratigraphic studies of Mercurian geology, using images acquired by Mariner 10, have continued. Analyses of magnetic field data were completed, and new theories have emerged concerning the formation history of Mercury.

With the increase of interest, we felt the time was propitious for a conference on the planet Mercury, and an accompanying book. Our inquiries brought a strong response from the scientific community. To quote one of the responses we received: "We desperately need this conference!" The Mercury Conference, sponsored by the Division for Planetary Sciences of the American Astronomical Society and International Astronomical Union Commission 16, was held 6–9 August 1986 in Tucson, Arizona.

Most University of Arizona Space Science Series books follow a landmark event in the study of a solar-system object, or significant advances in the study of a phenomenon. One of our objectives for this conference and book was to stimulate interaction among scientists of different disciplines in order to coordinate future groundbased and spacebased studies of Mercury, which is so difficult to study due to its close proximity to the Sun. With this in mind, we allowed chapter authors the freedom to adjust their manuscripts to include ideas that developed after the conference.

The production of this book was supported by the National Aeronautics and Space Administration Planetary Geology program. The advice of Tom Gehrels throughout the development of the conference and book was invaluable, as was the participation of Melanie Magisos who worked tirelessly on the conference and book. We warmly thank them both.

Faith Vilas
Clark R. Chapman
Mildred S. Matthews

MERCURY: INTRODUCTION TO AN END-MEMBER PLANET

CLARK R. CHAPMAN
Planetary Science Institute

This Chapter introduces the major themes of this book. In many ways, Mercury is an extreme planet, and thus it provides a unique benchmark for testing our theories about the origin and evolution of other (particularly terrestrial) planets. Emphasis is given to synthesizing and critiqueing the book's chapters on the planet's origin, its metal-rich composition, its thermal and geophysical evolution, and its cratering history; these topics are complex and controversial, and this book contains a variety of new perspectives on them. Mercury's geology, atmosphere and magnetosphere are discussed more briefly. We also place the study of Mercury in its historical context and in the context of the spacecraft exploration program, both past and future. This Introduction is not intended to be a comprehensive summary of the contents of the book; readers wanting an overview of Mercurian science should supplement this chapter by reading the abstracts of the book's chapters.

I. INTRODUCTION

In a solar system of just nine planets, each planet is important. Mercury is especially so for reasons elaborated on in this book. It is the closest planet to the Sun and it is made of the densest materials; both dynamically and compositionally, it is an "end-member" planet. It exists in the most intense solar radiation environment, and it is most affected by solar tides. The diurnal temperature range of its surface is the most extreme in the solar system. Of all the planets, its present geochemistry may be the most modified, by catastrophic processes, from its primordial state. As one of very few terrestrial planets, Mercury plays an important role in comparative planetological studies of processes relevant to our own planet Earth (ranging from cratering histories to magnetic field structures).

In both the popular imagination and in planetary science, Mercury has been rather ignored. It is a small planet, difficult to see in the sky. Its superficially Moon-like characteristics tended to obscure its important differences from our well-studied nearest neighbor in space. After Mariner 9's spectacular revelations about Mars, Mariner 10's reconnaissance of Mercury seemed a bit anticlimactic. In the late 1970s, follow-on planetary missions were planned at the next level of sophistication, requiring orbiters and landers; but spacecraft trajectory studies at first appeared to show that such missions to Mercury were impossible without development of new technology: Mercury was simply too deep in the Sun's gravity well to be reached by conventional ballistic rockets. These considerations helped relegate Mercury to secondary status, along with its well-studied look-alike—the Moon—when COMPLEX (the planetary committee of the National Academy of Science's Space Science Board) established planetary exploration objectives (COMPLEX 1977). Later, when NASA's Solar System Exploration Committee recommended an implementation of the COMPLEX strategy in the form of a set of prospective planetary missions for the rest of the century (Solar System Exploration Committee 1983), Mercury remained an also-ran, and the effect rippled into European plans for Mercury exploration, as well.

A decade after the last conference on Mercury (Comparisons of the Moon and Mercury 1977), perceptions about Mercury had changed, providing impetus to the 1986 Mercury Conference, from which this book is an outgrowth. For one thing, many significant planetological questions and enigmas were raised by Mariner 10 that continued to intrigue Mercury aficionados, even though the wider scientific community remained largely unaware of them. Also, groundbased observations—both by ever-improving radar facilities and with optical telescopes—were continuing to add to the database about Mercury. And there were surprises, such as the discovery of prominent emission lines from Mercury's virtually negligible atmosphere which has opened up a whole new topic for groundbased research. Possibly the most significant changes in our scientific conception of Mercury during the past decade have been inspired by the advancement of planetary science in general, both by active spacecraft exploration of Mars and the outer solar system and from continued theoretical modeling of planetary origin and evolution. Viewed from the context of modern planetary science, Mercury's relevance and significance has been enhanced from the mid-1970s' perspective. Now with the remarkable discovery at the Jet Propulsion Laboratory (see Yen 1985; also see Chapter by Stern and Vilas) of elementary ballistic trajectories to Mercury, involving multiple encounters with Venus and Mercury itself, we can make practical plans to place scientific payloads in circular orbits around Mercury using available launch vehicles.

The Challenger disaster and resulting restructuring of American and international space programs may provide a window for wider interest in spacecraft missions to the planet Mercury. A recent study by the National Academy of Sciences ("Space Science in the Twenty-First Century") considered a Mercury mission as having one of the broadest appeals in the space sciences: astrophysicists are interested in relativity tests, solar physicists in a near-Sun platform, planetologists in many of the problems addressed in this book, and space physicists (including those in the newly revamped NASA Office of Solar-Terrestrial Relationships) in the interactions of Mercury with its space environment. A Mercury mission has recently been promoted by a team of 42 European scientists (Neukum et al. 1985) and, although that mission was not selected for Phase A study by the European Space Agency, Mercury exploration clearly provides the potential for international, as well as interdisciplinary, collaboration. As this book appears, COMPLEX is completing a revision of its decade-old strategy for the exploration of the inner planets; that study will undoubtedly revise the now-obsolete considerations that relegated Mercury missions to the graveyard, and there is already thinking in NASA that a Mercury orbiter might make sense as a near-term Planetary Observer mission.

Before beginning to address the chief scientific themes of this book, and Mercury's contributions to understanding the other planets in the solar system, let us consider the historical roots of our subject.

II. THE HISTORY OF VISUAL MAPPING OF MERCURY*

I will not claim that much was added to our knowledge of Mercury from pre-Space Age telescopic observations of Mercury: but for the coming of space research, we would know little indeed. Good and useful maps were made of the Moon and Mars in the nineteenth and first half of the twentieth centuries (cf. Antoniadi 1930; Moore 1984); Antoniadi's nomenclature for Mars is retained even today. Mapping Mercury, however, is much more difficult. The first serious attempts to see detail were made in the early 19th century, at Lilienthal, by J. H. Schroeter and his assistant K. L. Harding. Schroeter was convinced that he had recorded 20 km high mountains, plus a certain amount of albedo variation on the disk, and from this work F. W. Bessel derived a rotation period of $24^h 0^m 53^s$, with an axial inclination of 70°. William Herschel was as skeptical about the Mercurian mountains as he had been about those of Venus; he never saw any detail on Mercury.

Various observations were made during the latter half of the last century. Vague bright and dark features were reported between 1867 and 1886 by L.

---

*This section is written by Patrick Moore.

Prince, J. Birmingham, H. C. Vogel, and É. L. Trouvelot. In 1892, Trouvelot showed some terminator deformations and cusp blunting, probably due to defective seeing conditions. In 1881, W. F. Denning made a series of sketches with his 24 cm reflector, claiming that patches on Mercury were "easy" and pronounced enough to suggest an analogy with those of Mars; the rotation period which he derived was 25 hr.

The first map of Mercury to be worthy of serious discussion was due to G. V. Schiaparelli between 1881 and 1889; he used 22 cm and 49 cm refractors, and always observed when Mercury was high above the horizon, in broad daylight; his work made him confident that the Mercurian rotation was synchronous with Mercury's 88-day orbital period.

At Flagstaff, Percival Lowell began to observe Mercury in 1896 with his 61 cm refractor. From his observations of 1896–7, he produced a map (this map, and several others referred to in this chapter, are reproduced in a book by Strom [1987b]). He accepted the synchronous rotation, dismissed the idea of surface water or organic material, and concluded that Mercury's surface was "covered with long, narrow markings best explained as the results of cooling." In his map, he named 78 features, either dark linear streaks or dark patches, with romantic names (e.g., Psychopompos, Lichanos Hyperbolaeon, Keryx and Parmese Meson). Lowell was a good mathematician, a great organizer and benefactor of astronomy (but for him, there would be no Lowell Observatory), a brilliant writer, and apparently a first-class speaker as well. The one thing he was not, unfortunately, was a good observer, and we have to concede that his linear features on Mercury do not exist.

Can we say the same about the work of T. J. J. See? See must have been a strange man, and to say that he was unpopular with his colleagues would be an understatement. (Of him, A. E. Douglass wrote, "I have never had such aversion to a man or beast or reptile or anything disgusting as I have to him.") He was forced to leave Lowell Observatory; in 1901 he was at the U.S. Naval Observatory, using its 66 cm refractor. He reported craters on Mars and also on Mercury (see Gordon 1983). His most interesting drawing was made in June 1901 under excellent conditions, though the apparent diameter of Mercury was then only 6.6 seconds of arc. Apparently he did not concentrate upon albedo features, but on craters, of which he reported many. See has had many detractors, but also supporters (Young 1978; Baum 1979). I am skeptical, if only because on the occasions when I have used large refractors to study Mercury, I have been unable to see anything definite. It is rather tempting to suggest that one large crater shown by See is the 625-km crater we now call Beethoven, but I am inclined to believe that if See truly saw any craters, then Antoniadi would have seen them too.

During subsequent decades, many observers studied Mercury, especially in France (the most extensive observations were by V. Fournier between 1909 and 1941). In 1934 Antoniadi published his important book about Mercury (Antoniadi 1934). Using the Meudon 83-cm refractor, he made numerous observations, always during daylight, and produced a map. He was convinced that the rotation was synchronous, and also that the Mercurian atmosphere was dense enough to support fine dust, so that the local veilings (see Fig. 1) were "more frequent and more obscuring than those of Mars." On his map, he named the main dark feature Solitudo Hermae Trismegisti; he called it "a vast shading, which I discovered on 17 August 1927 and which I have since seen almost continuously, even with very bad images." There was also the Solitudo Criophori, "a very dark feature 2500 km long, curved with its point or horn directed toward the north-east." Antoniadi believed that Schiaparelli had discovered S. Criophori, lettering it *mb*. The maps by Antoniadi and Schiaparelli have other features in common (Schiaparelli's *q* is Antoniadi's S. Iovis; his *p* is S. Maiae; his linear feature *rq* is H. Vallis, and so on), though Schiaparelli did not record S. Hermae Trismegisti.

From 1942–1944, Pic du Midi astronomers produced a map based upon photographs with the Pic refractor. It may be that we can identify Antoniadi's S. Criophori, S. Atlantis and S. Lycaonis. Further observations by A. Dollfus in 1950 also showed dark patches. Maps attempted by observers using smaller telescopes are, I fear, decidedly suspect; that by G. Wegner during the period 1956–1960 (see Sandner 1963) is typical.

The last really serious maps, before Mariner 10 gave us knowledge of what Mercury's surface is really like, were those of Chapman (1968) and of Camichel and Dollfus (1968). By then the rotation period was known to be 58.6 days. Chapman's map was derived from 130 representative drawings made between 1882 and 1963; he concluded that "this chart probably represents accurately the gross surface features of Mercury." The Camichel/Dollfus map was also based on historical drawings and photographs, reinterpreted with the correct rotation period, and it closely resembles Chapman's map.

What can we learn from all this? Obviously, there is no chance of seeing much detail on Mercury from beneath the Earth's atmosphere, and the one point at issue is whether any genuine features were seen at all. There have been suggestions that Antoniadi was unconsciously prejudiced by Schiaparelli's map, but this I reject. I knew Antoniadi, and he was not the type of observer to be prejudiced in any way. Moreover, Cruikshank and Chapman (1967) have pointed out that in the Antoniadi drawings which lack the Solitudo Criophori, the feature would not have been on the visible disk at the time of observation (see Fig. 1). I feel it is fair to say that even Chapman's map, based upon computer analysis of the best drawings by the most skilled observers over a long period of time, is not a sufficient basis for retaining any nomenclature from the previous maps. My own conclusion is that although it is probable that a few real albedo features were glimpsed occasionally, the errors in observation were so unavoidably large that, without the spacecraft, we would never have learned anything definite about Mercury's features. The visual observers did all that they could; the fact that they failed was no fault of theirs.

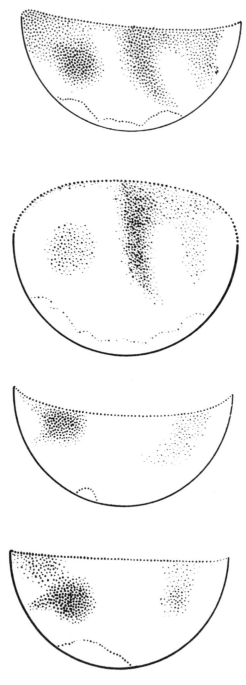

Fig. 1. Four sketches of Mercury by E. M. Antoniadi when Mercury was at eastern elongations in Aug. 1924, June 1927, Oct. 1927 and Sept. 1929. The so-called Sinus Criophori, the large dark equatorial feature in the two right-hand drawings, seemed to be veiled in those to the left. However, the latter show a different face of the planet, about 140° away, not the same as Antoniadi assumed.

III. RECENT HISTORY OF MERCURY STUDIES

Mercury is always difficult to observe with Earth-based telescopes, because it never strays as far as 30° from the Sun. Various erroneous conclusions were reached as new astrophysical measurement techniques were applied to Mercury in the middle of the 20th century. For example, early polarimetric measurements yielded the false conclusion that Mercury has a substantial atmosphere (see Sec. II.B of the Chapter by Veverka et al.; also Chapter by Hunten et al.). The difficulties of mapping Mercury, summarized above by Moore, were complicated by an accidental stroboscopic effect: Mercury's favorable eastern (or western) elongations occur for several years in a row every three synodic periods (350 days), which closely equals six 58.65-day rotation periods (352 days). Therefore, observers saw the same features in the same positions on Mercury's disk at favorable apparitions for several years in a row. It was just such observations that justified the fallacious conclusion that Mercury keeps one face always to the Sun, and rotates synchronously. So Mercury reminded us that mere consistency with the data is insufficient proof of a hypothesis.

The first hint that something was awry came in 1962, when thermal radio emissions were detected from Mercury. Given the geometry, much of the emission had to be coming from the dark side, implying a temperature far higher than would have been possible on an airless planet with a perpetually dark side. In 1965, Pettengill and Dyce (1965) used the Arecibo radar to measure the limb-to-limb Doppler frequency spread of returned echoes from Mercury. It was larger than expected; soon refinements proved that Mercury rotates in exactly two-thirds of its 88-day yr. Subsequent analysis of spin evolution, given the strong solar tides and the large eccentricity of Mercury's orbit, showed that such a two-thirds spin-orbit coupling is readily understood (see Chapter by Peale).

Some efforts were made to obtain groundbased photographs of Mercury to produce a reliable map. Real features were recorded (Smith and Reese 1968), but they were fuzzy and had no discernible planetological significance; the new photographs were used to augment the Camichel-Dollfus map (Murray et al. 1972). Later in the 1960s, observers concentrated on measuring the global photometric and radiometric properties of Mercury's surface. In a general sense, Mercury appears to be much like the Moon (see Chapter by Veverka et al.), except for its proximity to the Sun and unusual rotational geometry. For example, it was pointed out by Soter and Ulrichs (1967) that there are two "hot poles", with the Sun overhead at perihelion, which receive more than twice the solar energy received at longitudes 90° away. At a hot pole, the noontime temperature may exceed 700 K, but drops toward 100 K at night.

By the time Mariner 10 had passed Venus, and was on its way to what would be the first of its three encounters with Mercury (29 March 1974), old notions about Mercury had been shed, and planetologists were awaiting the new results with an open mind. As with other spacecraft of its generation, major emphasis was placed on Mariner 10's imaging system to reconnoiter the planet's surface topography. Some thought had been given to improving the resolution by using a film system instead of a vidicon; that was not done, but the camera was improved, nevertheless, and high-resolution images were obtained. Mariner 10 carried an array of particles-and-fields instruments, as well. Long after the mission was planned, it was discovered that the spacecraft could be brought back to Mercury for second and third encounters (21 September 1974 and 16 March 1975). A popular account of Mariner 10 is given by Murray and Burgess (1977). Initial scientific results were published in *Science* (1974, Vol. 185) and more complete analyses in *Journal of Geophysical Research* (1975, Vol. 80). A post-Mariner 10 Mercury conference was held in Pasadena, 10–16 June, 1975; resulting papers appeared in a special issue of *Icarus* (August 1976, Vol. 28, No. 4). An important review of the status of Mercury research as of a decade ago is that of Gault et al. (1977).

IV. THE ORIGIN OF MERCURY AND ITS COMPOSITION

The origin of Mercury is the least well-understood topic discussed in this book. Chapters dealing with other topics are well-organized reviews of the observable properties of Mercury and its inferred history since the end of the late heavy bombardment. However, we have a plethora of individualistic discussions of Mercury's bulk composition and of the various physical scenarios that try to describe how this iron-rich planet came to be. Here, I will attempt to place some order on this confusion. We should not be surprised, however, that the earliest stages of solar system history are so poorly understood, especially concerning a planet from which we have no samples to measure ages or chemistry. However, we know enough about Mercury already from Mariner 10's reconnaissance to raise a variety of speculations, some of which have profound implications for the origin and early history of all the terrestrial planets, including the Earth. Studies of Mercury have accelerated the trend of the past decade away from orderly views of the development of the terrestrial planets toward a chaotic history dominated by random giant impacts. I begin this section by considering Mercury's unusual chemical composition and conclude with an evaluation of physical models for the origin of Mercury.

Although Johannes Kepler proposed that Mercury had the highest bulk density of all the planets, reliable evidence about Mercury's high density has not been available for long. For example, the 1926 edition of Russell, Dugan, and Stewart's classic *Astronomy* text lists Mercury's density as "3.8 times that of water,—intermediate between the densities of Mars and the Moon, as seems reasonable." Later this century, it gradually became clear that Mercury's density is nearly equal to the Earth's and, correcting for the Earth's much greater gravitational compression, Mercury must be made of unusually dense materials. As Goettel notes (see his Chapter), there is no rigorous way of determining what heavy elements are responsible for the high bulk density. But, given cosmic abundances, Mercury must have a large proportion of iron (Urey 1952). During recent decades, this has been the chief constraint on scenarios for the planet's origin.

One approach to understanding the high density has been to invoke Mercury's proximity to the early Sun's heat. Bullen (1952), echoed a decade later by Ringwood (1966), proposed vaporization of much of the silicate portion of Mercury during the Sun's formation. Sagan (1974) succinctly summarized this view: "Mercury may have originally had amounts of silicate and iron-nickel quite similar to the primitive Earth, and the planet would have been more intensely bombarded by solar electromagnetic and corpuscular radiation in the earliest phases of the evolution of the Sun. This radiation may have been intense enough to vaporize and drive off much of the upper silicate mantle of Mercury."

Cameron's giant-gaseous-protoplanet theory has led to the most recent version of the vaporization model in which Mercury's composition is obtained from the high-temperature loss of everything except the least volatile elements (see Chapter by Cameron et al.). The chief question about Cameron's latest version of the vaporization model is whether or not the extensive atmosphere of rock vapor that is produced can actually be removed from Mercury: it seems that turbulent flow of the solar nebula wind past Mercury would have to entrain atmospheric parcels and carry them away with a very high efficiency. The unproven vaporization model nevertheless makes predictions: measurements of Mercury's chemical composition should identify the refractory-rich, volatile-poor signature of mantle vaporization, if it occurred. Even if the bulk composition of Mercury has not been affected by vaporization, it is conceivable that Mercury's immediate surface layer has been vaporized at some time in the past; this possibility may invalidate attempts to extrapolate observations of surface composition (cf. Chapter by Vilas) to the bulk planet.

Lewis (1972) proposed that the densities of the terrestrial planets could be explained by the condensation, under conditions of chemical equilibrium, of solids from a high-temperature, gaseous solar nebula of solar composition (hereafter called the "equilibrium condensation model"). In particular, Mercury formed at a distance so close to the Sun that only iron and refractory silicates condensed (temperatures > 1400 K). Venus and the Earth were diluted in iron by the much larger fractions of silicates that condensed at slightly lower temperatures. Implicit in the equilibrium condensation model are three important requirements: (a) accretion of grains and cooling of the nebula were slow relative to reaction rates, so that gas-solid equilibrium could be attained; (b) Mercury formed very quickly and/or the nebula was dispersed rapidly relative to nebular cooling, so that less than 15% of lower-temperature magnesium-bearing silicates condensed near Mercury and accreted onto it; and (c) there was very little subsequent mixing of materials from Venus' zone or beyond, otherwise the density contrast could not have been maintained. As we will see, the degree of interzone mixing depends on important details of planetary accretion models, such as the size-distribution of planetesimals.

Nobody, not even Lewis, ever believed in the strict, end-member version of the equilibrium condensation model. But during the period when Mercury was being studied in the aftermath of Mariner 10, most scientists accepted the basic idea, perhaps even too literally, and regarded deviations from the pure model to be relatively modest. The history of the equilibrium condensation model is explored, at varying lengths, in several Chapters in this book, including those by Goettel, Lewis, Cameron et al. and Wasson.

The equilibrium condensation model places rather stringent constraints on the abundances within Mercury of many other constituents important for considering the planet's subsequent geophysical and geochemical evolution. For example, boundary conditions for Mercury's thermal evolution are set by the model's predictions for abundances of radioactive elements: short-lived potassium-40 should be absent, but refractory elements uranium and thorium should be enriched. Also, because of its comparatively low condensation temperature, sulfur—which plays an important role in the cores of iron meteorite parent bodies and in the Earth itself—should be entirely absent within Mercury and its core. With no sulfur, Mercury's core may be expected to freeze solid early in the planet's history, which is inconsistent with inferences that an active dynamo still exists in the core and generates the planet's magnetic field (see below).

Later in the 1970s, various researchers investigated supplementary or alternative processes for iron/silicate fractionation as it became clear (see, e.g., Barshay 1981) that extreme segregation of Mercury from planetesimals of lower-temperature compositions was not possible; this work has been extended by Lewis (see his Chapter), who confirms Barshay's essential conclusion. Weidenschilling (1978) reviewed various mechanical fractionation mechanisms that depend on physical differences between metal and silicates (e.g., differential response of iron and silicates to impact comminution and to aerodynamic sorting); he proposed that Mercury formed iron rich due to such processes. In his model, Mercury is special in this regard due to the higher gas density and shorter dynamical time scales in the innermost part of the solar nebula.

As Wasson (see his Chapter) points out, there is circumstantial meteoritical evidence for mechanical iron/silicate fractionation: chemically similar EH and EL chondrites differ in metal content by a factor of two. If such processes, enhanced by factors of several, occurred near Mercury's location, they could explain the planet's density without resorting to small differences in condensation temperature between iron and silicates. Naturally, physical fractionation of silicates from iron should not affect the chemistry of either phase; therefore, the composition of Mercury's mantle could be volatile-rich and oxidized, despite the planet's high iron content (cf. Goettel's Chapter).

Wasson, however, considers that Mercury may be reduced and related to the enstatite (E) chondrites (see below).

The major question posed by Mercury is whether its metal-rich composition is symptomatic of its being a refractory planet, formed at the hottest end of the sequence of planetary distances (and where the solar nebula was most dense), or whether it is an "accident" that has nothing (or little) to do with its proximity to the Sun. If it is of refractory composition, as formerly assumed, then it is an end-point in a sequence of planetary compositions that seem to be controlled by primordial conditions in the nebula: from refractory Mercury, through volatile-rich Mars and carbonaceous water-rich asteroids, to icy bodies in the outer solar system. The alternative view is that, whatever initial conditions prevailed in the solar nebula, subsequent evolutionary processes have dominated planetary traits.

The elegance of Lewis' original proposal of equilibrium condensation now seems much too simple to be true. Other authors have tried to fit the range of planetary and meteoritical compositions into a less constrained sequence ranging from most reduced near the Sun to most oxidized farthest from the Sun (see Wasson's Chapter); this effort is hindered by lack of secure knowledge about the formation distances, or even present locations, of the meteorite parent bodies. Wasson would have Mercury made from materials chemically similar to enstatite chondrites, but mechanically fractionated in order to enhance the proportion of iron. The most extreme opposing view is suggested by recent work of Wetherill (see his Chapter): Mercury's composition might have little to do with its location near the Sun; indeed the planet may originally have formed at a different location, perhaps well beyond 1 AU. The most startling aspect of Wetherill's work is his suggestion of an extraordinary metal/silicate fractionation mechanism for Mercury: it was stripped practically to its core by one or more giant collisions.

Wetherill's work is an outgrowth of 2-dimensional numerical studies by Cox and Lewis (1980) and 3-dimensional studies by Wetherill (1980a, 1985a) of the dynamical evolution of some hundreds of large planetesimals as they accumulate into the terrestrial planets. These types of studies have been limited by computer capabilities, so it is important to understand whether the starting conditions for simulating this final stage of planetary accumulation truly reflect nature, or are compelled by the computer limitations. The earlier stages of planetary accumulation, during which countless km-scale planetesimals accumulate into larger bodies, can be modeled only in a statistical way. Results of such statistical simulations by Greenberg et al. (1978) suggested that a very few planetesimals might begin to grow to large sizes while most of the mass of the planetesimal population remained in small bodies; such a size distribution, which would be difficult to model, differs from the starting conditions for Wetherill's late-stage studies in which the mass is dominated by some hundreds of bodies near the large end of the size distribution. However, the Greenberg et al. simulations had artifacts (resulting, for example, in ar-

tificially short time scales) due to faulty algorithms for mass-binning and gravitational stirring (Patterson et al. 1988). Revisions of the Greenberg et al. code
by Patterson, Spaute and others and new work by Stewart and Wetherill
(1988) are clarifying the degree to which "run away" planet growth occurs
and the form of the planetesimal size distribution during the middle stages of
accumulation. It is important to learn if the behavior is compatible with Wetherill's initial conditions for the late stage. (A logical extreme, for which, however, there is now no modeling evidence, is that small planetesimals accumulated into just four large embryos in the terrestrial planet zone, which grew to
the planets we know today.) While we cannot yet be sure about the middle
stages of planetesimal growth, Wetherill's modeling of the resulting evolution
of these larger planetesimals commands our attention.

What happens is that quite a few of the bodies approach planetary dimensions, and they interact gravitationally with each other, resulting in a scattering of their orbits. From time to time they collide, sometimes accumulating
and at other times shattering each other. According to Wetherill's results, at
the end there is only approximate retention of any initial compositional gradients in the inner solar system after the dynamical and collisional evolution is
complete. In his scenario, proto-Mercury may have originated anywhere between its present location and beyond the orbit of Mars. It may have been
impacted by another body of nearly planetary dimensions (or by a series of
somewhat smaller impacts), resulting in the blasting away of much of its
silicate mantle. Equivalent giant impacts may have had profound effects on
other planets, as well; for example, a currently popular idea is that the Moon
originated from the impact of a Mars-sized planet with the Earth.

According to Wetherill's scenario, Mercury is metal-rich not because it is
near the Sun but rather because it is, as an accidental result of its stochastic
growth, so small. At any time during the late stages of terrestrial planet
growth, there were the largest, full-grown planets, and a distribution of smaller bodies. Once a planet became as large as the Earth, the largest possible
impact could, at most, create a comparatively small Moon, but could not strip
away a significant fraction of the mantle. Protoplanets much smaller than
Mercury were either so small that they were readily destroyed by high-velocity impacts with smaller bodies, or they would become incorporated into
the largest planets. Mercury happened to be at an intermediate size where
probable collisions could strip away much of its rocky mantle but could not
utterly destroy the planet.

Wetherill would be one of the last to insist that his advances in calculating the planetary accretion end-game necessarily prove that giant impacts
dominated the early history of the terrestrial planets or that primordial compositional trends were necessarily erased by orbital chaos. We really do not
know how the size distributions and velocity distributions evolved as planetesimals grew from km-scale to the sizes Wetherill has used as a starting point
(in fact, the hypothesis that km-scale planetesimals ever existed, as Goldreich

and Ward (1973) envisioned, is far from secure). Nor, despite the recent studies using hydrodynamic codes (see Chapter by Cameron et al.), is it at all clear just what happens when planetary-size bodies collide with each other: perhaps the Moon can be "splashed off" the Earth, and perhaps proto-Mercury's mantle can be stripped away, but the calculations have not yet reached the maturity required for confidence.

There are other reasons, however, besides the recent numerical modeling to favor the Wetherill scenario over equilibrium condensation, or at least to favor some intermediate model (cf. Chapter by Goettel). As I describe below, some other features of Mercury's geophysical evolution are difficult to understand from the perspective of equilibrium condensation and may be more compatible with Wetherill's view. The authors of this book's chapters on composition and origin take a variety of advocacy positions concerning the most probable mode of origin for Mercury, and they variously dismiss some of the other models. In my view, however, we know too little about primordial conditions in the solar nebula and about processes of planetary accretion, and we know too little about Mercury, to take a firm stand about any of the models. Fortunately, most of them may be testable, with little ambiguity, by practical spacecraft studies of Mercury.

V. THE CRATERING HISTORY OF MERCURY

A fundamental conclusion of the first decade of spacecraft studies of the terrestrial planets is that the first 0.5 Gyr of their history has been greatly obscured by the "late heavy bombardment." (This term generally refers to a high cratering rate that was, at least on the Moon, still continuing as late as 4.2 to 3.9 Gyr ago, as long as 600 Myr after the origin of the solar system. It does not necessarily imply a higher bombardment rate than during the preceding periods.) The bombardment has been proven only for the Moon, for which age dating of returned rock samples permitted the establishment of a relationship between crater density and age. Most of the lunar basins and other large craters date from around 4 Gyr ago. However, considerations of the dynamical scattering of cometary and asteroidal projectiles by the Earth and other planets prove that any heliocentric projectile population that could have struck the Moon at that time (and it is difficult to believe that a geocentric projectile population existed at that time), must have been scattered about the early solar system and struck the Earth, Mars, Venus and Mercury, as well. Indeed, both Mars and Mercury have basins and heavily cratered terrains similar to those on the lunar highlands, and it is widely believed that those features formed contemporaneously with the late heavy bombardment on the Moon. What is secure, however, is that the visible craters on those planets formed either at that time *or later,* but certainly not before.

What happened during the missing first 0.5 Gyr on Mercury? The eventual sweep up or dispersal of the last large planetesimals left the four terrestrial planets, plus the Earth's Moon, free from chances for any more planet-shattering impacts. But there was undoubtedly a continual rain of smaller remnant planetesimals, both from the terrestrial planet zone, and from those populations of more distant remnant planetesimal populations that still crater the planets today—the asteroids and comets. Although we have no direct observations relevant to the magnitude of bombardment by these now-decayed populations, they could have had a profound effect on the surface layers of the newly formed planets. Mercury's crust could well have been pummeled by impacts for hundreds of Myr, at a frequency well exceeding that recorded by the late heavy bombardment. There could have been sufficient mass in the early projectile populations to modify the chemical composition of the planet, or at least of its surficial layers (see Chapter by Lewis).

It is possible that some remnant population of planetesimals in Mercury's zone could have been swept up so slowly that their impacts on Mercury actually post-dated the period of late heavy bombardment. Chapman (1976) first questioned the Mariner 10 imaging team's extension of the lunar cratering time scale to Mercury and suggested that the projectile populations affecting Mercury might be quite different from those affecting other planets. He and his colleagues Davis and Weidenschilling later proposed that certain Mercury-zone planetesimals might, indeed, be Mercury-specific impactors and might be expected to last as a projectile population long past the late heavy bombardment. They termed these hypothetical bodies "vulcanoids"; the hypothesis is developed by Leake et al. (1987), who conclude that vulcanoids may have had some modest effects on the early visible cratering record of Mercury.

The essential points of the vulcanoid hypothesis are: (a) Most models for the origin of Mercury, including orderly planetesimal accretion scenarios as well as those involving giant impacts (see Wetherill's Chapter), could plausibly result in remnant vulcanoids orbiting interior to Mercury's orbit. (b) Because of secular evolution of the eccentricity of Mercury's orbit, only occasionally can Mercury sweep up vulcanoids near $a = 0.25$ or 0.26 AU; sweep-up time scales for such bodies would be of the order of 1 Gyr so vulcanoids could readily dominate the cratering history of Mercury well after the end of the late heavy bombardment. (c) Mercury is too small and too deep in the Sun's gravity well to scatter vulcanoids out to Venus or beyond, so vulcanoid bombardment would not be expected to be manifest on the Moon or other terrestrial bodies. (d) Surveys of heliocentric objects near the Sun have not yet placed useful observational constraints on the existence of a population of vulcanoids similar to the asteroid belt remaining to the present time, but more sensitive infrared searches are planned. (e) A major constraint on the vulcanoid hypothesis is that, presuming vulcanoids were ever formed, calculations show (Leake et al. 1987) that those in orbits near Mercury's would have collisionally destroyed each other over a period of ≤ 1 Gyr, even if they were made of strong metal, so they would not have lasted long past the late heavy bombardment. This conclusion pertains only to the limited collisional scenarios studied by Leake et al.; Wetherill (personal communication) believes the constraint may be relaxed according to the discussion in his Chapter. This conclusion would be refuted by the observational discovery of some vulcanoids remaining today in orbits near Mercury.

There is little reason to believe that the cratering history on Mercury has been affected by vulcanoids over the last 3 Gyr. However, it remains an open question as to whether vulcanoids were important, or dominant, during and immediately following the period of late heavy bombardment occurring throughout the inner solar system 4 Gyr ago. There is no reliable, independent way to identify whether craters were formed by vulcanoids, or by projectiles from the same population responsible for lunar bombardment. Leake et al. (1987) describe some modest differences in the crater distributions on the two bodies that suggest, but hardly prove, that different populations may have been involved.

Schultz (see his Chapter) has investigated cratering physics and concludes that size dependencies of certain crater morphological parameters on Mercury, the Moon and Mars may reflect impactor traits, such as velocity and bulk density. He finds consistency if high-velocity cometary impactors are responsible for a larger fraction of recent craters on Mercury than on the Moon and Mars (as expected; cf. Hartmann et al. 1981). However, some of the larger, older craters and basins are more readily interpreted as being due to low-velocity, low-density impactors in circular heliocentric orbits. Vulcanoids could be responsible for many of the large (100- to 300-km diameter) two-ring basins and central-pit craters that formed before the Caloris basin; they would, indeed, impact with relatively low velocities compared with other source populations, but would less likely have had low (e.g., comet-like) densities.

Schultz's efforts to take a new look at scaling relationships for crater morphologic parameters are somewhat speculative, and they diverge in some ways from the previous work by Schmidt, Holsapple and their colleagues (see, e.g., Holsapple and Schmidt 1979; Housen et al. 1983). Readers may contrast Schultz's work with that of Pike (see his Chapter), who tends to de-emphasize the role of impactor velocity in contributing to crater morphology. Ultimately, the work of both Schultz and Pike should improve our understanding of the cratering process and should establish a firm foundation for interplanetary comparisons of crater statistics.

Strom and Neukum (see their Chapter) compare the highland crater size-frequency distributions on Mercury, the Moon and Mars and conclude that they are probably due to the same population of late-heavy-bombardment impactors. They use as evidence a high degree of similarity in the shape of the crater distributions, after correcting for certain differences that can be ascribed to endogenic processes. Their argument is plausible, but not incontrovertible. Given the modest statistical significance of the crater counts at large diameters, the subjectivity of corrections for endogenic crater-erasure processes (e.g., eolian processes on Mars, volcanic intercrater plains-forming processes on Mercury), and the "free parameter" of impactor velocity and other scaling factors that permit the curves to be shifted in diameter (i.e., shifted laterally on the standard R-plot), I think it would be prudent to keep an open mind about whether a specific impactor population was responsible for all late heavy bombardment cratering in the inner solar system.

Nevertheless, certain earlier disputes about late heavy bombardment cratering have now been largely laid to rest. Strom has partially retreated from his earlier insistence that the highlands units are not saturated with craters; he now relies on several studies (most recently by Chapman and McKinnon 1986) which show that size distributions similar to those seen in the lunar highlands could result from production functions similar to the observed ones whether saturation has been reached or not. The other question is Hartmann's (1984) belief that the shallow slope of the small-diameter leg of the crater curve is mostly or wholly due to endogenic crater-erasure processes on all three bodies: Mercury, the Moon and Mars. As argued by Chapman and McKinnon and reiterated by Strom and Neukum (see their Chapter), the paucity in Mercurian craters <50 km diameter compared with the Moon is rather modest in spite of a *much larger fraction* of intercrater plains on Mercury than on the Moon. Thus, obliteration of craters by intercrater plains formation on either body cannot explain the *gross* shape of the small-diameter leg of the crater curve, so its shallow slope must chiefly be a reflection of the shape of the production function.

By lateral shifting of the crater curves in the 40- to 150-km diameter range so that the shapes overlap, Strom and Neukum have deduced the relative impact velocities for crater-forming projectiles hitting each body. Evidently, the late heavy bombardment impactors in the inner solar system had relatively low velocities that varied from body to body in a way that points to remnant planetesimals left over from the growth of Venus and the Earth as the probable impactors. However, the required velocities for Mercury are too high, they say, for a lower-velocity vulcanoid population to be the source of impactors for Mercury. A corollary of Strom and Neukum's conclusion is that the late heavy bombardment in the inner solar system is not related to any similar bombardment episode in the outer solar system. Strom and Neukum's results have motivated new research to investigate how a population of accretion remnants could be stored for 0.5 Gyr in orbits between Venus and the Earth and whether they could maintain, in spite of collisional fragmentation, the characteristic size distribution of the late heavy bombardment. Wetherill had suggested that tidal disruption of a large body passing close to the Earth or Venus might result in such a size distribution, but Mizuno and Boss (1988) have shown that such tidal disruption probably cannot occur. Therefore, there is new impetus to resolve a fundamental mystery about the early history of the terrestrial planets: what was the source of the late heavy bombardment?

VI. GEOPHYSICAL AND THERMAL HISTORY

Mercury's evolution as a planet provides a benchmark against which to test our ideas of planetary evolution generally. Several of Mercury's traits must be reconciled with each other and with our understanding of geophysical processes. These include: (a) the presence of a dipolar magnetic field; (b) an axial spin locked at two-thirds the orbital period; (c) the apparent expression of the late heavy bombardment and an inferred lunar-like cratering chronology; and (d) expression in surface geology of tectonic processes possibly related to global spin-down and global contraction. In this section I first discuss the history of thermal modeling. Then I amplify on the magnetic field, spin-down and associated global lineaments, and the lobate scarps.

Thermal Evolution

The initial discoveries of Mariner 10 and some subsequent analyses evoked surprise and evidence of contradiction. The magnetic field—evidently due to dynamo action in a molten core—had not been anticipated, in part due to Mercury's slow spin but also due to widespread expectations that Mercury's core should be frozen solid. If the cratered terrains on Mercury's surface were indeed due to the late heavy bombardment, dated on the Moon as occurring 4 Gyr ago, then core formation and much of the subsequent planetary cooling must have taken place within the first 0.5 Gyr (Solomon 1976; Solomon 1977; Toksöz and Johnson 1977); otherwise late core formation would have erased surface geology, or late planetary cooling would have led to much more extensive surface expression of global contraction than is observed. Just possibly, the magnetic field might be due to fossil remanent magnetism of the planet's interior.

A major problem was that in the then-popular model of equilibrium condensation, a still-molten core could not be maintained, especially if it formed early in Mercury's history; without the compositional constraint of equilibrium condensation, we can now appeal to sulfur in the core to lower its melting temperature and keep at least part of it molten and convecting. Some early thermal models delayed core formation and planetary contraction, which violated the standard cratering chronology, but might be accommodated by an extended period of vulcanoid bombardment. However, those models overlooked what is now believed to be the inevitability of early core formation and the efficiency with which subsolidus convection of the mantle cools the planet subsequent to core formation. The early thermal models are reviewed in the Chapter by Schubert et al.

We now believe there are abundant sources of heat early in Mercury's history so that it must be assumed that its core separated very early, if not contemporaneously with the planet's formation. Not counting other sources of early heat, one or more of which must have been active to differentiate many meteorite parent bodies (e.g., short-lived radioactivity and electromagnetic heating), the ambient temperature of Mercury during accretion and added accretionary heat should have been sufficient, augmented by the expected heating from long-lived radioactives. Using a range of parameters, including sulfur concentrations in the core of a few percent, Schubert et al. (see their Chapter) extend and refine the calculations of Stevenson et al. (1983), who demonstrated a thermal evolutionary history for Mercury that seems to meet most observational constraints. They calculate that Mercury's lithosphere would already be 50 km thick at the end of the late heavy bombardment, and about 150 km thick today. A solid inner core would form early in Mercury's history and steadily grow, relegating the sulfur to a molten outer core shell. This thermal history model predicts that Mercury must have globally contracted, due to inner core growth and mantle cooling, by 6 to 10 km in radius; this exceeds the observational limit of about 2 km established by Strom (1979; see also Chapter by Melosh and McKinnon) from the numbers and geometries of the lobate scarps (that limit pertains, however, only to the period subsequent to retention of observable topography; see below).

Magnetic Field

Connerney and Ness (see their Chapter) summarize what has been learned about the magnetic field of Mercury; it is weak compared with the Earth's field but otherwise (to within the wide uncertainties inherent in the Mariner 10 data) is dipolar and perpendicular to the ecliptic. Schubert et al. believe it is possible to generate a magnetic field by convection within the thin shell of the outer core. However, a problem is that their modeled strength for the modern Mercurian magnetic field greatly exceeds the measured field strength, suggesting that generation of the field may be by a different mechanism. Other researchers (cf. Chapter by Connerney and Ness) are less sanguine than Schubert et al. about the prospects for generating a field from a thin core shell. Most researchers agree that the earlier concept of a remanent field is untenable.

Spin Down and Global Lineaments

The Chapter by Peale discusses tidal spin down of Mercury from its presumably more rapid primordial spin, and capture into its present 59-day period. While such capture is nearly inevitable under certain simple conditions, Peale believes a rapidly spinning Mercury would have been slowed and captured into a higher-order spin resonance if it had an early molten core, due to additional dissipation between such a core and the mantle. Either (a) Mercury never had a spin period more rapid than 44 days (otherwise it would have been captured into the 44-day 2:1 resonance), (b) a molten core had not formed until after spin down, or (c) Peale's modeling is erroneous or incomplete. Option (b) seems unlikely, according to Schubert et al., who assume and believe that the core is "initially completely molten." Option (a) is not particularly satisfying; perhaps a rapid initial spin (often taken to be of order 20 hr) is less deterministically required by the new large impact scenarios (Chapter by Wetherill) than in the more orderly planet-growth scenarios previously accepted.

The duration of spin down is uncertain but is unlikely to exceed 1.5 Gyr and may be as short as 100 Myr for very low values of the dissipation parameter Q (Leake et al. 1987; Chapter by Melosh and McKinnon). Stresses in the lithosphere induced by such spin down should be reflected in the surface topography of the planet if the lithosphere was thinner than about 100 km during spin down and if that topography has been preserved. The conditions on lithosphere thickness are met by the Schubert et al. thermal models for the first 2 Gyr of Mercury's history. Observed lineament orientations on Mercury are roughly in accord with the predicted stresses, and are expressed chiefly in the older stratigraphic units (see Chapters by Melosh and McKinnon and by Thomas et al.). Thus, as Melosh and McKinnon conclude, there is a consistent picture of Mercury despinning during roughly its first 0.5 Gyr and preserving the tectonic expression of that part of its history.

Lobate Scarps and Global Contraction

The widespread network of lobate scarps is unique to Mercury. These thrust faults occurred relatively late in Mercury's history, commencing after the last intercrater plains were formed, just prior to the Caloris basin impact (Leake et al. 1987; Chapters by Melosh and McKinnon, by Thomas et al. and by Spudis and Guest), which occurred at about 3.85 Gyr (Chapter by Strom and Neukum). These scarps are ascribed to global contraction caused by solidification of the inner core and cooling of the mantle. As noted above, the observed scarps account for only a rather small fraction of the total contraction expected from thermal modeling. Much of the contraction would have predated the observable geologic record, according to Schubert et al.'s calculations, which seems to account for the discrepancy. But then there is no ready explanation for why the lobate scarps were not forming throughout Mercury's recorded geologic history, in particular during its early history. Most interpretations of Mercury's geologic history (cf. Chapters by Spudis and Guest and by Strom and Neukum) consider the period of 4.2 to 4.0 Gyr ago to be a period of crustal expansion with widespread volcanism, followed by the onset of global contraction, evidenced by lobate scarp formation and the closing of vents (which ended smooth plains volcanism). Since such a history is inconsistent with the models showing Mercury to be in global contraction throughout this period, we may have to appeal to a thin but rapidly thickening lithosphere, or some other factor, to reconcile the thermal modeling with the geologic observations.

VII. GEOLOGIC HISTORY OF MERCURY

In comparison with the Moon or Mars, it has been somewhat more difficult to synthesize the geologic history of Mercury. We have only the images
taken by Mariner 10, supplemented by groundbased radar (Chapters by Harmon and Campbell, and by Clark et al.). Much of the planet not imaged by
Mariner 10 remains *incognito*, and even the pictures we do have are of rather
modest resolution and taken with a limited range of solar illuminations.
Adding to the difficulty is the inherent lack of prominent albedo- or color-
contrasts on Mercury (those that do exist have been discussed by Rava and
Hapke [1987]).

Nevertheless, more than a decade of study seems to have resolved many
of the early debates (see Chapter by Spudis and Guest). Although Mercury
was originally thought to be rather lacking in basins, remnant basins have now
been recognized in numbers similar to those observed on the Moon. Most of
them are very old, however, and have been substantially degraded by intercrater plains formation and subsequent cratering. Intercrater plains are very
widespread on Mercury. Spudis and Guest lean toward attributing them to
abundant volcanism even though, in their degraded state, they lack clear volcanic morphology; Leake is more certain about the volcanic provenance of the
intercrater plains, and notes an association between later intercrater plains and
some volcanic landforms (Leake et al. 1987). The alternative idea is that intercrater plains are basin ejecta, but the plains lack well-defined spatial relationships to obvious source basins. (Debates about the origin of Mercurian
plains resemble debates about the origin of lunar intercrater plains prior to the
Apollo 16 mission, which firmly established the association of such lunar
plains with basin ejecta. Even on the Moon, the situation may be complex,
and there is no compelling reason to extend that particular lunar example to
Mercury.)

The impact that formed the Tolstoj basin began a period of cratering and
plains formation termed the Tolstojan stratigraphic system. The impact of the
Caloris basin is the dominant event in Mercury's history. Extensive smooth
plains formed during the Calorian period. Some of those plains units appear to
be related to Caloris, although Spudis and Guest point out that they are widespread across the observed portions of Mercury and new crater counts suggest
that many of the units formed significantly *after*, not contemporaneously with,
the impact that formed Caloris. Although such smooth plains have also been
described as basin ejecta, Spudis and Guest share the consensus view that they
are of volcanic origin. Subsequent to cessation of smooth plains formation,
little else has occurred on Mercury besides occasional cratering.

Synthesizing the perspective of Spudis and Guest with the cratering chronology developed by Strom and Neukum (see their Chapter), we can conclude
that the entire observable history of geological evolution of Mercury began
with ancient saturation cratering of basins about 4.25 Gyr ago and ended with

the final smooth plains emplacement about 3.8 Gyr ago. This period of still-visible endogenic history (tectonics and volcanism) is a slightly older and distinctly shorter history than that recorded on the Moon, where volcanism lasted for another 1 Gyr. (The absolute ages for Mercury could be appreciably younger if vulcanoids contributed to Mercurian cratering.) According to Strom and Neukum, it is possible that smooth plains volcanism was simply a continuation of intercrater plains volcanism, which finally terminated when the lithosphere thickened and became compressed by global shrinkage, shutting off vents to magma source regions in the mantle.

VIII. ATMOSPHERE AND MAGNETOSPHERE

The atmosphere and magnetosphere of Mercury are very tenuous and at least the magnetosphere is highly time variable. This makes study of their fascinating behavior rather difficult. The three slices cut by Mariner 10 through the magnetosphere of Mercury constitute a necessarily limited sampling of the rapidly varying phenomena within that large volume of space surrounding Mercury. Mariner 10 also measured the ultraviolet signatures of hydrogen, helium and oxygen in Mercury's atmosphere, but only very recently did groundbased observers discover the signature of somewhat more abundant sodium (see Chapter by Hunten et al.). That emissions from sodium, potassium and possibly other constituents may be monitored from the Earth provides hope for understanding the behavior of Mercury's atmosphere, but little time has yet elapsed since the discovery of sodium to carry out this work.

The discovery of emission from sodium and potassium by resonant scattering of sunlight has generated active debate about sources and sinks of the observed material (see Chapter by Hunten et al.). There is naturally some curiosity about whether we can eventually learn about the composition of the surface of Mercury, or about the nature of infalling meteoritic material, from studies of Mercury's atmosphere. The initial indication that the sodium-to-potassium ratio is an order of magnitude larger than found in meteorites or lunar samples is intriguing. Firm conclusions, however, will have to await some degree of resolution of the current debates about the processes that govern the atmosphere, including thermal escape, radiation pressure, photoionization and gas-surface interactions. New observations have revealed both time variability and spatial asymmetries in the sodium. There is hope that the models can be constrained by further telescopic data as well as additional analysis of Mariner 10 data in the light of the new findings.

Of particular interest is the possible role of Mercury's magnetic field in controlling either the supply to, or removal from, the atmosphere. For example, it is believed that the solar wind actually reaches the surface of Mercury at times of highest solar activity but stands off most of the time. Perhaps even more important is the rapid antisunward convection of the magnetosphere.

Mercury's magnetosphere plays an important role in the comparative study of planetary magnetospheres (see Chapter by Russell et al.). Mercury is a nearly atmosphereless planet; it has little or no ionosphere. Since the planet's surface is insulating, the dynamics of Mercury's magnetosphere must be quite different from the Earth's. There are other significant differences: Mercury has a magnetic moment 1,000 times weaker than the Earth's, it is close to the Sun, and its magnetosphere is much smaller than Earth's (only 5% in linear scale). Mercury occupies a much larger fraction of the volume of its magnetosphere than do the other planets. Russell et al. describe a variety of interesting phenomena measured during the three passes by Mariner 10 and highlight a number of outstanding questions.

Relativistic Jovian electrons may play an important role within Mercury's magnetosphere. They are known to spiral in toward the Earth, and Mariner 10 observed their characteristic signatures as it approached Mercury. A particular very high-energy electron burst was observed in 1974 that has been attributed to such a Jovian source. Jovian electrons, with their characteristic hard-energy spectrum, could dominate Mercury's magnetosphere at times since it is difficult to understand how they could be generated at Mercury (e.g., the recirculation mechanism operating in the Earth's magnetosphere is likely to be inoperative at Mercury due to differences in scale and stability between the two magnetospheres).

Several phenomena associated with Mercury's magnetosphere are potentially observable from the Earth, including the effects of direct solar wind interaction with the surface of Mercury, possible surface-heated auroral zones, and transient synchrotron-emitting radiation belts due to the Jovian electrons within the inner magnetosphere of Mercury. Naturally, a much wider variety of phenomena could be explored with a future orbiter mission to Mercury.

IX. CONCLUSION

I have only highlighted some of the interesting scientific themes raised during the last decade of research about Mercury and its relationship to other terrestrial bodies. This book also contains a wealth of basic data about Mercury (e.g. geologic maps, crater morphology data, radar profiles) that can form the basis for continuing research. While some of the ambiguities and contradictions in the literature a decade ago appear to be resolved as a result of strenuous attempts to understand them, Mercury has taught us often enough in the past to be hesitant about claiming that we finally understand the planet. Indeed, major discrepancies still remain between the inferred geologic history and the best calculations of the thermal evolution of the planet (e.g., why are there no lobate scarps dating from the Tolstojan and earlier periods when the rate of planetary contraction should have been even higher than during the time the observed scarps were formed?). Our understanding of other issues— the origin of Mercury, its chemical composition, and the behavior of its atmosphere—remains very immature. Our basic data base for Mercury is very incomplete compared with the other planets (for example, less than half the planet has been imaged at useful resolutions).

Stern and Vilas (see their Chapter) outline future directions for Mercury studies to take. Much more can be done with groundbased optical and radar telescopes. Soon there will be opportunities to observe Mercury from above the Earth's atmosphere, using sounding rockets and Earth-orbital telescopes. Serious work may soon commence to design a Mercury orbiter mission, perhaps with surface landers, to address the fundamental planetological questions. I believe that this book about the innermost planet of the solar system establishes a firm foundation for the next decade of research on Mercury.

Acknowledgments. I thank my colleagues at Planetary Science Institute (especially R. Binzel, D. Davis, D. Spaute, and S. Weidenschilling) as well as D. Hunten and G. Wetherill for helpful reviews. I also wish to thank all the chapter authors and contributors to the Mercury Conference for providing grist for my mill.

FUTURE OBSERVATIONS OF AND MISSIONS TO MERCURY

S. ALAN STERN
University of Colorado
and

FAITH VILAS
NASA Johnson Space Center

The continued and expanded study of Mercury is important to several aspects of planetary science. We first review the broad scientific objectives of such exploration and describe the methods by which such scientific objectives may be addressed. Groundbased optical, infrared and radar astronomy are discussed first, followed by Earth-orbital observations and in situ missions to Mercury. Several planned NASA missions, including the ASTRO Spacelab payload and the Hubble Space Telescope, have the potential for making important contributions to the study of Mercury. Sounding rockets can obtain ultraviolet spectroscopy when spacecraft lack such technical capabilities as extensive optical baffling. There are difficult performance requirements for getting spacecraft to Mercury, although technical solutions have been proposed to overcome these difficulties. A method that offers immediate potential to mount substantial Mercury missions with current launch vehicle inventory is the use of the multiple gravity-assist trajectories recently discovered by Yen. We discuss potential payloads for Mercury orbiters and the importance of eventually landing on the surface.

I. INTRODUCTION

Motivations for continued observation and exploration of Mercury are made clear by various authors in this book: Mercury's atmosphere displays unique characteristics important to the understanding of satellite and cometary atmospheres. Mercury's geology is relevant to the general processes which affect terrestrial and satellite bodies. Mercury's magnetosphere is complex and enigmatic: it represents the sole known example of a substantially magnetized small body and of a magnetosphere existing in the absence of an ionosphere. Additionally, Mercury's evolution represents an end-member case because it formed—or at least is now located—at a temperature extreme in the solar system.

Although Mercury is an important object for study, its location near the Sun makes it a difficult object either to reach or observe. Fielding orbiter and lander missions is difficult because of the tremendous launch and orbit insertion energy requirements imposed by Mercury's orbit close to the Sun. Additionally, the insolation environment imposes considerable constraints on spacecraft design, and hence mission cost. Mercury's physical proximity to the Sun also makes the planet difficult to observe from the Earth, since solar elongation angles in excess of 28 degrees are never obtained.

Due in part to the intrinsic difficulty of reaching Mercury, potential missions have received low priority in planetary exploration plans. Neither NASA's Solar System Exploration Committee nor the NRC's Space Science Board have formally recommended high priority new missions to Mercury.

Faced with combined technical and programmatic difficulties, researchers interested in Mercury may for some time be faced with the prospect of adapting instruments and spacecraft designed for other purposes to observe Mercury remotely. Here, we examine the near-term avenues presently available for observing Mercury, both from the Earth and from space. We also discuss the scientific rationale and technical concepts for missions to Mercury. Before addressing these subjects, however, we first review the broad scientific questions which future observations must address.

II. KEY SCIENTIFIC QUESTIONS AND EXPERIMENTS

As described in this book, our knowledge of Mercury today stands at a point somewhat similar to our state of knowledge of Mars prior to Mariner 9. Although Mariner 10's three successful flybys of Mercury provided sufficient information to develop pointed scientific questions, we still lack the data necessary to characterize the basic compositional make-up, phenomenological processes, and evolutionary history. Many of our questions require synoptic observations.

Simply put, the key scientific questions which must be addressed by the next generation of Mercury observations are:

1. What does the still unimaged hemisphere of Mercury look like and what are the inferred geomorphological and tectonic processes?
2. What is the chemical and mineralogical composition of the surface? What

are the textural properties of the surface? How do these vary among geologic units?

3. What is the full chemical composition of the atmosphere? How do the
composition and pressure vary with location on Mercury and orbital phase?

4. By what processes are the Mercurian atmosphere generated?

5. By what process is the Mercurian magnetosphere generated and how does
this magnetosphere interact with the time-dependent atmosphere and variable solar wind?

6. Does Mercury have a present-day liquid core and attendant dynamo? If so,
how much of the core is molten?

7. What are the global geophysical properties of Mercury (gravity field, heat
flow, and seismicity)?

8. What is the chronology of internal and external processes that have modified Mercury over time? Are there clues about how the planet formed?

Beyond these questions, Mercury observations also offer promise toward
the general understanding of planetary magnetospheric, cratering and exospheric processes. Further still, Mercury provides a unique location for tests
pertaining to general relativity and for the study of solar physics.

In order to address the scientific goals of future Mercury studies, multiple
observational techniques must be employed: Question (1) requires high-quality imaging and altimetry of the entire planet. Question (2) requires multispectral imaging, orbital X-ray and gamma-ray fluorescence measurements,
and surface geochemistry experiments. Questions (3) and (4) cannot be answered definitively without high-resolution synoptic spectroscopy (particularly in the ultraviolet), as well as in situ charged-particle measurements, and in
situ mass spectroscopy. Questions (5) and (6) await in situ magnetic field and
charged-particle environment observations extending over time scales at least
as long as the Mercurian year (88 days). Complete answers to Question (7)
require surface exploration. Question (8) involves a synthesis, and will therefore rely upon information obtained from all of the aforementioned techniques,
and others.

This recounting of the key scientific questions and the methods by which
these questions can be resolved provides two insights. First, without future
spacecraft missions to Mercury, even our first-order questions cannot be fully
answered. Second, however, synoptic Earth- and space-based remote sensing
can clearly still contribute important findings, particularly in terms of improved imaging, altimetry and spectroscopy. In the remainder of this chapter,
the specific projects and programs which can improve our knowledge of Mercury are reviewed.

III. FUTURE GROUNDBASED STUDIES

Mercury must always be observed from the Earth at small solar elongation angles. At its maximum, Mercury never strays farther than 28° from the

Sun. Such close proximity makes observations difficult at best. However, as evidenced by the recent discovery of sodium and potassium in Mercury's atmosphere by Potter and Morgan (1985a, 1986a), groundbased work can still make important contributions.

Among the priorities for future groundbased work are: (1) expanded atmospheric composition and abundance studies, particularly as a function of Mercury's orbital position; (2) surface composition studies by the technique of spectrophotometry; and (3) radar observations. In addition, related activities such as radar and spectrophotometric observations of asteroids, satellites, and the Moon as well as laboratory studies of meteorites will contribute to our understanding of Mercury by placing it in the larger context of planetary studies.

Several key projects remain concerning Mercury's atmosphere. Most important is the study of spatial and temporal variations in the atmospheric abundance of sodium, potassium and any other as-yet-undiscovered constituents of the Mercurian atmosphere. The measurement of such variations over Mercury's orbital and rotational periods is vital for understanding the mechanism(s) responsible for the generation of this tenuous atmosphere (cf. Chapter by Hunten et al., and references therein). Temporal variations in Mercury's atmosphere are particularly important to understand in light of the well-known convolution of true abundance and observed abundance caused by the high radial velocities of the planet in its elliptical orbit. In addition to spectroscopy, stellar occultation opportunities (Mink 1987) could potentially provide measurements of the vertical composition, pressure, and temperature structure of Mercury's atmosphere.

Groundbased spectrophotometric observations of Mercury's surface should unlock significant information about Mercury's formation conditions. Resolving the question of the presence of Fe^{2+} in the surface mineralogy (Chapter by Vilas) would provide information about volatiles in the surface materials, and by extension, address questions concerning the extent of the feeding zone which Mercury sampled during its formation. Mercury's silicate composition could be probed with a search for the Restrahlen bands at thermal infrared wavelengths.

New groundbased instrumentation allows telescopic observations to be made across extended spectral ranges, and facilitates compensation of the effects of high airmass and bright sky background. The correlation of compositional variations with geologic units identified in Mariner 10 images, and the extension of such research to the as-yet-unimaged hemisphere of Mercury, is of particular interest.

Groundbased observations are also useful for studying surface properties and composition. However, as with atmospheric studies, such observations are complicated by Mercury's small size and its proximity to the Sun. However, spectrophotometry (Chapter by Vilas) and polarimetry (Gehrels et al. 1987) each continue to be profitable. Of particular interest would be the correlation of compositional variations with the specific geologic units mapped

by Mariner 10 (Rava and Hapke 1987), and the extension of such research to
the prediction of the as yet unmapped portions of Mercury.

Radar and radio observations can provide information about topographic
slope and altitude, surface electric properties, surface thermal properties,
spin-axis orientation, and the ephemerides of Mercury (R. Landau, personal
communication). Indeed, it was by radar techniques that Mercury's 2:3
orbital-to-rotational resonance was discovered.

Radar work has demonstrated that Mercury's surface may be less lunar-
like than once suspected. In particular, some radar data have indicated that
Mercury is comparatively smooth from an rms-slope standpoint, but anomalously rough on small scales (1–10 cm). Studies by a variety of groups (see,
e.g., Chapters by Clark et al. and Harmon and Campbell) have demonstrated
that radar-derived roughness and slope data appear to correlate with terrain
units imaged by Mariner 10. Over the next decade, radar system improve-
ments at both Goldstone and Arecibo, as well as increased latitudinal coverage will significantly improve the radar data base on Mercury, and provide our
only means of actively probing the planet without mounting a space mission.

Radio and infrared observations can also contribute to studies of the thermal, mechanical and electric properties of the surface regolith.

IV. OBSERVING MERCURY REMOTELY FROM EARTH ORBIT AND DEEP SPACE

Remote observations of Mercury from space offer certain advantages
over Earth-based studies. They can achieve broader spectral coverage and
diffraction-limited resolution and also can circumvent solar-elongation difficulties with appropriate baffling systems. In this section we demonstrate the
desirability of adapting future satellite platforms for observations of Mercury
from Earth orbit.

It is important to recognize that a few "proof-of-concept" observations
of Mercury have already been carried out from Earth orbit. Included among
these are serendipitous observations by the Skylab ATM and Solar Maximum
Mission coronagraphs. Additionally, one simple Space Shuttle payload
(CHAMPS) had planned to observe Mercury during orbital twilight periods to
obtain low-resolution ultraviolet spectra; CHAMPS was destroyed with the
Orbiter Challenger on its tragic last mission.

While thermal and telescope baffling constraints have precluded observations of Mercury by recent orbiting observatories, including Copernicus,
IRAS and IUE, several possibilities exist for near-term space-based observations of Mercury. Among the possibilities, however, only one spacecraft is
actually planning to make observations—the Hubble Space Telescope (HST).
While HST was not originally believed capable of observing at small solar
elongation angles, a study (LMSC 1984) conducted by the manufacturer
(Lockheed Missiles and Space Company) has demonstrated that *imaging* ob-

servations can indeed be made during orbital twilight periods when the Sun is occulted by the disk of the Earth.

It is estimated that HST will be able to achieve 30 to 60 km resolution at Mercury. This will permit the as yet unmapped hemisphere to be imaged, thereby revealing the full pattern of global geology for the first time. At a spatial resolution of 30 to 60 km, basins, large craters, scarps and other large-scale constructs may be recognizable. The question of hemispheric geologic asymmetries may also be addressed. Sufficient observations are planned to map the entire planet (except for certain polar regions) at 1:150,000 scale. Because HST's imaging systems include extensive filter wheels, multispectral imaging will also be possible.

While Space Telescope can make images of Mercury, it cannot obtain Mercurian spectra. This limitation derives from engineering constraints on the spacecraft's navigational capabilities. In short, HST must navigate to and guide on Mercury by means of gyros because its star trackers cannot operate in close angular proximity to the Sun. Using gyros limits the ability of the Space Telescope to place accurately its extremely small spectroscopy slits on Mercury. By contrast, the imaging apertures are quite large. Thus, while one confidently expects to obtain albedo and large-scale topographic maps from HST, it is beyond HST's capability to obtain Mercury spectra.

Concerning ultraviolet observations, it is particularly important to note that the important spectral region from 1800 to 3200 Å remains *wholly* unexplored at Mercury. The Mariner 10 spacecraft did not include this capability (Broadfoot 1976). This portion of the ultraviolet is diagnostic both to atmospheric studies (Barth 1969) and to mineralogical investigations (Wagner et al. 1987). Ultraviolet spectra of Mercury cannot be obtained by Space Telescope for the technical reasons cited above.

High-resolution ultraviolet spectra of Mercury and its atmosphere are of considerable scientific import. They would (1) permit diagnostic surface mineralogy bands to be observed (including the iron-oxide absorptions at 2600 Å), (2) establish the presence or absence of currently undiscovered constituents of Mercury's atmosphere such as Mg and Mg^+, (3) permit the time-dependent measurement of atmospheric composition, temperature and pressure profiles, (4) document the ultraviolet albedo curve of Mercury, and (5) search for aurora. Given the importance of ultraviolet spectroscopy, we now examine how such data might be obtained in the essence of missions to Mercury.

One way ultraviolet observations could be made is by using the recently proposed Earth-orbiting Planetary Telescope mission. Such a satellite, as now envisioned, would employ extensive solar baffling specifically to permit synoptic, long-term imaging and spectroscopy of Mercury and Venus, comets approaching perihelion, and certain asteroids. Both visible and ultraviolet detectors have been suggested for the payload. However, this mission is not yet approved and is therefore, at best, years away.

Another proposal for obtaining exploratory high-resolution Mercury ultraviolet spectra involves one or more flights of an ultraviolet telescope/spectrometer aboard a sounding rocket (cf., Table 1). As has been pointed out (Stern et al. 1986), a typical 300 rocket flight observation using a 16-inch telescope with modern digital array detectors could obtain Å-class surface spectroscopy in the mid-ultraviolet. The detection and equivalent-width measurement of a variety of suspected but unobserved atmospheric atoms and ions, including magnesium, could be carried out by a sounding-rocket mission. Such an experiment would also provide the ultraviolet surface albedo curve to high accuracy (<1% rms).

Similar sounding rocket experiments have successfully obtained spectra of Venus, Jupiter and bright comets. The advantages of a rocket experiment (Fig. 1) include its low cost, its rapid-response time scale (12–18 months) and its independence from the oversubscribed Shuttle and ELV manifests. By combining a twilight solar occultation and optical baffling, extremely high off-axis light rejection capabilities can be obtained. After an initial "survey" flight, reflights could (a) employ eschelle resolution spectroscopy to measure resonance-line profiles, (b) perform surface spectroscopy of the opposite hemisphere, and (c) explore the orbital-phase dependence and time dependence of atmospheric composition and magnetospheric structures.

In addition to dedicated sounding rocket or satellite missions, ultraviolet spectra of Mercury could also be obtained by adapting the ASTRO Spacelab payload. Were the ASTRO observatory to incorporate a more extensive solar baffle, it could obtain a wealth of data about Mercury and its atmosphere. ASTRO carries three meter-class ultraviolet telescopes with imaging, spectroscopic and polarimetric capabilities.

It may also be possible to observe Mercury and its environment remotely if the proposed CRAF cometary spacecraft travels inside 1 AU (thereby being able to make Mercury observations at higher elongation angles than are available from Earth, hence reducing the need for extensive solar baffles).

TABLE I
Scientific Rationale for Observing Mercury's Ultraviolet Spectrum from a Sounding-Rocket Mission

1. **Obtain a Mid-UV Survey Spectrum of Mercury:**
   1800–3000 Å is completely unexplored

2. **Surface Science Objectives:**
   Albedo as $f(\lambda)$
   Diagnostic mineralogy, including silicate and FeO absorption bands

3. **Atmospheric Science Objectives:**
   Identification of emissions from potential neutral and ionic constituents which could include: Mg, Si, Al, Fe, S

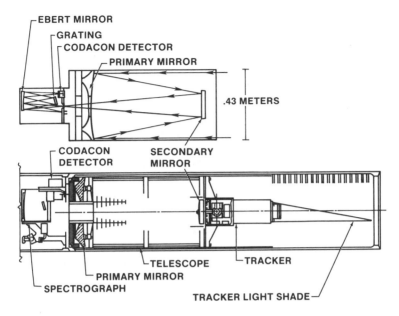

Fig. 1. Schematic diagram of proposed ultraviolet spectroscopic sounding rocket experiment for observing Mercury in 1800 to 3000 Å spectral range.

In a similar vein, we also point out that the Pioneer Venus Orbiter has been used on several occasions to observe bright objects (e.g., comets) unobservable from Earth due to poor geometry. Although this spacecraft is spin-stabilized, the Pioneer Venus Orbiter Ultraviolet Spectrometer (PVOUVS) experiment can detect Mercury Lyman-α emissions at the level detected by Mariner 10's ultraviolet spectrometer in a 12-hour integration (A. I. Stewart, personal communication). The routine monitoring of Mercury's hydrogen envelope (which is not possible from within the Earth's geocorona) would provide a unique data set in the absence of an *in situ* Mercury mission. The PVOUVS can also measure Mercury's ultraviolet albedo.

V. MISSIONS TO MERCURY

As we have pointed out, certain key questions about Mercury can be answered only by sending spacecraft to observe the planet closely and to sample its surface and its environment. Although the planetary science community has not formally endorsed missions for the study of Mercury in the past decade, the objectives for such a mission have been assessed (Jet Propulsion Laboratory Publ. 77-51, 1977).

Because the first-order questions about Mercury can only be fully answered by long-term measurements, flyby missions to Mercury are not scientifically preferable. We therefore restrict our attention in this chapter to orbiter and lander missions.

As noted, the launch and orbit insertion energy requirements necessary to mount orbiter and lander missions are severe (Yen 1985). This is due to a combination of Mercury's position deep in the solar gravitational well, the planet's small mass which reduces its gravitational braking potential, and its lack of a substantial atmosphere which could be used for aerobraking. Additionally, the spacecraft design requirements for Mercury missions are complicated by the hot near-Sun environment.

In the past, a number of technology-driven techniques for achieving the ΔV performance necessary to mount Mercury missions have been studied. Included in these studies were: extremely large launch vehicles (Friedman 1978), solar-sail propulsion systems (French and Wright 1986), and solar-electric propulsion systems (Friedlander and Feingold 1977). Unfortunately each of these techniques suffers a common weakness: they are technology driven. Neither Saturn-class launch vehicles, nor solar sails, nor solar-electric propulsion systems are available today; they are not even under development by NASA. Given Mercury's past lack of priority in the queue of potential planetary space exploration missions, we regard those missions requiring the development and flight test of new high-performance propulsion systems to be substantially less likely to occur. Since an available and less costly alternative has recently been discovered, we will not consider technology-driven approaches further; the interested reader is instead referred to the references cited above.

How might Mercury be reached using available launch vehicles and propulsion systems? Recently, Yen (1985) has discovered a family of multiple Venus-Mercury encounter trajectories which, through successive gravity assists, reduce mission performance requirements to levels deliverable by available systems, such as Titan-Centaur, Atlas-Centaur and Shuttle/TOS.

Venus gravity assists have long been known to reduce the launch requirements necessary to reach Mercury; in fact, this technique was employed by Mariner 10 to conduct its pioneering mission to Mercury in 1974. While such trajectories are useful for flybys, they are not adequate for orbiters and landers. This is because of the large orbit inject propulsion requirements at Mercury resulting from the high heliocentric approach velocity of the Venus/Mercury transfer orbits. Yen's contribution has been to find trajectories that employ repeated Mercury encounters to dramatically reduce the braking requirements for Mercury orbit injection.

For example, a single launch in July of 1994 using a Titan-Centaur combination could place a 1477-kg payload into orbit around Mercury. A schematic of a sample EV^2M^4 trajectory is shown in Fig. 2. Numerous launch windows capable of supporting the Yen-multiple gravity assist technique are available between 1991 and 2004: each opportunity offers several distinct en-route trajectories which trade transit time against injected payload mass (see Table II).

Given this remarkable set of Shuttle and ELV compatible trajectories to Mercury, it now becomes possible to "field" substantial Mercury missions almost as easily as Mars and Venus missions. We therefore next consider the candidate experiments which would be most valuable on a return to Mercury. Of course, many of the important Mercury observations are similar in nature to those proposed for the study of other solid bodies including Mars and asteroids, and most particularly the Moon.

While it is generally regarded important that a close correspondence between future lunar and Mercury payloads be maintained, the breadth of Mercury studies is greater than that of lunar studies. This is because Mercury has an atmosphere, which the Moon has not, and because Mercury is magnetized, which the Moon is not. Further, Mercury's close proximity to the Sun provides a unique vantage point for measurements of the possible time variations in the gravitational constant, for frame-dragging tests important to general relativity theory (Committee on Gravitational Physics, 1981), and for precise measurements of (or at least constraints on) the solar quadrupole moment (Committee on Solar Space Physics, 1980). Given these differences between the Mercury and lunar science objectives, it is clear that additional instruments, such as an

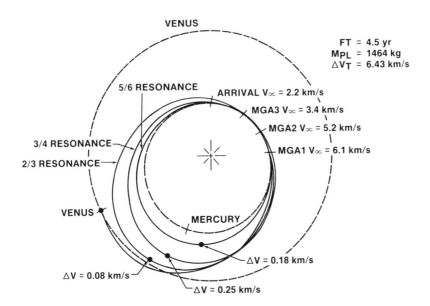

Fig. 2. Proposed EV^{2}M^4 gravity-assisted trajectory necessary to put a 1464-kg payload into 300 km circular orbit around Mercury. The payload, launched using a Titan 34D7/Centaur G', requires 4.5 yr flight time from the Earth to Mercury (figure from Yen 1986).

TABLE II
Sample Mercury Orbiter Mission Opportunities, 1990–2010[a]

Launch Date	Trajectory Type	Transit Time (yr)	Payload Configuration[b]	Payload Mass[c] (kg)
7/30/91	EV^{2}M^4	4.5	OSL^3	1347
7/23/94	EV^{2}M^4	5.6	OSL^5	1610
7/05/96	EVM^4	4.4	OS	499
7/10/96	EV^{2}M^4	5.3	OSL^3	1325
7/25/99	EV^{2}M^4	4.9	OSL^2	1027
9/07/02	EV^{2}M^4	4.5	OSL	895
6/30/04	EVM^4	4.3	OS	652
7/10/04	EV^{2}M^4	5.6	OSL^2	1053
8/05/05	EV^{2}M^4	5.6	OSL^4	1398
7/09/07	EV^{2}M^4	5.2	OSL	951

[a]Table adapted from Yen (1985).
[b]O = 600 kg orbits; S = 500 kg subsatellite; L = 200 kg lander. Exponent on L indicates the number of landers.
[c]Maximum mass carried into 300 km altitude circular orbit, after injecting a 500 kg subsatellite into a 300 km × 12 hr orbit.

ultraviolet spectrometer and precise radio-science instrumentation should be included on future missions to Mercury.

The key measurements which a spacecraft could make from orbit about Mercury are: multispectral and fluorescence spectroscopy composition maps, complete surface mapping and altimetry, magnetospheric studies, atmospheric composition and dynamics observations, dust environment studies, radiometry observations, and the determination of the first few gravitational harmonics (by radio tracking). Although a Mercury orbiter would also be useful to gravitational physics, these relativity experiments are best suited to a small stable-orbit free-flying subsatellite which can be accurately ranged upon from Earth (Vincent and Bender 1988).

To exploit most fully the capabilities of an orbiting spacecraft, a high inclination (e.g., polar) orbit is required. In combination with high inclination, a circular low-altitude orbit is preferable for surface composition and geoid mapping experiments. For magnetospheric and atmospheric studies, however, a high-altitude (apoapsis $\gtrsim 3\ R_M$) elliptical orbit is desired. This higher orbit is also preferable from the standpoint of spacecraft thermal design. A mission compromise would be to insert into high orbit for 88 to 176 days, and then to lower the orbit for the remainder of the mission.

In Table III, we list the components of a Mercury-orbiter payload designed by European investigators (Neukum 1985) to study surface geology and geochemistry, atmospheric composition and structure, the local particle and fields environment, and solid-body rotation dynamics. Experiments to address tests of relativity would augment this list (see Bender et al. 1986). We

TABLE III
Proposed Mercury Science Payload[a]

Instrument/Experiment	Scientific Objective	Est. Weight (kg)
Gamm-Ray and Solar Neutron Spectrometer	Chemical abundance of K, U, Th, O, Mg, Al, Si, Ca, Fe, possibly Na, Ti, Cr, Cl. Measurement of solar and cosmic gamma ray lines and continuum, solar neutrons	13
X-Ray Spectrometer	Chemical abundances of Mg, Al, Si, Fe	8
Visual and Infrared Mapping Spectrometer	Mineralogical composition: olivine, pyroxene, plagioclase, Fe and Ti abundance	10
Multispectral Imager	Surface structure/morphology, spectral/compositional mapping	6
Infrared Radiometer	Thermal properties: thermal conductivity and temperature gradient in the regolith, additional information on composition of surface materials	4
Microwave Detector	Heat flow in combination with IR radiometer: thermal conductivity, thermal gradient	13
Altimeter	Topography: necessary for gravimetry experiment	10
Gravity Experiment	Internal structure, general relativity, gravitation	—
Ultraviolet Spectrometer	Composition of atmosphere: He, O, Xe, Ar, Ne	6
Magnetometer	Determination of the 3 elements of the magnetic field at different latitudes, longitudes, altitudes	2.5

*Adapted from Mercury Polar Orbiter proposal to ESA (Neukum 1985).

note that the solar neutron spectrometer included for the purpose of surface compositional studies also has application to the study of solar flares and could also measure low-energy solar beta decay events unobservable from 1 AU (Cooper 1986). A detailed discussion of the science objectives of a Mercury Orbiter is given by Neukum (1985).

Finally, we point out the importance of landers and rovers on Mercury. The important measurements which landing vehicles could provide are: *in situ* surface composition, surface mechanical properties, internal activity (seismic and thermal), total atmospheric pressure and fractional abundance measurements, as well as local seismic profiling, and an assessment of the surface magnetic field and charged-particle environment. The emplacement of the lander's radio beacon on Mercury would also permit the amplitude and period of the planet's physical libration to be determined. By combining libration data, with orbital measurements of the gravitational harmonics and spin obliquity, it is possible to determine definitively the size of Mercury's molten core (see Peale's Chapter). However, given the historically high cost of planetary landers, an attractive alternative for Mercury exploration might be the emplacement of a penetrator network (Friedlander and Davis 1976). Because they are *impact* landing devices, penetrators are inherently simpler, lighter and less costly than soft landers. A network of three or more penetrators could provide a capable magnetic, seismic and heat-flow array for geophysical studies of Mercury's interior.

VI. SUMMARY

Mercury occupies an important place in planetary science. At present, however, our knowledge of Mercury is perhaps more primitive than of any other planet save Pluto. Mercury is difficult to observe because of its perpetual angular proximity to the Sun, and difficult to reach because of its position deep in the solar gravitational well. Still, its study from the Earth and space is important.

Valuable remote observations remain to be performed from the Earth and from Earth orbital spacecraft equipped with suitable optical baffles. For the near term, two key projects appear to carry the greatest potential for the improvement of our knowledge about Mercury from space; these are:

1. High resolution ultraviolet spectroscopy;
2. Large-scale global structural and multispectral mapping.

Because ultraviolet observations hold high promise for Mercury, low cost adaptations of existing space instruments (e.g., ASTRO) or the commission of a sounding rocket program to explore Mercury's ultraviolet spectrum appears to be a key future requirement. Additionally, continued radar/microwave experiments and long-term spectroscopic monitoring of Mercury's atmosphere by groundbased observers will continue to provide important data.

For the longer term, spacecraft missions to Mercury *must* be undertaken if we are to understand the planet more fully. Flyby missions do not appear to be capable of providing the synoptic observations required to complete the exploration of Mercury. An observer-class polar orbiter, however, can satisfy the key objectives of a moderately comprehensive exploration program. Such missions are feasible with the current stable of launch vehicles.

Acknowledgments. Conversations and critiques which improved our manuscript were provided by M. Davies, L. Esposito, R. Landau, D. Siskind, H. Smith, A. Friedlander and W. Mendell. It is a pleasure to acknowledge A. Alfaro's assistance in the preparation of this manuscript.

PHOTOMETRY AND POLARIMETRY OF MERCURY

J. VEVERKA and PAUL HELFENSTEIN
Cornell University

BRUCE HAPKE
University of Pittsburgh

JAY D. GOGUEN
Jet Propulsion Laboratory

Application of Hapke's (1986) photometric function to observations of Mercury and the Moon confirms the remarkable similarity of photometric properties. Mercury's average photometric behavior is matched best by that of lunar terrains commonly classified as average on the basis of their albedos. In detail, the Mercurian regolith appears to differ from its lunar counterpart by virtue of (1) slightly less backscattering particles, (2) a possibly higher state of compaction, and (3) an apparently stronger opposition surge. However, the average values of the large-scale roughness parameter $\bar{\theta}$ are very similar. An interesting (and probably significant) difference is that while the highland/mare albedo ratio is about 2 on the Moon, it is only about 1.4 on Mercury. Yet the brightest areas on Mercury are much brighter than their lunar counterparts, reaching reflectances of 0.45, compared to maximum values of < 0.3 on the Moon. Although the polarimetric properties of Mercury and the Moon are basically similar, Mercury shows slightly lower values of P_{max} (the maximum positive polarization observed), a difference that is consistent with either a slightly higher average reflectance for Mercury, or with a difference in regolith texture, or both. Dollfus and Auriere found that Mercury's polarization properties can be matched best by Apollo soil samples of higher than average albedo. Infrared, microwave, and radar measurements of Mercury indicate that the thermal and electrical characteristics of Mercury's regolith (specifically: thermal inertia, dielectric constant,

*loss tangent, radar reflectivity, and the range of variation of these parameters)
are nearly identical to those of the lunar regolith. Observationally there is a
critical need of disk-integrated and disk-resolved photometry of Mercury at
phase angles below 50°.*

In this chapter we review information about the photometric and polarimetric properties of Mercury's surface, concentrating on data concerning the
nature of the regolith and emphasizing comparisons with the Moon. Specifically, we investigate to what extent the photometric characteristics of regions
on Mercury are similar to those of the Moon. How similar are the average
albedos, and the distributions of reflectances? Our analyses for Mercury and
the Moon are carried out entirely in terms of Hapke's photometric function
(Hapke 1981, 1984, 1986). However, we compare our conclusions with those
reached earlier by Lumme and Bowell (1981b) and Lumme and Irvine (1982),
among others.

We also investigate the extent to which polarimetry sheds further light on
regolith characteristics. Dollfus and Auriere (1974) have argued that polarimetry indicates a slightly higher average albedo for Mercury than for the
Moon, and possibly suggests a slightly different particle size distribution
and/or texture for the Mercurian regolith. Inferences based on other remote
sensing techniques, i.e., infrared radiometry, microwave, radio measurements
and radar studies, are summarized briefly in the context of the photometric
and polarimetric results.

I. PHOTOMETRIC OBSERVATIONS OF MERCURY

A. Disk-Integrated Results

Summary of Observations.   Because of its proximity to the Sun, Mercury is a very difficult object to observe from Earth. Due to this factor, we still
rely on visual observations of this planet and have very little information
about its brightness at low phase angles (at low phase angles Mercury is near
superior conjunction and therefore behind the Sun as seen from Earth).

The first useful visual observations of Mercury's phase curve were made
in the 19th century by Zollner (1865) and Muller (1893). As reviewed by
Harris (1961), these measurements were superseded by the admirable efforts
of Danjon (1933,1949,1953) who obtained observations ranging in phase angle from 3° through 123°. Danjon also showed that in the region of overlap his
results agreed well with those obtained by Muller between 1878 and 1888.
His phase curve reproduced in Fig. 1 can be fitted by a cubic equation

$$\Delta m(\alpha) = 3.80 \left(\frac{\alpha}{100°}\right) - 2.73 \left(\frac{\alpha}{100°}\right)^2 + 2.00 \left(\frac{\alpha}{100°}\right)^3 \qquad (1)$$

where α is the phase angle expressed in degrees. Note that this is an arbitrary
magnitude system. Various attempts have been made to convert this magni-

tude system to the V system of the standard UBV set. One of the most detailed of these by de Vaucouleurs (1964) concluded that Danjon's

$$m(\alpha = 0°) = -0.21 \tag{2}$$

should be equated to

$$V(1,0) = -0.42 \tag{3}$$

where $V(1,0)$ is the conventional absolute magnitude reduced to $\alpha = 0°$ and planet-Sun and planet-Earth distances of 1 Astronomical Unit.

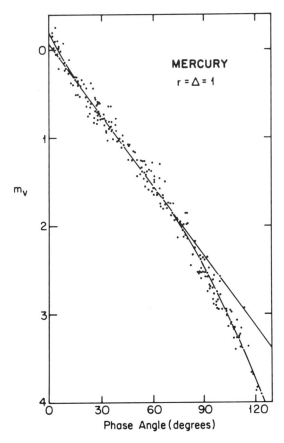

Fig. 1. Observed phase curve of Mercury from Danjon (1949). Shown are both Danjon's cubic fit to the data, and the best straight-line fit.

Of necessity, in what follows we make heavy use of Danjon's results. Therefore, a fundamental issue is: how reliable are visual estimates of the brightness of a planet? There are two important clues that Danjon's data were obtained with great care and are therefore worthy of detailed analysis. First, photoelectric, disk-integrated measurements of Mercury between phase angles of 58° and 115°, obtained by Irvine et al. (1968), agree very well with Danjon's data using the de Vaucouleurs conversion of Danjon's measurements to the *UBV* system (Fig. 2). Second, Danjon's results are consistent with those of Ratier (1972), who used a coronagraph to measure the brightness of Mercury during the superior conjunction of 15–16 Nov. 1969. In Danjon's magnitude system he found that at $\alpha = 0°58'$

$$m = -0.14 \pm 0.2 \tag{4}$$

to be compared with $m = -0.17$ derived by extrapolating Danjon's cubic formula to this phase angle. Clearly, the agreement is excellent. Converting to the V system, the apparent V magnitude at $\alpha = 1°$ is estimated by Ratier (1972) to be $m_v(\alpha = 1°) = -0.31 \pm 0.1$.

Early Interpretations.   The close similarity of the phase curves of Mercury and the Moon, as well as the almost identical average geometric and

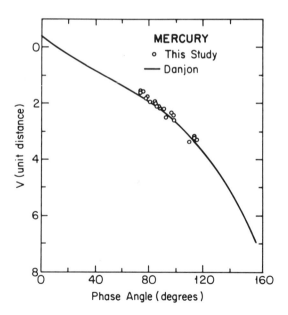

Fig. 2. Visual phase curve of Mercury. Observations of Danjon as reduced to UBV by de Vaucouleurs (1964) represented by solid line; open circles from Irvine et al. (1968). (Figure from Irvine et al. 1968.)

Bond albedos indicate that the surface of Mercury must be very similar to that of the Moon in terms of overall photometric characteristics. This realization is evident in all late 19th century and 20th century discussions of Mercury. One of the first attempts to model the photometric characteristics of Mercury was made by Hameen-Antilla et al. (1965). While the particular conclusions of that paper are of no more than historical interest today, the authors made a strong case that interpretations of phase curves such as those of the Moon and Mercury must include an accounting for the presence of topography and large-scale roughness. This argument had already been made very convincingly in the case of lunar photometry by Schoenberg (1929), but remained unappreciated for more than thirty years.

A more detailed attempt to model the phase curve of Mercury was made by Pikkarainen (1969). Pikkarainen dealt with large-scale roughness using the approach developed by Hameen-Antilla et al. (1965), in which an arbitrary fraction of the surface is assumed to be covered with paraboloidal holes (rimless craters) of a specific depth-to-diameter ratio (actually, this is a generalization of the hemispherical hole model used by Schoenberg in his lunar studies). Assuming that the photometric function is given by Lommel-Seeliger's law, with the phase function determined by laboratory measurements on dark basalt samples, Pikkarainen succeeded in obtaining respectable fits to both the lunar and Mercurian phase curves. He showed that Rougier's lunar phase curve could be matched by a model in which 20% of the Moon's surface is covered with holes whose depth-to-diameter ratio is 1:0, and that Danjon's Mercury observations implied a surface of which 50% was covered by holes whose depth was 0.25 of the diameter. Pikkarainen presented rationalizations to justify his conclusion that Mercury is more cratered than the Moon, but that its craters are more eroded. A retrospective verdict is that Pikkarainen's results are non-unique and model dependent; his rationalizations are inconclusive and almost certainly incorrect.

Lumme and Bowell (1981a,b) published a semi-empirical photometric theory which they applied to available disk-integrated observations of solar system objects. The theory attempts to account for the photometric effects of texture due to "porosity", and to surface "roughness". Lumme and Irvine (1982) applied the theory in detail to lunar data, but so far its only application to Mercury is that in Lumme and Bowell (1981b), where the authors analyze disk-integrated lunar and Mercurian data over a limited range of phase angles (less than 25°) and reach the following conclusions: "Data for the Moon, Mercury, and Deimos fall exactly in the asteroid domain . . . implying that both physical properties (roughness and volume density) and the optical properties (characterized by the asymmetry factor g) are similar." The authors conclude that the Moon, Mercury and the asteroids all have similar porosity and roughness properties, as well as similar single-particle phase functions. In their view, differences in reflectances, phase coefficients, etc., are predominantly caused by differences in the effective single particle scattering albedo of the regoliths. Implied in this interpretation is the idea that all these regoliths experienced approximately similar geologic processes, both in terms of determining microscopic texture, and in producing large-scale topography.

New Analysis in Terms of Hapke's Function.   An ultimate aim of planetary photometry is to identify geologically interpretable surface physical properties. Hapke's (1981,1984,1986) application of radiative transfer theory has yielded a successively refined suite of equations representing the most recent and rigorous effort to describe bidirectional reflectance in physically meaningful terms. In Hapke (1984), bidirectional reflectance is modeled in terms of five parameters: the average single scattering albedo of a particle w; first and second order coefficients b and c of a Legendre polynomial representation of the scattering phase function of an average particle; a semi-empirical regolith compaction parameter h describing the angular width of the opposition surge; and an average topographic slope angle $\bar{\theta}$ describing the macroscopic surface roughness at subresolution scales (Helfenstein 1987). For $\bar{\theta} = 0°$, the equations in Hapke (1984) reduce to the description formulated for a smooth, particulate surface in Hapke (1981).

Values of these parameters for Mercury ($w = 0.25$, $h = 0.4$, $b = 0.579$, $c = 0.367$, $\bar{\theta} = 20°$) derived by Hapke (1984) predict reflectances consistent with both disk-integrated data of Danjon (1949) and disk-resolved scans from calibrated Mariner 10 images. Hapke's values of w and h are identical to lunar values reported by Buratti (1985). Buratti did not derive a lunar value of $\bar{\theta}$, and used the Henyey-Greenstein asymmetry factor g in place of Hapke's b and c parameter. An equivalent Mercurian value of the asymmetry factor, $g = -b/3 = -0.19$ is quite similar to Buratti's lunar $g = -0.25$, especially in view of the probable uncertainty in the data sets from which these results were derived.

Hapke (1986) offers a physically motivated replacement for his earlier, semi-empirical description of the opposition surge. In the new description, the total amplitude of the opposition effect B_0 is shown to depend on w, b and c, and a new zero-phase amplitude term $S(0)$ related to the refractive indices and microstructure of average regolith particles. This approach differs from that in Hapke (1981,1984) where B_0 was strictly an empirical function of w. In addition, the new description provides a more rigorous definition of the compaction parameter h in terms of the compaction state of the surface, the rate of change of compaction with depth, and the particle size distribution in the regolith.

The improved ability of Hapke (1986) to isolate specific regolith physical properties from different aspects of photometric behavior yields a more complete and accurate framework for the comparison of planetary surfaces. The new $S(0)$ term, for example, may identify differences in average particle microstructure. It is essential to note that the parameters h which occur in both versions of Hapke's theory are defined differently in the two cases; therefore numerical values of the new Hapke compaction parameter cannot be compared directly with values obtained using the older theory. However, since the new *h* is derived more rigorously, its geological interpretation should be more meaningful. Fitting to observed data Hapke's earlier theory required that in choosing *w* a compromise be made between its function in describing the amplitude of the opposition surge and its broader role in describing multiple scattering and hence the overall brightness of the surface at any phase angle. Since *w* is less strongly coupled to the opposition surge in the new equation, it can be estimated more accurately. Improved estimates of *w* can, in turn, result in more accurate estimates of the remaining parameters.

In the present study, the model of Hapke (1986) was applied to disk-integrated observations of Mercury (Danjon 1949; Ratier 1972) and of the Moon (Lane and Irvine 1973; Shorthill et al. 1969). In each case, values of the parameters were sought to minimize the summed squares of differences between observed disk-integrated brightnesses and those predicted from Hapke's equation for corresponding illumination and viewing geometries (for an outline of the method, see Helfenstein and Veverka [1987]).

The reliability of results depends strongly on the absolute calibration of the data. For planetary objects absolute brightnesses involve assumptions about the absolute brightness of the Sun. Modern measurements of this parameter are only accurate to about 0.1 magnitude, and thus constitute a significant source of uncertainty. In the present study, Mercurian and lunar disk-integrated brightnesses were calibrated assuming a value of $M_\odot = -26.74$ in the *V* filter of the *UBV* system (Johnson 1965; Allen 1976).

Hapke parameters derived from our analysis of the disk-integrated phase curves of Mercury and the Moon are listed in Table I. The corresponding model phase curves are compared with observations in Fig. 3*a,b*.

TABLE 1
Photometric Parameters for Mercury and the Moon

	w	*h*	B_0	*b*	*c*	$\bar{\theta}$
Disk-integrated						
Mercury[a]						
Solution 1	0.20	0.11	2.4	0.20	0.18	21°
Solution 2	0.23	0.09	2.5	0.18	0.15	25°
Moon	0.21	0.07	2.0	0.29	0.39	20°
Disk-resolved (Lunar)						
Dark	0.13	0.05	2.6	0.27	0.20	12°
Average	0.23	0.07	2.2	0.34	0.31	19°
Bright	0.29	0.06	2.3	0.35	0.29	20°

[a]Solution 1: fitting *I/F* data on linear scale (cf. Fig. 3).
Solution 2: fitting on linear scale for $\alpha \leq 20°$; on magnitude scale at higher phase angles.

Note that in the case of Mercury two nearly identical solutions are given in Table I. Solution 1 was obtained by fitting the phase curve on a linear I/F scale as plotted in Fig. 3. Solution 2 involved fitting the near opposition portion ($\alpha \leqslant 20°$) on a linear scale and the remainder on a magnitude scale. The second approach retains maximum detail in the opposition portion, while reducing the sensitivity to errors at high phase angles where light levels are low. Obviously, for perfect data the results of the two solutions should be identical, and indeed they turn out to be very similar. From a practical aspect, the differences between the two solutions provide minimum estimates of uncertainties in the parameters derived. In what follows we adopt Solution 2 as the basis for our discussion of Mercury.

The new values of w, $\bar{\theta}$, b and c for Mercury can be compared directly with earlier determinations by Hapke (1984). Hapke's $\bar{\theta} = 20°$ estimate is similar to our values, as is his result of $w = 0.25$. The new values of b and c suggest a less backscattering single-particle phase function. This difference can be attributed directly to the different manner of treating the opposition surge in Hapke's 1984 model and in his 1986 work which forms the basis of our analysis.

A fundamental question concerns the degree to which both the particle phase function $p(\alpha)$ and $\bar{\theta}$ can be determined separately from disk-integrated data; both of these parameters influence the shape of the phase curve and to some extent can be traded off against each other. Although it is impossible to demonstrate that our solution is a unique solution, we can show readily that it not only fits the disk-integrated observations but also reproduces satisfactorily specific disk-resolved scans from Mariner 10 images (see Fig. 6 and discussion in Sec. I.C below).

Comparison of Fig. 3a and b confirms the overall similarity of lunar and Mercurian disk-integrated photometric behavior. Table I shows that both objects have nearly identical values of $\bar{\theta}$, and similar single-scattering albedos. Photometrically the two regoliths appear to differ primarily in their particle phase functions and opposition surges. Both are dominantly backward scattering; however, the asymmetry in backscattering is greater for the Moon. The lunar asymmetry factor derived from our results ($g = -0.1$) is identical to that reported by Lumme and Irvine (1982). The values of B_0 from Table I suggest that the Mercurian opposition effect may be slightly stronger than that for the Moon. The h parameter is significantly larger for Mercury, suggesting that the Mercurian regolith is more compacted, a difference which can be considered consistent with Mercury's higher gravity. Alternatively, a slightly different particle size distribution could account for the larger value of h. It is important to note, however, that the derived value of h is sensitive to small differences in the distribution of data points at small phase angles. If the single data point of Ratier (1972) is ignored, a value of $h = 0.08$ best matches the angular width of the opposition surge. Within observational uncertainty, this latter value is indistinguishable from the lunar case. A more reliable determination of h

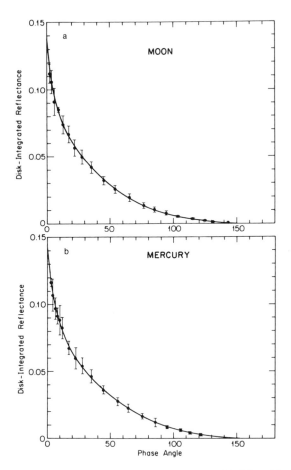

Fig. 3. (a) Hapke's function fitted to disk-integrated phase curve of the Moon. The lunar data are from Shorthill et al. (1969) and Lane and Irvine (1973). (b) Hapke's function fitted to Danjon's phase curve for Mercury.

cannot be made without more accurate low phase angle observations of Mercury.

The disk-integrated lunar data correspond to observations of the Earth-facing hemisphere, which is heterogeneously divided into highlands terrain materials and the lower-albedo maria. The photometric properties of Mercury are thus being compared to the behavior of an areal average of several diverse terrains. It is more useful to identify a particular lunar terrain that best characterizes average Mercurian photometric behavior (see below).

B. Disk-Resolved Observations

Pre-Spacecraft Observations of Mercury. Earth-based studies of the brightness distribution on Mercury have been very limited: typical Mercury images are only 6″ to 8″ across, and are seriously affected by smearing due to seeing. All such studies to date involve special efforts to allow for the effects of seeing. All agree that as far as one can tell, the brightness distribution across Mercury's disk is indistinguishable from that on the Moon at a particular phase angle.

The earliest study, based on Pic du Midi photographs, is that of Hameen-Antilla (1967), who concluded that in terms of surface photometry, Mercury and the Moon "both . . . have identical surface materials." A similar study by Veverka (1970), based on photographs obtained in 1967 at the McDonald Observatory by Morrison and Veverka, concluded that, at least in the phase angle range covered by the observations (70° to 122°), the brightness distribution on Mercury is closely similar to that of the Moon.

The most detailed study of this type is that of Beebe and Herzog (1975), who found that the brightness across Mercury images at phase angles of 31°, 55°, and 92°, could be fitted well using a lunar photometric law (Fig. 4). The

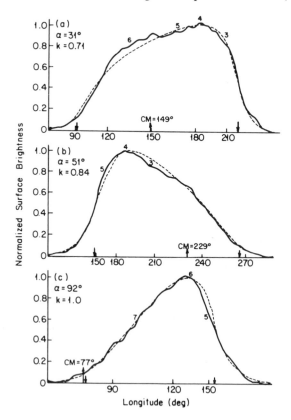

Fig. 4. Brightness scans across Mercury obtained by Beebe and Herzog (1975) compared with a lunar photometric function indicated by dashed line. The numbers identify individual albedo features on the planet.

study also produced lower limits on contrast ratios between bright and dark areas on the planet (greater than 10 to 20% at 630 nm).

Mariner 10 Mission Results.   Mariner 10 obtained some two thousand images of Mercury, but only a small fraction has been analyzed photometrically. The phase angle coverage is restricted to 80° to 110°. Thus, the spacecraft data contain valuable information on regional variations of photometric properties, and provide good constraints on certain parameters such as the large-scale roughness parameter $\bar{\theta}$. However, they provide no information on the phase curves of the surface materials near opposition.

The first conclusion from Mariner 10 is that Mercury's photometric characteristics are indeed very similar to those of the Moon, as anticipated from Earth-based observations. Second is that the range of albedos found on Mercury exceeds that on the Moon, apparently mostly due to the presence of unique bright patches in some large craters (see below). In their preliminary report, Murray et al. (1974) found that on Mercury the brightness distribution at a particular phase angle is "virtually identical with . . . lunar data," a conclusion borne out by the subsequent detailed analysis of Hapke et al. (1975). These authors found that the brightness distribution across Mercury at phase angles of 80° and 100° is closely similar to that observed on the Moon, a result in agreement with our analysis of Mariner 10 data at $\alpha = 77°$ (Fig. 5a). They estimated normal albedos (actually measured at 5° phase) to range from 0.09 to 0.45[a]. All of the highest albedo values (0.26 to 0.45) occurred in association with craters (see below). Hapke et al. found that the heavily cratered plains have an average albedo of about 0.17 (a value similar to the lunar 0.16), but that the darkest plains on Mercury are brighter than their lunar counterparts. While a representative albedo for lunar mare is about 0.09, the smooth plains on Mercury tend to have albedos of about 0.13. Thus, while the highland/mare albedo ratio is almost a factor of 2 on the Moon, it is only about 1.4 on Mercury. These conclusions are consistent with those reached on the basis of additional comparisons of Mariner 10 photometry of Mercury with lunar results by Hapke (1977).

Dzurisin (1977) identified morphologically and photometrically anomalous bright patches associated with certain large craters on Mercury, and estimated normal reflectances ranging from 0.39 to 0.45 (some 60% brighter than the brightest feature on the Moon, the crater Aristarchus which has a reflectance of 0.26). Dzurisin concluded that these bright patches constitute a morphologically and photometrically distinct class of surface features with seemingly no lunar counterparts. He speculated that the bright deposits may be due to physico-chemical activity, presumably referring to possible fumarolic activity along impact-induced fractures. The bright patches have drawn com-

---

[a]Using the parameter for Mercury given in Table I, we find that all values of "normal albedos" given in Hapke et al. (1975) should be increased by a factor of 1.35 to bring them from $\alpha = 5°$ to $\alpha = 0°$.

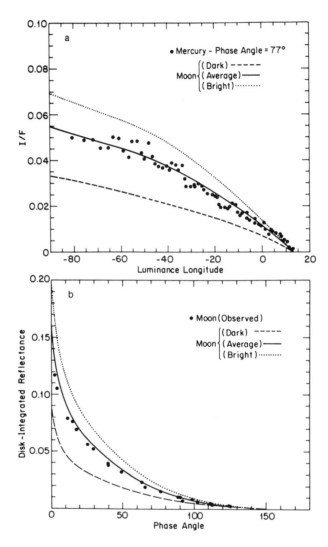

Fig. 5. (a) Comparison of the brightness distribution across Mercury at a phase angle of 77°, with lunar data for three classes of areas. The lunar curves are derived from fitting Hapke's function to the lunar data of Shorthill et al. (1969); the Mercury scan is from Hapke et al. (1975). (b) Comparison of the disk-integrated phase curve of the Moon from Lane and Irvine (1973) with synthetic curves generated for three types of lunar areas based on the work of Shorthill et al. (1969).

ments from Schultz (1977), who suggested that they "may represent intensely shock-altered crustal material that remain exposed on the floor, walls and central peaks. . .".

A Summary of Lunar Data.   Reliable disk-resolved photometric measurements of lunar terrains have been made by several investigators (see, e.g., Fedoretz 1952; van Diggelen 1959; Gehrels et al. 1964; Shorthill et al. 1969). In the detailed study of Shorthill et al. (1969), lunar terrains are divided into three classes on the basis of their normal albedos, A_n: dark terrains ($0.06 \leq A_n \leq 0.09$), average terrains ($0.10 \leq A_n \leq 0.12$), and bright terrains ($0.13 \leq A_n \leq 0.16$). Helfenstein and Veverka (1987) have derived Hapke photometric parameters for examples of these terrains. Some preliminary results are reported in Table I.

Intercomparisons of the lunar terrains show that they differ primarily in their single scattering albedos and macroscopic roughnesses. Average and bright lunar terrains appear to differ principally in terms of w. Dark (mare) materials have lower values of w, are significantly smoother, and possibly have a stronger opposition surge than average and bright lunar terrains. Mercurian values of the photometric parameters most closely match those of average albedo lunar materials. A demonstration of this fact is given in Fig. 5a, which shows a brightness scan at constant luminance latitude obtained from a calibrated Mariner 10 image, compared with model scans predicted using the lunar parameters given in Table I.

C. Summary of Photometric Results

Basic Photometric Quantities.   An early summary of Mercury's photometric properties is provided by Harris (1961). In Table II we summarize available data on the geometric albedo p, the phase integral q, and the spherical albedo pq, all in the V filter of the UBV system. In the case of Mercury, the Harris (1961) and de Vaucouleurs (1964) data are based on the observations of Danjon (1933,1949,1953). The Dollfus and Auriere (1974) value is an estimate derived from Mercury's polarization curve (cf. Sec. II). Note that Irvine et al. (1968) did not derive values of p or q independently of the Danjon data, because their observations were restricted to phase angles between 58° and 115°. Our values for the parameters in Table II are derived directly from our fits to the Danjon phase curve, using the de Vaucouleurs conversion of those data to the V system (Fig. 2). In the case of the Moon, we compare our results, obtained from the fit shown in Fig. 3a, with those of Harris (1961) and Lane and Irvine (1973). For Mercury, we quote numbers for each of the two solutions discussed above; for practical purposes the two approaches give identical values of p and q.

In Table III we summarize information on the radiometric Bond albedo of Mercury and Moon. We concentrate on values derived directly from photometry. In the case of the Moon, all previous work is superseded by that of

TABLE II
Basic Photometric Parameters

	p_v	q_v	A_v
Moon			
Harris (1961)	0.115	0.585	0.067
Lane and Irvine (1973)	0.113	0.611	0.069
Lumme and Irvine (1982)	0.152	0.476	0.072
This work	0.136	0.451	0.061
Mercury			
Harris (1961)	0.100	0.563	0.056
de Vaucouleurs (1964)	0.104	0.56	0.058
Dollfus and Auriere (1974)	0.130[a]	[0.56][b]	[0.073]
This work			
Solution 1	0.140	0.473	0.066
Solution 2	0.138	0.486	0.067

[a]Evaluated at $\alpha = 5°$.
[b]Assumed following Harris (1961) and de Vaucouleurs (1964).

Lane and Irvine (1973), who determined p and q as a function of wavelength, between 350 and 1000 nm, and used these values to calculate the radiometric Bond albedo shown in Table III. In the case of Mercury, we lack sufficient information to derive the wavelength dependence of q, and as far as p is concerned, an adequate first approximation is that the wavelength behavior is similar to that of the Moon (see Harris 1961; Chapter by Vilas). Thus, a reasonable first-order assumption is that

$$\left(\frac{A_B}{A_V} \right)_{\text{☿}} \simeq \left(\frac{A_B}{A_V} \right)_{\text{☾}} \tag{5}$$

For the Moon we use the data of Lane and Irvine (1973): $A_V = 0.069$ and $A_B = 0.123$. Taking our value of $A_V = 0.067$, for Mercury from Table II (Solu-

TABLE III
Estimates of Bond Albedo

	p_v	q_v	A_B
Moon[a]	0.113	0.611	0.123 ± 0.002
Mercury[b]	0.138	0.486	0.119

[a]Values from Lane and Irvine 1973.
[b]This work.

tion 2), we arrive at A_B = 0.119 for Mercury, using the relationship above (Table III).

Discussion of Photometric Results. While our Hapke parameters for Mercury are derived from disk-integrated data, it is easy to show that they also reproduce disk-resolved scans from Mariner 10 images (Fig. 6). This agreement between the model and disk-resolved data is a strong additional indication that $\bar{\theta}$ in the range of 20° to 25° is a correct average value for Mercury.

The application of Hapke's (1987) photometric function to Mercury and the Moon confirms their remarkably similar photometric properties. Lunar terrains classified by Shorthill et al. (1969) as average terrain materials characterize best Mercury's average photometric behavior (Table I; Fig. 5a,b). The Mercurian regolith appears to differ from its lunar counterpart by virtue of (1) slightly less backscattering particles, (2) a possibly higher state of compaction, and (3) an apparently stronger opposition surge. The average values of the large-scale roughness parameter $\bar{\theta}$ are similar.

An interesting (and probably significant) difference between Mercury and the Moon first noted by Hapke et al. (1975) is that while the highland/mare albedo ratio is about 2 on the Moon, it is only about 1.4 on Mercury. Yet the brightest areas on Mercury are much brighter than their lunar counterparts, reaching reflectances of 0.45, compared to maximum values of < 0.3 on the Moon.

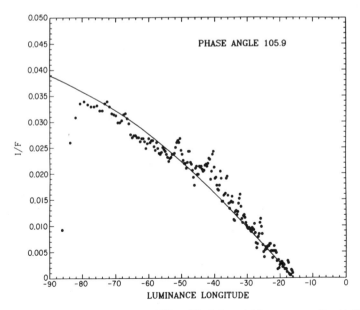

Fig. 6. Comparison of Mariner 10 data at 105°9 and Hapke's model for parameters listed in Table I (Solution 1). Comparable agreement between model and observation occurs at α = 77° to 110°.

II. POLARIZATION PROPERTIES OF MERCURY

Polarization measurements of Mercury began with the work of Bernard Lyot in 1929–30 and extended up until the Mariner 10 encounter in the early 1970s. A thorough review of all this work, except for that recently done by Gehrels et al. (1987), is provided by Dollfus and Auriere (1974). A major focus of polarimetric observations of the planet was the attempt to place limits on the density of any Mercurian atmosphere, but subsequent investigations have made it evident that any atmosphere that the planet may have is so tenuous as to be invisible to polarimetric techniques (see Sec. II.B below). Thus, today the importance of the accumulated polarimetric data is in terms of the information that they provide on the general properties of the regolith.

A. The Work of Lyot and More Recent Results

Lyot made his pioneering measurements using a visual polarimeter of his own design during the late 1920s. He found that Mercury's disk-averaged polarization is almost identical in its dependence on phase angle to that of the Moon, but did comment on the fact, substantiated by later observations, that P_{max} (the maximum positive polarization observed) is slightly lower for Mercury (Lyot 1929,1930). Such a difference would be consistent with either a slightly higher average reflectance for Mercury, or with a difference in regolith texture, or both. Lyot concluded that like the Moon, Mercury must be covered with a layer which in terms of both texture and optical properties must be similar to dark volcanic ash.

The work of Lyot was extended by Dollfus in France and by Gehrels and Coyne in the United States. Selected highly accurate measurements were also made by Ingersoll (1971), with the specific purpose of placing limits on the density of any putative Mercurian atmosphere.

The work of Dollfus (see summary in Dollfus and Auriere [1974] and Fig. 7) involved both visual measurements aimed at detecting polarization differences across Mercury's disk and photoelectric observations of disk-integrated light over a significant range of wavelengths (310 to 610 nm). Dollfus found evidence that darker areas on Mercury polarize light more strongly than brighter ones (Fig. 8). He estimated that at $\alpha = 5°$, the results were consistent with a reflectance of 0.144 for the brighter regions, and 0.122 for the darker areas. However, he stressed that due to the very limited effective resolution of such observations, this contrast difference (about 15%, compared with 50% for highland/mare contrasts on the Moon) must be considered a lower limit.

Dollfus and Auriere (1974) review evidence that the negative branch of the polarization curve is a sensitive indicator of surface texture. They make two important points: first, Mercury's negative branch is very similar to that of the Moon, indicating that, to first order, Mercury is covered with a regolith whose optical properties and texture are similar to those of the lunar surface. Their second point is that in spite of this overall similarity, Mercury's negative branch

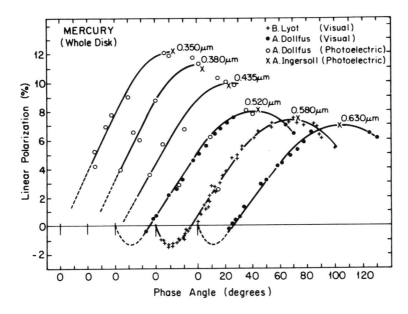

Fig. 7. Summary of polarization measurements of Mercury (whole disk) from Dollfus and Auriere (1974). The data of Gehrels et al. (1987) are not included.

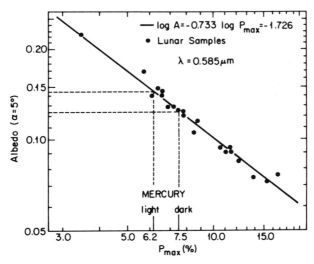

Fig. 8. Empirical relationship between the maximum linear polarization (P_{max}) and the albedo of
a surface used by Dollfus and Auriere to estimate the albedos of areas on Mercury. (Figure from
Dollfus and Auriere 1974.)

is slightly shallower than that of the Moon, but almost identical to some individual Apollo lunar soil samples such as 15291.55 (Fig. 9).

Dollfus and Auriere (1974) also discuss the wavelength dependence of P_{max} and conclude that in general P_{max} is lower for Mercury than for the Moon (consistent with Lyot's result) and that Mercury's polarization does not increase as steeply with decreasing wavelength as does that of the Moon (Fig. 10). These observational conclusions are very similar to those reached by Gehrels et al. (1987) on the basis of independent observations (Fig. 10). Dollfus and Auriere argue that the P_{max} observations for Mercury can be matched by a mixture of lunar soil samples having a reflectance of about 0.13 at $\alpha = 5°$. Thus, the polarization data would make Mercury's average reflectance about 20% higher than suggested by the photometric data of Danjon (1933,1949,1953).

In his summary of the results of his past polarimetric work on Mercury, Dollfus (1964) finds that the shape of the polarization curve, its wavelength dependence, and the P_{max}-albedo relationship are identical to those for lunar samples from the lightest (i.e., brightest) mare regolith material. Based on their attempt to interpret detailed differences between the observed polarization characteristics of Mercury and the Moon, Geake and Dollfus (1986) find that the average regolith grain size may be smaller. Using a plot of P_{min} (the minimum polarization observed) against V (their designation for the phase

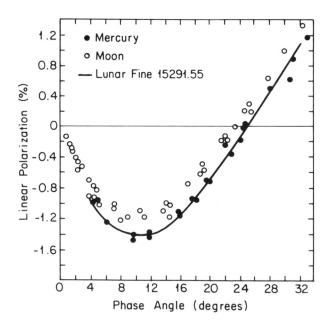

Fig. 9 Comparison of the negative branch of Mercury's polarization curve with data for the whole Moon and for Lunar Soil Sample 15291.55. (Figure from Dollfus and Auriere 1974.)

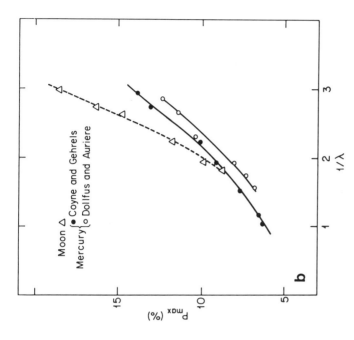

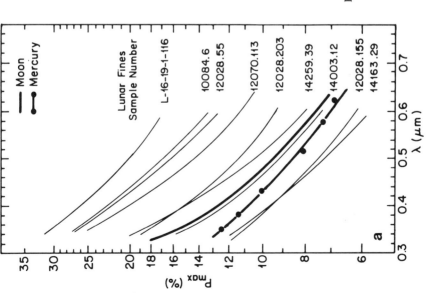

Fig. 10 The wavelength dependence of P_{max} for Mercury. (a) Dollfus and Auriere (1974) showed that for Mercury P_{max} increases less steeply with decreasing wavelength than is the case for the whole Moon, but that the observed trend is similar to that displayed by some lunar samples. (b) Corresponding data for Mercury and Moon from Gehrels et al. (1987). For comparison we have added the data of Dollfus and Auriere (1974) for Mercury.

angle at which the polarization goes from negative to positive—usually near 20–25°), they conclude that the point for Mercury falls in the same zone as for the lunar fines, although it is near the edge, which suggests possibly a larger proportion of the smallest particles.

The extensive polarization measurements by Coyne and Gehrels have only recently been reduced for publication (see, e.g., Gehrels et al. 1987). Their wavelength range is more extensive than those by Dollfus, and in general the agreement between the two data sets is good. The results concerning P_{max} as a function of wavelength are compared in Fig. 10. Gehrels et al. (1987) also noted that P_{max} appears to vary with time, an effect which is most likely an indication of longitudinal terrain differences on the planet.

B. Polarimetric Upper Limits on a Possible Atmosphere

Polarimetric searches for an atmosphere on Mercury go back to the time of Lyot. The basic idea is that at phase angles near 90° a Rayleigh component due to a thin atmosphere will have a positive polarization close to 100% which will be added to that of the light scattered from the surface. If one assumes that the ground polarizes much like the lunar surface, one can place a limit on any atmospheric component. Of course, one can do much better if one has measurements as a function of wavelength near P_{max}, since the contribution of any Rayleigh component would increase rapidly with decreasing wavelength. The methods are described in detail by Ingersoll (1971) and Dollfus and Auriere (1974), among others.

Since such upper limits are now only of historical interest, we restrict ourselves to summarizing published estimates in Table IV. It is clear that even the most optimistic estimates come nowhere near detecting gas concentrations as tenuous as those detected by the Mariner 10 Ultraviolet Spectrometer (Broadfoot et al. 1974) nor those indicated for the reported sodium and potassium clouds (Potter and Morgan 1985a, 1986a).

TABLE IV
Polarimetric Searches for a Thin Atmosphere on Mercury[a]

Polarization Pressure	Remarks	Reference
≤ 7 mbar	Upper limit	Lyot 1929
About 3 mbar	Assuming polarization of surface independent of λ. Later proved to be incorrect.	Dollfus 1950 O'Leary and Rea 1967
≤ 1 mbar	Assuming $\Delta P_{max}/\Delta\lambda$ does not decrease at short wavelengths.	Ingersoll 1971
≤ 0.25 mbar	"	Dollfus and Auriere 1974

[a]Table after Dollfus and Auriere 1974.

III. THERMAL INFRARED, MICROWAVE AND RADAR CONSTRAINTS

Infrared, microwave and radar measurements provide additional constraints on the physical properties of Mercury's regolith. Infrared measurements, like those at visible wavelengths, depend only on the regolith properties in the uppermost millimeter or so of the surface. The microwave and radar results refer to depths of roughly ten times the wavelength of observations, or about 10 cm to 1 m. The weight of the overlying regolith compresses the subsurface regolith, resulting in an increase in the bulk density and changes in the bulk electrical properties.

The Mariner 10 Infrared Radiometer obtained high spatial resolution (> 20 km) measurements of Mercury's thermal emission in two broadband channels centered at 11 μm (8.5–14 μm) and 45 μm (34–55 μm) (Chase et al. 1976). From independent observations of the same surface area viewed from different emission angles, the emissivity was found to vary as the cosine of the emission angle raised to the 0.19 ± 0.07 power. Assuming a bolometric emissivity of 0.90 at 0° emission angle, the surface thermal inertia discussed here in cgs units as cal cm^{-2} s$^{-1/2}$ K^{-1},

$$I = (K \rho C)^{1/2} \tag{6}$$

where K is the thermal conductivity, ρ is the density, and C is the specific heat, was determined by modeling the variation of the 45 μm flux with longitude on Mercury's night side. The spatial coverage consisted of a nearly equatorial swath a few tens of kilometers wide along a path from about (180°W, 20°N) to (350°W, 20°S). Between 295 and 340°W long, nighttime cooling was shown to be consistent with a homogeneous surface with $I = 0.0018$. A larger value, $I = 0.0026$, was found for 200 to 230°W longitude, and marked variations of I, ranging from 0.0016 to 0.0031, were found between 230 and 290°W longitude. According to Chase et al., these variations may be understood in terms of a uniform regolith with $I = 0.0016$ interrupted by outcrops of higher thermal inertia rocks, characterized by $I = 0.05$, covering at most some 8% of the surface. These values of the thermal inertia and its range of variation are similar to lunar values (see, e.g., Veverka et al. 1986).

The linear polarization of the thermal radiation is largely determined by the dielectric constant of the surface material. Landau (1975) measured the linear polarization at 3.5 μm of a portion of Mercury's disk and set a lower limit of 2.7 on the dielectric constant with a most likely value near 4.0. Cuzzi (1974) estimated the electrical properties of Mercury's subsurface at depths from a few centimeters to a few meters from microwave observations at several wavelengths. Interferometric measurements of the variation of the linear polarization across the disk at 3.71 cm were used to infer a dielectric constant of 2.4 ± 0.3 at about 50 cm depth. Using a self-consistent thermophysical model and measurements of the disk-integrated brightness temperatures at

TABLE V
Summary of Remote Sensing Data[a]

Quantity	Mercury	Moon
Visual geometric albedo (p_v)	0.138	0.136
Midnight temperature (K)	100	100
Thermal inertia (cal cm^{-2} s$^{-1/2}$ K^{-1})	630	850
Thermal skin depth (cm)	11[b]	10
RF skin depth (λ)	10	8–50
Radar cross section $(\times \pi r^2)$	0.06	0.07
Dielectric constant	2.7	2.9
Density of regolith (g cm^{-3})	1.6	1.8

[a]Sources: Pettengill 1969; Morrison 1970; Dollfus and Titulaer 1971; Goldstein 1971; Chase et al. 1974; Dollfus and Auriere 1974; Mendell and Low 1975.
[b]Lunar thermal skin depth scaled to Mercury.

0.31, 0.33, 6 and 18 cm, evidence was presented that the dielectric constant varies from about 1.8 in the upper few centimeters to about 2.9 at a few meters depth; the loss tangent increases from about 0.005 near the surface to about 0.009 at a few meters depth. These values of the dielectric constant and loss tangent, as well as their increase with depth, are comparable to those deduced for the lunar regolith (Mayer 1970). The higher value of the dielectric constant at the surface found by Landau (1975) can be attributed to the difference in scale (by orders of magnitude) between the infrared and microwave wavelengths. The void space between particles in the regolith affects the long-wavelength dielectric constant, but is negligible in the micrometer range (Campbell and Ulrichs 1969).

The radar reflectance at 12.6 cm of Mercury's surface is between 0.06 and 0.08, consistent with a dielectric constant of about 2.9 (Evans et al. 1966; Pettengill 1969). This reflectance is similar to that of the Moon (Dyce 1970; Hagfors 1970).

In summary (Table V), infrared, microwave and radar measurements of Mercury indicate that the thermal and electrical characteristics of Mercury's regolith (specifically: thermal inertia, dielectric constant, loss tangent, radar reflectivity, and the range of variation of these parameters) are nearly identical to those of the lunar regolith, indicating a strong similarity in composition and physical structure.

Acknowledgments.   We thank B. Buratti and E. Bowell for helpful comments and suggestions. This research was supported in part by a grant from the National Aeronautics and Space Administration.

SURFACE COMPOSITION OF MERCURY FROM REFLECTANCE SPECTROPHOTOMETRY

FAITH VILAS

NASA Johnson Space Center

Reflectance spectra of Mercury have been obtained periodically from 1963 through 1984. Since 1969, these observations were made in an effort to learn about the surface mineralogical composition of Mercury. Using the phases of the planet around maximum elongations, Mercury's 6°.1385/day rotational rate, and the theory of bidirectional reflectance spectroscopy, some spatial resolution across the planet has been obtained. A very shallow absorption feature, which has been attributed to Fe^{2+} in orthopyroxenes, is evident in two recent, high-quality spectra but is noticeably absent in a third. This difference cannot be explained by reflected light from different terrain. No noticeable spectral differences exist between the portion of Mercury photographed by Mariner 10 and the unimaged portion of the planet. All of the reflectance spectra display the same slope seen in lunar reflectance spectra, attributed to Fe- and Ti-bearing agglutinates in the lunar regolith. Gravitational focusing of meteoric material at Mercury's heliocentric distance and the extreme surface temperatures experienced by Mercury's sunward hemisphere both suggest that the regolith formation rate would be higher than the Moon's, and that agglutinates of unknown composition cause the spectral slope.

Electromagnetic radiation received from Mercury at the Earth emanates from two sources: (1) diffuse reflected sunlight in the near-ultraviolet, visible and near-infrared spectral range; and (2) thermally emitted radiation beginning at ~ 1.6 μm resulting from incident sunlight heating the planet's surface and reradiating at longer wavelengths. The method of determining the mineralogy of rock samples by observing how the crystal structure of certain minerals changes the spectrum of diffuse reflected light was pioneered during the late

1960s (see, e.g., Burns 1970; Adams and Filice 1967) and continues to be studied today (see, e.g., Pieters 1983). Concurrently, photoelectric photometry was developed during the 1960s and applied directly to telescopic observations of the planets. In the late 1960s, groundbased telescopes with narrowband filter photometers were turned toward the Moon and other solar system objects and used to measure how reflected sunlight is affected by the surface regolith. Samples returned from the various Apollo missions in the late 1960s and early 1970s confirmed the basaltic surface mineralogy suggested by the reflectance spectra, thus affirming the validity of this approach to determining the surface composition of a planetary body (see, e.g., Adams and McCord 1970). Today, reflectance spectrophotometry remains the dominant method of remotely sensing the surface mineralogical composition of solar system objects.

Reflectance spectrophotometry can be used as a probe of up to ~100 μm depths of surface regolith materials, depending upon the composition (and therefore crystal structure) and particle size of the material (Pieters 1983; Morris and Mendell 1984). Spectral reflectance data can potentially constrain our knowledge about Mercury's surface composition. Indeed, it could even constrain the primordial bulk chemistry of the planet; the assumption made here is that Mercury's surface was not altered or eradicated by a major event such as the volatilization of the outer layer(s) of the planet (Chapter by Cameron et al.), or catastrophic collision (Chapter by Wetherill). In this case, the presence, amount and form of FeO on the planet's surface would serve as an indication of the oxidation state and amount of volatiles in the outer portion of the planet, which is expected to contain silicates.

The primary indicator sought in these spectra is the presence and characteristics of an absorption feature centered near 1.0 μm caused by interelectronic transitions of Fe^{2+} in an octahedral site in olivine [$(Mg,Fe)_2SiO_4$] or pyroxene [$(Mg,Fe)SiO_3$] or both. Transition energy differences cause the absorption features to be centered at different spectral locations. If a regolith contains both olivine and pyroxene, the spectral width of the absorption band will be broader. These absorption features are common in reflectance spectra of the Moon and asteroids (see, e.g., McFadden et al. 1984; Adams and McCord 1970).

In the context of several models for the condensation of Mercury, the presence and amount of volatiles could show where, and under what circumstances, planetary formation occurred. Lewis (see his Chapter) discusses why various condensation and accretion models are inadequate alone to explain Mercury's high density, and how the amount of FeO might be an indicator of the formation process. Wasson (see his Chapter) proposes that the area near Mercury is the formation region for the enstatite chondrites, and would have little or no FeO. Evidence of FeO in the planet's surface material would suggest that the accretion phase of Mercury's formation sampled material from a greater range of heliocentric distances than covered by Mercury's narrow feeding zone. Thus, spectral reflectance studies of Mercury have emphasized the search for a shallow absorption feature centered near 0.9 μm, seen in some spectra but not in others.

This chapter discusses the basic procedure for reducing spectrophotometric data of Mercury; the controversies surrounding (and the implications of) the existing spectra of Mercury, as well as a methodology for defining the portion of Mercury's surface contributing the greatest amount of light to an individual spectrum, including its application to these spectra.

DATA ACQUISITION AND REDUCTION

As viewed from the Earth, Mercury has a maximum angular elongation of approximately 28° from the Sun. Due to the close visible proximity of the two objects, observations of Mercury must be conducted either during daylight or during twilight while the Sun is below the horizon but Mercury is still visible. This restriction introduces some unavoidable observing problems: careful attention must be paid to avoid scattered sunlight in the telescope tube and instrumentation. The limited dynamic range (operating range within which the detector produces a measureable 1:1 output signal for each input irradiance level) of earlier detectors caused saturation of the received planetary signal during daylight hours. As a result, all spectral reflectance observations reported here were made during twilight except for the 1984 CCD spectrograph measurements. Spectrophotometric observations must also be corrected for the attenuation of light by the Earth's atmosphere along the line of sight from the object to the telescope. The atmospheric thickness will vary from a minimum at the zenith to a maximum along the horizon. The general extinction formula

$$S_\lambda = S_{0\lambda} \, e^{-\tau_\lambda X} \tag{1}$$

describes the signal attenuation corresponding to the extinction τ_λ caused by the atmosphere at a given wavelength λ, and the atmospheric thickness at that location. The airmass X is a logarithmic definition of the atmospheric thickness defined as 1 at the zenith and increasing to infinity at a zenith angle of 90° (the horizon). The signal S_λ is the expected counts for that spectral interval at the airmass X, and $S_{0\lambda}$ represents the counts at an airmass equal to zero. Differential refraction (the dispersion of light at different wavelengths by the thick Earth atmosphere) becomes a problem at high airmass, especially for data acquired at the near-ultraviolet and blue wavelengths obtained with narrowband filter photometers. Active telescope guiding by the observer can compensate where necessary for the images positioned differently on the instrumentation due to refraction. Water present in the Earth's atmosphere (telluric H_2O) has absorption features centered near 0.73, 0.82 and 0.93 μm. The water content of the atmosphere can change throughout an observing session, and observations of Mercury at high airmass can aggravate the H_2O absorption. Removal of the effects of water absorption in the Earth's atmosphere from Mercurian reflectance spectra is probably the most difficult task in the data acquisition and reduction procedure. The presence of water absorption affects part of the spectrum covered by the Fe^{2+} silicate absorption feature.

Spectrophotometry of Mercury has been obtained in the visible and near-infrared spectral region with a variety of instruments, however, the data reduction procedure has generally remained the same. [Some exceptions for specific characteristics of certain instruments have also been included in the data reduction procedures of individual observers. For example, Chapman and Gaffey (1979) describe corrections for coincidence and misalignment of optics in the dual-beam photoelectric photometers which were used to take Mercury data (McCord and Adams 1972a,b; Vilas and McCord 1976). Vilas and Smith (1985) describe the uniformly illuminated "flat field" observations necessary to correct for variations in the spectral responsivity of individual pixels across a 2-dimensional CCD camera.]

For a given spectral interval $\Delta\lambda$ and integration time, the signal received from a solar system object during one observation S_λ is calculated as

$$S_\lambda = C_\lambda - B_\lambda - D_\lambda \qquad (2)$$

where C_λ is the raw photon count of the object plus the background sky, B_λ is the background sky count, and D_λ is the dark count (background count generated by the detector). Since photon counts comprise the data, the calculated signal-to-noise is based upon Poisson statistics. The spectral interval is physically defined by the attributes of the instrumentation (e.g., the passband of a narrowband filter, the grating resolution of a spectrograph).

Observations of a calibrated standard star are obtained across a large airmass interval which includes ideally the airmass interval covered by Mercury when observations are made of the planet. For a given spectral interval, the extinction coefficients are calculated using a least-squares fit to the logarithm of the counts vs airmass for all of the observations of the standard star during one observing session. The standard star observation closest in airmass to a specific Mercury observation is identified, and the extinction coefficients are used to adjust the standard star counts to those counts expected at the same airmass as the planet's observation as

$$\ln S_{\lambda C} = \ln S_\lambda + \tau_\lambda (X - X_C) \qquad (3)$$

where S_λ is the count for the wavelength interval at the airmass X, τ_λ is the calculated extinction coefficient and $S_{\lambda C}$ is the count of the standard star at the planet's airmass X_C. The inverse logarithm of this expression produces the corrected standard star value.

The corrected standard-star spectrum and the planet's spectrum are independently scaled to 1.00 at a specified wavelength. The reflectance value for

Mercury, corrected for the spectral signature of the reflected sunlight, is calculated as

$$\left(\frac{\text{Mercury}}{\text{standard star}}\right) \times \left(\frac{\text{standard star}}{\text{Sun}}\right) = \frac{\text{Mercury}}{\text{Sun}}. \tag{4}$$

The selection of appropriate standard stars is difficult. To date, the selection and calibration of standard stars for planetary spectral reflectance data has been accomplished in two different ways. One method has produced a net of bright standard stars located around the ecliptic (P. Owensby personal communication) observed using the same instrument used for the planetary observations. These stars have been calibrated using observations of Apollo lunar landing sites from which lunar soil samples have been returned to the Earth. Laboratory reflectance spectra of the lunar soil samples have then been used to correct the standard star/Apollo landing site spectra to become a standard star/Sun spectrum. This is the only method known of calibrating standard stars which closes the loop between observations of star and planetary object, and calculation of object/Sun, although there remain questions of how representative the returned samples are of the large site area measured telescopically.

Alternately, Hardorp (1980) has conducted a careful search for stars whose spectra are solar analogues, documenting in his publications when variations in absorption features exist. If, for a given spectral interval, the ratio of the scaled solar analogue star to the theoretical Sun is approximately 1.0, then the second step in Eq. (4) can be eliminated, i.e., Mercury/solar analogue star sufficiently represents Mercury/Sun. A reduction in the calculation steps should eliminate one source of calculated uncertainty. Various observers have tested the resulting reflectance spectra obtained using a solar analogue star. McFadden et al. (1984) compared the reflectance spectra of asteroid 2100 Ra-Shalom ratioed to solar analogue 61 Cygni B and to standard star α Equulei. They find that the spectra are consistent within the calculated errors, except for the near-ultraviolet wavelengths, and they endorse the continued use of both of these standard star sets. For the Eight-Color Asteroid Survey (ECAS), Tedesco et al. (1980) calculated the differences between ECAS photometry of stars that Hardorp designated as spectrally identical to the Sun, and stars he determined were close to solar analogues. Interpolating between the values for these filters to obtain higher spectral resolution data for these stars, Vilas and Smith (1985) note that the differences between most of the Hardorp solar analogue stars and the assumed true Sun are less than the scatter in their asteroid reflectance spectra. They conclude that object-to-solar analogue star ratios represent the object-to-Sun reflectance spectra adequately.

The advent of the use of spectrograph and detector combinations, such as CCD spectrographs, which provide comparatively high resolution from 0.95

to 1.00 μm, introduces a further problem. The spectra of some early-type stars (e.g., ε Aquarii) contain the hydrogen Paschen lines. These lines, which are smeared in the wider passband filters used in astronomical photometers, are resolved in the higher-resolution photometry. If the values of early-type standard star/Sun ratios used in narrowband filter photometry are interpolated for the purpose of correcting the higher-resolution spectra for reflected sunlight, the interpolated values will not have sufficient spectral resolution to correct for the structure introduced into the spectrum by the Paschen lines. Even if subsequent stellar calibrations account for the Paschen lines, ratios of solar system objects to early-type standard stars can introduce noise in the spectra. Changes in seeing, instrument flexure, or exact positioning of the target in a spectral slit have required subpixel interpolation of CCD reflectance spectra (see, e.g., Vilas 1985; Vilas and Smith 1985; Buie 1984). The 0.95 to 1.00-μm spectral range covers part of the spectral reflectance absorption feature caused by Fe^{2+} in olivine or pyroxene or a combination of both, crucial to studies of Mercury. Using a solar analogue star would prevent the introduction of unnecessary structure in the 0.95- to 1.0-μm spectral range, caused by the subpixel misalignment of two spectra, from degrading a high-resolution reflectance spectrum.

In this chapter, the spectral reflectance curves are all scaled to the value of 1.00 at 0.7 μm, a spectral range common to all of the data sets discussed, allowing the data to be intercompared. Other scientific literature on Mercury present data scaled differently, but the significance of the data is unaffected by the choice of scaling.

One other factor must be considered in reduction of Mercury reflectance data: the high surface temperatures of the illuminated side (590 − 725 K) cause the thermal component of the radiation to become greater than the reflected component beyond 1.6 μm. Corrections must be made to remove the thermal radiation component, in order to consider the spectral reflectance characteristics of Mercury at longer wavelengths where some silicate absorption features are prominent. Clark (1979) described one method of removing this component from circular-variable-filter (CVF) data of Mercury. These analyses are left to future observers and are not discussed further in this chapter.

CONSTRAINING THE SPATIAL RESOLUTION OF MERCURY'S SURFACE

The reflectance spectra obtained of Mercury to date spatially cover the integral illuminated portion of the planet presented to the Earth at the time and date of the observations. Some correlation of the integral planet spectrum with surface terrain is desired. The planet's orbital and rotational periods, the conditions of illumination, and the observational geometry can be used to derive some spatial resolution of the planet's surface. When Mercury is at maximum elongation, the phase (portion of the planet illuminated) dictates the longitudinal interval of the planet illuminated during one observing session (longitudinal dimension covered is equal to the difference between 180° and the phase angle). The rotational period of Mercury is taken to be equal to two-thirds of its orbital period or 58.6462 days.

Table I contains information about the locations and physical ephemerides of Mercury on all of the dates for which available spectral reflectance observations were made. An eastern elongation as seen from the Earth shows

TABLE I
Observational Data of Mercury

Observation Date (UT)	Phase Angle[a] (deg)	Bright Limb Longitude (deg)	Terminator Longitude (deg)	Hemisphere	Telescope (m)	Instrument[b]
16 Jun 63	−98.3	328.4	246.8	S	0.4	NFP
14 Aug 63	69.5	46.2	156.7	S	0.4	NFP
15 Aug 63	71.0	50.8	160.0	S	0.4	NFP
18 Aug 63	75.5	65.5	170.0	S	0.4	NFP
19 Aug 63	77.0	70.3	173.4	S	0.4	NFP
22 Aug 63	81.9	85.4	183.5	S	0.4	NFP
25 Aug 63	87.0	100.7	193.7	S	0.4	NFP
26 Aug 63	88.9	105.9	197.0	S	0.4	NFP
19 May 64	−116.0	269.6	205.6	S	0.4	NFP
20 May 64	−113.7	275.2	208.9	S	0.4	NFP
22 May 64	−109.2	286.1	215.4	S	0.4	NFP
15 May 65	−83.0	331.1	234.1	S	0.4	NFP
18 May 65	−77.1	345.3	242.4	S	0.4	NFP
17 Jun 69	−121.5	109.6	51.1	unknown	—	NFP
18 Jun 69	−118.8	115.2	54.0	unknown	—	NFP
26 Dec 69	70.7	156.3	265.6	S	1.5	NFP
13 Mar 72	89.1	358.5	89.4	N	0.9	NFP
29 Sep 74	73.8	107.1	213.3	S	0.9	NFP
5 Oct 74	86.4	138.2	231.8	S	0.9	NFP
6 Oct 74	88.9	143.6	234.7	S	0.9	NFP
7 Oct 74	91.5	149.0	237.5	S	0.9	NFP
8 Mar 75	−78.2	83.6	341.8	S	0.9	NFP
9 Mar 75	−76.6	88.6	345.2	S	0.9	NFP
10 Mar 75	−75.0	93.6	348.6	S	0.9	NFP
11 Mar 75	−73.4	98.6	352.0	S	0.9	NFP
21 Apr 76	77.4	167.9	270.5	N	2.2	CVF
22 Apr 76	81.3	172.4	271.1	N	2.2	CVF
23 Apr 76	85.3	177.2	271.9	N	2.2	CVF
24 Nov 84	71.8	146.9	255.1	S	1.0	CCD

[a] Positive phase angle represents eastern elongation; negative phase angle represents western elongation.
[b] NFP = Narrowband filter photometer; CVF = Circular variable filter photometer; CCD = CCD spectrograph.

the western side of Mercury to be illuminated; however, the eastern longitudes of the Mercurian hemisphere facing the Earth are illuminated. The reverse is true for a western elongation. Mercury is locked into a quasi-commensurability in which 54 Mercurian sidereal periods are equal to 13.00600 terrestrial tropical years. Thus, the Mercurian physical ephemeris repeats itself every 13 years. During one 13 year interval, northern hemisphere observations favor Mercurian central meridian longitudes of 90 and 270°, while the longitudes of 0 and 180° are poorly viewed. However, southern hemisphere observations cover equally well all central meridian longitudes except for 90 and 270°. This unusual observing circumstance may have contributed to the late discovery of Mercury's rotational period. Smith and Reese (1968) describe the implications of this phenomenon: ". . . had Schiaparelli or Antoniadi made their observations from Africa, Australia or South America, the true rotation period of Mercury might have been established long ago." Recent observations of the spectral reflectance of Mercury have been conducted at both northern and southern hemisphere observatories, providing relatively complete longitudinal coverage of the planet.

Photometric theory can be used to define a subsection of the illuminated portion of Mercury's disk which would contribute the greatest amount of reflected sunlight to a reflectance spectrum. Using Hapke's (1981) bidirectional reflectance formula for a planetary surface composed of closely packed particles of arbitrary shape (Hapke's Eq. 16), Vilas et al. (1984) calculated the distribution of bidirectional reflectance along the illuminated Mercurian surface for different phase angles, assuming a constant albedo planet. The brightest portion of the planet lies at the luminance equator on the planet's limb. As more of the planet's disk is illuminated, the distribution of brightness across the surface becomes more uniform. Figure 1 shows the longitude interval vs the date of observation for the reflectance spectra of Mercury. Insufficient spatial resolution exists for the Earth-based observations to distinguish among the surface terrain types seen in Mariner 10 imagery. In this discussion, the assumption is made that spectral differences on the planet's surface could at best be distinguished between the two dominant geologic terrain types, the intercrater plains and the smooth plains encompassing Caloris Basin (Trask and Guest 1975). Figure 2 shows a sketch of the distribution of these two terrain types as inferred for spectral reflectance studies from Mariner 10 imagery, groundbased radar and albedo data (Chapter by Clark et al.; Zohar and Goldstein 1972; Murray et al. 1972).

REFLECTANCE SPECTRA

Harris (1961) reported the earliest broadband UBVRI photoelectric measurements of the planet Mercury (see the Chapter by Veverka et al.). Both spectral range and spectral resolution were extended by Irvine et al. (1968) with observations made of Mercury taken around four maximum elongations of the

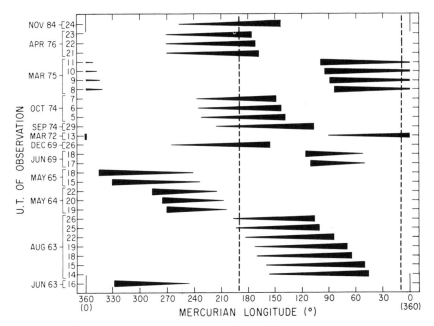

Fig. 1. Longitude intervals for observation dates of Mercurian reflectance spectra. The broad side is the bright limb tapering to a point for the terminator (indicating the greater contribution of light to the spectrum from the bright limb area). The hatched lines mark the 10° to 190° interval photographed by Mariner 10.

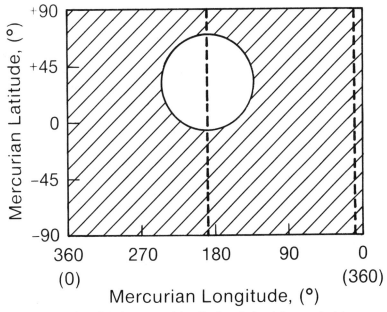

Fig. 2. The distribution of the intercrater plains (lined portion) and the smooth plains encompassing Caloris Basin (blank portion) on Mercury's surface as inferred for spectral reflectance studies.

planet during the years 1963 through 1965. These observations were made using 9 or 10 different narrowband interference filters having passbands centered from 3147 Å to 10635 Å. In this review, the corresponding relative intensity for each filter was calculated from the data tabulated by Irvine et al. (1968) using the relationship

$$I = 10^{-0.4 \text{ m}} \tag{5}$$

and scaling the intensities to 1.0 for a value interpolated for 0.7 μm from between the values of the 6264 Å and 7297 Å filters. Vilas et al. (1984) grouped these data around the maximum elongations of Mercury which occurred on 13 June 1963 (23°W), 24 August 1963 (27°E), 24 May 1964 (25°W), and 6 May 1965 (27°W). Figure 3 shows plots of the unweighted mean and standard deviations of the spectral reflectance values taken on dates around these elongations. Irvine et al. do not list errors for the daily Mercury data, and general errors for all observations made during monthly intervals (see Table IX in Irvine et al. 1968) are not included in the errors shown in Fig. 3.

McCord first looked at Mercury's reflectance spectrum with sufficient spectral resolution to see mineralogical absorption features in the late 1960s and early 1970s (McCord and Adams 1972a,b; Vilas et al. 1984) using a photometer with 22 narrowband interference filters covering a spectral range of 3190 to 10530 Å having passbands of ~250 Å. The calibration technique using areas on the Moon having known reflectance values was used in the reduction of spectral reflectance values taken near three maximum elongations of 23 June 1969 (23°W), 27 December 1969 (20°E), and 14 March 1972 (18°E). The data are scaled to 1.000 for the 6990 Å filter. Figure 4 shows a spectrum composed of the unweighted mean and standard deviation of 4 observations made on the nights of 17, 18 June 1969 from an unknown location, and the unweighted mean and error of 6 observations taken on 13 March 1972 from KPNO. The spectrum formed from the unweighted mean and error of 6 observations taken on 26 December 1969 from CTIO is included in Fig. 5.

During 1974 and 1975, Vilas and McCord (1976) obtained spectra on dates around the maximum elongations of 1 October 1974 (26°E) and 6 March 1975 (27°W) at CTIO using a photometer with 24 narrowband interference filters covering a spectral range of 3350 to 10640 Å. Vilas et al. (1984) re-reduced these spectra using improved standard star calibrations and grouped the spectra by phase angle. The weighted average and error of 37 observations obtained on 5, 6, 7 October 1974 formed one spectrum shown in Fig. 5. Eight observations of Mercury taken on 29 September 1974 and the weighted average and error of 56 observations obtained on 8, 9, 10, 11 March 1975, produced two additional spectra shown in Fig. 4. These data are scaled to 1.000 for the 6990 Å filter.

The spectral range was shifted to the near-infrared (0.65 − 2.5 μm), and spectral resolution was improved to 50 Å when McCord and Clark (1979)

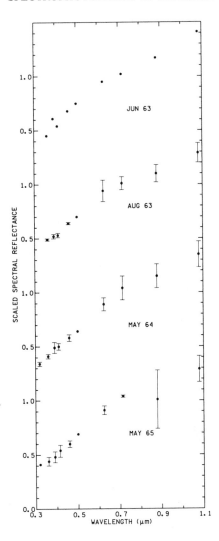

Fig. 3. Reflectance spectra obtained from 1963 to 1965.

acquired reflectance spectra of Mercury near the maximum elongation of 28 April 1976 (21°E) at Mauna Kea using a circular-variable-filter photometer with 120 passbands. A composite spectrum of 51 observations made on the dates of 21, 22, 23 April 1976 was produced (Fig. 6).

No further advances in instrumentation were applied to the problem of observing Mercury until 1984 when Vilas used a 2-dimensional CCD with a spectrograph to get 16 Å resolution data across a more limited visible and near-infrared spectral range of 0.5 to 1.0 μm. The multiplexing advantage of

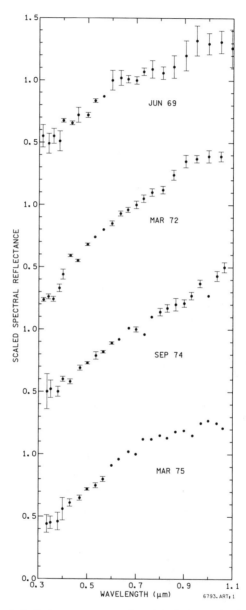

Fig. 4. Reflectance spectra covering solely the intercrater plains.

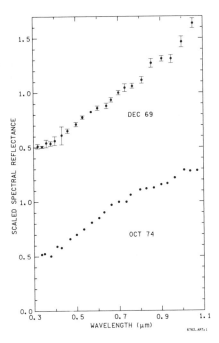

Fig. 5. Reflectance spectra obtained with a narrowband filter photometer covering the Mercurian surface area containing both intercrater plains and smooth plains.

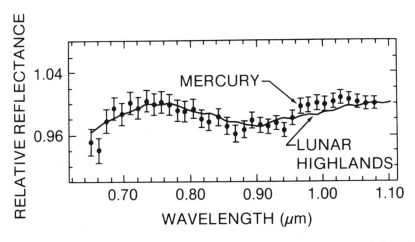

Fig. 6. McCord and Clark's (1979) April 1979 reflectance spectrum of Mercury (individual points) compared with a laboratory spectrum of Apollo 16 (lunar highlands) soil showing the 0.89 μm Fe^{2+} orthopyroxene absorption feature. The sloped continuum (reddening) seen in the other spectra has been removed. The spectrum covers the Mercurian surface area containing both smooth plains and intercrater plains.

[71]

this combination of instrument and detector allowed short exposures to be taken of the planet and sky concurrently, permitting observations to be made during the day when the planet was observed through a low airmass. A composite spectrum of 59 observations of Mercury taken on 24 November 1984 at CTIO near the maximum elongation of 25 November 1984 (22°E) covered an airmass range of 1.002 to 1.014 (Fig. 7). The greater dynamic range of the CCD accommodated the background sky brightness. The 2-dimensional CCD allowed the background sky to be mapped concurrently with the planet, thus accounting for temporal changes in sky conditions. The signal-to-noise ratio for this spectrum exceeds 100:1, and a better indicator of the quality of the spectrum is the scatter within the spectrum. Telluric H_2O absorption is also very apparent in the spectrum, and is attributable to changes in atmospheric water content during the 6.5 hour delay between the observations of Mercury

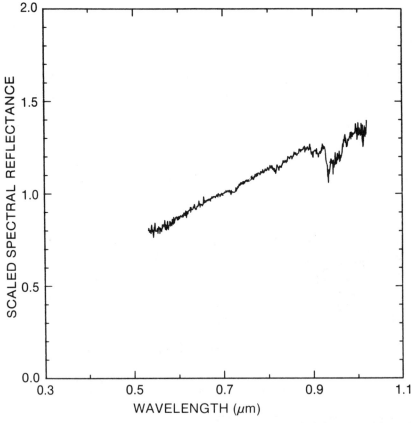

Fig. 7. Reflectance spectrum of the Mercurian surface area covering both intercrater plains and smooth plains, obtained in 1984 with a CCD spectrograph.

and the observations of the Hardorp solar analogue standard star, SAO 147237, used to make the reflectance spectrum.

A total of twelve composite reflectance spectra of Mercury have been produced from observations made around maximum elongations over the years of 1963 through 1984. Here, these spectra have been divided into three groups: early spectra, spectra covering solely intercrater plains and spectra covering terrain split roughly between intercrater plains and the smooth plains encompassing Caloris Basin.

Early Spectra

The spectra taken during 1963 to 1965 (Irvine et al. 1968) lack the spectral resolution to show detailed mineralogical information (Fig. 3). Vilas et al. (1984) drew some general conclusions about the surface compositional properties from these spectra. The most important feature of these spectra is that three of the spectra cover terrain unimaged by Mariner 10, while one spectrum covers terrain known to consist of primarily intercrater plains. The similarity between these four spectra suggests that there is no radical difference in mineralogical composition between the unimaged portion of Mercury and the area imaged by Mariner 10. At wavelengths >0.7 μm, these data exhibit increasing slope (reddening) with increasing phase angle, as seen for the Moon. Irvine et al. (1968) noted this same reddening effect when they grouped the spectra by phase angle, not by elongation date.

Intercrater Plains

Four of the Mercurian spectra (June 1969, March 1972, September 1974, March 1975) having adequate spectral resolution for surface compositional studies (Fig. 4) cover terrain in the 0–90° quadrant of the Mercurian surface, consisting structurally of predominantly intercrater plains. Most absorption features present in these spectra, however, can be attributed to the incomplete removal of telluric H_2O. McCord and Adams (1972a,b) first interpreted the dip at 0.95 μm in the March 1972 spectrum as a possible Fe^{2+} pyroxene absorption. The dip at 0.82 μm in this spectrum indicates, however, that the incomplete removal of telluric H_2O is the probable origin of this feature.

Caloris Basin and Smooth Plains

The smooth plains are considered to be younger than the intercrater plains, emplaced after the impact which formed Caloris Basin. A debate continues over whether the smooth plains are volcanic or impact-related in origin (see Chapter by Spudis and Guest). The spectra in Fig. 5 are those in the visible range covering roughly 50% intercrater plains and 50% smooth plains. A portion of the composite near-infrared spectrum of McCord and Clark (1979) covering the same area is shown in Fig. 6. McCord and Adams (1972a,b) interpreted the dip at 0.95 μm in the December 1969 spectrum as due to pyroxene, but the spectrum shows absorptions near 0.73 and 0.82 μm which again suggests the incomplete removal of telluric H_2O.

The three spectra with sufficient signal-to-noise to merit serious consideration all cover approximately the same surface region (see Table I, Fig. 1). The October 1974 spectrum shows a shallow absorption feature beginning at 0.8 μm, centered near 0.9 μm, extending to 1.0 μm. The incomplete removal of telluric H_2O is also noticeable in the spectrum. Vilas et al. (1984) compared this spectrum with the April 1976 composite spectrum which shows a very weak absorption feature beginning at 0.78 μm and extending to 0.95 μm, centered at 0.89 μm. McCord and Clark (1979) noted an absorption band depth of 4%, and compared the spectrum with the spectrum of Apollo 16 lunar highlands soil containing a minor amount of pyroxenes contributing ~5.5% FeO to the composition of the soil.

In contrast to these two spectra, the composite spectrum obtained on 24 November 1984 shows only the incomplete removal of the O_2A band and telluric H_2O absorption features (Vilas 1985).

The absence of the absorption feature in the 1984 spectrum reopens the question of the existence of the weak feature seen rarely in the other Mercury spectra. There are two current opinions on the subject. One states that the proposed absorption feature does not exist, and that inaccuracies introduced into the formation of the October 1974 and April 1976 composite spectra produced the weak absorption features. All three spectra could be affected by the incomplete removal of telluric H_2O due to the delays between the times that Mercury was observed and the times that the standard stars used for calibration (α Lyrae in 1974, β Geminorum in 1976, SAO 147237 in 1984) were observed. (The thinner atmosphere at Mauna Kea's higher altitude reduces the amount of telluric H_2O which would have affected the 1976 spectrum.) Vilas (1985) has also suggested that averaging spectra obtained on more than one observation date could have introduced additional problems in the 1974 and 1976 spectra. The phase angle difference among these three spectra is within 20°. Using the results of laboratory studies conducted by Pieters (1983) on the effects of phase angle on the reflectance spectrum of the enstatite component of websterite (EN_{89}), Vilas has shown that the phase differences among these spectra could reduce the observed band depth by at most 6%. The complete disappearance of this feature could not be caused by differences in observational conditions alone.

An alternate opinion is that, for reasons still not understood, the absorption feature appears under some conditions affecting Mercury and disappears with changes in these conditions. Earth-based observing conditions do not mask a feature, so physical reasons on the planet's surface must be examined. The surface temperature of the sunward portion of Mercury varies between 590 and 725 K depending upon location on the planet's surface and the planet's orbital position. Laboratory studies by Singer and Roush (1985) show that for temperatures up to 448 K the orthopyroxene absorption band located near 0.9 μm is enhanced in width and strength; the short-wavelength edge of the feature is constant in spectral location, the long-wavelength edge increases in location, and although the spectral position of the band minimum does not change, the depth increases up to 15%. At higher temperatures, the spectral position of the band minimum increases (Sung et al. 1977). For these Mercury reflectance spectra, the short-wavelength edge of the 0.9 μm absorption feature should not be obscured by the telluric H_2O absorption. The issue of the tenuous existence of this absorption feature remains unresolved.

What do these spectra imply for the surface composition of Mercury? The slope (reddening) present in all of the Mercurian spectra, and its similar behavior with changes in phase angle to the slopes seen in lunar spectra, suggest that similar material (either compositionally or in physical state) to the lunar regolith exists on Mercury. The slope in the spectra of different lunar terrains (maria, highlands) is caused by Fe- or Ti-bearing agglutinitic glasses, produced by micrometeoroid bombardment, in the surface regolith (Adams and McCord 1973). Metallic iron particles were also found with the lunar agglutinate samples. The environment of Mercury suggests that the regolith could contain agglutinates formed in a similar manner. The composition of these glasses remains unknown at this time. Cintala (1981) has shown that the melt production rate at the Mercurian surface is almost three times greater than at the hottest lunar point, due to the increased surface temperatures at Mercury and gravitational focusing of meteoric material by the Sun at Mercury's heliocentric distance. The flux of meteoric particles is at least eight times greater at Mercury than at the Moon (Leinert et al. 1981), suggesting that a much greater regolith formation rate is operating on the Mercurian surface, although that does not necessarily increase the fraction of agglutinates expected. However, the higher impact velocities might increase the fraction of agglutinates. The composition of possible Mercurian agglutinates remains unknown.

The weak absorption feature has been attributed to Fe^{2+} in orthopyroxenes (McCord and Adams 1972*a,b;* McCord and Clark 1979) in the surface material. If the 0.9 μm pyroxene absorption feature is not present, as the most recent composite spectrum suggests, then information about the surface composition can be inferred only from its conspicuous absence. The lack of an absorption feature suggests that the surface of Mercury is highly reduced. Since condensation sequences proposed for the origin of the solar system generally have increasing FeO with decreasing temperature, and increasing volatile content with decreasing condensation temperature, a lack of volatiles on Mercury's surface could be inferred from the lack of FeO.

Future spectral reflectance observations of Mercury in the visible and near-infrared spectral range could resolve the question of the existence of the Fe^{2+} absorption feature more firmly. Valuable data on the mineralogical composition can also be obtained in the thermal infrared and near-ultraviolet spectral range from Earth-based studies. The difficulties with groundbased observations remain, although improved instrumentation and observations of Mercury above the Earth's atmosphere from space will enhance future observing efforts.

GOLDSTONE RADAR OBSERVATIONS OF MERCURY

P. E. CLARK
Jet Propulsion Laboratory

M. A. LEAKE
Valdosta State College

and

R. F. JURGENS
Jet Propulsion Laboratory

Radar observations of Mercury were made during the past two decades at the Goldstone radar facility. Correlations of these observations with geologic maps are presented in this chapter. Topographic profiles indicate that Mercurian craters are rather shallow. Some topographic features are seen on the side of Mercury not imaged by Mariner 10. There are global correlations between topography and radar roughness. Mercury's surface may be rougher on a 1-cm scale than on a 10-cm scale, in comparison with the Moon.

This chapter summarized geologic studies of Mercury based on ground-based radar data obtained in the 1970s and 1980s with the Goldstone/JPL Solar System Radar facility at Fort Irwin, California. These data (Table I) are summarized here for the first time, along with a brief description of the data reduction procedure and some analyses of the data.

The Mariner 10 mission has provided most of the basis for our still limited understanding of Mercury. Unfortunately, all three flybys encountered the same hemisphere of the planet. The effective coverage for geologic interpreta-

TABLE 1.
Groundbased Radar Observations of Mercury

Type	Form	References
12.5-cm polarized	reflectivity images	Goldstein (1971)
		Zohar et al. (1974a,b)
		Clark et al. (1984a)
		Clark (1987)
12.5 cm-polarized	topography profiles	Goldstein (1971)
		Zohar et al. (1974a,b)
		Clark et al. (1984b)
		Clark (1987)
3.8-cm polar-ized/depolarized	bulk scattering	Clark et al. (1986)
3.8-cm polarized	topography profiles	Clark and Slade, personal communication

tion is still less because many parts of the observed hemisphere were imaged at high Sun angles and, therefore, lack shadows which are necessary for studies of geologic structure. Supplementary groundbased data, including radar observations taken at Goldstone and at Arecibo (see Chapter by Harmon and Campbell), augment our understanding of Mercury. At Mercury's closest approach to the Earth, surface features down to 10 km in size can be resolved using radar techniques. This chapter discusses Goldstone radar observations made at 12.5 cm (S-band) and 3.5 cm (X-band) wavelengths.

I. DERIVATION OF GEOLOGIC INFORMATION FROM RADAR DATA

Many useful global- and regional-scale physical properties of Mercury have been derived from groundbased radar observations (Pettengill and Dyce 1965; Pettengill 1978; Jurgens 1974; Zohar et al. 1980; Jurgens 1982). This

TABLE II
Physical Parameters Derivable from Radar Data

Type	Property	Feature Scale
Figure	shape	global
Figure	spin axis	global
Figure	rotation rate	global
Surface	topography (profiles)	km
Surface	topography (reflectivity, polarized component)	km
Surface	roughness (rms slope)	dm to m
Surface	roughness (reflectivity, depolarized component)	mm to dm

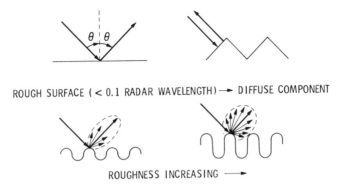

SMOOTH SURFACE (> 10 RADAR WAVELENGTH) → SPECULAR COMPONENT

ROUGH SURFACE (< 0.1 RADAR WAVELENGTH) → DIFFUSE COMPONENT

ROUGHNESS INCREASING →

MODELLING ROUGHNESS BY ASSUMING A SURFACE COMPOSED OF
BLOCKS (WITH HEIGHT H) WITH DEFINED SPACING (RESULTING IN θ)

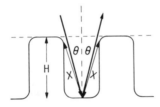

Fig. 1. A diagrammatic view of interaction of radar waves with smooth and rough planetary
surfaces.

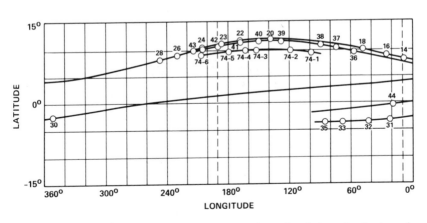

Fig. 2. Tracks of reduced Goldstone Solar System radar observations in imaged and unimaged
hemispheres of Mercury made to date.

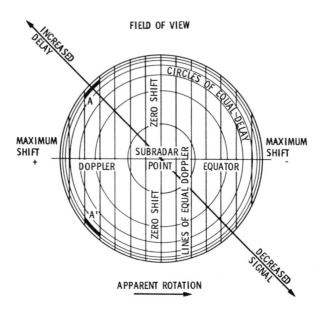

Fig. 3. Interaction of radar beam with planetary surface resulting in delay/Doppler transformation of the returning signal.

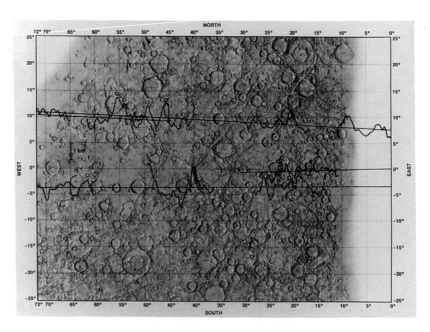

Fig. 4. Plot of Goldstone topographic profiles on portion of 1:15000000 shaded relief map corresponding to H-6 quadrangle. Scale of profiles is indicated on map. Footprint size is discussed in text.

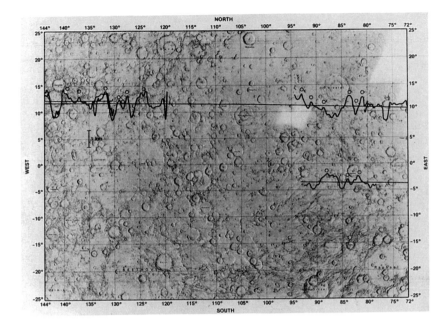

Fig 5. As in Fig. 4 for the H-7 quadrangle. Feature recognition in some portions of this map is reduced by high Sun angle imaging.

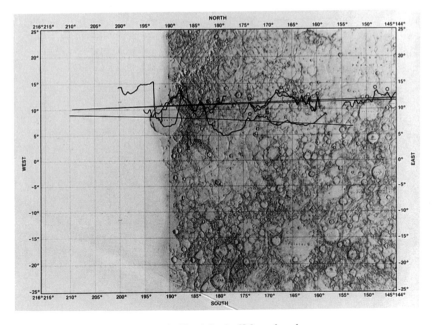

Fig 6. As in Fig. 4 for the H-8 quadrangle.

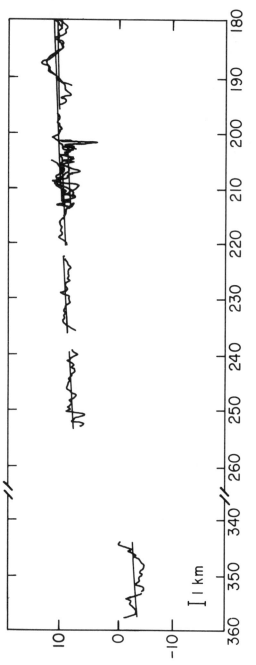

Fig 7. Plot of Goldstone topographic profiles on portion in visually unimaged hemisphere.

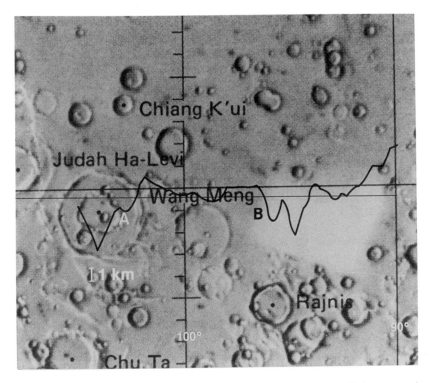

Fig 8. S-band topographic profile for crater Wang Meng in H-7 quadrangle. Profile is superposed on portion of 1:5000000 shaded relief map. Goldstone track is indicated by a straight line through profile. Vertical scale is indicated by a bar, and the footprint is indicated by an open rectangle.

chapter addresses the radar characterization of surface structure at smaller scales ranging from tens of km down to mm (see Table II). Some basic assumptions (Fig. 1) are made in modeling radar data to obtain surface parameters (Pettengill 1978; Jurgens 1982). Most of the returned signal is due to scattering from the first solid surface encountered by the radar wave. We can distinguish between two types of such surface scattering, depending on the roughness of the surface structure compared with the wavelength of the radar.

Specular reflection occurs if the surface is smooth on the scale of the radar wavelength (cm to dm). The specular component is generated by reflections from slopes of surface features that meet two criteria, which together specify that the slopes are perpendicular to the radar beam: (1) their strikes (or orientations) are concentric about the subradar point; and (2) their dips (or slope angles) approximately equal the incidence angle of the radar waves, which, in turn, equals the number of radial degrees away from the subradar point. According to the laws of reflection and Fresnel scattering, such spec-

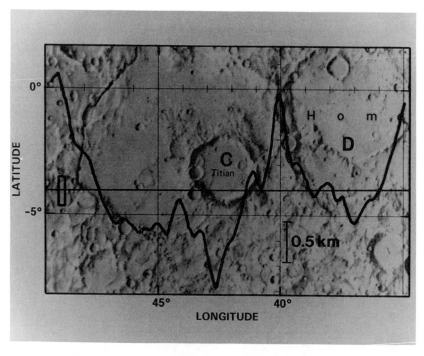

Fig 9. As in Fig. 8, topographic profile for craters in H-6 quadrangle.

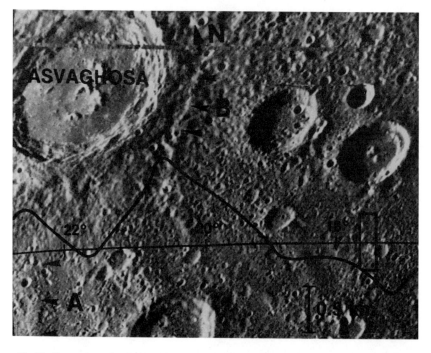

Fig 10. Comparison of typical landform and radar-derived topographic profile in H-6 quadrangle. Profile is superposed on Mariner 10 image of part of Santa Maria Scarp, plotted as in Fig. 8.

ular reflection is reversed in phase by 180°; such reversed circular polarization is termed polarized (Pettengill 1978).

If the surface is rough relative to radar wavelengths, a portion of the returned radar signal is diffusely scattered according to Lambertian scattering laws (Pettengill 1978). Such scattering scrambles the sense of polarization by multiple scattering, reradiation from sharp edges, and internal reflections from random dielectric layer blocks such as rocks and boulders. The diffuse component consists of a polarized part (first reflections) plus a depolarized part. The polarized component of diffuse scattering is often obscured by the quasi-specular reflections from smooth surfaces near the angle of incidence. Because most real surfaces are a combination of specularly and diffusely scattering elements, a surface may be modeled approximately by the superposition of two components: rough patches on smooth slopes.

Although radar waves sent to Mercury are tuned to a single frequency, the returned echoes are spread in frequency by Doppler effects. Data acquired in the form of frequency spectra are called continuous wave (or CW) data. If the returned echoes are separated into time delay gates (ranging) as well as frequencies, delay/Doppler 2-dimensional arrays can be produced, which then can be transformed into planetary coordinates to produce reflectivity

7-C   Enlarged view of the northeast region of the H-7 photomosaic

Fig 11. As in Fig. 10 for the H-7 quadrangle. Profile is superposed on Mariner 10 image of an area dominated by long bright linear streaks which appear to be subdued ridges.

maps. We describe below how such data may yield elevation plots along the
apparent path of the subradar point.

II. EXPERIMENTAL METHODOLOGY AND APPROACH

Radar data of Mercury were acquired at Goldstone during approximately
50 observation periods in the 1970s. Some of these data were published at the
time (Zohar and Goldstein 1974), but many were not reduced or published
until recently (Clark et al. 1984; Clark et al. 1985). Figure 2 shows the tracks

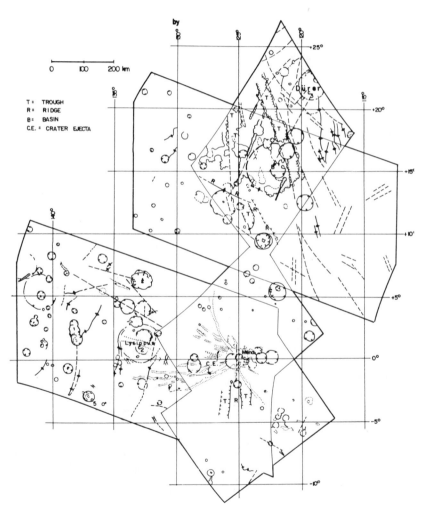

Fig 12. Morphological map of portion of H-7 quadrangle (Thompson et al. 1986). Linear and
curvilinear features with associated relief were mapped from Mariner 10 stereo pairs.

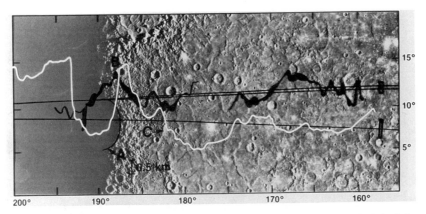

Fig 13. As in Fig. 10 for the H-8 quadrangle. Profile is superposed on 1:5000000 shaded relief map in area dominated by Caloris-related structures. Note drop in elevation between 170 and 175°, indicating the presence of an inter-ring trough. Note Caloris-radial scarps at B and C.

of all Goldstone radar data from which geologically useful products have been derived to date. These are 12.5 cm (2388 Hz or S-band) polarized data; that is, the receiver was set for circularly polarized echoes in the opposite sense from the transmitted beam (thus primarily quasi-specular reflection).

Figure 3 illustrates the geometry of a delay-Doppler map (cf. Pettengill 1978; Jurgens 1982). The first part of the planet from which echoes are received is the subradar point; later echoes come from concentric rings farther back towards the silhouette of the disk. The Doppler frequencies map in parallel stripes, due to approach (by one side of the disk) and recession (of the other) caused by apparent rotation of the planet. There is a two-fold ambiguity

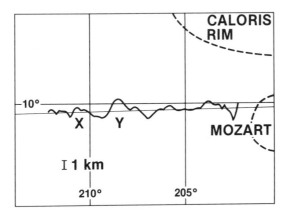

Fig 14. S-band profile in unimaged portion of Caloris region east of Fig. 13. Note scarps X and Y which are similar to Caloris radial scarps in Fig. 13.

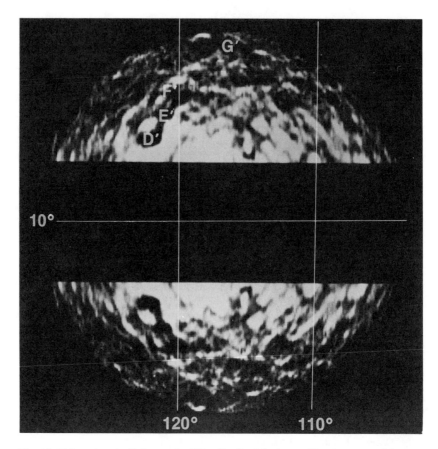

Fig. 15. H-7 quadrangle 12.5 cm radar (S-band) reflectivity image. Note the structural features which are observable in both this image and the visual image with overlapping coverage (Fig. 16): crater chain marked D', E' and F', and lineaments marked G'.

between delay-Doppler coordinates and latitude-longitude: as shown by A and A' in Fig. 3, each delay-Doppler cell north of the Doppler equator has a corresponding cell to the south. Elimination of the ambiguity requires use of radar interferometric techniques such as those applied to Venus (Jurgens et al. 1980) but not yet to Mercury.

In the reflectivity images shown here, the longitude resolution is approximately constant at 10 km, while latitude resolution varies from ~65 km at the Doppler equator to 5 km at the top or bottom of the image. The statistical certainty of the reflectivity images decreases radially outward from the sub-radar point due to the angular backscattering properties associated with the quasi-specular scattering mechanism and due to the increased resolution associated with the delay-Doppler coordinate system (see Fig. 3). In topography profiles, latitude resolution is 65 km, but longitude resolution varies from ~10 to 20 km due to smoothing applied during profile extraction. The altitude

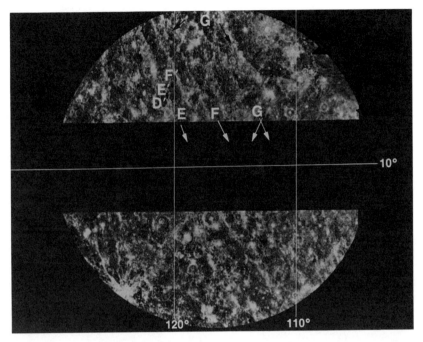

Fig. 16. Portion of photomosaic covered by Fig. 15 radar image. Note features D', E', F' and G', as described in Fig. 15.

resolution is roughly 150 m, but degrades with distance from the subradar point.

JPL ephemeris routines using the IAU coordinate system (Standish et al. 1976) were used to determine the delay-Doppler coordinate frame and the planetary coordinate mapping transformation matrix for each observation. Although a recent ephemeris was used for the 1974 data, the reduction of the 1972 and 1973 data did incorporate Mariner 10 geometric data; all are believed to be consistent within the spatial resolution of the data.

Delay-Doppler imaging of the small-scale roughness associated with the depolarized backscattering is difficult because of the smaller effective radar cross section. Such images would be very useful for comparing Mercurian craters with well-characterized lunar craters. Alternately, we have tried to characterize the roughness associated with the quasi-specular scattering component using Hagfors' (1968) model. This model assumes a Gaussian height distribution and an exponential auto-correlation function whose effective roughness is specified by a parameter C. In this case, C is the inverse of the n mean squared slope associated with structure sizes larger than the wavelength.

A two-stage process was used to unravel the effects of topography from orbital elements and to define the range and Doppler frequency centers for each image. An initial estimate of the range to the subradar point was made by forming an array of echo power points as a function of range, and by fitting them to models of the range resolving function of the radar and a scattering law for a rough spherical planet. This yields the echo amplitude, range to first contact (the subradar point), and an estimate of the Hagfors C parameter. These estimates are refined using the 2-dimensional delay-Doppler array and a model employing the 2-dimensional instrument resolving function convolved with a spherical model of the planet. The resulting fit forms the reference range and Doppler coordinates for the image.

The background noise must be subtracted from the data. All images include at least four range gates recorded before the first real signal; the average noise in those gates is subtracted from the remaining data. Data frames are normalized to each other by (1) multiplying by the ratio of the number of spectra in a given frame to the average number of spectra per frame, and (2)

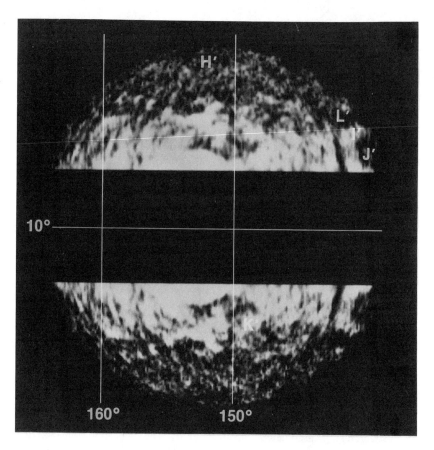

Fig. 17. As in Fig. 15 for the H-8 quadrangle. Note the structural features which appear both here and in the visual image (Fig. 18) of the same area: curvilinear features I′ and L′, craters J′ and K′. Note distinct appearance of smooth plains marked H′.

multiplying the result by the ratio of the inverse square of the noise to the inverse square of the average noise per frame. The mean scattering law (derived as explained above) is then removed by dividing the reflectivity by the square root of the mean scattering at each delay-Doppler coordinate. The square-root normalization leaves an impression of the quasi-specular scattering law while reducing the dynamic range of the images so that features are readily seen. The data are then transformed into the planetary coordinate frame. Separate frames can be averaged to form a sharp image by removing the effects of planetary motion.

We have derived topography profiles using the method described below. Mercury's apparent rotation is so rapid that the time required for a surface element to pass through one Doppler resolution cell is the same order as one round-trip period (about 10 minutes). The fitting procedure assumes that the delay-Doppler templates for a spherical planet match those of the actual planet if the model profile is delayed to match the first contact with the surface at a specific Doppler frequency. Profiles of delay measurements are formed for each frequency, relative to a spherical model, and are transformed into altitudes. The altitudes are less accurate away from the subradar point due to greater noise, however, averages in longitude bins help reduce that noise.

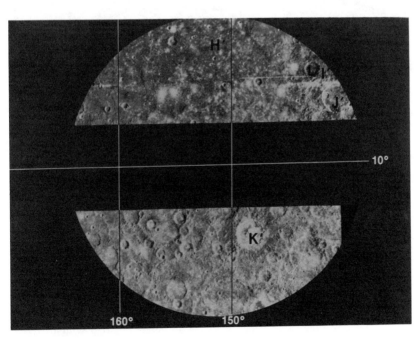

Fig. 18. Portion of photomosaic covered by Fig. 17 radar image. Note features H', I', J', K', L' as described in Fig. 17.

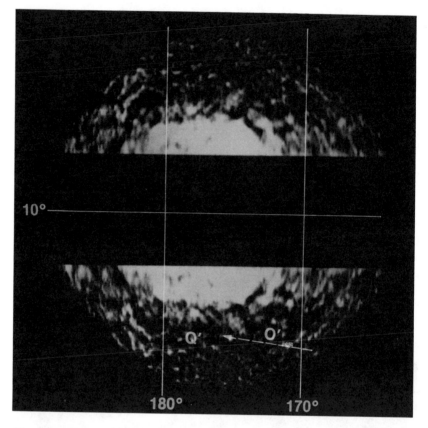

Fig. 19. As in Fig. 15 for part of the H-8 quadrangle. Note the relatively featureless smooth plains at Q'.

Radar profiles, reflectivity images and Hagfors C parameter were plotted on 1:5000000 shaded relief maps or photomosaics derived from Mariner 10 images. A longitude correction ($0°.5$ W) was applied to compensate for differences between IAU and USGS coordinate systems (Davies 1976). Surface features greater than 10 km in scale are being studied to address such major questions as: (a) characteristic roughness or topography trends of the major terranes and structures; (b) relationships between radar-derived and visually-derived surface properties; (c) the nature of the unimaged hemisphere; and (d) the significance of preferred alignments and other statistical properties of observable features.

III. GEOLOGIC ANALYSIS

Radar coverage of Mercury includes an equatorial belt which extends from ~15°N to 5°S. Coverage is more extensive in the hemisphere imaged by Mariner 10 than in the unimaged hemisphere.

Topographic Profiles

Altimetry profiles have been used to study geologic structures on Mercury, especially craters. For the larger craters studied rim heights are low and floor depths are shallow. Excursions in elevation beyond 2 km are rare (see Figs. 4–9). The profiles yield minimum depths because the radar footprint extends at least 50 km from N to S, averaging elevations across the crater. The profiles show maximum depths most accurately for carters larger than 50 km diameter, especially if the tracks pass nearly centrally through the crater. The depth-to-diameter (d/D) ratio of Titian (although off-center from the track) and the unnamed basin to the west is about 1:75 (Figs. 4 and 9). Lysippus (Fig. 5) is another good example of a transected crater; it has a d/D ratio of 1:75, whereas Mozart (Fig. 6) is about 1:90. Note Wang Meng in Fig. 8. Equivalent data for lunar craters of similar diameter (Apollo Lunar Sounder or Laser Altimeter maps; Elachi et al. 1976; Wollenhaupt and Sjogren 1977) show d/D ratios about 3 times greater than for Mercury (the 30 × 30 km footprints for the lunar data are comparable to the 12.5 × 50 km Mercury data). For example, Neper, Wyld, Maraldi and Kastner (90 to 120 km diameter craters) have d/D of about 1:30. The difference can be ascribed to Mercu-

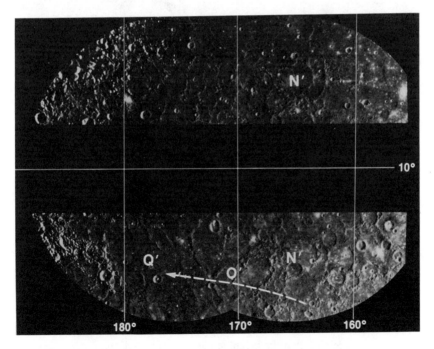

Fig. 20. Photomosaic partly covered by Fig. 19 radar image. Portion of the photomosaic of the H-8 quadrangle which lies in the area of coverage of the reflectivity image shown in Fig. 19 is the circular portion to the left. Note feature marked Q' as described for Fig. 19.

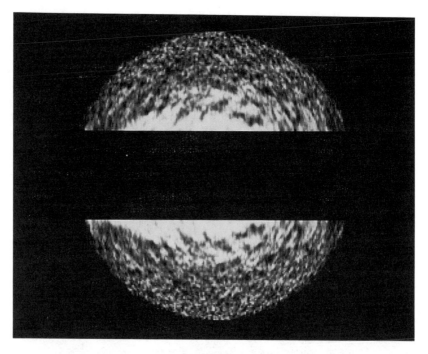

Fig. 21. As in Fig 15 for the unimaged hemisphere, which is directly southwest of Caloris. Note the overall similarity of the features observable here and those which can be observed in Fig. 19 which is centered about 20° due east, and to the southeast of Caloris.

ry's greater gravity (see also Chapter by Pike). It is uncertain to what degree craters might be modified by factors such as lava filling or viscous relaxation of Mercury's warmer, more plastic crust (Schaber et al. 1977).

Topographic profiles show many examples of linear features of tectonic origin. The H-6 quadrangle (Fig.4) is dominated by rugged highlands, intercrater plains and well-defined scarp systems. Figures 4 and 10 show profiles of the Santa Maria scarp system, which appears to consist of a series of asymmetric ridges in an en echelon pattern. A profile of Haystack Vallis yields a depth of 1.5 to 2 km.

Quadrangle H-7 (Fig. 5) contains both intercrater plains and smooth plains. Radar profiles partly compensate for the high Sun angles of Mariner 10 pictures of this region. Small (0.5 to 1 km) elevation increases occur where radar profiles cross some prominent bright linear streaks, suggesting that the streaks are subdued ridges (Fig. 11). Figure 12 shows a map of one such area made using Mariner 10 stereo pairs by Thompson et al. (1986), who confirmed the impression that the bright streaks are elevated, with a few cases of trough-like depressions between pairs of bright ridges.

The H-8 quadrangle (Fig. 6) is dominated by smooth plains and by struc-

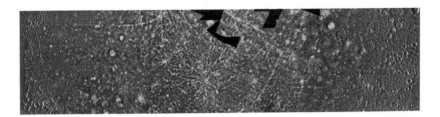

Fig. 22. Mercator projection photomosaic of images of the equatorial region on Mercury. Note how the character of the features changes depending on the Sun angle (higher near center, and lower nearer edges) in the photography. Most prominently displayed here are the long, bright, orthogonal, vertically and horizontally trending features in the H-7 quadrangle. These features are not clearly equivalent to crater rays.

tures both radial and concentric to Caloris basin. Well-formed radial scarps transect the profiles between 175° and 185° long (Figs. 6 and 13); a drop of 1 to 2 km is shown on one profile between 170° and 175°. This may be an inter-ring depression related to Caloris, but it is also near the ancient multiring basin Tir Planitia, mapped by Spudis and Guest (see their Chapter). Relief in the H-8 quadrangle is less than 2 km with rms slopes of only 4° to 5°, whereas quadrangles to the east, which are less dominated by smooth plains, have relief up to 4 or 5 km and rms slopes of 6 to 8 degrees.

Figure 7 shows radar profiles in the unimaged hemisphere extending southwest of Caloris. There may be scarp-like features between 200° and 210° long, where radial scarps from Caloris would be expected (Figs. 7 and 14). Terrain between 220° and 240° and between 280° and 300° long appears similar to the H-8 quadrangle smooth plains.

Radar Reflectivities

Despite the two-fold ambiguity, radar reflectivity images (Clark et al. 1984; Zohar and Goldstein 1974a) show interesting features. Bright curvilinear or textural features appear to be correlated with observed structures, as seen in the portions of the photomosaic which accompany the reflectivity images (Figs. 15–21). Craters and crater chains are readily interpretable. Figures 15 and 16 show striking circular features associated with the crater chain marked D′, E′ and F′. There is good correspondence for craters J′ and K′ in Figs. 17 and 18.

Smooth plains show as darker, relatively featureless areas on radar reflectivity maps, e.g., Budh Planitia (H′ in Figs. 17 and 18) and Tir Planitia (Q′ in Figs. 19 and 20). Some linear features appear on radar reflectivity images, e.g., the NE trends associated with Durer (G′ in Figs. 15 and 16) and an EW trend (marked I′ in Figs. 17 and 18). Figure 21 shows the unimaged area southwest of Caloris. Its character is similar to that of Fig. 20, which is just to the east; thus, we infer that it also consists predominantly of smooth plains, with linear structures related to Caloris.

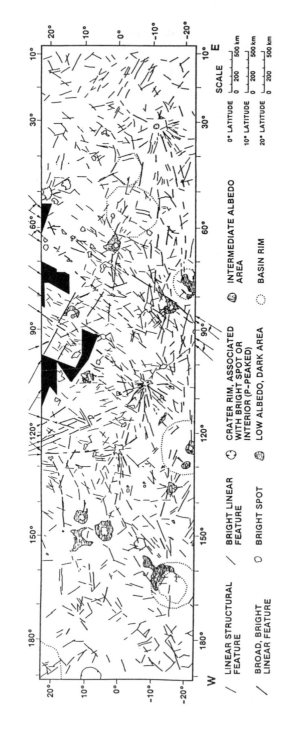

Fig. 23. Plot of linear albedo and structural features observed in Fig. 22 photomosaic. Note roughly orthogonal character of bright linear features in the region of highest Sun angle (figure courtesy of M. A. Leake).

/ LINEAR STRUCTURAL FEATURE

/ BRIGHT LINEAR FEATURE

⊙ CRATER RIM, ASSOCIATED WITH BRIGHT SPOT OR INTERIOR (P-PEAKED)

⊛ INTERMEDIATE ALBEDO AREA

/ BROAD, BRIGHT LINEAR FEATURE

○ BRIGHT SPOT

⊛ LOW ALBEDO, DARK AREA

⋯ BASIN RIM

SCALE

0 200 500 km 0° LATITUDE

0 200 500 km 10° LATITUDE

0 200 500 km 20° LATITUDE

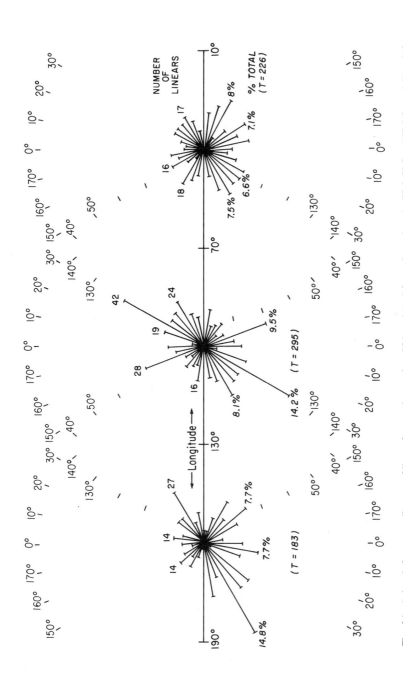

Fig. 24. Azimuth frequency diagram of linear features longer than 95 km plotted for each quadrangle (H-6, H-7 and H-8 from right to left, respectively). Note distinctive distribution of linear albedo features in the H-7 quadrangle which may correspond to the proposed lineament system as discussed in the text. Note the similarity in distribution of linear structural features in quadrangles H-6 and H-8.

Linear Features

For the purposes of comparing radar data with a Mariner 10 photomosaic of the equatorial regions (Fig. 22; R. G. Strom, personal communication), we have mapped (Fig. 23) and characterized (Fig. 24) the linear features in that region. The observed regularities in azimuth and spacing of bright albedo streaks and linear structures suggest a pattern of nearly orthogonal fractures or stresses, which could be remnants of the proposed global lineament system (Pechmann and Melosh 1979; see Chapter by Melosh and McKinnon). The numerous features striking N 30° E and N 20° W in the Beethoven quadrangle area reflect the contributions of high-albedo streaks. Such trends are less prominent in the Kuiper area (which displays trends open to the north) and Tolstoj (which displays trends radial and concentric to Caloris, N 60° E, N 40° W). Close examination of albedo streaks mapped from the photomosaic show that they are often positive relief features (polygonal crater rims, hilly and lineated terrain, ridges and scarps), which is unexpected since they had been

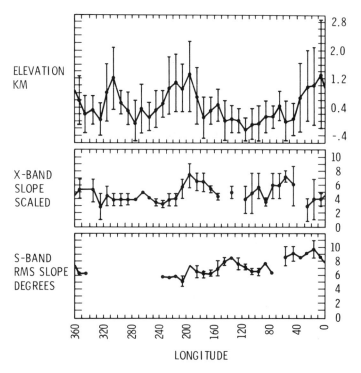

Fig. 25. Comparison of topography, scattering properties and geologic map classification with 10° longitude bins in the equatorial region. These profiles are derived from all available Gold-stone, Arecibo, and Haystack topography and rms slope data. Error bars are equivalent to standard deviation. X-band slope values are relative variations which are derived from reflec-tivity data (Ingalls and Rainville 1972).

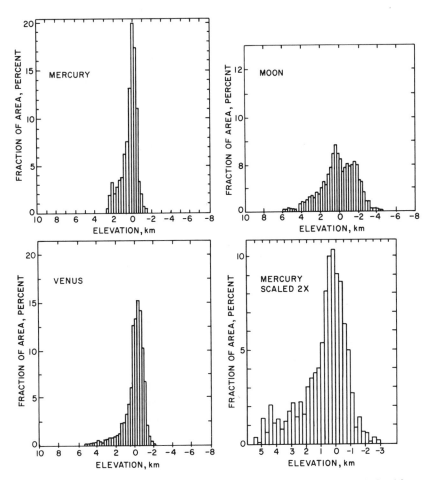

Fig. 26. Comparison of topography for terrestrial planets. These histograms are calculated from available topography data for each of the planets. Note similarity between histograms for Venus and Mercury. Two elevation histograms exist for the Moon. One histogram has un-normalized topography data. In the other, lunar elevations are gravity scaled for comparison with Mercurian elevations: lunar elevations are decreased by a factor of 2 because the Moon's gravitational field is less than half that of Mercury.

considered analogous to lunar rays (Strom 1984; Allen 1977). This finding could link the lineament system to the proposed episode of crustal expansion accompanying core formation (Solomon 1977,1978; Strom 1984). Impact events may reactivate old structural features. Radar characterization of the global lineament system has not been completed.

Comparison of Roughness at Various Scales

We have assembled equatorial data from Goldstone (Goldstein 1971; Zohar and Goldstein 1974; Zohar and Goldstein 1974, unpublished; Clark et al. 1984) and Arecibo (Harmon et al 1986), as well as rms slope estimates from Goldstone S-band (Jurgens 1974) and early Haystack 3.5 cm (X-band) data (Ingalls and Rainville 1972) derived from the Hagfors C parameter. These estimates pertain to slopes at all scales larger than roughly 10 wavelengths, averaged across fields of view approximately 20° across. Figure 25 shows relationships (in 10°-longitude bins) between topography (km-scale) and rms slopes at two wavelengths corresponding roughly to meter and 20-cm scales for X- and S-band, respectively. Uncertainties are generally greatest in the unimaged hemisphere and for the topographic data.

Within the equatorial zone, high topographic regions seem to be associated with orbital antipodes. The X-band rms slopes may show a hemispheric asymmetry; there appears to be a correlation with the topography from at least 140° to 300° long associated with the distribution of smooth plains, which seem to occur preferentially in the imaged portions. The X-band radar cross sections for both the polarized and depolarized components were derived from Goldstone data by Clark et al. (1986). The polarized cross section veries from 0.05 to 0.06, while the depolarized component varies from 0.01 to 0.015. The depolarization ratio is 0.17 near 155° long and 0.14 near 275°. The estimate of 0.08 for the depolarization ratio at S-band (Goldstein 1971) is about half what we find at X-band (see also Ingalls and Rainville 1972), indicating that an enhanced diffuse scattering component exists in the cm- to mm-size range. This enhancement in small-scale roughness may be greater for Mercury than for the Moon.

Figure 26 shows topography histograms for the Moon, Venus and Mercury (the latter shown twice, once rescaled, to account for gravity scaling, for comparison with the Moon). Lunar data are from laser altimetry observations and Venus data from Pioneer Venus altimetry. Whereas the lunar topography appears bimodal, Mercury's topography seems unimodal.

Acknowledgments   We wish to thank R. M. Goldstein for his pioneering work in obtaining Mercury radar observations. We are very grateful for the substantive discussions which we have had with G. Schaber, M. Strobell, R. Strom, M. Kobrick, W. Sjogren, G. Trager, C. Wood and F. Vilas. We particularly wish to recognize the significant contributions to software and hardware development necessary for this work made by G. Downs and S. Johnson. This work was performed under NASA contract at the California Institute of Technology Jet Propulsion Laboratory.

RADAR OBSERVATIONS OF MERCURY

JOHN K. HARMON and DONALD B. CAMPBELL
National Astronomy and Ionosphere Center

Earth-based radar observations have provided information on the equatorial topography, surface scattering properties, and rotation of Mercury. The recent observational program at Arecibo Observatory has been devoted primarily to the measurement of altimetric profiles using the delay-Doppler technique. These profiles, which were derived from observations made over a 6 yr period, provide fairly extensive coverage over a restricted equatorial band. The data have sufficient resolution and accuracy to permit the identification of radar signatures for features as small as 50-km diameter craters and km high arcuate scarps. More importantly, they have been used to identify large-scale topographic features such as smooth plains subsidence zones and major highlands regions. In this sense the radar data complement spacecraft images, as quantitative altimetry from the latter is limited to shadow measurements of small, high-relief features such as crater rims. Radar data such as that obtained at Arecibo also provide most of what little information is available on the unimaged hemisphere. Measurements of radar cross section, Doppler spectrum shape, and depolarization provide information on the dielectric properties and roughness of the surface. These results can be readily compared with similar measurements for other planets.

Radar probing of Mercury, Venus and Mars began in the early 1960s. The early Mercury observations, like those of her sister planets, were mainly continuous-wave (CW) measurements of radar cross section and Doppler spread. Although these observations provided some information on surface conditions, their most notable contribution was the discovery of Mercury's nonsynchronous rotation (Pettengill and Dyce 1965).

With the early 1970s came more detailed studies of the planet's radar scattering properties (Goldstein 1971) as well as the first attempts to measure Mercurian topography using pulsed or coded transmissions (Smith et al. 1970; Ingalls and Rainville 1972). The latter suggested that Mercury has a somewhat more subdued topographic relief than is the case for the equatorial zones of Venus and Mars. The radar altimetry measurements of Zohar and Goldstein (1974a), the last to be published prior to Mariner 10, revealed topographic detail in the form of hills and valleys with 1 to 2 km relief and features resembling large craters.

The Mariner 10 encounters of 1974–75 brought the first great advance in our knowledge of Mercury and, with it, an understandable falloff in interest in Earth-based observations of the planet. The spacecraft results were by no means exhaustive, however. Mariner 10 carried no altimeter and quantitative altimetry was limited to shadow measurements of high-relief features such as crater rims and scarps. Moreover, many of the images were obtained at unfavorable illumination angles and one entire hemisphere was left unimaged. The prospect of doing useful, complementary radar work was greatly enhanced with the installation, in the mid-1970s, of a sensitive S-band radar on the upgraded Arecibo telescope. Starting in 1978, a regular program of radar observations of Mercury was undertaken at Arecibo, primarily for the purpose of obtaining accurate measurements of surface topography. The results of the altimetry measurements obtained over the period between 1978 and 1984 were recently reported by Harmon et al. (1986). This chapter presents several of the more important findings from that paper along with a summary of the current state of our knowledge of the radar scattering properties of the Mercurian surface.

I. ALTIMETRY MEASUREMENT TECHNIQUE

The standard technique for spatially resolving planetary radar echoes is the so called delay-Doppler method, which combines pulsing or coding of the transmitted wave with Fourier analysis of the coherently detected echo. The technique has been used very successfully on Venus to obtain high-resolution maps of radar reflectivity. Similar attempts to map Mercury have met with much less success due to the weaker echoes for that planet. Thus the emphasis of the Mercury program at Arecibo has been to use the delay-Doppler method to measure altitudes along the Doppler equator rather than to map radar reflectivity.

The altimetry estimation method used at Arecibo is essentially that described in detail by Ingalls and Rainville (1972) and Shapiro (1972). Here one fits numerically computed templates of echo power vs delay to the delay-Doppler data array to estimate the time delay to the leading edge of the planet at each Doppler longitude. From these delays one subtracts the computed delays to a reference sphere of given radius. The residual delays can then be expressed as altitudes relative to an assumed datum ("sea level"). Each altitude datum point in the Arecibo profiles corresponds to a spatial resolution cell (radar footprint) measuring 0.15 degrees in longitude by 2.5 degrees in latitude (6×100 km). The typical altitude accuracy (for a surface which is flat over this cell) is 100 m.

This estimation method is feasible only for the strong echo from the subradar region. For example, a single day's observing session at Arecibo yields an altitude profile spanning roughly 14 degrees of longitude along the Doppler equator. Using Mercury's 5 deg/day rotation, one can construct longer profiles by connecting the individual profiles obtained from observations made on several successive days. Altitude profiles traversing up to 90 degrees of longitude have been obtained at Arecibo by this means. Obviously, to obtain even a modest degree of planetary coverage by this method entails a major observational effort. The data presented by Harmon et al. (1986) were derived from approximately 150 days of observations spread over a 6 yr period.

The primary limitation of Earth-based radar altimetry is that it is restricted to the narrow equatorial zone defined by the possible subradar tracks. Mercury's 7° orbital inclination to the ecliptic renders a $\pm 12°$ equatorial band accessible to the radar. The most interesting known feature lying outside this zone is the Caloris basin. Fortunately, the accessible band does encompass representative examples of most of the important terrain types found on the planet.

II. EQUATORIAL GLOBAL TOPOGRAPHY

To show the equatorial topography on a global scale, we have plotted all of the Arecibo profiles from 1978 to 1982 on a 0-360° longitude scale in Fig. 1. The data are plotted on an absolute altitude scale where the zero-altitude datum is defined by a 2439.0-km radius reference sphere. In Fig. 2, we show a histogram giving the distribution of altitudes from Fig. 1.

The mean of the altitudes in Fig. 1 is $+0.7$ km (2439.7 km mean radius). This result is consistent with the mean equatorial radius reported from earlier radar observations (Ash et al. 1971) and radio occultations (Fjeldbo et al. 1976). The zero altitude datum corresponds to the typical elevation of Mercurian lowlands and is close to the most probable altitude ($+0.3$ km) as given by the peak of the histogram.

The extreme range of measured Mercurian altitudes is 7 km, as measured from the lowest crater floors to the high plateau near the 0°W longitude meridian. For comparison, the Moon has approximately 10 km of peak-to-peak relief about its center of mass (Kaula et al. 1974) or about 7 km of relief (excluding crater floors) about its center of figure (Brown et al. 1974). The typical elevation difference between Mercurian highlands and lowlands is about 3 km, which corresponds to the approximate equivalent width of the altitude distribution in Fig. 2. Like the Moon, Mercury's altitude distribution is weakly bimodal and skewed toward positive altitudes.

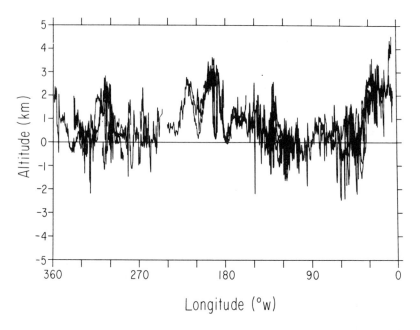

Fig. 1. Combined plot of all Arecibo radar profiles of Mercury from 1978 to 1982, showing absolute altitudes relative to the 2439.0 km radius reference sphere. Latitudes of profiles on this figure range from 12°N to 5°S. All profiles are simply superposed with no latitude averaging (figure from Harmon et al. 1986; copyright Amer. Geophys. Union).

The equatorial zone shows two major topographic highs. The first of these is roughly centered at 10°W longitude and crosses the Mariner 10 eastern terminator. This plateau-like feature has an abrupt drop off on its western side which is associated with an extensive system of faults (see Sec. III.B). The second major high area covers a broad region south of the Caloris basin between 160°W and 240°W. This region contains two local topographic lows at 180°W and 210°W which correspond to smooth plains regions. A third, less extensive highlands area can be seen in the more northerly profiles near 310°W in the unimaged hemisphere.

Mercury's resonant spin state indicates that the long axis of the planet's dynamical figure is aligned with the perihelion subsolar points at 0°W and 180°W longitude. The Arecibo results in Fig. 1 show that Mercury's topographic figure is roughly aligned with its dynamical figure, although the two largest bulges appear to be more closely aligned along a 10–190°W longitude axis. Goldreich and Peale (1966) have shown that a difference between the equatorial moments of inertia of only about 0.01% (assuming an ellipsoidal figure) would suffice to ensure a high probability of capture into the 3:2 resonance state. This corresponds to variations in the dynamical figure of about 100 m. Hence, it is conceivable that the dynamical figure of Mercury is dominated by a long-wavelength component of uncompensated topography associ-

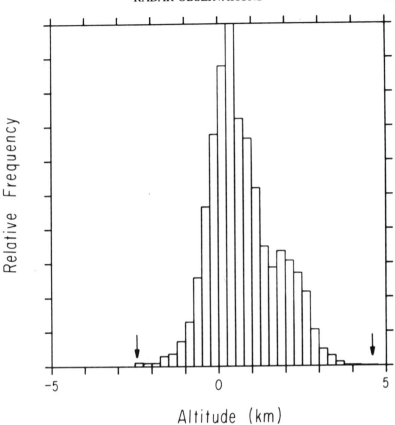

Fig. 2. Histogram of the altitudes shown in Fig. 1. The histogram is normalized by the number of altitude data points within each altitude bin, not by area. The lowest and highest altitudes measured (−2.4, +4.6 km) are denoted by arrows (figure from Harmon et al. 1986; copyright Amer. Geophys. Union).

ated with the observed bulges. An alternative explanation for the 3:2 spin-orbit resonance is that there is a lunar-like mascon associated with the smooth plains in or around Caloris. Melosh and Dzurisin (1978) have argued that a positive gravity anomaly associated with 400 m of uncompensated material in the circum-Caloris smooth plains would suffice to control the planet's dynamical figure. The Arecibo altitude profiles over the circum-Caloris plains are consistent with a subsidence of these plains under an emplaced load (see Sec. IV.A) which, if partially uncompensated, could strongly influence Mercury's dynamical figure.

III. TOPOGRAPHIC FEATURES: THE IMAGED HEMISPHERE

Mariner 10 imaged the hemisphere of Mercury extending from 10°W to 190°W longitude. Much of the structure in the radar altimetry profiles can be identified with specific features in the USGS maps of this hemisphere. In this section we present several of the more important results obtained from a comparison of the Arecibo altimetry with the Mariner 10 images and image-derived maps.

A. Crater Depths

Shadow measurements from Mariner 10 images have yielded depth estimates for Mercurian craters with diameters in the range 1 to 170 km (Gault et al. 1975; Malin and Dzurisin 1977,1978). These studies have revealed a strong similarity between the crater depth/diameter relations for Mercury and the Moon, although large Mercurian craters tend to be shallower than lunar craters of comparable size. Both planets show a flattening or turnover of the depth-vs-diameter curve for diameters > 10 to 20 km.

Radar altimetry offers an alternative crater depth measurement technique, one which is not restricted to structures near the Mariner 10 terminator. However, the large size of the radar footprint limits reliable depth estimates to craters larger than about 40 km in diameter, well beyond the turnover point in

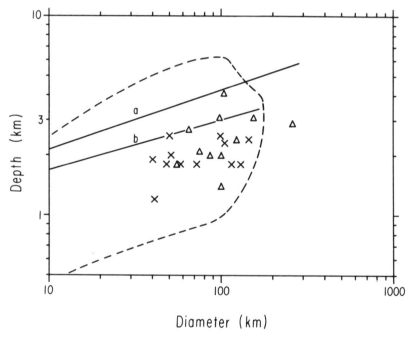

Fig. 3. Depth vs diameter for fresh (triangles) and degraded (crosses) Mercurian craters as measured by radar altimetry. The dotted line shows the approximate range of shadow-derived values for fresh and degraded Mercurian craters (Malin and Dzurisin 1977). The straight lines are the fitted power-law depth/diameter relations for fresh craters on (a) the Moon (Pike 1974), and (b) Mercury (Malin and Dzurisin 1977) (figure from Harmon et al. 1986; copyright Amer. Geophys. Union).

the depth/diameter relation. The accuracies of the radar depth estimates for large craters such as Mozart and Handel are comparable to or better than the 15 to 20% accuracy which has been claimed for shadow-derived depths. For smaller craters, the radar depth estimates tend to be less reliable due to the coarse latitude resolution, a point which is discussed in some detail by Harmon et al. (1986).

Figure 3 shows the radar-derived depth estimates vs diameter for 23 craters in the diameter range 40 to 260 km. These data are tabulated crater by crater in Harmon et al. (1986). Only craters in the imaged hemisphere have been included, although the depths shown in Fig. 3 are consistent with the depths of several of the more obvious impact features seen in radar profiles from the unimaged hemisphere. The craters in Fig. 3 are identified as being either fresh (USGS classes C3–C5) or degraded (USGS classes C1–C2). Included in Fig. 3 are the shadow-derived depth data of Malin and Dzurisin (1977); their power-law relation for fresh Mercurian craters is plotted along with an envelope encompassing their range of values for both fresh and degraded craters. Also shown in Fig. 3 is the power-law depth/diameter relation for fresh lunar craters (Pike 1974). The radar-derived depths fall well within the range of the shadow depths, although the radar depths of fresh craters lie, on the average, 17% below the fresh-crater depth/diameter curve of Malin and Dzurisin (1977). Since the radar and shadow depth data sets have no craters in common, it was not possible to determine if there is, in fact, a systematic disagreement between the two techniques. It is clear, however, that the radar depths in Fig. 3 lend strong support to the assertion that large, fresh craters on Mercury are shallower than those on the Moon.

B. A Major Fault System

Figure 4 shows the Arecibo altitude profiles obtained over the USGS H-6 quadrangle along with a map showing the location of the radar ground tracks. The most distinctive large-scale topographic feature in this region is the marked drop in mean elevation which can be seen between 30°W and 40°W longitude. The most impressive part of this drop occurs just west of Handel crater, where the elevation changes by 3 km within 1.5 degrees of longitude (70 km). To the north, a shallower slope occurs across and to the west of Yeats crater. To the south, the high eastern rim and asymmetric profile of Homer basin suggest that this basin also straddles the slope. Inspection of Mariner 10 images and of the geologic map of DeHon et al. (1981) shows that this regional slope occurs in an area with several west-facing intracrater fault scarps. These scarps, and some associated intercrater lineaments, are shown on the schematic map in Fig. 5. The intercrater features are identified as ridges by DeHon et al. (1981), although they have a highlighted appearance in the Mariner 10 images which would seem to be more suggestive of west-facing scarps. In any event, it is clear that the 3 km drop seen in the radar altimetry is associated with an extensive system of faults which trends north-south over

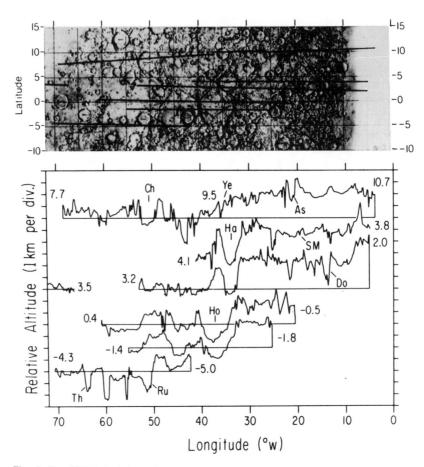

Fig. 4. *Top:* USGS shaded-relief map of the H-6 quadrangle, with dark lines to indicate the subradar tracks. *Bottom:* The corresponding altitude profiles for H-6. The vertical scale is relative altitude only. The zero-altitude datum is indicated by horizontal lines for those profiles for which accurate absolute altitudes were available. Numbers next to profiles indicate latitude. Abbreviations denote identified features: As, Asvaghosa; Ch, Chaikovskij; Do, Donne; Ha, Handel; Ho, Homer; Ru, Rudaki; SM, Santa Maria Rupes; Th, Thakur; Ye, Yeats (figure from Harmon et al. 1986; copyright Amer. Geophys. Union).

the northern half of the H-6 quadrangle. It is interesting to note that while this fault system and the arcuate scarps (see next section) have entirely different altimetric signatures, they both give rise to intracrater scarps which are similar in morphology.

C. Arcuate Scarps

The most distinctive tectonic features identified in the spacecraft images are the arcuate scarps, which are generally believed to be thrust faults driven by planetary contraction. Shadow measurements of these features give typical

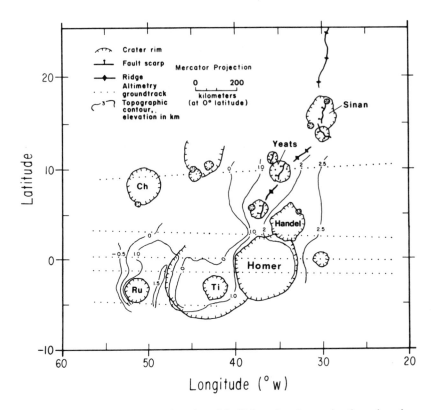

Fig. 5. Schematic map of the central portion of the H-6 quadrangle covering the region where radar altimetry shows a large west-facing downslope. Approximate topographic contours based on the radar data are shown. The indicated ridges and intracrater scarps are from the geologic map of DeHon et al. (1981). The names of several craters are abbreviated: Ru, Rudaki; Ch, Chaikovskij; Ti, Titian (figure from Harmon et al. 1986; copyright Amer. Geophys. Union).

heights in the range 500 to 1000 m, although a height of 3 km has been estimated for a portion of Discovery Rupes (Dzurisin 1978; Strom et al. 1975).

Three of the arcuate scarps shown on the tectonic maps of Dzurisin (1976,1978) and Strom et al. (1975) have been identified in radar profiles. All three are located in the same region of the H-6 quadrangle (Fig. 4); we display them on an expanded scale in Fig. 6. These scarps are delineated by shadow and, as they are near the Mariner 10 eastern terminator (illumination from the west), they were mapped as east-facing dips. The radar signatures of the three scarps are similar in that each appears to be a ridge-like feature with a height of 700 m and an across-strike width of roughly 70 km. In each case, the eastern slope of the radar feature agrees well with the location of the image shadow. The two scarps in the lower profile of Fig. 6 have slightly asymmetric profiles which could be consistent with overthrusting from the west. The scarp

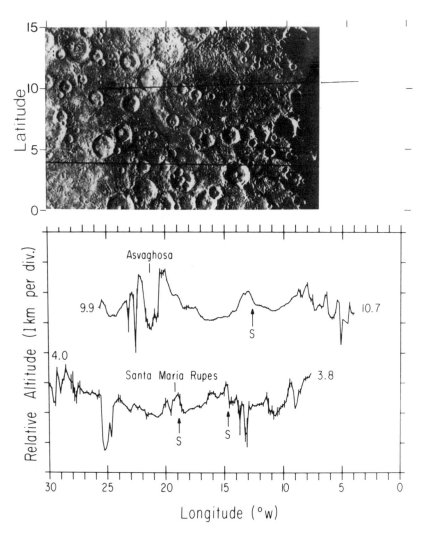

Fig. 6. Altitude profiles selected from the H-6 quadrangle (see Fig. 4) showing topography across the three mapped arcuate scarps discussed in the text. Arrows (S) denote the locations of the downthrown (shadow) sides of the scarps as determined from USGS maps and Mariner 10 images. Vertical bars indicate the ± 1 standard deviation altitude errors. The subradar tracks for the two profiles are shown on the Mariner 10 image at the top (figure from Harmon et al. 1986; copyright Amer. Geophys. Union).

east of Asvaghosa crater, on the other hand, has a more symmetric and rounded profile. All three scarps have significant downslopes on their western sides
and hence are more ridge-like than one would infer from images obtained at a
single illumination aspect.

D. A Large Basin (H-7 Quadrangle)

Most of the Arecibo radar coverage on the USGS H-7 map is concentrated in the western half of the quadrangle in an area which is dominated by a
large, degraded impact basin. In their survey of large Mercurian basins,
Schaber et al. (1977) listed this basin as an 839-km diameter, single-rim structure centered at 130°W, 1°8N. It is the second largest Mercurian basin in their
survey after Caloris. More recently, Spudis and Strobell (1984) claim to have
identified a multiple ring structure for this basin.

In Fig. 7 we show four of the altitude profiles across this basin, along
with markers indicating the approximate basin edges as given by the USGS
shaded-relief map and Schaber et al. (1977). The profiles show that the topography across this basin is complex and strongly latitude dependent. Some
portions of the basin floor appear to have been significantly altered by post-
basin impacts. Other portions of the basin appear topographically smooth,
possibly indicating a smooth plains fill. The southernmost profile at 4°5S is the
simplest of the four profiles in Fig. 7. It shows 1.2 km of rather smooth,
down-bowed relief, an upraised rim in the east, and a western rim which
coincides with a scarp visible in Mariner 10 images. The most northerly profile in Fig. 7 shows a very prominent basin rim in the northwest along with
two smooth, down-bowed sections of basin floor in the northwest and northeast. The two northern profiles in Fig. 7 show very little topographic expression across the northeast rim of the basin, whereas the next profile to the
south may show some basin rim structure on the eastern side. The radar data
show that, overall, the interior of the basin is not significantly lower than the
level of the adjacent terrain. This suggests that the basin has been severely
modified by post-impact processes such as isostatic relaxation, impact cratering or volcanic infilling. In addition, the interior of the basin may have experienced some local subsidence of smooth plains fill.

E. Smooth Plains

Not surprisingly, regions on Mercury which have been mapped as smooth
plains tend to have relatively smooth altitude profiles. The largest mapped
expanses of smooth plains lie within the H-8 quadrangle (Schaber and Mc-
Cauley 1980), and our profiles in this quadrangle (Fig. 8) have a generally
smoother appearance than do profiles in, for example, the H-6 quadrangle
(Fig. 6). Although smooth, the profiles in Fig. 8 do show some very distinctive large-scale undulations. The Tir Planitia smooth plains appear to be
strongly down-bowed. The lowest parts of these plains lie 1 to 1.5 km below
the terrain to the east (which includes a mix of Caloris ejecta, smooth plains

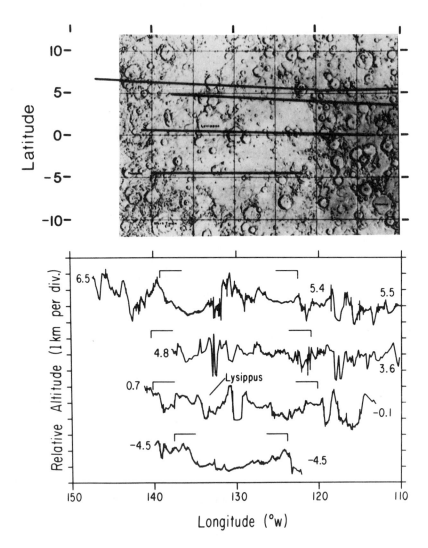

Fig. 7. Altitude profiles selected from the H-7 quadrangle showing topography across the large basin centered at 130°W, 1°8N. Brackets denote the approximate locations of the basin rim as given by the USGS shaded-relief map and Schaber et al. (1977). Vertical bars indicate ±1 standard deviation altitude errors. The subradar tracks for these profiles are shown on the USGS shaded-relief map at the top (figure from Harmon et al. 1986; copyright Amer. Geophys. Union).

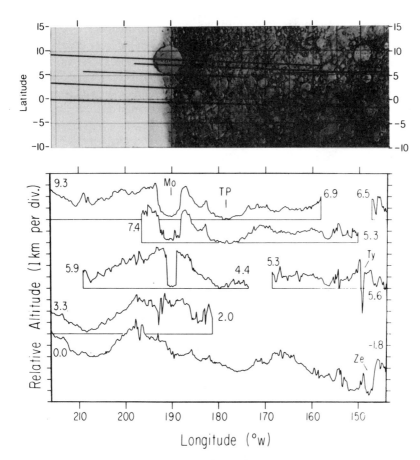

Fig. 8. *Top:* USGS shaded-relief map of the H-8 quadrangle, with subradar tracks indicated. Note that the Mariner 10 terminator is at 190°W, with terrain to the west unimaged. *Bottom:* The altitude profiles for H-8. The display format follows that of Fig. 4: Mo, Mozart; TP, Tir Planitia; Ty, Tyagaraja; Ze, Zeami (figure from Harmon et al. 1986; copyright Amer. Geophys. Union).

and intercrater plains) and the Mozart ejecta blanket to the west. The extension of these plains into the unimaged hemisphere is discussed in the next section.

IV. TOPOGRAPHIC FEATURES: UNIMAGED HEMISPHERE

Mercury showed the same face to the Sun at each of the three Mariner 10 encounters, leaving half of the planet unimaged. The unimaged hemisphere extends from 190°W to 10°W longitude. Radar has provided most of what little information is available for this side of the planet.

A. The Circum-Caloris Smooth Plains

The Mariner 10 western terminator runs through the large crater Mozart. As can be seen from Fig. 8, the profiles in the unimaged terrain west of Mozart have the same smooth, down-bowed appearance as the profiles over the imaged plains of Tir Planitia to the east of Mozart. This indicates that the circum-Caloris smooth plains extend well into the unimaged hemisphere, at least to the western edge of the H-8 quadrangle.

Harmon et al. (1986) suggested subsidence under an emplaced load as the most likely explanation for the down-bowing of the smooth plains in Fig. 8. The distinctive topography and the existence of mare-like ridges in Tir Planitia suggest that the circum-Caloris region has been subjected to a lithospheric loading and flexure process similar to that for the lunar maria. In this sense, the radar topography offers indirect support for a volcanic origin for the Mercurian smooth plains. However, certain objections to the volcanic hypothesis remain unanswered, and the question of the origin of the smooth plains must still be considered open.

B. Other Features

In Fig. 9, we show the Arecibo profiles in the region 250–360°W longitude, which includes the western half of the H-9 and the entirety of the H-10

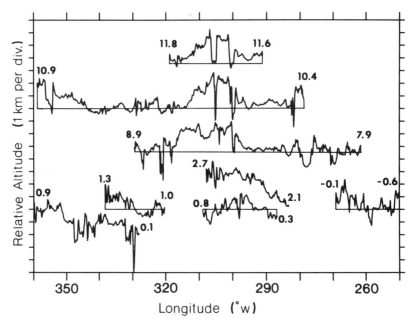

Fig. 9. Altitude profiles for the longitude range 250 to 360°W. These longitudes include the H-10 quadrangle and the western half of H-9. This entire region lies in the unimaged hemisphere. The display format follows that of Figs. 4 and 8.

quadrangles. Here we note a variety of terrain features. The most striking large-scale topographic feature is the highlands region between 295–315°W. The 2 to 3 km height of these highlands above the adjacent plains is comparable with the height of the large plateau dominating the eastern half of the H-6 quadrangle. The eastern slope of the H-6 plateau can in fact be seen near 350°W at the western edge of the H-10 quadrangle (Fig. 9). Relatively smooth terrain, possibly intercrater plains, can be seen extending over 280-295°W and in a northerly profile between 330 and 355°W. A number of features can be seen in Fig. 9 which are suggestive of large impact craters. The largest of these is a 2.5 km deep crater at 279°W, 8°.5N. The size and structure of this crater suggests that it may be a ring basin of the same class as Renoir and Rodin. Two other large craters can be seen atop the highlands at 300°W and 306°W, and a third crater can be seen at 356°W, 11°N.

The topography of the small portion of the unimaged hemisphere which we have studied does not appear to be markedly different from that of the imaged hemisphere. Also, no evidence has been found for the existence of another Caloris-scale impact structure in the equatorial zone of this hemisphere.

V. RADAR SCATTERING PROPERTIES

Most of the recent radar studies at Arecibo have concentrated on surface topography. However, work has been done, both at Arecibo and other facilities, which bears on the radar scattering properties of the Mercurian surface. These studies can be used to place some constraints on the roughness and dielectric properties of the upper surface layers of the planet.

A. Radar Cross Section and Quasi-Specular Roughness

Some useful information can be derived from simple CW observations of the planet. If the echo is predominantly quasi-specular, then the total radar cross section (normalized by the projected area of the planet) provides a first-order estimate of the Fresnel reflection coefficient of the surface. Recent (unpublished) CW observations at Arecibo and earlier observations at Goldstone (Goldstein 1971) have estimated the average radar cross section of Mercury as 0.06 at 13-cm wavelength. For comparison, solid rock surfaces have reflection coefficients in the range 0.15 to 0.25, while rock powders have reflectivities of 0.03 to 0.06 (Campbell and Ulrichs 1969; Evans 1969). One can conclude, therefore, that the upper meter or so of the Mercurian surface is dominated by relatively porous soils or regolith rather than by consolidated rock. The same conclusion holds for the Moon, which has a comparable radar cross section (0.06) at 23-cm wavelength (Hagfors 1967; Evans 1969). Venus, by contrast, has an average reflectivity of 0.11 at 13-cm wavelength (Carpenter 1966), which suggests that this planet has more exposed bedrock than does Mercury. The total radar cross section of Mercury varies over the range 0.04 to 0.08, depending on the location of the subradar point on the planet. This is much smaller than the range of cross sections (0.04–0.15) measured for Mars.

The central portion of the radar Doppler spectrum of a terrestrial planet is typically dominated by a narrow peak associated with quasi-specular scatter from smooth surface undulations with horizontal scales greater than the observing wavelength. The width of this peak can be used to estimate the rms slope of these undulations. This type of analysis has been done most extensively for Mars, where rms slopes have been found to vary between $0°.25$ and $10°$. On the average, Mercury spectra are broader and the inferred rms slopes are larger than those obtained for Venus and Mars (Pettengill 1978). At Arecibo, we have used the delay-Doppler data to estimate specific cross section as a function of incidence angle along the Doppler equator. Such a technique is more sensitive to spatial inhomogeneities in scattering than is an analysis based on CW-derived Doppler spectra. We find the smoothest and most homogeneous surfaces on Mercury to be in the smooth plains, for which we measure rms slopes of about $4°$. However, the smooth plains are considerably rougher than many of the plains regions on Mars, where rms slopes can be 1° or less. Presumably this reflects the fact that Mercury has not been subject to the same erosional/depositional processes which have altered the Martian surface. Outside the smooth plains, the Mercurian Doppler spectra tend to be broader and to be extremely irregular in a way which is indicative of highly inhomogeneous scattering. In many cases, these scattering variations can be identified with known features such as large craters.

B. Diffuse Scattering and Radar Polarization

Studies of the wings of the Doppler spectrum and of the degree of depolarization of the echo can provide information on the relative importance of diffuse scattering by wavelength-scale surface structure. We have estimated that diffuse scatter accounts for approximately 27% of the total radar cross section of Mercury, a percentage which is comparable to that estimated for the Moon and Venus (Harmon and Ostro 1985). Volcanic plains regions on Mars have been found to show much higher percentages of diffuse scatter as well as near-complete depolarization of the diffuse echo, conditions which are indicative of extremely chaotic small-scale surface texture. Goldstein (1971) noted two depolarization enhancements arising from rough regions in the unimaged hemisphere of Mercury. Recent dual-polarization CW observations at Arecibo have revealed some bumps and asymmetries in the Mercury spectra, including Goldstein's feature near 230°W longitude. These features are very subdued compared to the Martian features, however, and in no case have we found depolarization to exceed 40% for any portion of the diffuse echo of the planet. If Mercury does possess volcanic plains, they must have a smoother small-scale texture than those on Mars.

VI. CONCLUSION

Earth-based radar has proven to be an important tool for the remote sensing of Mercury and may be one of the few sources of new observational data on the planet for the foreseeable future. Data coverage over the accessible subradar zone remains incomplete, and additional observations are planned at Arecibo to fill in some of the gaps. Some upgrades to the Arecibo telescope and radar system are being considered which would improve the quality of the altimetry data, especially for observations made at larger geocentric distances, and which would justify a more serious effort at reflectivity mapping of the planet.

It is obvious that the next great advance in Mercury studies must await another spacecraft, preferably an orbiter. The Arecibo altimetry data provide a good sample of the sort of topographic features one would expect to measure with an orbiting altimeter. Among the most interesting results of the radar observations are the discoveries of large-scale topographic signatures for such features as the circum-Caloris smooth plains and the H-6 fault zone. While some preliminary speculation on the tectonics of such regions based on the radar results may be in order, a fuller understanding of Mercurian tectonics will only come from a comparison of altimetry with gravimetric data and higher-quality imagery.

Acknowledgments. We wish to thank I. I. Shapiro, D. L. Bindschadler, J. W. Head, T. Forni and J. Chandler for their assistance in various aspects of this work. We also wish to thank the staff of the Arecibo Observatory for their support of the Mercury program, especially R. Velez, A. Crespo, A. Hine and G. Giles. We thank the American Geophysical Union for permission to reproduce material first appearing in the *Journal of Geophysical Research*. The S-band radar program at Arecibo is supported by a grant from the National Aeronautics and Space Administration.

STRATIGRAPHY AND GEOLOGIC HISTORY OF MERCURY

PAUL D. SPUDIS
U.S. Geological Survey

and

JOHN E. GUEST
University of London Observatory

The history of Mercury can be reconstructed through photogeological analysis of global geologic relations of rock-stratigraphic units. Mercurian stratigraphy is subdivided into five time-stratigraphic systems based on mappable basin and crater deposits. These systems are the pre-Tolstojan (oldest), Tolstojan, Calorian, Mansurian, and Kuiperian (youngest). The pre-Tolstojan system includes crater and multiring basin deposits and extensive intercrater plains materials that were emplaced before the Tolstoj basin impact; at least some of these plains may be of volcanic origin. The Tolstojan system, defined at its base by Tolstoj basin deposits, includes deposits of that basin and other basin, crater and plains materials. The Calorian system comprises Caloris basin materials, which constitute an extensive stratigraphic datum, and subsequent widespread smooth plains materials. The global extent, depositional settings and age relations of Mercurian smooth plains indicate that they are predominantly of volcanic origin. Widespread thrust faults, indicating a period of global compression, finished their development in mid-Calorian time. Smaller-scale regional compression is evidenced by numerous wrinkle ridges within the smooth plains materials. If Mercury's impact-flux history is similar to the Moon's, most of these major geologic events were probably completed within the first 1 to 1.5 Gyr of Mercurian history. The Mansurian and Kuiperian systems, defined by the

[118]

craters Mansur and Kuiper, include only impact crater deposits; no evidence for regional volcanic or tectonic activity of this age has been recognized. Thus, Mercury appears to have completed most of its geologic evolution early in its history. This primitiveness makes Mercury an important object for the study of the comparative geologic evolution of the terrestrial planets.

I. GEOLOGIC MAPPING OF MERCURY

The geologic map is an important weapon in the armory of a geologist tackling the problems of understanding the stratigraphy and surface history of a planet. The work of W. Smith and others in the early 19th Century first demonstrated the value of geologic mapping in the interpretation of a planetary surface. They recognized that the different types of rocks could be organized into identifiable 3-dimensional units, or strata, and that the distribution and geometrical form of these could be expressed in map form by drawing together a large number of individual observations into an understandable synopsis. Relative ages of a succession of rocks—their order of emplacement—could be determined from the geometrical relations between defined strata. They also recognized that strata of different ages contained assemblages of distinctive fossils that allow worldwide time correlations between different rock units. The science of stratigraphy was thus developed in which rock units could be studied and placed in time sequence to enable geologists to interpret the geologic history and evolution of the Earth's surface.

Studies of the stratigraphy and the geologic histories of planetary bodies other than the Earth have, perforce, relied to a large extent on photogeological techniques. The Moon was the first body to receive such attention. In a paper that was to be a landmark in the history of lunar studies, Shoemaker and Hackman (1962) showed that the Moon could be mapped geologically by using telescopic observations with resolutions no better than 0.5 km and that the stratigraphy and thus surface history could be determined. Since that time, the geologic history of the Moon has been studied in considerable detail and geologic mapping has been aided and tested by the study of rocks from Apollo and Soviet landings on the near side of the Moon. Analyses of lunar samples largely confirmed the stratigraphic sequence established photogeologically. The techniques of lunar geologic mapping developed since the early 1960s and their stratigraphic implications have been discussed extensively by Mutch (1970), Wilhelms (1970,1984,1987a) and Wilhelms and McCauley (1971). The extension of these techniques for use on other planetary bodies with different surfaces and conditions has been reviewed by Wilhelms (1972,1987b).

The Moon is a relatively easy body to map geologically as it has no atmosphere and thus was molded by fewer and less complex processes than those which operate on planets like the Earth. Individual lunar rock bodies of different origins and ages can be identified by photogeological techniques with relative confidence because each unit has a distinctive morphology.

Without the aid of fossils, relative ages of units must be determined by super-position and cross-cutting relations between the units. On a planet such as Mars, which has a much more complex history involving many different pro-cesses, geologic mapping *sensu stricto* is more difficult. Here, mapped units of distinctive morphology may in some cases represent *surfaces* of different origins and ages, rather than the rock units into which the landforms are cut (Milton 1974). On other planetary bodies such as Ganymede, the geologic map may be more akin to a tectonic map until further data on the nature of the crustal materials become available.

Before the three Mariner 10 flybys in 1974 and 1975, very little was known of the geology of Mercury. This innermost planet lies within 28 an-gular degrees of the Sun and thus is difficult to study with Earth-based tele-scopes, which can obtain resolutions of no better than 300 km. Astronomical observations show faint albedo markings (see, e.g., Murray et al. 1972). Po-larimetric and photometric observations (Lyot 1929; Hameen-Anttila et al. 1970; McCord and Adams 1972a; Dollfus and Auriere 1974) indicated that the surface texture of Mercury is similar to that of the Moon; this similarity implies that the surface has been broken by impact cratering.

The Mariner 10 images provided the first data that could be used in a truly geological investigation of the surface (Murray et al. 1974). The first flyby was an equatorial pass at about 700 km distance at closest approach on the dark side of the planet. The second was a south polar pass at about 50,000 km. The third and final flyby was over the northern hemisphere at just under 400 km. The most valuable material was supplied by the first two flybys which produced over 2000 useful pictures, the best resolution being near 100 m for a few photographs taken at closest approach. Less than 50% of the surface of the planet was observed by Mariner 10, and the best resolution of the pictures that cover most of this area was at 1 to 1.5 km. Mariner 10 provided images of Mercury that are comparable to Earth-based photographs of the Moon. Because this was a flyby mission, the pictures suffer from the same problems as those taken of the Moon with a telescope on Earth. The images become progressively foreshortened toward the limbs, thus making interpretation less reliable as foreshortening becomes more severe. In addi-tion, for each of the flybys, the lighting geometry was virtually the same. This means that close to the terminator small-scale topography is enhanced and the effects of albedo are relatively weak, whereas the reverse is the case towards the limbs. Nevertheless, Mariner 10 provided excellent imagery for pho-togeological studies, although the variable lighting geometry from one area to another somewhat reduces the consistency of global mapping. A bonus was that the second flyby coverage overlapped in places with that from the first flyby, providing stereoscopic pairs that can be used for qualitative examina-tion of surface morphology. The recovery of images taken during the third flyby, which were intended to cover specific areas at very high resolution, was hampered by technical problems on Earth. Because of the consequently limited coverage, these images are less valuable to geologic studies than had been hoped.

As expected, the virtual lack of an atmosphere and the similarity in many respects of Mercurian to lunar processes made Mercury ideal for photogeologic mapping of the type already carried out for the lunar surface. Although the image resolution is less than that obtained by many of the Lunar Orbiter and Apollo missions, the success of lunar geologic mapping based on telescopic observations promised similar success in applying the same techniques to Mariner 10 images of Mercury.

The first geologic map of Mercury based on Mariner 10 images was prepared by Trask and Guest (1975), who used mainly first-encounter pictures. This map was termed a geologic/terrain map because the mapping scale, image resolution, and variable lighting conditions commonly made it necessary to combine several rock units of different lithologies and ages as "terrain units" on the basis of distinctive landforms. Despite these handicaps, this mapping provided a basis for discussing the broad geologic history of the planet and for subsequent mapping. Geologic mapping of the southern hemisphere based on the second-encounter pictures was later carried out by Guest and O'Donnell (1977) and O'Donnell (1980).

The next stage in the geologic mapping of Mercury was the preparation of 1:5,000,000 scale quadrangle sheets published by the U.S. Geological Survey (Holt 1978). So far, seven such sheets have been produced showing the geology in more detail than had been possible at the scale used by Trask and Guest (1975). This more detailed mapping has been reassessed with new global mapping by Spudis (in preparation). It is the new synoptic map that is the basis for part of this chapter.

The production of a new global map was facilitated by the development of a Mercurian time-stratigraphic system (Spudis 1985) that is based on the rock-stratigraphic classification developed during quadrangle mapping and the definition and subdivision of the Caloris Group by McCauley et al. (1981). The formal time-stratigraphic scheme was defined on the basis of recognizable crater and basin deposits, as will be described in detail below. This new system has enabled a planetwide correlation of geologic events on Mercury, and it also readily facilitates interplanetary correlations with similar systems developed for the Moon (see, e.g., Wilhelms 1984) and Mars (Scott and Carr 1978).

II. GEOLOGIC TERRAINS

Mercury's surface may be divided into several morphologic and physiographic categories, including basins, craters, plains and tectonic features. The geomorphology of Mercury has been reviewed by Gault et al. (1977) and in detail by Strom (1979,1984). In the following section, we review briefly the surface features of Mercury, with the specific aim of categorizing them into *geologic provinces* (McCauley and Wilhelms 1971) that will form the basis for the planetwide stratigraphy to be described below.

A. Cratered Terrain and Intercrater Plains

During the initial Mariner 10 encounters, it was recognized that Mercury is a heavily cratered body (Murray et al. 1974; Gault et al. 1975); thus comparisons naturally centered on lunar similarities. In the post-Apollo period, an outline of lunar history was perceived to be well understood (Wilhelms 1984), and many studies of Mercurian craters made detailed comparisons to the populations and densities observed in the heavily cratered lunar highlands (see, e.g., Murray et al. 1974; Gault et al. 1975; Strom 1977).

It was soon apparent that although Mercury displays regions of high crater density, its most densely cratered surfaces are not as heavily cratered as the lunar highlands (Fig. 1). This difference is due primarily to large exposures of intercrater plains, which also are present in the lunar highlands, but in restricted quantities (Strom 1977). At no place on Mercury do we find heavily cratered regions of overlapping large craters and basins like those that characterize the lunar highlands. The first-order implications of this observation are that Mercury has undergone some type of early resurfacing and that this resurfacing was more intense than comparable activity on the Moon.

The intercrater plains were first recognized, described, and mapped as a terrain unit by Trask and Guest (1975). The intercrater plains were defined as the level to gently rolling terrain between and around the cratered terrain. Superposed on the plains are numerous craters < 10 km in diameter. Many of these small craters are elongate and shallow or are open on one side; crater clusters as well as crater chains are common, and thus their morphology and distribution is similar to that of secondary impact craters. Trask and Guest (1975) believed that the only likely sources for these craters were the large craters and basins of the heavily cratered terrain. However, they realized that the crater rims of the heavily cratered terrain are narrow and that there is little evidence of surrounding continuous ejecta; but since Mercurian crater ejecta sheets are emplaced close to the crater (Gault et al. 1975), they argued that the crater ejecta would become more readily degraded than the more extensive lunar ejecta sheets. On the basis of these observations, Trask and Guest (1975) postulated that the majority of the intercrater plains predate the heavily cratered terrain. Murray et al. (1975) went further by suggesting that a global volcanic event that supposedly formed the intercrater plains wiped out most of the early craters.

Further study and the availability of the second-encounter images of the southern hemisphere showed that the intercrater plains have a more complex stratigraphy in relation to the heavily cratered terrain. Malin (1976*a*), Guest and O'Donnell (1977) and Strom (1977) found a number of old craters that had been embayed or covered by intercrater plains materials. Thus, some of the highland craters must predate parts of the intercrater plains. In a detailed study, Leake (1982) attempted to distinguish intercrater plains of different ages on the basis of degree of degradation of craters superposed or embayed by the intercrater plains. She found that these materials were emplaced over a time period represented by Class 5 through Class 3 craters[a] and that the area covered by these plains materials generally decreases with decreasing age. The secondary craters superposed on the intercrater plains apparently were derived mainly from Class 1 to 3 craters and Class 4 basins of the heavily cratered terrain.

Interpreting the origin of the intercrater plains is controversial. They were considered by Wilhelms (1976) and Oberbeck et al. (1977) to be old basin-ejecta material. Supporting this hypothesis is the intimate association of the lineated terrain of Trask and Guest (1975) with intercrater plains. The lineated terrain modifies both the heavily cratered terrain and the intercrater plains, and it consists of areas cut by parallel straight valleys that are as much as 10 km across and have scalloped margins. It is similar to the lunar Imbrium sculpture formed by secondary impact of ejecta from the Imbrium impact basin. Therefore, the Mercurian lineated terrain is probably related to large basins and it could be argued that the intercrater plains themselves are different facies of basin ejecta. However, Strom (1979) points out that there is an apparent lack of source basins to account for such a widespread unit.

Several workers have suggested a volcanic origin for the intercrater plains, although landforms of unambiguous volcanic origin have not been found. Malin (1978) identified some domical features that may be volcanic, although subsequent study by Spudis (1984) and Spudis and Prosser (1984) suggests that many of these features are massifs associated with ancient multiring basins, discussed below (Sec. II.B). Dzurisin (1978) recognized linear ridges that might be the sites of extrusive activity on the intercrater plains. In addition, morphological studies of craters by Cintala et al. (1977) suggest that in physical properties, the intercrater plains materials are more like the lunar maria and Mercurian smooth plains than they are like the megaregolith of the lunar highlands. Although the Mariner 10 images have allowed the recognition of a stratigraphy within the intercrater plains terrain, the origin of these materials remains ambiguous.

It may be that heavily cratered terrain exists on Mercury, but that it is largely buried by extensive intercrater plains materials. Recently, a large population of multiring basins, all of which predate the Mercurian intercrater plains, has been discovered (Spudis 1984; Spudis and Strobell 1984). Many of these basins have been partly mapped by several workers, including Trask and

---

[a]These numbered crater classes (1-5) based on morphologic degradation are used by Leake (1982) and others from the University of Arizona's Lunar and Planetary Laboratory and are the exact opposites of the U.S. Geological Survey's classes 1-5, based on stratigraphic position. Thus, a Univ. of Ariz. class 5 crater (very degraded) corresponds to a USGS class 1 crater (very old).

Fig. 1. Comparison of heavily cratered terrain on Mercury and the Moon. (a) Mercury. Regions between tracts of overlapping craters appear smooth at low resolution (intercrater plains). Large-crater population ($D > 300$ km) appears to be mostly absent. Mariner 10 incoming mosaic. (b) Moon. In the heavily cratered terrain northeast and east of Mare Marginis, note the large, overlapping craters, as large as basin scale (AS16-3029).

Strom (1976), Schaber et al. (1977), De Hon (1978), Frey and Lowry (1979) and Croft (1979). A typical ancient multiring basin is Eitoku-Milton, south of Tolstoj shown in Fig. 2 (Spudis 1984; Spudis and Prosser 1984). Ancient basins on Mercury are delineated and mapped by a combination of criteria, including arcuate scarps and ridges, circular alignments of massifs, isolated massifs, and localized concentrations of smooth plains within otherwise heavily cratered terrain. At least 15 ancient basins, randomly distributed, have been mapped in the area of the Mercurian hemisphere imaged by Mariner 10. Their presence suggests that heavily cratered terrain of the lunar type does exist on Mercury, but that it has been largely obliterated by planetwide deposition of the intercrater plains (see Sec. III.A below).

Fig. 1-b.

B. Basins Younger than the Intercrater Plains

Several large impact basins that postdate the intercrater plains have been recognized on Mercury as evidenced by the superposition of their ejecta and secondary craters on the intercrater plains. The oldest of these basins is Dostoevskij ($-44°$, $177°$) shown in Fig. 3a. Dostoevskij is very degraded; only one obvious ring (about 400 km in diameter) is preserved. Partial preservation of a radially textured ejecta blanket is evident, and long chains of secondary craters, radial to the center of Dostoevskij, are preserved in some regions. Crater-density data suggest that Dostoevskij is one of the oldest basins on the planet (Spudis and Prosser 1984).

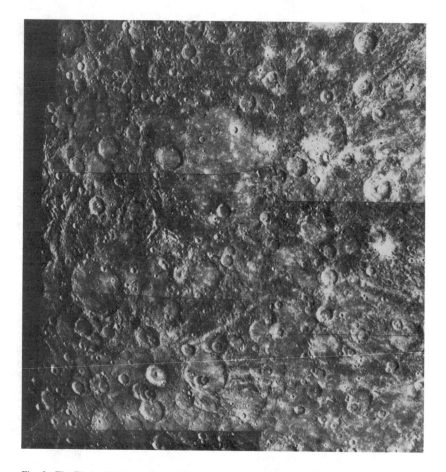

Fig. 2. The Eitoku-Milton basin. (a) Highland region, south of Tolstoj basin. (b) Overlay showing basin rings of Tolstoj (short dashes) and underlying Eitoku-Milton basin (long dashes).
Note arcuate, scarp-like rim portion (A), circular ridge beneath intercrater plains (B) and isolated massif (C). (North at top in each figure.)

Tolstoj ($-16°$, $164°$) is a true multiring basin, displaying at least two and
possibly as many as four, concentric rings (Figs. 2 and 3b). Tolstoj displays a
well-preserved, radially lineated ejecta blanket extending outward as much as
about one-basin diameter (500 km). The interior of Tolstoj is flooded with
smooth plains that clearly postdate the basin deposits. The Tolstoj ejecta
blanket is not radially symmetric with respect to the basin rim, as it appears to
be absent in the northern and western sectors of the basin. This suggested to
Schaber and McCauley (1980) that Tolstoj ejecta in this sector had been buried
by intercrater plains materials. It is suggested here instead that Tolstoj postdates the intercrater plains and displays natural ejecta asymmetry. Such asym-

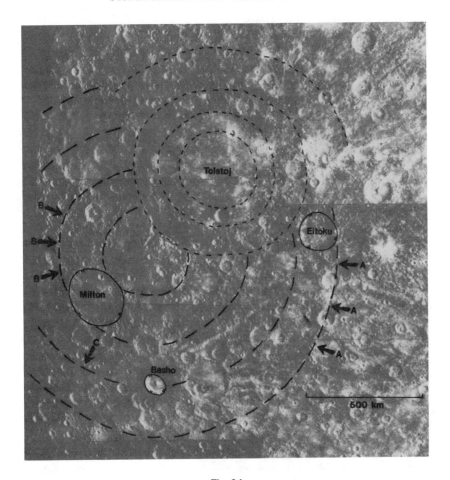

Fig. 2-b.

metry is observed in most mappable ejecta blankets of lunar basins (Wilhelms 1984) and may indicate basin formation by oblique impact (Gault and Wedekind 1978).

The Beethoven basin ($-20°$, $124°$) was imaged by Mariner 10 under high Sun illumination and its morphology and stratigraphy are difficult to observe. It appears to have only one ring, a subdued massif-like rim about 625 km in diameter. Even under these unfavorable lighting conditions, however, Beethoven displays an impressive, well-lineated ejecta blanket that extends as far as 500 km from its rim. The Beethoven ejecta blanket is also asymmetric with respect to its rim, but large expanses of younger, smooth plains occur in the

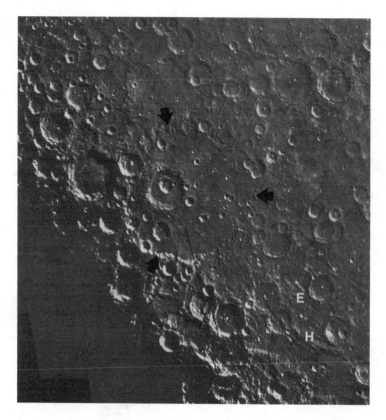

Fig. 3. Basins that postdate the intercrater plains. (a) Dostoevskij basin, H-12 quadrangle. Although heavily degraded, remnants of main rim (arrows) and radially textured ejecta (E) are preserved. Hero Rupes (H) is a lobate scarp that cuts Dostoevskij ejecta. Superposed crater density suggests that this basin is of late pre-Tolstojan age. (North at top; width of scene about 750 km.) (b) Tolstoj basin. Note strong radial texture to northeast and southeast of basin (Tolstoj ejecta). Interior of basin is flooded by smooth plains; some preplains (but postbasin) craters are visible on floor of Tolstoj (arrows). (North at top; width of scene about 1300 km.)

areas where its ejecta appear to be absent. The subdued rim of Beethoven has suggested to some workers that it is an extremely old feature (Scott and King 1987). In contrast, Spudis and Prosser (1984) found a surprisingly low crater density on the ejecta blanket; Beethoven probably postdates Tolstoj and may even be slightly younger than the Caloris basin.

The Caloris basin lies on the terminator as imaged by Mariner 10, thus only half of its circumference is visible (Fig. 4). Several different terrain units were identified by Trask and Guest (1975) as being associated with this impact basin. Surrounding the basin is the Caloris mountain terrain consisting of a ring of mountains with a diameter of about 1300 km. The mountains consist of smooth-surfaced blocks rising some 1 or 2 km above the surrounding ter-

Fig. 3-b.

Fig. 4. Caloris basin. Main rim, defined by rugged massifs, is 1340 km in diameter. Radially textured ejecta (Van Eyck Formation) is well developed to northeast of the basin rim (VE). Hummocky-to-knobby ejecta (Odin Formation) are prevalent due east of basin rim (O). Smooth plains (SP) appear to embay the Caloris ejecta in several regions. The basin floor displays radial and concentric ridges and a crudely polygonal fracture pattern. (North at top; width of scene about 700 km.)

rain and having a width of between 100 and 150 km. They are broken up into individual massifs that are typically 30 to 50 km long; the inner edge of the unit is marked by basin-facing scarps. Much of this massif material probably consists of bedrock uplifted from deep within Mercury.

Trask and Guest (1975) also recognized intermontane plains that occur in depressions between the mountains. This unit is distinguished from smooth plains elsewhere on the planet by its more rugged surface morphology and its position within the Caloris mountains. The materials are interpreted as a mixture of fallback material and impact melt.

Trask and Guest identified two other units that they considered to be ejecta from Caloris. The first, termed the Caloris lineated terrain, extends for about 1000 km out from the foot of a weak discontinuous scarp on the outer edge of the Caloris mountains. It consists of long hilly ridges and grooves subradial to the Caloris basin and is best expressed northeast of Caloris in the region of Van Eyck crater. In morphology, it is similar to the lunar Imbrium sculpture and is interpreted as ejecta from secondary craters of the Caloris basin.

The second unit forms a broad annulus about 800 km from the Caloris mountains, termed by Trask and Guest (1975) the hummocky plains. They consist of low, closely-spaced-to-scattered hills about 0.3 to 1 km across and from tens of meters to a few hundred meters high. In some places, the hills are aligned concentrically with the rim of Caloris, giving the plains a corrugated appearance. The outer boundary of this unit is not always easy to map as it tends to be gradational with the (younger) smooth plains materials that occur in the same region. Both the hummocky plains and the lineated terrain are considered to be facies of Caloris ejecta.

The floor of the Caloris basin consists of an extensive plains unit included by Trask and Guest (1975) as part of the smooth plains unit that is also found elsewhere on the planet. However, Schaber and McCauley (1980) and Guest and Greeley (1983) mapped this as a separate unit as it shows more intense secondary deformation in the form of sinuous ridges and fractures, giving the plains a grossly polygonal pattern. The origin of this material is uncertain but it may be volcanic, possibly formed by the release of magma as part of the impact event, or alternatively, a thick impact-melt sheet produced during basin formation.

C. Plains Materials

Widespread regions of Mercury are covered by relatively flat, sparsely cratered plains materials (Trask and Guest 1975; Strom et al. 1975). Immediately after the Mariner 10 data acquisition, Murray et al. (1974) drew the obvious parallel between the Caloris basin and its surrounding smooth plains and the lunar Imbrium basin and its surrounding maria. An obvious difference between the Mercurian smooth plains and the lunar maria, however, is that the plains on Mercury differ little in albedo from the average heavily cratered highlands and intercrater plains. This single characteristic has sparked considerable controversy.

The smooth plains are flat-to-gently-rolling, possess numerous lunar-mare-type wrinkle ridges, and are most strikingly exposed in a broad annulus around the Caloris basin (Fig. 4). They tend to fill depressions that range in size from regional troughs to crater floors. In high-resolution photographs, no unequivocal volcanic features, including flow lobes, leveed channels, and domes or cones, are visible.

The key issue concerning the smooth plains on Mercury is their origin. The Mariner 10 imaging team (Murray et al. 1974,1975; Strom et al. 1975) favored a volcanic origin, primarily on the basis of analogy to lunar basins and maria. This conclusion was challenged by Wilhelms (1976), who noted (correctly) that the arguments used by the Mariner team for a volcanic origin for the smooth plains were similar to arguments designed to advocate a volcanic origin for the lunar Cayley plains, subsequently shown by Apollo 16 mission results to be of impact-ejecta origin. The Cayley plains surround the Imbrium basin in the way that the Mercurian plains surround Caloris, although these plains cover more area than the lunar Cayley plains. Wilhelms (1976) speculated that the Mercurian plains are an ejecta facies of the Caloris basin, similar to the lunar Cayley plains, but much more voluminous. The debate on the origin of Mercury's smooth plains has continued in subsequent years, the volcanic hypothesis gradually becoming generally accepted, mostly by default, as little clear-cut evidence is available directly from the Mariner 10 data. Curiously, one key piece of evidence for the origin of the smooth plains has not been adequately discussed in the literature: the relative crater densities of the smooth plains and the Caloris ejecta and other Caloris-related features. This is discussed more fully in Sec. II.D below.

It was recognized during the systematic mapping of Mercury at 1:5,000,000 scale that some smooth plains appear to be transitional in age between those surrounding the Caloris basin and the older, intercrater plains. Those plains are not as extensive as the younger smooth plains, but their existence suggests that some mechanism of plains deposition operated throughout Mercury's early history.

D. Craters

Mercurian craters have all the morphological elements of lunar craters: the smaller craters are bowl shaped while with increasing size they develop scalloped rims, central peaks, and terraces on the inner walls (see Figs. 5-9). The ejecta sheets have a hummocky radial facies and swarms of secondary impact craters. The fresher craters of all sizes have dark or bright halos and well-developed ray systems. Although Mercurian and lunar craters are superficially similar, those on Mercury show subtle differences, especially in the effective range of ejecta. The continuous ejecta and fields of secondary craters are far less extensive (by a factor of about 0.65) for a given rim diameter (Gault et al.

Fig. 5. Kuiperian crater Basho (−32°, 170.5°; 70 km diameter). Note fresh-appearing, smooth floor, crisp rim topography, dark inner halo, and rays (FDS 0166846). (North at top.)

Fig. 6. Mansurian crater Zola (50°.5, 178°; 60 km diameter). A few superposed craters are seen, but rim and ejecta appear fairly fresh; no rays are preserved (FDS 99). (North at top.)

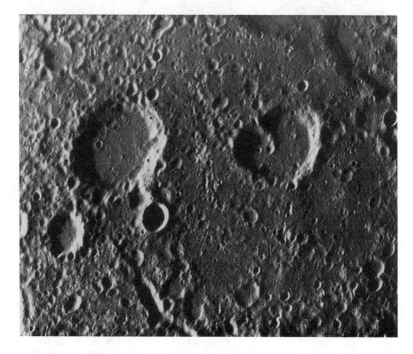

Fig. 7. Two unnamed Calorian-age craters. Crater at left ($-8°1$, $19°2$; 45 km diameter) and double crater at right ($-8°1$, $17°2$; 30 km diameter) show degraded rims and smooth plains fill. Both craters show that, at this age, wall terraces are largely obliterated (FDS 27460). (North at top.)

1975) than those of comparable lunar craters. This difference is considered by Gault et al. to result from the 2.5 times higher gravitational field on Mercury compared with that on the Moon.

As on the Moon, impact craters have been progressively degraded by subsequent impacts. There is thus a complete range of crater-degradational morphologies. The freshest craters have ray systems and a crisp morphology. Secondary craters are well developed and the continuous ejecta shows radial lineaments. Somewhat older craters have a similar form, but like their counterparts on the Moon, have lost their ray systems. With further degradation, the craters lose their crisp morphology, and features on the continuous ejecta become more blurred until only the raised rim near the crater remains recognizable, although it has many superposed smaller impact craters. Degradation continues until only a weak circular ring of hills is visible. This sequence of events may be interpreted by cross-cutting relations between craters in different stages of degradation.

The effect of degradation by secondary craters is much more marked on Mercury than it is on the Moon. This difference results from the much tighter

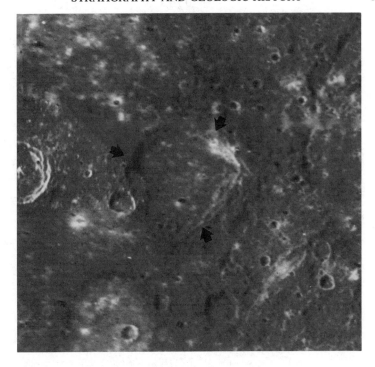

Fig. 8. Tolstojan crater Rudaki ($-3°5$, $51°5$; 120 km diameter). At this degradational state, crater ejecta are completely obliterated and numerous superposed craters are evident. Smooth plains fill is typical for craters of this age. (Portion of H-6 photomosiac; north at top.)

clustering of secondary impact craters, which can age an adjacent crater to a much greater extent than would be the case on the Moon. Additionally, the higher surface gravity of Mercury causes ballistically emplaced material to land at much higher velocities on the Mercurian than on the lunar surface (Scott 1977), thereby increasing the erosive power of secondary impacts. This proximity aging may make the earlier crater appear older than it actually is.

Because craters become progressively degraded with time, the degree of degradation gives a rough indication of the crater's relative age. Techniques for using the morphology of craters as a dating tool were first established for the Moon by Pohn and Offield (1970). Their method depends on the degree of degradation of certain morphologic components of craters and assumes that impact craters in a given size range were initially of the same form. The components they used were rays, secondary craters, ejecta facies, rim sharpness, inner terraces and central peaks. On the assumption that craters of similar size and morphology are roughly the same age, it was possible to place constraints on the ages of other underlying or overlying units.

A scheme similar to that of Pohn and Offield for the Moon was established for Mercury by N.J. Trask (see, McCauley et al. 1981). He defined five classes

Fig. 9. Unnamed pre-Tolstojan crater ($-21°$, $28°$; 110 km diameter). Crater is almost completely obliterated; only the rim crest is partly preserved. Intercrater plains overlie the rim (FDS 27428). (North at top.)

of craters from C_5 as the freshest to C_1 as the most degraded (see footnote in Sec. II.A). Examples were given for each of these classes for two sizes of crater, one between 20 and 100 km in diameter and the other larger than 100 km. Typical examples of Mercurian crater-degradational types are shown in Figs. 5–9.

This scheme was used extensively by geological mappers of the 1:5,000,000 scale quadrangles, and the system has proved useful in correlating ages in areas distant from suitable major impact basins. However, some difficulties have been encountered. For example, in the Shakespeare quadrangle (Guest and Greeley 1983), it was considered better to restrict Class 5 craters to those with dark or light halos, and/or rays. It was found that by including those craters that were fresh appearing but without rays within Class 5, virtually no Class 4 craters remained. As a result, the 150 km diameter crater Verdi was mapped as a C_4 crater despite its being the type example of Trask for a C_5 class of crater. It was recognized, however, that albedo patterns are less easy to distinguish close to the terminator, and crater form less easy to distinguish towards the limb. In consequence, no scheme can be applied with rigorous consistency until all of Mercury is imaged at different lighting angles.

A similar problem was encountered by Spudis and Prosser (1984) during the mapping of the Michelangelo quadrangle. They found that Dostoevskij (see Sec. II.B and Fig. 3) possessed a crater density indicating that it is one of the oldest basins on the planet. In contrast, Dostoevskij is the type example of a large C_3 (middle-age) crater in the Trask system (McCauley et al. 1981). Spudis and Prosser (1984) suggested that the stratigraphic significance of crater degradation is only approximate and that a variety of evidence, including regional geologic setting, proximity to other units, crater density, and type of postcrater modification, must be used in concert to establish the relative ages of Mercurian features.

III. STRATIGRAPHY

In the earliest stages of planetary geologic mapping, the first steps usually taken are terrain reconnaissance and classification. Later, when the relations between photogeologically defined units become clearer, it is desirable to establish a formal chronostratigraphic (i.e., time-stratigraphic) classification scheme in order to correlate widely-separated geologic units and provide a framework for interplanetary correlations of geologic time. Shoemaker and Hackman (1962) presented a chronostratigraphic classification for the Moon in which chronostratigraphic systems were defined on the basis of superposition relations of observable rock-stratigraphic units (Wilhelms 1970,1987a). This method, although modified by more detailed knowledge of lunar geology, is still in use today (Wilhelms 1987a).

The geology of Mercury was initially mapped with physiographic/terrain units, although an attempt was made to indicate their lithology and relative stratigraphic order (Trask and Guest 1975). Geologic mapping for the 1:5,000,000 scale quadrangle maps of the Mercurian hemisphere imaged by Mariner 10 relies on an informal rock-stratigraphic classification of crater, basin and plains units arranged in order of relative age (Holt 1978). After mapping was begun, McCauley et al. (1981) proposed a series of formal rock-stratigraphic units associated with the Caloris basin as the first step in formulating a stratigraphy for Mercury. Spudis (1985) described the use of the Caloris group of McCauley et al. (1981), together with newly defined rock-stratigraphic units, to develop a formal Mercurian chronostratigraphy (Table I).

The chronostratigraphic scheme developed for Mercury is very similar to the system used for the Moon. In Table I, the approximate ages of, and lunar equivalents to, the Mercurian systems are given, although exact correlations of the systems and times are not intended. This aspect of the history of Mercury is discussed below in Sec. IV.B.

A. pre-Tolstojan System

The two oldest Mercurian chronostratigraphic systems are divided by the deposits of the Tolstoj basin (see Sec. II.B). The informal pre-Tolstojan sys-

TABLE I
Mercurian Chronostratigraphic Scheme

System	Major Units	Approx. Age of Base of System[a]	Lunar Counterpart[b]
Kuiperian	crater materials	1.0 Gyr	Copernican
Mansurian	crater materials	3.0–3.5 Gyr	Eratosthenian
Calorian	Caloris Group; plains, crater, small-basin materials	3.9 Gyr	Imbrian
Tolstojan	Goya Formation; crater, small-basin, plains materials	3.9–4.0 Gyr	Nectarian
pre-Tolstojan	Intercrater plains, multiring basin, crater materials	pre–4.0 Gyr	pre-Nectarian

[a]Approximate ages based on the assumption of a lunar-type impact flux history on Mercury.
[b]Included for reference only; no implication of exact time correlation is intended.

tem includes all deposits formed before the impact that created the Tolstoj basin. The oldest recognizable pre-Tolstojan units are the remnants of ancient multiring basins (Fig. 2). The currently recognized population of pre-Tolstojan multiring basins is listed in Table II (designated pT). These basins are randomly distributed over the Mercurian surface (Fig. 10) and probably represent the remnants of a lunar-like, heavily cratered terrain. Some basin sites have served as depositional environments for later smooth plains units (see also Leake 1982), and many ring remnants are recognized by the deflection of later tectonic features into circular or arcuate patterns. Thus, this population of pre-Tolstojan basins forms the broad-scale structural framework for the subsequent geologic evolution of Mercury's surface.

Because of the degraded nature of these ancient basins, both the ring diameters and the original basin configuration are difficult to ascertain. However, in all of the basins, at least one ring appears to have more topographic or structural expression than others. These conspicuous rings are italicized in Table II and are considered to be the rings that were originally the basin rims, i.e., the structural equivalents of the main massif ring of Caloris (1340 km in diameter) or the Cordillera ring of the lunar Orientale basin. Diameters of the main rim are used in the crater-frequency curve of Fig. 11. This curve shows two parallel production functions. For the lower-size ranges, representative cratered terrain in the H-12 quadrangle was used. The crater curve at the larger diameters shows a distinct "knee" at about 500 km (Fig. 11). At diameters larger than 500 km, the function is again in production. These relations suggest that the population of large ancient basins represents only the large members of a heavily cratered surface, of which most craters smaller than about 500 km were obliterated.

TABLE II
Mercurian Multiring Basins

Basin[a]	Center	Age[b]	Ring Diameters (km)[c]	Comments[d]
CALORIS	30°,195°	C	(630),900,1340,(2050),(2700),(3700)	
TOLSTOJ	−16°,164°	T	(260),330,510,(720)	
VAN EYCK	44°,159°	T	150,285,(450),520	
SHAKESPEARE	49°,151°	pT	(200),420,680	
SOBKOU	34°,132°	pT	490,850,1420	Partly mapped by Croft (1979)
BRAHAMS-ZOLA	59°,172°	pT	340,620,840,(1080)	
HIROSHIGE-MAHLER	−16°,23°	pT	150,355,(700)	Partly mapped by Leake (1982)
MENA-THEOPHANES	−1°,129°	pT	260,475,770,1200	Partly mapped by Schaber et al. (1977)
TIR	6°,168°	pT	380,660,950,1250	
ANDAL-COLERIDGE	−43°,49°	pT	(420),700,1030,1300,1750	
MATISSE-REPIN	−24°,75°	pT	410,850,1250,(1550),(1990)	Partly mapped by Croft (1979)
VINCENTE-YAKOVLEV	−52°,162°	pT	360,725,950,1250,(1700)	Spudis and Prosser (1984)
EITOKU-MILTON	−23°,171°	pT	280,590,850,1180	Spudis and Prosser (1984)
Borealis	73°,53°	pT	860,1530,(2230)	Partly mapped by Trask and Strom (1976)
Derzhavin-Sor Juana	51°,27°	pT	*560,740,890*	Partly mapped by Schaber et al. (1977)
Budh	17°,151°	pT	580,850,1140	
Ibsen-Petrarch	−31°,30°	pT	425,640,930,1175	
Hawthorne-Riemenschneider	−56°,105°	pT	270,500,780,1050	Spudis and Prosser (1984)
(Gluck-Holbein)	35°,19°	pT	240,500,950	
(Chong-Gauguin)	57°,106°	pT	220,350,580,940	
(Donne-Moliere)	4°,10°	pT	*375,700,(825),1060,1500*	
(Bartok-Ives)	−33°,115°	pT	*480,790,1175,(1500)*	Spudis and Prosser (1984)
(Sadi-Scopas)	−82°5,44°	pT	*360,600,930,(1310)*	

[a] Basin names in capital letters definitely exist, names in upper-and-lowercase probably exist and names in parentheses possibly exist. Double names are applied to degraded, nearly obliterated basins following the practice devised for naming degraded basins on the Moon (Wilhelms and El-Baz 1977; Wilhelms 1987a).

[b] Relative age; C for Calorian; T for Tolstojan; pT for pre-Tolstojan.

[c] Italicized diameters correspond to physiographically most prominent ring (basin rim); diameters in parentheses reflect uncertain measurement due to discontinuous rings.

[d] References to previous studies that originally recognized degraded basins.

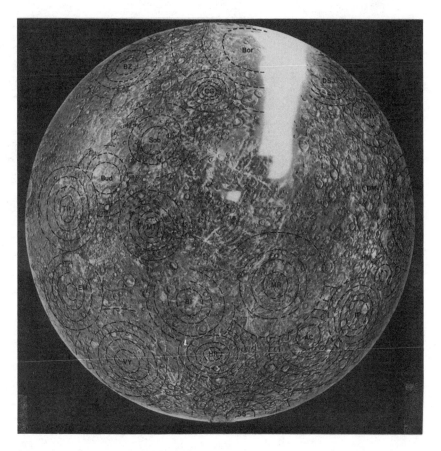

Fig. 10. Map showing the distribution of pre-Tolstojan basin rings. All basins shown here under-
lie the intercrater plains. Basins are listed in Table II. Base is Lambert Equal-Area projection,
centered on 0°, 100°, showing hemisphere of Mercury imaged by Mariner 10. (North at top.)

Apparently, this ancient cratered surface on Mercury was largely erased
by the widespread intercrater plains materials (Fig. 12). Trask and Guest
(1975) considered that the intercrater plains represent an ancient primordial
surface, mainly because so many secondary craters are superposed on the
intercrater plains. Malin (1976a) observed that, in several regions, large craters can be detected underneath intercrater plains. Strom (1977) noted that the
production of similar, although less extensive, highland plains on the Moon
had likewise obliterated the smaller-crater populations in these regions. Not-
ing the global distribution of the Mercurian plains, he suggested that widespread volcanic resurfacing provides the best explanation for the origin of the
intercrater plains.

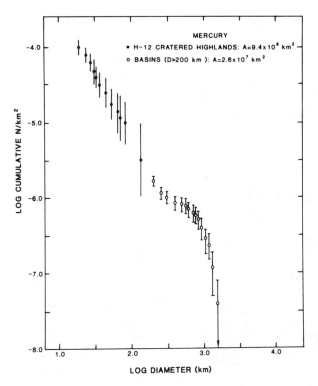

Fig. 11. Frequency-distribution of large craters and basins on Mercury. Smaller-diameter population (black dots) is a representative section of heavily-cratered terrain in the H-12 (Michelangelo) quadrangle. Larger-diameter population (open dots) includes ancient, multiring basins (Fig. 10; Table II). Bend in curve occurs between diameters of 300 and 500 km; this suggests that the intercrater plains largely obliterated a pre-existing, heavily cratered surface. Only the largest craters of this older population can be discerned.

Whatever process is invoked to explain the intercrater plains, it must be capable of explaining their global distribution and their efficacy in obliterating all craters smaller than 300 to 500 km. While no diagnostic volcanic land-forms are associated with the intercrater plains, neither are there any obvious source basins to provide ballistically emplaced (Cayley-type) debris. We think that the global distribution of the intercrater plains is a compelling argument for volcanic activity being at least partly responsible for their emplacement. Because the intercrater plains were emplaced early in the planet's history, during the time of heavy impact bombardment, it is likely that these plains have been extensively brecciated and probably do not retain any remnant of original surface morphology.

In addition to ancient multiring basins and the intercrater plains, several large basins formed in the pre-Tolstojan period after most of the intercrater

Fig. 12. Intercrater plains south of Homer basin ($-1°$, $37°$). Rough, undulating surface is due mostly to numerous superposed secondary craters. (Portion of H-6 photomosaic; north at top; width of scene 680 km.)

plains had been emplaced. These basins include Dostoevskij (see Sec. II.C and Fig. 3), Shakespeare ($49°$, $151°$), and Homer ($-2°$, $38°$; Fig. 12). A particularly interesting pre-Tolstojan basin is Surikov ($-37°$, $125°$). The name was originally applied only to the peak ring of this basin. Subsequent geologic mapping (Spudis and Prosser 1984) demonstrated that Surikov is a three-ring structure, but that the outer rings are much more subdued than the innermost, peak-ring structure. This observation suggests that some type of structural rejuvenation may be partly responsible for the present morphology of Surikov and, by extension, that such a process may be operative in the production of basin morphology elsewhere on Mercury.

B. Tolstojan System

The base of the formally named Tolstojan system is marked by the base of the distinctive lineated terrain unit composed of deposits of the Tolstoj basin impact (Fig. 2). This basal, lineated unit was named by Spudis (1985) the Goya Formation, after the unrelated pre-Tolstojan crater Goya ($-7°$, 152°). The Goya Formation consists of coarsely lineated to hummocky material that extends as far as one basin diameter (500 km) from the rim of the Tolstoj basin (Fig. 13). As previously discussed, the Goya Formation appears to be largely absent from the western and northwestern exterior of the basin; its total areal extent is about 7.8×10^5 km^2. The crater density on the Goya Formation serves as the reference crater density for the base of the Tolstojan system; for craters with diameter > 20 km, the reference crater density is $8.5 \pm 1.4 \times 10^{-5}$ km^{-2}. All surfaces with crater densities higher than this value are pre-Tolstojan in age.

Tolstoj displays four rings, with varying expressions of continuity and

Fig. 13. Geologic sketch map of the Tolstoj basin. Tolstoj ejecta blanket (Goya Formation; G) is well exposed to northeast, southeast and south of the basin. Tolstoj displays four rings 260, 330, 510 (basin rim) and 720 km diameter (dashes). (North at top; width of scene about 1200 km.)

morphologic prominence (Fig. 13). The main basin rim consists of prominent massifs and scarps that form a ring 510 km in diameter. Other rings are more discontinuous, but may be mapped at 260, 330, and 720 km in diameter (Table II). On the floor of the Tolstoj basin are several "Archimedean" craters, i.e., craters that have formed on the basin floor before deposition of smooth plains materials in the basin interior. This stratigraphic relation and the lower crater density on the smooth plains indicates that the plains were not emplaced at the time of the Tolstoj impact, but substantially postdate the basin. Wilhelms (1976) argued that the plains fill of Tolstoj is a distal ejecta deposit of the Caloris basin; however, no Caloris basin smooth plains are observed superposed on the Goya Formation. A more likely interpretation is that the smooth plains are volcanic lavas that flooded the Tolstoj basin some time after the basin impact.

After the Tolstoj impact, numerous large craters and basins formed, including Scarlatti (40°, 100°), Durer (22°, 119°) and Renoir (−18°, 52°). All of these features are double-ring basins, which are randomly distributed around the hemisphere of Mercury imaged by Mariner 10. In addition to these small basins and craters, smooth plains materials were emplaced in late Tolstojan time, some of which are still exposed. Typically, these plains morphologically resemble the later smooth plains units, but they possess a substantially higher crater density. The Tolstojan plains generally crop out around the margins of regional deposits of younger, post-Caloris basin plains; a reasonable inference is that the present surface exposure of the older smooth plains is only a fraction of their total areal extent, the rest being buried by younger units. This observation suggests that emplacement of plains materials may have extended over most of the early history of Mercury.

The Beethoven basin (−20°, 124°) probably formed toward the end of the Tolstojan Period, based on its crater density (see Sec. II.B). Beethoven possesses a well-preserved, lineated ejecta blanket that is well exposed east and south of the basin rim (Fig. 14), but appears to be absent west of the basin. Because of the relative youth of Beethoven, this relation is probably best explained by a natural ejecta asymmetry, as in the case of Tolstoj. Beethoven displays only one ring, about 625 km in diameter; because Mariner 10 images of this basin were taken under very high Sun illumination, other rings probably exist that cannot be discerned on available images. The floor of Beethoven is covered by smooth plains that possess a considerably higher crater density than do the Caloris smooth plains (Fig. 14). These plains materials appear to have completely buried any expected interior ring of Beethoven; this implies that they are fairly thick in the basin interior. We consider it likely that these plains were formed by massive flood lava eruptions, immediately following the Beethoven impact. An alternative hypothesis is that these plains represent the original floor material of Beethoven, that an interior ring(s) never formed, and that the volume of impact melt generated by the Beethoven impact was several orders of magnitude greater than that generated

Fig. 14. Mosiac of part of the Beethoven basin (rim 625 km diameter; dashed). Basin floor is filled with older, cratered smooth plains. Lineated ejecta blanket (E) extends outward up to one basin diameter southeast of rim. (North at top.)

during the impact of other Mercurian basins. We consider this complex explanation less likely than flooding by lavas to form the plains.

C. Calorian System

Caloris Basin.   The impact that formed the Caloris basin (Fig. 4) was a watershed event in the geologic history of Mercury. The extensive ejecta deposits of Caloris are presumed to have formed instantaneously within the Mercurian geologic record. These deposits therefore form a useful marker horizon and can be used to divide the geologic time scale in the same way that the ejecta of the Imbrium basin has been used on the Moon (Shoemaker and

Hackman 1962). The Caloris deposits were used as a stratigraphic marker by Guest and Gault (1976), Guest and O'Donnell (1977) and Leake (1982), who referred to stratigraphic units as pre- and post-Caloris.

The products of the Caloris impact have been formalized into a rock-stratigraphic Caloris Group, consisting of several formations (McCauley et al. 1981). This nomenclature was developed during the 1:5,000,000 scale geologic mapping of the Tolstoj (H-8) quadrangle (Schaber and McCauley 1980) and the adjacent Shakespeare (H-3) quadrangle[a] (Guest and Greeley 1983). The units designated as formations were essentially the morphological units recognized by Trask and Guest (1975), each of which was considered an individual rock unit. The mountain material was named the Caloris Montes Formation, the intermontane plains the Nervo Formation, the hummocky plains the Odin Formation, and the lineated plains were included in the Van Eyck Formation. In addition, clusters of secondary craters from Caloris were recognized by Schaber and McCauley (1980) and were included as a facies of the Van Eyck Formation.

All the formations of the Caloris Group are interpreted to represent deposits formed by the impact that formed the Caloris basin. These materials include basin impact melt, clastic ejecta, locally reworked material, and secondary crater material. The base of the Caloris Group defines the base of the Calorian system (Spudis 1985); its reference crater density ($D > 20$ km) is $5.8 \pm 1.3 \times 10^{-5}$ km^{-2}.

The Caloris basin (Fig. 4) has been the subject of several detailed geologic studies (Strom et al. 1975; McCauley 1977; McCauley et al. 1981) and is only briefly described here. (See Figs. 15-18 for rock-stratigraphic formations of the Caloris Group [McCauley et al. 1981].) The main basin rim is 1340 km in diameter; it consists of rugged massifs arranged into a concentric pattern (Fig. 15). The Caloris rim is here considered to be the structural equivalent of the main rings of lunar basins (e.g., Cordillera of Orientale; Apennines of Imbrium). Mantling the intermassif areas near the Caloris rim is the Nervo Formation (Fig. 15), which forms an undulating-to-smooth unit that is interpreted by McCauley et al. (1981) as fallback ejecta. Much of the Nervo Formation may consist of impact melt, ejected from the excavation cavity of the basin; it resembles small melt ponds observed in the ejecta of the lunar Orientale basin. The Odin Formation (Fig. 16) consists of knobby, plains-like deposits that are widely exposed directly east of the basin rim. The Odin Formation morphologically resembles knobby basin deposits found around lunar basins, such as the Alpes Formation of the Imbrium basin (McCauley et al. 1981). Large tracts of the Odin Formation apparently are partly buried by later smooth plains materials (Fig. 16).

[a]In the Shakespeare quadrangle map (Guest and Greeley 1983), the Odin Formation—mantling deposit (com) and the Van Eyck Formation (cvl) have been wrongly labeled in the description of units, although they are correctly designated in the accompanying text.

Fig. 15. Caloris Montes Formation (M), forming the southern rim of the Caloris basin. Nervo Formation (N) occurs between massifs. Lineated facies of Van Eyck Formation (VEL) at lower right (FDS 111). (North at top.)

Fig. 16. Knobby deposits of the Odin Formation (O). Caloris deposits are partly mantled in this region by smooth plains (P) (FDS 73). (North at top.)

The radially lineated ejecta blanket of the Caloris basin is expressed by the Van Eyck Formation, lineated facies (see Figs. 15 and 17). This unit is equivalent to the lunar Fra Mauro Formation of the Imbrium basin. The Van Eyck Formation, lineated facies, is not widely exposed around the Caloris basin, and apparently was largely buried by subsequent smooth plains materials (Fig. 15). The other facies of the Van Eyck Formation, secondary craters, is well exposed in the highlands southeast of Tir Planitia (Fig. 18). Caloris basin secondary craters consist of groups or chains of large (10–20 km), overlapping, irregular craters. No recognizable Caloris secondaries have been found on the smooth plains surrounding the basin, indicating that the plains postdate all recognized basin secondaries.

An unusual terrain, consisting of hilly and furrowed material, occurs near the antipode of the Caloris basin around $-25°$, $20°$. Schultz and Gault (1975) and Hughes et al. (1977) have provided compelling arguments that this terrain was produced by the global focusing of seismic waves associated with the Caloris impact. Although these materials are not a member of the Caloris Group (McCauley et al. 1981), we accept the interpretation of their origin by Caloris basin seismic waves, and we therefore assume that these hilly and furrowed units on the opposite side of Mercury from Caloris demarcate the base of the Calorian System in their region of occurrence.

Fig. 17. Nervo Formation (N) and lineated facies of Van Eyck Formation (VEL). The large crater is Nervo, 50 km in diameter (FDS 103). (North at top.)

Fig. 18. Secondary crater facies of Van Eyck Formation (arrows) (Portion of H-8 photomosaic; north at top.)

The ring system of the Caloris basin has been the subject of some controversy. The only obvious ring is the 1340 km main basin rim, defined by the Caloris Montes Formation. However, the Caloris basin is centered within a broad, regional concentric pattern that is formed by wrinkle ridges, scarps and physiographic boundaries, both inside and outside the Caloris basin (Fig. 19a). The smooth plains/highlands contact in the H-8 quadrangle forms a major physiographic boundary concentric with the Caloris basin, and some major dorsa (e.g., Schiaparelli dorsum) are also concentric with the basin rim. Experience with the geologic mapping of the lunar basins (e.g. Imbrium; see Spudis 1986) suggests that regional mapping of concentric structures can aid

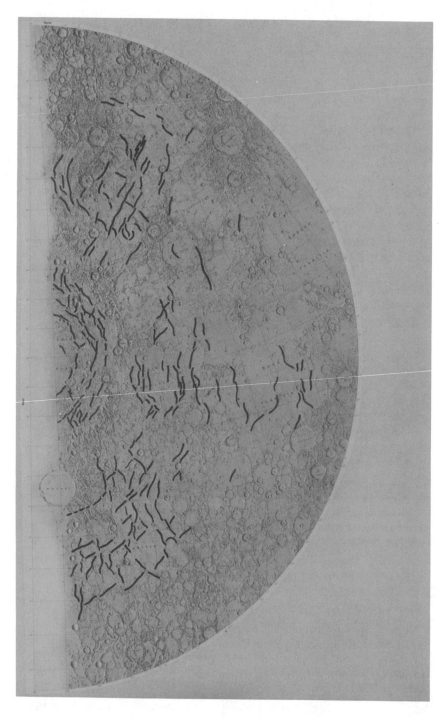

Fig. 19. Basin-centered view of Caloris basin showing distribution of basin radial and concentric structures. Stereographic projection by U.S. Geological Survey. (North at top.) (a) Map of

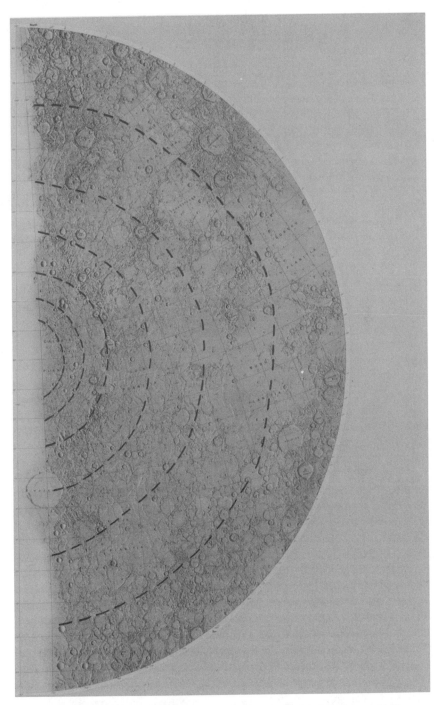

major ridges within Caloris Planitia and smooth plains exterior to Caloris basin, and (b) Ring interpretation map, based on (a). See Table II for ring diameters.

in the delineation of subdued basin rings. On this principle, we present a regional map of structures around the Caloris basin (Fig. 19a) and a ring interpretation map based on these features (Fig. 19b). In this reconstruction, Caloris displays six rings, ranging from 630 km to 3700 km in diameter (Fig. 19b; Table II). Caloris bears a striking similarity to the lunar Imbrium basin, also a six-ring basin, containing two rings interior and three rings exterior to the main basin rim (Schultz and Spudis 1985; Spudis 1986); both basins are of similar size and morphology. In addition, both Imbrium and Caloris display large tracts of smooth plains in a broad, exterior arc external to the basin rim (Oceanus Procellarum around Imbrium; Tir, Odin and Susei Planitia around Caloris).

The striking parallels between the lunar Imbrium and Mercurian Caloris basins indicate that a similar sequence of events was responsible for the development of both. This similarity, in addition to other relations of the Mercurian smooth plains discussed below, suggests that the bulk of the smooth plains of Mercury are of volcanic origin and are related to the Caloris basin impact only in the same sense that the lunar maria are related to their containing basins, i.e., as younger geologic units deposited in structural and topographic lows.

Calorian plains materials. In the above discussion on the Caloris Group, the floor material of the Caloris basin (see Figs. 4 and 20) was specifically excluded. McCauley et al. (1981) considered that this material was of uncertain relevance to the basin. The Caloris floor material displays numerous wrinkle ridges (indicative of surface compression) and graben-like troughs (indicative of surface tension; Fig. 20). The wrinkle ridges are cut by (predate) the fractures; the implication of these observations is that the floor of the Caloris basin originally subsided, and then uplifted, possibly as a result of up-doming (rebound) in the basin interior (Strom et al. 1975; Dzurisin 1978).

This hypothesis for the history of the Caloris basin floor does not address the ultimate origin of the basin floor plains. We suggest, on the basis of the general appearance of undeformed portions of the basin floor (Fig. 20), that the original basin floor of Caloris was flooded by smooth plains volcanic lavas soon after the Caloris impact; the subsequent deformation described above is probably a result of regional, postbasin tectonic activity. The age of the Caloris basin-fill material appears to be slightly younger than Caloris ejecta (discussed below).

Aside from the intercrater plains, the Calorian-age smooth plains are the most widespread plains materials on Mercury (Fig. 21); the smooth plains cover about 10.4×10^6 km^2, or almost 40% of the total area imaged by Mariner 10. In contrast to some earlier opinions, the Calorian-age smooth plains are distributed all over the hemisphere of Mercury imaged by Mariner 10. Moreover, more than 90% of the regional exposures of smooth plains on Mercury are associated with basin depositional environments (cf. Figs. 10, 19b and 21). Smooth plains also occur as fill materials for both smaller double-ring basins and large craters.

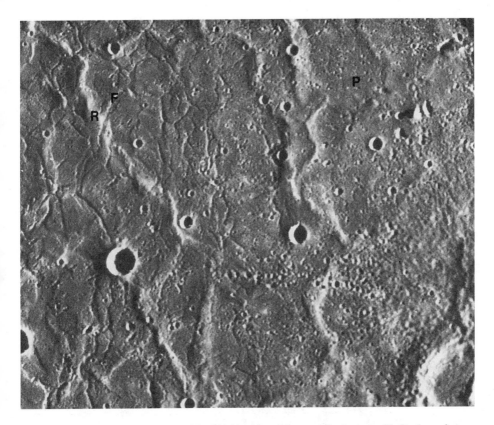

Fig. 20. Caloris basin floor materials. Wrinkle ridges (R) are cut by fractures (F). Portions of floor appear similar to smooth plains (P), suggesting that Caloris basin fill consists of smooth plains material, tectonically deformed (FDS 106). (North at top; width of scene 250 km.)

For the purpose of this study, we have made new crater counts for a variety of Mercurian geologic units (Table III). Results indicate that both the Caloris basin floor material and the smooth plains exterior to and concentric with the Caloris basin are younger than the basin ejecta. These data support the geological arguments in the previous section that the Mercurian smooth plains are not genetically related to Caloris basin ejecta, but were emplaced (apparently within a limited time) well after the Caloris impact.

In the debate regarding the origin of Mercury's smooth plains, several geological constraints may now be placed on the generating mechanism. First, the smooth plains have a planetwide distribution (Fig. 21), and are not solely confined to the distal margins of the Caloris basin. Second, the total volume of the smooth plains, whose estimation must now be greatly increased over the large estimates of Strom et al. (1975) and Trask and Strom (1976), is much greater than that which could be reasonably expected from the Caloris

TABLE III
Crater Densities for Some Mercurian Geologic Units.

Unit	Region[a]	Area (10³ km²)	Number (D ≥ 20 km)	Density (10⁻⁵ km⁻²)[b]
Smooth plains	H-3	494	12	2.4 ± 0.7
Smooth plains	H-8	528	15	2.8 ± 0.7
Smooth plains	H-3; H-8	1022	27	2.6 ± 0.5
Caloris floor	H-3	280	11	3.9 ± 1.2
Bach basin	H-12	120	5	4.2 ± 1.9
Caloris basin	H-3	360	21	5.8 ± 1.3
Beethoven basin	H-7; H-12	440	31	7.0 ± 1.3
Tolstoj basin	H-8	413	35	8.5 ± 1.4
Dostoevskij basin	H-12	360	45	12.5 ± 1.9
Surikov basin	H-12	120	19	15.8 ± 3.6

[a]Region refers to USGS quadrangle maps where crater counts were done.
[b]Crater density plus one standard deviation.

Smooth Plains
(Calorian-Tolstojan)

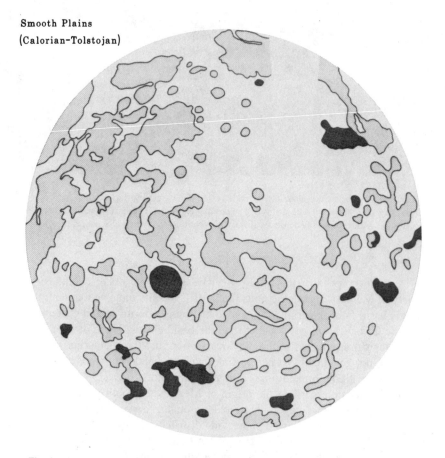

Fig. 21. Distribution of Calorian (light grey) and Calorian and/or Tolstojan (dark grey) smooth plains. Smooth plains cover about 10.4×10^6 km² or 40% of the part of Mercury photographed by Mariner 10. Base is a Lambert equal-area projection, centered on 0°, 100°. (North at top.)

impact, or several Caloris-sized impacts. Third, crater-density data (Table III) indicate that major expanses of the circum-Caloris smooth plains substantially postdate all recognized major Mercurian basins.

By analogy with the lunar maria, these constraints support the hypothesis that the Mercurian smooth plains were emplaced as volcanic lavas, created during partial melting of Mercury's mantle. The long-standing reluctance on the part of some investigators to accept the volcanic hypothesis for smooth plains emplacement appears to be motivated partly by the lack of diagnostic volcanic morphologies in the smooth plains and partly by the lack of any color or albedo contrast between the plains and the cratered terrain on Mercury. Schultz (1977) and Malin (1978) have pointed out that the apparent absence of volcanic landforms is largely due to the poor coverage and resolution of Mariner 10 data; similar coverage of the Moon would likewise reveal few, if any, volcanic features in the lunar maria. The lack of albedo contrast between Mercurian plains and cratered terrain may simply reflect a generally lower FeO, and possibly TiO_2, content of Mercurian vs lunar lavas (Hapke et al. 1975; Strom 1984). The typical albedo of the circum-Caloris smooth plains is about 0.12 to 0.13 (Hapke et al. 1975). This value is identical to that of a regional light plains deposit on the Moon, the Apennine Bench Formation. Geologic and remote-sensing studies have shown that these lunar plains are probably composed of extrusive basaltic rocks (Spudis and Hawke 1986) that are higher in Al_2O_3 and lower in FeO than typical lunar mare basalts. Thus, the lack of an albedo contrast between the Mercurian smooth plains and highlands does not necessarily support an impact-ejecta origin for the plains.

Although these arguments are indirect, we consider that the evidence for a volcanic origin for the Mercurian smooth plains is compelling; similar conclusions were reached by Kiefer and Murray (1987). The age and distribution of the smooth plains suggest that Mercury underwent large-scale volcanic resurfacing after the Caloris basin impact.

Tectonic features.   The tectonic features and history of Mercury are discussed in detail elsewhere in this book (see the chapter by Melosh and McKinnon) and will not be dealt with directly here. However, the lobate scarps of Mercury are of great importance to the geologic evolution of the planet, and stratigraphy can place some constraints on their time of formation.

The lobate scarps are widely distributed over Mercury (Strom et al. 1975). They consist of sinuous to arcuate scarps that transect pre-existing plains and craters (Fig. 22). They are most convincingly interpreted as thrust faults, indicating a period of global compression (Strom et al. 1975; Solomon 1977). All these scarps cut intercrater plains materials, which suggests that they began to form after the intercrater plains materials were emplaced in pre-Tolstojan time. Additionally, Dostoevskij basin ejecta (Fig. 3) is cut by a lobate scarp, Hero Rupes; this suggests that scarp formation began some time in the late pre-Tolstojan period. The lobate scarps typically transect smooth

Fig. 22. View of Discovery Rupes, one of a global system of lobate scarps. Crater offset (arrow) suggests these scarps are thrust faults. (North at top; portion of H-11 photomosaic.)

plains materials (early Calorian age) on the floors of craters (Fig. 22). Strom (1984) notes that superposed craters are of the Class 1 and 2 degradational types (Kuiperian and Mansurian age). These observations suggest that lobate-scarp formation was confined to an interval beginning roughly in the late pre-Tolstojan Period and ending in the middle to late Calorian Period.

In addition to the lobate scarps, numerous wrinkle ridges occur in the smooth plains materials (see Figs. 4 and 20). These ridges probably were formed by local to regional surface compression caused by lithospheric loading by dense stacks of volcanic lavas, as suggested for those of the lunar maria (Solomon and Head 1980).

Calorian craters and basins. Although Caloris was the last major basin-forming impact on Mercury, numerous smaller two-ring basins formed during the Calorian Period and are randomly distributed over the planet. Such basins include Strindberg (53°, 136°), Rodin (21°, 18°), Michelangelo ($-35°$, 109°) and Bach ($-69°$, 101°; Fig. 23). These two-ring basins are well preserved, display in most cases an exceptionally complete peak ring, and are commonly filled with smooth plains materials. In addition to these small basins, numerous craters of Calorian age are widely distributed over the planet. Many of these craters are partly to completely flooded by smooth plains materials.

D. Mansurian System

The Calorian Period witnessed what was virtually the last major, regional cycles of geologic activity on Mercury. After the final deposition of smooth plains materials, only impact craters have formed on the Mercurian surface. Rocks formed in the long interval of time between the smooth plains emplacement and the present are divided into two chronostratigraphic systems. The lower contact of deposits from the crater Mansur (48°, 163°), shown in Fig. 24, defines the base of the Mansurian System. This system, like the lunar Eratosthenian System (Table I), includes materials of slightly degraded, but still relatively fresh craters, some of which contain minor plains materials. All of these Mansurian-age plains are confined to crater floors and they most likely consist of crater impact-melt sheets and fallback ejecta deposits. Mansurian-age craters possess no rays, but fine-scale structures in their ejecta deposits are largely preserved (Fig. 24). All craters of this age are randomly distributed over the planet and no regional deposits of plains of Mansurian age have been recognized.

E. Kuiperian System

The end of the Mansurian system is loosely defined by superposed, rayed craters, typified by the deposits of the crater Kuiper ($-11°$, 31°.5) shown in

Fig. 23. Calorian age two-ring basin Bach (−69°, 101°; 210 km diameter). Inner peak ring (103 km diameter) is exceptionally well preserved and basin is flooded by younger smooth plains materials. (North at top; portion of H-15 photomosaic.)

Fig. 25. The Kuiperian System includes the formation of all fresh, rayed craters on Mercury, extending to the present; it is composed wholly of crater deposits, almost all, of which are only slightly degraded if at all and still possess bright ray systems (Fig. 25). This system is thus analogous to the lunar Copernican System (Table I). Several unusual high-albedo markings with swirl-like appearance have been recognized near the crater Handel (4°, 34°) by Schultz and Srnka (1980); they suggested these swirls are geologically young (thus Kuiperian in age) and were produced during the impact of a cometary coma by scouring the Mercurian surface with hot gases and high-energy plasmas. Other than this unusual feature, no regional deposits of Kuiperian age have been recognized.

Fig. 24. Mansur crater (48°, 163°; 75 km diameter). The base of deposits from this crater define the top of the Calorian and bottom of the Mansurian Systems. (North at top; portion of H-3 photomosaic.)

Fig. 25. Kuiper crater (−11°, 31°5; 60 km diameter). The base of deposits from this crater define the top of the Mansurian and bottom of the Kuiperian Systems (FDS 27304) (North at top.)

IV. GEOLOGIC HISTORY OF MERCURY

The chronostratigraphic scheme described above has been used to make a global geologic map of Mercury at a scale of 1:10,000,000 (Spudis in preparation). We have used this geologic map to prepare a series of paleogeologic maps, showing our best estimate of the changing appearance of Mercury through time (see Plates 1-6 in the color section). The five maps show the distribution of geologic units in middle pre-Tolstojan (Plate 2), early Tolstojan (Plate 3), early Calorian (Plate 4), middle to late Calorian (Plate 5), and Kuiperian times (Plate 6). The last map (Plate 6) shows the current surface geology of Mercury and is an abstract of the new global map at 1:10,000,000 scale.

A. Summary of Mercury's Geologic Evolution

The earliest decipherable geologic event in Mercury's history was the formation of its crust. On the basis of analogy with lunar history, Warren (1985) considers that Mercury underwent early global crustal melting. This process, called the "magmasphere" stage by Warren (1985), would have been similar to the lunar "magma ocean", whereby large-scale melting of at least the outer few hundred kilometers of the planet would tend to concentrate low-density plagioclase feldspar into the uppermost part of the Mercurian crust. If this process operated during Mercury's early history, its crust is composed largely of anorthositic rocks (anorthosites, anorthositic norites and anorthositic gabbros). This is consistent with the limited full-disk spectra of Mercury obtained by McCord and Clark (1979), which suggest that Mercury's surface is compositionally similar to Apollo 16 highland soils. However, these data are of such limited quality that few constraints can be placed on Mercury's surface composition.

The earliest cratering record of Mercury has been largely destroyed by the deposition of the intercrater plains. However, the largest impact features of this period (multiring basins) have been partly preserved and suggest that Mercury's surface may have originally resembled the cratered lunar highlands (Color Plate 2). Sometime during the heavy bombardment, the emplacement of massive quantities of intercrater plains materials largely obliterated the older crater population. The global distribution of the intercrater plains suggests that they may be at least partly volcanic in origin, although subsequent cratering probably reduced the original surface to breccia and primary surface morphologies were probably destroyed.

The impact that formed the Tolstoj basin marked the beginning of the Tolstojan Period (Color Plate 3), still a time of high impact rates. Although some flood lavas may have erupted during this time, their preservation is sporadic (Fig. 14) and they may be largely covered by subsequent lavas. The Beethoven basin probably formed near the end of the Tolstojan Period.

The Caloris impact formed the largest well-preserved basin on Mercury's surface (Color Plate 4) and provided an extensive stratigraphic datum on the planet. Catastrophic seismic vibrations from the Caloris impact probably formed the hilly and furrowed terrain on the opposite side of the planet. Some finite, but probably short, time after the Caloris impact, came massive extrusions of flood lavas to form the Mercurian smooth plains (Color Plate 5). A rapidly declining cratering rate has produced minimal changes to Mercury's surface (Color Plate 6) since the final emplacement of the smooth plains. This low cratering rate presently continues to produce regolith on all Mercurian surface units.

B. Interplanetary Comparisons

The chronostratigraphic systems used in the geologic mapping of Mercury are similar to those used for the Moon (Table I). On both bodies, almost all geologic activity occurred early in planetary history—during pre-Tolstojan to early Calorian time on Mercury and pre-Nectarian to early Imbrian time on the Moon (Wilhelms 1984;1987a). A significant difference between the two planets is the continued period of mare deposition on the Moon, which may have continued into Copernican time (Schultz and Spudis 1983). Such a wide range in age for plains material is not evident on Mercury, at least within the hemisphere imaged by Mariner 10.

Previous studies of the Mercurian basin population have emphasized the apparent deficiency of basins on Mercury relative to the Moon (Malin 1976b; Schaber et al. 1977; Frey and Lowry 1979). Wood and Head (1976) considered that Mercury has a high basin density, but they included many complex craters and protobasins in their sample; the first 19 entries in their table of Mercurian basins are all smaller than 200 km in diameter. The cumulative basin density on Mercury ($D \geq 200$ km) is reported by Schaber et al. (1977) to be only 37% of that on the Moon. The Moon has at least 62 basins, $D \geq 200$ km (Schaber et al. 1977), resulting in an average basin density of $1.72 \pm 0.13 \times 10^{-6}$ km^{-2}. Only about 35% of Mercury's surface was photographed by Mariner 10 with lighting conditions adequate (Strom 1979) to recognize ancient basin structures. The degraded basins recognized by systematic geologic mapping (Sec. III.A; Table II) suggest that Mercury has at least 50 two-ring and multiring basins ($D \geq 200$ km) within this area. Thus, the average density of Mercurian basins is $1.92 \pm 0.14 \times 10^{-6}$ km^{-2}. This value suggests that Mercury is not deficient in multiring basins and may possibly have more per unit area than the Moon. The presence of a large population of basins that predate the Mercurian intercrater plains suggests that basin production was a continuous process during the early history of the planets.

The relative ages of some comparable Mercurian and lunar geologic units may be plotted against crater densities to estimate the cratering rates in the early histories of the two planets (Fig. 26). Although we have radiometric ages for only a few lunar geologic units (Fig. 26b), it is clear that the general shapes of the two curves may be interpreted to be similar. Deposition of the smooth plains materials on Mercury was essentially that planet's last global

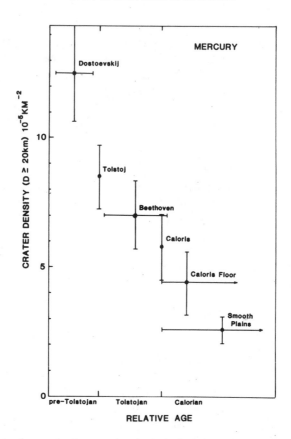

Fig. 26. Plots of crater density on various geological units on Mercury and the Moon versus stratigraphically defined relative ages. *Mercury*: Tolstoj and Caloris have no horizontal error bars because they define beginning of stratigraphic systems. Note that both Caloris basin fill and exterior smooth plains postdate Caloris basin ejecta. See Table III. *Moon*: Nectaris and Imbrium have no horizontal error bars because they define beginning of stratigraphic systems. Crater densities for lunar basins from Wilhelms (1987a); data for average maria from Basaltic Volcanism Study Project (1981). Note that for comparable stratigraphic positions, the lunar units display somewhat lower absolute crater densities than Mercurian units. Absolute ages in billion (10^9) years.

geologic event and only impact craters have formed since then. Crater densities on the Mercurian smooth plains are comparable to the crater density on the lunar Imbrium basin deposits (Fig. 26; see also Murray et al. 1974). This may imply that the last major geologic activity occurred on Mercury roughly 3.8 Gyr ago (Basaltic Volcanism Study Project 1981), although it is by no means certain that absolute ages may be directly compared between the two bodies (Chapman 1976). In any event, it appears that Mercury, like the Moon, was most active geologically early in its history although the comparative

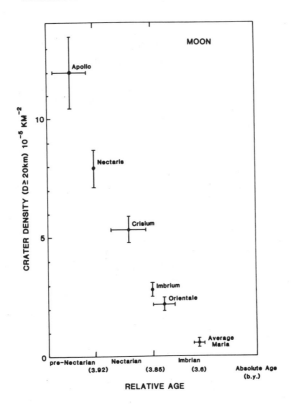

timing of the decline in impact-cratering rate and somewhat later cessation of volcanism must await radiometric dating of samples collected from the surface. Whatever the timing, no relatively uncratered units, such as are widespread on Mars (Scott and Carr 1978; Basaltic Volcanism Study Project 1981), so far have been observed. Thus, there has been no extensive resurfacing of Mercury in its more recent history. The question remains, however, why the Moon and Mercury, with such different internal structures, had such apparently similar surface histories.

V. CONCLUSION

The preceding discussion on the geologic evolution of Mercury is based entirely on the data returned more than a decade ago by the Mariner 10 mission. Roughly 40% of the surface of the planet has been examined at resolutions comparable to that of Earth-based telescopic observations of the Moon. Our knowledge of Mercury is similar to our knowledge of the Moon in about 1965, after the first three Ranger missions. We now perceive that Mercury is a planet that underwent intensive, early impact bombardment and widespread resurfacing by volcanic lavas, which essentially ended as much as 3 Gyr ago.

If we should have learned any single principle in the past 25 years of planetary exploration, it is that the universe is very perverse about rapidly disclosing its secrets and that today's conceptual models tend to be naive. One only has to look at the experience with knowledge of Mars in 1970 to appreciate the dangers of presuming to understand a planet without systematic global mapping (Hartmann and Raper 1974). We have little doubt that our model for the geologic history of Mercury is incomplete and major surprises may well be waiting to be discovered by a future mission to Mercury. It is our hope that the stratigraphic framework developed from the Mariner 10 data will be useful to future workers, who will have an abundance of global data to decipher the complex geologic history of Mercury.

Acknowledgments.   We thank A. Dial for providing unpublished crater-count data in the H-12 quadrangle and D.E. Wilhelms for his crater-count data on lunar basins. W.S. Kiefer, M.A. Leake, E.C. Morris and D.E. Wilhelms provided constructive reviews of this chapter. The geologic mapping of Mercury was supported by the NASA Planetary Geology Program.

GEOMORPHOLOGY OF IMPACT CRATERS ON MERCURY

RICHARD J. PIKE
U.S. Geological Survey

Morphologic complexity of impact craters on Mercury increases systematically with their diameter D. *At least 15 crater attributes are strongly size dependent: depth* d, *rim height* h, *rim-wall width, peak and floor diameter, peak and rim-wall complexity, ring frequency and spacing, and presence/absence of a bowl-shaped interior, central peak, flat floor, scalloped rim crest, slump deposits, and rim-wall terraces. The 447 crater sample divides into 7 groups that resemble lunar classes, although size ranges are more like those on Mars: simple (0.225 km ≤ D ≤ 14.4 km), modified-simple (4.6 km ≤ D ≤ 12.2 km), imma-ture-complex (9.5 km ≤ D ≤ 29.1 km), and mature-complex (30.0 km ≤ D ≤ 160 km) craters, protobasins (72 km ≤ D ≤ 165 km), two-ring basins (132 km ≤ D ≤ 310 km), and multiring basins (285 km ≤ D ≤ 1600 km). Linear log-log regressions of* d *on* D *for simple* (n = 104) *and complex* (n = 127) *craters resolve ambiguities in past Mercurian* d/D. *The equations intersect at* D = 4.7 ± 0.6 km, *half the previous value. Terrain-related contrasts in* d/D *exist only for modified-simple craters, which are shallower on smooth plains. Comparable* D *from* d/D *fits for the Earth, Mars and the Moon [1.9, 3.1, and 8.6 (mare) and 10.9 km (upland)] all differ. Linear fits to* h/D *on Mercury intersect at* D = 11 km, *less than the old value (16 km) and, importantly, much more than the diam-eter of the inflection in* d/D. *The diameter of the transition from simple to com-plex craters,* D_t, *on Mercury is 10.3 ± 4 km, much less than the 16 km deter-mined from prior work. It was recalculated from new morphologic observations and* d/D *and* h/D *data* (n = 10). *The new* D_t *and comparable values for the Earth, Mars and the Moon (3.1, 5.8, and 19 km) scale strongly (inversely) with surface gravity* g *and even more so with* g *and the approach velocity of asteroids and short-period comets* g/V_∞. *Thus, Mercurian and lunar craters differ signifi-cantly in size dependence of form. Mercury's protobasins, large craters with both a central peak and an inner ring, differ significantly from two-ring basins as well as from craters: (1) both interior features in protobasins are propor-*

tionally smaller ($D_{cp} = 0.11D$, $D_i = 0.4D$); *and (2) peak and ring diameters within individual protobasins are complementary. Mean protobasin diameters (n = 20), close to Mars' but not the Moon's, also scale with g, but less so with* g/V_∞. *Two-ring basins more closely resemble multiring basins than protobasins: their mean inner-ring/rim-crest ratio is 0.5 rather than 0.4. The 0.5D ratio is a member of the* $2^{0.5}D$ *spacing interval for multiring basins. The average diameter of Mercury's 31 two-ring basins, closest to that for Mars, scales well with either g or with* g/V_∞. *At least 23 old and fragmented multiring basins have been recognized on Mercury. They have the same* $2^{0.5}D$ *spacing between adjacent ring positions, both inside and outside the main ring, as those on the Moon, Mars, and to a lesser extent Earth and the Galilean satellites. Neither average nor onset sizes of multiring basins on Mercury and the three other planets scale with g or* g/V_∞ *to the* -1.0 *value. This study concludes that surface gravity, substrate rheology and impactor velocity decrease in importance with increasing size of the impact, with g the last to disappear. Results here uphold the 1975 findings of D. E. Gault and others that the transition from simple to complex craters strongly reflects g. The* D_t *of four rocky bodies appears to scale better with g to the* -1.0 *than to* -0.25. *Substrate differences, evident on Mercury only in the morphology of small craters, are less important.* D_t *and possibly* $D_{h/D}$ *are the only diameter thresholds whose scaling is improved by adding* V_∞. *Implications for geologic process include both craters and basins: (1) although crater-interior complexity developed largely through gravity-driven failure of the rim, d/D and h/D data for Mercury and the three other planets indicate that inertially-driven uplift of the crater bowl (peak recoil) played the major role in initiating this collapse, (2) the similar, discrete spacing of basin rings on the four planets and the apparent absence of external controls on the largest impacts suggest that basin formation is dominated by some combination of energy-scaled and hydrodynamic-periodic processes, which determine ring position if not necessarily the mode of ring emplacement.*

I. INTRODUCTION

Mercury has again become "the forgotten planet" of the solar system. Its Moonlike surface, imaged well over a decade ago by the Mariner 10 spacecraft (Murray et al. 1974), now appears prosaic compared to the tectonic terrains of Venus shown by the Venera orbiters, the diverse topography of Mars seen in the Viking pictures, or the bizarre landscapes of some of the outer satellites revealed by the Voyagers. However, Mercury is not simply another Moon. Its surface features furnish essential pieces to the interpretive puzzle posed by contrasts in the geomorphology of the inner planets (see, e.g., Strom 1984, p. 17, 54).

This chapter seeks to identify possible influences on the impact-cratering process from the morphology of fresh-looking Mercurian craters 1 km to 1500 km across. Here, "crater" denotes young structures formed by primary impact, including all the two-ring and multiring landforms commonly called basins. Degraded craters (minus basins) and those formed by secondary impact raise problems other than those addressed here, and are not considered.

Size-dependent differences in crater shape, when correlated with those else-where, both clearly establish the uniqueness of Mercury among other planets and satellites, and indicate constraints on cratering problems.

Crater Morphology

Impact craters are the premier surface features on solid planets and satel-lites across the solar system, from Mercury to Uranus (see, e.g., Shoemaker 1977; Beatty et al. 1987). They are rare only on the geodynamically active Earth, Io and Venus. Although hypervelocity impact is thought to be an essen-tially similar process everywhere and over a wide range of energy, the result-ing landforms are not alike in morphology. Impact craters vary systematically in many aspects of form according to their planetary host (Gilbert 1893; Hartmann 1972), size (Gilbert 1893), location on a planet (MacDonald 1931; Dence 1972) and relative age (Baldwin 1949).

Contrasts in crater shape reflect both internal and external controls. The key internal variable is magnitude of the impact itself. The mix of physical processes and their relative importance almost certainly changes with increas-ing energy of the event (Baldwin 1963,p.184). This is especially likely in the very large impacts that form multiring basins.

Several external influences affect the impact process (Quaide et al. 1965; Gault et al. 1975). They may include density, velocity and angle of impact of the projectile; strength, structure and physical state of the target material; planetary surface gravity g; atmospheric density; thermal history; and post-impact erosion, sedimentation, isostasy and magmatism. It has long been known that some of these influences differ considerably with magnitude of the impact (see, e.g., Heide 1964). Most recent observations suggest that the relative importance of external variables on crater shape also differs from planet to planet (Wood and Head 1976; Pike 1980a). The identification of specific factors that affect impact cratering on each body, a major goal of planetology, requires complementary methods of analysis and interpretation. One of these methods is geomorphology (see, e.g., Carr et al. 1984,p.3).

The geomorphic approach characterizes common aspects of crater shape, compares the resulting descriptors for craters on a planet and among planets, and then develops explanatory models for key similarities and differences (Gault et al. 1975; Head 1976; Schultz 1976a; Malin and Dzurisin 1977; Pike 1977a; Smith and Hartnell 1978; Wood et al. 1978; Mouginis-Mark 1979; Cintala 1979; Dence and Grieve 1979; Croft 1981a; Ivanov et al. 1986). De-scriptive variables for the geomorphic study of craters need to be quantitative and widely applicable. The transition of internal morphology from *simple* (bowl shaped) to *complex* (displaying central peaks, terraces and other fea-tures) with increasing crater size serves this purpose particularly well (Quaide et al. 1965; Dence 1972). Other quantitative transitions in morphology mark the diameter of onset of paired and multiple concentric rings in large craters and basins (Hartmann 1972).

The transition from simple to complex craters is a major geomorphic threshold that can be recognized on at least 16 impacted bodies larger than Amalthea: the Earth, the Moon, Mars, Mercury, Venus, Ganymede, Callisto, Tethys, Enceladus, Iapetus, Rhea, Dione, Mimas, Titania. Umbriel and Ariel. The mean crater diameter at which the simple-to-complex transition occurs D_t has been determined with an accuracy and precision of about 35% for the first four planets named above and estimated less reliably for some of the satellites (Hartmann 1972; Pike 1980a,b; Basilevsky 1981; Basilevsky and Ivanov 1982). The case of Venus remains indeterminate (Ivanov et al. 1986). The value of D_t is not constant, but differs from body to body. A parallel, less well-defined set of values has been determined for the onset of paired concentric rings in large craters and small basins (Hartmann 1972; Basilevsky and Ivanov 1982).

The Problem

Impact craters and basins on Mercury pose something of an enigma because opinions differ over the role of external influences, particularly surface gravity g and impact velocity V on their morphology (Murray et al. 1974; Gault et al. 1975). Evidence has been adduced to support diametrically opposed views on the importance of g and V (see reviews by Strom 1979,1984). The problem is particularly evident in the anomalous position of Mercury on the plot of D_t against g (Fig. 1). One explanation for the different D_t values invokes a dependence on target strength and an inverse dependence on g (Pike 1980a,b). Data from prior work show that the transition on Mercury occurs at a much larger crater size ($\sim$ 16 km diameter) than would be anticipated ($\sim$ 9 km) from its observed g and presumed target type, crystalline silicate rock.

This discrepancy between expected and observed D_t is so great that some crater size/shape data for Mercury resemble those for the Moon, which has half Mercury's g. Thus, the role of gravity in both creating basin rings (Head 1978; Boyce 1980) and shaping crater interiors on Mercury and other bodies (Cintala et al. 1977; Malin and Dzurisin 1978; Wood et al. 1978) has been questioned. Impact velocity has been proposed as an alternative control (Cintala et al. 1977; Chapter by Schultz). However, the radial extent of ejecta and secondary-impact craters around large craters on Mercury compared to that on the Moon is consistent with the higher g of Mercury, at least in the origin of secondary ballistic landforms (Gault et al. 1975). Still other arguments favor gravity scaling of both the formation of basin rings (Wood and Head 1976; Basilevsky 1981) and the shaping of crater interiors (Gault et al. 1975; Smith 1976; Smith and Hartnell 1978; Pike 1980a).

The effects of external influences on the morphology of multiring basins have been difficult to assess, mainly because quantitative data for basins on the four terrestrial planets are few (see, e.g., Hartmann 1972; Wood and Head 1976; Croft 1979). New measurements of basin-ring diameters by Spudis (1984) remedy this problem for Mercury. Now, both onset size for basin oc-

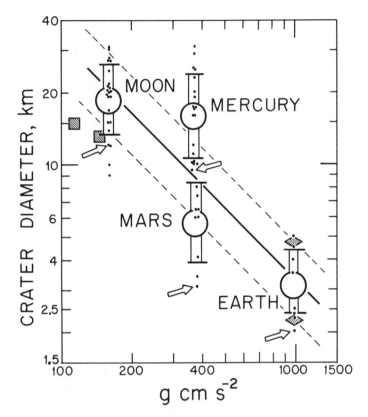

Fig. 1. Discordance of Mercury (old data) and Mars with the Earth and Moon (cf. Fig. 24). Transitions from fresh simple to complex craters are interpreted in terms of planetary surface gravity g and target characteristics. Each small dot ($n = 59$) is a statistically derived crater diameter that describes an aspect of the crater interior, such as inflection of depth/diameter data (arrows), onset of central peaks, etc. (Pike 1980a,Table 3). Large circles are geometric means D_t for the 8,10,18 and 23 diameters (small dots); bars are one-sigma limits. Solid line represents least-squares fit to the four D_t values: log D_t = log 3300 − 1.00 log g, interpreted as indicating targets of average strength or degree of stratification. Strong or weak targets suggested by different D_t values for terrestrial craters are subdivided by substrate: upper diamond, D_t for craters in crystalline rock; lower diamond, D_t for craters in sedimentary rock. Dashed lines extrapolate these two D_t values to lower g. Lower dashes suggest weak or strongly layered substrates, e.g., Mars and Callisto (upper square: presence of central peaks only) and Ganymede (lower square). Upper dashes are interpreted as resistant or homogeneous targets. Mercury lies well above "hard-target" trend.

currences and radial spacing of the ring diameters D may be able to provide clues to any role of g or V in the formation of multiring basins and thereby yield constraints on geophysical models proposed to explain them. Here, I try to ascertain whether or not the oft-cited $2^{0.5}D$ spacing interval between rings (Fielder 1963a) is valid for Mercury, how the verdict compares with those for the other terrestrial planets (Pike and Spudis 1985,1987), and thus if ring spacing merits consideration as a basic ingredient in basin-forming hypotheses.

II. BACKGROUND

Ambiguity in the relative roles of g, V and other variables in cratering on Mercury stems partly from uncertainties in identifying key geomorphic contrasts and the crater sizes at which they occur. Some of these morphologic uncertainties are evident in prior work on the interior form of Mercurian impact craters. The parameters of interest are those that compose the D_t value and other crater-size thresholds. The D_t constituents include the crater depth/diameter d/D ratio, the rim height/diameter h/D ratio, central-peak and floor morphology, and various features attributed to failure of the crater rim. A second important diameter transition on Mercury, that of large craters to small (two-ring) basins (Gault et al. 1975), is not yet well defined statistically; and a third, the diameter transition from two-ring basins to those with three or more rings, was virtually unknown until this study (cf. Wood and Head 1976; Croft 1979). The following section summarizes problems with existing data on the morphology of Mercurian craters.

Previous Work

The first study of Mercury (Murray et al. 1974) found its complex craters markedly shallower than those on the Moon, a result that has not changed. However, shortcomings are evident in all published d/D analyses for craters on Mercury. A common problem is failure to distinguish between simple and complex morphologies before summarizing the data by regression analysis. Thus, an inflected d/D distribution (e.g., Figs. 11,12, below) may not necessarily separate two key types of craters, particularly as the two linear branches of the log d/log D distribution usually overlap (Pike 1977b; Malin and Dzurisin 1978).

Four attempts have been made to derive a functional d/D relation for fresh craters on Mercury: Gault et al. (1975); Malin and Dzurisin (1977); Mouginis-Mark and Wilson (1979), and Pike and Clow (1982). (The radar altimetry date of Harmon et al. [1986] did not resolve craters small enough to straddle the simple-to-complex transition.) Mouginis-Mark and Wilson obtained their depths from photoclinometry (Mouginis-Mark and Wilson 1979,1981), while the other three studies used the shadow-length technique. The four results are inconsistent. The initial d/D study in 1975 found that the two log-linear regressions intersected at a crater diameter of about 9.3 km. Revision of this figure to 9.8 km in the 1977 paper was not significant, but the 1982 shadow-length results suggested that the intersection diameter is only 6 to 8 km, and the photoclinometric work estimates it at 13 km. These obvious discrepancies have remained unresolved, and their sources unidentified.

Problems with prior d/D results clearly lie with the crater depths; measurements of crater rim (and floor) diameter, by comparison, are fairly accurate. Several d values tabulated in Malin and Dzurisin (1977, 1978) are unrealistically high or low, given the observed scatter of d/D data on Mercury and experience with measurements of other planets (Pike 1977a, 1980a). These values may have resulted from inaccurate crater locations, which have been improved in subsequent iterations of the Mercury control net (Davies and Batson 1975; U.S. Geol. Survey 1979). Comparisons, made for this study, of image coordinates generated from camera-pointing data and listed in the Supplementary Experimental Data Record, SEDR (Image Processing Laboratory 1976) with actual picture coordinates obtained directly from the map (U.S. Geol. Survey 1979) through photointerpretation reveal many discrepancies. Some of these are serious: e.g., footprints for Mariner 10 frames FDS # 0027405 and # 0166751, both containing the crater Sadi (No. 282 in Appendix B), are displaced westward by about 4°.

Incorrect values of d in Malin and Dzurisin (1977, 1978) also arise from illumination angles that are unsuited to the morphology of the crater. One crater listed in Malin and Dzurisin (1978, Table 1) as 32 km in diameter (and hence unquestionably of complex morphology; see Fig. 11 below) and 5.54 km deep has a shadow angle of 32° on frame FDS # 0000222. The shadow corresponding to such a steep angle is so short that its tip must fall on the inner wall, not out on the crater floor. The 5.5 km depth cannot be correct. In another case (the crater Brahms; see Fig. 5b,c below), the depth value listed is actually the relative height of the central peak.

The rim-height/diameter results of Cintala (1979), although appearing reasonable, also warrant a second look. Again, in the absence of any simple/complex distinction in overall morphology, significance of the inflection in the h/D distribution must remain moot. Second, the measured craters number only 30, too small a sample given the inherent difficulties in obtaining reliable h values from the shadow-length technique at extremely low Sun angles. The data exhibit such great scatter that the h/D fits are not robust: subtracting a single point (Cintala's No. 29) changes the slope of his steeper h/D least-squares fit ($n = 21$) from 1.01 to only 0.83.

Hale and Head's (1980a) analysis of central peaks in craters on Mercury needs to be reexamined for two reasons. First, their least-squares fit to peak-base diameter D_{cp} and crater diameter D presumed (without proof) an arithmetic-linear relation rather than the logarithmic-linear relation normally encountered in analyses of crater dimensions (Baldwin 1949, 1963). Whether both variables (and nearly all other crater variables: see any size-frequency distribution) are in fact distributed normally in the log x domain, as I contend, remains to be formally tested. The D_{cp} measurements, inherently among the least accurate and precise data on crater shape, are important because they permit large craters and small basins to be compared in plan, quantitatively (Pike 1982a, 1983a). The failure to use logarithmic scales also attends some widely cited equations for basin-ring diameters (Head 1978).

Second, Hale and Head (1980a) concluded that central peaks in Mercurian, as well as lunar, craters do not increase significantly in morphologic complexity with size of the host crater. Their claim is at odds with other descriptive data on Mercurian central peaks (Smith and Hartnell 1978) as well as a large body of prior observations on the Moon (see, e.g., Hartmann and Wood 1971; Schultz 1976a).

Previous results of size-morphology analysis on Mercurian craters from nonmetric data also are inconsistent. Three independent studies on the presence or absence of interior features, namely bowl shape, central peaks, minor wall failure (slump deposits beneath scalloped rim crests), and major wall failure (terraces) yielded three different frequency-of-occurrence curves (cf. Cintala et al. 1977; Malin and Dzurisin 1978; Smith and Hartnell 1978). Attributes such as these, which are essential to a reliable value of D_t, need to be systematically reexamined. Moreover, data of Smith and Hartnell (1978, Appendix II) on the complexity of peaks and terraces have never been given the detailed analysis warranted. (Their paper also must be read carefully, because of serious caption errors in their Figs. 14 and 15, Table III, and Appendix II.)

Ambiguity extends also to Mercury's crater-to-basin transition. The diameter for this morphologic threshold has been rather vaguely defined, ranging somewhere between 70 km and 200 km (cf. Gault et al. 1975; Wood and Head 1976; Hale and Head 1980a; Pike 1982a). This diameter needs to be determined with greater precision. More importantly, the physiographic character of the transition is debatable. Hale and Head (1979a, 1980a) challenged the traditional view, carried over from lunar work, that central peaks in Mercurian craters expand gradually into interior basin rings (Gault et al. 1975). They suggested rather that the change from peaks to inner rings may reflect an abrupt shift in formational mechanism, a difference of opinion that needs to be examined. The nature of the crater-to-basin transition is further confused by the nonmonotonic relation between diameters of inner and main rings described for the analogous lunar case by Hale and Grieve (1982, Fig. 11).

Anomalies appear in ratios of ring diameter for two-ring basins and large craters on Mercury. The data of Wood and Head (1976) show many low values for the ratio of inner D_i to outer D ring diameters, which typically lie close to 0.5 for the Moon, Mars and Mercury. The plot of Croft's (1979) basin-ring data by McKinnon (1981) confirms the low ratios and shows further that they correlate preferentially with the smallest basins. This size-dependent relation, which has been examined in detail for the Moon (Pike 1983a), warrants further study on Mercury, where the more numerous two-ring structures provide better statistics. The D_i/D contrast has implications for ring-ratio models that have been developed in the course of explaining the origin of two-ring and multiring basins (see, e.g., Head 1978; Croft 1979).

The transition from two-ring to multiring basins on Mercury is not known at all (Wood and Head 1976). Few multiring basins were even identified there (Croft 1979) until recently (Spudis 1984; Spudis and Prosser 1984), let alone analyzed for onset size or spacing of their rings. Prior work suggests that adjacent-ring spacing for multiring basins on Mercury is half that for mature (large) two-ring basins, in agreement with the earliest results for lunar basins (Hartmann and Kuiper 1962). Preliminary analyses of Mercurian basin rings (Croft 1979; Pike 1982a) need to be updated to include the latest basin discoveries and ring measurements.

Present Work

This chapter summarizes most of the available information on morphology of the crater-to-basin continuum on Mercury and compares the analytical results with those for the other three terrestrial planets. It fully exploits two new sets of data: the d/D measurements and morphologic observations made expressly for this study, and the recent basin-ring measurements of Spudis (1984). The work reexamines various aspects of crater and basin morphology over the entire diameter range first observed on Mercury (Murray et al. 1974) to bring both the observational data base and morphologic analyses up to the standard of those for the Earth, Moon and Mars. Specific objectives that were attained fall into nine categories:

1. A new value for D_t and re-evaluation of the possible g or V dependence of both crater and basin transitions;
2. Integration of a new and reliable d/D data set with a new and uniform set of size-morphology observations (the oft-made distinction between crater "morphology" and "morphometry" is artificial and counterproductive);
3. The incorporation into these new data of some prior sets of observations that are judged comparable in accuracy and precision;
4. A classification of craters and basins on Mercury, but focusing on constituent morphologic elements and transitional features as much as on the formal taxonomic groups;
5. Tests for dependence of crater morphology on target terrain;
6. Reevaluation of some size-dependent characteristics of central peaks in craters and inner rings in basins;
7. Tests for systematics in the spacing of basin rings;
8. Comparison of size-dependent morphologic thresholds for craters on Mercury with those on other bodies, with special reference to scaling by g and $V;$
9. Application of the results to constrain models for the onset of complex crater interiors and the spacing of concentric rings.

The scope of this study is primarily empirical. While physical theory and experimental work contribute complementally to solutions for cratering problems, the approach here emphasizes measurement, classification, cross-comparison, and analysis of craters according to their surface dimensions. The element of geologic history (see the Chapter by Spudis and Guest) is ignored, because time-dependent contrasts in crater shape do not contribute centrally to solving the issues addressed. Morphologic changes in craters with increasing relative age (Malin and Dzurisin 1977; Trask 1981) are not examined; nor is the detailed chronology of geologic units from crater size-frequency distributions (e.g., Leake 1981) addressed. Finally, description of crater classes or individual features as "mature" or "immature" does not connote time, but rather the completeness of morphologic development.

Some procedural standards are introduced here to reduce subjectivity and to improve the statistical presentation over past studies. Much of the analysis is by regression of paired variables, almost entirely in the log-log domain (Baldwin 1949). Results are carried to three decimal places to minimize round-off error. Each regression equation has a goodness-of-fit test of its significance and an error envelope calculated at the 95% confidence interval (CI) (Davis 1986,pp.66-67,176-186; Snedecor and Cochran 1967, pp.135-140). The CIs are calculated in the log-log domain but are expressed in arithmetic terms (the conversion is described in Croxton et al. [1967,pp.435-441]). The 95% CI values, calculated for Y_c of log-log regressions *at the midpoint of X,* also are close to the standard error of the estimate S_{yx}. All regressions of ring diameters for two-ring and multiring craters and basins carry weights that better reflect the varied quality of ring measurements. Where two crater subsets are compared, a statistical test for similarity of their regression lines is carried out through an analysis of covariance (Snedecor and Cochran 1967,pp.432-436). Finally, the slope of a regression line can be compared with another value by a *t*-test (Lentner 1972,pp.227-228).

The result of the work described in this chapter is a diameter-dependent and largely quantitative synthesis of crater systematics on Mercury. The integrated model of crater morphology and geometry builds on the work of previous investigators and extends the findings first outlined in Pike and Clow (1982) and in Pike and Spudis (1984). Various processes and influences that may be responsible for the morphology of Mercury's complex craters and basin rings are discussed in light of this empirical synthesis, as well as parallel results from the study of impact structures elsewhere.

III. THE NEW DATA

Observations and measurements were made of 447 impact craters on Mercury. Of these, 316 were selected for the study of size/morphology, including d/D, 57 are craters suitable for a restudy of h/D (Table I, Appendix B), and 74 are basins or large craters that have at least two concentric rings

TABLE I
Distribution of Crater Sample[a]

316-Crater Subset (depths)

Quadrant on Mercury	Interior Morphology					
	Simple (culled)	Simple (selected)	Modified-Simple	Immature-Complex	Mature-Complex	All
NE	4	4	1	14	10	33
SE	10	14	5	15	19	63
NW	40	73	10	31	19	173
SW	12	13	3	9	10	47
All four	66	104	19	69	58	316
Smooth plains	33	58	12	30	10	143
Cratered terrain	33	46	7	39	48	173

57-Crater Subset (rim heights)

	Interior Morphology		
Quadrant	Simple	Complex	All
NE	—	3	3
SE	—	—	—
NW	21	21	42
SW	11	1	12
All	32	25	57

[a]Supporting data may be found in Appendix B.

(Tables II-IV). Most data on multiring basins are from Spudis (1984). These three sets of new data were supplemented by six separate sets of observations made of Mercurian craters previously, by other workers, as indicated.

Fresh-looking craters were chosen for analysis. No formal classification of relative age or state of degradation was applied as in Malin and Dzurisin (1977), Wood et al. (1978), Smith and Hartnell (1978), Leake (1981), and Trask (1981). However, the sample was restricted to craters with a crisp overall and, where possible, detailed morphologies, continuous rims, and where resolution permitted, some indication of textured ejecta or satellitic craters (cf. Pike 1980a,p.2173-2175). These constraints were relaxed for the largest craters, which were judged to have lost little of their initially great depth to infilling. Ringed basins of any relative age or state of preservation were accepted, because the basin measurements of interest, rim-crest diameters of rings, probably change little with degradation.

Suitability for depth measurement by the shadow-length technique restricted the main sample to 316 craters located close to the terminator (see

Table I and Appendix B). Although craters at all latitudes on both east and west edges of the illuminated disk are included, most of the sample is in the northwest quadrant of the imaged hemisphere (map quadrangles H-3 and H-8), where the best high-resolution images are located. After freshness of morphology, the main criterion for the d/D sample was shadow position within the crater. The depth value most characteristic of a bowl-shaped crater is the maximum, usually at the exact center of the rounded crater bottom. Thus, in craters with featureless interiors, the shadow tip should fall close to the center ($n = 170$). The depth value most descriptive of a crater with a more complex interior is an average value that represents the entire flat or hummocky floor. Thus in craters without bowl-shaped interiors, the shadow tip should fall on some part of the quasi-level floor, away from the sloping walls, and not distorted by any central mound or peak ($n = 146$).

The 316 craters were subdivided on the basis of the topography that surrounds them. I have followed a previous convention (Cintala et al. 1977; Smith and Hartnell 1978) of dividing Mercury into two or more major and contrasting geomorphic realms, here cratered terrain (CT) and smooth plains (SP), in an attempt to mimic the upland/maria dichotomy on the Moon and thus possibly distinguish substrate properties that might correlate with differences in crater shape (Pike 1980*a*). The basis of the two-part classification is the generalized geologic terrain map by Trask and Guest (1975): My "smooth plains" is theirs; my "cratered terrain" is all other units (cf. Cintala et al. 1977). Final assignment of a crater to its background topography was made from Mariner 10 images and photomosaics.

Diameter, Depth and Rim Height

Mean rim-to-rim crater diameters and rim-to-bottom shadow lengths, maximum for bowl-shaped interiors and mean for complex interiors, were measured directly on Mariner 10 images, both contact prints and enlargements (Appendix A). No data were taken from the 1975 third-encounter pictures. All estimates were made to 0.01 mm precision with a 7X Bausch & Lomb optical comparator (Measuring Magnifier, No. 81-34-35) equipped with a reticle calibrated in 0.1 mm divisions. Replication of measurements was very good. Internal checks on diameters and shadow lengths were obtained by both deliberate and accidental duplications of measurements, on the same images at both similar and dissimilar scales, and on different images. The quality of the measurements varies with that of the image, as well as with image scale. Few data were available for Mercurian craters under 1 km across, which is equivalent to only a few pixels on the largest-scale images.

As a further check on measurements for the sizes of small and nearly circular craters imaged at low view angles, three diameters were calculated: E-W, in the direction of maximum foreshortening, and in the direction of no foreshortening (Appendix A). Generally the median value is the one reported in Appendix B. The three diameters were averaged for large complex craters with irregular outlines in plan to obtain a more representative value. Position of the shadow tip in each crater was identified without photometric analysis, which has successfully solved this problem on Mars. Close correspondence between Martian crater depths obtained both with the photometric technique (Pike and Arthur 1979) and without (Pike 1980a) gives assurance that visual discrimination of shadow umbra from penumbra in the Mercurian work has not introduced much error into the new depth estimates.

Crater depths and diameters were obtained from raw measurements through equations that accommodate any spacecraft altitude or look angle, planetary location, or degree of foreshortening encountered in the Mariner 10 images (Appendix A). There are 15 input quantities, 4 of which are held constant here: vidicon width, camera focal length, radius of Mercury and solar latitude. Image width has only two values, one for contact prints and one for enlargements. Besides the crater diameters (three) and length of the shadow, the algorithm requires values for colatitude and longitude of a point on the shadow-casting rim, available to the required accuracy from mapped results of the latest Mercury control net (U.S. Geological Survey 1979). The remaining four variables: longitude and colatitude of the Sun and longitude and altitude of the spacecraft, were taken from the SEDR (Image Processing Laboratory 1976). After calculating spacecraft range (Eq. A6, Appendix A), five angles relating elements of the spacecraft-Sun-crater rim geometry (Eqs. A1,A2,A5,A7,A8) and magnification of both vidicon and image (Eq. A11), the algorithm computes three diameters of the crater, the length of the shadow, and the depth of the crater, both on the image and on the surface of Mercury (Eqs. A3,A4,A9,A10).

The d/D results are given in full for all 316 craters in Appendix B. The smallest crater for which a depth (at $d = 0.044$ km, also the shallowest) was measured is 0.225 km in diameter. It is located on smooth plains at $+30°5$, $162°9$ W, just north of Odin and Budh planitia. The largest crater measured (Boccaccio, at $-81°$, $41°5$ W) is 160 km across ($d = 4.0$ km). The deepest crater (Ictinus, at $-78°$, $159°$ W) extends 4.8 km from floor to rim crest. Both large craters are situated in cratered terrain.

Of the 170 bowl-shaped craters for which depths were calculated, only 104 were retained in the subsequent analysis (all equations in Table V). The 66 culls are (slightly) overly deep or shallow craters wherein the shadow tip was displaced from the exact center of the crater by more than 5% of the crater diameter, and whose shadow-casting angle either exceeded 26° or fell short of 18°. These criteria were determined from an empirical analysis of the inverse relation between shadow length/crater diameter (median 0.50, range 0.34 to 0.63) vs shadow-casting angle (median $21°4$, range 15° to 30°) and its related effects on crater depth ($n = 170$). The smaller the shadow angle, the shallower the crater appears, and *vice versa*). The remaining 104 d/D values are of exceptionally high quality, because the shadow tip is positioned almost exactly in the center of these craters. An analysis of covariance shows that retained and discarded craters differ significantly only in dispersion, as expected.

The algorithm developed for the d/D work (Appendix A) was also used to calculate relative height h of the rim crest above the surrounding terrain for 57 craters on Mercury (Appendix B). The sample included 28 of the 30 craters previously measured by Cintala (1979). I observed no external shadow for his no. 11, and could not satisfactorily locate the position of the shadow tip in the uneven terrain surrounding his no. 24. Discrepancies between the two data sets are rare and minor. I found only two of Cintala's diameters, nos. 6 and 8, and two of his rim heights, nos. 2 and 5, to be less than would normally be expected from the scatter of the 28 pairs of values in each case. Good rim heights can be obtained from the shadow-length technique only for craters on relatively smooth surfaces located near the terminator. Because errors in shadow angle, in position of the shadow-casting rim and the shadow tip, and in separating umbra from penumbra are inherently more severe for these low angles, the height data (Appendix B) are substantially less precise than the depth data (e.g., lowest r values in Table V).

Morphologic Identifications

Presence or absence of six internal morphologic features was ascertained for each crater where a depth or rim height was calculated: bowl shape, flat floor, scalloped rim, slump block, terrace and central peak. Reliable observations were restricted to the 316 craters used for d/D analysis, because shadow angles of the 57 craters measured for h/D were so low that most of the crater interiors were concealed (Appendix B). The latter 57 craters were classified provisionally, as simple ($n = 32$) or complex ($n = 25$), so that rim height could be compared with interior morphology as closely as possible under such difficult circumstances. In practice, this assignment was arbitrary only for the 11 h/D craters smaller than the largest simple crater in the 316-crater sample (14.4 km) but larger than the smallest complex crater (9.5 km): the six smaller craters were designated simple, the five larger complex (each subset has about the same median h).

Presence/absence observations have limitations that do not apply to the comparatively straightforward d/D and h/D calculations. These limitations particularly need to be assessed for craters that may have a transitional morphology, which I have defined as internal features immaturely, i.e. incompletely, developed. The following evaluations for each of the six interior features are not evidence that the observations are unreliable, but rather describe problems routinely encountered in assembling the data set. These remarks are offered as guidelines and possible criteria for rejecting craters from a sample rather than risk misidentifying a feature and perhaps incorrectly classifying a crater. Such operational questions, which arise less frequently in lunar or Martian data where image resolution is consistently higher, assume real importance for Mercury and the satellites of the outer planets, where resolution is lower.

A bowl-shaped crater interior is a rounded and smooth surface devoid of

any of the major morphologic features described below (Fig. 2). At high resolution some stratification in the upper wall and texture, probably the result of
mass-wasting, may be present on the lower wall. Characteristically, the rim
crest is smooth, unbroken and highly circular in plan. Isolated and obvious
secondary-impact craters, which usually are less crisp in appearance and less
symmetric in plan, but which otherwise have a bowl-shaped interior, have
been excluded from this study.

Small central peaks at very low resolution cannot be confidently distinguished from a small "floor" of impact fallback or sheet-slump debris or
from one or more lumps or hummocks of slumped rim debris that have collected in the lowest, central part of the crater (Figs. 3 and 4). This is not a
trivial distinction, as it bears on interpretation of the transition from simple to
complex craters (Fig. 1): addressing the question of whether peak formation
or rim failure occurs first in small craters depends critically on accurate observations of these features. Even at high resolution it is possible to confuse a
peak with a slump block in a small crater. Because both slump deposits and
peaks can occur in the same small crater (Fig. 4), *small peaks may lie hidden
beneath the slump materials*. Fallback from the impact also can conceal nas-
cent peaks. Finally, an especially small peak cannot be seen if the interior
shadow is long enough to conceal it, although in most craters on Mercury a
peak has sufficient relief to show above a long shadow (Fig. 5).

Three older data sets on the detailed morphology of central peaks were
restudied to supplement the new observations on basic presence/absence in

 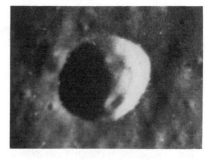

Fig. 2. (left) Simple impact crater 10.8 km across and 2.07 km deep, on Mercurian cratered
terrain at $-03°6$, $28°7W$, on Mariner 10 frame FDS 0027475. Elements of simple morphology
include smooth and circular rim crest, featureless bowl-shaped interior and commensurate 1/5
depth/diameter (d/D) ratio (cf. Figs. 10-12), minor wall texturing.

Fig. 3. (right) Modified-simple crater 9.7 km across and 1.4 km deep, on Mercurian smooth
plains at $-06°0$, $172°5W$, on Mariner 10 frame FDS 0000061. Elements of the modified-
simple morphology include minor departures of rim crest from circularity, small slump deposits
at foot of interior wall and commensurate $< 1/5$ d/D, and rudimentary flat crater bottom (cf.
Figs. 10-12).

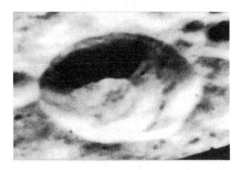

Fig. 4. Immature-complex crater 20 km across and 1.82 km deep, on Mercurian cratered terrain at $-12°2$, $26°6W$, on Mariner 10 frame FDS 0027473. Elements of the immature-complex morphology include departures of rim crest from circularity, concentric rim crest topography, large slump deposits or rudimentary terraces on inner wall and flattish floor, low d/D, and frequently (although not evident here) a small central peak (cf. Figs. 10-12).

Appendix B. Central peak diameter D_{cp} was not remeasured for complex craters on Mercury (Figs. 4 and 5), although it was for protobasins (Fig. 6). Rather, 138 of the 140 values plotted against crater diameter, 15 km $\leq D \leq$ 175 km, by Hale and Head (1980a,Fig.2) were adopted for analysis. Values of D and D_{cp} were scaled directly off a prepublication enlargement of their diagram (I could only resolve 138 points). The technique of measurement is described fully by Hale and Head (1980a). I also reexamined their data on peak complexity (but not "geometry") for the same 140 craters (their Fig. 5A). Finally, the detailed observations on central peak complexity by Smith and Hartnell (1978, their Appendix II) for 129 craters were examined to ascertain more quantitatively any size dependence in peak morphology (with due caution, as I found some errors in their raw data).

The identification of flat floors involves problems of crater taxonomy and the effects of image resolution. First, flat floors can be confused with central peaks or slump debris at very low resolution, often just at the crater sizes where such distinctions are important (Figs. 3 and 4). Second, "flat floor" is a much-abused term whose meaning and significance often are glossed over. It is useful to try to distinguish among flat floors that may look much the same at low resolution but that look distinctly different at high resolution, that have different origins, and tend to occur systematically in craters of different sizes. At least five types of primary flat floors can be recognized (cf. Schultz 1976a): ejecta fallback, minor slump deposits (Fig. 3), major slump deposits (Fig. 4), trough surrounding peak in smaller craters, and melt-veneered annulus in larger craters (Figs. 5 and 6). Equally important is the small flat floor formed by ballistic ejecta from a nearby crater or post-impact volcanic fill (Leake 1981), and thus not a primary feature at all, but rather an attribute that could lead to misclassification of the crater as being more complex than it really is.

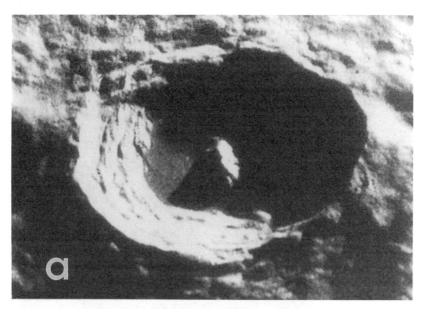

Flat floor diameter F of complex craters (Figs. 4-6) was not measured afresh here, but instead 61 of the 63 values plotted against crater diameter by Malin and Dzurisin (1978, their Fig. 5) were accepted from their Table I as sufficiently accurate for restudy here. These data are so much more reliable than their d measurements, that I did not remeasure them. However, F values for the two smallest craters were discarded as being indeterminate; neither one has a measurable flat floor. The measurement technique is described by Malin and Dzurisin (1978).

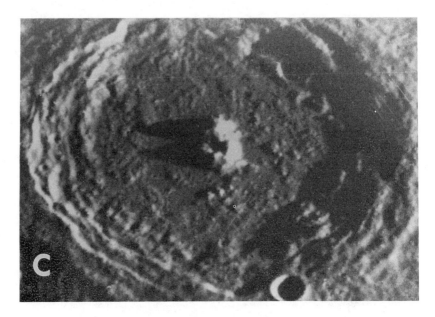

Fig. 5. Mature-complex craters on Mercury. (a) Small example 36 km across and 2.13 km deep,
    on cratered terrain at −51°8, 20°0W, on Mariner 10 frame FDS 0027466. Elements of mature-
    complex morphology in smaller craters (cf. Fig. 5b) include pronounced rim-crest acircularity
    and concentricity, strong terracing, distinct flat floor and well-developed single central peak,
    and low d/D (cf. Figs. 10-12). (b) Large crater (Brahms) 97 km across and > 3.15 km deep
    (shadows cast by rim crest do not reach the floor in this and in craters and basins in Figs. 6-9)
    on smooth plains at 58°7, 174°2 W, on Mariner 10 photomosaic 3-B. Overhead view. Elements
    of mature-complex morphology in larger craters (cf. Fig. 5a) include highly irregular rim crest,
    elaborate terracing, broad and hummocky flat floor, compound or complex central peak, and
    low d/D (cf. Figs. 10-12). (c) Close-up high-oblique view of detail in large crater Brahms (Fig.
    5b) on Mariner 10 frame FDS 0000080, showing effects of increased resolution: intricate
    terracing, detailed elements of floor texture, and the large central peak.

The scalloped rim crest is perhaps the least ambiguous of all the morphologic attributes of fresh craters on Mercury (Figs. 3-5). Well described for
clearly depicted small craters on the Moon such as Dawes (Settle and Head
1979), especially in contrast to terraces in larger craters, scalloped rims have
been studied in craters on Mercury for possible variation of their frequency
with crater size, planet and substrate type (Cintala et al. 1977). The cuspate
outline almost always is matched by a corresponding deposit of slump material lower down on the wall, or at the bottom of the crater (Figs. 3 and 4). The
rounded deposits appear to be material of low cohesion. Although scalloped
rims and slump deposits seem to be replaced in larger craters by terraces and
more coarsely scalloped rim outlines (Figs. 5 and 6), the finer-scale features
actually persist to much larger crater sizes on Mercury than usually is recognized (Fig. 7; cf. Schultz 1976a, p.24).

Fig. 6. Protobasins on Mercury. (a) Verdi, 135 km across and > 2.8 km deep, at +65°, 165°W, on Mariner 10 photomosaic 3-B. Elements of protobasin morphology are rudimentary in this example, which clearly is transitional between large complex craters (Fig. 5b,c) and well-formed protobasins (Fig. 6b). Central peak is large and morphologically complex; the inconspicuous inner ring is very fragmentary. Verdi occupies an anomalous position in Fig. 18 below (see also Table II). (b) Van Gogh, 98 km across at −76°, 136°W, and Bernini, 165 km across at −79°, 135°W, both on Mariner 10 photomosaic 15-B. Well-formed elements of protobasin morphology include craterlike rim, undersized inner ring ($\sim 0.4\ D_i/D$) and central peak ($\sim 0.13\ D_{cp}/D$), and clear physiographic subordination of inner to main ring (cf. Fig. 18). Only half of the inner ring of Van Gogh is evident (Table II).

The downslope masses of material that often correspond 1 : 1 to scallops in cuspate crater rim-crests are somewhat less easily identified than the crenulations themselves. Variously termed "scallops" (Head 1976), hummocks and slump masses, blocks or deposits, these features range considerably in appearance, from very small individual lumps at the foot of the crater wall (Fig. 3) to masses of arcuate, strongly textured material (the "swirl texture" of Smith and Hartnell [1978]) that occupy the entire crater interior (Fig. 4). In small complex craters on Mercury, slump masses can appear to coalesce to form what at low resolution can be mistaken for rudimentary terraces (Fig. 4). Slump masses probably persist to the largest crater sizes, although their presence is overwhelmed by that of terraces (Figs. 5-7; Schultz 1976a,p.24).

Terraces, arcuate and continuous slump features on crater walls, can be difficult to identify in small Mercurian craters for reasons noted above. They are larger and longer than slump blocks. Varying criteria for distinguishing slump blocks from terraces (cf. Figs. 4 and 5) can lead to difficulties in comparing independent studies of crater morphology (Cintala et al. 1977; Malin and Dzurisin 1978; Smith and Hartnell 1978). In large craters at good resolution, terraces appear to be composed of more coherent material than slump blocks, and have the stair-step morphology that is characteristic of (multiple) rotational landslips (Figs. 5 and 6). The topographically highest terraces on a crater wall correspond to scalloped, cuspate segments of the rim. There is some tendency for terraces to increase in width with diameter of the host crater.

Two older data sets were incorporated into the analysis of terraced crater walls on Mercury. Width of the terraced rim wall W of complex craters (Figs. 4-6), the complement of F (see above), was not remeasured here. Rather, 61 of the 63 values plotted against crater size by Malin and Dzurisin (1978,Fig. 6) were taken from their Table 1 (see F, above), omitting the same two small craters. Additionally, the detailed "wall complexity" assessments of Smith and Hartnell (1978, Appendix II) for 147 complex craters on Mercury were examined here (for the first time) to gain a clearer picture of crater-size dependence in the development of wall failure.

Basin Measurements

Mercury has many more basins than was initially recognized (Figs. 7 and 8), but they are very old (Fig. 9). The number of ancient circular structures with two or more concentric rings has grown steadily (Murray et al. 1974; Strom et al. 1975a,b; Malin 1976b; Wood and Head 1976; Schaber et al. 1977; DeHon 1978; Strom 1979; Frey and Lowry 1979; Croft 1979). Recent photogeologic analysis has identified still more two-ring (Pike 1982a) and multiring basins on Mercury (Spudis 1984; Spudis and Prosser 1984; Chapter by Spudis and Guest). Multiple basin rings, which coincide with occurrences of smooth plains materials (Spudis 1984; Chapter by Spudis and Guest), also are evident on topographic profiles from Earth-based radar (Spudis and

Fig. 7. Two-ring basins on Mercury. (a) Bach, 210 km across, at $-69°$, 103°W, on Mariner 10 photomosaic H-15C. Elements of two-ring basin morphology include overall circularity in plan, albeit with some asymmetry, equally conspicuous inner and outer rings, $\sim 0.5\ D_i/D$, with no central peak (see Table III and Fig. 18). (b) Ahmad Baba, 132 km across at $+59°$, 127°W, and Strindberg, 190 km across at $+54°$, 134°W, both on Mariner 10 photomosaic 3-C. Inner rings are one-half the diameter of outer ones, but are less complete and less conspicuous than that of Bach (see Fig. 7a, Table III and Fig. 18).

Fig. 8. Young multiring basin Caloris, 1340 km across at the main ring, centered at $+30°$, 195°W, Mariner 10 computer photomosaic 3-F (stereographic projection). Elements of multiring basin morphology include overall plan circularity, highly variable completeness and physiographic strength of concentric rings or ring arcs, with one usually more conspicuous than others, and approximate spacing of rings by $2^{0.5}D$ increment (see Table IV and Figs. 19-23,25). Some rings of the large older Tir basin (Fig. 9 below) lie to the NE, and some of the small old basin Van Eyck (Table IV) lie to the SE.

TABLE II
Ring Diameters of Protobasins on Mercury

Name	Center Coordinates (deg)	Observed Diameter[a]			W[b]
		Rim	Inner Ring	Base of Central Peak	
—	−28, 20W	72	36	7	(2)
—	+01, 17W	76	25	15	(3)
—	+13, 27W	95	32	15	(2)
Asvaghosa	+11, 21W	86	29	13	(3)
Bernini	−79, 135W	165	65	10	(3)
Boethius	−01, 74W	122	50	12	(3)
Brunelleschi[c]	−09, 23W	128	34	17	(3)
Chong Chol	+47, 116W	138	58	11	(2)
Equiano	−39, 31W	105	46	18	(2)
Hawthorne	−51, 116W	125	55	25	(2)
Hitomaro	−16, 16W	115	40	14	(3)
Jo' Kai	+73, 136W	86	38	12	(3)
Lu Hsun	00, 24W	101	41	22	(2)
Mansur	+48, 163W	95	38	12	(2)
Scarlatti	+40, 100W	152	69	6	(3)
Sinan	+16, 30W	136	60	7	(3)
Ts'ai Wen Chi	+23, 23W	123	54	14	(3)
Van Gogh	−76, 136W	98	37	18	(2)
Verdi	+65, 169W	140	50	35	(1)
Zeami	−03, 147W	130	44	12	(1)

[a]Source: measured or remeasured by author on Mariner 10 photomosaics, after lists of Croft (1979,Table 5.2) and Wood and Head (1976,Table 2). Values for rim and inner ring are means; values for peak base are maxima (Hale and Grieve 1982).
[b]Weight values (see text; cf. Table IV) reflect quality of observations (3 = strongest).
[c]Unusually small inner ring possibly is an open or expanded cluster of central-peak elements (Pike 1983a).

Strobell 1984). The current total of large impact structures having at least two measurable ring diameters (Figs. 6-9) stands at 74 (Tables II-IV). This list is not exhaustive. For example, the large craters Lermontov and Giotto may have inner rings as well as central peaks (De Hon et al. 1981), but the high-Sun Mariner 10 images preclude good estimates of their dimensions. For further reading on the identification of large basins, see Hartmann (1981) and Pike and Spudis (1987). Basin depths and the relative heights of rings are generally unavailable save for occasional features measured from radar altimetry (Harmon et al. 1986) because shadow lengths on Mariner 10 images are too short.

The diameters of large, multiring basins (Table IV) were measured by fitting circles (Fig. 9c) to the segments and fragments of ring crests (Fig. 9b) mapped on computer mosaics (Fig. 9a) in the photographic edition of the Mercury Atlas (Davies et al. 1978), although individual Mariner 10 images

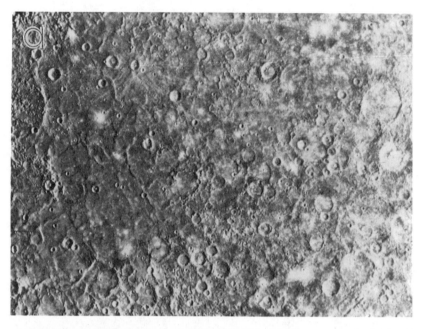

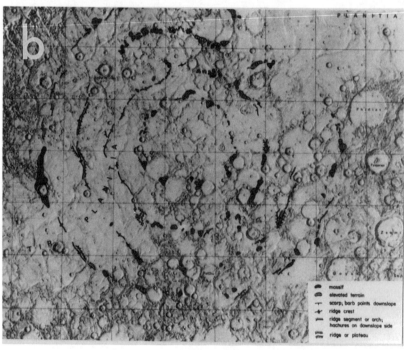

PLANITIA

PLANITIA

PLANITIA

T I B

Puduan

Copmo

Zeam

G o v

🦪 massif

🦪 elevated terrain

⊢ scarp; barb points downslope

⊹ ridge crest

ᴍᴍᴍ ridge segment or arch;
hachures on downslope side

ᴍᴍ ridge or plateau

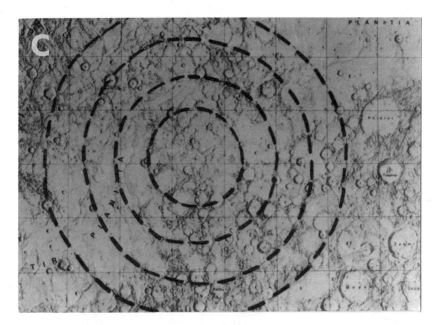

Fig. 9. Old multiring basin Tir, showing mapping of subdued rings. (a) Site of Tir basin, 1250 km across at the main ring, at +06°, 168°W, on partial photomosaic of the Tolstoj quadrangle. (Ring elements are easier to identify on enlargements of full-sized mosaics.) Rim of Caloris basin is in upper left. Scene is 2000 km across with north at top. (b) Terrain map of basin-associated elements of the Tir basin of Spudis (1984), on shaded relief map of part of the Tolstoj quadrangle. Location of multiple rings are indicated by massifs, mare-ridge alignments and topographic highs. (c) Four-ring interpretation of the Tir basin derived from terrain map in (b), after Spudis (1984). Largest ring is 1250 km across (see Table IV and Figs. 19-23,25). Base map same as in (b). Panels a,b and c are from Pike and Spudis (1987).

were used in some cases. Most of the new data come from Spudis (1984), who has applied the detailed mapping techniques of Wilhelms and El Baz (1977) and Schultz et al. (1982) in identifying degraded basins on the Moon and Mars.

Small basin diameters (Fig. 7, Tables II and III) were measured along four or more azimuths, by dividers. Many basin rings on Mercury are badly fragmented and evident only as isolated and low, yet concentrically disposed, topographic hummocks and massifs (Fig. 9). In the smallest basins (actually large craters) the inner ring often is only a partial circle or arc of hummocks, and it is possible in some cases (e.g., Verdi: 140 km across; see Fig. 6a) to confuse a true ring (Figs. 6b and 7) with an annulus of central-peak elements loosely disposed about the main mass of the peak (Fig. 5b,c). Where a central peak was present, the diameter of the base D_{cp} was measured by fitting it with a circumscribed circle so that the data could be compared with those of Hale and Head (1980a).

TABLE III
Ring Diameters of Two-Ring Basins on Mercury

Name	Center Coordinates (deg)	Diameter[a] Rim	Diameter[a] Inner Ring	W[b]
Ahmad Baba	+59, 127W	132	60	(3)
Al-Hamadhani	+39, 90W	155	75	(3)
Bach	−69, 103W	210	103	(3)
Botticelli	+64, 110W	148	70	(2)
Cervantes	−75, 122W	203	106	(3)
Chekhov	−36, 62W	205	100	(3)
Dorsum Schiaparelli Basin	+17, 166W	225	120	(2)
Durer	+22, 119W	190	95	(3)
Handel	+04, 34W	153	68	(2)
Hitomaro Basin	−15, 15W	255	125	(3)
Homer	−01, 37W	310	172	(2)
Ma Chih-Yuan	−59, 78W	190	90	(3)
Mark Twain	−11, 138W	150	80	(2)
Mendes-Pinto	−61, 20W	195	100	(2)
Michaelangelo	−45, 110W	225	105	(3)
Mozart	+08, 191W	255	130	(3)
North Pole Basin	+85, 171W	170	85	(2)
Pushkin	−65, 24W	235	125	(3)
Renoir	−18, 52W	225	110	(3)
Rodin	+22, 18W	235	117	(3)
Shelley-Delacroix Basin	−48, 136W	222	105	(2)
Sotatsu	−48, 19W	160	74	(2)
South of Moliere	+10, 16W	295	145	(1)
Strindberg	+54, 134W	190	90	(2)
Surikov	−37, 125W	235	110	(2)
Valmiki	−23, 142W	210	105	(2)
Vivaldi	+14, 86W	205	100	(3)
Vyasa	+49, 81W	280	145	(1)
Wang Meng	+09, 104W	165	80	(3)
Wren	+25, 36W	208	120	(2)
—	+48 169W	175	80	(2)

[a]Measured or remeasured by author on Mariner 10 photomosaics, after lists of Croft (1979,Table 5.2) and Wood and Head (1976,Table 2). Values for rim and inner ring are means.
[b]Weight values (see text; cf. Table IV) reflect quality of observations (3 = best).

Each ring diameter (both collectively for two-ring basins) is assigned a relative weight of 1 (weakest), 2, or 3 (strongest) in subsequent analysis. This value reflects both the likelihood that the ring exists and accuracy with which it could be measured, and thus roughly reflects the ring's physiographic prominence or "strength"—height, breadth and continuity (Pike and Spudis 1987). Moreover, five ring positions in Table IV contain two split segments (see Wilhelms et al. 1977), whose individual diameters are footnoted (Pike and Spudis 1987). (Spudis and I recognized split rings *before* the ranking step, explained later.) Geometric mean diameters of split rings were calculated from the constituent diameters using the 1, 2 and 3 weights. Split basin rings have the weight of the weaker constituent.

IV. THE SIZE:SHAPE PROGRESSION

Fresh-looking impact craters on Mercury vary widely in morphology over the observed diameter range of 0.2 to 1600 km (Murray et al. 1974; Gault et al. 1975). As on the Moon (Gilbert 1893), the Earth (Dence 1972) and other bodies, physiographic complexity increases with crater size. At least 15 systematic, size-dependent contrasts in crater shape, including d/D, are examined in this section. They are sufficiently clear-cut on Mercury so that the continuum of unmodified craters can be subdivided into fairly discrete classes, each of which reflects dominance of a different combination of formational processes.

Seven distinct classes are recognizable: simple craters (Fig. 2), modified-simple craters (Fig. 3), immature-complex craters (Fig. 4), mature-complex craters (Fig. 5), protobasins (peak-plus-ring craters: Fig. 6), two-ring basins (Fig. 7), and multiring basins (Figs. 8 and 9). Six Mercurian categories have counterparts formally or informally recognized on the Moon and Mars (Schultz 1976a; Cintala et al. 1976). Three of my four classes resemble those found by Wood and Andersson (1978) to include most fresh lunar craters (e.g., ALC, TRI, and TYC, but not BIO), and the three classes of basins or large craters are those identified for the Moon by Hartmann and Wood (1971). Modified-simple craters, as a class, may be recognized here for the first time.

Simple Craters

Simple craters on Mercury between 225 m and 14.4 km in diameter display virtually perfect correspondence of a featureless bowl-shaped interior to a d/D of almost exactly 1/5 (Figs. 2, 10 and 11). These 104 craters probably are equivalent to slump-modified transient cavities plus the uplifted rim. For the evolution of this type of geometric model, see Gault et al. (1968), Grieve et al. (1981) and Grieve and Garvin (1984). Slope of the least-squares regression to the d/D data (variables here and in subsequent fits are in km),

$$d = 0.199\, D^{0.995} \tag{1}$$

is isometric (1.0 in log-log space) and scatter of the depth data is remarkably small (goodness-of-fit parameter R^2 is 0.99; see Table V and Fig. 12; complete data for all equations and results for test of significance are given in Table V). These craters correspond morphologically to the lunar type ALC (Albategnius-C) of Wood and Andersson (1978).

The 1/5 d/D relation for simple craters on Mercury is much the same as those for comparable craters on the Moon and Mars (Pike 1980a,b) with the

TABLE IV

Ring Diameters of Multiring Basins on Mercury

Basin[a]	Center Coordinates (deg)	Mean Observed Diameter by Ring Ranks I-VII (km)[b,c]							
		"A"	I	II	III	IV	V	VI	VII
Borealis[d]	+73, 53W	—	—	(860)	—	1530	[2230]	—	—
Gluck-Holbein	+35, 19W	—	—	(240)	—	500	—	(950)	—
Derzhavin-Sor Juana	+51, 27W	—	—	—	—	560	(810)	—	—
Sobkou	+34, 132W	—	—	(490)	—	850	[1385]	—	—
Brahms-Zola	+59, 172W	—	—	325	—	620	(840)	[1080]	—
Donne-Moliere	+04, 10W	—	375	—	(785)	1060	(1500)	—	—
Hiroshige-Mahler	−16, 23W	—	—	150	—	355	—	(700)	—
Mena-Theophanes	−01, 129W	—	(260)	—	[475]	770	1200	—	—
Tir	+06, 168W	—	380	660	950	1250	—	—	—
Budh	+17, 151W	—	—	—	580	850	(1140)	—	—
Ibsen-Petrarch	−31, 30W	—	—	—	425	640	(930)	[1175]	—
Andal-Coleridge	−43, 49W	—	[420]	(700)	1030	1320	(1750)	—	—
Matisse-Repin	−24, 75W	—	—	(410)	—	850	1250	[1785]	—
Bartok-Ives	−33, 115W	—	(480)	—	790	1175	[1500]	—	—

Name	Lat., Long.								
Hawthorne-Riemenschneider	−56, 105W	—	—	270	—	530	780	1050	—
Vincente-Yakolev	−52, 162W	—	—	[360]	—	725	950	[1550]	—
Eitoku-Milton	−23, 171W	(290)	—	590	(850)	1180	—	—	—
Sadi-Scopas	−83, 44W	—	360	—	(600)	930	[1310]	—	—
Tolstoj	−16, 164W	—	—	[260]	330	510	(720)	—	[3700]
Caloris	+30, 195W	—	—	[630]	(900)	1340	(2050)	[2700]	—
Chong-Gauguin	+57, 106W	220	350	—	(940)	—	—	—	—
Shakespeare	+49, 151W	—	—	[200]	[580]	420	680	—	—
Van Eyck	+44, 159W	—	—	150	—	285	[450]	[520]	—

a Many names are provisional and do not constitute official nomenclature.

b Source: measured by Spudis (1984, Table 1); assignment of rings to seven ranks and one provisional rank, by author, explained in text; weight values: [ ] = 1 (weakest), ( ) = 2, all others = 3 (see text); cf. Tables II and III.

c Underscored diameters are weighted geometric means of two closely spaced ("split") rings (see text), as follows (in km):

Mean Diameter	Constituent Diameters	
(810)	[740],	890
[1385]	[1280],	1420
(785)	700,	(825)
[1785]	(1550),	[1900]
[1550]	(1250),	(1700)

d See also Spudis and Strobell (1984).

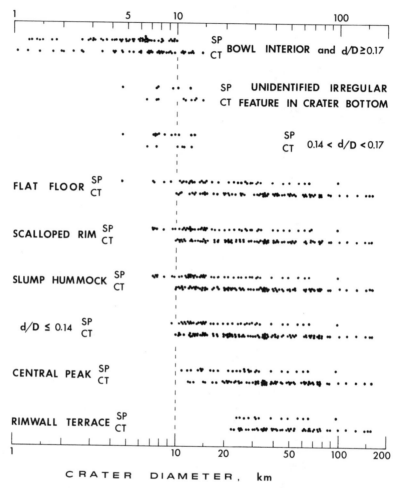

Fig. 10. Size dependence of interior morphology, from new observations on 316 fresh impact craters on Mercury (see Table II and Appendix B). Depth/diameter d/D and seven other interior features, independent of crater taxonomy, are plotted by diameter and topographic background. SP: smooth plains; CT: cratered terrain. Transition from simple to complex craters is centered near dashed vertical line (cf. Fig. 11 and Table X). Complex features onset at different crater sizes (see text), but persist to largest craters. Relative order of onset is read from top to bottom of diagram. Minor wall failure (flat floor, scalloped rim, slump hummocks) precedes appearance of central peak and wall terraces, and also occurs earlier (i.e., in smaller craters) on smooth plains.

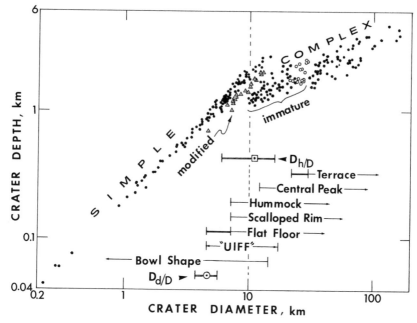

Fig. 11. The simple-to-complex transition on Mercury with depth d as a function of rim diameter D for 316 fresh craters from new shadow-length measurements (Table I and Appendix B): 104 simple craters (dots in 1.0-sloping array, left); 19 modified-simple craters (triangles); and 127 complex craters in 0.4-sloping array (immature: dots to left of circles; mature: dots to right of circles; 16 taxonomically equivocal craters with rudimentary terraces: circles). Regression equations are in Table V, and confidence bands in Fig. 12 below. Barred square: mean and standard deviation for diameter of intersection of Eqs. (2) and (5) in text (Table V). Barred circle: same statistics for Eqs. (1) and (17) in text (Table V). Diameter ranges are those of seven attributes in Fig. 10 above; uncertainty in onset of flat floors and terraces are shown by bars. Four crater types shown here correspond to presence and/or onset of specific morphologic features. The vertical line at 10 km D lies close to the center of the simple/complex transition in craters on Mercury (Table X). The change in d/D precedes that in h/D.

important exception that the largest simple crater on Mercury is about 14 km across (Fig. 10), whereas on the Moon it is about 21 km and on Mars it is about 10 km across. Simple craters on Earth are a little shallower, even with allowances for degradation, and do not exceed 4 km in diameter (Pike 1980a; Grieve et al. 1981).

The variation of rim height h with increasing diameter (Fig. 13) does not differ appreciably from that first observed by Cintala (1979). The h/D relation for Mercury lies above that for other planets. The 32 putative bowl-shaped craters for which values of h were measured (2.4 km $\leq D \leq$ 12 km) follow the near-isometric trend

$$h = 0.052 \, D^{0.930}. \tag{2}$$

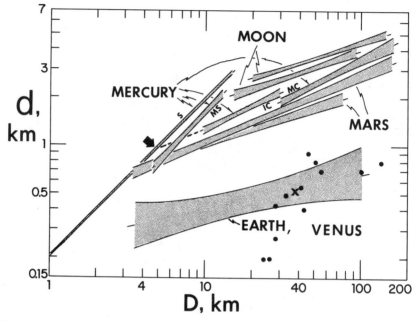

Fig. 12. Depth/diameter *d/D* on Mercury. Results confirm four crater classes and are consistent with *g* scaling of complex craters on all terrestrial planets (cf. Figs. 24,26,28 and Table XI). New *d/D* data (see Fig. 11 above) and those for four other planets, are shown by regression equations and 95% confidence interval (CI) envelopes. Mercury's immature-complex (IC) craters differ significantly (at 1% level) from the mature-complex (MC) craters. The four classes of fresh craters on Mercury (Table V) are compared with complex craters only on other planets (simple craters are similar): Moon, upper curve: uplands, $n = 37$ and lower curve: mare, $n = 20$; Mars, upper curve: cratered terrain, $n = 54$ and lower curve: plains, $n = 51$; and Earth, $n = 12$ (all three after Pike 1980a). Arrow marks 4.7 km intersection diameter of Eqs. (1) and (17) for Mercury. Dots: 12 fresh impact craters on Venus (Ivanov et al. 1986,Table 4). Location of the centroid X within terrestrial field supports *g* scaling of complex-crater depth first indicated for Venus in Pike (1980a) from early data.

Despite the great scatter ($R^2 = 0.55$), this regression is significant (at the 1% level). Thus, rim crests of simple craters on Mercury may be markedly higher than on other bodies (Cintala 1979) provided the *h* measurements are accurate. The issue probably cannot be resolved from Mariner 10 data.

Terrain-related differences in morphology and *d/D* for simple craters on Mercury are negligible (Fig. 10). An analysis of covariance reveals no difference, at any level of significance, in the two regression equations (CT, SP in Table V). However, the largest simple crater in the cratered terrain ($n = 46$) is 14.4 km across, whereas the largest one on the smooth plains ($n = 58$) is only 10.0 km in diameter. The CT/SP contrast in maximum crater size parallels that for uplands/maria on the Moon and for crystalline rock/sedimentary

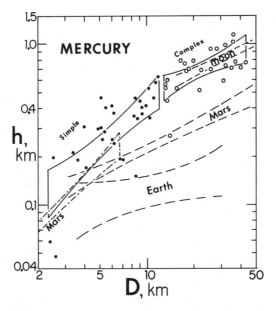

Fig. 13. Change in rim height *h* of Mercurian craters at a diameter *D* of 11 km (cf. Fig. 12). Anomalously high rims of both simple and complex craters (cf. Cintala 1979) are shown by *d*/*D* for 57 fresh craters, from new shadow-length measurements (Table I and Appendix B). Dots: 32 simple craters; circles: 25 complex craters (9 largest have mature morphology). Solid envelope curves define 95% confidence interval (CI) for corresponding linear regressions (Table V). Dashed CI curves are comparable for *complex* craters on the Moon, Mars and the Earth, and dash-dotted envelope defines CI for simple craters on Mars (Pike and Davis 1984); *h*/*D* results for *simple* craters on the Earth and the Moon (not shown) are similar.

rock targets on Earth. That observed for highlands/plains on Mars is the inverse (Pike 1980*a*).

Modified-Simple Craters

Modified-simple craters on Mercury differ from simple ones only in that they are significantly, although not much, shallower (*d*/*D* typically a little less than 1/6), and have one or more small interior features, almost always at the crater bottom (Figs. 3, 10 and 11). The exact identity of the feature(s) often is unclear (Appendix B). Where these small features can be identified, however, they are never central peaks, but rather indications of minor or incipient wall failure: slump blocks, scalloped rim crests or small flat floors. Modified-simple craters on Mercury are shallower and morphologically more complicated than lunar craters of the BIO (Biot) type of Wood and Andersson (1978), and thus do not correspond to them; nor are they like the larger SOS (Sosigenes) class craters of Wood and Andersson (1978).

These 19 craters, wherein a slight diminution in depth is commensurate with an equally modest increase in interior complexity, mark the first and most elementary departure from the simple morphology on Mercury. They measure 4.6 km to 12.7 km across, the size range in which evidence of minor rim collapse was identified by both Malin and Dzurisin (1978) and Smith and Hartnell (1978). This D interval lies wholly within the diameter range of simple craters. The d/D relation for modified-simple craters,

$$d = 0.151 \, D^{1.029} \tag{3}$$

also exhibits a slope of nearly 1.0 and little dispersion (R^2, Table V; Fig. 12). An analysis of covariance for Eqs. (1) and (3) shows that simple and modified-simple craters are different at the 1% level of significance.

Terrain-related differences for modified-simple craters on Mercury are also minor but significant. First, the 7 located in cratered terrain are slightly deeper than the 12 in smooth plains, which parallels lunar and Martian observations (Pike 1980a). An analysis of covariance demonstrates that the least-squares fits for the Mercurian CT and SP subsets (Table V) differ at the 5% level, even in this small sample. Second, observations for both "unidentified irregular features in crater bottom" ($n = 15$) and "intermediate d/D" ($n = 19$) in Fig. 10 suggest—although data in this D-range often are questionable (Appendix B)—that wall failure develops in smaller craters on smooth plains (median $D = 8$ km) than it does on cratered terrain (median $D = 10$ to 11 km). The difference could reflect a sampling deficiency, in that Fig. 10 shows unidentified features in the only three CT craters in this size range. Moreover, none of these 15 craters has the scalloped rim that is commonly diagnostic of minor wall failure.

Immature-Complex Craters

Two classes of complex impact craters can be recognized on Mercury: immature and mature. The two differ subtly, yet recognizably, in almost all morphologic respects. This and the ensuing sections document these contrasts. Statistically significant differences exist in crater depth (Fig. 12), presence/absence of wall terraces (Fig. 14 and 15b), and floor diameter and rim width (Fig. 17). The two crater classes do not differ significantly with respect to central peak base diameter (Fig. 16) or central peak complexity (Fig. 15a), which increase progressively throughout the D-range of all complex craters.

The 69 craters classified here as immature complex (Fig. 4) are much like the lunar TRI-class (Triesnecker) craters of Wood and Andersson (1978); the 58 craters classified as mature complex (Fig. 5) correspond largely to lunar craters of their class TYC (Tycho). The size boundary between the two classes on Mercury seems to fall sharply at 30 km diameter, although the taxonomic ambiguities discussed below could put it at 21 to 22 km. On the Moon the boundary is much more diffuse, spanning crater diameters of about 30 km and 50 km (Wood and Andersson 1978).

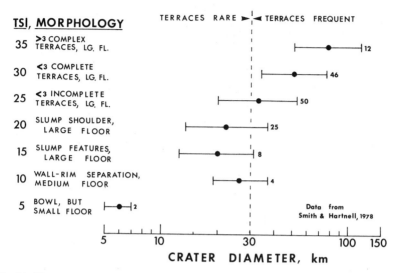

Fig. 14. The mature/immature-complex crater distinction (cf. Fig. 12): correlation of terrace frequency with terrace complexity on Mercury. Size-dependent increase of wall complexity is shown for 147 craters (data from Smith and Hartnell 1978, their Appendix II). Barred dots give geometric mean and standard deviation; small numbers are sample sizes. The vertical line at 30 km diameter separating immature- from mature-complex craters (cf. Figs. 11,12) also divides most terraced craters of Smith and Hartnell from nonterraced craters. For explanation of "terrace shape index" number (TSI) and fuller morphologic descriptions, see Smith and Hartnell (1978).

Immature-complex craters on Mercury are larger (9.5 km $\leq D \leq$ 29 km), much shallower for their size, and morphologically more complicated than both classes of simple craters (Figs. 4, 10 and 11). Diameter ranges of all three classes overlap (Fig. 12). In sum, minor wall failure is the dominant geomorphic signature of immature-complex craters: scalloped rim crests and their accompanying slump deposits are more evident here than in any other craters. The introduction of central peaks is a secondary distinction of this class. Finally, some aspects of morphology increase in complexity with crater size, whereas others remain about the same. All of these observations are elaborated below.

The increasing complexity with size of craters in this class is paralleled by a 50% decrease in the depth/diameter ratio, from about 0.13 at 10 km diameter to 0.07 at 30 km D. The allometric (not 1.0 in log-log space) slope of the least-squares fit to the 69 pairs of d/D data,

$$d = 0.410 \, D^{0.490} \tag{4}$$

is half that for both types of simple craters. The equation is not continuous with that for the larger complex craters on Mercury (Eq. 8, below), differing

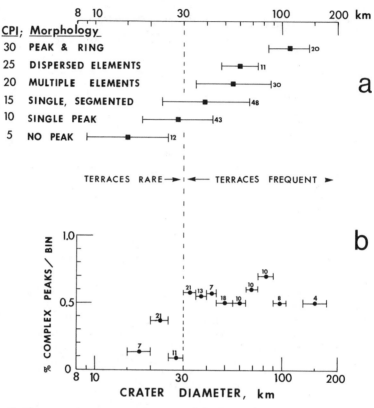

Fig. 15. The mature/immature-complex crater distinction: central-peak complexity markedly increases at the 30 km diameter threshold separating the two classes on Mercury. (a) Size-dependent increase of peak complexity (CPI) in 164 craters and protobasins (data from Smith and Hartnell [1978, their Appendix II]; see also Table II). Barred dots give geometric mean and one-sigma value; small numbers are sample sizes. Craters with no peak or only a single peak lie to left of the vertical line at 30 km diameter (cf. Figs.11,12). For explanation of "central peak index" number (CPI) and fuller morphologic descriptions, see Smith and Hartnell (1978). (b) Dramatic size-dependent increase, at 30 km diameter, in percentage of 138 complex craters on Mercury that have central peaks classified as "complex", as opposed to "simple", by Hale and Head (1980a,Fig.5A). Vertical axis shows percentage of craters that have "complex" peaks for each of 12 diameter bins. Barred dots give geometric means and one-sigma values for the 12 bins; small numbers are sample sizes.

from it significantly at the 1% level. It also is much steeper than the comparable fit for class TRI on the Moon (cf. Wood and Andersson 1978, their Table 1, and Fig.4).

The variation of rim height with increasing diameter differs little from that documented by Cintala (1979); complex craters on Mercury have much higher rims than might be expected (Fig. 13), given the intermediate d/D of Mercury *vis-a-vis* those on the Moon and Mars. However, least-squares re-

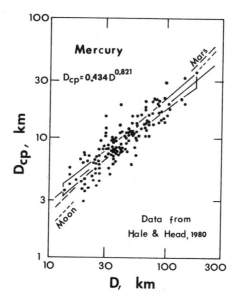

Fig. 16. Similarity of central-peak size on three planets. Reanalysis in log-log space of the relation between base diameter of the central peak, D_{cp}, and rim diameter D for 138 complex craters on Mercury (data from Hale and Head 1980a,Fig.2). Confidence interval (calculated for this work) is at 95%. Analogous equations for Mars (Hale 1983) and the Moon (Hale and Head 1980b) reveal that mean D_{cp}/D ratios for Mercury are like those elsewhere even though the planet's linear equation (Table V) slopes at less than 1.0.

gression for the 25 assumed complex craters for which h values were obtained (13 km $\leq D \leq$ 43 km, and thus mostly immature-complex craters),

$$h = 0.150\, D^{0.487} \tag{5}$$

virtually coincides with that for the Moon (Pike and Davis 1984). It does not lie above the lunar least-squares fit as did that by Cintala (1979). Whereas Cintala's simple/complex crater equations intersect at 16.3 km D and 0.27 km h, those here (Eqs. 2 and 5) cross at 11 km D ± 5 km and 0.48 km h ± 0.15 km. The difference between the h/D results for simple and complex craters is real, despite the great scatter (Table V), because Eqs. (2) and (5) differ significantly at the 1% level. Thus, the puzzle of anomalously high h/D values for Mercurian craters persists.

Two aspects of the h/D inflection may be critical to a causal interpretation of the simple-to-complex transition. First, the h/D break coincides with onset of central peaks at 12 km crater diameter (Figs. 10 and 11). More importantly, the h/D inflection occurs at a much larger D than that of d/D (Figs. 11 and 12).

Wall features attributable to slump become more varied as the size of immature-complex craters increases from 9.5 km to 29 km (Fig. 14). Smith and Hartnell (1978) studied features on the walls of Mercurian craters and devised a "terrace-shape index" number (TSI) that increases with geomorphic complexity. According to their data (Smith and Hartnell 1978, their Appendix II), median diameter is 17 km for craters bearing a TSI of 20, but reaches 26 km for craters with a TSI of 35 (Fig. 14).

My data (Appendix B) show that unambiguous wall terraces on Mercury first appear at a diameter of 22 km in perhaps 16 of the largest immature-complex craters. At this crater size most terraces are rudimentary and poorly formed, and the diameter transition from slump deposit to terrace is ambiguous and difficult to compare from study to study (cf. Cintala et al. 1977; Malin and Dzurisin 1978; Smith and Hartnell 1978). These 16 craters would have been classified mature complex, but the high d/D values that put them in the immature complex d/D category were judged more diagnostic than the presence of primitive terraces.

Central peaks are not observed in the smallest immature-complex craters (about 10 km D), but appear at about 12 km D, near intersection of the h/D equations, and gradually become more frequent until they are universally present in the largest ones (about 30 km D; Fig. 10). The 12 km onset D accords with previous observations (Gault et al. 1975; Cintala et al. 1977; Malin and Dzurisin 1978; Smith and Hartnell 1978). Peaks tend to be single, small edifices. According to data from Fig. 5A of Hale and Head (1980a), the percentage of craters 9.5 km to 29 km across that have complex rather than simple peaks is low, by their classification, only about 10% to 35%, compared with 25% to 80% for mature-complex craters (Fig. 15). However, this trend does not apply within the class itself. I could detect no substantial increase in peak complexity with crater size (Fig. 15), within the immature-complex subset, in the data of either Smith and Hartnell (1978, their Appendix II) or Hale and Head (1980a,Fig.5A).

The base diameters of central peaks within immature-complex craters occupy about 24% to 27% of rim diameter (Fig. 16), somewhat more than they do in mature craters, according to measurements by Hale and Head (1980a,Fig.2) on those 33 (of the 138) craters that are less than 30 km in diameter. This percentage is about the same as that on Mars (Hale 1983) and Earth (Pike 1985), but it is higher than that observed on the Moon (Hale and Grieve 1982). The D_{cp}/D data for Mercurian craters under 30 km across are so scattered that a separate least-squares fit for the 33 *presumed* immature craters (not shown here) is barely significant at the 5% level (R^2 is only 0.14), according to an F-test for goodness of fit (Davis 1986,p.182-184). This equation does not differ (at the 1% level) from a D_{cp}/D regression for the 105 craters *over* 30 km across. Thus, the combined fit (Fig. 16 and Table V) for all 138 complex craters is more representative.

The two classes of complex craters on Mercury differ more with respect to size of the interior floor. Diameter of the flat floor F and width of the rim W do vary isometrically with crater size in the immature-complex class (Fig. 17). Floors occupy an invariant fraction, about half, of the crater diameter. Within the diameter range occupied by immature-complex craters, floor and wall widths for 16 of the 61 craters from Table 1 of Malin and Dzurisin (1978) describe (necessarily) complementary regressions.

$$F = 0.50 \, D^{0.99} \tag{6}$$

and

$$W = 0.23 \, D^{1.02} \tag{7}$$

that both slope at unity ($R^2 = 0.66$ and 0.67, respectively). Both of these regressions differ significantly (1% level) from the allometric plots (nonunity slope) for mature complex craters, (see Table V and Fig. 17: Malin and Dzurisin 1978,Table 2). The two contrasts suggest that 30 km D indeed separates two relatively distinct morphologic classes on Mercury.

Strong terrain-related contrasts for immature-complex craters on Mercury, *as a class,* remain unproven, where sufficient data exist to provide a meaningful test. First, the 30 craters on smooth plains occupy nearly the same diameter range as the 39 on cratered terrain (onset D for both is 10 km). The SP craters have a median D of 14.5 km, whereas it is 17 km for CT craters, a small difference. Removing the 16 large craters that show rudimentary terraces (see above) does not materially change these values. Second, unlike modified-simple craters, immature-complex craters on both types of terrain follow much the same depth/diameter relation. Analysis of covariance reveals no significant difference in d/D for log-log regressions of the two subsets (see Table V). Finally, the various interior features are not developed preferentially on either type of surface *within the diameter range* of this class (Fig. 10).

Terrain-related contrasts in crater morphology on Mercury do exist, however, if the taxonomic groups are ignored. Figure 10 shows clearly that the three chief effects of incipient rim failure—a flat floor, a scalloped rim crest, and slump blocks—all are observed in craters over 7 km across on the smooth plains, whereas they do not appear on cratered terrain until craters exceed 10 km. This nicely complements the persistence of simple craters to larger diameters (by 5 km) on cratered terrain (Fig. 10). All of the smooth-plains craters that have these features and are under 10 km across are classified as modified simple. This diameter-related difference may be real, despite the difficulty in identifying features in such small craters on Mariner 10 images. If real, the D-difference is only 3 km; perhaps too much should not be read into it, particularly in the absence of accompanying d/D differences and diameter-related contrasts for central peaks and terraces.

TABLE V
Least-Squares Fits to Crater Dimensions[a,b]

Text Eq.	Variables (Y/X)	Class on Mercury	TB[c]	n	r[d]	Slope ± CI	Y-cept	$\bar{X}$[e]	$Y_c/\bar{X} \pm$ CI[f]
1	d/D (selected)	Simple	—	104	0.997	0.995 ± 0.015	0.199	1.8	0.198 ± 0.004
—	d/D (culled)		—	66	0.989	0.996 ± 0.037	0.198	3.0	0.197 ± 0.006
—	Depth/diameter	Simple	CT	46	0.985	0.982 ± 0.030	0.202	3.9	0.197 ± 0.005
—			SP	58	0.998	1.003 ± 0.018	0.198	1.5	0.198 ± 0.005
2	Height/diameter	Simple	—	32	0.739	0.930 ± 0.316	0.052	5.3	0.046 ± 0.007
3	Depth/diameter	Mod. Sim.	—	19	0.975	1.029 ± 0.121	0.151	7.7	0.160 ± 0.005
—			CT	7	0.982	0.905 ± 0.200	0.206	9.0	0.167 ± 0.007
—			SP	12	0.980	1.051 ± 0.149	0.141	6.9	0.156 ± 0.008
17	Depth/diameter	Complex	—	127	0.898	0.418 ± 0.036	0.492	39.0	0.58 ± 0.002
5	Height/diameter	Complex	—	25	0.640	0.487 ± 0.252	0.150	23.6	0.030 ± 0.003
9	Peak/Rim Diam.[g]	Complex	—	138	0.875	0.818 ± 0.077	0.439	51.2	0.214 ± 0.009

#	Variable	Comp.	Topo[c]	n	r	b ± CI	a	X̄	Y at X̄
4	Depth/diameter	Imm. Comp.	—	69	0.766	0.490 ± 0.100	0.410	16.6	0.098 ± 0.003
—			CT	39	0.775	0.520 ± 0.141	0.373	16.8	0.096 ± 0.004
—			SP	30	0.760	0.463 ± 0.153	0.446	16.6	0.099 ± 0.005
8	Depth/diameter	Mat. Comp.	—	58	0.823	0.496 ± 0.090	0.353	69.4	0.042 ± 0.002
—			CT	48	0.828	0.509 ± 0.103	0.335	69.4	0.042 ± 0.002
—			SP	10	0.784	0.424 ± 0.273	0.476	54.9	0.047 ± 0.005
6	Floor/Rim Diam.[h]	Imm. Comp.	—	16	0.813	0.999 ± 0.410	0.504	22.2	0.503 ± 0.035
10		Mat. Comp.	—	45	0.990	1.262 ± 0.055	0.203	69.0	0.616 ± 0.017
7	Rim Width/Rim Diam.[h]	Imm. Comp.	—	16	0.816	1.015 ± 0.412	0.232	22.2	0.244 ± 0.017
11	Mat. Comp.	—	45	0.904	0.600 ± 0.087	1.003	69.0	0.184 ± 0.008

[a]Linear equations of form $Y = aX^b$, where X and Y are variables in col. 2, a = Y-cept (at $X = 1$ km), and b = regression coefficient, or slope.
[b]Dispersion values, CI, are error estimates calculated at the 95% confidence interval throughout.
[c]Topographic background (cratered terrain or smooth plains).
[d]Results of a goodness-of-fit test, a t-test on (slope/sample standard deviation of slope) with n-2 degrees of freedom, indicate that all regressions are significant at the 1% level (Snedecor and Cochran 1967,p.138; Davis 1986,p.67).
[e]Geometric midpoint of X.
[f]Calculated Y at $\bar{X}$.
[g]Data from Hale and Head (1980a,Fig. 2).
[h]Data from Malin and Dzurisin (1978,Table 1).

Mature-Complex Craters

Mature-complex craters on Mercury differ from immature ones mainly in their larger size (30 km $\leq D \leq$ 175 km), the presence of well-formed terraces, their floor and rim-width geometry, and the morphologic diversity of their central peaks. They are also slightly shallower for their size and convey the somewhat misleading impression of greater uniformity of appearance (Figs. 5, 10 and 11). All interior features described previously are present (Fig. 5). Rim-crest scallops and slump deposits that are not terraces also persist to the largest crater sizes.

Virtually all geomorphic aspects of mature complex craters change with increasing diameter. The d/D ratio, although slightly lower overall than that for immature complex craters, similarly diminishes by 50% from 0.06 at $D = $ 30 km to 0.03 in the larger craters (Fig. 11), as revealed by the < 1.0 slope of the least-squares fit to the 58 pairs of d/D values,

$$d = 0.353\ D^{0.496}. \tag{8}$$

The equation, which is discontinuous with that for smaller complex craters (Eq. 4) and differs from it at the 1% level, also has a markedly steeper slope than that for lunar class TYC craters (Wood and Andersson 1978).

The size of central peaks is one attribute that changes little with crater size. Base diameters of central peaks in mature-complex craters occupy roughly 17% to 25% of the rim diameter, according to a least-squares fit (Fig. 16) to the 138 pairs of D_{cp}/D data of Hale and Head (1980a, their Fig.2) in the log domain,

$$D_{cp} = 0.44\ D^{0.82}. \tag{9}$$

The relation is the same for both mature and immature craters (above). The reason for the low 0.8 slope, which according to t-tests at the 1% level is both significant and differs from 1.0, is unclear, although it could reflect the poor resolution of Mariner 10 images used for the smallest craters. Slopes for analogous D_{cp}/D relations for the Moon ($n = 175$) and Mars ($n = 1672$) by the same principal author are essentially isometric: 1.05 and 0.99, respectively (Hale and Grieve 1982; Hale 1983). However, importantly, the mean D_{cp}/D fractions are not very different on all three planets (Fig. 16).

Central peaks in mature-complex craters not only are more complicated in form than those in immature-complex craters, but intricacy *increases within the class* until peaks in the largest craters are closed or open clusters of peak elements. According to the data of Hale and Head (1980a, their Fig.5A), the percentage of craters 30 km $\leq D \leq$ 160 km that have complex rather than simple peaks is high, between 50% and 80% (cf. 10% to 35% for immature-complex craters), and the percentage increases systematically with crater size

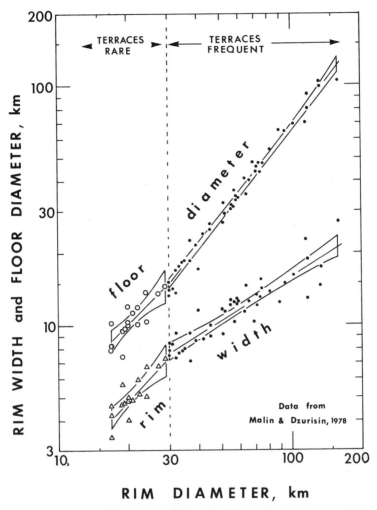

Fig. 17. The mature/immature-complex crater distinction: morphology-dependent relation of floor diameter F and width of the rim wall W to rim-crest diameter D is given for 61 complex craters on Mercury (data from Malin and Dzurisin 1978, Table 1). Division of data according to the immature (circles, triangles)/mature (dots) categories introduced here in Figs. 11 and 12 above yields significantly (at 1% level) different linear relations, not evident in original analysis. Onset of maturely formed terraces (cf. Fig. 14) and central peaks (cf. Fig. 15b) at 30 km D correlates with these contrasts in rim/floor geometry.

(Fig. 15b). Data of Smith and Hartnell (1978, their Appendix II) also show
that morphologically more complex peaks occur within larger craters (Fig.
15a), but the contrast is less dramatic. According to values of their "central
peak index" number (CPI), the median diameter for craters with single peaks
is 35 km, whereas it is almost 60 km for craters with dispersed peak elements.
Moreover, median crater size increases, within the mature-complex class
alone, with increasing values of the CPI.

Width of the rim wall and diameter of the floor vary allometrically with
crater size in mature-complex craters, whereas they vary isometrically for the
smaller complex craters (Fig. 17). Malin and Dzurisin (1978, their Figs.5,6)
first defined the complementary relation between rim width W and floor diameter F for 63 large craters on Mercury, wherein floor size increases with crater
diameter at the expense of the rim wall. Closer examination of both relations
shows that they inflect. The distinction made here between immature- and
mature-complex craters reveals an even stronger size dependence in the allometric changes of rim width and floor diameter with increasing crater size
(Fig. 17). The revised log-log regressions to the 45 pairs of values for the
mature craters are

$$F = 0.203 \, D^{1.262} \tag{10}$$

and

$$W = 1.003 \, D^{0.600}. \tag{11}$$

Equation (10) is much like that for large craters on the Moon (Pike 1977a;
Wood and Andersson 1978). The slopes, which depart significantly from 1.0
(Table V) at the 1% level, demonstrate the extent of these class-related differences, especially for width of the rim wall (cf. Eqs. 6 and 7). The inflections in the F/D and W/D distributions appear to coincide with onset of well-
formed rim terraces in large mature craters, a relation that is not yet well
defined for lunar craters of classes TRI and TYC (Wood and Andersson 1978,
their Table 1 and Fig.8).

Data from Smith and Hartnell (1978, their Appendix II) demonstrate
further that the number and intricacy of terraces on walls within large craters
also increase markedly and steadily with diameter (Fig. 14). Median diameter
for craters assigned their TSI value of 25 is 40 km, whereas D is 80 km for a
TSI of 35. This increasing complexity parallels the increasing relative width
of the floor (F/D) in mature-complex craters (cf. Eqs. 6 and 10). Moreover,
diameters of central-peak bases in these craters occupy about 17% to 25% of
rim diameter (Fig. 16). The slight decrease in D_{cp}/D values, but commensurate increase in F/D values (Fig. 17), for mature-complex craters with increasing diameter may be consistent with modestly increased flooding of the larger
crater floors, perhaps by impact melt or volcanics.

Major terrain-related contrasts within mature-complex craters on Mercury have not been found. Although craters on the smooth plains are smaller than those on cratered terrain (median/maximum D: 47/52 km and 96/160 km, respectively; see also Table V), this difference probably reflects *dwindling numbers of large impacts on the younger plains surface over time* rather than a substrate effect. The onset diameter is the same (30 km; or 21 km on smooth plains and 24 km on cratered terrain with the 16 large immature-complex craters added). Depth/diameter results (Table V) are virtually identical, even though craters on the smooth plains number only 10; analysis of covariance for the d/D regressions for CT and SP reveals no significant differences. Adding the 16 large immature-complex craters does not alter the d/D results. The morphologic observations made here (Fig. 10) also support the conclusion reached earlier (see, e.g., Cintala et al. 1977; Malin and Dzurisin 1978; Smith and Hartnell 1978) that *large* craters on Mercury do not seem to have responded to differences in terrain or substrate characteristics, if such differences exist.

Protobasins

Protobasins on Mercury are transitional landforms, 72 km (crater at $-28°$, 20°W) $\leq D \leq$ 165 km (Bernini), that have both a central peak and an interior ring but otherwise look like large complex craters (Table II and Fig. 6; see also Gault et al. 1975; Stuart-Alexander and Howard 1970; Hartmann and Wood 1971; Hale and Head 1979a; Leake 1981,p.74, her Table 5). They are much more common on Mercury than on the Moon or Mars. Many more surely exist on the unobserved 60% of the planet. In size, the 20 protobasins (up from 12 in Pike 1982a) slightly overlap two-ring basins, but are wholly contained within the diameter range of mature-complex craters (Fig. 18). Topographic prominence of the inner ring is conspicuously subordinate to that of the rim, even in large protobasins such as Verdi (140 km, a marginal example; (see Fig. 6) and Bernini (the largest), which has a nearly complete inner ring.

The term protobasin was introduced to improve nomenclature (Pike 1983a). Sometimes termed "central-peak basins" (Wood and Head 1976), protobasins too closely resemble mature-complex craters to warrant the "basin" distinction, which should be restricted to the larger, less equivocal, two-ring and multiring structures (Figs. 7-9). The need for a qualified term for these transitional structures, which could just as easily be called "peak-and-ring craters" or "supercraters," is met by protobasin. "Proto-" denies them full basin status by suggesting immaturity, while "-basin" implies presence of the inner ring and thus sets them apart from large complex craters. The choice of nomenclature depends upon whether "basin" should be an inclusive or exclusive term. The latter philosophy (e.g., "basins are special"), favored here, has a lengthy history well-grounded in observation (Baldwin 1949, pp.200-202).

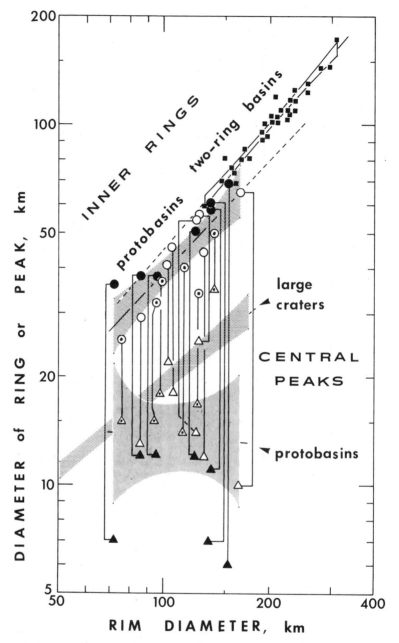

Fig. 18. Morphologic transition from central-peak craters to two-ring basins on Mercury, in plan. Regression of peaks and rings are from both old and new measurements. The distinctive, hybrid morphology of protobasins is evident in two respects: First, both peaks (triangles) and inner rings (circles) of protobasins ($n = 20$; Table II) are smaller than they are in large complex craters (peaks: longest shading; $n = 138$) or two-ring basins (rings: squares; $n = 31$; Table III).

The plan-view geometry of protobasins on Mercury (and elsewhere Pike, 1982a, 1983a) is unique in two important respects. Both of these involve complementary relations of the central peak and the inner ring, and both were first identified in lunar protobasins, from many fewer data, by Hale and Head (1979a). Their significance is examined in the section on morphologic transitions.

First, both the inner ring and the central peak in protobasins are significantly smaller than they are in mature two-ring basins and complex craters, respectively (Figs. 18 and 19). A least-squares fit to inner ring diameter D_i and that of the rim D for 20 protobasins (Table VI)

$$D_i = 0.316 \, D^{1.044} \qquad (12)$$

shows that the inner ring is less than ⅖ its size in larger, two-ring basins (Eq. 13, below, and Fig. 19). Equation (12) and ensuing ring fits are weighted (see Tables II-IV). This log-log regression is very similar, except for the diameter range, to that recently discovered for protobasins on the Moon (Pike 1983a). Not only do the D_i/D regressions for protobasins and two-ring basins differ at the 1% level of significance, according to an analysis of covariance, but a two-sided normal test reveals that mean ring ratios of two-ring basins (0.49 ± 0.01) and protobasins (0.39 ± 0.03) on Mercury (also the Moon and Mars) differ significantly at the 1% level (Fig. 19). This difference indicates that there are two distinct types of two-ring structures. They must not be combined into one data set when examining statistical properties of basin rings.

Central peaks of protobasins typically are only about 2/3 as broad as those of mature complex-craters (cf. Eq. 9). The difference is indicated by a least-squares fit to their D_{cp} and D (Fig. 18) as well as centroid (geometric mean) protobasin values of D_{cp} and D, 13 km and 120 km, respectively. The regression (not given here or in Table V) has a negative slope, which is not statistically significant, even at the 10% level. The poor goodness-of-fit (R^2 is only 0.04) may in part be due to later embayment of central peaks, presumably infilling by ejecta and/or lava. However, analysis of covariance for this expression and Eq. (9) reveals that they *do* differ significantly (at the 1% level) with respect to residual variance and elevation (Y_c).

Two (long) upper dashed lines and shaded areas give least-squares fits and 95% confidence intervals for inner D_i and main D rings of protobasins and two-ring basins (Table VI). Two lower (very short) dashed lines and shaded areas give regressions for base diameter of central peaks D_{cp} and main rim D of large craters (Table V) and protobasins (latter fit, not given here, is not significant; it is replaced by a centroid value). Second, peak-and-ring relations for nearly 2/3 of the protobasins are interdependent: seven out of 11 *rings above* the regression line (solid circles) are in protobasins that also have *peaks below* the centroid (7 out of 10: solid triangles), and vice-versa (6 out of 9 rings lie *below* fit: dotted circles, and 6 out of 10 peaks are *above* the centroid: dotted triangles). These complementary relations suggest a link between processes that form central peaks and inner rings in protobasins.

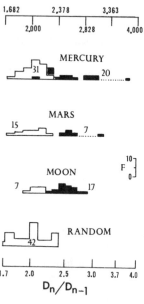

Fig. 19. Vivid contrast in plan-view geometry between two-ring basins and protobasins. Ratios of paired ring diameters ($D_n = D$ = outer, $D_{n-1} = D_i$ = inner) for Mercury, Mars and the Moon (Mercury D/D_i from Tables II and III). The 94 ratios fall into two groups: those for two-ring basins (open histograms, left, $n = 53$) and those for protobasins (filled histograms, right, $n = 44$). Ratios for two-ring basins lie wholly within the same $2^{0.5}D$ interval, 1.682 to 2.378, occupied by alternate ranks of multiring basins (cf. Fig. 23 below), and cluster around the spacing-model value of 2.00 for alternate-ranked rings. Ratios for protobasins do not occupy a $2^{0.5}D$ interval that is related to the spacing of two-ring or multiring basins, but rather cluster around a value of ~ 2.6 that is unique to protobasins and differs significantly from those for two-ring and multiring basins (Table IX). Random uniform distribution is that described in Fig. 23 below. (Figure after Pike and Spudis 1987.)

The second unique characteristic of protobasin geometry on Mercury, and elsewhere (Hale and Head 1979a; Pike 1983a), is the inverse relation between sizes of central peaks and inner rings (Fig. 18). Diameters of the peak and inner ring, already undersized in comparison with "normal" craters and basins, are themselves complementary. This relation is evident for 13 out of the 20 protobasins on Mercury: seven of the 11 rings lying *above* Eq. (12) for D_i/D (solid circles in Fig. 18) occur in protobasins whose accompanying peaks lie *below* (seven out of 10; solid triangles) the centroid value (in the absence of a significant linear fit) for D_{cp}/D (Fig. 18). The converse also applies: six of the nine rings lying *below* Eq. 12 (dotted circles in Fig. 18) occur in protobasins whose peaks lie *above* (six out of 10; dotted triangles) the D_{cp}/D centroid. This is a remarkably systematic relation, given the intrinsically imprecise, and probably inaccurate, nature of most D_{cp} estimates.

Size-dependent contrasts in morphology within the 20 protobasins thus far recognized on Mercury are inconspicuous. Some minor variation is evident in that the inner rings tend to be less complete in the smaller examples, which are otherwise indistinguishable from large craters. Flat floors of protobasins appear to be broader with increasing size, essentially continuing the trend described for large craters. Mercury's protobasins were not examined for terrain-related differences. which have been shown (above) to affect only small craters.

Two-Ring Basins

Mercury has at least 31 two-ring (or "peak-ring": Wood and Head 1976) basins between diameters of 132 km (Ahmad Baba) and 310 km (Homer). This is many more than any other body, despite only 40% of the surface having been observed (Table III, Figs. 7 and 18). The descriptive term two-ring basin is preferable to the interpretive term peak-ring basin, which unjustifiably implies that the central peak in large craters expands to become an inner ring in small basins. (Coexistence of both peaks and inner rings within protobasins warrants such caution.) In diameter, this class just overlaps mature craters and protobasins (Fig. 18) and barely overlaps multiring basins (see Fig. 25, below). These are true basins, wherein the inner ring is both proportionally larger in diameter and topographically more prominent than it is within protobasins. Moreover, the main ring (or "topographic rim": Wilhelms et al. 1977) is less craterlike and less conspicuous than it is in protobasins.

Near equivalence of physiographic prominence of both rings, particularly on Mercury (Murray et al. 1974; Wood 1980), is one of the two major traits that distinguish these mature basins from protobasins (cf. Figs. 6 and 7). The other characteristic (Murray et al. 1974; Head 1978) is a ratio of about 0.5 between diameters of the inner D_i and outer D rings, which is shown here to average 0.49 ± 0.01 at the geometric mean basin size of 202 km (Table VI). The value of 0.5, which is uncontaminated by the much lower ratios for protobasins ($0.38 - 0.40$), is derived here from a linear least-squares fit to the 31 pairs of ring diameters (Tables III and VI),

$$D_i = 0.247\, D^{1.130} \qquad (13)$$

(which is not isometric, at 1% level of significance). The $\sim 0.5\, D_i/D$ ratio for two-ring basins on the Moon (Fig. 19) has been known since the initial work by Hartmann and Kuiper (1962).

The morphologic and geometric contrasts between protobasins and two-ring basins (Fig. 19) argue conclusively against combining the two classes of structures when examining the numerical characteristics of basin rings. This inadvertent practice has led to misleading graphical and statistical conclusions (see, e.g., Head's [1977] Fig. 5; Head's [1978] Fig. 2; Croft's [1979] Figs. 5-16; Boyce's [1980] Figs. 1 and 2; Croft's [1981b] Fig. 1; Hale and Grieve's

TABLE VI

Ring Spacing in Protobasins and Two-Ring Basins from Least-Squares Fits to Diameters[a]

D_i/D Ring Groups[c]	Results of Regression Analysis[b]					Statistics of Ring Spacing		
	n	r[d]	b, Slope	a, Y_c at $X = 1$ km (km)	$\bar{X}_g$[e]	Y_c, 95% CI at $\bar{X}_g$[e] (km)	Observed $Y_c/\bar{X}_g$, 95% CI[f]	Observed x 95% CI[g]
MOON								
2-Ring basins	7	0.70	0.88 ± 1.04	1.005	342	165 ± 9	0.484 ± 0.026	1.874 +0.215 −0.196
Protobasins	17	0.96	0.913 ± 0.141	0.640	222	88 ± 3	0.400 +0.014 −0.013	1.280 +0.091 −0.082
MERCURY								
2-Ring basins	31	0.98	1.130 ± 0.096	0.247	202	100 ± 2	0.494 ± 0.009	1.952 +0.074 −0.072
Protobasins	20	0.85	1.044 ± 0.378	0.316	109	42 ± 3	0.387 +0.030 −0.028	1.198 +0.194 −0.167

MARS

2-Ring basins	15	0.99	1.018 ± 0.078	0.454	152	75 ± 3	$0.496 \begin{array}{l} + 0.021 \\ - 0.020 \end{array}$	$1.968 \begin{array}{l} + 0.170 \\ - 0.155 \end{array}$
Protobasins	7	0.95	0.920 ± 0.366	0.545	99	37 ± 5	$0.377 \begin{array}{l} + 0.049 \\ - 0.043 \end{array}$	$1.137 \begin{array}{l} + 0.315 \\ - 0.245 \end{array}$

[a] Source: calculated from weighted inner D_i and outer D ring diameters in Tables II and III only (no multiring basins), and from Table II of Pike and Spudis (1987).

[b] Linear equations of form $Y = aX^b$, where $Y =$ observed D_i, $X =$ observed D, $Y_c =$ calculated D_i, and ± values of b estimate error at the 95% confidence interval; values of Y_c and $\bar{X}_g$ given here in the alog domain.

[c] Protobasins are large craters with an inner ring and a central peak; 2-ring basins have no central peak.

[d] All regressions are significant at the 1% level by a goodness-of-fit test, except that for lunar 2-ring basins (5% level).

[e] $\bar{X}_g =$ geometric midpoint of observed D for each subset of paired rings.

[f] Y_c calculated at $\bar{X}_g$; model $Y_c / \bar{X}_g$ for all subsets shown here $= 0.500$.

[g] $_X = $ (alog [log($Y_c / \bar{X}_g$) + 3 log $2^{0.5}$])^2, where ± x values are at the 95% confidence interval; model $\bar{x} = 2.000$ (Fielder [1963a] for multiring basins; cf. Table VII).

[1982] Fig. 11). Specifically, combining protobasins with two-ring basins conceals some remarkable consistency in the ring spacing of two-ring basins.

For example, ring-ratio diagrams for Mercury by Wood and Head (1976, Fig.14) and by McKinnon (1981,Fig.5, using data from Croft [1979]) show, respectively, median values of ~ 2.0 (D/D_i) and ~ 0.5 (D_i/D). These values are important because they are related to those for the larger, multiring basins. However, both diagrams also show substantial deviations from the median values of 2.0 and 0.5. In both cases, deviations vanish when the 9 and 13 protobasins in the respective diagrams are removed (cf. Fig. 19). Respective median ratios (D measurements by Wood and Head and Croft, not those in Tables II and III above) then change from 2.1 ($n = 33$) to 2.0 ($n = 24$) and from 0.49 ($n = 37$) to 0.50 ($n = 24$), which are closer to the model values discussed here (below).

Two minor morphologic trends may make two-ring basins on Mercury appear as transitional rather than as discrete morphologic features. First, D_i/D increases slightly with basin size (Eq. 13) (see also, McKinnon's [1981] Fig. 5, less the 13 protobasins). However, this tendency (Fig. 18) may only reflect the undersized inner ring of the smallest basin, Ahmad Baba (132 km), and the unusually large inner ring (or perhaps the tectonically modified outer ring) of the largest basin, Homer (310 km). The other change is subtler and possibly not real. "Strength" of the outer ring relative to that of the inner may decrease as basin size increases. In small basins such as Ahmad Baba and Strindberg (190 km), the outer ring often is the more conspicuous on Mariner 10 images. In larger basins, including Renoir (225 km) and Rodin (235 km), the inner ring is at least as prominent as the outer one. However, differences in relative basin age and degree of post-impact degradation, especially flooding, obscure the trend (e.g., Homer: $D = 310$ km; unnamed 295-km basin at 10°N, 16°W.

Multiring Basins

Basins on Mercury that have at least three measurable rings vary in size from about 285 km (Van Eyck) to 1530 km (Borealis) across at the strongest observed ring (Table IV, Figs. 8 and 9). The number of such basins on the imaged hemisphere now far exceeds the four noted by Croft (1979) and may be as high as 23 (Spudis 1984). Even the most recent count (Table IV, revised from Pike and Spudis 1984), which includes 87 individual rings (five are "split" rings), must be conservative: the scarcity of basins 285 km to 500 km in diameter, especially in the area of high-Sun Mariner 10 coverage, indicates others yet to be discovered. Recent mapping has upgraded the status of some basins: Shakespeare and Van Eyck, once thought to possess only two rings, now appear to be genuine multiring structures. Generally no more than four or five rings are readily observed.

Multiring basins on Mercury have two salient characteristics. One is the inconspicuous geomorphic expression of their partial and degraded rings (Fig.

9; Wood and Head 1976; De Hon 1978; Croft 1979; Spudis 1984; Spudis and Prosser 1984). The second is radial spacing of rings, or incomplete ring fragments and arcs, at an interval that increases incrementally outward by approximately $2^{0.5}D$. This multiplier was first recognized by Fielder (1963a) to describe the ring spacing of four of the Moon's better-preserved basins.

The degree of spatial orderliness among basin rings remains controversial. It has been evident, ever since lunar basin rings were first measured (Hartmann and Kuiper 1962; Hartmann and Wood 1971), that radial spacing of rings on the Moon, Mars and Mercury (mostly two-ring basins in the latter two cases; Wood and Head 1976) is not random. The most frequently cited spacing constants are $2^{0.5}D$ and $2D$. With the recognition of many newly discovered multiring basins on Mercury, the spacing issue now must be investigated on that planet too.

The problem, does a radial spacing constant exist and if so what is it, may be addressed in three ways (Pike and Spudis 1984,1985; Pike et al. 1985). Results of all three approaches are expressed in terms of the *spacing increment* **x** where $x^{0.5}D$ is the radial interval between each basin ring and the next larger ring (Tables VI-IX). In the following three sections, I summarize the analysis and results for Mercury only (for Mars and the Moon, see Pike and Spudis [1985,1987]).

Groups of Ranked Rings   The first approach compares the size of each basin ring with that of the main ring (Figs. 20, 21 and Table VII). Ring spacing is recognized by graphical analysis. The first step after mapping and measuring all rings for a basin is selecting the physiographically strongest ring. Plotting diameters of all other rings against that of the strongest in each basin. in the log domain, separates all but 3 of the 87 Mercurian rings into 5 linear trends. The trends emerged by clustering the points in Fig. 20a *so as to maximize discreteness of a minimum number of linear D_n/D_m groups sloping near 1.0, with no more than one ring per basin in each group.* The 5 trends are subparallel and spaced at roughly equal increments (Fig. 20a). Rings are not evenly distributed among the 5 groups, but vary from 7 to 18 per group. Some rings fall between the better-defined trends, and in 5 cases, 2 closely-spaced ("split") rings from the same basin occupy a position normally filled by only one.

The 5 distributions in Fig. 20a, plus a sixth formed by plotting the strongest ring against itself, constitute an empirical geometric model for ring spacing on Mercury (Fig. 20b). If the 6 groups are numbered I through VI, in order of increasing ring size, the strongest and most frequently occurring ring, the topographic rim, ranks as IV. Provisional taxonomy followed here for the other ring groups is: (I) inner, (II) peak (sic; see above discussion of two-ring basins), (III) intermediate, (V) outer-1, and (VI) outer-2. The equivalent ring diameters are $D_{I...VI}$. Two other possible groups of rings, one a rank below I (here designated ring A) and the other a rank above VI (ring VII), are represented in Fig. 20b by two points and a single point, respectively.

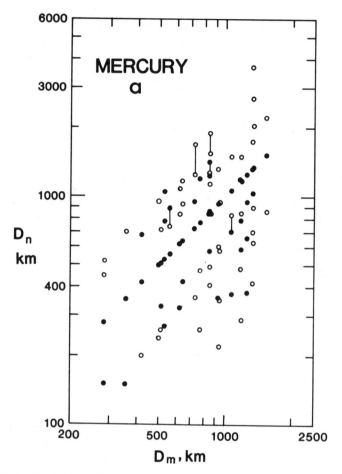

Fig. 20. Ranking the rings of basins on Mercury. (a) Ring-diameter plot for 23 multiring basins. *Unranked* observations: i.e., main ring identified but no assumptions yet made for relative positions of other rings. Dots: diameter D_n of each ring in a basin plotted against that of main ring D_m (Table IV). Circles: diameters of partial arcs and less-certain rings. Short vertical lines join two closely spaced rings or arcs, presumably split segments of one ring (determined from photogeologic analysis; see text), which most commonly lie outside the main ring. Several distributions sloping at 1.0 can be identified. Discreteness of linear groupings is less evident here than for the Moon, but more so than for Mars (Pike and Spudis 1987). Subsequent ranking (Fig. 20b) is determined by three criteria: (1) only one ring or split ring *in each basin* can belong to each group of rings; (2) points are assigned to ranks *so as to maximize discreteness of*

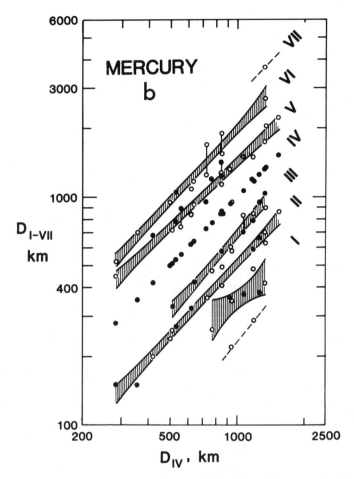

linear D_n/D_m groups sloping at or near 1.0.; and (3) the number of ring groups is minimized. (b) Resultant ring ranks and correlations for Mercury. Dots: basin-ring diameters, $D_{I...VI}$ ($n = 87$), as a function of main-ring diameter D_{IV} plotted in Fig. 20a. Shaded pairs of curves enclose 95% confidence intervals for linear regressions to 5 of the 6 major clusters of rings (rank VII represented by one point only; upper dashed line not a calculated fit). Two points (lower dashed line) belong to a rank below I. All adjacent equations differ significantly at the 1% level. Least-squares fits (Table VII) slope at about unity and are spaced at about $2^{0.5}D$ intervals. Intervals III-IV and IV-V are slightly wider, as on Mars and the Moon (cf. Fig. 21). Partial arcs and less-certain rings (circles) are weighted less (Table IV) in correlations. Split segments occupying same rank were averaged for the regressions.

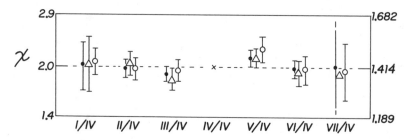

Fig. 21. Agreement of observed with model spacing for multiring basins: ranked ring groups on Mercury (Fig. 20 and Table VII), Mars and the Moon from regression analysis (after Pike and Spudis 1984). Horizontal axis: rank-grouped ring *positions* wherein diameters are paired in least-squares fits (except IV/IV). Vertical axis (left): Intervals of mean radial separation are converted to x domain, between $D_{I...VII}$ and D_{IV} observed for 7 groups of basin rings on Mercury (triangles; from Table VII), the Moon (dots) and Mars (circles). Values of **x** and 95% confidence interval (error bars) are calculated in log domain (see Table VII, footnote). Vertical axis (right): raw-ratio equivalents of **x** for adjacent ring positions (cf. Fig. 23, top scale, and Table IX, first two entries). Averaged spacings of observed ring groups cluster about model value ($\bar{x} = 2.0$), and avoid intervals between groups (x = 1.4, 2.8). Rings III and V lie farther from model value than other rings (cf. Fig. 20). Best fit to $2.0^{0.5}D$ model is **x** = 2.0; poorest possible agreement is for extreme values of **x,** 1.414 and 2.828, which are equivalent to distances midway between "model" d_n/D_{IV} equations spaced exactly at $2.0^{0.5}D$ intervals. (Note that Figs. 21, 22 and 23 all show in a different way the same result.)

The 6 linear groups in Fig. 20b have familiar lunar equivalents. Mercury groups II through VI are essentially the five clusters of lunar rings evident in Fig. 27 of Hartmann and Wood (1971). The same 5 groups also seem to be the classes of lunar rings identified by Wilhelms et al. (1977), with their "inner" ring group equivalent to Mercury groups II and III. Ring groups II, III and IV correspond, respectively, to the lunar central-peak ring, intermediate ring and outer ring of Head (1977). Mercury ring groups I through V also correspond to the 6 lunar classes of Croft (1981*b*), with his 2 types of "intermediate" rings equivalent to Mercury group III. The sixth ring on Mercury is an outer arc whose lunar equivalent is only slightly less common.

Ring-spacing intervals on Mercury correspond well to the $2^{0.5}D$ model, although variances within the 5 groups in Fig. 20b (see also Clow and Pike 1982), as well as the fragmental nature of the ring record, preclude treating the $2^{0.5}D$ spacing as a deterministic "law" or "rule" (Fig. 21). Rings II and VI are very close to the model $2^{0.5}D$ spacing, and ring I deviates from it only modestly. The strongest departures of observed from hypothetical spacings are symmetric about the topographic rim: Rings III and V are each displaced from Ring IV by more than the $2^{0.5}D$ increment (Fig. 21). Their spacing is approximately $2.1^{0.5}D$ (Table VII). This small but systematic dependence of ring spacing on ring rank is evident by:

1. Fitting each ring group (weighted data; see Table IV) with a linear regression in the log-log domain (Table VII, Columns 1-5);

TABLE VII

Ring Spacing in Multiring Basins on Mercury, from Least-Squares Fits to Diameters Grouped by Rank

| | Results of Regression Analysis[a] | | | | | Statistics of Ring Spacing | | | | |
| | | | | | | $Y_c/\bar{X}_g$ 95% CI | | | | |
Ring Groups[b]	n	r[c]	b, Slope	a, Y_c at X = 1km (km)	Y_c 95% CI at $\bar{X}_g$[d] (km)	Model	Observed	Form[e]	Observed x 95% CI[f]	
"A"/IVg	2	—	—	—	—	0.250	0.24 —	x/8	1.8	—
I/IV	7	0.81	0.798 ± 0.601	1.437	359 +72 −34	0.354	0.356 +0.037 −0.034	$x^{0.5}/4$	2.028 +0.448 −0.366	
II/IV	15	0.99	1.065 ± 0.082	0.332	334 +15 −14	0.500	0.506 ± 0.022	x/4	2.047 +0.185 −0.170	
III/IV	12	0.99	1.142 ± 0.122	0.261	559 +24 −23	0.707	0.676 +0.029 −0.028	$x^{0.5}/2$	1.828 +0.161 −0.149	
IV/IV	23	—	(1.000)	(1.000)	762	1.000	(1.000)	x/2	(2.000)	
V/IV	18	0.98	0.934 ± 0.089	2.235	963 +35 −34	1.414	1.458 +0.053 −0.051	$x^{0.5}$	2.124 +0.157 −0.147	
VI/IV	9	0.99	1.065 ± 0.100	1.286	1208 +58 −55	2.000	1.955 +0.094 −0.090	x	1.912 +0.189 −0.172	
VII/IVg	1	—	—	—	—	2.828	2.8 —	$2x^{0.5}$	1.9	—

[a] Linear equations of form $Y = aX^b$, where Y = observed $D_{I...VII}$, X = observed D_{IV}, Y_c = calculated $D_{I...VII}$, and ± values of b estimate error at the 95% confidence interval; values of Y_c and $\bar{X}_g$ are given here in the alog domain.

[b] Source: calculated by author from weighted data in Table IV only; no two-ring basins or protobasins; provisional rank "A" lies inside rank I.

[c] All regressions are significant at 1% level, by a goodness-of-fit test.

[d] $\bar{X}_g$ = geometric midpoint of observed D_{IV} for each ring group.

[e] Expresses spacing between each equation and that for the main ring (IV/IV) in terms of adjacent $2^{0.5}$ groups, through a variable x raised to power of the constant 0.5.

[f] $x = (a\log[\log(Y_c/\bar{X}_g) + (5 - R)\log 2^{0.5}])^2$, where R = rank of rings for which Y_c is calculated; ± x values estimate error at the 95% confidence interval; model $\bar{x}$ for all ring groups = 2.000 (Fielder 1963a).

[g] No least-squares fit; $Y_c/\bar{X}_g$ and x are geometric means or single values only.

2. Calculating the geometric midpoint $\bar{X}_g$ of observed D_{IV} for each ring group and values of Y_c for $D_{I...VI}/D_{IV}$;
3. Then expressing the ratio $Y_c/\bar{X}_g$ as a function of $\mathbf{x}^{0.5}$, where the model value of the spacing increment $\bar{\mathbf{x}} = 2.0$ (Table VII). Differences between observed values of $\mathbf{x}$ for this ratio and the closest fractions or multiples of $2^{0.5}D$ are small, and range from 0.028 (D_I; $n = 7$) to 0.172 (D_{III}; $n = 12$). Percentage deviations for multiring basins on the Moon and Mars have both a similar range and a similar, symmetric, correlation with ring rank (especially ranks III and V) (Fig. 21; Pike and Spudis 1984,1987).

The rings of two-ring and multiring basins on Mercury share the same fundamental spacing. The equation describing D_{II}/D_{IV} for the "peak" (sic) ring within 15 multiring basins is

$$D_{II} = 0.322 \, D_{IV}^{1.065} \tag{14}$$

The average D_{II}/D_{IV} *ratio*, 0.506 ± 0.022, is statistically indistinguishable from the $0.494 \, D_i/D$ for the 31 two-ring basins described by Eq. (13). Although the two regression *equations* differ at the 1% level, for general descriptive purposes their combined 46 data points can be represented by the single expression,

$$D_{II} = 0.401 \, D_{IV}^{1.039} \tag{15}$$

for which the average D_{II}/D_{IV} ratio is 0.506 ± 0.011. The lunar case is similar (Pike 1983a).

The inner ring of two-ring basins is *not* equivalent to the "intermediate" (sic) ring of multiring basins. Equation (15) has a slope of slightly more than 1.0. The two original regressions (Eqs. 13 and 14) have slopes that exceed unity by more than 0.04, but not enough for inner rings of two-ring basins (Eq. 13) to align with "intermediate" rings of multiring basins (III), the next-highest group in Fig. 20b. Its equation is:

$$D_{III} = 0.261 \, D_{IV}^{1.142} \tag{16}$$

Average D_{III}/D_{IV} for the 12 intermediate rings is 0.68, far exceeding (by a factor of $2^{0.5}D$) the 0.50 ratio of all D_{II}/D_{IV} distributions.

Adjacent rings observed for a basin on Mercury do not always occupy adjacent groups in Fig. 20b. For example, the 4 rings of the provisional Vincente-Yakolev basin fall into groups II, IV, V and VI, but not group III. This does not mean that the $2^{0.5}D$ spacing does not apply to this basin, because all its other rings belong to various $2^{0.5}D$ groups, but rather that the D_{III} *position* is unoccupied. Thus not all adjacent *observed* rings, e.g., rings II and IV of Vincente-Yakolev are necessarily spaced at the fundamental $2^{0.5}D$ interval.

Rather, they are spaced at some integer multiple or fraction of it, such as 2.0 D or 0.5 D. The "missing" ring III of Vincente-Yakolev may have formed but is now buried, or it may never have formed. It follows further that the 0.5 D_i/D spacing frequently documented for two-ring basins (e.g., Eq. 13; Wood and Head 1976) may be a variant on the $2^{0.5}D$ spacing wherein a potential ring position always remains unfilled in small basins. The 0.5 D interval of two-ring basins thus probably arose from the same fundamental mechanism(s) that formed the multiple rings of larger basins.

Rings of Single Basins. Ring spacing also can be examined basin-by-basin if enough ($\geq$ 4; Pike et al. 1985) rings are present (Fig. 22 and Table

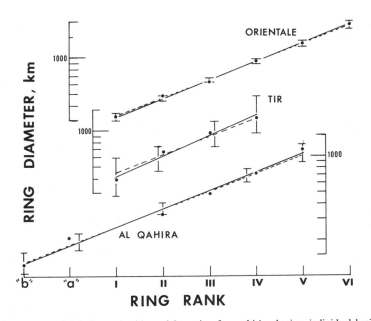

Fig. 22. Agreement of observed with model spacing for multiring basins: individual basins. $2^{0.5}D$-sloping relations between log ring diameter D and ring rank n (arbitrary units) are given for representative multiring basins on Mercury (Tir, Fig. 9), the Moon and Mars. Mercury data are from Table IV, the other from Pike et al. (1985). Solid lines are linear least-squares fits of the form $\log D_n = \log D_{IV} + (n - 4) \log b$, where n is any ring rank, D_{IV} is the diameter of the main basin rim, and $\log b$ is the slope of the regression line. Ring diameters are weighted by quality of observations (see text and Table IV). Error bars define 95% confidence interval. Calculated slope b yields average spacing increment between adjacent rings of each basin. All values of b are significant at 1% level. Dashed lines: spacing model, where a$\log b$ = 1.414 ($\bar{x}$ = 2.000). Values of b for the three basins are: 1.482 $\pm$ 0.286 (x = 2.198 $\pm$ 0.862) for Tir, 1.420 $\pm$ 0.036 (x = 2.016 $\pm$ 0.101) for Orientale (Moon), and 1.420 $\pm$ 0.074 (x = 2.017 $\pm$ 0.210) for Al Qahira (Mars). Table VIII gives results for the other 22 multiring basins on Mercury. Average values of x for Mercury and the other two planets ($\geq$ 4 rings per basin) lie close to the model value of 2.0 (Table VIII). (Note that Figs. 21, 22 and 23 all show in a different way the same result.)

VIII). Values of the spacing increment **x** were estimated for the resulting 16 basins on Mercury (67 rings) from a functional dependence of log II ring diameter Y upon radial-ring position (rank) X. The linear least-squares fits, after Clow and Pike (1982) from Fielder (1963a), are of the form $\log D_n = \log D_{IV} + (n - 4) \log b$, where D_n is the diameter of a ring of any rank n and D_{IV} is diameter of the main ring (km). Ring ranks n are similarly spaced integers in arbitrary units, and log b is the slope. Again, diameters were weighted 1, 2 or 3, by quality (Table IV). Figure 22 shows examples of the observations for Tir

TABLE VIII
Least-Squares Fits to Basin-Ring Rank and Diameter[a]

Basin[b]	Number of Rings	r[c]	D_{IV}* (km)	Slope, b	x, 95% CI[d]
MERCURY					
Borealis	3	0.997	1567	1.358	1.843 +1.519 −0.833
Gluck-Holbein	3	0.999	487	1.411	1.990 +0.929 −0.633
Sobkou	3	0.984	900	1.374	1.889 +6.100 −1.442
Brahms-Zola	4	0.999	609	1.362	1.854 +0.169 −0.155
Donne-Moliere	4	0.999	1066	1.414	1.999 +0.113 −0.107
Hiroshige-Mahler	3	0.997	336	1.475	2.176 +2.265 −1.110
Mena-Theophanes	4	0.996	788	1.466	2.150 +0.504 −0.409
Tir	4	0.988	1333	1.482	2.198 +1.024 −0.699
Budh	3	0.997	827	1.407	1.979 +1.980 −0.990
Ibsen-Petrarch	4	0.994	624	1.430	2.044 +0.564 −0.442
Andal-Coleridge	5	0.986	1329	1.396	1.948 +0.440 −0.359
Matisse-Repin	4	0.999	857	1.448	2.096 +0.069 −0.067
Bartok-Ives	4	0.993	1125	1.341	1.800 +0.426 −0.345
Hawthorne-Riemenschneider	4	0.999	537	1.410	1.987 +0.155 −0.144
Vincente-Yakolev	4	0.992	710	1.412	1.993 +0.598 −0.460
Eitoku-Milton	4	0.999	1189	1.420	2.017 +0.088 −0.085
Sadi-Scopas	4	0.993	909	1.374	1.888 +0.469 −0.376
Tolstoij	4	0.993	499	1.439	2.072 +0.615 −0.474
Caloris	6	0.997	1330	1.438	2.067 +0.171 −0.158
Chong-Gauguin	4	0.992	900	1.407	1.979 +0.615 −0.469
Shakespeare	3	0.996	438	1.511	2.282 +3.512 −1.383
Van Eyck	4	0.995	286	1.372	1.884 +0.382 −0.317
SUMMARIES FOR THREE BODIES[e]					
Moon (15 basins)	73	—	—	1.424	2.029 +0.220 −0.192
Mercury (16 basins)	67	—	—	1.414	2.000 +0.394 −0.312
Mars (19 basins)	93	—	—	1.417	2.008 +0.260 −0.217

[a]From weighted data in Table IV; regression analysis is in semi-log domain (see text); values of D_{IV}* and b are shown here in alog domain.
[b]Many names are provisional and do not constitute official nomenclature.
[c]All regressions are significant at the 1% level by a goodness-of-fit test.
[d]$x = b^2$; error estimates are at 95% confidence interval.
[e]Means weighted by number of rings (≥ 4); Moon and Mars data from Pike and Spudis (1987).

basin on Mercury (four rings, $D_{IV} = 1250$ km) and one each on the Moon and Mars.

A fit of the equation to each basin yielded the slope $\log b$ and an estimate of Y at $n = 4$: diameter of the main ring, $D_{IV}*$ (Table VIII). Respective values for Tir are 1.482 and 1333 km. All 22 slope values are significant at the 1% level (t-test). The mean spacing increment for adjacent (ranked) rings within a basin, x, is the square of a $\log$ slope, b^2, $2.198^{+1.024}_{-0.699}$ for Tir. Rings III and V, which lie farther from ring IV than the nominal 2.0 spacing (see last section) but were used here so as not to exclude more basins, did not push the fits toward high slopes. The mean excess of slope over 2.0, 0.012, is so small it may not be statistically significant. Values of x for individual basins on Mercury range from 1.800 (Bartok-Ives) to 2.282 (Shakespeare).

The mean value of x for Mercury, $2.000^{+0.394}_{-0.312}$, corresponds to the model spacing increment of 2.0. Comparable values for the Moon and Mars are 2.029 and 2.008, $n = 15$ and 19 (Pike et al. 1985). Dispersion for ring spacing on Mercury is somewhat greater than that derived from analysis of grouped basins above (Table VII). Nonetheless, this 95% confidence interval, 1.688 to 2.394, lies well within one $2.0^{0.5}D$ interval, 1.414 to 2.829 (defined by distances midway between $\log (D_n/D_{n-1})$ model ring-spacing values 1.0, 1.414, 2.0, 2.828, etc.; Fig. 21). Whether the observed dispersion is sufficiently small to differ significantly from dispersion that might arise from random processes (Clow and Pike 1982) could be tested (see next section). However, this is unnecessary. A t-test for the correspondence of $\log b$ and $\log 2.0^{0.5}$ (Lentner 1972, p. 227) reveals no difference between the two, for all 22 basins, at the 1% level of significance.

Ring-Diameter Ratios. The third parameter of ring spacing is the ratio of diameters of adjacent *observed* basin rings D_n/D_{n-1} where n is ring rank (Fig. 23, Table IX). This was the first (Hartmann and Kuiper 1962) and most commonly applied approach to the problem (Fielder 1963a; Hartmann and Wood 1971; K. A. Howard et al. 1974; Wood and Head 1976; Pike and Spudis 1985). The 87 ranked Mercurian rings (Table IV) include both *adjacent ranked* and *alternate ranked* rings. Ratios excluding rings ranked III and V, but including rings of two-ring basins, have a mean *spacing increment* $\bar{x}$ of about 2.0 (Table IX). Corresponding *raw* ratios average about either 1.4 or 2.0 (Figs. 19 and 23). Ratios including rings ranked III and V, as for the least-squares groups, above, have a larger increment $\bar{x}$ averaging 2.1. The F-test indicates, at the 5% level of significance, that these two spacings did not arise by chance.

Ratios for Mercurian basins fall into the two clusters of values found by Hartmann and Kuiper (1962) for multiring and double-ring basins on the Moon. The first cluster, centered at about 1.4 to 1.5, usually describes adjacent ranks ($n = 46$); the second, centered at about 2.0, usually is for alternate ranks ($n = 42$; Fig. 23). To supplement the sparse observed alternate-rank

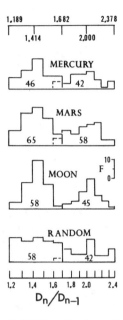

Fig. 23. Agreement of observed with model spacing for multiring basins: the ratios are of diameters (where n = any rank) in multiring basins on Mercury, the Moon and Mars (after Pike and Spudis 1985,1987). Most of the 314 observed ratios (88 Mercury ratios from data in Table IV) fall into two groups (Table IX): those for *adjacent ranked* rings (histograms on left, n = 169; cf. Fig. 21) and those for *alternate-ranked* rings (histograms on right, n = 145). Most ratios in each group lie wholly within one of two $2^{0.5}D$ intervals (cf. Fig. 21): either 1.189 to 1.682 or 1.682 to 2.378 (upper scale), and cluster around the respective spacing-model values of 1.414 for adjacent-ranked rings and 2.000 for alternate-ranked rings (calculated $\bar{x}$ = 2.0 in *both* cases; see Fig. 21). Alternate-ranked rings have the same ratios as two-ring basins, but not protobasins (cf. Fig. 19). Two frequency distributions of ratios were drawn at random for the same two $2^{0.5}D$ intervals, 1.189 to 1.682 and 1.682 to 2.378, to test the groups of observed ratios for nonrandomness (see results in text and Table IX). Dispersion of the n = 42 sample is identical to that of the other, despite the "peak" at D_n/D_{n-1} = 2.05. (Note that Figs. 21,22 and 23 all show in a different way the same result.)

ratios (n = 18), I generated a second set of 24 ratios by deliberately skipping one *observed* ring in calculating each ratio (Fig. 23 and Table IX). Additionally, the three subsets of ratios were grouped according to inclusion or exclusion of rings III or V, now known to be systematically displaced. Finally, ratios for protobasins and two-ring basins were added for comparison.

Mean and standard deviation were calculated, in the log (D_n/D_{n-1}) domain (basin-ring *ratios* also are skewed strongly toward high values), for each of the 7 groups of ratios observed for basins and protobasins on Mercury (Table IX). Four other groups had < 10 ratios and were not tested. All means and standard deviations were transformed to the domain of the spacing increment **x** so that (1) statistics for both adjacent *and* alternate ranks are directly comparable (Table IX); and (2) the results are comparable with those from the first two methods of analysis. Calculations for ratios from multiring basins were weighted 1 to 5 by combining weights assigned to individual rings (Table IV).

The 67 observed ratios (x equivalents) in the 4 multiring-basin categories for Mercury (Table IX) cluster around the model spacing increment of 2.0. The weighted mean is $1.997^{+0.319}_{-0.274}$ (comparable values for the Moon and Mars are 2.091 and 2.027; $n = 82$ and 118). The combined weighted mean for all three planets is $2.039^{+0.383}_{-0.284}$; the net excess of observed $\bar{x}$ over model $\bar{x}$, 0.040, probably is statistically insignificant.

Differences in mean spacing among "close", "wide", and "neither close nor wide" ratios for multiring basins on Mercury, although not tested for statistical significance, are strong enough to show up clearly in averaged x values of these groups: respectively, $1.888^{+0.359}_{-0.298}$ ($n = 12$ ratios), $2.097^{+0.327}_{-0.283}$ ($n = 30$), and $1.961^{+0.339}_{-0.288}$ ($n = 13$). Comparable values for the Moon and Mars and for weighted three-planet means are given in Pike and Spudis (1985,1987). Averaged x values for Mercury's protobasins and two-ring basins (Fig. 19) are $3.294^{+0.416}_{-0.358}$ ($n = 20$) and $2.048^{+0.259}_{-0.230}$ ($n = 31$). Two-ring basins have the same fundamental spacing as multiring basins, but protobasins do not (Table IX). The large difference between protobasins and all other basin-ring categories, which is significant at the 1% level, by a two-side normal test, confirms that protobasins are not basins in the strict sense.

The observed mean values of x, although close to 2.0, support the spacing model only when a second condition is met: dispersion about the mean must occupy limits too narrow to have occurred by chance. This requirement is necessary because $\bar{x}$ for randomly chosen ratios within one $2.0^{0.5}$ interval *also* is 2.0 (Table IX). Accordingly, I compared, by an F-test at the 5% level of significance (Natrella 1963), the standard deviation of each subset in Table IX with a value of this statistic from one of two sets of ratios, assuming a *uniform* distribution function, drawn from a table of random numbers (RAND Corp. 1955), one set ($n = 58$) corresponding to ring ratios for adjacent ranks (1.189 to 1.682), the other ($n = 42$) to ratios for alternate ranks (1.682 to 2.378). (The intervals are defined by distances midway between two log (D_n/D_{n-1}) model ring-spacing values, respectively 1.0, 1.414 and 2.0, and 1.414, 2.0 and 2.828.) For Mercury (also the Moon and Mars; Pike and Spudis 1985), the spread of observed basin-ring ratios about the mean *is* significantly less than that for the randomly drawn ratios, i.e., dispersion that might arise from random processes. This difference, with qualifications noted below, indicates that the $2.0^{0.5}D$ ring spacing on Mercury is real.

Results of the F-test for basins on Mercury are systematic by ring-sample size (Table IX). For the two subsets of ≥ 20 ratios, the observed dispersion is significantly less than random dispersion (i.e., a "pass"). Of the three subsets containing ≤ 19 ratios, only one passes. Combined results for the Moon, Mars and Mercury (Pike and Spudis 1987) are, respectively, 6 passes out of 6

TABLE IX
Ratios of Ring Diameters for Basins and Protobasins on Mercury:
Statistics of Central Tendency and Dispersion[a]

Groups of Ring Ratios[c]	Mean, ± One-Sigma[b], (Number of Ratios) x equivalents[d] are unbracketed raw ratios are [bracketed]
ADJACENT RING RANKS	
Randomly chosen values from 1.189 to 1.682	2.069 +0.475 −0.387 (105) [1.438 +0.157 −0.141]
II/I, VII/VI (spacing neither wide nor close)	—[e] —
III/II, VI/V (close spacing)	1.888 +0.359 −0.298 (12) [1.374 +0.125 −0.113]
IV/III, V/IV (wide spacing)	2.097 +0.327 −0.283 (30)[f] [1.448 +0.109 −0.101]
ALTERNATE RING RANKS: NO OBSERVED RING SKIPPED[g]	
Randomly-chosen values from 1.682 to 2.378	2.020 +0.460 −0.375 (42) [2.010 +0.217 −0.196]
IV/II, VI/IV (spacing neither wide nor close)	1.961 +0.339 −0.288 (13) [1.981 +0.164 −0.152]
III/I, VII/V (close spacing)	—[e] —

($n \geq 20$ ratios) and 4 passes out of 12 ($n \leq 19$ ratios). Clearly, when the sample is sufficiently large, dispersion of observed ring ratios is significantly less than that ascribed to chance. Most subsets failing the F-test probably are just too small to be representative samples.

Summary of Ring Spacing. Three analyses of a new data set suggest strongly that the spacing of basin rings on Mercury is nonrandom. One spacing interval, $(2.0 \pm 0.3)^{0.5}D$, dominates the ring geometry of two-ring and multiring basins and a secondary interval, $(2.1 \pm 0.2)^{0.5}D$, applies to rings immediately flanking the main ring in a multiring basin. Spacing of both intervals is the same both inside and outside the main basin rings. Such "exterior" rings, once thought to be rare, are shown here to be nearly as common as interior rings. Protobasins have an entirely disparate ring spacing, commensurate with their small size, their morphologic immaturity, and the presence of a central peak.

TABLE IX (*Continued*)

Groups of Ring Ratios[c]	Mean, ± One-Sigma[b], (Number of Ratios) x equivalents[d] are unbracketed raw ratios are [bracketed]
ALTERNATE RING RANKS: ONE OBSERVED RING SKIPPED	
IV/II, VI/IV (spacing neither wide nor close)	$1.897 +0.237 -0.211$ (12)[f] $[1.948 +0.118 -0.111]$
III/I, VII/V (close spacing)	—[e] —
Rings V/III (wide spacing)	—[e] —
TWO-RING BASINS AND PROTOBASINS[h]	
D/D_i for two-ring basins	$2.048 +0.259 -0.230$ (31)[f] $[2.024 +0.124 -0.117]$
D/D_i for protobasins	$3.294 +1.155 -0.855$ (20)[i] $[2.567 +0.416 -0.358]$

[a]Raw data in Tables II–IV.
[b]Calculations in log domain; ratios weighted 1 to 5 according to quality of data (see text).
[c]Include ratios formed from two rings ranked below I ("A").
[d]$x = (Y_c/\bar{X}_g)^2$ for adjacent ring ranks, and $(Y_c/\bar{X}_g)^2/2$ for alternate ring ranks and for 2-ring basins and protobasins.
[e]< 10 ratios.
[f]Dispersion of observed ratios is significantly less than that of ratios selected randomly within one $2^{0.5}$ interval, according to one-sided F-test carried out in the log domain at the 5% level.
[g]No wide spacings (rings V/III).
[h]D equivalent to ring IV, D_i to ring II; protobasins have a central peak in addition to the inner ring.
[i]Mean D/D_i values for protobasins differ significantly from those for two-ring basins, by a two-sided normal test carried out in the log domain at the 5% level.

V. MORPHOLOGIC TRANSITIONS

The size-dependent array of crater and basin morphology thus far observed on Mercury is not a perfect continuum, even though large craters in any one class tend to resemble small ones in the next larger size class (Murray et al. 1974; Gault et al. 1975). Rather, the continuum is interrupted by several transitions from one crater class, or dominant assemblage of impact-generated landforms, to another. Crisp morphologic contrasts do not separate all crater classes. There are three principal transitions, each of which occupies a dis-

TABLE X
Constituents of a Revised D_t for Craters on Mercury[a]

Morphologic Attribute	Statistic	Sample Size	Diam.[b] (km)
Depth/diameter	Intersection of curves	316	4.7
Rim height/diameter	Intersection of curves	57	11.0
Occurrence of simple craters	Largest D	170	14.5
Occurrence of complex craters	Smallest D	146	9.5
Occurrence of modified-simple craters	Median D	19	8.5
Bowl shape to flat floor	Median D of overlap	316	9.4
Bowl shape to scalloped rim	"	"	10.0
Bowl shape to floor-wall hummocks	"	"	10.0
Bowl shape to central peaks	"	"	13.0
Bowl shape to terraces	"	"	17.8

[a]Geometric mean D_t is 10.3 km +4.4 −3.1 km (one-sigma); comparable means for the Moon (19 km), Mars, (5.8 km) and Earth (3.1 km) derived in Pike (1980a,Table 3).
[b]Source: this chapter.

crete range of crater diameters on Mercury: simple to complex craters, at an average rim diameter of about 10 km (Table X, and Fig. 24); complex craters to small, two-ring basins at about 110 km (Figs. 25 and 26); and two-ring to multiring basins at perhaps 400 km (Fig. 25).

The three transitions are observed on other planets and satellites throughout the solar system, albeit at different threshold sizes (Table XI below; Hartmann 1972; Wood and Head 1976; Croft 1979; Hale and Head 1979a; Pike 1980a; Basilevsky 1981). Unique crater morphologies distinguish the simple/complex and crater/basin transitions. Each of the three changes in morphology probably reflects dominance of a different set of cratering processes, as target materials over increasingly large areas respond to higher levels of impact energy.

Simple-to-Complex Craters

The morphologic transition from simple to complex craters on Mercury occurs at a nominal diameter D_t of 10.3 km (Fig. 24). This revised diameter is significantly less than the 16 km derived previously (Pike 1980a,Table 2). Even the upper one-sigma limit for the new interval estimate, 7.2 km to 14.7 km, is less than the old 16 km point estimate. The new value corroborates the initial conclusions by Gault et al. (1975) that the simple/complex contrast in crater shape occurs at a comparatively small crater size on Mercury, and is related inversely to planetary surface gravity (Fig. 24).

The new D_t was calculated by the same technique used previously (Pike 1980a). D_t is the geometric mean of diameters that mark major size-dependent contrasts in crater shape, in this case, 10 (Table X). All 10 aspects of the

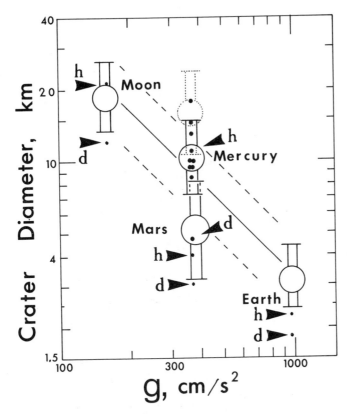

Fig. 24. Two new size-dependent relations. First: consistency of Mercury with the Earth and the Moon in the g-scaled simple-to-complex transition of crater morphology (cf. Fig. 1). The revised inverse function of D_t (large open circles) with g for four terrestrial planets, incorporates new data for Mercury. Plain dots: 10 newly-calculated variables (Table X). Dotted symbol: position of Mercury in Fig. 1 from the old data (sloping lines are the same as in Fig. 1). Position of remaining anomalous body, Mars, could reflect an especially "weak" layered or volatile-laden upper crust. The second size-dependent relation is the preferential grouping of "dry" planets, Moon and Mercury apart from "wet" planets, the Earth and Mars. Arrowed dots are intersection diameters of h/D and d/D regressions, which also reveal the systematic diminution of crater depth before (i.e., in smaller craters) rim height on the terrestrial planets. Sources of the data are given in text.

change from bowl-shaped to terraced-wall craters were measured from the results of this study. The smallest constituent diameter is 4.7 km. It marks the intersection (diameter at 4.7 km, depth at 0.94 km) of the d/D fits for the 104 simple (Eq. 1) and that for the 127 complex (69 immature plus 58 mature) craters,

$$d = 0.492 \, D^{0.418} \tag{17}$$

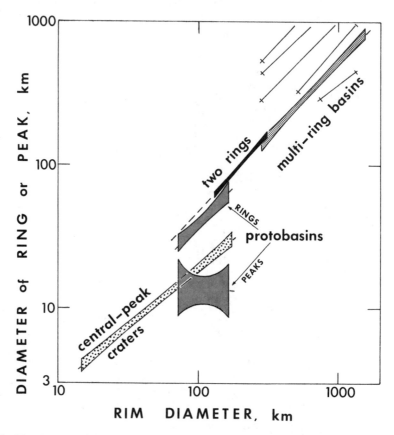

Fig. 25. Morphologic transition for Mercurian crater features observed in plan: large central-peak craters (cf. Fig. 16), through protobasins and two-ring basins (cf. Fig. 18), to multiring basins (emphasizing rank II; Fig. 20). Shaded areas are 95% confidence intervals from linear regression (Tables V-VII); lines are fits for multiring-basin rings of rank other than II (Table VII). Transitions on the Moon and Mars are similar (Pike 1982a). The chief elements of these size-dependent changes are: (1) the central peak in large complex craters is replaced by an undersized peak and inner ring in protobasins; (2) the latter two features are replaced by one full-sized ($0.5\,D$) inner ring in two-ring basins; (3) two rings are replaced by multiple rings spaced at exactly half the $0.5\,D$ interval.

(Table V; also see Pike and Clow 1984). This expression is as steep as that for Mars (Pike 1980a), but much steeper than those for the Moon (Pike 1977a) or Earth (Pike 1980a). The 19 modified-simple craters could have been added to Eq. (17) without materially changing the results. The largest constituent diameter of D_t in Table X is 17.8 km, midpoint of the diameter interval that separates the smallest terraced craters from the largest nonterraced craters.

External influences.   Morphologic effects related to contrasts in to-

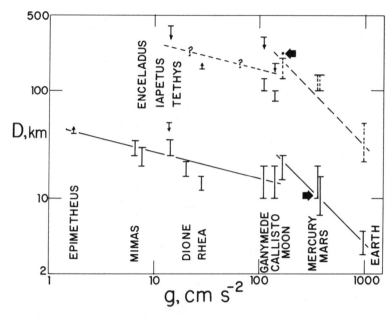

Fig. 26. Compound gravity-scaled model for size-dependent changes in crater morphology in the solar system. Terrestrial planets have a −1.0 dependence of crater shape log D on log g, whereas it is only −0.25 on the outer satellites. A D_t/g diagram is given for both the simple-to-complex crater transition and the crater-to-basin transition (after Basilevsky and Ivanov 1982; Pike 1982*b*). Lower bars: approximate one-sigma values for onset of central peaks in complex craters; upper bars: approximate one-sigma values for onset of inner rings in protobasins and two-ring basins. Upper arrow: revised mean for the Moon, from data in Table XI; lower arrow: revised mean for Mercury (data in Fig. 10). Both revisions improve the linearity of inverse monotonic trends for the four terrestrial bodies. An alternative model, whereby changes in crater shape on all bodies scale at −0.25, requires craters on the Earth to differ in their shape, and possibly in their mode of formation, from those elsewhere.

pography or substrate on Mercury (CT/SP) were judged neither large enough nor consistent enough to warrant separate D_t calculations for the two different groups of terrains. The CT/SP contrasts are real, but so minor that they affect only small craters. Terrain or substrate effects on the simple-to-complex transition are much stronger on the Earth, the Moon or Mars (cf. Pike 1980*a*).

On Mercury, only the *initiation* of the transition reflects differences in substrate. This sensitivity of crater shape to CT/SP contrasts may not have been detected previously because the effect diminishes so rapidly with increasing crater size. Three observations (Fig. 10) indicate that CT/SP differences affect only the first manifestations of crater complexity, over the 6- to 10-km diameter range: (1) persistence of simple craters to larger sizes on cratered terrain; (2) slightly greater depth of modified-simple craters on cratered terrain; and (3) development of wall-failure characteristics—flat floors, slump blocks, scalloped rims and unidentified irregular features—in smaller

SP than CT craters. Interior features in larger, complex, craters 10 to 30 km across reveal no such sensitivity to substrate type. Elements of the complexity include onset size of central peaks and terraces (which are really members of the same morphologic continuum as slump blocks and scalloped rim crests in the smaller craters), and low d/D values.

Interplanetary differences in target characteristics are more significant than intraplanetary contrasts in two respects. First, the revised D_t value for Mercury contradicts previous conclusions that Mercury's craters resemble those of the Moon with respect to interior form and its relation to surface gravity (see, e.g., Wood et al. 1978). Figure 24 shows not only that Mercury's D_t lies midway between that of the Moon and that of Mars, but also that Mars, not Mercury, is the anomaly when $\log D_t$ is plotted as a function of $\log g$. With respect to overall morphology, however, Mercury's craters clearly are Moon-like. None of the morphologic features unique to Martian craters is observed on Mercury.

Second, Mercury's revised D_t is consistent with at least two different scaling relations that reflect the predominant influence of gravity on planetary surface features (Figs. 24 and 26). In each case, however, other effects modulate or even obscure the role of gravity. According to the first model, wherein craters on silicate-rock planets follow a steep log-log D_t/g trend (about -1.0) and those on primarily icy bodies follow a shallower trend (about -0.25: Basilevsky 1981; Pike 1980a), crater morphology may depend strongly on the rheology of target materials (Fig. 26).

The inverse dependency of D_t on g alone for just the four terrestrial planets is monotonic (Fig. 24). The scatter, mainly the low D_t of Mars, is reduced further while preserving the -1.0 relation by introducing V, the calculated approach velocity of asteroids or other bolides at each body (Hartmann 1977). Seven of the nine relations thus created are essentially log-log linear with D_t (Fig. 27). These four-planet relations are consistent with the hypothesis that gravity and velocity may be two of the strongest controls on the crater size at which the simple-to-complex change occurs, although they by no means prove the point. Combining g with velocity does not explain the strong target dependence of D_t for terrestrial complex craters (Dence 1972; Grieve et al. 1981; Pike 1985).

The alternative interpretation (Fig. 26), wherein all observed bodies but one describe a single, shallow D_t/g trend (about -0.25), requires that impacts forming complex craters on Earth (and, by inference, on Venus) differ somewhat from those elsewhere (Pike 1982b). For example, if even large impactors fragment in Earth's thick atmosphere before impact, in opposition to the conventional view (see, e.g., Heide 1964), the resulting craters may be larger and shallower per unit impact energy than craters on relatively airless bodies (Passey and Melosh 1980; Melosh 1981). Such a difference might also explain the target-related contrasts in D_t in that impacts by fragmented, slower bolides would be more sensitive to differences in target strength. Impact craters on

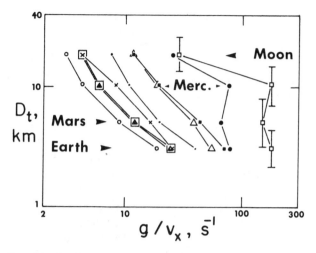

Fig. 27.   Scaling of the simple-to-complex transition in crater morphology with gravity and bolide velocity (cf. Fig. 24). Log D_t values for the four terrestrial planets are given as a function of log surface gravity g and log of 9 different estimates for velocity V of impacting bodies in the solar system. All D_t:g/V relations but two show a nearly linear inverse monotonic trend that suggests a simple functional dependence. Velocity estimates from Hartmann et al. (1981,Tables 8.5.1-3) are the following. Circles: rms approach V of parabolic comets; small dots: modal impact V of Earth-crossing asteroids; crosses: rms approach V of observed parabolic comets. "Adopted" approach V are from Hartmann (1977, his Table I) as follows. Solid squares: asteroids and short-period comets (used for all curves in Fig. 28 below); large open squares: long-period comets; small open squares: local planetesimals in heliocentric circular orbits. Calculated impact V are from Hartmann (1977, his Table III) as follows. Large dots: local planetesimals; open triangles: asteroids; small closed triangles: long-period comets.

Venus may provide a test for these speculations. Preliminary d/D results (Ivanov et al. 1986) suggest *complex* craters on Venus are as shallow as those of the Earth (Fig. 12). However, much depends on the validity of some recent views (see, e.g., Melosh 1981) on bolide fragmentation (cf. Heide 1964).

Competing Models.   The simple-to-complex transition is most conspicuous between simple and immature-complex craters. Explaining this morphologic change is not the same as determining the influences upon it. The transition on Mercury is a size-dependent phenomenon with the following three features:

1. Modified-simple craters show evidence of incipient rim collapse—in small craters on smooth plains and in slightly larger and deeper craters on cratered terrain—with commensurate minor deposition of slump material in the bottom of the crater;
2. Immature-complex craters display recognizable rim failure and central-peak emplacement that is sufficient to markedly reduce crater depth,

through deposition of slump materials and uplift of the crater bottom. This trend toward modification by slumping and related processes continues in mature-complex craters, which are larger than D_t in the strict sense;

3. Mature-complex craters show massive wall failure accompanied by distinct, coherent terraces as well as amorphous slump deposits, plus deposits of impact melt on the floor.

The problem of whether wall collapse or central-peak rebound is the driving mechanism of crater modification on the planets has long remained a "chicken-and-egg" dilemma (Gilbert 1893; Shoemaker 1959; Quaide et al. 1965; Head 1976; Melosh 1977a; Settle and Head 1979; Pike 1980b,1983b; Melosh 1983). Terrestrial evidence seems to favor peak rebound as the initiating mechanism (Milton and Roddy 1972; Ullrich 1976; Roddy 1979; Pike 1980b). For rebound not to apply on all other planets means that extraterrestrial complex craters somehow form differently, even though they appear much the same as those on Earth. The gravity scaling of morphologic differences could suggest either alternative, although a stronger effect of g might be expected for the gravity-driven wall collapse than the presumably more inertially-induced dynamic rebound.

Some of the morphologic data presented here can be interpreted either way. On one hand, Fig. 10 shows that wall failure definitely is *observed* in smaller craters more often than a central peak, thus suggesting that walls failed before peaks formed. The well-known sequence of emplacement of central peaks *before* terrace formation in complex craters on Mercury (Figs. 10 and 11) and elsewhere (Pike 1983b), once thought to conflict with this interpretation, does not, because *rim terracing (on Mercury at* D > 20 km$)$ is continuous with the minor wall failure observed in smaller craters.

However, the occurrence of wall failure before peak emplacement is not evidence for collapse as the driving mechanism. The rudimentary wall failure observed in modified-simple craters and the clear cases of collapse seen in immature-complex craters could just as easily have been induced by incipient peak recoil that was sufficient to bring on wall collapse but was not strong enough to thrust a central peak through the overlying slump deposits in the crater bottom. Evidence for the latter is twofold: first, the onset of central peaks coincides with the diameter at which the h/D regressions inflect (Fig. 11). However, this coincidence on Mercury is perhaps too tentative, given the vague h/D data, to establish firmly causality between peak emplacement and rim collapse.

Second, and more importantly, the d/D distribution for simple and complex craters inflects at a much smaller diameter, 4.7 ± 5 km, than does that for h/D, 11 ± 5 km (cf. Figs. 11 and 13). This large difference suggests that substantial filling and/or recoil of the crater bowl are underway well before (i.e., in smaller craters) the crater rim begins to lose much relief (in larger craters), presumably through collapse and slumping. Moreover, this d/D-before-h/D order exists on all four terrestrial planets (Fig. 24). The supporting data come from Pike (1977a), Wood and Andersson (1978), Pike (1980a), Pike and Davis (1984) and Table X. I interpret this consistent size-dependent contrast as perhaps the strongest and least ambiguous morphologic evidence yet adduced for peak recoil as the mechanism driving the simple-to-complex transition. Some final thoughts are presented below, in Sec. VI.

Craters to Basins

The transition from large central-peak craters to mature two-ring basins (Fig. 25) may be the strongest discontinuity in the size-shape progression on the terrestrial planets. Prior analysis of this transition has focused on a possible genetic link between central peaks and interior rings (Wood and Head 1976; Head 1977,1978; Croft 1979; Hale and Head 1979a,1980a,b; Hale and Grieve 1982; Pike 1982a,1983a). Study of the peak-and-ring geometry here, as exemplified by the particularly abundant data for Mercury (Fig. 18), supports an interpretation favoring a substantial discontinuity in topographic form and perhaps process(es) of large-crater modification (Hale and Head 1979a,1980b). However, my data and analysis indicate no major change in the fraction of energy expended in forming the two central features.

Magnitude of the central disturbance, as reflected by plan size of the peak and inner ring, increases monotonically with size of the host crater or basin. A similar and parallel pair of log-log correlations, between central peak and crater size D_{cp}/D and between inner ring and basin size D_i/D for two-ring basins, is observed on Mercury (Fig. 18) and five other bodies: the Earth, the Moon, Mars, Ganymede, Callisto (Pike 1982a); observations are not yet available for Venus. The resulting pairs of least-squares fits exclude protobasins, which are now known to have a different inner ring geometry (Fig. 19). The respective log-log curves do *not* intersect on the graph (cf. Hale and Head 1979a,1980a,b; Hale and Grieve 1982). I interpret the parallel log-log D_{cp}/D and D_i/D trends in Fig. 18 to indicate that "normal" full-sized central peaks and inner rings both form at a level of energy that increases at a linear rate with magnitude of the impact.

On all five bodies, the inner ring of each two-ring basin is about half the size of the main ring, whereas diameters of central-peak bases measure barely a quarter the diameter of their mature-crater hosts (Pike 1982a). The Y-intercepts D_{cp} of least-squares fits at a mature-crater diameter D of 50 km (in the middle of the distribution) occupy a narrow range, from 10 to 13 km (10.5 km $\pm$ 0.5 km for Mercury). The Y-intercepts D_i at a basin diameter of 200 km also range narrowly, from 100 to 110 km (100 km $\pm$ 2 km for Mercury; Table III and Fig. 18). The D_i/D curve for two-ring basins on Mercury (Fig. 18 and Table VI) parallels other log-log monotonic relations between the main and auxiliary rings of multiring basins (Fig. 25 and Table VII).

Diameter overlap of the peak and ring (D_{cp}/D and D_i/D) data marks the transition from craters to basins on Mercury (Fig. 25) and elsewhere (Pike 1982a). The range of rim diameter that brackets occurrence of the smallest inner ring of *two-ring* basins and the largest central peak of mature craters

TABLE XI
Morphologic Transitions in Craters and Basins[a]

Variable	Body			
	Moon	Mercury	Mars	Earth
Intersection of depth/diameter equations: depth, $d_{d/D}$ (km)	$2.2 \begin{smallmatrix}+0.5\\-0.4\end{smallmatrix}$[c]	$0.93 \begin{smallmatrix}+0.12\\-0.10\end{smallmatrix}$	$0.65 \begin{smallmatrix}+0.16\\-0.13\end{smallmatrix}$	$0.27 \begin{smallmatrix}+0.11\\-0.09\end{smallmatrix}$
Intersection of depth/diameter equations: diameter, $D_{d/D}$ (km)	$10.9 \begin{smallmatrix}+2.3\\-1.7\end{smallmatrix}$[c]	$4.7 \begin{smallmatrix}+0.7\\-0.5\end{smallmatrix}$	$3.1 \begin{smallmatrix}+0.7\\-0.5\end{smallmatrix}$	$1.9 \begin{smallmatrix}+0.6\\-0.4\end{smallmatrix}$
Diameter of simple-to-complex transition, D_t (km)[b]	$18.7 \begin{smallmatrix}+7.7\\-5.4\end{smallmatrix}$	$10.3 \begin{smallmatrix}+4.4\\-3.1\end{smallmatrix}$	$5.1 \begin{smallmatrix}+3.1\\-1.9\end{smallmatrix}$d	$3.1 \begin{smallmatrix}+1.3\\-0.9\end{smallmatrix}$
Size of protobasins D_{pp}, geometric mean diameter (km)	$220 \begin{smallmatrix}+60\\-45\end{smallmatrix}$	$110 \begin{smallmatrix}+30\\-25\end{smallmatrix}$	$120 \begin{smallmatrix}+45\\-35\end{smallmatrix}$	$32 \begin{smallmatrix}+35\\-15\end{smallmatrix}$
Size of two-ring basins D_{tr}, geometric mean diameter (km)	$340 \begin{smallmatrix}+80\\-55\end{smallmatrix}$	$200 \begin{smallmatrix}+50\\-35\end{smallmatrix}$	$140 \begin{smallmatrix}+110\\-60\end{smallmatrix}$	$\left[>36 \begin{smallmatrix}+45\\-15\end{smallmatrix} \right]$[e]

Onset of ≥3 rings D_{to} geom. mean diam. of main rings for five smallest basins (km)	390^{+55}_{-45}	400^{+110}_{-90}	300^{+110}_{-80}	—
Size of multiring basins D_m, geometric mean diameter of main ring (km)	550^{+240}_{-170}	760^{+430}_{-280}	620^{+490}_{-270}	—
Surface gravity g (g/cm²)	162	370	375	981
Approach velocity of asteroids and short-period comets V_∞ (km s⁻¹)[f]	14	19	8.6	14
g/V_∞ (s⁻¹)	12	20	44	70

[a] See Fig. 28; geometric means and one-sigma values; some data from Pike and Spudis (1987).

[b] Moon and Earth data from Pike (1980a, Table 3).

[c] Upland craters only; comparable values for mare craters are: $D = 8.6^{+1.9}_{-1.5}$, and $d = 1.7^{+0.6}_{-0.4}$ km.

[d] D_t revised by adding h/D and d_a/D data from Pike and Davis (1984).

[e] Least reliable value in table.

[f] Values adopted by Hartmann (1977, Table I).

always includes some protobasins (Fig. 18; Pike 1983a). The ranges of D_{cp}/D and D_i/D overlap are approximate: Mercury, 132 to 160 km; Earth, 15 to 25 km; Mars, 52 to 90 km; for Ganymede possibly 70 to 120 km, and for Callisto perhaps 80 to 100 km (Pike 1982a). No such crater/basin overlap exists for the Moon, the largest central peak crater being 200 km across and the smallest two-ring basin 320 km across (Pike 1983a): lunar protobasins overlap both classes. Some of these threshold diameters on the four terrestrial planets (Table XI) correlate inversely with g and with g/V_∞ (see Fig. 28, below).

The geometry of protobasins is uniquely transitional. Protobasins differ from complex craters and two-ring basins in that (1) the relation between size of the inner ring and size of the accompanying central peak is complementary; and (2) either one or both of these features is smaller than it would be had it appeared alone in a crater or basin of that diameter. The correlation, described fully in the last section, is especially evident on Mercury (Fig. 18). Similar, if less clearly defined (e.g., from fewer data), complementary relations occur on the Moon (Hale and Head 1979a; Pike 1983a) and on Mars (Pike 1982a); Ganymede and Callisto may have no protobasins.

The origin of the peak/ring transition must remain speculative in the absence of well-constrained theory of impact at such high energy. However, protobasins may offer clues. One possibility is that at some physical (and undefined) threshold, inner rings replace central peaks as the more efficient structural expression for a constant fraction of cratering energy. However, a qualitative change in the form of the central feature indicates no change in the energy fraction itself; where both a peak and an inner ring form in the same protobasin, one or both features are smaller than normal. This prompts the further speculation that the energy fraction available for the central disturbance might be shared between the peak and the inner ring rather than be increased to accommodate both features (Hale and Grieve 1982; Pike 1982a).

A final, and important, point: the appearance of *both* a peak *and* an inner ring in the *same* protobasin indicates that *the ring cannot be simply a larger version of a central peak*. Thus, coexistence of the two features suggests that they might not even form by the same mechanism. On the other hand, their complemental geometry indicates concurrently that emplacement may be controlled by interacting, and hence related, processes. The dilemma cannot be resolved here. The possibility remains that the crater-to-basin transition may be marked by a real change in process, as suggested by Hale and Head (1979a,1980b), although not for all the reasons they cited. Discussion of alternative ring-forming processes is deferred to the next section.

Two-Ring Basins to Multiring Basins

The transition from two-ring to multiring basins on Mercury (and other terrestrial bodies; see, e.g., Wood 1980; Croft 1981b) is perhaps the simplest in terms of morphologic expression (Fig. 25). It is marked by: (1) an increase in the number of concentric rings; (2) a change in the observed ratio between adjacent rings from exclusively $2D$ to $2^{0.5}D$ and its multiples (most often $2D$); (3) development of rings outside the main, or topographic, ring; and (4) a marked diminution in ring prominence with increasing radial distance (Table IV). On Mercury these changes are observed first in basins about 300 to 400 km across at the main ring: Homer, at 310 km, is the largest two-ring basin; Van Eyck (285 km), Hiroshige-Mahler (355 km), and Shakespeare (420 km) are the smallest multiring basins.

I interpret the transition largely as (1) the extension of an already well-developed structure and landform—the concentric inner ring spaced at a 2D ratio, into a larger size range, and (2) multiplication of auxiliary inner and outer rings at exactly half the 2D spacing, in response to increasing energy of impact (Fig. 25). This change contrasts with the transition from simple to complex craters (Figs. 10 and 11), where central features develop in response to either or both inertial/dynamic or gravitational instability of the crater. The basin-to-basin transition also differs from that of large craters to small basins, where an increasingly (with crater size) unstable landform, the central peak, is replaced by a different and (only temporarily) more stable one, the inner ring at $D_i/D = 0.4$ (Figs. 18 and 25). Any g or V scaling of the appearance of multiple rings is much less evident than that for crater complexity or twin rings (see Fig. 28 below, the last two plots).

Multiring basins are more complicated than two-ring basins in that many of the rings form outside the main, or topographic ring (Table IV and Fig. 25). Ring proliferation suggests that operation of the ring-locating mechanism might not necessarily be confined wholly within a deep transient cavity in the largest impacts. Thus, among other things, the transient cavity (or the excavation cavity) may be compound in profile, with a deep inner bowl and a shallow outer shelf (Pike 1980b; Grieve et al. 1981). The multiplicity of rings, together with persistence of the comparatively regular $2^{0.5}D$ spacing of rings, despite radial decay of ring prominence, suggests further that rings or at least their positions develop through a periodic (wave or resonance) phenomenon (Chadderton et al. 1969; Pike and Spudis 1987). It is further possible that the *ring-spacing* mechanism operates in addition to, rather than in place of, other processes proposed for *forming* rings in multiring basins (see discussion in Pike and Spudis 1987).

The formation of concentric-ring basins remains one of the main unsolved problems of planetology, on Mercury as elsewhere. Aside from their obvious impact origin, no agreement has been reached on the exact means of ring spacing and emplacement (Baldwin 1974; Croft 1981a,b,; McKinnon 1981). Contending ring mechanisms, reviewed by Pike and Spudis (1987), include various "collapse" and "mega-terrace" models, the "nested-crater" hypothesis, the "rock-tsunami" and "oscillating-peak" schemes, and "elastic-plate tectonism." The data reported here preclude none of these alternatives as processes that could form *some* basin rings on Mercury.

Both the rhythmically spaced deformation implicit in the observed $2.0^{0.5}D$ spacing on Mercury (as well as on the Moon and Mars; Pike and Spudis 1985,1987; Pike et al. 1985) and its roughly equal prevalence both inside and outside the main ring suggest a process that must be able to (1) operate independently of broad-scale target properties, impact velocity and impactor composition; and (2) locate as few as two, but up to six or seven similarly spaced ring *positions* in one basin. The first requirement poses almost insurmountable difficulties for the nested-crater and elastic-plate schemes, while the second condition argues strongly against collapse and mega-terrace models and to some extent the oscillating-peak hypothesis.

This process of elimination leaves some type of periodic (wave) mechanism, although not specifically the rock-tsunami scheme or the oscillating-peak hypothesis (see Baldwin 1974), as the most attractive means for locating, and perhaps even partly emplacing, the rings of basins and protobasins on Mercury and the other terrestrial planets. Physically, the model remains undefined at this time. The process must also explain the prominence of ring IV and the association of rings with deep-seated structures. The issue is far from closed. The ring problem can be expected to yield only to a concerted approach by mathematical modeling, field study, photogeologic analysis, and perhaps laboratory experiment (see, e.g, Grieve et al. 1981; Grieve and Garvin 1984).

VI. DISCUSSION

The effects of two and perhaps three influences on the impact process may be assessed or interpreted from the size-dependent transitions in crater and basin morphology on four terrestrial planets. The data and analysis presented here in Tables XI and XII and in Fig. 28 suggest that these are, in order of diminishing importance: gravity, target rheology and velocity of the impactor. I have based the interpretations on some form of the size-dependent scaling established for terrestrial planets (Figs. 1, 24 and 26-28). Insufficient data exist for the outer satellites (Fig. 26). This four-planet comparison is possible only now that the data on Mercurian craters carry an accuracy comparable to those for craters on the other three bodies (Pike 1980a).

Seven quantitative morphologic characteristics of craters and basins listed in Table XI were examined in Fig. 28 in log-log space as a function of surface gravity g and both g and the approach velocity of asteroids and short-period comets g/V_∞ (Hartmann 1977). In order, the first five are: $d_{d/D}$ and $D_{d/D}$, respectively the depth and diameter at which d/D of Eqs. (1) and (17) for simple and complex craters intersect; D_t, the diameter of the simple-to-complex transition; and D_{tp} and D_{tt}, the geometric mean diameters for protobasins and two-ring basins, respectively. The latter two quantities are good

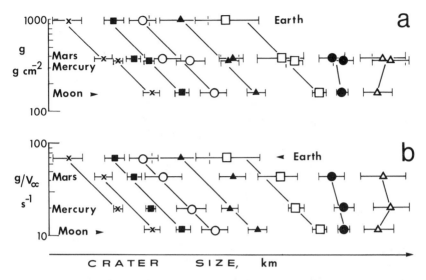

Fig. 28. Seven transitions in crater and basin morphology on the terrestrial planets: scaling by crater size (Table XI) as a function of log g (Fig. 28a) and log g/V_∞ (Fig. 28b). Symbols are geometric means; error bars are one-sigma values. Crosses: $d_{d/D}$, crater *depth* at crossing of Eqs. (1) and (17); solid squares: $D_{d/D}$, crater diameter at same intersection; circles: D_t (see Table X); solid triangles: D_{tp}, diameter of protobasins; open squares: D_{tt}, diameter of two-ring basins; large dots: D_{to}, onset of multiring basins (diameter of five smallest main rings); open triangles: D_{tm}, diameter of multiring basins. For clarity, horizontal axis shows only *relative* position of the 7 transitions; true diameters (one is depth) are given in Table XI. V_∞ values are Hartmann's (1977) estimates for asteroids and short-period comets (solid squares in Fig. 27 above). Slanted lines in the five relations on left are *not* statistical fits, but rather −1.0 index trends: models for comparison with observations. The agreement is good, with exceptions noted in text. Slanted lines in the two relations on the right *are* estimated fits to observations: correlation with the −1.0-sloping model is poor. Overall trend is one of decreasing influence of impactor velocity, target rheology, and g with increasing size of impact.

approximations for diameters documenting the lower size limit for small basins. Measures of central tendency D_{tp} and D_{tt} are statistically more robust than observations of "onset" diameter, which by virtue of being an extreme value is a less stable statistic. Despite this limitation, D_{to}, the geometric mean of the five smallest multiring-basin main rings, D_{IV}, and D_{tm}, the geometric mean of all basin main rings. also are included in an attempt to identify any scaling trends for these very large features.

Figure 28a shows that five of the seven quantities scale inversely and monotonically with g, displaying varying degrees of scatter. The two exceptions describe multiring basins. For craters, represented by the first three relations, the largest deviations from the −1.0 log-log trend may be ascribed to differences in target characteristics (Cintala and Mouginis-Mark 1980; Pike 1980a). The improved data for Mercury have materially reduced the scatter in all three $D : g$ relations for craters. The reduction of Mercury's D_t from 16 to 10.3 km brings it close to the -1.0 trend for terrestrial planets, much closer than Mars. Similarly, the new $D_{d/D}$ value for Mercury of 4.7 km lies almost exactly on the trend, leaving only Mars, falling well below the trend, as an anomaly. The two Martian anomalies could reflect an overall "wet" (Cintala and Mouginis-Mark 1980) or a "wet, layered" (Pike 1980a) condition of the target, which may encourage onset of the complex morphology at smaller crater sizes on that body. Presumably this would occur through closer coupling of impact energy to layered volatile-rich materials or through a reduction in overall target cohesion or both. However, the Earth also is a comparatively "wet" target.

The case of crater depth is less straightforward in Fig. 28 because the inverse dependence of crater $d_{d/D}$ on g suggests two trends. The Moon and Mercury have a higher value, whereas the Earth and Mars follow a lower, parallel trend. Again, the difference could reflect the influence of a volatile-rich, and weaker target on *both* the Earth and Mars, in contrast with the "dry" targets presented by the Moon and Mercury. The similarly dichotomous h/D relation discussed above (Fig. 24) suggests that rim height also may respond to this contrast.

Two-ring basins and protobasins also follow the single -1.0 $D : g$ trend (Basilevsky 1981; Pike 1982a) in Fig. 28. The sizes of both features, less a well-defined D_{tt} value for Earth where the largest two-ring basins almost certainly have not been observed, are consistent with the same strong inverse g dependence as for complex craters. The relation for protobasins is quite linear. The fact that neither protobasins nor two-ring basins on Mars fall below these trends, in contrast to craters, is consistent with diminution of the influence of target rheology with increasing impact energy and crater or basin size (Baldwin 1963,p.184).

The difference in depth $d_{d/D}$ in Fig. 28a is particularly important because crater complexity often has been interpreted as having developed from gravitational collapse of a transient cavity that has exceeded some critical depth (Quaide et al. 1965; Melosh 1977a, 1983). If a "wetter" or otherwise weaker target could reasonably be expected to sustain only a lower crater wall before collapsing, then a gravity-driven model is consistent with the $d_{d/D}$ plot. However, this consistency does not preclude an alternative model (Roddy 1979; Pike 1980b), which was developed in response to the observation that centripetal rim collapse is not a convincing mechanism for producing the central uplift in terrestrial meteorite craters (Milton and Roddy 1972). Namely, crater complexity, while obviously involving rim collapse as one element of the change in morphology in the late stage of the impact, is driven in its initial development by inertial, not gravitational, forces, whose most important result is uplift of the crater bottom into a central peak after full development of the transient cavity in the earliest stage of cratering. According to this model, collapse of the rim is a g-scaled event that occurs late in the impact, but the collapse is thought to be *caused by* the central uplift (see, e.g., Pike 1980*b*). The *d*/*D* and *h*/*D* data (Fig. 24) are particularly persuasive evidence for this view.

The conflict between inertial and *g*-driven models for crater complexity may be more illusory than real (Pike 1983*b*; Melosh 1983). Because rim slumping is much less complicated theoretically (Quaide et al. 1965; Melosh 1977*a*; McKinnon 1978) than peak rebound (Harlow and Shannon 1967; Ullrich 1976), slumping has been the more frequently modeled of the two, and thus seems to have gained more acceptance. Recoil is far less amenable to tests and physical analysis. Modeling, however, does not in and of itself bestow credibility or preclude other alternatives.

The two contending views may be resolved by invoking (1) an inertial mechanism, recoil, for the central uplift, reflected in diminution of depth in small craters; and (2) a *g*-driven mechanism, collapse, for wall failure, reflected in diminution of rim height in craters 1.32 to 2.3 times larger (Fig. 24). Possibly the effects of the two are mixed during the course of an impact, rather than restricted entirely to one stage or another. Thus, on Mercury, wall failure may be observed in smaller craters than may central uplift (subject to the caveat on buried small peaks described earlier) (Fig. 10).

The velocities of impacting asteroids and comets differ according to their source location in the solar system (Wetherill 1975; Hartmann 1977; Hartmann et al. 1981). Such contrasts, although subject to much uncertainty, may be evident in observations on crater shape (Wood and Head 1976; Cintala 1979; Chapter by Schultz). Figure 28b, a plot of the five crater attributes against g/V_∞, V_∞ being the approach velocity of asteroids and short-period comets (Hartmann 1977, Table I), is an attempt to discover possible effects on crater morphology. The velocity V_∞ was chosen over 7 other V values (Hartmann 1977, his Tables I,III; Hartmann et al. 1981, their Tables 8.5.1-2) as most likely representing impacting bodies, but the other 8 V estimates also yield -1.0-sloping plots against D when combined with g (Fig. 27). I emphasize that *no differences in crater shape correlate with* V_∞ *itself,* but rather with V_∞ *in combination with* g. Figure 28 shows that any influence of V_∞ is difficult to separate from that of g because the $D{:}g$ and the $D{:}g/V_\infty$ relations are similar in overall form.

Perhaps the most convincing case for any substantial influence of bolide velocity on crater shape, from the observations presented here, is the plot of D_t against g/V_∞ in Fig. 28b. All four planets lie close to a -1.0 log-log trend, and deviate from it much less than do the D_t/g observations. However, this especially linear -1.0 correlation does not persist for other crater variables or for basins. The $D_{d/D}{:}g/V_\infty$ plot, for example, excludes Mercury, not Mars— as above, from its otherwise linear trend. Instead, Mercury's *d*/*D* intersection diameter documented by the new results in Eqs. (1) and (17) lies 2 km below that suggested by the -1 trend. A plot of $D_{h/D}{:}g/V_\infty$ (not shown; cf. Fig. 24) is linear, with virtually no scatter, but the slope of -1.3 differs significantly from unity. Large craters and small basins show even more discordance in $D : g/V_\infty$ space: Martian protobasins lie 50 km *above* the -1.0 trend for three planets, and the data for two-ring basins do not lie as close to the -1.0 trend as do the $D : g$ data.

The $d_{d/D} : g/V_\infty$ results, in particular, suggest that g/V_∞ scaling may not be as consistent across the four terrestrial planets as that for g alone. Figure 28b shows that the Moon and Mars follow one inverse trend, whereas Mercury and the Earth follow a parallel, lower trend. This is an unlikely pairing for these four planets, given that Mars and the Earth are known to have been "wet" targets and the Moon and Mercury "dry". Such a pairing does correspond to differences in the respective mean densities of the bodies, but any connection with processes affecting crater depth remains unknown.

Multiring basins differ from craters and smaller basins on the four terrestrial planets in that their morphologies do not appear to scale with gravity, target rheology or impactor velocity. Onset size of multiple (≥ 3) rings shows at best only a weak inverse correlation with gravity, and the mean size of basin main rings shows, if anything, a slight positive dependence on g (Fig. 28a,b). Thus the overall picture, summarized in Table XII and Fig. 28, is one of a systematically declining influence of external factors on crater and basin morphology with increasing crater and basin size, and hence energy of impact (Baldwin 1963,p.184; Dence and Grieve 1979).

Figure 28 indicates further that the size-dependent order of declining influence *overall,* velocity first, then target properties, and finally gravity, is the same as their relative importance *for small craters*. This internal consistency suggests that multiring basins on terrestrial planets are much more the product of impact-energy-scaled phenomena than outside influences such as crustal layering or thickness, or even such an important control as gravity. Thus, conditions of high impact energy and consequentially low target strength *over a large target area* might well favor the operation of a wave or

TABLE XII
Possible Influences on Morphologic Transitions[a]

	Variable[b]		
Transition	Surface Gravity	Target Rheology	Impactor Velocity
Simple-to-complex craters	*	*	?
Craters to basins	*	?	—
Multiple rings, onset	?	—	—
Multiple rings, mean basin size	—	—	—

[a]Summary interpretation of Fig. 28 (see Table XI).
[b]Explanation of symbols: (*): variable has some effect; (?): variable may have some effect; (—): effect not discerned.

resonance phenomenon to produce the statistically constant spacing of multiple basin rings and ring fragments (Pike and Spudis 1984,1987).

The view that the structural and topographic effects of very large impacts differ in important ways from those of smaller impacts because of magnitude-dependent changes in physical processes is hardly new (Baldwin 1949, 1963, 1974). In the largest impacts, the enormous forces generated may simply overwhelm all other influences that would have an effect at smaller energy levels, and create morphologic features that *predominantly* reflect an inertially-driven, hydrodynamic regime where strength of materials over a large target area is essentially zero. The data presented here on the craters and basins of Mercury are consistent with such a view.

VII. SUMMARY AND CONCLUSIONS

New observations and analyses of the surface form of 447 impact craters on Mercury constrain their geomorphic interpretation and that of fresh primary craters on other planets. The findings in this chapter include both new results and confirmation or elaboration of those reached by others. They fall into two groups: conclusions unique to Mercury, and those that apply to all terrestrial planets and satellites. The following summary ranges from crater taxonomy and numerical observations on morphologic features to speculations on processes responsible for basin-ring geometry.

Mainly Mercury

1. The morphologic complexity of Mercury's impact craters increases systematically with the diameter. At least 15 features of the crater interior are strongly size dependent: depth, rim height, rim width, peak and floor diameter, peak and rim-wall complexity, frequency and spacing of concentric rings, and the presence or absence of a bowl-shaped interior, flat floor, central peak, scalloped rim crest, slump deposits, and rim-wall terraces.

2. Seven classes of craters are recognized on Mercury, in order of increasing diameter D and morphologic complexity: simple, modified-simple (a new category), immature-complex, and mature-complex craters; protobasins; and two-ring and multiring basins.

3. Three size-dependent transitions in crater morphology on Mercury are identified and quantified: D_t: simple to complex craters at 10.3 km $\pm$ 4 km diameter; D_{tp}: complex craters to double-ring basins at about 110 km $\pm$ 25 km D; and D_{to}: the onset of $\geq$ three rings at about 400 km $\pm$ 100 km D.

4. Mercury's new D_t (10.3 km) is significantly smaller than the old value (16 km). The number of its morphologic constituents is increased to 10,

including measurable variations in depth, rim height, floor and rim-wall width, and rim-wall and central-peak intricacy.

5. The new depth/diameter d/D relation for fresh craters on Mercury differs substantially from previous results. The log-log equations for simple and complex craters intersect at a low D of 4.7 km and $d = 0.94$ km. Slopes of the complex-crater fits are much steeper on Mercury than they are on the Moon or Earth, a still-unexplained difference.

6. Putative differences in substrate affect crater depth in only one class on Mercury: modified-simple craters on cratered terrain are slightly, but significantly, deeper than those on smooth plains.

7. Complex craters on Mercury comprise two distinct types (cf. the Moon; Wood and Andersson 1978): immature-complex craters, < 30 km across, are characterized by a higher d/D ratio, shallow wall-failure, smaller interior floors, and the onset of morphologically simple central peaks. Mature-complex craters, > 30 km across, have a lower d/D ratio, relatively well-formed rim terraces, complicated central peaks, and broader flat floors.

8. Central peaks in complex craters on Mercury increase markedly in complexity with crater size, in opposition to the conclusions of Hale and Head (1980*a*).

9. Crater morphology on Mercury shows only weak dependence on topographic background, commonly interpreted as indicating substrate rheology. Flat floors, scalloped rims, slump blocks and largest bowl-shaped interior—all attributes that reflect the initiation of wall failure—appear in craters 3 to 5 km smaller in diameter on smooth plains than they do on cratered terrain. No such correlation exists for onset diameters of craters with central peaks, terraces and low d/D, all of which mark major morphologic changes that may be less sensitive to substrate contrasts than minor slope failure. This difference (with conclusion (6), above) indicates that substrate contrasts are minor and affect only small craters.

10. New values of rim height for simple and, especially, complex craters on Mercury remain anomalously high, as first reported by Cintala (1979), although the intersection diameter for the regression lines is now 11 km rather than 16 km.

11. The size-dependent variations in crater shape identified here indicate two dominant mechanisms explaining the onset of crater complexity on Mercury. Both inertial forces (central-peak recoil) and gravity (wall failure) share in development of the complex morphology. However, the inflection of d/D well before (in smaller craters than) h/D indicates that the initiating mechanism is recoil of the crater bowl, not slumping of the rim.

12. The two unique attributes of protobasins, wherein both the inner ring and the central peak are smaller than they would be had each formed singly in a crater or basin and the diameters of both of these inner features are complementary, are best developed on Mercury.

All Terrestrial Planets

13. New observations force changes in the taxonomy of impact craters: "modified-simple," "immature-complex" and "mature-complex" classes reflect size-dependent differences better than do existing terms. "Protobasin" and "two-ring basin" replace "central-peak basin" and "peak-ring basin," respectively, as both more descriptive and freer of genetic connotations.

14. Mercurian craters fall about midway between Martian and lunar craters with respect to the 10 size-dependent attributes of the crater interior that define D_t. In other morphologic respects, however, Mercurian craters resemble those of the Moon, not Mars.

15. Mercury's new D_t of 10.3 km is consistent with either monotonic-inverse gravity g or gravity/bolide velocity g/V_∞ scaling of D_t on the terrestrial planets. However, D_t and $D_{h/D}$ are the only transition parameters that may scale better with g/V_∞ than with g. Crater diameter and depth at the intersection of d/D equations both scale more consistently with g alone.

16. Interplanetary effects of gross target rheology are evident in D_t constituents. Both crater diameter at intersection of the h/D regressions and depth at that of the d/D fits divide planets into two groups, when plotted against g (but *not* g/V_∞): the "wet" planets Earth and Mars (lower $D_{h/D}$ and $d_{d/D}$) and the "dry" planets Moon and Mercury.

17. The new data reveal that Mars, not Mercury, now is the anomaly with respect to the inverse D_t/g relation and other D/g and $D : g/V_\infty$ relations for craters.

18. Mercury's *simple* craters are similar in depth to those on the Moon, Mars, and to a lesser extent, the Earth ($d/D = 1/5$), but do differ in their maximum size (14 km D). *Complex* craters on Mercury are significantly shallower than those on the Moon but deeper than those on Mars; terrestrial complex craters and probably those on Venus are much shallower.

19. Terrain dependency in crater d/D is much more evident on the other three terrestrial planets than it is on Mercury. This suggests that physical properties of the Mercurian upper crust are more uniform than those elsewhere.

20. Ring spacing of protobasins on Mercury, the Moon and Mars ($D_i/D = 0.40 \pm 0.03$) is unique. It differs significantly from that for two-ring basins ($D_i/D = 0.49 \pm 0.01$) and multiring basins (alternate rings: $D_{n-2}/D_n = 0.50 \pm 0.03$). This contrast invalidates prior conclusions from statistical and graphical analyses of basin-ring spacing, which were derived from samples incorporating protobasins with basins.

21. Average diameters of protobasins and two-ring basins are consistent with inverse monotonic g scaling of the crater-to-basin transition.

22. The 2.0 D spacing of rings in two-ring basins is a subset of the $2^{0.5}D$

spacing identified for multiring basins. Inner rings of two-ring basins are geometrically similar to rings that lie within the main ring of multiring basins and are spaced at twice the $2^{0.5}D$ interval.

23. The transition from two-ring to multiring basins does not scale with gravity, nor does the mean observed diameter of multiring basins (main ring) appear to scale with g or g/V_∞.

24. The average spacing observed between adjacent rings of multiring basins on Mercury is similar to that $2^{0.5}D$ identified for basins on the Moon, Mars, the Earth, and to some extent on the outer-planet satellites Ganymede and Rhea. Statistical tests indicate that this target-invariant interval did not arise by chance. Its exact mode of origin and significance remain unexplained, although available data suggest that the impact itself, rather than target properties, is the primary control on ring spacing and perhaps ring emplacement.

25. Rings flanking the main ring of multiring basins on Mercury, as on Mars and the Moon, are spaced at an interval somewhat greater than $2^{0.5}D$: $2.1^{0.5}D$. This systematic difference, which is not an observational artifact, probably relates to the physiographic prominence of the main ring.

26. Concentric-basin rings on Mercury and elsewhere are nearly as prevalent outside the main ring as within. Moreover, the mean spacing interval, $2.0^{0.5}D$, is the same both inside and outside the main ring, as is that for the rings spaced at $2.1^{0.5}D$. Thus, the ring-locating mechanism functioned independently of the size of the transient basin cavity.

27. The means by which basin ring topography and relief were emplaced are not necessarily those by which ring spacing was determined.

28. The overall picture of crater formation, with increasing energy of impact, is one in which physical processes are dominated less by gravity and material strength and more by hydrodynamic effects. At the crater-to-basin transition, hydrodynamic processes affect such large areas of zero-strength target material that simple periodic phenomena begin to appear. Above the two-ring-to-multiring transition, these wave effects are so well established over such a very large area that multiple rings are able to form.

29. Although not excluded as processes that *form* some rings in multiring basins, mechanisms such as "mega-terracing," "elastic-plate tectonism," and "nested-crater excavation" can be ruled out as ring-*locating* processes, which must both yield rhythmically spaced deformation and be equally effective both inside and outside the main ring. Not precluded by the new data is the class of ring-locating hypotheses that invoke hydrodynamic processes (specifically wave or resonance phenomena) operating over a large area.

30. Contrasting target rheology is not observed to affect the shape of impact craters in morphologic transitions at and beyond the appearance of protobasins D_{tp}. This suggests that upper-crustal contrasts on none of the

rocky bodies are sufficient systematically to modulate crater structure and morphology in large impact events.

31. Any effects of impactor velocity on the morphologic transitions for craters on silicate planets are difficult to distinguish from those of gravity, given the velocity estimates currently available for asteroids and comets.
32. The importance of three external influences on crater and basin shape: gravity, target rheology and impactor velocity, appears to decline systematically with increasing magnitude of the impact. The effects of bolide velocity are the first to disappear and those of gravity the last.

Acknowledgments. This chapter is dedicated to Ralph Belknap Baldwin—mentor, colleague and inspiration, on the occasion of his receipt of the G. K. Gilbert Award from the Planetary Division of the Geological Society of America. Ralph had many of the right answers long before any of us even knew there were questions to be asked.

I thank G. Clow for the FORTRAN tutorial, his least-squares algorithm, help with the statistics and vector algebra, and ongoing systems management; P. Spudis for his measurements of Mercurian basins, innumerable helpful discussions and morale boosts; A. Woronow for his excellent guidance on statistics; M. Davies for the Mariner 10 films; P. Spudis, A. Woronow, M. Leake and A. Dial for helpful reviews; G. Schaber for his encouragement; J. Boyce for funding the work and for his patience; and the Branch of Astrogeology for support over the last two decades: *Ave atque Vale.*

APPENDIX A

Calculation of Crater Dimensions From Oblique Images

GARY D. CLOW and RICHARD J. PIKE

Horizontal and vertical dimensions of impact craters imaged by spacecraft can be estimated by analyzing the Sun-spacecraft-planet geometry that existed when the images were obtained. For most flyby missions such as Mariner 10, spacecraft altitude at the time of image collection greatly exceeds the radius of the planet. The resulting oblique viewing angles induce severe image distortion. Thus, the geometric quantities that define positions of the Sun, the spacecraft, and a point on the planet's surface (Fig. A1) vary in a nonlinear fashion across an image and must be redetermined for each planetary feature of interest. The general method described here for estimating crater dimensions from such redeterminations can be applied to any planet or satellite. We have used the procedure to generate new measurements of diameter and depth or rim height for 373 craters on Mercury (see Appendix B).

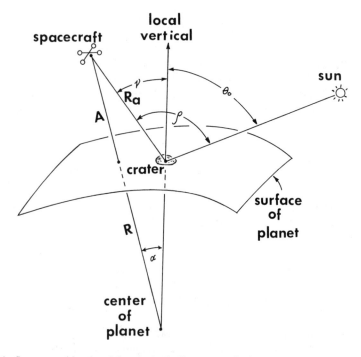

Fig. A1. Seven quantities that define the basic Sun-spacecraft-planet geometry: spacecraft altitude A, planetary radius R, spacecraft range R_a, viewing angle γ, phase angle ρ, solar zenith
angle θ_0, and angle α. Note that $\rho \neq \gamma + \theta_0$.

Crater diameter can be estimated by finding the distance between two
opposing points on the rim crest. Let the first point P of such a pair be located
at latitude ϕ and longitude λ in the planet's geocentric coordinate system. The
corresponding (planetary surface) coordinates of the spacecraft and the Sun are
(ϕ_s, λ_s) and $(\phi_\odot, \lambda_\odot)$, respectively. The azimuth of the spacecraft relative to
point P (Fig. A2) is

$$\psi_s = \cos^{-1}\left[\frac{\tan(\phi_s - \phi)}{\tan \alpha}\right] \tag{A1}$$

where the angle α, shown in Fig. A1, is given by

$$\alpha = \cos^{-1}\left[\cos \phi \cos \phi_s \cos(\lambda - \lambda_s) + \sin \phi \sin \phi_s\right]. \tag{A2}$$

Let the second point on the crater rim Q be located at an azimuth ψ relative to
P. When viewed obliquely from the spacecraft, foreshortening of the circular
crater will cause P and Q to "appear" to be separated by a distance

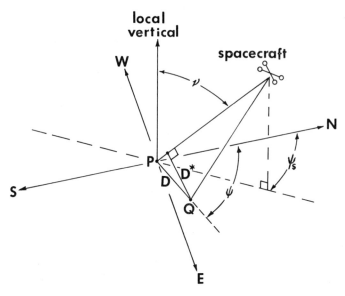

Fig. A2. Estimating horizontal distances from the locations of the spacecraft and a point Q relative to that of point P on the planetary surface, with reference to azimuth, ψ_s and ψ, respectively. D is the true horizontal distance between P and Q, and D^* is the "apparent", or foreshortened, distance viewed from the spacecraft.

$$D^* = D\left\{\left(\frac{1 + \cos\gamma}{2}\right) - \left(\frac{1 - \cos\gamma}{2}\right)\cos[2(\psi - \psi_s)]\right\}. \quad \text{(A3)}$$

Rearranging terms, the true horizontal distance between P and Q is

$$D = \frac{D^*}{\cos^2\left(\frac{\gamma}{2}\right) - \sin^2\left(\frac{\gamma}{2}\right)\cos[2(\psi - \psi_s)]}. \quad \text{(A4)}$$

The view angle γ, between the spacecraft and the local (planetary surface) vertical must be known to evaluate Eq. (A4). The view angle can be found from the planet's radius R, spacecraft altitude A, the distance between the spacecraft and a point on the planet's surface (spacecraft range) R_a, and angle α

$$\gamma = \sin^{-1}\left[\frac{(R + A)\sin\alpha}{R_a}\right]. \quad \text{(A5)}$$

The range R_a between the spacecraft and point P is given by

$$R_a = [R^2 + (R + A)^2 - 2R(R + A)\cos\alpha]^{1/2}. \quad \text{(A6)}$$

Equation (A4) predicts correctly that no foreshortening ($D^* = D$) will occur when the azimuth of Q differs from that of the spacecraft by 90°. Maximum foreshortening occurs when ψ and ψ_s differ by 0° or 180°, in which case $D^* = D \cos \gamma$. Thus, crater diameters derived from several pairs of opposing points along different azimuths can be compared as an internal check on consistency of the calculations. Each crater diameter on Mercury (see Appendix B) is a median value of three found from pairs of points measured (1) along the direction of maximum foreshortening, (2) along that of minimum foreshortening, and (3) along an east/west line through the crater center.

Depths for craters with bowl-shaped interiors can be estimated from the length of a shadow cast by the crater rim crest, provided the shadow reaches the center of the crater. Two angles are required for this calculation, the solar zenith angle θ_o and the phase angle ρ (the angle between the Sun and the spacecraft). The solar zenith angle is given by

$$\theta_o = \cos^{-1} [\cos \phi \cos (\lambda - \lambda_\odot)] \tag{A7}$$

and the phase angle is

$$\rho = \cos^{-1} \left[\left(\frac{R + A}{R_a} \right) \cos \phi_s \cos (\lambda_\odot - \lambda_s) - \left(\frac{R}{R_a} \right) \cos \theta_o \right]. \tag{A8}$$

If L^* is the *apparent* length of the shadow (viewed obliquely from the spacecraft; Fig. A3), the tip of the shadow lies below the crater rim at a distance

$$d = \frac{L^* \cos \theta_o}{\sin \rho}. \tag{A9}$$

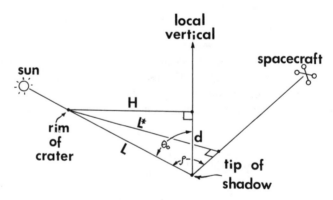

Fig. A3. Estimating vertical distances from the relation between crater depth d and the apparent shadow length L^*. True length of the shadow is L; H is the horizontal component of true shadow length.

This depth will be the correct (maximum) value for a bowl-shaped crater if the shadow tip falls near the lowest point on the crater bottom, which generally is at the crater center. Thus, the *horizontal* extent of the shadow

$$H = \frac{L^* \sin \theta_o}{\sin \rho} \qquad (A10)$$

should be close to half the crater diameter estimated from Eq. (A4). For depth/diameter analysis of bowl-shaped Mercurian craters, we used only those 104 craters in Appendix B whose shadow tips fell within 0.05 D of the crater center.

Depths for craters with flat interior floors can be estimated from any shadow whose tip falls on the floor. Many of these craters have such irregular rim crests that an averaged shadow length is required for a representative depth. Equation (A9) applies equally to shadows cast outside a crater, and hence can be used to estimate height of the rim above the surrounding terrain.

Crater 10 in Appendix B, located at image coordinates $X = 190$, $Y = 570$ pixels on Mariner 10 Frame FDS 0027475 (Fig. A4), illustrates our procedure for measuring the dimensions of a morphologically simple, bowl-shaped crater. The shadow-casting point on the rim crest of Crater 10 lies at latitude ϕ $= -3°6$ and longitude $\lambda = 28°7$ (U.S. Geological Survey 1979). Generally estimated to the nearest $0°1$, ϕ and λ for some very small neighboring craters were extrapolated to $0°01$ using Mariner 10 images. The best estimate of the spacecraft position comes from the Mariner 10 Supplementary Experimental Data Record or SEDR (IPL 1976). The coordinates of the spacecraft when Frame 0027475 was recorded were latitude $\phi_s = -21°21$, longitude $\lambda_s = 346°21$, and the altitude was 12,770 km. According to the SEDR, longitude

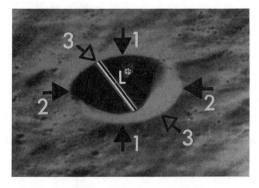

Fig. A4. Examples of measurements for a bowl-shaped Mercurian crater: diameter 1: direction of maximum foreshortening; diameter 2: direction of no foreshortening; diameter 3: east-west; L^*: apparent length of shadow. Crater No. 10 (see Appendix B) is shown on a portion of Mariner 10 Frame FDS 0027475. Diameter of rim crest is 10.8 km.

TABLE AI
Calculated Quantities for Crater 10

α	=	$44°86$
γ	=	$52°12$
ρ	=	$118°09$
θ_o	=	$70°81$
ψ_s	=	$108°60$
R_a	=	13,588 km

of the Sun was $\lambda_\odot = 99°47$. Because the obliquity of Mercury very likely is $<$ 1° (Klaasen 1976), most probably 0° (Davies and Batson 1975), the solar latitude $\phi_\odot$ for all Mercury images also is assumed (Image Processing Laboratory 1976) to be zero. The radius of Mercury is estimated from two independent sources to be $R = 2439$ km (Ash et al. 1971; H. T. Howard et al. 1974). Table AI lists the spacecraft range and the various angles, derived from these quantities, that are needed to compute D and d for Crater 10.

The apparent distances between opposing rim crest points were measured on a 131 mm-wide positive film print. Precision of D^*_p (and L^*_p) measurements varied with crater size and image resolution, from 0.1 to 0.01 mm. Focal length f of the Mariner 10 camera was 1.50 m and width of the vidicon was 12.35 mm. Thus, apparent lengths D^* on the surface of Mercury were reduced by a factor of

$$M = \left(\frac{131 \text{ mm}}{12.35 \text{ mm}}\right)\left(\frac{f}{R_a}\right) = 1.171 \times 10^{-6} \qquad (A11)$$

to those on the paper print. Table AII lists the measured apparent lengths D^*_p on the print, their corresponding apparent lengths on Mercury's surface D^*, and the estimated actual crater diameter D, for three measured directions. The three values of D differ by $< 1.5\%$, demonstrating, in this example, the consistency of the measurements and their resulting diameters. Our best estimate for the diameter of Crater 10 is 10.8 km.

TABLE AII
Measured and Calculated Diameters for Crater 10

Measurement Direction	ψ	D^*_p (mm)	D^* (km)	D (km)
East-West	90°	8.35	7.13	10.92
Maximum Foreshortening	ψ_s	7.75	6.62	10.78
No Foreshortening	$\psi_s \pm 90°$	12.60	10.76	10.76

An apparent shadow length L^*_p of 6.5 mm was measured for Crater 10 on the print, which corresponds to $L^* = 5.55$ km on Mercury. From Eq. (A9), crater depth is estimated at 2.07 km. The horizontal extent of the shadow is 5.94 km, and Crater 10 thus satisfies our $(0.45 \leq L/D \leq 0.55)$ condition for location of the shadow tip (5.94 km/10.8 km = 0.55).

APPENDIX B

Catalog of Mercurian Impact Craters[a]

LAT.	LONG.	TB	DIAM.	DEPTH	B	P	F	R	S	T	NB[b]	FDS No.	X	Y	D
							SIMPLE MORPHOLOGY								
+14.6	30.7	CT	12.100	2.200	*						1	0027241	540	510	*
+09.9	27.6	CT	5.890	1.050	*						2	0027453	265	360	
+08.9	31.9	CT	10.740	2.100	*						3	0027241	470	695	*
+08.1	27.2	CT	5.130	1.000	*						4	0027444	190	075	
+06.8	27.0	CT	7.000	1.380							5	0027444	215	290	
+06.0	27.6	CT	8.500	1.630	*						6	0027444	110	405	
+03.8	33.3	SP	6.400	1.320	*						7	0027445	270	645	*
+03.6	32.4	CT	11.700	2.280	*	?					8	0027445	375	645	*
-02.6	28.2	CT	7.900	1.490							9	0027475	300	235	*
-03.6	28.7	CT	10.800	2.070	*						10	0027475	190	570	*
-07.2	28.4	CT	6.910	1.250	*						11	0027474	270	375	*
-08.3	31.0	CT	6.700	1.220	*						12	0027436	295	405	*
-08.3	30.9	CT	8.800	1.660	*						13	0027436	340	405	*
-09.6	30.5	CT	7.700	1.360	*						14	0027436	365	595	*
-11.3	28.4	CT	6.700	1.220	*						15	0027473	360	205	*
-14.1	31.4	CT	8.500	1.690	*						16	0027432	160	185	*
-15.7	30.6	CT	5.900	1.080	*						17	0027432	250	405	*
-18.2	28.6	CT	5.870	0.990	*						18	0027366	080	010	*
-18.6	41.0	CT	9.270	1.900	*						19	0027477	240	460	
-18.7	39.5	CT	6.100	1.310	*						20	0027477	445	450	

APPENDIX B *(Continued)*

LAT.	LONG.	TB	DIAM.	DEPTH	B	P	F	R	S	T	NB^b	FDS No.	X	Y	D
−20.9	29.8	CT	5.100	0.910	*						21	0027471	150	350	*
−23.9	36.0	SP	5.500	1.060	*						22	0027424	050	225	
−24.5	32.0	CT	7.700	1.560	*						23	0027424	370	200	*
−26.0	30.1	CT	9.150	1.790	*						24	0027470	070	320	
−38.9	38.4	CT	9.200	1.860	*						25	0027420	685	310	*
−58.5	50.1	CT	10.840	2.350	*						26	0027399	510	665	
−65.4	66.0	CT	7.700	1.660	*						27	0166680	325	130	
−66.9	98.2	CT	12.460	2.590	*						28	0166740	535	305	*
−67.2	54.8	CT	10.300	1.650	*						29	0166680	705	025	
−67.6	74.7	CT	6.450	1.340	*						30	0166680	310	460	
−68.5	75.6	CT	9.100	1.800	*						31	0166680	360	535	
−72.4	83.2	CT	10.800	1.830	*						32	0166681	490	575	
+73.0	110.7	CT	11.100	1.830	*						33	0000085	580	330	
+70.0	120.0	CT	14.360	2.630	*						34	0000089	395	655	*
+66.5	149.7	SP	7.210	1.170	*						35	0000162	495	650	
+61.3	140.5	CT	4.600	0.960	*						36	0000093	785	665	*
+60.6	139.1	CT	2.600	0.530	*						37	0000093	630	620	*
+60.5	141.0	CT	2.750	0.610	*						38	0000093	730	575	
+60.2	135.5	CT	2.000	0.400	*						39	0000093	415	635	*
+60.2	134.5	CT	2.600	0.540	*						40	0000093	340	665	*
+60.0	137.5	CT	4.000	0.780	*						41	0000093	510	575	*
+59.2	137.4	CT	6.550	1.340	*						42	0000093	415	515	*
+58.1	140.2	CT	4.050	0.790	*						43	0000093	475	340	*
+56.3	141.0	CT	4.050	0.820	*						44	0000093	385	125	
+56.1	146.4	SP	8.130	1.630	*						45	0000093	725	005	*

APPENDIX B (Continued)

LAT.	LONG.	TB	DIAM.	DEPTH	B	P	F	R	S	T	NBᵇ	FDS No.	X	Y	D
+54.4	157.8	SP	9.260	1.810	*						46	0000163	460	480	
+53.15	154.5	CT	7.000	1.440	*						47	0000163	300	445	
+52.9	154.0	CT	4.160	0.820	*						48	0000096	305	505	*
+52.7	147.6	SP	9.500	2.050	*						49	0000159	365	425	*
+51.7	154.5	SP	2.120	0.470	*						50	0000096	270	360	
+51.3	151.0	SP	5.370	1.180	*						51	0000101	560	530	*
+51.1	153.2	SP	4.430	0.860	*						52	0000096	135	300	*
+50.5	154.7	SP	2.580	0.480	*						53	0000096	225	210	*
+50.4	154.7	SP	5.350	1.120	*						54	0000096	225	200	*
+50.3	144.5	CT	4.500	0.950	*						55	0000097	755	095	
+50.0	153.5	SP	2.840	0.560	*						56	0000096	130	205	*
+47.8	160.0	CT	10.710	1.890	*						57	0000163	420	060	*
+47.8	156.8	SP	6.100	1.240	*						58	0000163	270	095	*
+43.5	159.5	SP	3.020	0.560	*						59	0000104	155	670	*
+42.3	160.1	SP	3.010	0.530	*						60	0000104	190	435	*
+41.6	160.1	SP	9.240	2.000	*						61	0000104	145	420	*
+41.4	160.1	SP	6.000	1.130	*						62	0000104	130	395	*
+40.9	160.1	SP	6.400	1.190	*						63	0000104	110	355	*
+34.20	156.9	SP	7.000	1.460	*						64	0000186	615	665	
+31.75	167.12	SP	5.910	1.080	*						65	0000042	810	640	*
+31.7	159.8	SP	7.400	1.520	*						66	0000186	700	490	
+31.32	163.32	SP	0.300	0.060	*						67	0000043	680	680	*
+31.2	166.0	SP	6.600	1.320	*						68	0000190	345	325	*
+31.0	162.0	SP	3.450	0.680	*						69	0000043	030	530	*
+30.95	168.5	SP	1.900	0.350	*						70	0000042	720	140	

APPENDIX B (*Continued*)

LAT.	LONG.	TB	DIAM.	DEPTH	B	P	F	R	S	T	NB[b]	FDS No.	X	Y	D
+30.88	162.06	SP	1.280	0.260	*						71	0000043	290	490	*
+30.82	161.64	SP	1.240	0.270	*						72	0000043	080	480	*
+30.8	162.5	SP	0.900	0.180	*						73	0000043	440	410	*
+30.75	162.75	SP	0.800	0.130	*						74	0000043	120	395	
+30.75	161.45	SP	0.380	0.076	*						75	0000043	035	425	*
+30.75	162.5	SP	3.400	0.630	*						76	0000043	535	325	
+30.5	162.1	SP	0.280	0.060	*						77	0000043	180	210	*
+30.4	162.9	SP	0.225	0.044	*						78	0000043	530	070	*
+30.4	157.8	SP	10.000	2.220	*						79	0000186	600	425	
+30.3	164.1	SP	9.100	1.870	*						80	0000190	245	250	*
+30.27	162.7	SP	1.430	0.300	*						81	0000043	420	030	*
+30.25	162.7	SP	1.390	0.280	*						82	0000043	425	070	*
+30.1	164.9	SP	8.900	1.740	*						83	0000108	650	475	*
+29.6	165.3	SP	2.900	0.550	*						84	0000108	735	370	*
+29.5	165.2	SP	4.500	0.870	*						85	0000108	710	350	*
+29.1	162.1	SP	4.800	1.020	*						86	0000108	380	360	*
+28.7	165.9	SP	6.350	1.270	*						87	0000108	725	140	*
+28.1	162.3	SP	5.050	1.060	*						88	0000108	380	240	*
+28.0	162.4	SP	8.000	1.770	*						89	0000108	410	220	*
+27.7	165.7	SP	6.900	1.400	*						90	0000191	330	600	*
+27.6	163.5	SP	8.000	1.660	*						91	0000108	525	145	*
+27.0	168.4	SP	5.860	1.060	*						92	0000073	055	390	*
+26.7	160.0	SP	7.220	1.550	*						93	0000108	110	120	
+25.8	171.3	SP	2.540	0.440	*						94	0000072	750	630	
+25.3	168.6	SP	5.000	0.950	*						95	0000072	245	625	*

LAT.	LONG.	TB	DIAM.	DEPTH	B	P	F	R	S	T	NB[b]	FDS No.	X	Y	D
+25.0	155.0	SP	4.650	1.120	*						96	0000047	510	080	
+24.7	163.9	SP	7.260	1.630	*						97	0000191	220	410	*
+24.2	167.8	SP	9.990	1.790	*						98	0000191	405	350	*
+24.2	168.2	SP	4.600	0.860	*						99	0000072	040	310	*
+22.8	168.1	SP	3.950	0.730	*						100	0000071	180	580	*
+22.1	156.2	SP	1.840	0.400	*						101	0000048	675	170	
+20.9	170.4	SP	9.000	1.780	*		*				102	0000071	700	150	
+19.2	155.4	SP	3.900	0.870	*						103	0000049	240	260	
+19.1	155.5	SP	4.350	0.920	*						104	0000049	290	190	
+17.0	167.4	CT	3.440	0.720	*						105	0000069	200	510	*
+15.0	171.0	SP	11.300	2.090	*						106	0000226	330	500	
+14.4	166.8	SP	6.730	1.420	*						107	0000068	090	580	*
+14.4	169.9	SP	5.000	0.960	*						108	0000068	790	435	
+13.7	159.0	SP	4.520	1.020	*						109	0000051	780	670	
+13.3	171.7	SP	7.600	1.380	*						110	0000116	160	565	
+13.0	156.5	SP	2.200	0.460	*						111	0000051	060	540	
+12.9	159.2	CT	2.800	0.600	*						112	0000051	760	260	
+12.8	167.3	CT	5.800	1.170	*						113	0000068	180	110	*
+12.7	168.7	SP	6.610	1.300	*						114	0000068	540	055	*
+11.5	164.0	CT	9.000	2.070	*						115	0000117	675	605	*
+10.7	164.2	CT	6.800	1.450	*						116	0000117	705	505	*
+10.7	169.5	SP	6.180	1.200	*						117	0000067	760	655	*
+10.6	164.8	SP	4.600	0.910	*						118	0000117	760	490	
+10.4	159.9	CT	1.800	0.370	*						119	0000052	655	440	
+10.3	174.4	SP	8.300	1.380	*						120	0000116	490	200	

APPENDIX B (Continued)

LAT.	LONG.	TB	DIAM.	DEPTH	B	P	F	R	S	T	NBᵇ	FDS No.	X	Y	D
+10.3	168.1	SP	4.120	0.830	*						121	0000067	415	255	*
+10.0	158.0	CT	6.600	1.440	*						122	0000052	045	445	
+09.2	158.5	SP	8.800	1.780	*		?				123	0000052	190	040	
+09.2	159.3	SP	4.860	1.040	*						124	0000052	490	045	
+07.7	167.0	SP	3.700	0.750	*						125	0000066	100	540	*
+07.5	162.0	SP	1.130	0.220	*						126	0000053	810	530	
+07.4	169.0	SP	6.430	1.290	*						127	0000066	570	280	*
+07.4	169.4	SP	3.390	0.620	*						128	0000066	750	290	*
+07.0	161.0	SP	5.900	1.240	*						129	0000053	500	380	
+06.7	167.0	SP	1.700	0.330	*						130	0000066	065	095	*
+06.7	169.3	SP	9.540	1.920	*						131	0000066	670	100	*
+05.8	169.6	SP	1.820	0.360	*						132	0000065	720	590	*
+05.7	168.7	SP	4.860	0.950	*						133	0000065	280	630	*
+05.7	169.1	SP	3.110	0.610	*						134	0000065	555	650	*
+05.4	168.6	SP	6.420	1.260	*						135	0000065	390	640	*
+04.4	163.7	SP	1.090	0.210	*						136	0000054	645	640	
+04.2	162.7	CT	8.220	1.910	*	?	?				137	0000054	270	560	
+03.8	163.0	CT	1.330	0.270	*						138	0000054	350	440	
+03.8	168.9	CT	2.290	0.450	*						139	0000065	420	220	*
+03.5	163.6	CT	5.320	1.200	*						140	0000054	570	290	*
+02.9	172.0	SP	10.000	1.730	*						141	0000120	260	620	*
+00.7	165.9	CT	1.040	0.210	*						142	0000055	290	600	*
+00.6	165.6	CT	8.000	1.690	*						143	0000055	245	585	*
+00.4	165.8	CT	1.170	0.230	*						144	0000055	285	515	
+00.1	165.4	CT	1.950	0.400	*						145	0000055	115	380	*

APPENDIX B *(Continued)*

LAT.	LONG.	TB	DIAM.	DEPTH	B	P	F	R	S	T	NB[b]	FDS No.	X	Y	D
−00.2	166.8	CT	1.430	0.300	*						146	0000055	630	290	*
−00.3	166.8	CT	6.880	1.360	*						147	0000055	650	170	*
−00.6	170.8	CT	4.050	0.780	*						148	0000063	535	680	
−00.7	167.4	SP	5.320	1.080	*						149	0000055	830	055	*
−01.1	169.2	CT	5.600	1.230	*						150	0000063	055	605	
−03.4	169.4	CT	11.400	2.540	*						151	0000056	330	525	
−03.6	169.5	CT	1.210	0.230	*						152	0000056	310	675	*
−04.0	168.8	CT	1.730	0.330	*						153	0000056	135	385	*
−04.5	170.4	CT	2.700	0.530	*						154	0000056	650	230	
−04.5	169.6	CT	4.600	0.880	*						155	0000056	395	245	*
−04.8	170.6	SP	1.490	0.250	*						156	0000056	780	055	*
−05.0	170.5	SP	3.200	0.630	*						157	0000061	125	585	
−05.0	172.1	CT	3.700	0.670	*						158	0000061	515	635	
−05.1	172.1	CT	2.370	0.440	*						159	0000061	515	550	
−05.6	170.9	SP	3.350	0.590	*						160	0000062	310	550	
−05.8	172.9	SP	10.100	1.740	*						161	0000061	785	415	
−07.0	168.4	CT	12.400	2.440	*						162	0000244	465	350	*
−08.2	173.4	CT	6.750	1.120	*						163	0000058	665	440	
−08.3	172.7	CT	4.270	0.710	*						164	0000058	360	405	
−09.0	167.9	CT	8.000	1.450	*						165	0000125	465	620	*
−09.5	172.5	CT	3.260	0.580	*						166	0000058	290	135	*
−10.2	167.3	CT	7.600	1.670	*						167	0000125	410	500	*
−12.0	166.2	CT	9.700	1.850	*						168	0000125	290	370	*
−12.2	167.3	CT	11.000	2.210	*						169	0000125	410	135	*
−14.8	166.7	SP	7.700	1.400	*						170	0000125	360	135	*

APPENDIX B (Continued)

LAT.	LONG.	TB	DIAM.	DEPTH	B	P	F	R	S	T	NB[b]	FDS No.	X	Y
MODIFIED SIMPLE MORPHOLOGY														
+31.1	30.9	SP	10.200	1.670	?						171	0027325	680	260
−11.5	28.3	CT	7.650	1.370	?						172	0027473	375	295
−11.9	28.3	CT	6.600	1.120	?						173	0027473	350	420
−11.9	29.7	CT	7.500	1.270	?			?	?		174	0027473	090	440
−21.3	28.5	CT	10.000	1.540	?			?	*		175	0027471	370	270
−25.6	30.0	CT	10.750	1.800		?	?	*	*		176	0027470	060	190
+57.2	145.6	SP	7.500	1.240	*	?			?		177	0000093	760	115
+42.3	161.0	SP	12.730	2.030		?	*	?	*		178	0000104	230	500
+10.6	160.0	SP	7.600	1.290			?	*	*		179	0000104	080	310
+31.9	171.1	SP	7.440	1.140			*	*	*		180	0000075	420	030
+31.6	165.1	SP	7.300	1.200			*	*	*		181	0000108	760	635
+31.3	163.0	SP	12.100	1.930			*	*	*		182	0000108	540	620
+29.6	171.6	SP	4.630	0.690			*	?	?		183	0000074	635	215
+24.6	168.3	SP	9.200	1.460		?	*	?	?		184	0000072	060	400
+22.8	154.5	SP	8.380	1.370	*			?	?		185	0000048	290	565
+19.0	167.0	SP	7.150	1.000			*	*	*	*	186	0000070	075	325
−06.0	172.5	SP	9.700	1.400		?	?	?	?		187	0000061	650	390
−63.2	123.8	CT	12.200	2.050	*		?	?	?		188	0166742	360	415
−66.4	130.7	CT	11.000	1.780		?	?	*	?	*	189	0166742	730	520
IMMATURE COMPLEX MORPHOLOGY														
+75.0	41.8	CT	26.000	1.720	?	?	?	?	?	?	190	0000160	420	640
+73.0	99.5	SP	25.000	2.370	*	*	*	*	*	*	191	0000085	440	435

APPENDIX B (*Continued*)

LAT.	LONG.	TB	DIAM.	DEPTH	B	P	F	R	S	T	NBᵇ	FDS No.	X	Y
+71.5	79.4	SP	24.000	1.710	*	*	*	*	*	*	192	0000085	150	680
+13.6	18.6	CT	20.100	1.400	*	*	*	*	*		193	0027456	300	495
+13.5	19.8	CT	25.700	1.640	*	*	*	*	*	*	194	0027456	100	500
+13.4	25.1	CT	12.160	1.250	*	*	*	*	*		195	0027457	110	520
+10.5	27.2	CT	18.000	2.050	*	*	?	*	*		196	0027453	355	225
+10.0	25.4	CT	10.400	1.190			*	*	*		197	0027458	070	150
+09.9	23.2	CT	14.980	1.330	*	*	*	*	*		198	0027458	435	135
+07.0	22.3	CT	10.000	1.190		?	*	*	*		199	0027444	780	295
+05.9	21.3	CT	14.500	1.360	?	?	*	*	*		200	0027443	045	560
+05.1	23.7	CT	11.200	1.370	?	?	?	*	*		201	0027444	585	480
+04.0	20.2	CT	13.600	1.430			?	*	*		202	0027448	280	460
+03.7	22.8	CT	12.400	1.210	*	*	*	*	*		203	0027444	660	670
−00.4	20.2	CT	19.760	2.170		*	*	*	*		204	0027459	360	300
−02.1	20.8	CT	25.500	2.150	?	?	*	*	*	*	205	0027459	185	655
−02.1	28.8	CT	15.000	1.860			*	*	*		206	0027475	260	140
−09.3	22.0	SP	13.800	1.350			*	*	*		207	0027435	300	350
−12.2	26.6	CT	20.000	1.820		?	*	*	*		208	0027473	750	420
−17.2	28.0	CT	14.000	1.250		*	*	*	*		209	0027472	420	560
−22.1	30.4	CT	12.800	1.510		?	*	*	*		210	0027471	075	575
−23.4	27.0	CT	23.100	2.250		*	*	*	*		211	0027428	585	550
−24.0	17.4	CT	21.000	1.790	*	*	?	*	*	?	212	0027462	320	370
−27.2	29.0	CT	10.600	1.290			*	*	*		213	0027470	270	600
−27.6	28.4	CT	17.100	1.550		*	*	*	*	?	214	0027470	450	600
−31.0	27.6	SP	12.000	1.710	?	?	?	*	*		215	0027469	510	100
−45.7	18.1	CT	27.600	2.000	*	*	?	*	*	*	216	0027465	530	470

APPENDIX B   (Continued)

LAT.	LONG.	TB	DIAM.	DEPTH	B	P	F	R	S	T	NB[b]	FDS No.	X	Y
−45.9	23.7	CT	28.300	2.490		*	?	*	*	*	217	0027419	735	550
−46.0	25.9	CT	24.400	1.790		*	?	*	*	?	218	0027419	575	600
+69.7	153.0	CT	13.040	1.270		*	*	*	*		219	0000088	360	245
+68.4	133.4	CT	21.080	2.300		*	*	*	*	?	220	0000089	650	350
+67.0	148.2	CT	21.000	1.770		*	*	*	*	?	221	0000162	480	680
+61.7	154.0	CT	10.420	1.370			*	*	*		222	0000092	125	510
+59.4	159.4	CT	14.500	2.040		?	?	*	*		223	0000092	375	145
+57.4	144.3	SP	11.500	1.390		*	*	*	*		224	0000093	700	155
+53.7	153.1	CT	15.360	1.650		*	*	*	*		225	0000096	310	610
+52.1	153.5	SP	9.500	1.240		*	*	*	*		226	0000096	240	405
+44.4	174.6	SP	26.850	1.730		*	*	*	*	?	227	0000078	550	300
+30.6	168.2	SP	12.000	1.440		?	*	*	*		228	0000074	150	590
+23.6	179.8	SP	29.100	2.450		*	*	*	*	?	229	0000126	220	250
+22.7	177.4	SP	17.200	1.470		*	*	*	*		230	0000111	340	650
+20.5	169.8	SP	14.500	1.460		*	?	*	*		231	0000112	490	460
+20.3	175.6	SP	10.420	1.240		?	*	?	?	?	232	0000111	330	390
+19.3	173.9	SP	20.910	1.530		*	*	*	*		233	0000191	235	000
+19.2	179.0	CT	23.200	1.690		*	*	*	*		234	0000111	515	190
+18.8	167.7	SP	12.000	1.330		*	*	*	*		235	0000070	240	250
+18.0	175.4	SP	11.000	1.120		?	*	*	?		236	0000111	080	075
+15.8	169.0	SP	15.000	1.820		*	*	*	*		237	0000069	530	120
+14.1	175.2	SP	13.500	1.640		?	?	*	*		238	0000116	635	650
+14.0	181.5	SP	19.690	1.630		?	?	*	*		239	0000115	085	520
+13.9	168.4	SP	14.600	1.950		*	*	*	*		240	0000068	470	350
+13.8	173.4	SP	11.400	1.280		?	?	*	*		241	0000116	440	630

APPENDIX B (Continued)

LAT.	LONG.	TB	DIAM.	DEPTH	B	P	F	R	S	T	NBᵇ	FDS No.	X	Y
+11.8	172.5	SP	15.200	1.600	?	*		*	*		242	0000116	310	395
+11.2	171.5	SP	12.500	1.480		*		*	*		243	0000116	185	360
+10.3	176.1	SP	28.500	2.400	*			*	*	*	244	0000116	720	200
+09.5	168.1	SP	10.700	1.360		*	*	*	*		245	0000067	390	090
+06.3	159.9	SP	18.300	2.270	*	*	*	*	*		246	0000053	225	100
+00.8	174.9	SP	16.500	1.490	*	*	*	*	*		247	0000120	615	405
+00.5	174.5	SP	14.200	1.330	*	*	*	*	*		248	0000120	560	390
+00.4	170.6	SP	12.800	1.700		?	?		*		249	0000064	640	140
−04.9	179.0	SP	26.500	1.770	*	*	*	?	*	*	250	0000231	220	635
−10.3	175.0	CT	24.800	2.130	*	*	*	*	*	*	251	0000124	130	480
−10.9	176.0	CT	22.000	1.740	*	*	*	*	*	*	252	0000124	220	410
−17.9	177.0	SP	22.500	1.790	?	?	?	*	*	?	253	0000231	100	105
−66.6	132.4	CT	12.200	1.220	?	?	?		*		254	0166742	755	580
−67.6	146.3	CT	13.000	1.590	?	?	?	?			255	0166749	575	220
−69.8	148.8	CT	21.000	1.580				*	*		256	0166749	735	150
−72.3	146.4	CT	17.000	1.630				*	*		257	0166748	750	485
−72.3	133.4	CT	18.000	1.750					*		258	0166748	610	190

MATURE COMPLEX MORPHOLOGY

LAT.	LONG.	TB	DIAM.	DEPTH	B	P	F	R	S	T	NBᵇ	FDS No.	X	Y
+81.0	88.2	CT	44.400	2.530	*	*	*	*	*	*	259	0000156	730	540
+72.9	55.5	SP	61.000	3.100	*	*	*	*	*	*	260	0000156	375	685
+51.1	15.0	CT	59.500	1.950	*	?	?	?	?	?	261	0027377	420	080
+21.5	11.4	CT	148.000	3.690	*	*	*	*	*	*	262	0027323	265	550
+15.2	22.6	CT	30.000	1.860	*	*	*	*	*	*	263	0027457	660	160

LAT.	LONG.	TB	DIAM.	DEPTH	B	P	F	R	S	T	NB^b	FDS No.	X	Y
+09.1	17.7	CT	35.000	1.810		*	*	*	*	*	264	0027452	720	285
+05.6	15.0	CT	61.000	2.440		*	*	*	*	*	265	0027447	210	250
+02.7	14.9	CT	88.000	3.910		*	*	*	*	*	266	0027357	350	140
+01.0	16.2	CT	48.100	1.850		*	*	*	*	*	267	0027438	645	120
+00.02	17.3	CT	76.500	2.650		*	*	*	*	*	268	0027438	525	245
−09.7	13.1	CT	66.500	2.300		*	*	*	*	*	269	0027361	505	165
−13.1	14.2	CT	62.000	2.490		*	*	*	*	*	270	0027361	480	410
−23.0	20.9	CT	33.900	2.080		*	*	*	*	*	271	0027427	100	525
−25.8	13.8	CT	45.500	1.910		*	*	*	*	?	272	0027299	590	055
−44.4	14.0	CT	75.000	2.370		*	*	*	*	*	273	0027393	730	030
−51.8	20.0	CT	36.000	2.130		*	*	*	*	*	274	0027466	460	320
−54.5	21.2	CT	42.200	2.430		*	*	*	*	*	275	0027397	360	230
−58.6	24.8	CT	32.000	1.570		*	*	*	*	*	276	0027397	225	480
−61.5	24.8	CT	31.000	1.560		*	*	*	*	*	277	0027398	735	405
−63.0	24.0	CT	86.000	3.890		*	*	*	*	*	278	0027401	170	150
−64.1	37.0	CT	53.000	3.310		*	*	*	*	*	279	0027402	320	240
−65.7	34.2	CT	70.700	3.770		*	*	*	*	*	280	0027402	490	330
−75.9	50.1	SP?	31.400	1.900		*	*	*	*	*	281	0166687	490	380
−78.6	60.8	CT	62.200	2.480		*	*	*	*	*	282	0027405	215	550
−79.4	87.5	CT	41.000	2.540		*	*	*	*	?	283	0166752	370	090
−80.6	96.2	CT	51.000	2.600		*	*	*	*	*	284	0166752	460	225
−81.0	41.5	CT	160.400	4.000		*	*	*	*	*	285	0027405	630	530
−81.3	77.0	CT	36.000	2.410		*	*	*	*	*	286	0166689	370	100
−82.2	74.0	CT	39.000	2.300		*	*	*	*	*	287	0166689	440	125
+80.4	120.0	CT	35.000	2.000		*	*	*	*	*	288	0000156	780	410

LAT.	LONG.	TB	DIAM.	DEPTH	B	P	F	R	S	T	NB[b]	FDS No.	X	Y
+77.7	140.0	CT	40.100	1.970	*	*	*	*	*	*	289	0000084	500	550
+76.3	156.0	CT	34.000	2.050	*	*	*	*	*	?	290	0000165	275	335
+72.9	154.5	CT	32.400	1.750	*	*	*	*	*	*	291	0000084	560	040
+69.0	175.7	CT	53.300	2.550	*	*	*	*	*	*	292	0000166	555	570
+58.7	174.2	SP	97.200	3.150	*	*	*	*	*	*	293	0000080	340	540
+57.3	170.4	SP	32.800	2.310	*	*	*	*	*	*	294	0000167	380	575
+49.8	176.3	SP	64.900	2.560	*	*	*	*	*	*	295	0000079	700	180
+48.8	179.1	SP	3.880	2.800	*	*	*	*	*	*	296	0000098	385	435
+43.8	176.9	CT	41.500	2.700	*	*	*	*	*	*	297	0000103	420	480
+43.0	175.0	CT	34.800	2.090	*	*	*	*	*	*	298	0000103	220	370
+42.8	178.0	CT	66.650	3.350	*	*	*	*	*	*	299	0000103	560	340
+36.3	176.7	CT	49.000	2.980	*	*	*	*	*	*	300	0000194	380	410
+32.1	175.3	CT	35.000	2.330	*	*	*	*	*	*	301	0000107	720	540
+31.0	174.6	CT	75.000	3.520	*	*	*	*	*	*	302	0000107	690	350
+25.4	179.3	SP	55.700	2.640	*	*	*	*	*	*	303	0000126	230	490
+11.5	181.2	SP	49.000	2.270	*	*	*	*	*	*	304	0000115	085	215
+00.9	181.0	CT	39.000	2.500	*	*	*	*	*	*	305	0000119	060	360
+00.5	176.9	SP	31.000	1.680	*	*	*	*	*	*	306	0000120	000	350
−04.5	179.4	SP	45.000	2.470	*	*	*	*	*	*	307	0000231	270	660
−35.8	182.0	CT	100.200	3.330	*	*	*	*	*	*	308	0000310	465	240
−39.9	180.7	CT	156.000	4.430	?	*	*	*	*	*	309	0000232	325	640
−40.8	178.3	CT	69.600	2.270	*	*	*	*	*	*	310	0000232	200	625
−45.9	183.3	CT	74.700	3.600	*	?	?	?	?	?	311	0000310	415	075
−53.5	177.5	CT	113.000	3.690	*	*	*	*	*	*	312	0000232	140	410
−72.2	151.3	CT	77.000	2.560	*	*	*	*	*	*	313	0166755	325	300

APPENDIX B (Continued)

LAT.	LONG.	TB	DIAM.	DEPTH	B	P	F	R	S	T	NB[b]	FDS No.	X	Y
−75.4	161.6	CT	66.140	3.200		*	*	*	*	*	314	0166754	580	660
−78.1	143.2	CT	50.000	2.680		*	*	*	*	*	315	0166754	485	245
−78.0	159.0	CT	129.500	4.780		?	*	*	*	*	316	0166754	725	385

SIMPLE MORPHOLOGY (UNDIFFERENTIATED)

LAT.	LONG.	TB	DIAM.	HEIGHT	B	P	F	R	S	T	NB[b]	FDS No.	X	Y
+86.7	90.0		9.300	0.400	?						317	0000164	315	445
+85.3	128.4		11.100	0.460	?						318	0000164	290	365
+85.2	133.5		8.350	0.350	?						319	0000164	300	355
+76.8	168.2		6.200	0.240	?						320	0000164	395	030
+67.8	183.1		9.100	0.440	?						321	0000166	690	505
+63.1	184.7		2.500	0.195	?						322	0000090	125	325
+62.6	182.9		2.380	0.058	?						323	0000081	665	010
+62.4	185.3		5.080	0.300	?						324	0000090	165	365
+59.5	183.8		4.900	0.300	?						325	0000166	775	065
+42.8	185.6		10.500	0.350				?			326	0000102	325	225
+40.1	184.9		5.100	0.460	?						327	0000197	275	020
+39.3	186.3		8.500	0.460	?						328	0000198	380	485
+32.8	185.2		10.000	0.410	?						329	0000198	370	180
+29.9	186.6		5.200	0.260	?						330	0000199	495	550
+29.0	184.7		6.000	0.410	?						331	0000199	415	470
+28.9	186.5		12.000	0.620	?						332	0000199	495	480
+28.1	186.2		8.600	0.340	?						333	0000199	485	420
+26.5	185.6		6.800	0.190	?						334	0000126	705	595
+25.5	186.4		5.400	0.280	?						335	0000126	780	465
+23.7	185.8		6.000	0.250	?						336	0000126	720	245

APPENDIX B (*Continued*)

LAT.	LONG.	TB	DIAM.	HEIGHT	B	P	F	R	S	T	NB[b]	FDS No.	X	Y
+00.9	186.9		11.600	0.580	?						337	0000119	595	350
−00.5	186.8		3.400	0.210	?						338	0000119	565	190
−00.9	187.0		3.600	0.190	?						339	0000119	600	160
−01.2	186.4		10.300	0.530							340	0000119	535	120
−02.0	186.7		4.000	0.310	?						341	0000119	565	045
−08.7	186.0		4.100	0.170	?						342	0000123	280	620
−08.9	185.4		7.000	0.190	?						343	0000123	200	615
−09.7	187.8		8.540	0.150	?			?			344	0000123	430	530
−09.9	187.8		2.600	0.047	?						345	0000123	440	510
−10.9	186.2		5.500	0.400	?						346	0000123	295	430
−11.2	188.2		9.400	0.370							347	0000123	445	405
−15.5	187.2		6.190	0.370	?						348	0000123	330	050

COMPLEX MORPHOLOGY (UNDIFFERENTIATED)

LAT.	LONG.	TB	DIAM.	HEIGHT	B	P	F	R	S	T	NB[b]	FDS No.	X	Y
+84.8	24.5		35.000	0.710							349	0000164	340	545
+81.3	35.5		25.200	0.940							350	0000164	165	590
+75.5	24.4		31.000	0.590							351	0000164	055	665
+86.6	122.2		25.700	0.700							352	0000164	340	410
+83.8	150.0		32.000	0.740		*					353	0000164	310	305
+79.4	154.8		36.000	1.050		*					354	0000164	225	160
+64.4	182.1		14.000	0.270					?		355	0000081	490	275
+60.2	186.3		21.500	0.520				?			356	0000090	230	075
+54.1	179.1		13.100	0.540		?					357	0000094	325	620
+51.0	183.0		39.900	0.720		*				*	358	0000094	560	510
+46.1	188.3		43.000	0.770				?			359	0000197	370	350
+45.9	184.9		36.500	1.160							360	0000102	285	640
+44.4	186.4		37.000	0.780							361	0000102	335	450
+41.2	186.8		18.400	0.690							362	0000198	400	590
+38.2	186.2		22.150	0.690							363	0000198	370	415

APPENDIX B (Continued)

LAT.	LONG.	TB	DIAM.	HEIGHT	B	P	F	R	S	T	NB	FDS No.	X	Y
+31.8	182.2		16.500	0.790				?			364	0000199	290	655
+31.2	185.7		13.300	0.590	?						365	0000199	455	610
+31.0	185.0		13.500	0.440	?						366	0000199	420	595
+31.0	184.0		15.500	0.530	?			?			367	0000199	375	590
+26.5	184.4		32.000	0.980		*					368	0000126	580	575
+24.4	186.0		28.150	0.930				?			369	0000126	710	320
+24.3	184.2		20.000	0.780				?			370	0000126	555	320
+23.8	184.3		18.000	0.680				?			371	0000126	570	275
+23.2	185.4		17.500	0.750				?			372	0000126	675	180
-10.9	184.0		13.000	0.560							373	0000123	090	420

[a]Definitions of column headings:

LAT. Latitude of point on crater rim casting shadow, in degrees north or south of equator;

LONG. Longitude of point on crater rim casting shadow, in degrees west of prime meridian;

TB Topographic background; CT, cratered terrain; SP, smooth plains;

DIAM. Crater diameter, in km; mean derived from three measurements;

DEPTH Crater depth, in km; maximum for bowl-shaped interiors, mean for flat floors (replaced by HEIGHT in last part of listing).

HEIGHT Height of crater rim, in km, above surrounding topography.

B Bowl-shaped interior? (Presence denoted by asterisk, *; uncertain observations queried: ?).

P Central peak?   "

F Flat floor?   "

R Scalloped rim crest?   "

S "Slump block(s)"?   "

T Terraced wall?

NB Serial number, for reference.

FDS No. Flight data subsystem number, uniquely identifying Mariner 10 image frames.

X Pixel number of crater center on image, horizontal direction, from left to right.

Y Pixel number of crater center on image, vertical direction, from top to bottom.

D Simple crater with shadow angle between 18° and 26°, and a shadow length/rim diameter ratio between 0.45 and 0.55 (denoted by *).

[b]Crater names: 192: Myron; 259: Despréz; 260: Tung Yüan; 262: Melville; 266: Donne; 269: Dvořák; 273: Rilke; 278: Tsurayuki; 282: Sadi; 285: Boccaccio; 292: Martial; 293: Brahms; 295: Zola; 299: Nervo; 302: March; 309: Liang K'ai; 312: Dowland; 313: Dickens; 314: Martí; 316: Ictinus.

CRATERING ON MERCURY: A RELOOK

PETER H. SCHULTZ
Brown University

Impact craters on Mars and the icy satellites exhibit significant differences in shape, morphology and collective statistics from craters on the Moon. Contrasting crustal compositions (water/ice) provide one plausible explanation for such differences. But only Mercury permits calibrating the effects of impact velocity and gravity on cratering in a silicate crust depleted of volatiles. Different models of crater modification may account for the observed contrasts in crater morphologies on volatile-rich and volatile-poor planets and satellites. Results from laboratory experiments, however, indicate that crater shapes intrinsically become flatter as the time for energy/momentum transfer increases, provided that a critical transfer time is exceeded. If expressed in terms of crater diameter, transient crater aspect ratios (diameter/depth) should be constant up to a transition diameter dependent on $v^{1.17}g^{-1}$, above which craters become flatter with increasing size. These results predict that shallower craters on Mars relative to Mercury at the same diameter may reflect the low rms impact velocities at Mars, rather than the effects of buried volatiles. Changes in the size-frequency distributions of craters on a given planet through time, therefore, need not indicate changes in the size distribution of the impactors, but only a change in the velocity distribution. It is further suggested that elements of transient excavation craters produced by large, low-velocity impactors may be preserved even after wall collapse and floor uplift, in contrast with the same size craters produced by small, high-velocity impactors. One possible element includes the depressions within the central peaks of craters that increase in relative frequency from Mercury, to the Moon, to Mars. This different perspective may permit deriving information about planetary bombardment history and demonstrates the importance of the Mercurian cratering record for distinguishing the effects of contrasting combinations of impactors and targets.

I. INTRODUCTION

The heavily cratered and barren surface of Mercury contrasts with the complex geologic histories of Mars and Venus or the exotic histories of the satellites of Jupiter, Saturn and Uranus. Yet, Mercury provides a unique planetary-scale laboratory for understanding impact processes with surface gravity similar to Mars but a Moon-like vacuum environment, a volatile-free crust, and impact velocities exceeding 80 km s^{-1}. Because the surface of Mercury preserves a record of the inner solar system which is largely re-processed or hidden on the Earth and Venus, it holds fundamental clues for understanding impact mechanics.

Prior to the Mariner 10 encounters with Mercury in 1974, planetary scientists had a detailed look at cratering processes on the Moon through Apollo and a somewhat blurry-eyed view of Mars from Mariner 9. Since this encounter, an Apollo Synthesis program provided a detailed multidisciplinary analysis of the Moon; three international colloquia have focused on the results of the Viking mission to Mars; Pioneer-Venus and Venera have exposed the surface of Venus; and Voyager has recorded the cratering history of the outer solar system. Just as importantly, perspectives on impact cratering processes have been broadened by sophisticated applications of finite-element codes, the establishment of the NASA-Ames Vertical Gun Range as a national facility, provocative evidence for links between impacts and life extinctions as well as impacts and material launched from other planets, and synergistic conferences (e.g., Impact and Explosion Cratering, 1976, Flagstaff; Comparisons of the Moon and Mercury, 1977, Houston; Multi-Ring Impact Basins, 1980, Houston; Origin of the Moon, 1984, Hawaii; and the Consequences of Large-Body Impacts on the Earth, 1981, Snowbird, Utah). With the changes in interest, the significance of the Mercurian cratering record has been largely neglected but not lessened.

The following discussion re-examines the cratering record of Mercury in terms of a post-Mariner 10 view of impact processes. Rather than restating and judging past conflicts, this review attempts to place them in the context of different cratering models that are often implicitly accepted and defended. As a result, considerable emphasis is placed on cratering phenomenology that might be expressed on Mercury and other planets. Results from small-scale impact experiments further provide a new view of planetary-scale cratering processes that could not have been tested without the laboratory of Mercury.

The cratering process can be separated into three stages: compression, excavation and modification (see Gault et al. 1968). Different approaches must be used in order to resolve the effects of each stage on the final crater morphology. For example, laboratory experiments typically use sand with low shear strength (and cohesion) in order to explore the excavation process. The use of sand is justified because passage of the impact shock wave "preconditions" the target from competent rock to a strengthless pulverized state. Sand provides a questionable model, however, for simulating early-time processes due to possible complicating effects of porosity including grain disruption and pore space collapse. Moreover, laboratory impact velocities are generally insufficient to permit examining the effects of internal energy losses by shock-induced vaporization and melting. A sand target also poorly simulates the post-impact modification stage of large-scale cratering. Such shortcomings do not render laboratory experiments useless, just as unknown material properties under high pressure or large finite-element grid sizes do not render numerical codes useless. This preamble is necessary because the interpretations of impact crater morphology often incorporate subtle but major assumptions about each stage of crater formation. Some of these assumptions seem unsupported in the context of our current understanding, yet often contain an element of unexplored truth.

The following discussion is divided into several sections. First, the efficiency of impact cratering at laboratory and planetary scales is compared (Sec. II). Second, a holistic view of crater growth and scaling is reviewed in order to place conflicting theories in perspective (Sec. III). An overview of recent experimental data is included and provides a basis for more in-depth discussion of the cratering process. Possible observable contrasts among craters on different planets are explored in terms of crater formation and modification in Sec. IV. Next, comparison of crater morphology and statistics on different planets illustrates the importance of Mercury for resolving fundamental aspects of the cratering process (Sec. V). And last, in Sec. VI, the effects of higher-impact velocities and the increased gravity field on ejecta emplacement on Mercury relative to the Moon are reviewed. This multi-perspective approach converges on a somewhat different view of impact cratering processes and planetary bombardment history in general.

II. CRATER EXCAVATION

Cratering Efficiency

A complete description of cratering efficiency should parallel the approach of Gault et al. (1975) where the kinetic energy available for crater excavation is reduced by energy loss for heating, comminution, plastic and/or viscous deformation, gravity, and seismic effects. Empirical scaling relations for impacts into low-strength targets presumably incorporate to first order the effects of deformation, gravity and residual seismic energy. Consequently, another approach is to express cratering efficiency as an empirically derived relation proposed by Holsapple and Schmidt (1982) but modified to include effects not fully simulated at laboratory scales by multiplying cratering efficiency by a series of functions defined to be dimensionless:

$$\left(\frac{M}{m}\right)_P = \left(\frac{M}{m}\right)_L f_i(r,\mathrm{v},\theta)\, f(r/R)$$

$$= k\pi_2^{-\alpha}\, f_i(r,\mathrm{v},\theta)\, f(r/R) \tag{1}$$

where $(M/m)_p$ is cratering efficiency at planetary scales; $(M/m)_L$ is the efficiency observed in the laboratory; $f_i(r,v,\theta)$ includes a variety of correction parameters dependent on impactor size r, velocity v, and impact angle θ; $f(r/R)$ provides correction for very large projectiles relative to the planet radius R; k is a constant incorporating dimensional terms and π_2 is the dimensionless gravity-scaled parameter incorporating the effects of momentum and energy (Holsapple and Schmidt 1982);

$$\pi_2 = 3.22 \, \frac{g \, r}{v^2} \tag{2}$$

where g is the gravitational acceleration. The correction factor $f_i(r,v,\theta)$ can be expressed as:

$$f_i(r,v,\theta) = f_\theta(r,v) \, f_m(r,v,\theta) \, f_v(r,v,\theta) \, f_c(r,v,\theta) \, f(r/R) \tag{3}$$

where $f(r,v)$ are dimensionless partitioning functions incorporating the effects of impact angle f_θ, melting f_m, vaporization f_v, and comminution f_c.

Figure 1 illustrates how well the dimensionless π_2 parameter appears to accommodate the data for a given impact angle and target. Subtle but significant departures from a single power-law relation, however, occur if a single

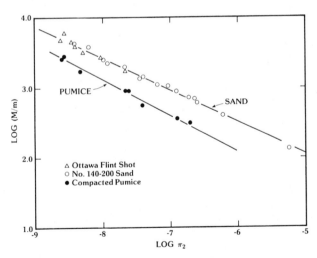

Fig. 1. Cratering efficiency (defined as the ratio of the displaced crater mass M to the projectile mass m) as a function of the dimensionless scaling parameter, π_2, given by Eq. (2). Impacts by aluminum and lexan projectiles into compacted pumice and sand targets with velocities from less than 200 m s^{-1} to over 6 km s^{-1} appear to follow a single power-law relation for a given target as first proposed by Schmidt and Holsapple (1980). The difference between sand and pumice targets reflects in part the difference in target cohesion and internal angle of friction (figure from Schultz and Gault 1985a).

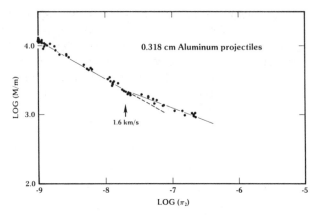

Fig. 2. Change in crater scaling as a function of velocity (constant projectile type and size) for impacts into No. 24 sand. A statistically significant change in the power-law relation between cratering efficiency and the π_2 parameter reflects a transition in physical processes occurring at early times. Small projectiles may disrupt an individual sand grain prior to coupling with the bulk target; large projectiles appear to undergo a transition related to the time required for transferring energy and momentum to the target. The transition generally depends on projectile size. Consequently, scaling relations derived from an average of different size projectiles or from an inappropriate early-time process may produce significant errors when extrapolated to planetary scales (large diameters and velocities) (figure from Schultz and Gault 1983a).

projectile size is examined as shown in Fig. 2 (Gault and Wedekind 1977; Schultz and Gault 1983a). Onsets of such departures occur at different values of π_2 for different size and density projectiles and are believed to indicate a different cratering regime as discussed in more detail below. Extrapolations of either an averaged power-law exponent for the two regions or an inappropriate exponent from one regime to another will result in significant errors. Nevertheless, the usefulness of using Eq. (1) in a given cratering regime can be demonstrated in Fig. 3 where impactor densities range from 10^{-3} to 7 g cm^{-3} (see Schultz and Gault 1985a).

The effect of impact angle was extensively studied by Gault et al. (1972), Gault (1973), and Gault and Wedekind (1978). Gault et al. estimated that the displaced mass for impacts into granite decreases as $\sin^2\phi$ with increasing angle from the vertical ($\phi = 90 - \theta$), but for impacts into sand, it decreases as $\sin\phi$. Gault and Wedekind (1978) further explored in detail the effects of oblique impacts on crater shape and ricochet for particulate targets. More recent studies suggest that impact-angle effects may depend on r and v as well (Schultz and Gault 1986a).

Gault and Heitowit (1963) and Cintala and Grieve (1984) suggest that internal energy losses (melting, vaporization, comminution) decrease energy available for crater excavation. Most studies suggest that relative energy losses due to melting and vaporization depend on impact velocities, i.e., are largely independent of projectile size (see, e.g., O'Keefe and Ahrens 1977):

$$f_m(r,v,\theta) \sim f_m^{-1}(v) \sim v^{-2} \quad (\theta = 90°)$$
$$f_v(r,v,\theta) \sim f_v^{-1}(v) \sim v^{-2} \quad (\theta = 90°) \tag{4}$$

Laboratory experiments typically do not attain velocities high enough to assess the role of melting and vaporization on gravity-controlled growth. Consequently, small quantities of dry-ice powder have been placed at the point of impact within a sand target in order to simulate first-order effects of vaporization (Schultz and Gault 1986, 1985b; Schultz and Crawford 1987). Preliminary results reveal the importance of impact angle on the vaporization process. At 15° from the horizontal, 6 km s⁻¹ impacts produce a vaporized cloud estimated to contain about 30% of the initial impactor kinetic energy. In contrast, vertical impacts (90°) result in nearly an order-of-magnitude reduction in the energy lost by vaporization. Ironically, cratering efficiency was found to increase with increased vaporization. This enhancement is believed to reflect large back pressures within the early-time sand cavity that amplify, rather than reduce, the radial flow field. Further studies are underway to explore the relevance of these results for vaporization/melting at planetary scales. Gault et al. (1975) noted that the fraction of kinetic energy converted to heat remains relatively constant for constant kinetic energy and increased impact velocity for velocities greater than 15 km s⁻¹. Although it is possible that the effects on cratering on different planets may not be evident, the effects on Eq. (2) could be significant.

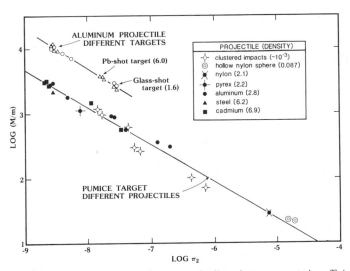

Fig. 3. The effect of contrasts in density between projectile and target on cratering efficiencies. The π_2 parameter is shown to accommodate a wide range of impactor/target density ratios from 10^{-3} to 7 (figure adapted from Schultz and Gault 1985).

Laboratory experiments using low-strength granular targets simulate conditions following shock-induced comminution. Consequently the factor $f_c(r, v, \theta)$ as defined in Eq. (3) must be greater than unity. Gault (1973), Gault et al. (1975), Cintala and Hörz (1984), and others observed that comminution energy directly reflects the volume and size distribution of shock-induced crushed and broken material. Gault et al. (1975) suggested that the comminution energy E_c can be expressed as:

$$E_c = kD_s^{3 - \omega\delta} \tag{5}$$

where D_s is the diameter of shock disruption zone and where ω and δ (>0) are exponents incorporating the size distribution of crushed material. Gault and Heitowit estimated that 20 to 30% of the original impactor energy is directly lost by shock comminution. Schultz and Mendell (1978) and O'Keefe and Ahrens (1987) noted that energy losses by comminution may increase further in large craters where shock broken material is subjected to shear and distortional stresses over long periods of time prior to excavation from the cavity.

The term $f(r/R)$ in Eq. (1) is unique to impacts that are large with respect to planet dimensions. Impacts into low-strength (pumice) planar and hemispherical targets at the NASA-Ames Vertical Gun Range reveal that this term may be greater than unity (Schultz et al. 1986). Cratering efficiency referenced to the pre-impact chord through the hemisphere closely matches data for the planar case. Consequently, cratering efficiency for global-scale basin impacts may have to be increased by the displaced mass above the chord.

The previous discussions have implicitly assumed that a single scaling relation between cratering efficiency and impactor variables (Eqs. 1 and 2) applies over all values of r and v for a given target. Gault and Wedekind (1977), however, documented a significant change in the scaled diameter as a function of kinetic energy for a given projectile size and target. Schultz and Gault (1985a) subsequently showed that similar changes in cratering efficiency occur as a function of the π_2 parameter for a variety of impactors (aluminum, pyrex, lexan) of different sizes as illustrated in Figs. 1–3. Preliminary analysis suggested that this change reflected the onset of projectile deformation, but more recent studies suggest that it is related to the rate at which energy and momentum are transferred to the target (Schultz and Gault 1987). Although further discussion is deferred to the following section about crater growth, it is important to emphasize that the observed change in cratering efficiency can have significant effects when extrapolated to planetary scales.

Crater Growth and Scaling

The growth of the impact cavity can be classified as either proportional or nonproportional (Fig. 4). Proportional crater growth depicts a crater that enlarges equidimensionally (diameter, depth, shape) throughout time (Fig. 4a). If this growth is the same for all size impacts, then proportional crater scaling

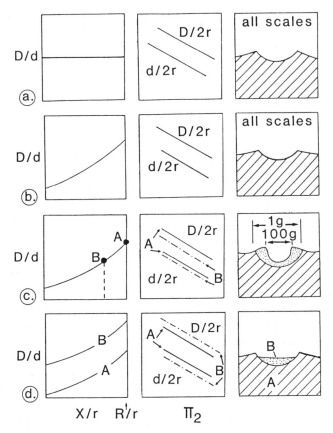

Fig. 4. Comparisons among different implicit models of changing crater shape during formation (left), scaling of crater dimensions (middle) and a cartoon of the effects of growth/scaling model on final crater shape (right). Transient crater shape is expressed as the ratio of the crater diameter D to depth d and is shown as a function of the radius of the crater X at a given time referenced to the final crater radius (left). Final crater dimensions are scaled to the projectile radius and are expressed as functions of the π_2 parameter (Eq. 2). Proportional crater growth and scaling (Fig. 4a) incorporate an implicit assumption that the cratering flow field is independent of the stage in crater growth and the scale (i.e., magnitude of the event). Nonproportional growth describes a crater whose diameter and depth grow differently during formation (Figs. 4b–d). If the cratering flow field is assumed to be independent of event scale, then final crater shape is invariant and the crater scaling remains proportional (Fig. 4b). If the cratering flow field is independent of event scale but a parameter controlling the final crater size depends on scale (Fig. 4c), then crater scaling is modified, i.e., modified proportional scaling. For illustration, A and B indicate the cases where higher gravity (B) limits the stage in crater growth (relative to A) but does not modify the cratering flow field. Consequently, the shape of the transient crater prior to collapse at small scales (1 g) is shallower than at large scales (100 g). If the cratering flow field depends on event scale, then both crater growth and crater scaling will be affected (Fig. 4d). All four descriptions of crater growth are explicitly or implicitly used in the current literature, but each can have different effects on interpretations of crater morphology and statistics.

also applies. In proportional crater growth and scaling, a crater of any size will appear similar at the same fractional stage of formation (e.g., with respect to time of formation or final crater size). For example, a growing crater can be viewed simply as a series of concentric hemispheres, or hemiparaboloids. If an experiment (laboratory or computer simulation) applies at one scale, then the results can be applied at all others. Proportional crater growth is a gross oversimplification of the cratering process but provides a useful order-of-magnitude estimate of crater diameter, depth and volume.

Nonproportional crater growth describes a crater that changes in shape during formation (Fig. 4b) and can result in three different types of scaling relations: proportional scaling, modified proportional scaling and nonproportional scaling. Nonproportional crater growth has been documented in laboratory (Gault et al. 1968; Stöffler et al. 1975; Gault and Sonett 1982) and computer (Orphal 1977; Schultz et al. 1981) experiments involving "normal" impactors.

During the compression stage, crater depth increases more rapidly than the diameter as the impactor penetrates the target; consequently, the instantaneous crater aspect ratio (diameter D to depth d) may initially be smaller than unity. Once the shock wave becomes detached from the front face of the projectile and moves rapidly through the target, crater growth enters the excavation stage and is controlled by the changing flow field in response to rarefaction waves created by the free-surface boundary. During the early stages of excavation in uniform materials, both diameter and depth grow at the same rate, thereby maintaining a constant aspect ratio. During the later stages of excavation, a maximum crater depth is attained while the diameter continues to grow. As a result, the crater aspect ratio increases, but the aspect ratio of the final transient cavity is the same regardless of the scale of the event. Proportional crater scaling with nonproportional crater growth is often implicitly used (see, e.g., Gault 1974; Schmidt and Holsapple 1982a).

Scale-dependent crater excavation models fall into two broad classes: modified proportional and nonproportional scaling. Modified proportional scaling includes models where the final crater aspect ratio (but not the cratering flow field) depends on the magnitude of the event. For example, the excavation flow field established by early-time conditions eventually evolves similarly at any scale, but gravity arrests the final stages, thereby affecting the final crater aspect ratio (Fig. 4c). Thus final excavation crater shape is viewed as a snapshot of an earlier stage. This view of crater growth is implicitly used in computer or analytic models where ballistic extrapolations are used (Orphal et al. 1980). Arresting the diameter of impacts (whether by gravity or strength) generally results in small values of D/d (see Schultz et al. 1981). The final crater shape may not, however, resemble this transient shape owing to collapse and floor uplift.

Nonproportional scaling with nonproportional growth includes models where *both* the cratering field and excavation dimensions depend on event scale

(Fig. 4d). On the basis of a largely intuitive model, Baldwin (1963) suggested impactor size might contribute to the observed flattening of lunar craters and basins with increasing diameters. Since then, various proposals for changes in crater scaling with size have been made and include the effects of increased material strength with depth (Head 1976; Hodges and Wilhelms 1978), some critical level of kinetic energy (Dence and Grieve 1979), and contrasting populations of large, low-density objects and smaller higher-density objects uniformly distributed throughout the solar system (Pike 1980a,b). Recent analyses of laboratory data involving a wide variety of targets and projectiles reveal a systematic change in crater shape that depends on impactor velocity *and* size (Schultz and Gault 1985a,1986b,1987). Below a critical value of the ratio between impactor size and velocity (i.e., time), the crater aspect ratio is approximately constant, whereas above this value, craters become shallower. Consequently, craters may become shallower on a given planet with a given average impact velocity. Such results predict that not only could there be a velocity effect on crater shape from planet to planet, but also a systematic change in crater shape with size on a given planet, thereby supporting Baldwin's suggestion. A more detailed review of this process will be considered in the next section.

Crater growth and scaling of crater dimensions include a variety of obvious and not-so-obvious assumptions about crater excavation. Although all views existed prior to the Mariner 10 mission, more than twenty cratered planetary objects have been added to the inventory. The widespread use of the cratering record as a chronometer, as a probe of planetary crusts, and as a possible signature of different impacting objects necessitates reviewing the evidence and consequences of nonproportional crater growth and scaling. The following discussion summarizes recent results from laboratory experiments.

Experimental Evidence for Nonproportional Growth/Scaling

Impacts into strength-controlled ductile targets such as aluminum have long been known to produce crater profiles that depend differently on impact velocity in different velocity regimes and at different impact angles (see, e.g., Summers 1959). At velocities less than the target sound speed c_t, crater aspect ratio is largely controlled by impact velocity: scaled penetration depth (scaled to projectile diameter) increasing with velocity. Impacts with velocities near c_t (transition regime) produce a nearly constant scaled penetration depth. At hypervelocities, scaled penetration depths once again increase with impact velocity but with a dependence different from the low-velocity regime. The onset of the transition and fluid impact regions could be clearly identified in the crater aspect ratio as well as the scaled penetration depths and energy scaling relations. Re-evaluation of published data for the fluid impact regime has revealed, however, that the crater aspect ratio decreases with increasing impact velocity for a given target and given projectile size and shape (Schultz and Gault 1987). For a given projectile and target, the crater aspect ratio was found empirically to increase with the time t_p required for the projectile to penetrate the target that could be approximated as $2r/v$ where $2r$ is the projectile diameter (spheres) and v is the impact velocity. Gault and Wedekind (1977) also documented a change in the scaling relation between crater diameter and impact energy for impacts into sand targets. Schultz and Gault (1985a, 1987) subsequently found that this diameter-energy scaling relation also could be related to t_p. The following discussion reviews these results in more detail.

Figure 5a shows the relation between $2r/v$ and crater aspect ratio for hypervelocity impacts into a variety of targets. For a given projectile and target, the crater aspect ratio increases with increasing values of $2r/v$. The different data sets merge empirically when $2r/v$ is multiplied by a series of terms incorporating the physical properties of the projectile (subscript p) and target (subscript t) as shown in Figure 5b:

$$D/d = k(\eta t_p)^\gamma \tag{6a}$$

where

$$t_p = \frac{2\,r_e}{v} \tag{6b}$$

$$\eta = \left(\frac{2r}{l}\right)\left(\frac{\delta_t}{\delta_p}\right)^{1/3}\left(\frac{c_p\,\delta_p}{c_t\,\delta_t}\right)\left[1 + \left(\frac{\delta_p}{\delta_t}\right)^{1/2}\right] \tag{6c}$$

$$k \sim S \tag{6d}$$

and where δ indicates density; r_e, the effective projectile radius defined as $(m/\delta_p)^{1/3}$ for a projectile mass m and density δ_p; c the sound speed; and S is an independent dimensionless strength parameter depending on the target yield strength. When ηt_p falls below a critical value defined here as $(\eta t)_c$, the crater aspect ratio becomes approximately constant (Figure 5b).

The derived empirical expression for impacts into ductile targets also applies to impacts into granular targets provided that the impact velocity exceeds the target sound speed (~ 130 m s^{-1}) and that the projectile size exceeds the size a of individual grains in the target. Impactors with velocities less than the target sound speed burrow through the target as penetration is largely controlled by frictional drag. High-velocity impactors with $2r/a \leq 2$ will disrupt an individual sand grain at contact, thereby changing the early-time transfer of energy to the target. Both cases produce aspect ratios that increase with impact velocity (Fig. 6a) but with different physical causes. Between these extremes, crater aspect ratios increase either with decreasing impact velocities for a given size projectile (Fig. 6a) or with increasing values of t_p (Figure 4b) for different projectile properties, a result consistent with the

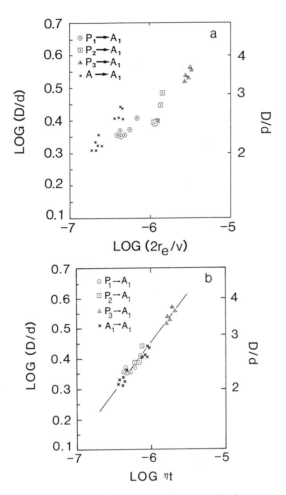

Fig. 5. Crater shape as a function of time for projectile penetration into nonporous ductile targets. Different combinations of projectiles and targets (a) consistently indicate an increase in the diameter/depth D/d ratio with increasing values of the effective projectile diameter $2r_e$ divided by the impact velocity, where P refers to polyethylene and A to aluminum. Different combinations of projectile and target, however, are displaced with respect to each other. All impact velocities exceed the target sound speed where the cratering process can be viewed as a fluid-like process. Multiplying the quantity $2r_e/v$ by a series of projectile/target physical properties merge the diverse data sets (b). These parameters are the density δ of target (subscript t) and projectile p; sound speech c; actual projectile diameter $2r$ and length l; and target strength (not included here since this is a single target type). The combination of terms can be physically related to the approximate time for transfer of energy and momentum to the target. The data sets refer to 7–9 km s^{-1} polyethylene into 2024 aluminum targets (Le Comte and Schaal 1966), 7–7.8 km s^{-1} high-density polyethylene with $2r/l = 1.8$ into 2024-T4 aluminum (Denardo 1962), 7.8–11.3 km s^{-1} high-density polyethylene with $2r/l = 3$ into 2024-T4 aluminum (Denardo 1966), 6.9–8.5 km s^{-1} 2017-T4 aluminum spheres into 2024-T4 aluminum targets (Denardo and Nysmith 1966).

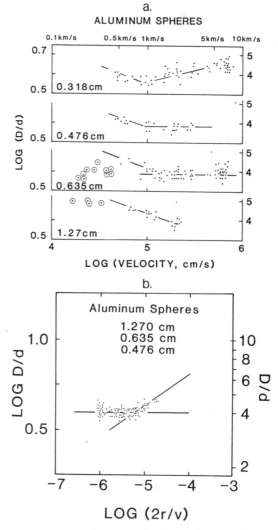

Fig. 6. The effect of impactor velocity and size on crater shape for low-strength sand targets. The impactors are aluminum spheres and the target is No. 24 sand. Crater shape depends on impact velocity for different size projectiles, but the specific dependence varies with projectile size (a). Projectiles less than 0.4 cm approach the size of individual sand grains (0.07 cm) and result in an increase in D/d with increasing velocities. Larger projectiles result in a decrease in D/d with an increase in velocity up to a critical velocity that depends on projectile size; crater shape is approximately constant at higher velocities. Circled data indicate impact velocities less than 3 times the speed of sound in sand. (b) combines the data in (a) with the variables $2r/v$ for projectiles >0.5 cm and impact velocities ≥ 400 m s^{-1}. Crater shape in sand becomes flatter with increasing values of impactor penetration time ($2r/v$) beyond common critical value.

form of Eq. (6) for ductile targets. Below a critical value of t_p, the crater aspect ratio is nearly constant provided that the value of $2r/a \gg 1$. The dimensionless strength factor S for granular targets now depends on target cohesion and internal friction angle.

Different projectile types impacting sand targets result in systematic offsets of the critical value of t_p (Fig. 7a), but the data merge if the physical properties in Eq. (6) are explicitly included (Fig. 7b). Even very low density impactors created by a cloud of hypervelocity pyrex fragments (Fig. 7c) can be accommodated where the penetration time refers to be time required for all fragments to impact and is given by Eq. (6) with $2r_e$ defined by the actual cluster dimensions with $c_p\delta_p$ now referring to properties of the projectile fragments.

The physical mechanism behind these empirical results can be related to the time required for energy transfer and the time for the impact shock to pass through the target. For very short shock transit times t_s, particles in the target within the shock are initially directed radially away from the point of effective energy transfer, but are then redirected by rarefaction waves influenced by the presence of the free surface (see Gault et al. 1968). The integrated path of such particles determines the cratering flow field and ultimately defines the crater shape. The flow field from impacts with very short energy transfer times (and short shock transit times) closely resemble point-source explosions (see, e.g., Dienes and Walsh 1970). As the time for energy transfer relative to the total impactor energy increases, the initial source will not be point-like except at very large distances. For very long shock transit times, gravity can begin to affect the post-shock rarefaction flow, in addition to modifying the ballistic paths of ejecta.

The interaction time t_i required to transfer impactor energy to the target is, to first order, proportional to the time required for the shock to engulf the projectile

$$t_i = \frac{2r_e}{U_p} \tag{7}$$

where U_p is the projectile shock velocity moving away from the impactor-target interface. An approximation for this interaction time has been given by Eichelberger and Gehring (1962):

$$t_i = \frac{2r_e}{v}\left[1 + \left(\frac{\delta_p}{\delta_t}\right)^{1/2}\right] \tag{8}$$

which can be shown to approximate Eq. (7) adequately for impact velocities near 6 km s^{-1}, but ranges from a factor of 0.6 to 0.7 too low for velocities higher than 15 km s^{-1}. Equation (6) results from multiplying Eq. (8) by a coupling factor including a projectile shape factor, the acoustic resistance be-

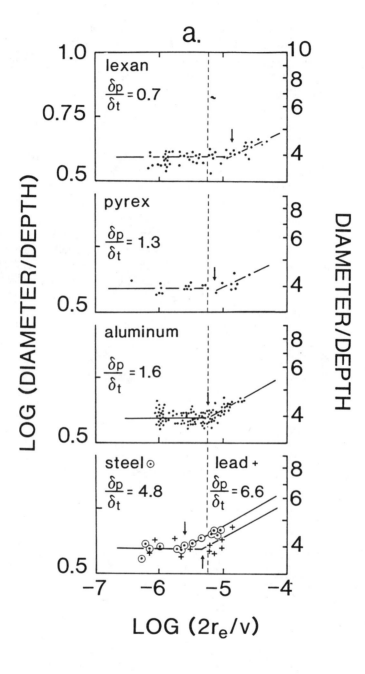

a.

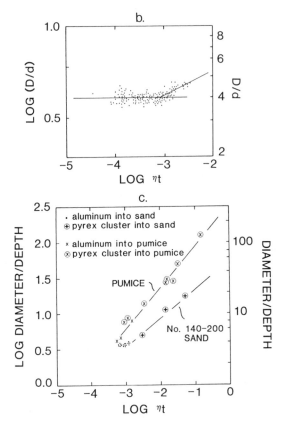

Fig. 7. The effect of different projectile types on crater shape for impacts into No. 24 sand. The dependence between crater shape and projectile penetration time ($2r/v$) found for aluminum impactors (Fig. 6a) also applies to impactors with different densities and strengths (Fig. 7a). Multiplying $2r/v$ by the same set of dimensionless parameters (excluding target strength) used for strength-limited impacts (Fig. 5b) brings the diverse data sets together (Fig. 7b). Impacts by very low-density clustered impactors can be similarly accommodated (Fig. 7c).

tween two dissimilar targets ($c_p\delta_p/c_t\delta_t$), and a mass balance factor represented by $(\delta_t/\delta_p)^{1/3}$ for equal distances travelled in projectile and target. Because the interaction time also controls the rise and decay of the shock wave, ηt_p in Eq. (6) also provides a first-order approximation for the shock transit time in the target.

   The possible effect of gravity on the cratering flow field will be proportional to a characteristic gravity interaction time t_g defined as c_t/g where g is the gravitational acceleration. The ratio between the shock transit time and the time for gravity to interact provides a physical basis for defining a dimensionless interaction time parameter t_p^*:

$$t_p^* = \eta \, \frac{t_p}{t_g} \tag{9}$$

where ηt_p is given by Eq. (6). Equation (9) permits scaling the laboratory results to different impact conditions.

Although the relative dimensions of diameter and depth change, the power-law relation between cratering efficiency (displaced target mass M relative to impactor mass m) and the π_2 parameter discussed by Holsapple and Schmidt (1982) remains unchanged within cratering regimes characterizing different D/d regimes. A constant D/d means that the scaled diameter or depth ($D/2r$ or $d/2r$) can be expressed as a function of the π_2 parameter. If a critical value of ηt_p^* is exceeded, then both the scaled diameter and depth change as $2r/v$ while cratering efficiency remains expressed as a single power-law function of π_2. The scaled diameter and depth can be derived by combining expressions for cratering efficiency and crater aspect ratio with crater shape approximated by a paraboloid:

$$\frac{D}{2r} = \left[\frac{4}{3} \frac{M}{m} \frac{\delta_p}{\delta_t} \left(t_p^* \right)^\gamma \right]^{1/3} \tag{10a}$$

and

$$\frac{d}{2r} = k \, \frac{D}{2r} \left(\frac{1}{t_p^*} \right)^\gamma \tag{10b}$$

where

$$\frac{M}{m} = k' \pi_2^{-\alpha'} . \tag{10c}$$

Figure 8 illustrates diagrammetrically the relationship between crater aspect ratio and both the independent target variables t_p^* and excavation crater diameter D_t, i.e., the diameter prior to or without rim collapse and floor uplift. The transition from a constant aspect ratio to a t_p^*-dependent value now can be expressed in terms of crater diameter D_t, impactor velocity, and material properties by combining Eqs. (6), (9) and (10):

$$D_t \sim \frac{v^{1 + \alpha/3}}{g} . \tag{11}$$

In summary, experimental data indicate that crater growth is nonproportional and that the ratio of final excavation dimensions depends on projectile penetration time for a given impactor-target combination. Most hypervelocity laboratory impacts have penetration times on the order of microseconds,

CRATER ASPECT RATIO

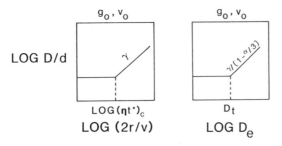

Fig. 8. Crater aspect ratio (diameter/depth) expressed in terms of time of energy/momentum transfer (Eq. 6 in text) or excavation crater diameter D_e. For a given impact velocity v_o and gravity field g, the critical transfer time parameter $(\eta t^*)_c$ correlates with a transition diameter D_t.

whereas a large-scale planetary impact resulting in a 100 km crater has a penetration time longer than 0.1 seconds. Consequently, planetary-scale craters may be significantly shallower (i.e., larger D/d) than laboratory-scale, gravity-controlled craters. Moreover, different rms (root mean square) impact velocities due to different orbital velocities and masses (as well as different impactor sources) on different planets will result in different crater aspect ratios for the same size crater. A subsection in Sec. III below will explore specific applications of these results to Mercury, Mars and the Moon.

Comparisons of Crater Excavation Models

The various proposals for crater growth and scaling need to be placed in the context of possible observations of planetary impact craters. This can be done by exploring explicit implications of each model for cratering efficiency, crater aspect and projectile size as a function of crater diameter. Cratering efficiency includes the total displaced mass by the impact, and is a measurable quantity if topographic data are available (with assumptions about post-cratering modification). Crater aspect is generally represented by depth plotted against diameter, and each planet exhibits a departure from a linear interdependence over a range of transition diameters (Pike 1974,1980a). Consequently, an understanding of crater excavation models provides a starting point for interpreting both the transition diameters on different planets and the general relation between depth and diameter at different scales. The relation between projectile size and crater diameter provides a link between the size distribution of astronomically observed bodies and the size distribution of craters. Figure 9 will be used to explore diagrammetrically the implications of different models of crater excavation on observable planetary dimensions. The following notation is used in Fig. 9: upper-case letters (A, B) refer to a specific set of independent variables affecting cratering efficiency with the power-law

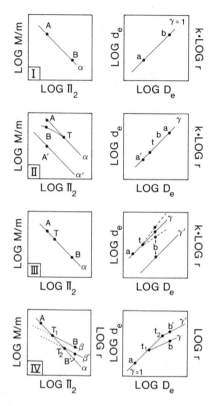

Fig. 9. The effect of different models of crater growth and scaling on the relation between cratering efficiency (displaced crater mass M normalized by the impactor mass m) and the π_2 parameter; excavation crater depth d_e and diameter D_e; and impactor radius r and excavation diameter. Model I presents the simplest case of proportional crater scaling where different impact velocities (A, B) fall on a single power-law dependence α for cratering efficiency, and cannot be discerned (a, b) from relation between crater depth and diameter with a single power-law dependence ($\gamma = 1$). Model II also shows proportional growth and scaling but includes two parallel dependences (α, α') between M/m and π_2 due to internal energy losses. If internal energy losses occur at a critical velocity (e.g., onset of disruption or melting) at T, then a different slope β may occur. For the assumed scaling/growth model, these differences are not recovera..e from the relation between crater depth and diameter. Model III corresponds to nonproportional growth with modified proportional scaling such that a transition at t may occur in crater shape but not cratering efficiency at T; see text for specifics. Model IV considers nonproportional growth and scaling as indicated by laboratory experiments. Transitions in scaling of cratering efficiency at T_1, T_2 also occur in the relationship between crater depth and diameter at t_1, t_2. If post-excavation modifications such as rim collapse and floor uplift can be estimated, then independent variables in π_2 may be deciphered from crater shape. Derivation of impactor size r from crater diameter depends on the assumed model of crater formation.

exponent indicated by α or β; lower-case letters refer to the corresponding values for crater depth and diameter with the exponent indicated by γ. A complete tie between the impact process and observable quantities also must include processes that modify crater dimensions, whether immediately afterwards or over significant time. This aspect of the problem is deferred to the next section.

The simplest cratering model considers proportional crater scaling regardless of crater growth (Fig. 9 model I) where different impact (r, v, δ_p) and target (δ_t) parameters map on a single straight power-law slope for cratering efficiency, crater shape and impactor size. Consequently, a very high impact velocity (point A) maps on the same power-law slope (α) as a low-velocity impact (point B) when relating cratering efficiency to the π_2 parameter. Crater depth, in this model, increases linearly with crater diameter:

$$d_e \sim D_e^\omega \qquad (12)$$

where $\omega = 1$. Impactor size increases with crater diameter with an exponent (ξ):

$$r_p \sim D_e^{\xi} \qquad (13)$$

where $\xi > 1$. Consequently, Fig. 9 illustrates both d_e and r_p in the same graph but the reader should be aware that the relative scales are different. Since complete similitude is assumed, impact parameters cannot be uniquely derived from the dimensions of the excavation cavity: different impact velocities plot on a single power-law relation between crater depth (or projectile size) and diameter. The derived scaling laws of Schmidt and Holsapple (1982a) exemplify such a model with Eq. (2) indicating $\xi = 1.2$. Similarly, if the energy-scaling relations derived by Gault (1974) are applied without inclusion of internal energy losses, $\omega = 1$ and $\xi = 1.19$.

A slightly more complete cratering model incorporates proportional crater growth but allows for nonproportional crater scaling due to inefficient transfer of impactor kinetic energy to crater excavation due to large internal energy losses (vaporization, melting), low impact angle, or large gravitational fields (Figure 9, model II). Efficient coupling between impact energy and target (e.g., vertical impact) corresponds to curve α' in Figure 9, model II for a given set of variables r, v, g. If cratering efficiency decreases with increasing impact velocity due to internal energy losses (e.g., the onset of melting at transition point T) then a change in the cratering efficiency exponent β results, but such processes by definition do not affect the relationship between depth and diameter. In its simplest form, the cratering flow field remains unchanged for different scale events; consequently, cratering efficiency—but not the final crater profile—is affected. Because crater growth is assumed to be proportional, the offset scaling relations α, α' follow a single power-law exponent γ for diameter vs depth.

Concern for the effects of energy losses has been one underlying reason for the use of impactor energy as the relevant independent variable (Charters and Summers 1959; Gault 1974). The Pi Theorem provides a highly useful independent variable (Chabai 1977; Schmidt and Holsapple 1980) by expressing crater scaling in dimensionless quantities as previously illustrated in Figs. 1–3. Energy partitioning and its effect on crater dimensions remain, however, a fundamental issue. The pioneering experimental studies by Gault illustrate this approach and include estimates of energy partitioning (Gault and Heitowit 1963), gravity (Gault and Wedekind 1977), and impact angle (Gault and Wedekind 1978; Gault and Schultz 1986). As shown in Fig. 9, impactor parameters a, b, T, a' may not be distinguished in crater profiles but might be identified in cratering efficiency either by a vertical shift α, α' or a different slope β. Additionally, planimetric crater shape and the distribution of ejecta provide important clues (Gault and Wedekind 1978).

An alternative to the preceding view maintains that cratering efficiency follows a single power law of the π_2 parameter (or energy) over all scales, but that crater growth is modified (Fig. 9, model III): i.e., proportional crater scaling but nonproportional crater growth. In this model, the effect of different impact parameters g, v, δ_p/δ_t, c_p/c_t remains unresolved in terms of cratering efficiency (points A and B) but can be distinguished in the crater profile. Several examples of such an approach exist in the literature. Preliminary experiments by Schmidt and Holsapple (1981) indicated that craters subjected to high gravitational accelerations have shallower excavation cavities (low g therefore corresponding to curve γ; high g to curve γ' in model III) although subsequent analysis did not confirm this result (Schmidt and Holsapple 1982b). Schultz et al. (1981) noted that gravity arrests the growth of the excavation crater at earlier stages. Since observations (Gault et al. 1968) and theoretical calculations (Orphal 1977; Orphal et al. 1981) reveal early hemispherical growth and late lateral growth, the effect of gravity should be the reverse of the previous scenario; high g corresponding to γ and low g corresponding to γ'.

Because gravity becomes more important with increasing scale, the relation between crater diameter and depth (or r) may change above a transition diameter. The two alternative scenarios are indicated by the two different slopes for diameters larger than t. The possible effects of impact velocity and density also fall in this class of models. High-velocity impacts have been proposed to resemble shallow-depth-of-burst explosions (see, e.g., Roddy 1977) or to redistribute residual energy for excavation due to internal heating (Cintala and Grieve 1984), thereby producing shallower craters (low v, γ; high v, γ'). Roddy (1968) proposed that low-density impactors produce anomalously shallow craters and might contribute to the range in planetary crater profiles. Schultz and Gault (1983a,1985a) confirmed this effect experimentally, whereas O'Keefe and Ahrens (1982) demonstrated it theoretically. Only minor departures from the cratering efficiency curve occur (Fig. 2) but major changes in the crater profile result (Fig. 7c). Because such changes require exceedingly low densities (Orphal et al. 1980; Schultz and Gault 1985a), the applicability to most planetary problems may be limited to problems such as disruption of projectiles during atmospheric entry (see, e.g., Melosh 1981).

Target properties also can have important effects on crater growth. Explosion cratering (Fortson and Brown 1958) and impacts (Quaide and Oberbeck 1968) revealed that a slight increase in target strength with depth can result in a mounded, flattened or terraced crater profile. Analogous changes in strength due to planetary structure have been used to account for changes in crater shape with size (Head 1976; Hodges and Wilhelms 1978). Such analogies, however, presume that the minor pre-impact strength contrasts in sand targets resemble the post-impact shock-disrupted material at large scale. This assumption remains debatable (see Gault et al. 1968) since possible effects of increasing overburden pressure or temperature on penetration depth or projectiles has been largely unexplored. Equation (10a) indicates that increased density with depth (regardless of strength) could theoretically play a role, but the cube-root dependence on this term limits its effects.

The fourth cratering scenario considers both crater growth and scaling as nonproportional. Two possible examples are shown in Fig. 9, model IV. For a given impactor size (and gravity field), the onset of significant impact melting (point T_1) may produce a change in power-law dependence β between π_2 and cratering efficiency, as in model II. If there is a corresponding change in crater profile (as in model III), then the onset t_1 or t_2 of shallower craters γ or γ' may reflect impact velocity. Gault and Heitowit (1963) note that, above 15 km s^{-1}, internal energy losses reach a maximum ($\sim$30–40%). Consequently, cratering efficiency never becomes independent of π_2, and crater depth never reaches a maximum. This example is hypothetical since experimental data have yet to establish unequivocally a change in the cratering efficiency exponent or a corresponding crater shallowing due to impact melting or vaporization. Instead, hypervelocity impacts into porous targets produce a relative increase in cratering efficiency with increasing velocity B to T_1 to A for a given projectile size with a corresponding decrease a to t_1 to b in relative depth (Gault and Wedekind 1977; Schultz and Gault 1985a,1987). This paradox is at least partly related to the relative size of projectile and grain size in the target (Schultz and Gault 1987). Although impact vaporization also increases cratering efficiency and D/d (Schultz and Gault 1985b), this effect may be related to large vapor back pressures within the cavity. The relevance of this observation to large-scale cratering remains unclear but intriguing.

The second example for case IV applies the evidence from laboratory experiments (discussed above) that D/d increases with t_p^* above a transition diameter D_t. Experimental data also indicate a change in scaling between cratering efficiency and the π_2 parameter (Fig. 2; Schultz and Gault 1982). Figure 9 represents these observations by the two different exponents α and β for a given ratio of velocity/gravity with a transition at T_1 (curve B-T_1-A) or α

and β' for a different ratio with a transition at T_2 (curve $B'\text{-}T_2\text{-}A$). These changes in cratering efficiency scaling are reflected by transitions (t_1, t_2) from a constant diameter-to-depth ratio ($\gamma = 1$) to a ratio that increases with increasing crater diameter ($\gamma = \gamma' \neq 1$). This cratering model does not require unusual impactor or target properties but only a sufficiently large impacting body for given values of impact velocity and gravity.

In summary, Fig. 9 reveals several different scenarios for crater scaling that have been used either explicitly or implicitly in the past. Because many studies use crater diameter as a characteristic dimension, different scenarios for crater growth and scaling will result in different inferences about impactor sizes derived from changes in crater size-frequency distributions and about the significance of substrates on the transitions in the diameter-depth relations. The observed crater dimensions may not correspond to the excavation craters shown in Fig. 9, however, due to a variety of modifying processes. The next step is to explore the implications of crater modification on the various observable quantities.

III. CRATER MODIFICATION

The observed crater diameter on a planet does not correspond to the diameter of the excavation cavity due to post-impact modification (see, e.g., Shoemaker 1962; Quaide et al. 1965; Mackin 1969; Gault et al. 1975; Grieve et al. 1977; McKinnon 1978; Croft 1981a). The observed crater diameter D_o is simply related to the gravity-limited excavation diameter D_e:

$$D_o = D_e + \Delta D \tag{14}$$

where ΔD represents the amount of enlargement by a variety of possible short-term and long-term processes. Similarly, crater depth achieved during crater formation may not reflect the final crater depth following crater enlargement and floor uplift. Figure 10 illustrates the observed relation between crater diameter and depth for the Moon (Pike 1980b). If proportional crater growth and scaling (models I and II, Fig. 9) represent the excavation stage of cratering correctly, then the observations require significant modification by crater enlargement and uplift. If nonproportional growth and scaling (model IV, Fig. 9) apply, then crater modification could play only a secondary role. The following discussion, therefore, reviews the possible effects of crater modification in order to get from Fig. 9 to Fig. 10. The causes and effects of crater enlargement are first considered independently of processes affecting crater depth and applied to selected models of Fig. 9. Different degrees of crater floor shallowing are then invoked in order to match the observations. Inversion of this approach provides a reference for estimating properties of the impactor to observed crater dimensions.

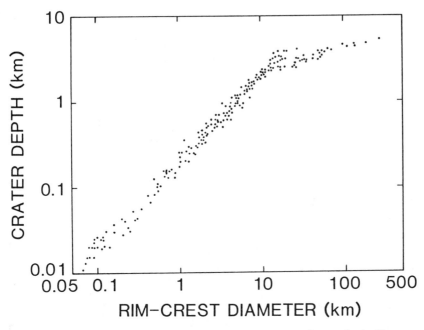

Fig. 10. The observed relationship between crater depth and rim-crest diameter for the Moon as given by Pike (1980*b*).

Rim/Wall Slumping

Short-term crater modification includes wall failure induced by simple mechanical collapse, i.e., slumping by dynamic collapse and by dynamic floor rebound. Crater wall failure in lunar craters is clearly expressed by terraces and jumbled hillocks at their base (see, e.g., Quaide et al. 1965; Howard 1974; Schultz 1976*a*). The embayment of wall slumps by once-molten floor material and the cascades of inferred impact melt across the wall terraces indicate that wall failure occurred during or immediately after crater formation (Schultz 1976*a;* Hawke and Head 1977). This process becomes important for craters above a transitional size D_t on different planets, reflecting at least in part the gravitational field as reviewed in the Chapter by Pike.

Figure 11 relates the observed (modified) crater diameter to the original excavation diameter (left), excavation depth and projectile size (middle), and the final observed crater depth (right) according to different models (A, B, C, D) of crater excavation and modification. Qualitative observations of wall scallops around lunar craters suggest that craters near the transition size may be enlarged to a greater proportion than are large craters much greater than the transition size. A simplified model relates ΔD to D_e by a power law:

$$\Delta D = kD_e^{\omega}. \tag{15}$$

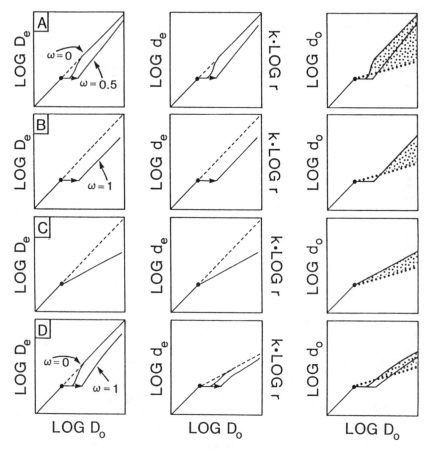

Fig. 11. Possible effects of crater modification processes on observed crater shape for different models of crater formation. The relation between the transient excavation crater diameter D_e and observed diameter D_0 is shown at left with the effect on the depth-diameter relations without floor uplift (middle). Dashed lines depict the assumed model of crater excavation; solid lines, the effect of modification; and dotted lines, the observed relation between depth and diameter on a given planet. The sequence at right illustrates the degree of uplift or floor filling required (stipple pattern) in order to fit observations. Craters that form with proportional growth and scaling may be enlarged by processes dependent on crater size: $D_0 = D_e + \Delta D$ where $\Delta D \sim kD_e^\omega$ with $\omega = 0$ and 0.5 (Model A), and $\omega = 1$ (Model B). Model A depicts one extreme where the width of wall failure is independent of crater size, whereas Model B includes the assumption that wall failure is directly proportional to the diameter (or depth) of the excavation crater. Model C considers the effect of dynamic collapse of the crater rim, where the width reflects the difference between the limit of shock disruption and gravity-limited crater growth. Model D depicts nonproportional crater growth with the effects of two values of ω.

If the *ad hoc* assumption is made that slump blocks have a constant size regardless of crater size, then $\omega = 0$ and D_o becomes the same as D_e for craters slightly larger than D_t (Fig. 11, model A). The resulting dependence between transient crater depth d_e and observed diameter exhibits a departure only near D_t; any size-dependent change in the final observed crater shape must be the result of floor uplift.

If the observed slump zone defines or is simply proportional to the degree of crater enlargement (i.e., $0 < \omega \leq 1$), then some fraction of the wall width W specifies the relationship between the excavation diameter and final observed diameter. Empirical expressions for W (see, e.g., Pike 1980a) permit exploring this possibility (Fig. 12). The observed final crater diameter should nearly parallel the excavation diameter in diameter-depth plots for craters slightly larger than the transition diameter, contrary to observations (Fig. 10). Consequently, considerable crater infill and/or floor uplift, as well as crater enlargement, seem required to produce the observed shallowing of very large craters. As indicated diagrammetrically in Fig. 11a (far right), the effect of such shallowing processes must increase with increasing crater size. Grieve et al. (1981) calculated that observed stratigraphic uplift beneath terrestrial craters is proportional to $0.06 \, D^{1.1}$. Such a dependence would not significantly change the power-law dependence between d_e and D_o, thereby not accounting for the difference between the assumed crater excavation/modification model and observations (the stippled zone). Consequently, either the assumed model of proportional crater growth/scaling is incorrect or structural uplift in terrestrial craters is only one aspect of crater shallowing processes on the planets.

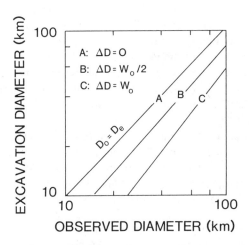

Fig. 12. Effect of crater enlargement on the relation between the excavation and observed crater diameter on the Moon, if the amount of crater enlargement ΔD is assumed to be proportional to the observed width W_0 of the wall slump zone.

Plastic Rim Collapse

The first model did not incorporate a reason for crater modification but only the assumption that observations of crater elements can be used to reconstruct the precollapse excavation crater. The next model incorporates a physical basis for crater modification. If it is assumed that shape of the excavation cavity is independent of scale, then the lithostatic overburden of the transient crater rim and presumably the limit of wall collapse increases linearly with crater size. Model B, therefore, incorporates a specific form of Eq. (15) where $\omega = 1$:

$$\Delta D = (1 + k)D_e. \tag{16}$$

The excavation diameter, impactor size and excavation depth for large craters then simply parallel extrapolations from small simple craters. This model, however, also predicts (and requires) simultaneous floor uplift which should be proportional to the lithostatic rim overburden of the transient crater.

Although the preceding model is intuitively reasonable, Melosh (1977a) and McKinnon (1978) demonstrated through a simple mechanical model that even large (15 km diameter) lunar craters should not exhibit wall failure unless highly unusual values of shear strength (30 bar) and internal friction (only a few degrees) are assumed. Melosh (1982a) subsequently specified the limit for crater collapse in terms of a mechanical model that satisfied the constraints for initiating wall failure. The model proposes that residual energy from the passing shock wave is trapped as acoustic intergrain vibrations. This noise reduces the strength of the shattered material surrounding the cavity to values below a critical yield strength, specified as the cohesion, relative to the gravity. Once failure is initiated, the collapsing cavity behaves as a Bingham fluid and can develop a variety of morphologies depending on the effective viscosity of the impact-fluidized material. The fluid behavior continues until energy losses reduce the acoustic vibrations (and thus the low cohesion) below the critical value. The onset of collapse occurs when crater diameter exceeds about 14 $C/g\delta_t$ where C is the effective cohesion of the material surrounding the crater. Model B is representative of numerous proposals where gravitational forces initiate important changes in crater morphology (Shoemaker 1962; Quaide et al. 1965; Gault et al. 1968; Mackin 1969; Gault et al. 1975; Dence et al. 1977; Grieve et al. 1981; Melosh 1982a; Shoemaker 1983). As in model A, however, crater floor uplift proportional to the excavation crater depth, which is proportional to crater diameter in this model, cannot alone account for the difference between the model and observations.

Dynamic Rim Collapse

Ivanov (1976), Schultz and Mendell (1978), Schultz et al. (1981) and Croft (1981a) explored possible implications of the difference between the limit of shock damage D_s and the limit of excavation permitted by gravity D_g.

Within D_g, material is ejected from the cavity along ballistic trajectories; outside D_g but within D_s, material follows ballistic trajectories until gravity prevents escape. Thus the potential for collapse exists in the fractured zone outside the gravity-limited cavity but inside the limit of shock damage. Moreover, the difference between D_s and D_g increases with crater scale. Croft (1985) explored in detail the implications of the conceptual model for complex craters that are summarized in model C (Fig. 11). The relation between the observed final diameter and excavation diameter is approximately expressed as $D_0 = D_e^{0.85}$, whereas the aspect ratio incorporates both the effects of crater enlargement and floor infill by the collapsing cavity. An important feature of this model (as well as model B) is that the transient excavation cavity is not preserved in the final crater morphology. As shown in Fig. 11c (at right), the difference between the model and observations is small.

Nonproportional Growth/Scaling

The preceding scenarios relied on crater widening and floor uplift to produce large, shallow craters. Nonproportional crater growth can produce a similar result with the degree of modification depending on the specific model for crater excavation (D, Fig. 11). Although the discussion with Fig. 7, (model IV) offered possible examples of nonproportional crater growth and scaling depending on specific impactor or target parameters, here implications of a progressive change in crater depth as a function of diameter on a given planet must be included. Settle and Head (1976,1979) and Malin and Dzurisin (1978) quantitatively reconstructed the rim of the excavation crater from lunar topographic data. Although rim collapse could account for crater shallowing in craters <50 km diameter, additional processes seemed necessary for larger sizes. Both groups and other researchers (see, e.g., Hodges and Wilhelms 1978; Pike 1983b) favored models of nonproportional crater excavation although the processes responsible were different. As shown in Fig. 11d (right), a qualitative match between the model and observations is possible. The next subsection will consider in more detail the specific implications of such models (including the experimental impact data).

Planetary Applications

While data for lunar diameters and depths (Fig. 10) provided a context for exploring possible causes for changes in crater shapes on a given planet, data for different planets (Fig. 13) provide a context for applying and testing the various models examined in Fig. 11. Each of the various models shown in Figs. 9 and 11 has at least some element of merit and not all are mutually exclusive. For example, nonproportional growth and scaling scenarios still must meet constraints on the failure criteria, and must invoke floor uplift in order to match observed relations between depth and diameter for large craters. Nevertheless, the choice of the dominant processes or models can have significant implications for recovering the original impactor populations,

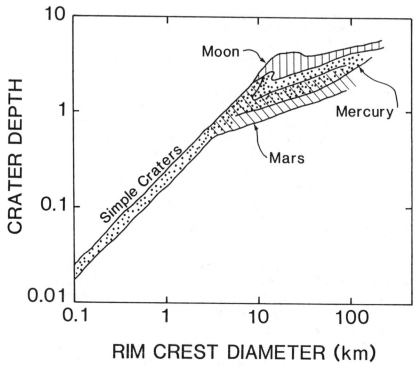

Fig. 13. Stylized comparison of depth vs diameter data for Mercury (see Chapter by Pike), the Moon (Pike 1980*b*) and Mars (Pike 1980*a*).

thereby affecting inferences about cratering chronologies and the evolution of early solar system objects. If plastic rim/wall failure occurs, then Fig. 11 indicates that impactor size and crater diameter are directly related but with different correction factors above and below the onset of failure. Dynamic collapse and nonproportional scaling incorporate different scaling relations for small and large craters; therefore, recovery of impactor sizes (for a given impact velocity) requires information about these scaling relations and the transition. Table I provides a reference for rms impact velocities by different objects on a given planet and their contribution to the total crater population based on data from Hartmann et al. (1981).

The diameters at which the observed crater depths depart from a nearly constant fraction of diameter depend on the proposed processes of crater formation and modification. Figure 14 illustrates different predictions resulting from the classifications shown in Fig. 11. Model A incorporates two alternative proposals. Head (1976) suggested that the transition diameter D_t depends on the depth of the megaregolith and is largely independent of gravity. Cintala et al. (1977) documented statistically that substrate contrasts on the

TABLE I
Impact Velocities (km s^{-1})[a]
(Fraction of Craters on Given Planet)

	Parabolic Comets	Periodic Comets	Asteroids
Mercury	87 (41%)	44 (3%)	34 (56%)
Venus	62 (12%)	44 (15%)	27 (73%)
Earth	53 (7%)	41 (26%)	25 (67%)
Moon	52 (10%)	39 (30%)	22 (60%)
Mars	42 (<3%)	21 (25%)	19 (72%)

[a]Data derived from Hartmann et al. (1981) with the relative proportions of lunar impacts by parabolic comets to periodic comets to Earth-crossing asteroids impacting the Moon assumed to be 1 : 3 : 6.

Moon (mare vs highlands) and Mercury (plains vs highlands) at least partly control the style of wall failure (i.e., the relative numbers of scalloped and terraced craters). If wall failure and the diameter/depth transition are expressions of the same process, then this proposal suggests that crustal structure is the controlling variable (Fig. 14, A'). Malin and Dzurisin (1978) further challenged the conclusions of Gault et al. (1975) that gravity controls D_t and proposed that shock decompression of the crater floor and ejecta loading of the rim more likely initiate mass movement and the onset of shallower craters since both lunar and Mercurian craters exhibit ratios of their transition diameters well below that predicted for just gravity. Since shock decompression should be related to kinetic energy, Malin and Dzurisin as well as Cintala et al. (1976) proposed impact velocity as the controlling variable with only a weak dependence on g. Subsequent analyses by Pike (1980b), however, indicated a signature of g on the terrestrial planets, with the possible exception of Mars.

Models depicting shallower craters due to gravity-initiated wall failure include those by Shoemaker (1962), Quaide et al. (1965) and Melosh (1982a). Figure 14(B) incorporates the formalism by Melosh where the onset of crater shallowing depends on $C/\delta_t g$. Consequently, small values of target density and gravity or large values of cohesion (internal friction) would appear to result in large transition diameters. If pre-impact competence is at all related to the post-shock state of extensive fracturing and acoustic vibrations, then this model seems to contradict the observations by Cintala et al. (1977) and Pike (1980b) that highland craters have slightly larger D/d and D_t. The evolving state of the Melosh model and the possible offsetting effect of δ_t, however, preclude using this observation as a test for the validity of the plastic rim collapse model. At present, the Melosh model does not explicitly link impact velocity with the onset of wall failure and D_t.

If results of laboratory experiments for porous and nonporous ductile

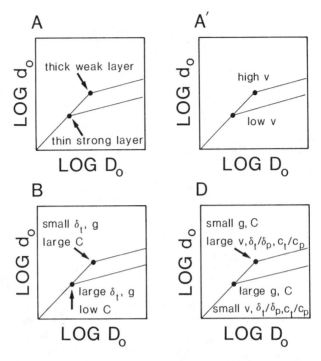

Fig. 14. Possible variables affecting the transition from craters with a constant diameter-to-depth value to craters that become progressively flatter with increasing diameter. Models A and A' correspond to the crater growth modification models of Fig. 11A and propose that crater shape changes as a consequence of a single variable, either crustal structure A or impact velocity A'. Model B includes specific predictions for onset of craters rim/wall failure controlled by target density δ_t, cohesion C, and gravity g as proposed by Melosh (1982a). Model D considers predictions based on nonproportional crater growth and scaling where projectile density δ_p as well as target/projectile sound speeds (c_t/c_p) are introduced.

targets (Schultz and Gault 1987) can be extended to planetary scales, then Eqs. (10) and (11) indicated that the onset of nonproportional scaling occurs when the cratering diameter exceeds D_t:

$$D_t \sim \frac{v^{1.17}}{g} \left(\frac{\delta_t}{\delta_p}\right)^{0.22} \left(\frac{c_t^2}{c_p}\right)^{0.83} \left[1 + \left(\frac{\delta_p}{\delta_t}\right)^{1/2}\right]^{-0.83} S^{-0.83} \qquad (17)$$

where a value of $\alpha \approx 0.50$ has been adopted. Extrapolation of the experimental data to lunar impact conditions ($\delta_t = \delta_p$, $c_p = c_t$, $v = 22$ km s^{-1}) predicts a 12 km transition diameter on the Moon. This extrapolation includes a correction for over-predicting the shock velocity from the approximation given by Eq. (8). The observed apparent diameter (referenced to the pre-impact surface) occurs between 10 and 15 km corresponding to rim-crest diameters between 12 and 18 km (Pike 1977*a*). More generally, Fig. 14 (D) shows that relatively large transition diameters should occur for high-velocity, low-density impacts into a given substrate and gravitational field *g*. Consequently, the transition diameter for diameter-depth should occur for larger craters on Mercury than on Mars if crustal properties are about the same. If gravity is introduced, then impacts on low-*g* planets should shift the transition to larger diameters. The Moon, therefore, should have a larger transition diameter, but the shift will be offset by the higher rms impact velocity for Mercury relative to the Moon. The transition from proportional to nonproportional scaling also depends on projectile and target properties. On a given planet (given *g*, rms v, δ_p, c_p), the laboratory data further indicate that regions of low average strength (low internal friction) should have larger transition diameters.

Equation (17) provides a basis for quantitative comparisons among the planets. If the Moon is used for reference, Table II reveals that transition diameters are reasonably matched for rms impact velocities predicted from orbital dynamics (Hartmann et al. 1981) without invoking dramatically different surface properties. Because both excavation and modification should be constrained by post-shock conditions, the latter result should not be surprising and is generally supported by the mechanical analysis of Melosh (1982*a*). Table II also shows that the possible ranges in transition diameters (controlled by impact velocity only) are in reasonable agreement with observations summarized in Pike (1980*b*).

From the discussion of crater excavation and modification (Figs. 9, 11 and 14), it becomes clear that recovery of original impactor parameters or, conversely, definitive testing of impact cratering models can be exceedingly

TABLE II
Transition Diameters (km)

	Crater Shape[a]				Morphology[b]	Statistics[c]
	Observed		Calculated			
	D_t	ΔD_t	D_t	ΔD_t		
Mercury	5.6	7–15	12	7–22	10 (14)	3–6
Venus	—	—	2.7	2–5	—	—
Earth	2	2–4	2.5	2–4	3 (4)	—
Moon	10	10–20	(10)	10–19	19 (21)	7–8
Mars	4	4–10	3.8	3–5	5 (10)	2

[a]Diameter/depth transitions from intercept of empirical fits for small and large crater data; range in values ΔD_t from data given by Pike (1980*b*; Chapter by Pike). Calculated values are based on the mean impact velocities for asteroidal impacts with the range (ΔD_t) from the range in rms velocities for different types of impactors. Calculations are based on impact velocities given in Table I and were normalized to 22 km s^{-1} for the Moon.
[b]Mean diameter and largest simple crater (parentheses) from Pike (1977*a*).
[c]Inflection in crater statistics from -3 to -2 slope for larger post-mare craters.

difficult from least-square fits of diameter-depth data within a factor of 2 to 3 of the transition diameter. The material dependent variations ΔD on a given planet may mask scaling changes until the effects of ΔD become less important (model A, B, D in Fig. 14). This effect should be most severe on the Moon where the relatively large D_t leaves little room before the onset of more complex modification processes at basin scales. It should be the least severe for Mars where D_t occurs at small diameters. As a consequence, the power-law ω increase in crater depth with diameter as defined in Eq. (12) should systematically increase from the Moon to Mercury to Mars as larger complex craters (but smaller than basin sizes) dominate the data set. Table III permits comparison of derived least-squares fits by Pike (1980*a*; Chapter by Pike) for all complex craters and just the larger examples. From Eq. (17), the value of ω is about 0.7 for $\gamma = 0.3$. The lower observed value of $\omega = 0.496$ could easily result from the additional effects of structural floor uplift.

Nonproportional growth and nonproportional scaling as applied from the experimental results have some similarities to the model proposed by Pike (1980*b*); however, there are fundamental differences. His model proposed proportionally shallower excavation depths with crater size, but he emphasized the role of both gravity and target properties to account for observed variations in planetary crater morphologies and depth-diameter transition. Although the Moon, Mars and the Earth were shown to exhibit an inverse relation between gravity and D_t, the transition diameter was significantly larger on Mercury than expected and slightly lower on Mars. He attributed this difference to *in situ* properties of certain planetary crusts: "hard" substrates on Mercury and "soft" for Mars. Through analogy with shallow depth-of-burst explosion craters and terrestrial analogues (see, e.g., Roddy 1968), Pike

TABLE III
Comparison of Diameter-Depth Relations for Complex Craters ($d = K D^{\omega}$)

Planet	ω	Diameter Range (km)	N	Reference
Moon				
undivided	0.301	10	33	Pike (1977*a*)
mare	0.332	10	24	Pike (1980*a*)
uplands	0.313	10	47	Pike (1980*a*)
TYC	0.411	30–170	32	Wood and Anderson (1978)
Mercury	0.260	>10	99	Malin and Dzurisin (1978)
	0.496	30–175	58	Pike (Chapter by Pike)
Mars				
undivided	0.415	>3.5	105	Pike (1980*a*)
plains	0.334	4–90	51	Pike (1980*a*)
cratered terrain	0.423	3.5–170	54	Pike (1980*a*)

(1980*b*) interpreted the onset of shallower craters as a reflection of large very low-density objects beginning to outnumber smaller, high-density objects throughout the solar system. He recognized, however, that some energy-threshold phenomena also might apply (see, e.g., Baldwin 1963).

The nonproportional growth model proposed here suggests that craters begin to become shallower when a dimensionless shock transit time parameter exceeds a critical value. For a given impact velocity, impactor size, gravity and internal friction largely control this critical value with pre-impact material properties only playing a contributing role. A population of large, low-density projectiles need not be invoked to account for the diameter-depth transition. Moreover, since internal friction refers to the post-shock mechanical properties of the target, "hard" vs "soft" rocks are believed to be less important during the excavation stage. Comparisons between the observations and predictions (Tables I and II) suggest that the average upper Martian crust closely resembles the bulk physical properties of the lunar and Mercurian crust. Crater rim collapse also depends on post-shock mechanical properties after the excavation flow field has been arrested. The specific style of subsequent crater modification (elastic rebound or fluid oscillation), however, may reflect absolute impact velocity as discussed below.

Further support for nonproportional crater scaling comes from changes in the displaced mass ratio as a function of crater diameter. Expressions derived from impact experiments (Eq. 10) indicate that displaced crater volumes should depart from a simple cubic relation with crater diameter on a given planet (i.e., given g and impactor/target properties). Croft (1978) confirmed that crater volume varies as $D^{3.00}$ for 47 craters smaller than 13 km, in agreement with proportional crater growth and scaling for craters smaller than the transition diameter. Croft (1978) and Hale and Grieve (1982), however, found that crater volume increases as $D^{2.31}$ and $D^{2.37}$, respectively, for approximately 20 craters between 17 km and 150 km. If the crater aspect ratio increases at $t^{0.3}$, then Eq. (10) can be used to demonstrate that crater volume should increase as $D^{2.67}$ if no floor uplift occurs. If structural floor uplift observed in terrestrial craters (Grieve et al. 1981) is included, then it can be shown that observed crater volume increases as $D^{2.43}$ for craters between 60 and 100 km, a result that closely matches the observations.

In summary, deriving impactor parameters from crater dimensions hinges on the chosen interpretation of cratering models. Both Mars and Mercury have essentially the same gravitational field but differences in crater profiles at the same diameter. Traditionally, this paradox has been interpreted as expressions of contrasting target strengths. If, however, crater profile data from laboratory experiments can be extended to planetary scales, then pre-impact target strengths are not as important, and the observed differences reflect a contrast in average impact velocities. Are there other possible signatures? The next sections reconsider certain aspects of interior crater morphology with this question in mind.

IV. CRATER MORPHOLOGY

The remarkable overall similarity of craters and basins on the Moon, Mercury and Mars underscores the similarity in the impact process over a wide range of impact velocities and gravitational fields. Equally important, however, are the subtle differences that hold clues for slight differences in processes, changes in dimensional scaling, impactor properties, and contrasts in planetary compositions. For the purposes of this chapter, several morphologic terms need definition. *Simple craters* refer to circular craters with minimal obvious modification. Although frequently cited as a "bowl shape," this is a misnomer for pristine craters, which typically exhibit a talus wall slope meeting a flat floor, thereby creating a truncated cone rather than a bowl. *Complex craters* refer to craters with a scalloped crater rim plan and include examples with evidence for *en masse* wall failure, e.g., slump terraces and debris-covered floors.

The distinction between simple and complex craters typically provides a first-order description of the general morphology but may carry with it an implicit connotation of a principal process, wall failure. If the goal is to understand the impact process, then a more useful classification divides a crater into concentric zones including the floor, wall and rim, each of which may reflect a response to different processes but does not presuppose that these processes are interrelated. For example, large secondary craters on the Moon contain large central mounds on their floors without evidence for extensive wall failure. The goal is to identify common features in each zone for comparison rather than characterizing an entire crater by one morphologic element. The crater rim zone is defined as the region outside the abrupt break-in-slope to the crater interior. In such a definition, the rim zone includes (but does not necessarily coincide with): the topographic rim crest; a concentric inner annulus of hummocky, fractured, blocky and smooth-ponded terrains; the continuous ejecta deposits; and the discontinuous ejecta deposits including secondary craters. The crater wall zone forms an annulus between the break-in slope from the rim and the generally flat, low-lying interior. The wall zone typically contains multiple slump terraces along the upper wall and hummocky or dissected debris near the base. The transition from the wall zone to the interior is typically abrupt owing to embayment of interior materials, but cascades of material across the wall onto the low-lying interior can be found. The crater floor zone comprises the low central portions of the crater, the outer edge delineated by the break-in slope from the wall. It includes a variety of textured terrains (fractured, hummocky, smooth) and features. Minor (hummocks) and major central relief (peaks, rings and pits) are defined as elements of the floor zone.

Ringed Peaks and Central Pits

Can the interior central peak morphology or dimensions provide a measure of impact conditions? Gault et al. (1975) and Melosh (1982a) emphasize the role of gravity-controlled collapse of the transient crater for central peak formation, thereby minimizing the effect of impact velocity. Analyses of terrestrial impact structures and explosion craters, however, have led others to conclude that central peaks result from dynamic rebound in response to interactions between the shock and rarefaction waves beneath the target (Milton et al. 1972; Wilshire et al. 1972; Dence et al. 1977); consequently, central peaks may be related to impactor kinetic energy. Although various models have proposed a progression from central peaks to a ringed arrangement of peaks, i.e., peak rings (Hartmann and Wood 1971; Head 1978; Hodges and Wilhelms 1978), subsequent studies seem to converge on a different mechanism to account for the more complex peak assemblages (Hale and Head 1980c). Measurements of central peak heights, volumes, and basal diameters on different planets generally support the latter alternative (Head 1977; Pike 1977a; Malin and Dzurisin 1978; Hale and Head 1979b,1980c; Hale and Grieve 1982). Above about 50 km, floor roughening in a ring pattern (Croft 1981c; Hale and Grieve 1982) and a corresponding departure in the power-law relation between peak volume and crater volume (Hale and Grieve 1982) suggest that a lateral redistribution of floor material and a different style of peak formation may begin. Lunar craters as small as 20 km in diameter, however, also contain peaks with central pits (e.g., Fig. 15a) and ringed central peak complexes (Schultz 1976a,b) which resemble the much more numerous Martian central peak pits. Lunar central peak depressions are not believed to be statistical flukes because they occur within numerous relatively young postmare craters and because they exhibit a nearly linear relation with crater size (Schultz 1976b). Although these structures are best expressed in the maria, they also are common in the highlands. Mercury exhibits similar but still fewer central structures (Fig. 15b). Because craters with central pits and pit-like structures are ubiquitous on Ganymede (Croft 1983), numerous on Mars (see, e.g., Wood et al. 1978; Hodges et al. 1980; Hale 1983), and largely forgotten on the Moon and Mercury (but see Croft 1981c; Pike 1983b), their mode of formation has been typically associated with the presence of water or water ice. Milton et al. (1972) associated a similar but differentially eroded pit within a terrestrial impact crater (Gosses Bluff, Australia) as an expression of uplifted water-saturated sedimentary strata.

Thus, there are several possible variables that could control the dimensions of central peaks, ringed peaks and pits, kinetic energy, impact velocity, material properties. However, a transition in the ratio between projectile size and velocity, as well as impact energy, can produce a transition in peak dimensions near a particular crater diameter on a given planet. A factor of three in impact velocity produces a factor of three in $2r/v$, possibly expressed as an 80% range in the average transition diameter and otherwise seemingly anomalous ring-like peak patterns at smaller crater diameters. Consequently, impactor penetration time needs to be added to the list of variables affecting central peak morphology. Since previous studies have focused on central peak dimen-

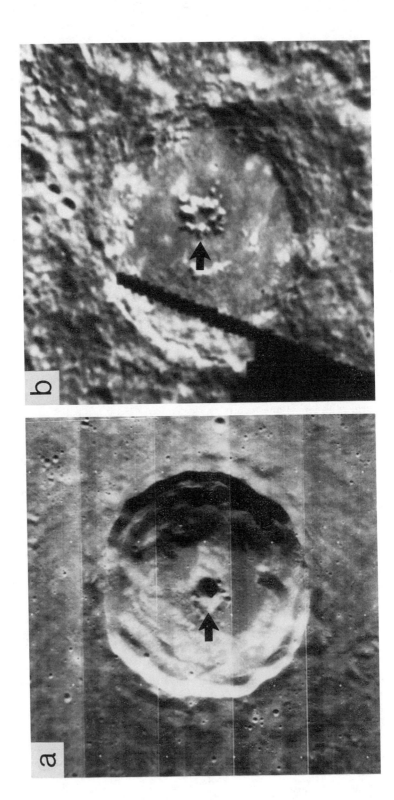

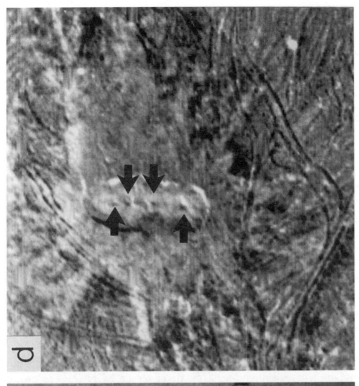

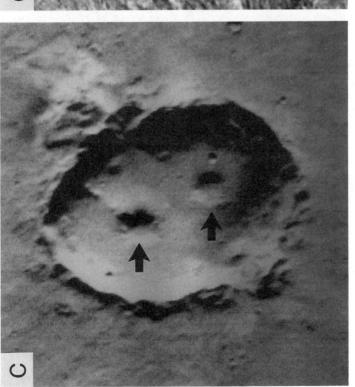

Fig. 15. Examples of craters on the Moon (a), Mercury (b), Mars (c), and Ganymede (d) with central depressions or pits (arrows) within the central peak complex. Figure 15a shows the 30-km diameter lunar crater Timocharis, whereas Fig. 15b shows the 130-km diameter Mercurian crater Hawthorne. Figures 15c and 15d illustrate overlapping craters on Mars and Ganymede, respectively, where central pits and ejecta ridges appear to indicate nearly simultaneous impacts. Such examples support the interpretation that central depressions reflect imprints of impactors requiring relatively long times to transfer their momentum/energy to the target. Planets with lower rms impact velocities require larger impactors to produce the same size crater, thereby increasing the relative size of the pit and frequency of occurrence on Mars and Ganymede. Paired craters on Mars are about 30 km wide; the example on Ganymede is about 40 km × 100 km. Lunar Orbiter IV-121-H3 (Fig. 15a) Mariner 10-H12 photomosaic (Fig. 15b), Viking 618A50 (Fig. 15c), Voyager 955J1 (Fig. 13d).

sions and occurrence, the following discussion explores the significance of ringed peaks and pits. Five observations need to be explored: the relative occurrence of ringed-peaks and pits on different planets, the diameter at which these structures begin to occur, the crater diameter range over which they occur, the relation between ringed-peak/pit diameter and crater diameter, and their relation to two-ringed impact basins and central peaks. These observations will be placed in the context of possible controlling variables of gravity, kinetic energy, impact velocity and the early-time coupling between impactor and target.

Figure 16 shows the relation between ringed-peak (or pit) diameter and crater diameter for Mercury, the Moon and Mars. If the dimensions of the interior ring structures are only controlled by the gravitational potential energy (e.g., rim collapse), then the diameter of the ring for a given crater diameter should increase with increasing gravity: Moon (162 cm s^{-2}) to Mercury or Mars (~380 cm s^{-2}). Figure 16a shows instead that ringed-peak/pit diameter decreases from the Moon to Mercury but increases from the Moon to Mars. It is often assumed that the presence of ice/water reduces the effective viscosity of the Martian crust, thereby accounting for the smaller onset diameters of complex crater morphologies and larger peak complexes than on the Moon. Gravity control of a less viscous substrate on Mars would account qualitatively for a larger ringed-peak/pit diameter on Mars relative to the Moon or Mercury, but not the relation between the Moon and Mercury.

The observed relation between ringed-peaks/pits and crater diameter (Fig. 14a) provides a basis for testing the possible roles of impact energy (impactor size and velocity), impact velocity only, and impactor size through crater-scaling formulas and predicted values for different planets (Table I). Because the observed crater diameter represents a modification of the excavation diameter used in most scaling laws, a correction to the observed relations in Fig. 16a may be necessary. As discussed above (Fig. 12), such a correction depends on the assumed crater growth/scaling model.

If the assumption is made that the diameter of the ringed-peak/pit is related to kinetic energy by a simple power law, then the observed relation (Fig. 16), kinetic energy, and scaling laws (Eq. 2) can be combined to give an explicit relationship between impact variables and observations. Specifically, this exercise results in the following expressions:

$$d_{pt} \sim (KE)^{0.292} \tag{18a}$$

$$d_{pt} \sim v^{-0.23} \, g^{0.18} \, \delta_p^{-0.29} \, \delta_t^{-0.35} \, D^{1.05} . \tag{18b}$$

For asteroidal impacts on the Moon (22 km s^{-1}), Mercury (34 km s^{-1}) and Mars (19 km s^{-1}) and the same ringed-peak and pit diameter (controlled by energy), craters on the Moon should be about 1.2 times larger than on Mars and about 1.3 times larger than on Mercury. These predictions can be made

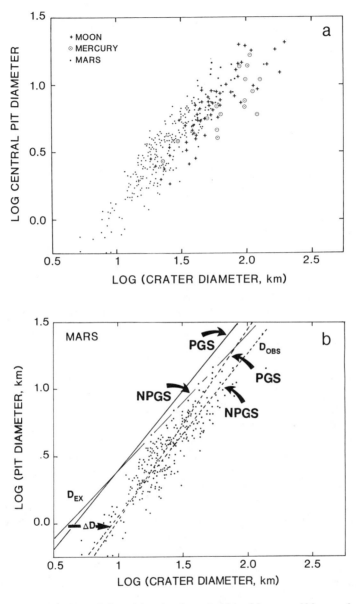

Fig. 16. Diameters of central pits and ringed peaks on the Moon, Mercury and Mars as a function of rim-to-rim crater diameter. Central pits/ringed-peaks increase in diameter for a given crater diameter from Mercury to the Moon to Mars (a), perhaps reflecting the relative size of the impactors for the expected rms impact velocities. In order to examine this hypothesis, crater diameters must be corrected for possible enlargement ΔD as shown diagrammetrically in Fig. 11. Two different models of crater growth/scaling for Martian data with $\Delta D \sim D_e$: proportional crater growth, PGS, and nonproportional growth/scaling, NPGS, are explicitly included in (b).

consistent with Fig. 16 if the density of Mercurian impactors is less than lunar/Martian impactors and if surface properties of Mars increase the size of ringed-peaks and pits at a given crater diameter. Cometary impacts at Mercury can meet the first condition and the presence of water/ice in the Martian crust can satisfy the second. Because comets significantly contribute to the Mercurian impact flux (see Hartmann et al. 1981), however, the higher energy for a given crater diameter should result in a proportionally larger number of ringed-peaks and pits.

A dependence between impact velocity alone (not size and velocity as in KE) and ringed-peak and pit diameter provides predictions more consistent with the observed dimensions:

$$d_{pt} \sim v^{-0.35} \, g^{0.18} \left(\frac{\delta_t}{\delta_p}\right)^{0.35} D^{1.05} . \tag{19}$$

For the expected relative rms impact velocities and the same ringed-peak/pit diameter, this assumption predicts Mercurian craters 1.00 times larger than lunar craters and 1.21 times larger than Martian craters with the same impactor/target properties. A slightly lower density for Mercurian impactors reflecting the predominant comet contribution brings these values closer to observations. As in the case of impact energy, however, the higher impact velocity at Mercury and the Moon would be expected to produce a larger number of examples, unless crater/ice in the Martian substrate is now used to enhance the expression without changing the relative dimensions.

If either impactor energy or velocity is the sole controlling parameter for ringed-peak/pit development, then several predictions can be made. First, the number of craters with such structures relative to the total number of craters at a given size would increase from Mars to the Moon to Mercury. Second, the onset and most frequent occurrence would increase from Mercury to the Moon to Mars. And third, central peaks, ringed-peaks/pits, and two-ring basins would not be expected to form at the same diameter. All three predictions are inconsistent with observations. First, the number of craters with ringed-peak complexes or pits decrease from Mars to the Moon to Mercury. Second, the onset diameter and modal diameter of occurrence decrease from Mercury to the Moon to Mars. And third, central peaks, ringed-peaks and pits, and two-ringed basins all can occur at the same diameter on the Moon, Mercury and Mars. Added to these observational constraints on occurrence, ringed-peaks/pits increase in diameter with crater diameter on all three planets. Consequently, the last alternative considers the possibility that the dimensions of such structures may be an imprint of coupling between impactor and target.

Ahrens and O'Keefe (1977) show that the far-field peak shock pressure decays with a power law of the distance from impact scaled to the projectile radius. The power-law exponent and the multiplicative constant is velocity dependent: -1.45 to -1.97 and -2.97 for the exponent and 1.4, 2.5 and 236 for the constant for impacts of 5, 15 and 45 km s^{-1} by anorthosite into anorthosite, respectively. A maximum pressure, therefore, can be related to multiples of the projectile size and can be expressed in terms of the crater diameter as in Fig. 9. Peak pressures exceeding 300 kbars and 100 kbars remain within 3 to 6 projectile radii for both 5 and 15 km s^{-1} impacts, but expand to more than 22 and 37 radii for a 45 km s^{-1} impact, respectively. High-velocity impacts, therefore, are characterized by high peak pressures engulfing a larger distance from the projectile and, if not offset by scaling relations, a larger fraction of the excavation cavity. Low-velocity impacts ($<$15 km s^{-1}) are characterized by a limited zone of high-peak pressures and compression approximately proportional to the projectile size. This difference between low- and high-velocity impacts may produce a difference in the structure of the central peak complex: low-velocity impacts retaining an imprint of the compressed/decompressed zone while high-velocity impacts do not. If this hypothesis is correct, then the diameter of central ringed-peak and pits may provide a measure of the impactor diameter for low-velocity impacts.

There is observational evidence that central peaks with pits and central pits reflect a decompressed zone surrounding the projectile. First, geologic analysis of certain complex terrestrial impact craters indicates a restricted central breccia column within the Decaturville, Missouri (Offield and Pohn 1977), Gosses Bluff, Australia (Milton et al. 1972), and Sierra Madera (Wilshire et al. 1972) impact structures. Second, numerical codes characteristically develop a central compressed and depressed zone surrounding the projectile (see, e.g., Bryan et al. 1978). Typically, this profile disappears during the ballistic "coast" phase as the calculation is carried to very late times (see, e.g., Orphal et al. 1980), but at least a transient stage exists that resembles the proposed model. Such calculations typically assume relatively small, high-velocity impacts that would not be expected to preserve the central compressed zone. And third, Fig. 15 shows two central pits in overlapping craters on Mars, whereas Fig. 15d illustrates four separate central pits in overlapping craters on Ganymede. The existence of a median ridge of ejecta extending in both directions away from the impacts strongly resembles experimentally produced simultaneous impacts (Oberbeck and Morrison 1974). The twin impacts were sufficiently close to prevent the formation of a wall/rim between them. Overlapping impacts of different ages would retain the rim/wall between the younger and older events (see Schultz 1976a); preservation of the ejecta facies and central pits precludes erosion destroying the rim/walls. If the central pits were the result of centripetal rim collapse, then a more complex single central peak structure would be expected.

Scaling relations can be used to relate the projectile size to the crater diameter and the resulting trends compared with observations. The first comparison presumes a model of proportional growth and scaling; the second considers implications of nonproportional growth.

If the Schmidt-Holsapple scaling relation (Eq. 2) is used without invoking a change in crater shape with scale or additional crater modification, then the distance to a given limit of shock compression (as expressed by the diameter of the central ringed-peak/pit) is given as:

$$d_{pt} \sim r_p \sim v^{-0.4} \, g^{0.2} \left(\frac{\delta_t}{\delta_p}\right)^{0.4} D^{1.2} \, . \tag{20}$$

This relation predicts Mercurian craters 1.00 and 1.2 times the size of lunar and Martian craters, respectively, for the same size ringed-peak/pit diameter for asteroidal impact velocities. Equation (20) predicts that d_{pt} should increase with the excavation crater diameter raised to a 1.2 power. Crater enlargement, however, increases this exponent near the simple-complex transition diameter (see Fig. 11). Only at the larger diameters can the excavation center diameter approach a linear relation with the post-collapse crater diameter. Figure 16b shows that exponents of 1.4 for small (5 km) and 1.2 for larger craters appear too high for the observational data unless other processes are invoked (e.g., a nonlinear relation between d_{pt} and r_p).

Equation (10) provides a model for nonproportional growth based on the shock transit time:

$$d_{pt} \sim r_p \sim v^{-0.23} \, g^{0.05} \, f(\delta,c) \, D^{1.05}$$

$$f(\delta,c) = \left(\frac{\delta_p}{\delta_t}\right)^{0.23} \left(\frac{c_p \, \delta_p}{c_t \, \delta_t}\right)^{0.123} \left[1 + \left(\frac{\delta_p}{\delta_t}\right)^{1/2}\right]^{0.123} \tag{21}$$

where the transition diameter (Eq. 11) and previous empirically derived values for the exponents have been introduced. The same impactors on all planets result in Mercurian craters 1.06 and 1.14 times the size of lunar and Martian craters, respectively. As shown in Fig. 16b, crater enlargement increases the diameter exponent at small sizes but approaches the derived value of 1.05 at large sizes in general agreement with observations. Hale (1983) independently derived a pit-crater diameter exponent of 1.01 for craters of all sizes.

The model of nonproportional crater growth/scaling implies that the excavation crater profile is in part preserved. The model of central ringed-peak and pit formation being considered here implicitly indicates that the onset of nonproportional scaling and the minimum size of craters with peaked-rings/pits should both be related to a critical value of t_p^*. Equation (11) provides a basis for explicitly predicting the relative onset diameter for this type of central peak morphology if

$$D_t \sim \frac{v^{1.17}}{g} \left(\frac{\delta_t}{\delta_p}\right)^{0.222} \left(\frac{c_t}{c_p}\right)^{0.833} \left[1 + \left(\frac{\delta_p}{\delta_t}\right)^{1/2}\right]^{-0.833} \, . \tag{22}$$

Equation (21) indicates that ringed-peak/pit diameter and crater diameter should exhibit a very weak dependence on gravity, whereas the onset diameter (Eq. 22) should be very sensitive to both gravity and velocity. For present estimates of asteroidal rms impact velocities (Table I), the onset diameter for ringed-peaks/pits on Mercury should be a factor of 1.4 times smaller than on the Moon but 2.0 times larger than on Mars. Observations indicate a minimum crater diameter for central pits to first appear near 20 km on Mercury, 18 km on the Moon and 5 km on Mars. The larger observed minimum size for Mercury may reflect insufficient resolution for characterizing the smaller central peak morphology or a higher average impact velocity.

Regardless of the scaling relations (Eqs. 20 or 21), the different sizes of the projectiles relative to the crater diameter for different velocities are within a factor of two, whereas the zone exceeding peak pressures of 100 kbar surpasses a factor of 7 for the same diameter crater. Consequently, material readjustment within the excavation cavity may be different. It is speculated that low-velocity impacts ($v < 15$ km s^{-1}) can preserve a marker of a given pressure limit proportional to projectile size. Higher-velocity impacts result in

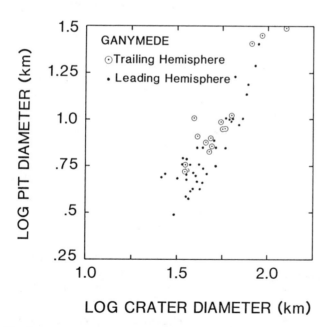

Fig. 17. The diameters of central pits on Ganymede are systematically larger on the trailing hemisphere (near 270°) relative to the leading hemisphere (near 90°) for a given crater diameter. Data for the trailing hemisphere were limited to 310-340° longitudes due to available coverage and resolution, whereas data for the leading hemisphere were restricted to 115-144°. The observed difference would be expected if central depressions are proportional to the size of the impactors for a given impact velocity.

a larger fraction of the crater subjected to high-peak pressures that respond by fluid collapse perhaps along the models of Melosh (1982a) or Croft (1985).

A possible test for the proposed model is the size distribution of central pits on Ganymede. Because the rotation of Ganymede is presently dynamically locked to its revolution around Jupiter, heliocentric objects impact the orbit-leading side with rms velocities of 27 km s^{-1} and the trailing side with velocities about 14 km s^{-1} (Shoemaker and Wolfe 1982). Diameters of central pits in well-preserved craters of a given diameter on the trailing hemisphere (near 90° longitude) are systematically larger than pits in craters on the leading hemisphere (near 270°) as shown in Fig. 17. The ratio of pit diameters on the trailing and leading hemispheres would indicate a factor of about two in impact velocities (from Eq. 21). Pit diameters for the leading hemisphere approximate the lunar data, thereby indicating impact velocities of about 22 km s^{-1} whereas data for the trailing hemisphere more closely match the inferred lower-velocity impacts in the Martian data, i.e., about 10 to 15 km s^{-1}. Analysis of craters of all ages did not reveal this systematic difference, perhaps due to the effects of global reorientation (Murchie and Head 1986).

In summary, the observed empirical data for ringed peak complexes and central peaks with pits on the Moon, Mercury, Mars and Ganymede can be reproduced by a relatively simple conceptual model based on the impactor penetration time (shock transit time) where ringed-peaks/pits reflect a unique record of impacts with comparatively large values of $2r/v$. If complemented by recent analysis of laboratory experimental data, then the relative number, relative size, and onset crater diameter for ringed-peaks and pits in craters on Mercury, the Moon and Mars can be understood. Mercurian impacts are dominated by high-velocity objects. Lunar and Martian impacts exhibit both populations with different proportions: a significant fraction ($\sim$65%) of Martian impacts probably occur at low-impact velocities ($v \le 15$ km s^{-1}) and as represented by the relative proportion of craters with ringed peaks and pits. The model does not need to invoke unique properties of the Martian crust, as consistent with observations by Hodges (1978) and Hale (1983), indicating that Martian pit craters are largely independent of latitude and terrain.

Two-Ringed Basins

As craters exceed a certain size on Mercury and other planets, they exhibit a characteristic two-ring pattern: the inner ring expressed as a ring of central peaks and the outer ring, as the crater rim crest. The origin of such a characteristic geometry has been attributed to: a discontinuity in pre-impact crustal structure (Wilhelms et al. 1977; Hodges and Wilhelms 1978); a simple progression in scale from central peak structure in smaller craters (Head 1977); an uplifted remnant of an inflection in the excavation crater profile related to the crater formation process rather than crustal structure (Schultz et al. 1981); some unspecified process related to impact velocity (Wood and Head 1976; Hale and Head 1980c); and as one of a continuum of morphologies reflecting the degree of gravity-induced uplift/collapse of the transient crater cavity (Gault et al. 1975; Dence et al. 1977; Murray 1980; Melosh 1982a). Mercury can provide important clues for the origin of the central ring structures since they are much more frequent than on the Moon, they develop craters with smaller rim-crest diameters, and they are relatively well preserved.

Interior ring structures in craters and basins on Mercury curiously occur in the same diameter craters where central peaks and central pits occur (Fig.

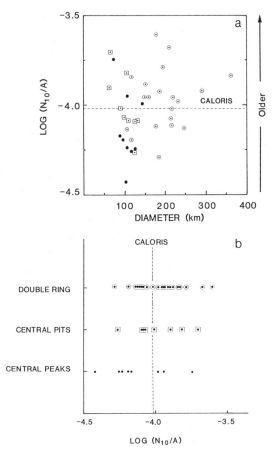

Fig. 18. Occurrence of different morphologies of central crater relief on Mercury as functions of time and crater size: central peak craters (dots), central ringed-peak and pits (squares), and two-ringed basins (circles). The relative age of different craters expressed by the areal density of craters> 10 km (N_{10}/A) is shown in (a). Two-ring impact basins characterize all impact structures> 150 km. All morphologies occur in craters <150 km, but with central peak structures characterizing the younger impacts. In (b), the occurrence of different central relief morphologies as a function of relative age only is shown. The approximate time of formation of the Caloris impact basin is indicated by the dashed line.

18a). The areal density of craters larger than 10 km (N_{10}/A) superposing these craters/basins reveals that large-diameter central rings (two-ring basins) dominate the central relief pattern in craters/basins larger than 150 km in diameter and are generally older than 50 to 150 km diameter craters with central peaks or peaks with pits (Fig. 18b). Moreover, craters with central peak/pits generally are older than craters with central peaks. Consequently, a progression in central peak morphology with time could be inferred. Such a progression qualitatively might indicate time-varying crustal properties with lateral heterogeneities accounting for the observed occurrence of craters with a central ring, peak-pit, and peak at the same diameter in post-Caloris time (Fig. 18a). Figure 19 reveals, however, that the ring-crater diameter power-law relation is nearly the same for Mercury, the Moon, Mars and Ganymede. It is unlikely that crustal structure is the same in these very different planets. The onset of two-ring craters/basins does not appear to be just gravity- (Mars vs Mercury) or strength- (Mars vs Ganymede) controlled. Another mechanism may play a role.

Two-ring impact basins on Mercury exhibit a much more well-defined inner ring typically expressed as an inward scarp. This morphology contrasts with the typical interior rings of lunar and Martian two-ringed basins, with some exceptions (e.g., the lunar Moscoviense basin), and more closely resembles the central-pit structures of much smaller Martian craters. The following discussion, therefore, considers the possible implications of the observations (Figs. 16 and 19) in the framework of the model of central ringed-peak/pit leading to Eq. (21). Since gravity should affect primarily the onset of occurrence for a given impact velocity (Eq. 22) and not the relative size of such structures (Eq. 21), the similar ring-crater diameter distribution for different planets could reflect impacts with relatively limited and similar velocity distributions. Reasonable agreement with Eq. 21 with observations is possible for very low-velocity ($\sim$5–7 km s^{-1}) objects with comet-like densities (0.5–0.8 g cm s^{-3}) and sound velocities (0.3 km s^{-1}). Higher-velocity objects are possible if the large size of the interior structure results in modification (e.g., enlargement by collapse) that precludes a simple extrapolation from the smaller ringed-peak/pit central structures. Regardless, such objects might indicate primitive accretion relicts in co-orbiting heliocentric circular orbits (see Hartmann 1977; Leake et al. 1987). The observed onset diameter of craters/basins with large rings, however, progress from 50 km on Mars to 110 km on Mercury and to 140 km on the Moon. This sequence closely resembles the predicted pattern for impacting objects with velocities typifying heliocentric co-orbits for each planet: 50 km for Mars, 110 km for Mercury (reference diameter) and 165 km for the Moon. Figure 18 also includes craters on Ganymede that exhibit anomalously large central pits (Croft 1983). The slightly larger pit diameter and smaller onset diameter would be consistent with Jupiter-orbiting objects impacting Ganymede with slightly lower impact velocities ($\sim$3 km s^{-1}).

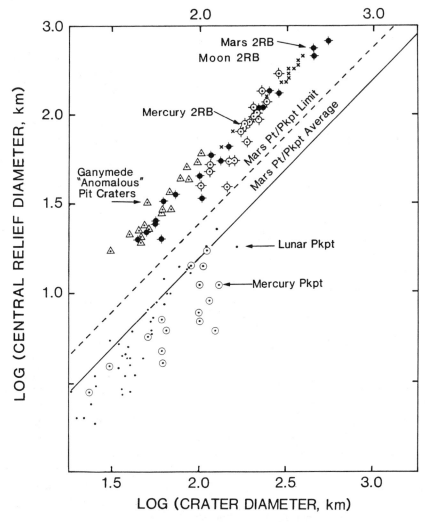

Fig. 19. Dimensions of central ringed-peaks/pits (Pt/Pkpt) and interior basin rings of two-ring basins (2RB) as a function of crater diameter for the Moon, Mercury, Mars and Ganymede. Two-ringed impact basins and "anomalous pit craters" on Ganymede (Croft 1983) do not exhibit the systematic differences in size observed for ringed-peak/pits in smaller craters but fall on nearly the same linear trend. This observation may indicate a different style of formation of rings, a common style of modification of the central relief, and/or formation by large, very low-velocity co-orbiting objects early in the geologic history for all bodies.

In summary, the cratering record on Mercury is consistent with a model of nonproportional crater growth and scaling that depends on shock transit time which depends on the projectile penetration time as first suspected by Baldwin (1963) and qualitatively described by subsequent studies as a consequence of impact energy or velocity (see, e.g., Malin and Dzurisin 1978; Pike 1980b; Hale and Grieve 1982). Craters with ringed peaks or pits may represent a distinctive morphology unique to impacts with relatively large shock transit times, which varies inversely with velocity. Although best expressed on Mars and the Galilean satellites Ganymede and Callisto, the presence of such morphologies on Mercury and the Moon lessens the role of water/ice in their formation, but perhaps not in expression. The proposed model does not preclude other models of central peak formation. Rather, oscillatory plastic flow as described qualitatively (see, e.g., Baldwin 1963, 1981; Murray 1980) and quantitatively (Melosh 1982a) requires a minimum impact velocity and energy when dynamic floor rebound destroys the original crater form.

V. CRATER STATISTICS

The cratering record of Mercury has been the focus of considerable debate, as reviewed by Woronow et al. (1982) and Strom and Neukum in their Chapter. Different slopes in the crater size-frequency distributions have been attributed to a variety of causes including different impactor populations, changing size distributions of a given population through time, broad-scale resurfacing processes, a balance between the rate of crater formation and destruction (i.e., equilibrium), and changing roles of secondary craters through time. Rather than reviewing the Mercurian cratering record specifically in these terms, the following discussion continues the holistic view of crater excavation and modification in order to explore possible signatures previously assigned to changes in the impactor population.

If a single size-frequency distribution of impactors is assumed, then previous discussions provide a basis for examining possible effects on the size-frequency distribution of the resulting craters. Figure 20 illustrates several options where the cumulative number N_{cd} of impactors is assumed to depend on projectile size r_p as $N_c \sim r_p^{-2.5}$. Use of the Schmidt-Holsapple scaling relation (Eq. 2) results in a cumulative size frequency of craters (N_{cd}) as $N_{cd} \sim D^{-\lambda}$, where $\lambda = 3.0$. Proportional crater scaling and simple wall failure, where $\Delta D \sim D_e^{\omega}$ and $0 < \omega \leq 1$, result in an offset of the crater production slope near the transition diameter where crater modification becomes important (Fig. 20a and b). Because the transition region is typically over a relatively narrow range in diameter compared to the statistical sample, proportional crater scaling ensures that any effect in the crater size-frequency distribution should not appreciably or permanently modify the simple relationship between impactor size and crater diameter. Figure 20c illustrates a further example where viscous relaxation has been included. Based on cen-

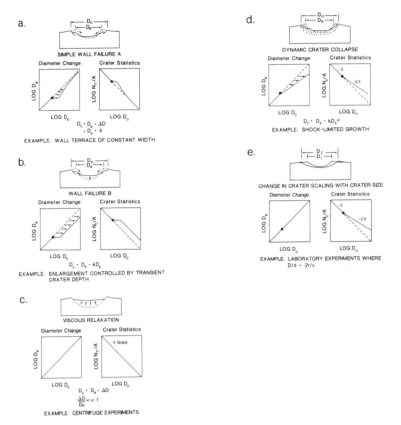

Fig. 20. Potential effects of different models of crater growth and/or scaling (Fig. 9), and modification (Fig. 11) on the statistics of observed crater diameters D_0 produced by impactors with a cumulative size-frequency distribution exponent of -2.5. Each illustration shows the effect of enlargement ΔD on the excavation crater diameter D_e as depicted in the cross section above. Plots a and d show models of proportional crater growth with different styles of modification, whereas plot e includes a model for nonproportional growth and/or scaling.

trifuge experiments by Wedekind et al. (1970) and data from Gault (personal communication), changes in crater depth by floor uplift do not affect crater diameter (see Fig. 21). Consequently, any significant change in the crater distribution must be the result of other processes and should not depend on gravity or impactor parameters for the same size distribution of objects. Any observed change with time can only reflect changes in crustal properties or the impactor size-frequency distribution.

The dynamic collapse model advocated by Croft (1981a, 1985) and others requires a change in scaling for craters larger than the transition diameter (Fig. 20d). For the assumed size-frequency distribution of impactors with a -2.5 power law, expressions derived by Croft (1985) indicate that the crater

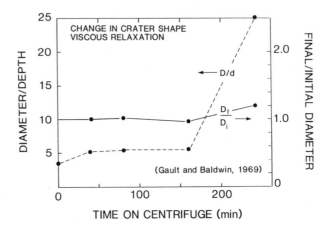

Fig. 21. Change in crater shape as a result of viscous relaxation induced on a centrifuge. Changes from the initial diameter-to-depth ratio of 4 : 1 largely reflected floor uplift rather than crater widening. Figure from Wedekind et al. (1970) and unpublished data from Gault (personal communication).

production slope for complex craters will be reduced with $\lambda \approx -2.1$. Since this model explicitly incorporates the limit of crater modification, the -2.1 power law represents the maximum value possible unless additional changes to the size distribution of impactors or target properties are involved. A value of $\lambda = -2.5$ for the cumulative size frequency of craters conversely indicates a cumulative size frequency of impactors with a -3.0 exponent.

If nonproportional growth/scaling applies, then the excavation cavity shape D/d changes with increasing diameter as discussed above, and the cumulative frequency distribution of craters should change near the transition diameter. Below the transition diameter, the projectile-scaled crater diameter depends on $\pi_2^{-\alpha'/3}(\delta_t/\delta_p)^{1/3}$; therefore, $N_{cd} \sim D_o^{-3}$. Above this transition, Eq. (10) applied and $N_{cd} \sim D_o^{-2.6}$. The transition diameter depends on impactor velocity and planetary gravity as well as physical contrasts between target and impactor. The transition diameter can vary over a factor of 2 due to possible ranges in impactor velocity even without changes in impactor density or target properties. Crater modification effects (Fig. 11) can further change λ to values approaching -2.

A given crater size-frequency distribution, therefore, will exhibit a single power-law dependence (with a minor shoulder near the onset of crater modification) if proportional crater scaling applies. A change in the scaling relation as described by Croft (1981a) or a transition to nonproportional crater scaling (Schultz and Gault 1987) will produce a change in the crater size-frequency power law even with a single population of impactors. Nonproportional crater scaling predicts a transition diameter dependent on $v^{1.17}/g$ (Eq. 22). Consequently, the transition from a -3.0 to -2.5 slope for a common impactor population varying principally in velocity (rather than bulk properties) should mirror this transition. On Mercury, the post-Caloris smooth plains exhibit a transition from -3.0 to -2.5 slope between 5 and 8 km (from published statistics in Schaber and McCauley 1980; Hartmann et al. 1981). On the Moon, the post-mare crater size-frequency distribution exhibits a similar transition between 2 and 3 km (see Hartmann et al. 1981). On Mars, the younger ridged plains exhibit essentially the same trends but shifted with the transition diameter between 0.5 and 1.0 km (Dial 1978). The relative transition diameter for the Moon : Mercury : Mars is therefore $9 : 3 : 1$. A dependence only on $v^{1.17}$ predicts 6:4:1 if it is assumed that both asteroidal and short-period comets dominate crater statistics for the Moon, parabolic impacts dominate for Mercury, and asteroids for Mars. If the possible physical differences between asteroids ($\delta_p = 2.5$g cm^{-3}, $c_p = 1.2$ km s^{-1}) and comets ($\delta_p = 1$ g cm^{-3}, $c_p = 1.2$ km s^{-1}) are included through Eq. (22), then the relative transition diameters become $10 : 6 : 1$. These results are in reasonable agreement with the observations considering other possible effects, such as secondary impact cratering.

Crater statistics for older and larger diameter craters show very similar distributions on the three terrestrial planets. This similarity has led Woronow (1978) and Woronow et al. (1982) to argue that the inner planets were impacted by the same family of bodies during late heavy bombardment. Such a family, however, exhibits lower cumulative slopes in the crater population that approach values expected for equilibrium cratering. The reader is referred to the Chapter by Strom and Neukum for further details on the controversy.

Figure 22 summarizes three ways to change the size-frequency distribution of impact craters on a given planet. The first involves a change in the size distribution of impactors due to either different provenances or different processes acting at different scales, e.g., fragmentation or secondary cratering. The second applies the effects on nonproportional scaling where a change in rms impact velocity results in a change in the transition diameter. A narrow shoulder (over a factor of two in diameter) exhibiting a lower slope is possible as a result of crater widening processes near the transition diameter. A third possible reason for a change in the crater production slope introduces a systematic size-dependent change in bulk density with size of the body. Such a change might reflect decreasing porosity with increasing size due to lithostatic pressure or accretionary/fragmentation processes. Although speculative, such an effect would be reflected in the crater size-frequency distribution regardless of the scaling law used. Conversely, an increase in target density with depth conceptually could produce a decrease in power-law production exponent. The latter option could be important only for the largest impacts on the terrestrial planets since depth-dependent density increases are gradual compared to other variables. It might yet play a role on the icy satellites and smaller objects.

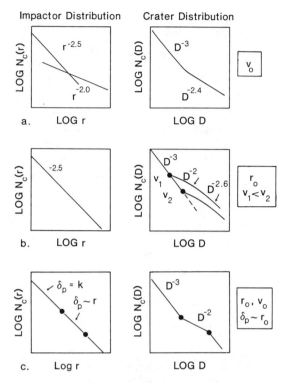

Fig. 22. Three possible origins for changes in the cumulative size-frequency distribution of craters $N_c(D)$ from a given distribution of objects $N_c(r)$ impacting the same reference area. For a given impact velocity (v_0), two different families of impactors with different size distributions can produce an inflection in crater statistics (a). If a change in the scaling relation for crater diameter occurs as a function of shock-transit time (impactor penetration time), then a single impactor size distribution can produce inflections in crater statistics depending on the characteristic impact velocity (b). An inflection in crater statistics from a single power-law distribution and impact velocity of objects is possible if the bulk density δ_p changes from a constant k to a density that increases with increasing size up to a limit due to gravitationally induced compaction.

In summary, interpretations of crater statistics cannot be completely decoupled from interpretations of crater excavation and modification. Changes in crater scaling relations can be just as important as changes in the size distribution or type of impactor. Conversely, if other observations could limit the options available for crater statistics, then perhaps new clues for the cratering process will emerge.

VI. EJECTA EMPLACEMENT

Could the higher gravity, high rms impact velocities at Mercury or its lithospheric history also leave potential signatures in the distribution and type of material excavated from craters and basins? Post-Mariner 10 analysis largely focused on gravitational effects on the secondary cratering process (Gault et al. 1975; Scott 1977), particularly with respect to the origin of the smooth plains (Wilhelms 1976; Oberbeck et al. 1977). This emphasis reflected analyses of samples returned from Apollo 14 and 16 (see, e.g., Warner 1972; Simonds et al. 1977; James 1980). The limited resolution available from Mariner 10 prevented exploring possible effects of impact velocity or impactor type. Nevertheless, clues do exit. The following discussion reviews the secondary cratering and ejecta emplacement processes with possible effects of the larger gravity, state of the lithosphere, and the higher impact velocity on Mercury relative to the Moon.

Secondary Cratering Processes

Gault et al. (1975) documented the clear effect of the increased gravity on Mercury (relative to the Moon) on the radial extent of ejecta deposits. The maximum extent of the continuous ejecta around well-preserved craters on Mercury extends only about 0.65 times that for lunar craters, whereas the areal density of secondary craters beyond the continuous ejecta is increased. Calculations by Gault et al. (1975) indicate that the areal density of secondaries increases from 5 times (near the crater) to as much as 10 to 20 times the density of lunar secondaries with impact velocities at the same relative distance from the primary increased by 50%. Consequently, the role of secondary cratering for ejecta emplacement on Mercury should be enhanced relative to the Moon. The Apollo 16 mission demonstrated the importance of the secondary cratering process for the origin of the smooth highland plains previously interpreted by many as ancient volcanic deposits. Studies synthesizing lunar samples, laboratory experiments, terrestrial field studies, and observations of the lunar surface built a strong case for the impact origin of the lunar smooth plains (Cberbeck et al. 1975) with equally important implications for the origin of the Mercurian smooth plains (Wilhelms 1976; Oberbeck et al. 1977). Strom et al. (1975b), Trask and Strom (1976), Strom (1977), and others have argued (and continue to argue) that in spite of the proposed physical mechanisms and Apollo 16 analogy, the smooth plains of Mercury developed as a consequence of volcanic processes more analogous to the emplacement of the lunar maria. The continued importance of secondary impact cratering for interpreting the origin of the Mercurian (as well as the lunar) smooth plains prompts a brief review of the controversy concerning ejecta emplacement processes.

Over the last decade, studies by Oberbeck (1971,1975), Oberbeck et al. (1975), and Hörz et al. (1983) have greatly influenced and increased our understanding of ejecta emplacement. Laboratory experiments demonstrated that impacts into granular targets can excavate more than 200 times the projectile mass. Consequently, at these scales, a ballistically transported fragment impacting this type of target would be virtually lost in locally excavated debris. In order to extend the laboratory results to planetary scales, Oberbeck et al. (1975) measured the diameter of Copernicus secondaries, assumed a D/d = 4 in order to estimate their displaced volume, and used energy-scaling relations from Earth-based explosions to derive adjusted values of cratering efficiency. As an example, continuous ejecta around a 25-km lunar diameter should contain 24 to 34% locally excavated debris and a 100 km crater, nearly 50%. With increasing distances on the Moon (or equivalently, greater gravity on Mercury), the secondary cratering process and locally derived debris increase in importance. At basin scales, the outward-moving inclined ejecta curtain incorporates the local debris, and the sustained momentum of the mixture produces an outward moving ejecta flow. Hörz et al. (1983) have successfully applied these results to the continuous ejecta deposits around the Ries Crater, Germany.

A different study of ejecta emplacement reconsidered several key assumptions (Schultz and Gault 1985a): (a) the physical state of the ejecta; (b) the relative dimensions of the ejecta curtain and contained ejecta; (c) the process of low-velocity (<2 km s^{-1}) impact cratering; and (d) the downrange dispersal of projectile material. Although the relative thickness of the ejecta curtain in the laboratory or at planetary scales may only be 5 to 10% of the crater diameter, the absolute thickness for a 100-km diameter crater or 1000-km diameter impact basin will be substantial: 5 to 10 km to 50 to 100 km, respectively. The weakness of naturally occurring materials increases with scale, and the average peak shock pressures of ejecta increase with absolute ballistic range (see, e.g., Schultz et al. 1981). Consequently, the ejecta curtain associated with the formation of large craters and basins will be composed of a wide range of fragment sizes with the greatest fraction occurring in the smallest sizes. The impact of such a debris curtain occurs over a significant time: 25 s at 0.3 R from the rim of a Copernicus-size impact. A snapshot of this process would more closely resemble impacts by an ensemble of debris rather than by a single large object. Consequently, Schultz and Gault (1985a) explored the effect of such a configuration on the secondary cratering process.

Clustered impacts into sand have cratering efficiencies 5 to 10 times less than the equivalent velocity and mass of a single impactor. This reduction in cratering efficiency depends on the cluster size, impact angle and target strength. The mass density of clustered impacts was found to have a marginal effect on cratering efficiency if the Schmidt-Holsapple scaling relation for low-strength targets is used. Oblique impacts ($<60°$ from the surface) produced shallow craters with extensive herringbone structure, a downrange ejecta fan, ridge-like rim and floor mounds—features characteristic of lunar secondaries. Targets with greater shear strength (e.g., compacted pumice) exhibited an even greater reduction in cratering efficiency. It is important to recognize that the secondary cratering process is a subsonic phenomenon: unless formation occurs completely in a low-strength material, analogies with single impacts into sand cannot apply. Important differences include much shallower depths of excavation, compression as well as excavation of the impacted target, and downrange ricochet and dispersal of the impactor. The net effect predicts near-rim continuous ejecta deposits of both lunar and Mercurian craters to be composed of 80 to 90% primary material, in general agreement with spectral observations of lunar craters (Pieters et al. 1985). At greater distances from the rim of the primary crater, ejecta ricochet out of secondary craters and are either deposited/mixed in the near-surface regolith or produce tertiary craters. Dispersion of the ricocheted material over broad areas downrange from the secondary craters is believed to play an important role in diluting the signature of primary material.

The distinction between the scaling of secondary cratering and ejecta emplacement models becomes most apparent at the largest scales. The model by Oberbeck et al. (1975) emphasizes the increasing contribution of locally excavated debris to the ejecta deposit with distance from the impact. The model by Schultz and Gault (1985a) proposes that the continuous ejecta deposits around craters and basins should contain a large fraction of primary material due to the process of emplacement. At greater distances, both models generally converge as ejecta disperse but with a few important differences. The Schultz/Gault model predicts a much lower cratering efficiency for a given mass, not only because of the proposed low density of the impacting cluster of debris but also because the low velocity of impact precludes extensive shock disruption of the target, thereby excluding the use of gravity-scaling relations. Although large secondaries may exceed the usually inferred onset diameter for gravity scaling, the process of shock preconditioning the target to an effectively strengthless state does not apply. The relatively low-impact velocities and the long interaction time indicate that clustered impactors forming a 10-km diameter secondary crater displace less (60%) of its mass, whereas a single impactor displaces nearly five times its mass. As further illustration, the formation of a 10-km diameter secondary on the Moon requires a 5 km solid block of ejecta, whereas a cluster of ejecta requires only that the diameter of the cluster approaches 5 to 8 km. Consequently, a much larger percentage of primary material (even at large distances) is possible. Possible signatures include fissured facies within large basin secondaries (Schultz 1981; Schultz and Gault 1985a). Subsequent studies by Hörz et al. (1983) summarize the Oberbeck et al. hypothesis in the context of the Ries crater, Germany, while recent analysis by Vickery (1986) provides support for the Schultz-Gault model. Studies of lunar spectra by Pieters et al. (1985) are consistent with the Schultz-Gault model of the near-rim continuous deposits, while derived mixing ratios for distant secondaries provide support for either model due to the ambiguity between mass mixing (reflecting cratering efficiency) and spatial mixing (reflecting not only cratering efficiency but also the downrange dispersal of the primary signature in the surface).

At basin scales, the ejecta emplacement process may be affected not only by the increase in impact velocity and size distribution of ejecta at the same relative distance from the crater (i.e., scaled to the basin radius) but also by the effect of surface curvature and the effective viscosity of the lithosphere. Impacts into planar targets produce a characteristic curtain inclined at a relatively constant angle with respect to the horizontal as it moves away from the crater rim. Impacts into hemispherical granular targets (Schultz et al. 1986*b*) demonstrated that during excavation the ejecta curtain angle is referenced to the local surface plane, but once the ejecta curtain becomes detached from the excavation crater, it also marches outward at a constant angle with respect to the surface tangent at the excavation crater rim (see Fig. 23). Due to surface curvature, the ejecta curtain meets the surface at increasingly steeper angles with increasing distance from the crater rim (Fig. 24), an effect that is enhanced if the uniformly downward-directed gravity field in the laboratory is theoretically corrected for a radially decreasing, inward-directed planetary

Fig. 23. The effect of surface curvature on ejecta curtain angle. Impacts into hemispheres of compacted pumice reveal that the ejecta curtain angle during excavation remains approximately constant with respect to the local surface tangent. After crater formation, the ejecta curtain becomes detached inertially and retains nearly a constant angle with respect to the surface tangent at the crater rim.

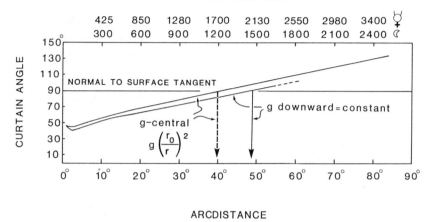

Fig. 24. Evolution of the ejecta curtain angle with respect to the surface tangent (a) at different distances from the rim as observed in the laboratory (gravitational field directed downward) and as applied to the Moon and Mercury (gravitational field directed to the center of mass). When the base of the ejecta curtain has advanced about 40° arc distance, the curtain forms a vertical wall of debris with respect to the surface.

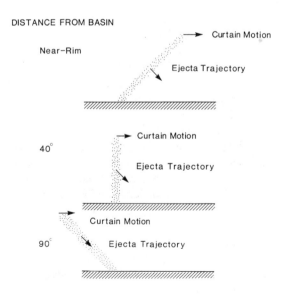

Fig. 25. The evolution shown in Fig. 24 illustrated in the frame of reference of the surface of a planet for different arc distances from the basin rim.

field (Fig. 25). At about 40° arc distance from the excavation cavity rim (~1700 km on Mercury), the ejecta curtain is oriented vertically with respect to the surface for initial ejection angles of 45°. If higher lithospheric temperatures existed at the time of impact (Schaber et al. 1977), basin impacts may have had steeper ejection angles analogous with impacts into low-viscosity material (see, e.g., Greeley et al. 1982). Ejecta within this curtain, however, continue on ballistic trajectories such that they impact at the original ejection angle, thereby increasing interactions between secondary ejecta produced by debris near the front of the curtain and still-incoming primary ejecta. One consequence of the changing ejecta curtain angle and the increased interaction time is the enhancement of ejecta flow within 2 radii of the rim.

Possible Signatures/Consequences of High Impact Velocities

Impacts by parabolic comets with rms velocities exceeding 80 km s^{-1} (see Table I) comprise a significant fraction of the crater population on Mercury. Because melting and vaporization relative to projectile mass increase as v^2, their effects on Mercury should be increased more than a factor of 10. How these effects will be expressed remains uncertain. Possible signatures include style of crater ray formation, regolith agglutinate production rates (and therefore crater ray destruction rates), rim and floor deposits of impact melt, crater aspect ratio, and local impact-induced magnetic fields.

High-velocity impacts producing primary craters on Mercury should increase the proportion of highly shocked materials for a crater of a given size. Although numerical codes comparing 5 and 15 km s^{-1} impacts of iron into anorthosite (Orphal et al. 1980) confirm this speculation, a wide range of shock levels are possible at any given distance from the crater rim (Schultz et al. 1981). The size distribution of ejecta only partly reflects the peak shock pressure history. Ahrens and O'Keefe (1978), for example, demonstrated that some high-velocity ejecta may be only slightly shocked, and Melosh (1984) proposed that near-surface spallation fragments increase in size with increased shock-transit time. Other studies suggest that comminution processes may become increasingly important with crater size (Schultz and Mendell 1978; O'Keefe and Ahrens 1987). If secondary craters reflect impacts by clusters of fragments, it may not be possible to characterize the characteristic fragment size from the size of the secondaries near the primary (Vickery 1986). Shoemaker (1962) and Vickery (1986) identified an apparent cutoff in ejecta with ejection velocities exceeding 1 km s^{-1} for Copernicus. This rapid fall off in areal density and median (not maximum) size of secondaries with derived ejection velocity may be related to the degree of shock comminution, thereby providing a possible measure of primary-impactor velocity. Other variables (strength of the substrate) and limited resolution for Mercury preclude exploring this possibility in detail.

Impact melt deposits can be tentatively identified on crater rims and floors (Hawke and Cintala 1977; Hawke et al. 1986), and dark halos surrounding fresh large craters qualitatively resemble glassy impact melts surrounding lunar craters (Hawke et al. 1986). If the relative fraction of energy partitioned to impact melting becomes constant for impact velocities >15 km s^{-1}, then the relative amount of impact melt on the floors and rims of Mercurian craters may not be that different from lunar craters. A few craters, however, have very distinctive offset halos in addition to limited bright-ray patterns that are unlike bright-rayed craters (Fig. 26). This pattern might reflect effects of surface scouring by impact vaporization or interactions between early-time ejecta and a cometary coma.

One consequence of higher-impact velocities producing the primary craters on Mercury would be the increased rate of agglutinate formation in the regolith (Cintala 1981). Since agglutinate formation has been generally attributed to the removal of albedo contrasts (see Pieters 1978), and since the impact flux may be twice as high on Mercury (Hartmann et al. 1981), the removal rate of crater rays and other albedo contrasts should be much higher on Mercury. Allen (1977) showed that the number density of rayed craters on Mercury is slightly lower than on the Moon, perhaps reflecting the increased

Fig. 26. Fresh Mercurian impact crater with mottled bright-ray pattern, offset halo, and darker deposits on one side (arrow). Albedo pattern may reflect combined effects of scouring by expanding impact vapor cloud and later deposition of impact melt glass. Bar scale indicates 20 km (figure from Mariner 10, H-12 photomosaic).

rate of regolith maturation. Bright rays survive between 1 and 2 Gyr around large lunar craters (50–100 km), although the specific survival rate depends on the contrast between local and nonlocal materials. This relatively short survival time places important limits on inferences drawn from bright crater rays. For example, Dobrovolskis (1981) explored the possibility that ray curvature and isolated ray arcs might reflect a time when the rotation of Mercury was much higher, thereby increasing the Coriolis term. The inferred high rate of ray removal, however, places the time for the proposed change in the rotation rate too late in Mercurian history (i.e., <0.5 Gyr); consequently, another mechanism seems necessary to produce the enigmatic pattern of ray arcs.

Distinctive swirl-like patterns on the Moon and Mercury are attributed to local regolith scouring and melting by higher-density gas/dust zones (10^{-10} g cm^{-3}) within the inner coma of comets (Schultz and Srnka 1980). On the Moon, such patterns are closely tied to strong magnetic anomalies (Hood et al. 1979). Schultz and Srnka (1980) proposed that a relatively long-lasting field produced by the compression of a cometary field was trapped in the lunar regolith during regolith melting and cooling. Spectral studies by Bell and

Fig. 27. Bright swirl pattern (arrows) on Mercury. Similar patterns on the Moon become more visible under large phase angles. They have been interpreted as alteration of uppermost regolith by high-velocity, low-density gas or dust or both in the inner coma of a comet (Schultz and Srnka 1980). Bar scale indicates 20 km (figure from Mariner 10, H-6 photomosaic).

Hawke (1981,1987) provided further support for this hypothesis. Very similar surface patterns found on Mercury (Fig. 27) may be associated with similar magnetic signatures. Hood et al. (1979) proposed an alternative hypothesis where the bright surface swirl patterns on the Moon are produced by a local pre-existing impact induced field that prevents surface darkening by solar protons. Hood and Vickery (1984) subsequently extended the model of Srnka (1977) to account for impact-induced remanence on the Moon. Recent laboratory experiments (Crawford and Schultz 1987) have produced inherent and enhanced ambient magnetic fields by low-angle impacts at modest velocities (<6 km s^{-1}), but further preliminary experiments also have produced enhanced magnetic fields by impacting plasmas on the surface. Regardless of the correct interpretation, the increased flux of high-velocity comets at Mercury could be expressed in a complex pattern of magnetic anomalies.

VII. CONCLUDING REMARKS

Recent focus on the outer planets and satellites perhaps has biased our view of planetary impact cratering. Crustal volatiles are often proposed to account for a wide variety of impact phenomena from increased crater collapse to interior peak morphology. Such proposals cannot be calibrated without a "dry" planet. Mercury provides an essential part of this calibration. In a similar way, our perspective of impact cratering has been influenced by experiments and analogies where the nearly instantaneous transfer of energy and momentum from impactor to target reinforces a point-source equivalence. But experiments and observational data provide a self-consistent but different viewpoint unique to the collisional process. Although further quantitative assessment must await future missions, impact craters appear to retain signatures of both the size and velocity of the impacting bodies through the onset of crater shallowing, the relation between crater depth and diameter, and the morphology of the central peak relief. This perspective lessens the role of contrasting crustal properties and increases the role of impactor provenance or placement in the solar system. The scaling relation between cratering efficiency and independent impactor/target variables need not change above the transition diameter, while the scaling relation between diameter and the independent variables continuously changes. Since crater diameter is the principal measurement for deriving chronologies on different planets, a change in crater scaling affects interpretations of surface ages and the population of impactors on different planets (or even a given planet through time). This situation could be viewed as an annoying complication; it also could be viewed as a means to decipher contrasting planetary bombardment histories.

THE CRATERING RECORD ON MERCURY AND THE ORIGIN OF IMPACTING OBJECTS

ROBERT G. STROM
University of Arizona

and

GERHARD NEUKUM
German Aerospace Research Establishment

The heavily cratered highlands of Mercury, Mars and the Moon all have similar crater size-frequency distributions. However, the Mercurian highlands show a marked paucity of craters < 50 km diameter compared to the lunar highlands. The paucity can be explained by crater obliteration by intercrater plains emplacement on Mercury and indicates that Mercury's intercrater plains were formed during the period of late heavy bombardment. They may range in age from about 4 to 4.2 Gyr. The post-Caloris crater population is similar to that of the lunar highlands over the same diameter range but has a much lower crater density than the lunar or Mercurian highlands, and a significantly higher density than the lunar maria. Unlike the lunar maria, Mercury's smooth plains were probably emplaced near the end of late heavy bombardment. Mercury's volcanic activity was very intense early in its history (much more so than the Moon) but ended sooner on Mercury (≈ 3.8 Gyr ago) than on the Moon. This is probably the consequence of the formation of Mercury's enormous iron core which caused extensive melting, global expansion and crustal tension during the period of late heavy bombardment, followed by cooling and global contraction to shut off magma sources early in its history. A comparison of the inner solar system cratering record with that at Jupiter shows that the heavily cratered surfaces of Ganymede and Callisto have a different crater population which cannot be explained by differences in crater scaling. The impact velocities needed to match the Callisto and lunar crater curves are completely unrealistic *for objects in heliocentric orbits that cross both Jupiter and the inner planets. Furthermore, the lateral displacements of the crater curves for the terrestrial planet highlands require impact velocity differences between Mercury, the Moon, and Mars that can only be explained by objects with small semimajor axes confined to the inner solar system. These results suggest that the objects responsible for the period of late heavy bombardment on the terrestrial planets were accretional remnants left over from the formation of the terrestrial planets and confined to the inner solar system. The cratering record in the outer solar system may have been produced largely by objects in planetocentric orbits.*

The cratering record on Mercury provides information on geologic processes, time scales and the origin of impacting objects. Mariner 10 images showed that, like the Moon and Mars, Mercury has a heavily cratered surface. This demonstrated that all of the inner planets, including the Earth and Venus, experienced a period of late heavy bombardment early in solar system history. Mercury's cratering record is important because it provides information on the impact history in the innermost region of the solar system.

Various aspects of the solar system cratering record, sometimes with widely disparate conclusions, have been discussed by Chapman and McKinnon (1986), Strom (1987a), Plescia and Boyce (1985), Shoemaker and Wolfe (1982), Neukum (1985), Hartmann (1984), Woronow (1978) and Horedt and Neukum (1984). These are by no means the only papers that discuss the cratering record of the solar system, but the reader is referred to these for different views on the origin of impacting objects and cratering processes. In particular, the reader is referred to the chapter in *Satellites* by Chapman and McKinnon (1986) for an excellent summary of the various viewpoints and the uncertainties in interpreting the cratering record.

This chapter is largely confined to a discussion of the terrestrial planet cratering record with emphasis on Mercury. First we review the geologic units on Mercury most relevant to the cratering record (Sec. I). In Sec. II the issue of equilibrium and saturation is reviewed and some new observations are presented. This is followed by a discussion of the Mercurian cratering record and its implications for geologic processes (Sec. III). Absolute time scales on the Moon and Mercury are considered in Sec. IV. The origin of the objects responsible for the period of late heavy bombardment in the inner solar system is discussed in Sec. V.

Two types of crater size-frequency distribution plots are used in this chapter: (1) the cumulative plot, and (2) the relative or "R" plot. In the cumulative size-distribution, the log cumulative crater frequency is plotted against the log crater diameter and is in the form $N \propto D^\alpha$ where N is the cumulative crater frequency, D is the crater diameter and α is the population or slope index. The "R" plot displays information on the differential size distribution, and is the ratio of the observed distribution to the function $dN \propto D^{-3} \, dD$. Because most large crater populations have slope or population indices within the range of ± 1 of the function D^{-3}, they plot as nonsloping or moderately sloping lines on these log/log plots. On an "R" plot, a horizontal line has a differential -3 slope index; one sloping down to the left at an angle of 45° has a differential -2 slope index and one sloping down to the right at 45° has a differential -4 slope index. The vertical position of the curve is a measure of crater density: the higher the curve, the greater the crater density. For a given single-sloped cumulative slope index, the equivalent differential slope index is decreased by 1, e.g., a cumulative -2 slope index is equivalent to a differential -3 slope index, a cumulative -1 is a differential -2, etc. The most frequently used plot in this chapter is the "R" plot, and therefore we will refer only to the differential population index.

I. GEOLOGIC UNITS

Like the Moon and Mars, Mercury's surface can be broadly divided into two physiographic provinces: highlands and lowland plains. The highlands of Mercury consist of heavily cratered areas interspersed with broad areas of gently rolling plains superposed with a high density of craters less than about 15 km diameter. These highland plains have been termed intercrater plains (Trask and Guest 1975). They occupy about 45% of the surface viewed by Mariner 10, and therefore constitute the major terrain type on Mercury. Many of the superposed small craters form chains or clusters suggestive of secondary impact craters. Both the Moon and Mars also have intercrater plains. On the Moon, however, the old (pre-Imbrium) intercrater plains are much less extensive than on Mercury and Mars.

The Mercurian lowland plains occur primarily within and surrounding the Caloris basin and in the north polar region. They have been termed smooth plains by Trask and Guest (1975). Smooth plains also occur on the floors of other large basins such as Tolstoij and Beethoven and as patches in the highlands. They resemble the lunar maria in both morphology and mode of occurrence. The albedo and color of the smooth plains differ significantly from those of the lunar maria (Strom 1984; Hapke et al. 1980; Hapke et al. 1975). The Caloris smooth plains are about 90% brighter than the lunar maria and the color differences are less than those within the lunar maria. The density of craters superposed on the smooth plains is considerably less than that on the intercrater plains indicating that they are younger. Their younger age is also confirmed by stratigraphic relationships (see chapter by Spudis and Guest).

Other geologic units occur on Mercury and are discussed in detail by Spudis and Guest in their chapter on the stratigraphy and geologic history. However, the Mercurian highlands with its extensive intercrater plains unit and the smooth plains are most relevant to the cratering record on Mercury. The origins of these two plains units are somewhat controversial. Both units have been ascribed to either volcanism or to basin-forming ejecta deposits. The evidence for these two very different origins has been reviewed by Strom (1984) and also in the chapter by Spudis and Guest. Current evidence seems to favor a volcanic origin for both plains units, but the evidence is considerably stronger for the smooth plains than for the intercrater plains. The reader is referred to the chapter by Spudis and Guest for a thorough discussion of this topic.

II. THE ISSUE OF SATURATION, STEADY STATE AND EQUILIBRIUM

There has long been a scientific debate in the field of impact cratering studies about the highest crater densities on the most ancient surfaces of the terrestrial planets and outer planet satellites. One school of thought believes these crater populations are saturated or in steady state (see, e.g., Hartmann 1984), while the other believes they are not (see, e.g., Woronow 1978). The concept of saturation is relatively simple. It merely means that a surface is so heavily cratered that each new crater destroys others in such a way that the crater density remains constant or nearly constant no matter how many craters are added. Thus, the surface is in a steady state or has reached equilibrium. Although this concept is simple, the process by which saturation occurs and the crater density or densities at which it takes place is quite complex.

The central issue for our purposes is *not* whether a surface is saturated, but *whether the observed crater size-frequency distribution is essentially the same as, or close to, the production size-frequency distribution.* A production size-frequency distribution is one that has retained its original form, and, therefore, reflects the projectile size-frequency distribution. In the absence of extensive erosion and deposition, a lightly cratered surface preserves the crater production population. Even on surfaces where craters have been destroyed by the cratering process, the production size-frequency distribution can still be preserved (Neukum and Dietzel 1971; Chapman and McKinnon 1986; Woronow 1977,1978). If one is dealing with a production population, then the projectile size-frequency distribution can be recovered by providing the correct values to parameters in an appropriate crater scaling law.

There are essentially two ways a production size-frequency distribution can be changed. One is by geologic (endogenic) processes such as erosion and deposition that destroy pre-existing craters, and the other is by the cratering process itself. Depositional and erosional processes destroy small craters more easily than large craters, resulting in a preferential loss of small craters relative to large ones. This results in an increase in the population index at smaller crater sizes, i.e., the slope index becomes less negative. At an extreme, all craters smaller than some threshold diameters may be destroyed. Another geologic process that can change the production population is the destruction of larger craters by viscous relaxation on icy surfaces. In this case, the population index decreases (becomes more negative) at larger crater diameters. Again, at the extreme, all craters larger than some threshold diameter may disappear.

The production size-frequency distribution can also change as saturation is approached and attained depending on the population index of the production population. For single-slope production size-frequency distributions with population indices more negative than -3, the population index increases (becomes less negative) as saturation is reached or approached. But this is not true for "shallower" initial size-frequency distributions. For example, production size frequencies with population indices equal to or less negative than about -2.5 retain their population indices at saturation (Woronow 1977). Neukum and Dietzel (1971) reached similar conclusions using an analytical approach.

Figures 1 and 2 show the crater size-frequency distributions for the lunar, Mercurian, and Martian highlands. They all display a complex curve of similar shape that cannot be represented by a single-slope population index. Between about 10 and 50 km diameter, the population index α is -2, at sizes between 50 and 150 km it is -3, at sizes 150 to 250 km it is about -4, and at diameters greater than about 300 km it is approximately -2, although the statistics are quite poor at these sizes.

The central question concerns the most heavily cratered surfaces in the solar system and whether or not they retain the signature of their production populations. There are two methods of addressing this problem: computer modeling and observation. The modeling approach concerns Monte Carlo computer simulations which crater a surface with size-frequency distributions of single or multiple population indices until that surface is saturated, i.e., the crater density stays nearly constant. Such simulations have been performed by Woronow (1977,1978) and by Chapman (Chapman and McKinnon 1986). The reader is referred to these publications for details and results of the simulations and thorough reviews of previous studies. The main difference between the simulation results is that Woronow found that saturation occurs at crater densities considerably higher than those observed on the most heavily cratered surfaces, while Chapman, using a somewhat different set of parameters, found that saturation occurs at lower crater densities closer to those observed. The difference between the two results appears to be due to the model parameters which govern the efficiency with which craters are destroyed. The values used by Woronow favor a lesser efficiency for destroying craters than those used by Chapman. The selection of these values is somewhat subjective. Chapman's values may be more realistic but the truth may lie somewhere between the two models. Probably all that can be said at this time is that the heavily cratered surfaces in the solar system may be relatively close to saturation density, but whether or not they have actually reached equilibrium is uncertain.

There are, however, certain results from these computer simulations that seem to be well founded. For a shallow-sloped (index > -3) production population of large dynamic range (large differences in crater sizes), the surface reaches only a state of quasi-equilibrium because crater obliteration is domi-

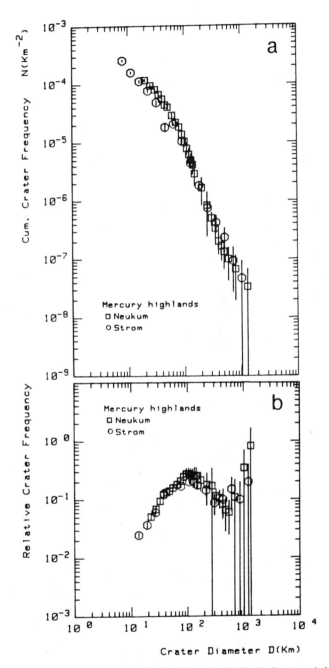

Fig. 1. Crater size-frequency distributions for the Mercurian highlands from two independent data sets of Neukum and Strom. The cumulative plot is shown in (a) and the relative plot in (b).

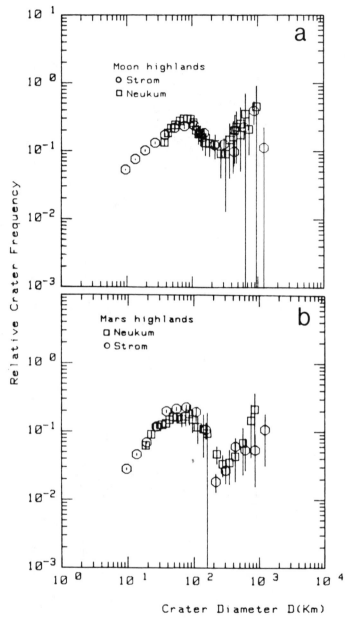

Fig. 2. "R" plots of the crater size-frequency distributions for the lunar highlands(a) and the Martian highlands (b) from two independent data sets of Neukum and Strom.

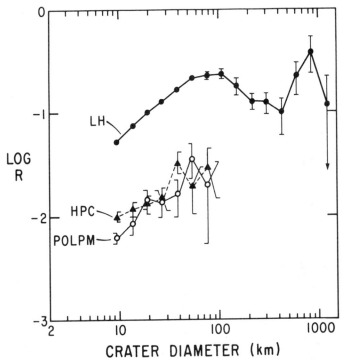

Fig. 3. Comparison of the crater size-frequency distributions of the lunar highlands (LH) with the post-Caloris crater population (HPC) and the post-Orientale population from which is subtracted the post-lunar mare crater population (POLPM). The shapes of the curves are similar over the same diameter range. See text for explanation.

nated by the largest craters (basins). There is not a unique crater density at which saturation occurs. It depends on the production function, and, in general, the shorter the dynamic range, the higher the density at which saturation occurs. One of the most important conclusions from the simulations is that the *observed size-frequency distributions on the most heavily cratered surfaces closely resemble their production size-frequency distributions, even if they are at or near saturation density.*

This latter result is supported by comparisons between crater populations on moderately cratered production surfaces and those on heavily cratered surfaces. Moderately cratered surfaces (surfaces with crater densities greater than that on the lunar maria and less than that in the highlands) show a crater size-frequency distribution that is essentially identical to that of the lunar highlands over the same diameter range of 8 to approximately 60 km. Such a population is found on the Orientale basin and continuous ejecta blanket (Fig. 3). This crater population represents the accumulation of impacts from the time of formation of the Orientale basin through the post-mare epoch up to the present

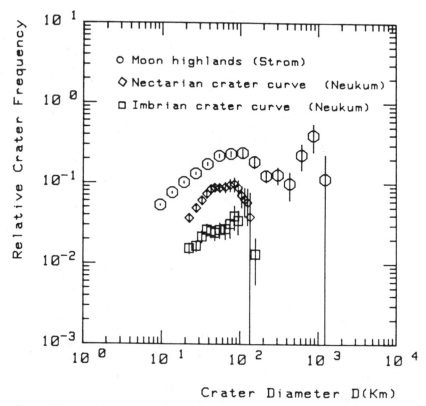

Fig. 4. "R" plot of the crater size-frequency distributions for the lunar highlands, and Nectarian- and Imbrian-aged craters. All size distributions have a similar shape but at different crater densities.

time. The Orientale impact is estimated to have occurred about 3.8 Gyr ago (Wilhelms 1984). The crater population formed between the Orientale impact and the emplacement of most of the lunar maria can be estimated by subtracting the post-mare population from the post-Orientale population normalized to the counting area on Orientale (Strom 1977). This post-Orientale/pre-mare population has a size-frequency distribution very similar to that of the lunar highlands (Fig. 3). The post-Caloris plains crater population on Mercury also has a crater density and size-frequency distribution very similar to the post-Orientale population (Fig. 3). Imbrian- and Nectarian-aged craters also have size-frequency distributions similar to that of the average lunar highlands (Fig. 4). The similarity between the production populations on moderately cratered production surfaces and the lunar highlands strongly suggests that the highlands crater population is basically the production population.

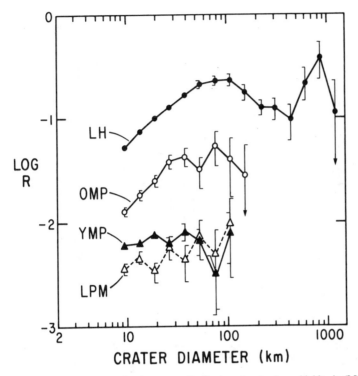

Fig. 5. "R" plot of the crater size-frequency distributions for the lunar highlands (LH), old Martian plains (OMP), young Martian plains (YMP), and the lunar post-mare (LPM) crater populations. The lunar highlands and old Martian plains crater populations are the same, but the young Martian plains and lunar post-mare populations are different at the chi-squared 99% confidence level.

On the Moon and Mercury, crater densities on these moderately cratered regions are relatively low or the counting areas are relatively small or both, so that the crater statistics are rather uncertain at diameters larger than about 40 km. Mars, however, has vast areas of plains which have a variety of ages. The number of craters between about 8 and 60 km diameter on these surfaces is large enough to provide good statistics with rather small uncertainties. On moderately cratered plains units the crater population is very similar to that of the lunar highlands (Fig. 5), again indicating the lunar highlands population is basically a production population.

The authors disagree between themselves about the crater population on the lightly cratered regions of the Moon (maria) and Mars (younger plains). One of us (G. Neukum) believes the size-frequency distribution is similar to the highlands population, while the other (R. Strom) believes it is significantly different. Neukum bases his belief on data consisting of craters interpreted to be of Erastothenian and Copernican age, while Strom includes all craters

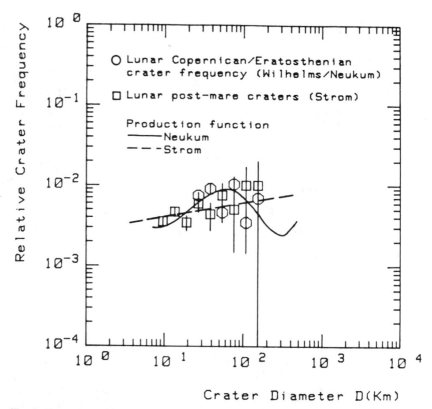

Fig. 6. Comparison of data sets for lunar post-mare and Eratosthenian plus Copernican craters. Also given are functions as advocated by Neukum and Strom, respectively. Here the authors disagree in interpretation. Neukum believes the production function is the same as for older populations as in Fig. 7 (see below), whereas Strom believes it has changed as in Fig. 5.

superimposed on the lunar maria and the younger Martian plains in his data. Chi-squared statistical tests of Strom's data indicate that the post-lunar mare population is different from that in the highlands at the 99% confidence level (Woronow et al. 1982; Strom 1977), and that the young post-Martian plains population is different at the 99.99% confidence level (Barlow 1987). These data are presented in Figs. 5 and 6, and the readers should judge for themselves the merits of the two interpretations. It should be noted, however, that this difference in interpretation is not central to the main issue of the heavily cratered terrains addressed in this chapter.

    The question of whether or not the highlands crater populations on the Moon, Mars and Mercury have been modified by geologic processes is another important issue. There is little doubt that compared to the Moon there has been a loss of craters less than about 50 km diameter by geologic pro-

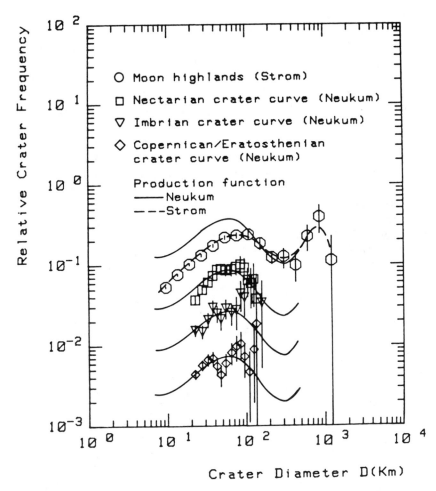

Fig. 7. Relative crater frequency plots of data from the lunar highlands, Nectarian-aged, Imbrian-aged, and Eratosthenian-aged surfaces. The highland data (Strom) are given in $\sqrt{2}$ intervals. The other data (Neukum) are given in much finer binning in order to bring out the structure of the distributions over the more limited size range better. These data have been smoothed by a floating average procedure to take out some of the statistical noise. Production functions are given in comparison. There is a slight disagreement of the authors in the interpretation of the data: Strom believes the lunar highland data totally reflect the production function for Imbrian to pre-Nectarian (highlands) ages, whereas Neukum believes that the production function as given here is not totally reflected in the highlands data but that the highlands distribution suffered some loss of craters (factor of 2 to 3) at sizes < 50 to 100 km from cratering and noncratering processes.

cesses on Mercury and Mars. This is discussed in Sec. III. The lunar highlands population at diameters greater than about 8 km must closely match the
production population because the population index between 8 and about 50
km diameter is essentially the same as that for the production populations on
moderately cratered areas of the Moon, Mars and Mercury. Between these
diameters, it is virtually identical to Neukum's "ideal production population."
This does not preclude the possibility that a portion of the smaller lunar crater
population has been erased by the cratering process, although the Monte Carlo
computer simulations by Woronow (1977) and Chapman and McKinnon
(1986) do not show this. Large basin-forming events will obliterate all smaller
craters over large areas. These areas are then recratered by smaller impacts
which build up the crater density. The production distribution for these small
craters is retained, although the crater density could be lower than that before
the basin-forming events. Neukum's ideal production population has attempted to take this into account by increasing the crater density at diameters
less than about 100 km while still preserving the observed production distribution (Fig. 7).

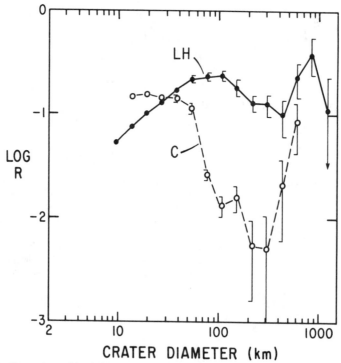

Fig. 8. Comparison of the Callisto (C) and lunar highlands (LH) crater size-frequency distributions. The crater populations are significantly different.

On Ganymede and Callisto the crater population shows a severe paucity of craters larger than about 60 km compared to the heavily cratered surfaces in the inner solar system (Fig. 8). This paucity of large craters has been attributed to obliteration of larger craters by viscous relaxation in ice (see, e.g., Passey and Shoemaker 1982; Shoemaker and Wolfe 1982). However, simulations by Woronow and Strom (1981), Chapman and McKinnon (1986) and Strom (1987a), and statistical tests by Gurnis (1987) show that this cannot be the case. These simulations reproduce the observed Callisto crater population using a lunar-highlands production function and incorporate diameter-dependent crater obliteration to simulate viscous relaxation. The resulting spatial distribution of craters is very nonuniform, exhibiting prominent holes where relatively recent large craters once existed. This is very different from what is observed on Callisto, where the spatial distribution of craters is quite uniform.

The main conclusion from computer modeling, simulations and observations of production populations on moderately cratered regions, is that the observed crater size-frequency distribution in the heavily cratered terrains on the inner planets (particularly the Moon) and at Jupiter are close to the initial production populations.

III. THE CRATERING RECORD AND GEOLOGIC PROCESSES

The age of Mercurian intercrater plains relative to the heavily cratered terrain has important implications for the early history of Mercury. The Mercurian highlands have a crater size-frequency distribution and density similar to that of the lunar and Martian highlands (Fig. 9). All show a highly structured curve with multiple distribution functions. The similar size-frequency distributions strongly suggest that the heavily cratered surfaces representing the period of late heavy bombardment in the inner solar system were the result of a single family of objects in heliocentric orbits. One difference between the lunar and Mercurian crater curves is that at diameters less than about 40 or 50 km diameter, Mercury shows a significant paucity of craters compared with the Moon (Fig. 9). Trask (1975) and Oberbeck et al. (1977) suggested that the Mercurian crater curve more closely represents the true crater production population analogous to a lunar highland curve from which widely distributed basin secondaries have been subtracted. However, it is much more likely that the paucity of craters on Mercury resulted from the obliteration of a fraction of craters by emplacement of intercrater plains (Strom 1977). Intercrater plains on Mercury are the major terrain type and cover a much larger fraction of the highlands than similar deposits cover on the Moon. They have surely been responsible for the obliteration of a portion of smaller craters since images show large craters embayed and partially buried by intercrater plains (Fig. 10). The Martian crater curve also shows a similar paucity of small craters relative to the Moon (Fig. 9) which must be due to plains formation in the Martian uplands and to eolian erosion. Furthermore, the post-Caloris crater

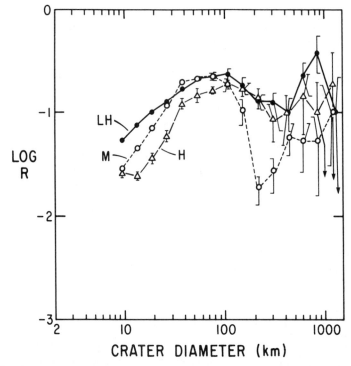

Fig. 9. "R" plot comparing the size-frequency distributions of the lunar highlands (LH), Mercurian (H), and Martian (M) highlands. The paucity of craters less than about 40 km on Mercury and Mars compared to the Moon is probably the result of crater obliteration by intercrater plains emplacement (see also Fig. 10).

size distribution must be in production and it is essentially identical to that of the lunar highlands, i.e., there is no paucity of smaller craters (Fig. 3). All of these observations strongly suggest that the paucity of smaller craters on Mercury's heavily cratered surface compared to the Moon is due to obliteration by intercrater plains emplacement. Intercrater plains show an abundance of craters less than about 15 km diameter that are largely secondary craters from craters in the heavily cratered terrain. Therefore, some of the intercrater plains are older than craters formed during the period of late heavy bombardment. This, together with the evident obliteration of craters by intercrater plains emplacements, strongly implies that intercrater plains formed *during* the period of late heavy bombardment. Leake (1982) and others have reached similar conclusions from stratigraphic studies. If the intercrater plains are volcanic in origin (Strom 1984; Leake 1982), then Mercury experienced much more volcanic activity during the period of late heavy bombardment than did the Moon.

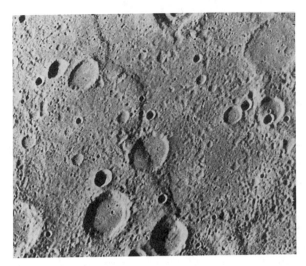

Fig. 10. Mariner 10 image of the intercrater plains in the Mercurian highlands. The large crater (70 km diameter) in the upper right-hand corner has been embayed by intercrater plains.

The crater population superposed on the smooth plains within and surrounding the Caloris basin has a size-frequency distribution that is essentially identical to that of the lunar highlands over the same diameter range of 8 to ~ 60 km (Fig. 3). The crater density of the post-Caloris population is about 10 times less than the Mercurian and lunar highlands, indicating that the smooth plains are relatively young. However, the post-Caloris population has a crater density about 5 times greater than the post-lunar mare crater population. It also has the same size distribution and crater density as the post-Orientale/pre-mare crater population (Fig. 11 in Strom 1977). The post-Orientale/pre-mare population is that which is superposed on the lunar Orientale basin and ejecta blanket from which is subtracted the post-mare population normalized to the area counted on Orientale. It therefore represents the time horizon between the Orientale impact and the emplacement of the lunar maria. The estimated age of the Orientale basin is about 3.8 Gyr (Wilhelms 1984). The similar shapes of the size-frequency distributions (lunar and Mercurian highlands, and post-Caloris and post-Orientale) indicate that the majority of Mercury's smooth plains were formed during, but near the end, of the period of late heavy bombardment.

If the intercrater and smooth plains are volcanic, then the cratering record suggests they were both emplaced early in Mercury's history during the period of late heavy bombardment. In fact, the formation of intercrater plains and smooth plains may not represent two distinct plains-forming episodes. Instead they may represent a more or less continuous period of plains formation lasting a shorter period of time than lunar volcanism. Volcanism may have lasted for a shorter period of time on Mercury than on the Moon because cooling of the lithosphere and the large iron core produced global compression which closed off magma conduits and inhibited surface volcanism earlier on Mercury than on the Moon.

IV. ABSOLUTE TIME SCALES ON THE MOON AND MERCURY

The only reliable way of determining absolute surface ages is by radiometric age dating of returned samples. To date, only samples from the Earth, the Moon and meteorites have been dated in this way. One class of meteorites, the SNC meteorites, probably originated from Mars, but the location on Mars from which these meteorites came is very uncertain.

Absolute ages of planetary surfaces can be derived from crater statistics provided the cratering rate as a function of time, or the integrated impact rate as a function of absolute surface age is known. The cratering rate depends on the origin (orbital elements) of the impacting objects. For example, objects in planetocentric orbits will have a much greater impact rate on satellites than objects in heliocentric orbits. The origin of the objects responsible for the period of late heavy bombardment in the inner solar system is discussed in the following section. From the similarities of the crater population at Mars, the Moon and Mercury and lateral shifts between the crater curves, it is concluded that the objects were in heliocentric orbits confined to the inner solar system. If this is the case, then the impact rate would have been roughly similar at the Moon and Mercury and the end of late heavy bombardment would have occurred at about the same time on each body. On Mars, the end of late heavy bombardment by accretional remnants could have been extended by about 1 Gyr, according to Wetherill (1977).

It is also possible that the end of late heavy bombardment could have been extended on Mercury compared with the Moon if there was a population of potential Mercury-specific impactors ("vulcanoids") in orbits nearer to the Sun than Mercury (Leake et al. 1987). However, searches from 1902 to 1981 for such objects that should still be present have proved negative (Leake et al. 1987). These searches have shown that there is no appreciable population of vulcanoids larger than 50 km diameter orbiting interior to Mercury, although smaller objects may be present. The authors believe that the cratering record representing the period of late heavy bombardment on Mercury cannot have been due to hypothetical vulcanoids because the size-frequency distribution is similar to that on the Moon and Mars, and because lateral shifts in crater curves between the Moon and Mercury indicate higher impact velocities on Mercury than the Moon. The similar crater curves indicate that the same family of objects was responsible for the period of late heavy bombardment on the Moon, Mercury and Mars. Whether vulcanoids would have the same size distribution is highly problematical. Vulcanoids would have much lower impact velocities on Mercury than objects impacting both Earth and Mars. Therefore, the ratio of Mercurian to lunar impact velocity should be less than 1 rather than 1.37 to 2.18 as observed (see Sec. V below).

Although we believe that the cratering rate and end of heavy bombardment at Mercury were similar to that at the Moon for the reasons given above, the reader should be aware that certain assumptions are required to derive the absolute ages that follow. The first assumption is stated above and the other is that the Caloris basin formed at the end of the period of late heavy bombardment 3.85 Gyr ago. If the age of the Caloris basin is not 3.85 Gyr, then the chronology and derivation of absolute ages is directly affected. For the assumed lunar-like time dependence of the Mercurian impact rate, there is about 100 Myr age difference for every factor of two in crater frequency. If, for example, the Caloris basin formed 3.95 rather than 3.85 Gyr ago, then all other crater retention ages would be shifted to older ages by 100 Myr. Stated another way, any crater frequency would have to be a factor of two lower to yield the same as for Caloris at 3.85 Gyr. If the impact probability decreased rapidly with decreasing cratering rate (a factor of $\sim$ 2 per 100 Myr) during the period of late heavy bombardment, then an uncertainty of about 100 Myr for the age of Caloris occurs. Hence, uncertainties greater than about 100 Myr in the derivation of absolute crater retention ages from this model appear unlikely.

If a surface is saturated many times over, then the crater frequencies cannot be used to determine absolute ages of these surfaces. However, we believe that the highland surfaces of the terrestrial planets are still in production although they may be close to saturation. It should be noted that if these surfaces were saturated many times over and the Caloris basin is about 3.8 Gyr old, then the highland surfaces would be as old or older than the age of the solar system.

The method to achieve relative and absolute age determination from crater statistics was developed for the Moon by different groups concentrating on different aspects (see, e.g., Öpik 1960; Shoemaker et al. 1962; Baldwin 1964,1971; Hartmann 1965,1966; Soderblom 1970; Neukum 1971,1981, 1982, 1983; Neukum and Wise 1976). In order to derive relative or absolute ages for a specific surface unit, it is necessary to verify by careful examination of the imagery and precise measurement that (1) a production density can be measured; (2) the structure to be dated is really one homogeneous geologic unit (no or negligible resurfacing in terms of destruction of craters of the population measured); (3) only superimposed craters and no relic or ghost craters from an underlying older unit are measured; (4) secondary craters and volcanic craters are eliminated; (5) the area of measurement is determined accurately; and (6) the size of craters is measured with highest possible precision since the exponential size-frequency distribution dependence amplifies small errors in diameter as large errors in crater density per unit area.

The relationship between a crater production size-frequency distribution and age (termed crater retention age) is as follows: The crater population on one planet approximately represents the mass-velocity distribution of the impactors responsible for the cratering record. Under the assumption of mean impact velocity, a meteorite population (i.e., the impactors) of the mass distribution $n(m,t)$ in the mass interval $(m, m+dm)$ causes a crater size-frequency distribution $n(D,t)$ in the crater diameter interval $(D, D + dD)$ for a specific exposure time t. The function $n(D,t)$ is termed differential distribution (number per unit area per diameter at time t). The relationship between differential distribution and differential cratering rate (number per unit area per diameter per time at time t) is given by:

$$n(D,t) = \int_o^t \varphi(D,t')dt' \tag{1}$$

where t is the exposure time with respect to the age of the crater population $(t > 0)$. The cumulative crater frequency $N(D,t)$ (number per unit area of all craters with diameters equal to or larger than D and which were formed during exposure time t) is given by

$$N(D,t) = \int_D^\infty n(D,t)dD' = \int_D^\infty \int_o^t \varphi(D',t')dD'dt'. \tag{2}$$

Note that this is in the continuous approximation; in reality, it is the sum of discrete numbers. The cumulative crater rate $\phi(D,t)$ is given by

$$\phi(D,t) = \int_o^t \phi(D,t')dt' \text{ or } \phi(D,t) = \partial N(D,t)/\partial t. \tag{3}$$

The function $\varphi(D,t)$ can be separated into a function $g(D,t)$ that reflects the underlying crater size distribution and into a function $f(t)$ that reflects the general functional dependence of cratering rate on time. Therefore one can write

$$\varphi(D,t) = g(D,t) \cdot f(t) \tag{4}$$

or

$$\phi(D,t) = \int_D^\infty g(D't)f(t)dD' = G(D,t)f(t). \tag{5}$$

If the size distribution is not directly dependent on time (i.e., the diameter distribution is the same over the whole exposure time), then the function $\varphi(D,t)$ can be separated in

$$\varphi(D,t) = g(D)f(t) \tag{6}$$

and

$$n(D,t) = g(D) \int_o^t f(t')dt' \tag{7}$$

or $n(D,t) = g(D)F(t)$ where $F(t) = \int_o^t f(t')dt'$. The cumulative crater frequency then is

$$N(D,t) = \int_D^\infty \int_o^t g(D')dD' \, f(t')dt' = G(D)DF(t) \tag{8}$$

which is a production distribution.

The production crater frequencies $n(D,t)$ or $N(D,t)$ show the same value, respectively, for areas exposed for an equal length of time on one planet for the same crater diameter or diameter interval regardless of location on the planet, provided the impactor flux is isotropic and target material compositions have no influence. In this case, we find the simple relationship

$$N_1(D,t_1)/N_2(D,t_2) = F(t_1)/F(t_2). \tag{9}$$

This means that the ratio of the crater production population frequencies on two areas (1 and 2) is directly proportional to the ratio of their functional dependence on time. From this relationship a relative age sequence of geologic units can be determined by direct comparison of their superimposed impact crater frequencies; i.e., the frequency of superimposed craters (per unit area) for a specific diameter or a specific diameter interval is a direct measure of relative age, termed relative crater retention age.

For direct extraction and comparison of relative ages of two or more units, it is necessary to measure in the same crater size range or, if that is not possible, to compare crater frequencies through application of the functional dependence of the production distribution (called standard or calibration distribution by Neukum and Wise [1976]) and to convert the measured frequencies in this way to the values which they would show at one and the same diameter if they could be measured there. The ideal production distributions for the Moon and Mercury were estimated over different crater sizes by a variety of authors and the results were synthesized by Neukum (1983) as

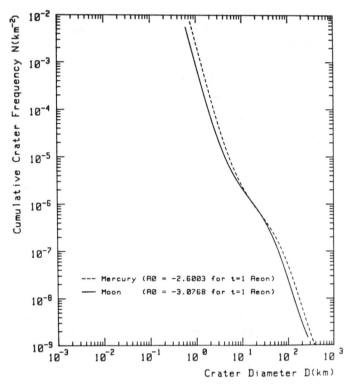

Fig. 11. Ideal cumulative crater size-frequency production distributions (polynomial of 11th degree) for the Moon and Mercury (figure after Neukum 1983).

shown in Fig. 11. The distributions are complex curves which cannot be approximated by simple power laws. The lunar and Mercurian production distributions as shown in Fig. 11 are directly related to each other and can be brought to coincide if one properly accounts for the difference in crater scaling due to factors like different impact velocity and gravity. The conversion from one planet to another along with the general relationships among the different terrestrial planets are summarized in Table 1.

The fact that two *different* values of crater frequency of the two planets at some diameter, or the *same* values at another diameter, may mean the same age of units on different planets is illustrated in Fig. 11 for the Moon and Mercury. The ideal cumulative production size-frequency distributions (calibration distributions) for both planets are given for the same age ($t = 1$ Gyr). Around $D = 10$ to 30 km, the size-frequency distributions fall almost upon each other, whereas the frequencies are quite different at smaller and larger diameters. This demonstrates that it is absolutely necessary to specify exactly the diameter or range of diameters in age determinations.

TABLE I
Conversion Factors for Crater Scaling on the Terrestrial Planets[a]

Object	V_{iP} (km s⁻¹)	$(2GM_P/R_P)^{1/2}$ (km s⁻¹)	g_P (cm s²)	$U_{\infty P}$ (km s⁻¹)	f_P	$F = f_P/f_M$	$\left(\dfrac{g_M}{g_P}\right)^{3/16}$	$\left(\dfrac{V_{ip}}{V_{iM}}\right)^{0.56}$	$D_P/D_M = \left(\dfrac{g_M}{g_P}\right)^{3/16} \times \left(\dfrac{V_{ip}}{V_{iM}}\right)^{0.56}$
Moon	14.1	2.38	162	13.90	1.03	1.00	1.000	1.00	1.00
Mercury	23.6	4.2	363	23.22	1.03	1.00	0.860	1.33	1.15
Venus	19.3	10.3	860	16.32	1.40	1.36	0.731	1.19	0.87
Earth	17.8	11.2	982	13.83	1.66	1.61	0.713	1.14	0.81
Mars	12.4	5.0	374	11.35	1.19	1.16	0.855	0.93	0.80

[a]Symbols have the following meaning:

D = crater diameter on the planet with gravitational acceleration.

$g_P = GM_P/R_P^2$ with M_P, R_P = planetary mass, planetary radius and $G = 6.67 \times 10^{-8}$ gravitational constant.

V_{iP} = impact velocity for projectiles of eccentricity $e = 0.6$.

$U_{\infty P}$ = relative velocity for projectiles on the border of gravitational sphere of action of planet.

$U_{\infty P} = (GM_\odot/1_{\odot P})^{1/2} (e^2 + e_P^2)^{1/2}$, $M_\odot$ = solar mass, e = eccentricity of projectile, e_P = eccentricity of planetary orbit, $1_{\odot P}$ = solar distance of planet.

$V_{iP} = (2GM_P/R_P + U_{\infty P}^2)^{1/2}$ = impact velocity = square root of the quantity consisting of square of escape velocity + square of relative velocity on the border of planet's gravitational sphere.

$f_P = (1 + 2GM_P/R_P U_{\infty P}^2)$ = focusing due to planet's gravitation.

Values for the Moon are indexed with M.

$D_P \propto V_{iP}^{0.56}/g_P^{3/16}$, $D_M \propto V_{iM}^{0.56}/g_M^{3/16}$, $D_P/D_M = (g_P/g_P)^{3/16} (V_{iP}/V_{iM})^{0.56}$.

$F = f_P/f_M = (1 + 2GM_P/R_P U_{\infty P})/(1 + 1GM_M/R_M U_{\infty M})$.

Subscript P = planet; subscript M = Moon.

The Mercurian cumulative crater production size-frequency distribution can be approximated by a polynomial of 11th degree as shown by Neukum (1983):

$$\text{Log}N = a_0 + a_1(\log D) + \cdots + a_{11}(\log D)^{11}$$

$$a_1 = -3.6712 \quad a_7 = 3.2493 \times 10^{-2}$$
$$a_2 = 0.2946 \quad a_8 = 1.1737 \times 10^{-2}$$
$$a_3 = 0.7630 \quad a_9 = -1.9272 \times 10^{-3}$$
$$a_4 = 0.1620 \quad a_{10} = -5.4447 \times 10^{-4}$$
$$a_5 = -0.2379 \quad a_{11} = 3.97 \times 10^{-5}$$
$$a_6 = -8 - 8.1361 \times 10^2. \tag{10}$$

The coefficient a_0 is variable and contains the age dependence of the distribution. Some specific values as for the curves in Fig. 13 below are

$$a_0 = -5.6003 \text{ for } t = 0.001 \text{ Gyr}$$
$$a_0 = -4.6003 \text{ for } t = 0.01 \text{ Gyr}$$
$$a_0 = -3.6003 \text{ for } t = 0.1 \text{ Gyr}$$
$$a_0 = -2.6003 \text{ for } t = 1.0 \text{ Gyr}$$
$$a_0 = -1.9461 \text{ for } t = 3.4 \text{ Gyr}$$
$$a_0 = -1.2669 \text{ for } t = 3.8 \text{ Gyr}$$
$$a_0 = -0.1409 \text{ for } t = 4.2 \text{ Gyr.} \tag{11}$$

A surface unit on Mercury that is exposed to meteorite bombardment will show more and more craters in all sizes according to the functional diameter-dependence of the production distribution and according to the functional dependence on time with respect to cratering rate. N is proportional to 10^{a_0} for a specific fixed diameter. Therefore one will find, as illustrated in Fig. 12, that the calibration distribution will shift in the log$N$ direction with exposure time (with *no* shift in diameter direction) by the same amount at all diameters, directly related to the functional dependence of cratering rate on time.

The cratering rate dependence on time with respect to the dependence of the accumulated cratering record as a function of age (i.e., the cratering chronology) cannot directly be determined for Mercury presently, since no absolute age information for surface rock units is available. Therefore it currently is only possible to estimate the Mercurian cratering chronology by extrapolating the lunar cratering chronology to Mercury under certain assumptions as discussed earlier. There exist essentially two different methods of extrapolating the lunar chronology to Mercury:

1. The cratering rate at Mercury compared to that at the Moon is estimated from present-day observations and theoretical considerations on asteroidal and cometary bodies and from relevant impact probabilities (Hartmann 1977; Basaltic Volcanism Study Project 1981). The functional dependence of the Mercurian cratering rate on time is assumed to be the same as for the

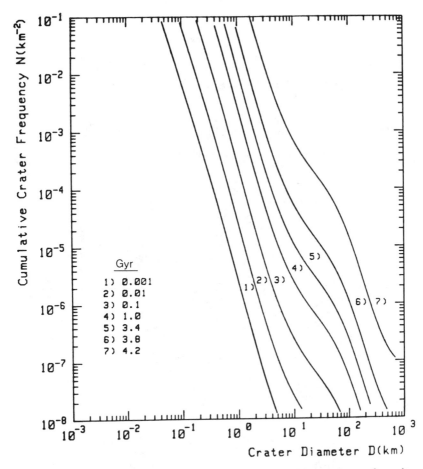

Fig. 12. Crater retention isochrons (polynomial of 11th degree) for Mercury. See text for explanation (figure after Neukum 1983).

Moon. Corrections for impact velocity, scaling and gravitational focusing effects are made similarly to our procedure in Table I.

2. Nothing is assumed about relative cratering rates. The functional dependence of the Mercurian cratering rate on time is assumed to be the same as for the Moon as in method (1). Specifically, it is assumed that the heavy bombardment that created the major basins on the Moon ended at the same time on Mercury; i.e., the age of the Imbrium or Orientale basin on the Moon (3.8 to 3.85 Gyr) is approximately the same as the age of the Caloris basin on Mercury (Neukum 1983). Caloris is set at 3.85 Gyr. Corrections for impact velocity, scaling and gravitational focusing effects are applied as given in Table I.

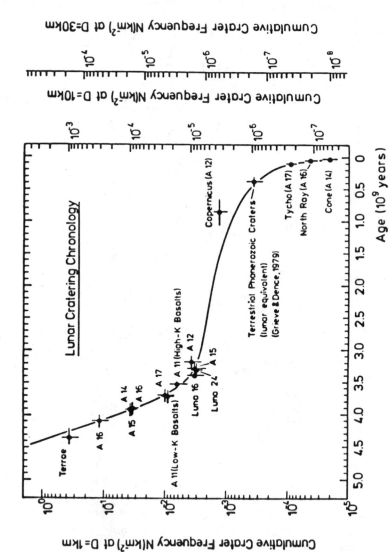

Fig. 13. Cumulative crater frequencies for various lunar surfaces dated from returned lunar samples.

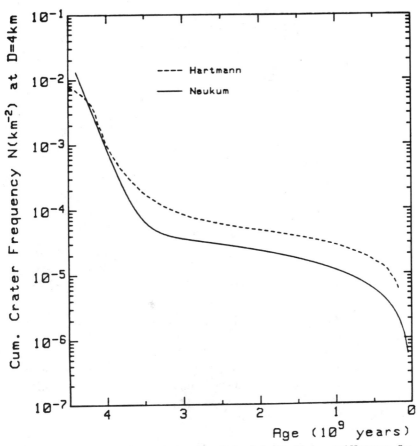

Fig. 14. Comparison of the lunar cratering chronologies derived by Neukum and Hartmann. See text for discussion.

The procedure described as method (2) is followed here.

Important in the derivation of the Mercurian cratering chronology is the quality of the lunar cratering chronology data. In our interpretation, we make use of the data by Neukum (1983) which are shown in Fig. 13. Hartmann's lunar data (Basaltic Volcanism Study Project 1981), shown in comparison in Fig. 14, lie systematically higher. The same is true for his Mercurian chronology curve. Application of Hartmann's curve to crater frequency data measured on Mercurian surface units would result in systematically younger ages. One major reason for the discrepancy between Hartmann's and Neukum's lunar chronologies is the difference in the data used for the determination of the chronology for ages < 1 Gyr. As discussed by Neukum (1983), Hartmann's data are based on a few scattered points with large error bars.

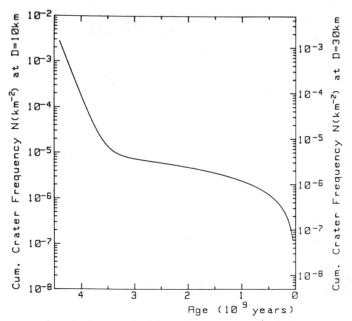

Fig. 15. Cratering chronology of Mercury derived by Neukum.

Our conversion of the lunar cratering chronology curve to the Mercurian case results in the dependence shown in Fig. 15. Numerically, the Mercurian cratering chronology in terms of cumulative crater frequencies at diameters D = 4 km, 10 km and 30 km reads:

$$N(D = 4\ km) = 1.37 \times 10^{-15}(e^{6.93t} - 1) + 2.10 \times 10^{-5}t \quad (12)$$

$$N(D = 10\ km) = 1.56 \times 10^{-16}(e^{6.93t} - 1) + 2.40 \times 10^{-6}t \quad (13)$$

$$N(D = 30\ km) = 3.16^{-17}(e^{6.93t} - 1) + 4.87 \times 10^{-7}t \quad (14)$$

where t is the age in units of 1 Gyr and N is the cumulative crater frequency per km^2.

The Mercurian cumulative crater frequencies at different diameters are related to each other through the Mercurian cumulative crater production size-frequency distribution as given in Figs. 11 and 12, and as approximated by the polynomial of 11th degree discussed above.

Absolute ages can be determined from crater statistics measured for specific Mercurian surface units in the following way:

1. Graphical determination: a value of N at a specific diameter is determined directly from the measured data or determined for a different reference

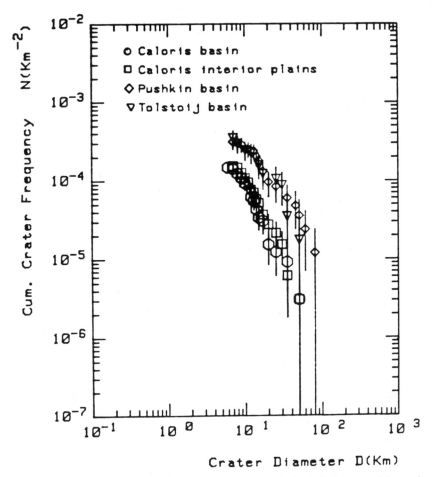

Fig. 16. Cumulative crater frequencies for Mercury's Caloris basin, Caloris interior plains, and the Pushkin and Tolstoij basins.

diameter (here 10 km or 30 km) through application of the Mercurian calibration distribution. The N value is introduced into the chronology graph (y-axis in Fig. 15) and the corresponding absolute crater retention age is read on the x-axis.

Numerical determination: a value of N for a specific diameter is computed by a least-squares fit of the calibration distribution to the measured data. This N value is introduced into the numerical expression of the Mercurian cratering chronology (as above) and the equation solved for t.

Crater frequency measurements for Mercurian highlands, basins and plains were conducted by Strom (1977), Leake (1982) and Neukum (1983)

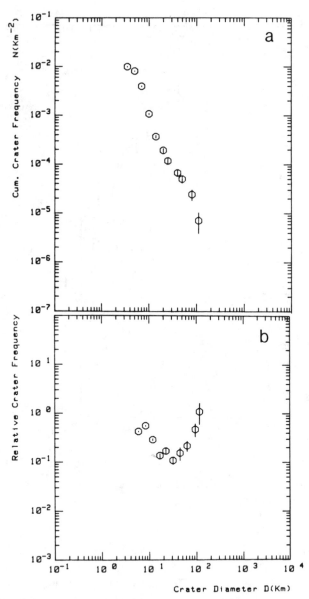

Fig. 17. Cumulative plot (a) and "R" plot (b) of the crater size-frequency distribution of the intercrater plains on Mercury.

TABLE II
Relative and Absolute Crater Retention Ages for Various Terrains on Mercury[a]

Area	Relative Crater Retention Age			Absolute Crater Retention Age (Gyr)
	$N(D = 4\text{km})\ \text{km}^{-2}$	$N(D = 10\text{km})\ \text{km}^{-2}$	$N(D = 30\text{km})\ \text{km}^{-2}$	
Highland	5.25×10^{-3}	5.99×10^{-4}	1.22×10^{-4}	4.18
Beethoven basin	1.34×10^{-3}	1.53×10^{-4}	3.10×10^{-5}	3.98
Chekhov basin	3.54×10^{-3}	4.04×10^{-4}	8.21×10^{-5}	4.12
Caloris basin	6.01×10^{-4}	6.85×10^{-5}	1.39×10^{-5}	3.85
Caloris mare (Interior)	6.01×10^{-4}	6.85×10^{-5}	1.39×10^{-5}	3.85
Dostoevskij basin	4.81×10^{-3}	5.49×10^{-4}	1.11×10^{-4}	4.17
Haydn basin	3.20×10^{-3}	3.65×10^{-4}	7.41×10^{-5}	4.11
Pushkin basin	3.02×10^{-3}	3.45×10^{-4}	7.00×10^{-5}	4.10
Raphael basin	3.23×10^{-3}	3.69×10^{-4}	7.49×10^{-5}	4.11
Tolstoij basin	2.33×10^{-3}	2.66×10^{-4}	5.39×10^{-5}	4.06
Intercrater plains				
Oldest	6.95×10^{-3}	7.93×10^{-4}	1.61×10^{-4}	4.22
Youngest	1.81×10^{-3}	2.06×10^{-4}	4.19×10^{-5}	4.02
Average	4.41×10^{-3}	5.03×10^{-4}	1.02×10^{-4}	4.15

[a]The errors of measurement are in all cases about $< \pm30\%$ in N.

among others. Examples for the basins Tolstoij, Pushkin and Caloris are presented in Fig. 16 and for the highlands in Fig. 1a (as cumulative crater frequencies). Data for Mercurian intercrater plains are shown in Fig. 17a (cumulative frequencies) and 17b (relative frequencies). Application of the Mercurian calibration curve (Fig. 11) to the cumulative crater frequencies yields the relative crater retention ages as listed in Table II (for $D = 4$, 10 and 30 km). Application of the Mercurian cratering chronology as in Fig. 15 (or the numerical expression for it discussed earlier) results in the respective absolute crater retention ages (Table II). With the age of Caloris basin set at 3.85 Gyr, the Mercurian highlands fall at ~ 4.2 Gyr on the average and the ages of the eight basins investigated fall between 4.2 Gyr and 3.85 Gyr.

The intercrater plains did not form at one point in time but during the whole period of heavy bombardment, the oldest plain having an average highland age, the youngest plain investigated having an age of ~ 4 Gyr. The fill of Caloris (smooth plains) has an age indistinguishable from that of the Caloris basin (rim and ejecta blanket). This means the volcanic filling of the basin occurred soon after the formation of Caloris.

V. IMPLICATIONS FOR THE ORIGIN OF IMPACTING OBJECTS

The heavily cratered highland surfaces on the Moon, Mars and Mercury represent the period of late heavy bombardment which ended, at least on the Moon, about 3.8 Gyr ago. Each of these surfaces displays similar crater size-frequency distributions indicating that they were impacted by the same family of objects (Fig. 9). Since these objects apparently spanned the entire inner solar system distance from Mercury to Mars, they were probably in heliocentric orbits.

At Jupiter, the cratering record is different (Strom et al. 1981). On Callisto and the heavily cratered areas of Ganymede, the crater size-frequency distribution shows a severe paucity of craters greater than 50 km diameter and an overabundance of craters less than about 30 km in diameter compared with that on the heavily cratered surfaces in the inner solar system (Fig. 18).

One could imagine that when crater scaling parameters are taken into account, the size-frequency distribution of the projectiles might be similar for the terrestrial planets and Jovian crater population. If the Ganymede or Callisto crater curve (Fig. 8) could be shifted a factor of 2 to larger diameters by differences in the scaling parameters, it would match more closely the crater curve of the terrestrial planet highlands (Neukum 1983, 1985; Horedt and Neukum 1984). Strom (1987a), however, employed a modified Holsapple-Schmidt (1982) crater scaling law to recover the size-frequency distribution of the impacting objects using present-day short-period comet impact velocities and found that the disparity between the projectile size distributions was even greater than that between the crater size distributions. Chapman and McKinnon (1986) reached similar conclusions.

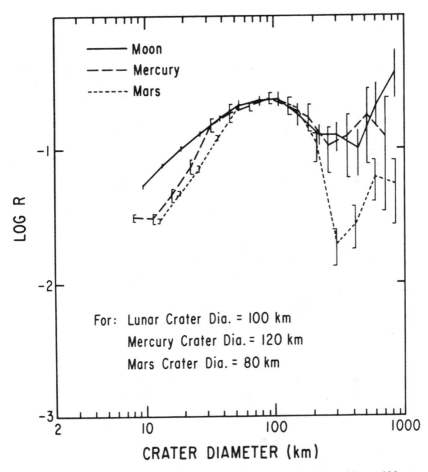

Fig. 18. The crater size-frequency distributions for the highlands of the Moon, Mars and Mercury (see Fig. 9) have been matched from about 40 km to 150 km diameter (the diameter range not affected by intercrater plains emplacement and having relatively good statistics). The lateral shifts in the curves require higher planet impact velocities with decreasing heliocentric distance (larger craters on Mercury and smaller ones on Mars compared to a given size crater on the Moon).

One can also analyze the problem dynamically by using the crater scaling law to solve for the impact velocity of a constant projectile diameter so that the crater curves more closely coincide. In this case, a given size projectile must make a crater twice as large on the Moon as on Callisto. The velocity V required to form a complex crater diameter D_r from a projectile diameter d is given by a modified Holsapple-Schmidt (1982) crater scaling law (Strom 1987a):

$$V = \left(\frac{D_q^{0.15} \, D_r^{0.85} \, g^{1/6}}{d^{1/1.2} \, Kc^{1/3}[1 - 0.095(1 - \sin A)]^{1/3}}\right)^3 \tag{15}$$

where D_q is the transition diameter from simple to complex craters, g is surface gravity, K is a coupling factor (~ 4.8), c is the ratio of projectile to target density, and A is the impact angle from the horizontal. The exponents of D_q and D_r are from Croft's (1985) empirical relationship D_e equals $D_q^{0.15} \, D_r^{0.85}$, where D_e is the excavation crater diameter. For simple bowl-shaped craters D_e is equal to D_r.

Since a high flux of comets early in solar system history has been proposed to account for the period of heavy bombardment throughout the solar system (Shoemaker and Wolfe 1982), we used a density of 1 g cm^{-3} for the projectile density. Table III lists the impact velocities required to form a 100 km crater on the Moon and a 50 km crater on Callisto for various comet diameters, i.e., the velocities required to shift the Callisto crater curve to match more closely that in the inner solar system. Also listed are the rms impact velocities for long-period and short-period comets on the Moon and Callisto. The required impact velocities are about 30 times greater on the Moon than on Callisto for a comet nucleus of similar size. This velocity difference is totally unrealistic for either short- or long-period comet orbits. The

TABLE III
Comet Impact Velocities to Form a 100 km Crater on the Moon
and a 50 km Crater on Callisto for a Given Size Nucleus

Comet diameter (km)	Moon impact velocity (km s^{-1})	Callisto impact velocity (km s^{-1})
7	113.4	3.75
8	81.2	2.69
9	60.5	2.00
10	46.5	1.54
11	36.6	1.21
12	29.5	0.98
13	24.1	0.80
14	20.0	0.66
15	16.9	0.56
16	14.4	0.48
17	12.3	0.41
18	10.7	0.35
19	9.3	0.31
	Long-period comets rms impact velocity (km s^{-1})	Short-period comets rms impact velocity (km s^{-1})
Moon	52	20
Callisto	26	14

maximum difference between the rms impact velocity at the Moon and Callisto (long-period comets at the Moon and short-period comets at Callisto) is 38 km s^{-1}, or only a factor of 3.7 compared to the factor of 30 required to account for the differences in the cratering records between the two satellites. In fact, heliocentric objects with any combination of orbital elements are incapable of generating the impact velocity difference required to account for the disparity in the cratering curves in the inner solar system and at Jupiter. This strongly suggests that the objects responsible for the period of late heavy bombardment in the inner solar system were confined to the inner solar system, and that the objects impacting the Galilean satellites were in planetocentric orbits or predated the late heavy bombardment.

Additional evidence for this comes from a comparison of the cratering curves representing the period of heavy bombardment on the Moon, Mercury and Mars. Figure 9 shows the crater size-frequency distributions for the heavily cratered highlands of the Moon, Mars and Mercury representing the period of late heavy bombardment. They all have similar shapes except at diameters less than about 40 km where intercrater plains emplacement and erosion have modified the curves for Mercury and Mars as discussed earlier. However, at diameters between about 40 km and 150 km, where the curves are probably unaffected by erosion and the statistics are good, the curves are laterally displaced with respect to each other. They are displaced in such a manner that higher impact velocities are required at planets with smaller heliocentric distances, i.e., larger craters on Mercury and smaller craters on Mars compared to a given size crater on the Moon. This is consistent with objects in heliocentric orbits. There are uncertainties in the amount of displacement giving the best match of the curves, but the outside limits are as follows: relative to a 100 km crater on the Moon, the displacement is from 0 to 20 km larger on Mercury and 20 to 30 km smaller on Mars. The best fit is shown in Fig. 18 where for a 100 km crater on the Moon, the crater size on Mercury is 120 km and on Mars 80 km.

Equation (15) can be used to calculate the impact velocities required to form a given size crater for a constant size projectile on the Moon, Mercury and Mars, and from these velocities determine the impact velocity ratios, Mercury/Moon and Mars/Moon. The outside limits on the displacements give a Mercury/Moon impact velocity ratio between 1.37 and 2.18, and a Mars/Moon impact velocity ratio between 0.48 and 0.68. The best fit shown in Fig. 18 gives a Mercury/Moon impact velocity ratio of 2.18 and a Mars/Moon impact velocity ratio of 0.68.

The orbital elements (semimajor axes and eccentricities) and impact velocity ratios of impacting objects in heliocentric orbits can be derived from the equations for impact velocity. In simplified form, the impact velocity V_{ip} for a heliocentric object with a planet is:

$$V_{ip} = (U^2 + V_e^2)^{1/2} \qquad (16)$$

where U is the encounter velocity of the object at infinity, and V_e is the escape velocity of the planet. For a heliocentric object impacting a satellite, the mean impact velocity V_{is} is:

$$V_{is} = (U^2 + V_{eo}^2 + V_{es}^2)^{1/2} \qquad (17)$$

where V_{eo} is the escape velocity of the planet at the orbit of the satellite, and V_{es} is the escape velocity of the satellite.

Figure 19 is a plot of the comparison of the impact-velocity ratios of Mercury/Moon (a), and Mars/Moon (b), vs impactor semimajor axes from

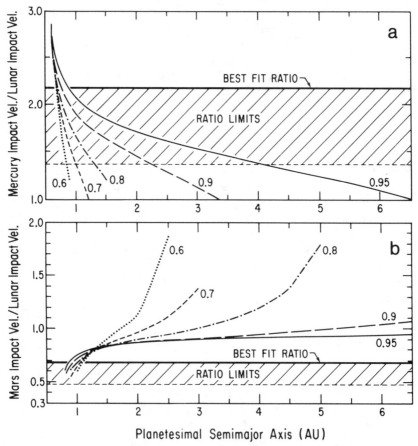

Fig. 19. Plot of the impact velocity ratios Mercury/Moon (a) and Mars/Moon (b) derived from matching the highlands crater curves (Fig. 18), vs impactor semimajor axes for eccentricities from 0.6 to 0.95. The hachured areas are the limiting impact velocity ratios for an acceptable curve fit, while the solid horizontal lines are the ratios derived from the best curve fit shown in Fig. 18. Only planetesimals with semimajor axes between about 0.8 and 1.2 AU lie within the same region of the impact velocity ratio limits. Jupiter crossers have semimajor axes >2.7 AU.

0.5 to 6.5 AU for eccentricities from 0.6 to 0.95. Also shown are the limiting impact-velocity ratios for an acceptable curve fit (hachured areas) and the ratios derived from the best curve fit shown in Fig. 18 (solid horizontal line). Only those parts of the eccentricity curves that lie within the hachured areas have impact-velocity ratios compatible with the observed shifts in the crater size-frequency distributions. The planetesimal orbits are further constrained by the need to have the curves lie within the same parts of the Mercury/Moon and Mars/Moon ratio limits. This occurs only for planetesimals with semimajor axes between about 0.8 and 1.2 AU. No matter what the eccentricity, planetesimals with semimajor axes greater than about 1.2 AU produce Mars/Moon impact velocity ratios that are too high to account for the shift between the Martian and lunar crater curves. The best fit of the crater curves is given by objects with semimajor axes of about 0.85 AU and eccentricities of about 0.95.

These results, together with those shown in Table III for the Moon and Callisto, strongly indicate that the objects responsible for the period of late heavy bombardment on the terrestrial planets were in orbits confined to the inner solar system. These objects are most likely accretional remnants left over from the accretion of the terrestrial planets. Wetherill (1977) has shown that accretional remnants could develop a long-lived tail consistent with the end of late heavy bombardment some 3.8 Gyr ago. This long-lived tail of accretional remnants could have had their orbits pumped up to large eccentricities by close approaches to the inner planets, which is consistent with the large eccentricities giving the best-fit impact-velocity ratios.

If the objects responsible for the period of late heavy bombardment of the terrestrial planets were confined to the inner solar system, then it follows that the period of late heavy bombardment of the outer planet satellites was the result of objects confined to that region of the solar system. Based on cumulative crater frequencies among the Saturnian satellites, Horedt and Neukum (1984) find that the objects responsible for the cratering record were more likely to have been in planetocentric orbits than in heliocentric orbits. Strom (1987a) reached a similar conclusion from differences of the crater populations among the outer planet satellites. Therefore, it is possible that the period of late heavy bombardment in the outer solar system was primarily the result of objects in planetocentric orbits. This conclusion should remain tentative until more detailed studies of the cratering record and celestial mechanical studies of planetocentric objects are concluded.

VI. SUMMARY

The cratering records in the lunar, Mercurian and Martian highlands all show similar size-frequency distributions which are characterized by highly structured curves with multiple distribution functions. On Mercury and Mars, there is a significant paucity of craters less than about 40 or 50 km diameter compared to the Moon. This paucity of craters is probably due to the obliteration of a fraction of craters by intercrater plains emplacement on Mercury and a combination of intercrater plains emplacement and atmospheric erosion and deposition on Mars. Intercrater plains are the major terrain type on Mercury, and both the cratering record and stratigraphic studies strongly indicate that they were emplaced over a period of time during the period of late heavy bombardment. Estimated absolute ages from crater densities suggest intercrater plains range from about 4.0 to 4.2 Gyr which coincides with the period of late heavy bombardment on the Moon. If intercrater plains are volcanic, then Mercury experienced much more volcanic activity during the period of late heavy bombardment than the Moon. This, however, is consistent with thermal history models that predict extensive lithospheric melting, global expansion and crustal extension resulting from the formation of Mercury's enormous iron core. Such conditions would provide a ready source of magma and an ideal tectonic setting for easy egress of lavas to the surface.

The Mercurian smooth plains within and surrounding the Caloris basin have a post-plains crater population that has a size-frequency distribution very similar to that of the lunar highlands over the same diameter range. The crater density is about an order of magnitude less than the Mercurian and lunar highlands, indicating that they are relatively young. The post-Caloris plains crater population, however, has a crater density identical to that superposed on the lunar Orientale basin and ejecta blanket which is about 5 times greater than that superposed on the lunar maria. This suggests that the Mercurian smooth plains were emplaced near the end of late heavy bombardment. They may be, on average, about 3.8 Gyr old, and therefore older than most of the lunar maria. Volcanism may have been shorter lived on Mercury than the Moon because the cooling of the lithosphere and the large iron core could have produced global compression that closed off magma conduits and inhibited volcanism earlier on Mercury than on the Moon. Mercury's cratering record and estimated ages from crater densities suggest that internal activity on Mercury was more extensive early in its history than on the Moon, but that it ended sooner.

The similarity of the crater size-frequency distributions for the heavily cratered surfaces in the inner solar system strongly suggests that they resulted from a single family of impacting objects in heliocentric orbits. The heavily cratered surfaces at Jupiter (Callisto and Ganymede) show a different crater population characterized by a paucity of large craters > 40 km diameter and an overabundance of small craters < 20 km diameter relative to that in the inner solar system. This difference cannot be explained by differences in crater scaling between Jupiter and the inner solar system. The impact velocities required to make the inner solar system and Callisto crater curves coincide are completely incompatible with impact velocities for objects in heliocentric orbits that cross Jupiter and the inner planets. This suggests that the objects responsible for the period of late heavy bombardment on the inner planets were confined to the inner solar system.

The inner planet highlands crater curves are laterally displaced with respect to each other in such a manner that higher impact velocities are required at planets with smaller heliocentric distances. The impact-velocity ratios Mercury/Moon and Mars/Moon derived from the Holsapple-Schmidt crater scaling law, and required to account for the lateral shifts, have fairly narrow ranges. Only objects with semimajor axes between about 0.8 and 1.2 AU are capable of producing the impact velocity ratios derived from the cratering record. The best curve fits require semimajor axes of about 0.85 AU and eccentricities of about 0.95. This strongly indicates that the objects responsible for the period of late heavy bombardment on the inner planets were confined to the inner solar system. These objects were probably accretional remnants left over from the formation of the inner planets. This implies that the outer solar system cratering record was the result of objects confined to that region, possibly in orbits that were mostly planetocentric.

THE TECTONICS OF MERCURY

H. J. MELOSH
University of Arizona

and

W. B. McKINNON
Washington University

Mercury is not by anyone's standards a tectonically active planet. However, this very lack of vigorous activity has preserved a record of the earliest tectonic events that occurred shortly after a brittle lithosphere developed, and thus pro- motes Mercurian tectonics to a position of great interest among students of the early development of planetary tectonic systems. The most ancient tectonic sys- tem is expressed by obscure lineaments and preferential orientations of later features such as crater rim segments and extrusive vents. A grid of NW-SE and NE-SW trending lineaments is well expressed in the equatorial regions but be- comes more oriented toward E-W directions near the poles. This grid is con- sistent with the fracture directions produced by despinning early in Mercury's history as the planet's original rapid rotation was slowed by solar tides and the equatorial bulge relaxed. Recent radar observations have confirmed the pres- ence of these lineament directions even on the unimaged hemisphere. The well- known lobate scarps are thrust faults produced by a slight ($\gtrsim$ 1 to 2 km) de- crease in Mercury's radius, probably due to cooling of the mantle and partial solidification of the core. Remnant despinning stresses may have been present at the time of scarp formation, but the case for preferred N-S orientation of the lobate scarps is weak. The remainder of Mercury's tectonic features are related to large impacts. In particular, the ring scarp around Caloris basin is probably due to collapse following the impact that excavated it. Grabens in the deposits filling Caloris are probably due to either later updoming that occurred during isostatic adjustment, or are perhaps a consequence of the loads induced by the *volcanic smooth plains that surrounded the basin. Mare ridges formed on these plains as the lithosphere sagged under their weight.*

I. INTRODUCTION

Mercury is not a tectonically active planet. Its dominant landforms are impact craters and volcanic plains. Tectonic features, where they occur, are subtle at the resolution of Mariner 10 and are often difficult to discern among other topographic forms on Mercury's rugged surface. Nevertheless, the tectonic features of Mercury are of great interest to planetary geologists precisely because Mercury does not exhibit vigorous tectonic activity. Mercury's surface preserves a record of tectonic events that occurred early in its history, a record that is either obliterated or overprinted and obscured on larger, more active planets such as the Earth or Mars.

Mercury's surface features preserve no record at all of the very earliest events in the planet's history. Processes such as accretion and differentiation can be deciphered only from the chemistry and isotopic ratios in minerals that make up Mercury's crust. This earliest history is unreadable because we lack samples of this crust. The presently known history of Mercury began when its lithosphere first became solid enough to retain topographic and structural features down to the present day. Curiously, the oldest record is probably not topographic, but structural; the ancient tectonic grid or lineament fabric, discussed in the next section, apparently predates even the most ancient and battered craterforms.

The preservation of tectonic history and, indeed, the detailed expression of tectonic events are thus tied to the existence of a lithosphere. All of the terrestrial planets possess lithospheres, which are the major unifying elements of planetary tectonics. Since several definitions of the term lithosphere are in current use, we must make it clear that in this chapter the term is used as Barrell (1914) originally defined it: the lithosphere is the outer portion of a planet that responds either elastically or fractures as a brittle solid under the action of applied stresses. The underlying asthenosphere is regarded as either inviscid or viscous, depending on the time scale of its deformation. The lithosphere may be entirely confined to the crust, or may include both crustal and mantle materials. The terms crust and mantle denote distinct chemical and petrological layers whose composition is only one factor (and not even the most important factor) in controlling the lithosphere's mechanical behavior.

The most important factors regulating the mechanical behavior of geologic materials are temperature and time. Hot rocks, in general, flow more readily than cold ones. The strain rate $\dot{\epsilon}$ of rocks subject to an applied deviatoric stress σ is given by an equation of form

$$\dot{\epsilon} = A\sigma^n e^{-gT_m/T} \tag{1}$$

(Weertman and Weertman 1975; Kirby 1983) where A, n and g are experimentally measured constants, n equals 3 to 5 for most crystalline substances, and g equals 27 to 30 for most non-icy geologic materials. T_m is the melting point of the solid. A 100°C change in temperature may thus increase the flow rate of typical silicate geologic materials by an order of magnitude.

The effect of time on the flow of rock materials is gauged by the Maxwell time τ_M, defined as the interval required for slow viscous flow (Eq. 1) to produce a strain equal to the initial elastic strain ϵ_{el}

$$\tau_M = \frac{\epsilon_{el}}{\dot{\epsilon}} \simeq \frac{1}{2A\mu\sigma^{n-1}} e^{gTm/T} \tag{2}$$

where μ is the elastic shear modulus. Using the definition of viscosity, $\eta = \sigma/2\dot{\epsilon}$, the Maxwell time can be written more conventionally as

$$\tau_M = \eta/\mu. \tag{3}$$

The Maxwell time is significant in that for time scales less than τ_M, elastic strain dominates and the material may be treated as an elastic solid (lithosphere), whereas on time scales longer than τ_M, flow dominates and the material may be treated as a viscous fluid (asthenosphere).

Because of the strong temperature dependence in Eqs. (1) and (2), the Maxwell time is a sensitive function of temperature. In general, rocks at absolute temperatures less than half their melting points (and at typical geologic stress levels of ~1 to 100 MPa) respond elastically over geologic time, whereas hotter rocks may flow appreciably.

Mercury's surface is sufficiently cold that its lithosphere has responded in an elastic/brittle fashion from the earliest times geologically recorded. Furthermore, tectonic evidence reviewed below shows that Mercury's lithosphere is rather thick: thicker than 100 km, compared with the ~25 to 50 km for old oceanic lithosphere on the Earth (Watts et al. 1980). The presence of this thick lithosphere is at least partly due to Mercury's small size and consequently low thermal gradient. Mercury may also possess a smaller average abundance of heat-producing radioactive elements than the Earth, suggested by Mercury's higher mean density and hence relative deficiency of silicate rocks compared with the Earth. Mercury's thick lithosphere, whatever its cause, has strongly influenced its response to both tectonic and impact stresses. The effective thickness of Mercury's lithosphere, like that of other planets, has probably also increased with time as the planet cooled and its surface heat flow declined. Thus, Mercury's early lithosphere may have been quite thin (although there is no evidence for this), whereas the present lithosphere could be considerably thicker than 100 km. Even so, this lithosphere is still a small fraction of Mercury's 2439 km mean radius so that it can be treated for most purposes as a thin elastic shell.

The important idea to keep in mind during this discussion of Mercury's tectonics is that throughout recorded geologic history, Mercury's mechanical structure has been that of a single thin elastic/brittle outer shell overlying a much more fluid interior. This thin shell is unable to resist long wavelength bending, like that associated with changes in the equatorial bulge. Large stresses may develop in it due to small changes in planetary radius. The lithosphere may be entirely penetrated by large impact events, leading to characteristic structures both near the impact site and at large distances from it.

II. MERCURY'S GLOBAL TECTONIC GRID

Mercury's most ancient tectonic feature is a fabric of fractures or weak zones that was impressed on the lithosphere before any presently recognizable topographic feature formed. The Mercurian grid, like the lunar grid (Strom 1964), is expressed only as lineaments or preferential orientations of much later features. The lineaments consist of ridges, troughs (excluding those radial to craters or basins) and linear segments of complex crater rims (Fig. 1). The orientations of these features are presumably determined by the ancient lines of weakness in the lithosphere.

The Mercurian grid was first recognized by Dzurisin (1976,1978), who noted three preferential lineament orientations: NE-SW, NW-SE and a weaker N-S set of directions. The NE-SW and NW-SE sets dominate near the equa-

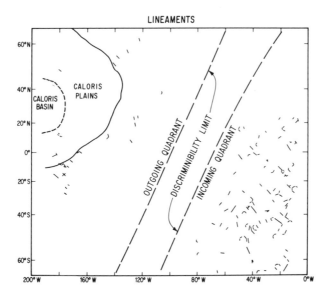

Fig. 1. Mercurian lineament map. The map includes linear scarps, ridges, troughs and linear segments of crater rims of inferred tectonic origin. The boundary of the Caloris plains is only drawn schematically on this Mercator projection. (Figure from Melosh and Dzurisin 1978a.)

tor, whereas orientations become more random as the south pole is approached (Fig. 2), although this may be partly due to the extreme lighting geometry near the pole. These observations are in qualitative agreement with similar work (Strom 1964) on the lunar grid. Recent remeasurement of the Mercurian grid (Chapter by Thomas, Masson and Fleitout) has confirmed Dzurisin's results and suggested that his N-S trend actually strikes more closely to N 20°E.

Lineaments are well developed in all ancient plains and heavily cratered regions of Mercury (Dzurisin 1978; Leake 1981). They have also been detected on portions of the planet not imaged by Mariner 10 in radar observations (Clark and Jurgens 1985a). Lineaments are rare on the younger Caloris plains, suggesting that most lineaments predate Caloris plains development.

The origin of the grid is commonly attributed to stresses that developed in Mercury's lithosphere due to tidal despinning, first analyzed by Burns (1976) and Melosh (1977b). It is generally accepted that Mercury's present 59 day resonant rotation period is the residual of an initially much faster rotation rate slowed by solar tides (Goldreich and Soter 1966).

Comparison of Mercury's angular momentum density to that of other planets (except Venus) suggests that its initial rotation period may have been roughly 20 hours (Kaula 1968; Burns 1975). Under such conditions, Mercury's hydrostatic polar flattening would have been about 1/160 (compared to the Earth's present value of 1/298). The spindown time τ from this initial state was derived by Goldreich and Soter (1966)

$$\tau = \frac{4}{9} \frac{CQ'R^6}{GM^2a^5} \; (\Omega_o - \Omega_f) \tag{4}$$

where Ω_o and Ω_f are the initial and final angular rotational velocities, respectively; C is Mercury's polar moment of inertia, a is Mercury's radius, R is Mercury's orbital semimajor axis, M is the Sun's mass, and G is the gravitational constant. The factor Q' is given by

$$Q' = Q \left(1 + \frac{19\bar{\mu}}{2g\rho a} \right) \tag{5}$$

where Q is the planetary dissipation factor, $\bar{\mu}$ is Mercury's mean rigidity, g is surface gravitational accleration, and ρ is mean density.

Goldreich and Soter (1966) concluded that Mercury could spin down to its present rotational state over the age of the solar system if Q is less than about 190. This can be compared with a Q of 13 they derived for the Earth, caused by dissipation in the oceans; a Q of ~30 for the Moon at a period of a month (Ferrari et al. 1980), possibly caused by fluid turbulence and dissipation at a core-mantle interface (Yoder 1981); and estimates for the Q of Mars of 50 to 150 (see, e.g., Veverka and Burns 1980). The first two mechanisms

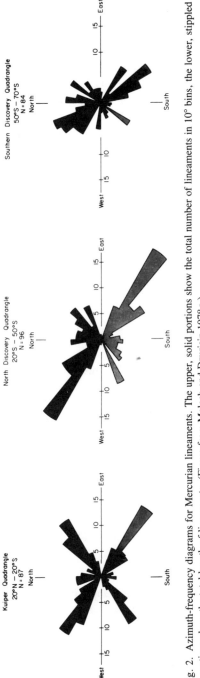

Fig. 2. Azimuth-frequency diagrams for Mercurian lineaments. The upper, solid portions show the total number of lineaments in 10° bins, the lower, stippled portions show the total length of lineaments. (Figure from Melosh and Dzurisin 1978a.)

involve dissipation at a solid-liquid interface and may be applicable to Mercury. In addition, Schubert et al. (see their chapter) have shown that the Maxwell time of the solid Fe-Ni inner core of Mercury may be comparable to or less than the present rotation period. This implies that Mercury's "solid body" Q may have also been rather low in the past and that Mercury's spin-down and thermal histories may have been intimately coupled. We conclude that Mercury's spindown time could have been considerably < 1 Gyr.

As Mercury despun, its equatorial bulge relaxed and the planet changed shape. If Mercury had a thin (< 100 km thick) lithosphere at the time of despinning, strains on the order of 1% would have accumulated in it. Stresses may have reached several kilobars, sufficient to produce extensive fracturing and faulting.

Vening-Meinesz (1947) presented expressions for the principal stresses in the thin elastic lithosphere of a planet stressed by the relaxation of an equatorial bulge

$$\sigma_{\theta\theta} = -\frac{5}{12}(m_o - m_f)\mu\left(\frac{1 + \nu}{5 + \nu}\right)(5 - 3\cos2\lambda) \qquad (6)$$

$$\sigma_{\phi\phi} = \frac{5}{12}(m_o - m_f)\mu\left(\frac{1 + \nu}{5 + \nu}\right)(1 + 9\cos2\lambda) \qquad (7)$$

where $\sigma_{\theta\theta}$ is the meridional (N-S) stress component and $\sigma_{\phi\phi}$ is the azimuthal (E-W) stress component. Note that $\sigma_{\phi\phi}$ is more compressional than $\sigma_{\theta\theta}$ at all latitudes except at the poles, where they are equal. The radial principal stress σ_{rr} is smaller than the horizontal stresses by the ratio of the lithospheric thickness to the planet's radius. The lithosphere is assumed to be thin, so that the radial principal stress can be neglected. In Eqs. (6) and (7), ν is Poisson's ratio and λ is the latitude at which the stresses are evaluated. The difference between the initial and final ratios of the centripetal acceleration at the equator to the gravitational acceleration, $m_o - m_f$, is given by

$$m_o - m_f = \frac{(\Omega_o^2 - \Omega_f^2)a^3}{GM_M} \qquad (8)$$

where M_M is Mercury's mass.

The fundamental assumption implicit in Eqs. (6) and (7) is that the planet was initially in a hydrostatic state (no deviatoric stresses were present). As the planet's surface cooled, the lithosphere became capable of supporting stresses for long intervals. Subsequent relaxation of the equatorial bulge generated stresses (Eq. 6 and Eq. 7). These equations are only valid as long as no fractures develop or other inelastic deformation occurs. The stress modifications produced by plastic flow are discussed by Melosh (1977b) and Pechmann and Melosh (1979).

The maximum stress difference achieved during despinning occurs at the equator and is given by

$$(\sigma_{\phi\phi} - \sigma_{\theta\theta})_{max} = 5(m_o - m_f)\mu \left(\frac{1 + \nu}{5 + \nu}\right) . \tag{9}$$

The stress differences predicted by this equation reach several kilobars for despinning from an initial 20 hr period, compared to typical rock crushing strengths of one kilobar or less, so that a pervasive fracture network should certainly have formed if the lithosphere was sufficiently cool when Mercury despun.

Melosh (1977b) extended these thin-lithosphere equations to the general case of a thick lithosphere and examined the tectonic consequences of despinning (Fig. 3). If the lithosphere is thicker than about 0.05 of the planet's radius (122 km on Mercury), a zone of N-S trending thrust faults forms near the equator, reaching a maximum latitudinal extent of about 25° north and south of the equator for lithosphere thicknesses between 0.5 and 0.8 of the planet's radius. A zone of strike-slip faults forms in intermediate latitudes for all lithosphere thicknesses. This zone includes the equator when the lithosphere is thinner than 0.05 of the planet's radius. East-west trending normal faults

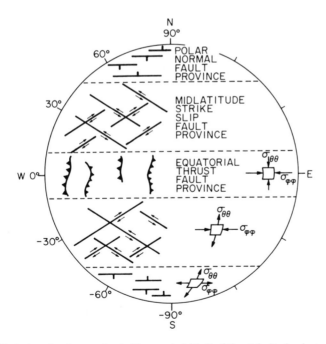

Fig. 3. Tectonics of a despun planet. The equatorial belt of thrust faults develops only if the lithosphere is thinner than 0.05 of the planet's radius. (Figure after Melosh 1977b.)

form poleward of about latitude ±55°, nearly independent of lithosphere thickness.

The NE-SW and NW-SE lineament trends within about 50° of the equator are consistent with tidal despinning for a lithosphere 100 km or less thick (the present lithosphere may, of course, be considerably thicker; the 100 km estimate is valid only at the time when most of the despinning took place). The rose diagrams (Fig. 2) also show that the acute angle between the two lineament directions is in the eastern and western quadrants, in agreement with the predicted dominance of E-W compression. The absence of any organized lineament directions in the polar region is a little more puzzling, although the recent work of Thomas et al. (see their chapter) shows some evidence for the expected E-W normal fault trend.

The N-S or N 20°E lineament trends are more difficult to explain on the basis of tidal despinning. Perhaps the lithosphere thickened sufficiently toward the end of despinning to induce some N-S thrust faulting, or perhaps contraction in addition to despinning resulted in N-S fracturing (Pechmann and Melosh 1979). This trend, however, is weak and not very consistent, so perhaps other, heretofore unrecognized, tectonic stresses were at work. Lineament directions are also unclear in the polar regions, where the despinning model makes a clear prediction of E-W normal faulting. Further study of these issues may have to await new data returned from spacecraft.

In summary, both theoretical and observational evidence support the idea that the global Mercurian lineament grid was predominantly due to tidal despinning early in the planet's history, although some difficulties remain unresolved. The paucity of lineaments on the young Caloris plains suggests that despinning was substantially complete before these units formed.

A. Lobate Scarps

One of the few geologic surprises of the Mariner 10 encounter with Mercury was the discovery of numerous lobate scarps (Strom et al. 1975). Figure 4 depicts Discovery scarp, one of the largest scarps on Mercury, which is up to 2 km high and over 500 km long. These scarps vary in length from 20 km to over 500 km and have heights from a few hundred meters to one or two kilometers. The scarp crests are rounded and the terrain behind the scarp varies from flat to gently dipping away from the scarp. Individual scarps often transect several types of terrain, including craters. Most scarps are lobate in plan on a scale of a few tens of kilometers, although members of one subclass, termed arcuate scarps by Dzurisin (1978), form relatively smooth arcs 100 to 600 km long and 500 to 1100 m high. The convex side of these arcuate scarps (as seen in plan) is always down, rather reminiscent of terrestrial island arcs. Radar altimetric profiles exist for a few examples (Harmon et al. 1985). Irregular scarps are entirely confined to the interiors of large craters. This last subclass may be either linear or highly sinuous in plan form. They are relatively small, less than about 100 km long and 400 m in height.

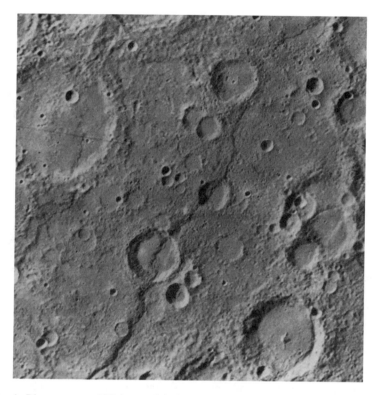

Fig. 4. Discovery scarp. This is one of the largest lobate scarps on Mercury, extending 500 km from end to end. It is up to 2 km high and transects two craters 35 and 55 km in diameter.

Lobate scarps are of young to intermediate age in Mercury's geologic history (Strom 1984). They are younger than the majority of large craters on Mercury, although young, medium size craters are superposed on the scarps in some places. No lobate scarp is embayed by intercrater plains, so they evidently postdate the formation of all intercrater plains (or nearly all [see Leake 1982]). A segment of Discovery scarp appears to be controlled by the Mercurian tectonic grid, so it is younger than grid formation. Some lobate scarps occur on the smooth plains surrounding Caloris, hence lobate scarp formation postdates Caloris itself and the subsequent extrusion of the smooth plains units.

Lobate scarps were almost immediately recognized to be thrust or reverse faults (Strom et al. 1975), partly on the basis of morphology, and partly because of transection relations between the 130 km long Vostok scarp and a 65 km diameter crater, Guido d'Arezzo. The northwest rim of Guido d'Arezzo is offset some 10 km by the scarp in a manner that strongly suggests crustal shortening by thrust faulting. If lobate scarps are the traces of thrust faults, Strom et al. (1975) estimated that there has been a net decrease in Mercury's surface area of about 6.3×10^4 to 1.3×10^5 km^2, equivalent to a 1 to 2 km decrease in the planet's radius, assuming dip angles for the faults of 45° and 25°, respectively. Strom et al. (1975) emphasized that these are minimum estimates because of probable undetected scarps; if dips of the fault planes are shallow at depth, which is common terrestrially, then the estimate of radius decrease is raised again.

The radius decrease implied by the formation of lobate scarps is probably due to the secular cooling of Mercury's mantle and the cooling and partial freezing of its core, following the heat pulse (and unrecorded expansion) consequent upon planetary differentiation and core formation. In agreement with the observations, thermal models (Solomon 1976, 1977) predict about 2 km shrinkage in Mercury's radius subsequent to core formation for secular cooling alone. Complete freezing of Mercury's core would result in an additional 8 km radius decrease (Solomon 1977). Considering the uncertainty in the minimum estimates of Strom et al. (1975), it is possible that a major fraction of the core has solidified during the past 4 Gyr. A relatively large solid inner core is an unavoidable outcome of all thermal models to date (see, e.g., Chapter by Schubert et al.).

The horizontal stresses that develop as a result of a change Δa in a planet's radius a are given by

$$\sigma_{\theta\theta} = \sigma_{\phi\phi} = 2\mu \left(\frac{1 + \nu}{1 - \nu} \right) \frac{\Delta a}{a} . \qquad (10)$$

A 1 km change in radius corresponds to stresses on the order of 1 kbar, adequate to initiate large-scale thrust faulting.

Lobate Scarp Orientation. Because the horizontal stresses induced by contraction (Eq. 10) are equal, the resulting thrust faults are not expected to show any preferred orientation. Whether or not this is in fact the case has been a hotly debated topic. Melosh and Dzurisin (1978a) noted an apparent N-S preference in lobate scarp orientation which they attributed to the predominance of E-W compression, a late inheritance from tidal despinning. Pechmann and Melosh (1979) showed that a combination of despinning and contraction proceeding at different rates could plausibly produce an early network of NW-SE and NE-SW lineaments followed by a later overprint of N-S trending thrust faults. However, Cordell and Strom (1977) argued that the apparent preferential orientation of scarp azimuths (Fig. 5) is an observational effect and that scarp azimuths are actually random. Figure 5 also shows that scarp azimuths tend to lie along the NW-SE and NE-SW grid directions.

Recent new work (Thomas et al. 1982; Chapter by Thomas et al.) has used the limited stereo coverage of Mercury in an attempt to overcome any observational bias in favor of scarps oriented perpendicular to the local Sun direction. Limited to the southern Mercurian quadrangles H-11 (Discovery), H-12 (Michelangelo) and H-15 (Bach), these stereoscopically identified

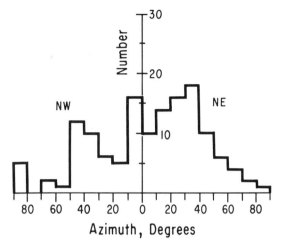

Fig. 5. Azimuth-frequency diagram of lobate scarps for all latitudes (−90° to ±60°). The origin of the apparent concentration of lobate scarp orientations between +45° and −45° is controversial. It may be either tectonic or due to lighting effects. (Figure after Cordell and Strom 1977.)

scarps appear to show a preferred orientation that is neither N-S nor E-W, but is radial to Caloris basin. Thomas et al. (see their Chapter) attribute this orientation to the far-field tectonic effects of the Caloris impact basin (Fleitout and Thomas 1982). Although this is an interesting new idea, the available stereo coverage is confined to the south polar regions where it is difficult to make a strong case for this proposal, since Caloris' effects are expected to be extremely small there. It would be more satisfying to demonstrate a strong preference for radial orientation close to Caloris where the basin's tectonic effects are expected to be largest.

B. Ridges

Dzurisin (1978) noted the occurrence of several linear ridges on the intercrater plains. These ridges, exemplified by Mirni ridge in Discovery quadrangle, range from 50 to 350 km long and are 100 to 1000 m high. They are typically rounded and convex upward in cross section. The ridges are roughly symmetric in transverse profile, so they are not closely related to lobate scarps (which are highly asymmetric, with steep slopes on one side and gentle backslopes on the other), although lobate scarps sometimes transform into ridges longitudinally. Ridge orientations generally follow the Mercurian grid directions. The ridges are similar in age to the lobate scarps, since they postdate the intercrater plains and are overlapped only by a few medium-sized craters. Little is known about the origin of these ridges. Dzurisin (1978) proposed that Mirni ridge is a dike-like volcanic extrusion that followed a pre-existing fracture of the global grid system. However, the morphology of Mirni and the other ridges is reminiscent of lunar and Martian mare ridges, so a compressional tectonic origin (see, e.g., Plescia and Golombek 1986) cannot be ruled out.

C. Additional Effects

Of necessity, the foregoing discussion presents a simplified picture of Mercury's evolution. There remain a number of potentially important physical effects that have not yet been examined in detail: insolation-derived lithosphere thickness variations, nonhydrostatic overburden stresses and viscous relaxation. We discuss each in turn.

As pointed out by McKinnon (1981), the permanent equator-to-pole subsurface temperature contrast on Mercury of ~350 K (see Morrison 1970; Cuzzi 1974) results in an elevation of lithospheric isotherms and thus thinning of the lithosphere at the equator. This thinning is potentially severe, as the bounding isotherm for the elastic lithosphere probably lies in the 600-900 K range (Watts et al. 1980), so reductions in lithosphere thickness by more than a factor of two with respect to the "cold" poles are conceivable. However, this effect is no doubt mitigated by asthenospheric advection of heat towards the poles. Mercury's equatorial "hot" poles, a consequence of its 3/2 spin-orbit resonance, are on average ~100 K warmer than equatorial longitudes 90° away, so these hot regions possess even thinner lithospheres. In terms of stresses due to despinning or radius changes, the thin lithosphere case (in the sense used previously) is instructive. The membrane strains imposed by the planetary shape change are the same as in the uniform lithosphere case, but the resulting meridional *force* gradient due to the nonuniformity causes a redistribution of strain energy. The outcome is, naturally enough, an enhancement of stress and strain in regions of thinner lithosphere. To first order, the predicted fault patterns do not change, only the magnitudes of the stresses. However, the sequence of pattern development (discussed in Melosh [1977*b*] and Pechmann and Melosh [1979]) may be affected as the plastic failure limit is approached. Even in this case, though, it is the equatorial region that fails first, nonuniform lithosphere or not. Lineaments may have preferentially formed in regions of thinner lithosphere, but Mariner 10 observations do not have sufficient coverage at proper solar incidence angles to test this idea adequately. Insolation-derived lithospheric thickness variations are clearly worthy of more study, however, whether on Mercury or on the icy satellites of the outer solar system.

Overburden stresses may be nonhydrostatic; in particular, ejecta and regolith layers and multilayered volcanic piles may have horizontal stresses at depth z that are smaller than the vertical stress $\rho g z$ (see, e.g., Haxby and Turcotte 1976; Golombek 1985). If the tectonic stresses are reduced in a fractured surface layer of high compressibility (as would describe a so-called megaregolith), the nonhydrostatic overburden stresses may be important. This is especially important if the faults originate in or at the base of the fractured layer. Here, the relative augmentation of the vertical compressive stress compared with the horizontal stresses can transform a region of thrust faults into one of strike-slip faults, and a region of strike-slip faults into one of normal faults; regions of normal faults remain unaffected. The effect on the despinning solution of Melosh (1977*b*) is to shift all the fault province boundaries in Fig. 3 equatorward. This might relax the constraint that Mercury's lithosphere was less than 100 km thick at the time of despinning, but whether this overburden effect is realized is problematic, due to competing thermal effects of either sign that arise in the deposition of volcanic and impact units. In addition, the Mercurian tectonic grid is an ancient feature, not simply the surface expression of deeper lithospheric stresses. As noted above, the stress levels involved in despinning are potentially quite large, so it is most likely that the lithosphere was thoroughly fractured. The NE-SW and NW-SE lineament trends are most likely the surviving expression of deeply penetrating strike-slip faults, an important exception to the conclusions of Golombek (1985).

Because of viscous relaxation, the rate at which stress accumulates is as important as the stress levels achievable in purely elastic situations. Membrane stresses caused by despinning or planetary volume change relax over a few lithospheric Maxwell times; topography relaxes over tens of Maxwell times or longer depending on crustal thickness and degree of isostatic compensation (Solomon et al. 1982). The composition and thickness of Mercury's crust are unknown, but regions on Mercury are up to 200 K warmer on average than corresponding regions on the Moon (cf. Morrison 1970; Keihm and Langseth 1975), so viscosities could be reduced by up to two orders of magnitude in comparison to the Moon if similar composition and heat flow are assumed. Schaber et al. (1977) argued that some Mercurian basins have been nearly compensated by this process and that viscous relaxation has reduced the total number of identifiable basins. On the other hand, there is the evidence that the lithosphere of the planet responded brittlely to despinning and global contraction stresses. Careful evaluation of these constraints in conjunction with plausible thermal and tidal histories may yield new information on Mercury's crustal properties.

III. TECTONIC FEATURES RELATED TO CALORIS BASIN

A. Ring Tectonics

Caloris basin is a 1300 km diameter impact structure that lies about 30° north of one of Mercury's hot poles (180° long). Caloris was on the terminator at the time of the Mariner 10 encounters, so that only one half of the basin was imaged. It is the largest impact structure known on Mercury, and is also one of the best preserved. A highly degraded basin named Beethoven in the southwestern portion of Beethoven quadrangle (H-7) is about half as large as Caloris, but few tectonic details can presently be discerned in it.

The term basin was first applied to large multiringed structures of impact origin by Hartmann and Kuiper (1962). Their definition emphasizes the presence of both radial structures (mainly troughs and ridges in the ejecta blanket) and two or more concentric rings. Although many theories have been proposed to account for the origin of the rings, current models recognize them as tectonic structures. The prominent asymmetric scarp surrounding multiring basins is thought to be the trace of an extensional ring fracture that formed outside the rim of the original crater. The crater rim itself and other interior rings with symmetric cross sections are due to crater excavation and collapse and are not included with tectonic features in this review (see Chapters by Pike and by Schultz).

Melosh and McKinnon (1978) and McKinnon and Melosh (1980) showed that the development of ring scarps depends upon the thickness of the lithosphere underlying the initial impact crater. If the crater fails to penetrate the lithosphere, no ring scarp forms. One or more ring scarps develop if the crater completely penetrates the lithosphere (Fig. 6). The scarps are a response to the drag of the asthenosphere as it flows inward to fill the crater cavity. The overlying elastic lithosphere is brought into extension by this flow and may fracture if it is weak enough. Solomon and Head (1980) showed that ring development on the Moon is correlated with lithosphere thickness, which they estimated from the tectonic pattern induced by mascon loading.

Multiring basins are common on the Earth's Moon, and on Ganymede and Callisto. A few very degraded basins appear to have developed on Mars. The Callisto basins, of which Valhalla is the prime example, differ from the lunar basins in having many more rings. This style is evidently a response to a very thin lithosphere (Melosh 1982). Mercury, on the other hand, has very few multiring basins. It can be argued that even Caloris itself is not a true multiring basin because only one asymmetric ring can be seen, which may be the rim of the crater that initiated the structure. Only a faint, incomplete ring scarp has been recognized to the northeast of the main scarp. This faint scarp lies between 100 and 160 km beyond the main scarp. If it is another ring, the ratio of its diameter to the diameter of the main ring falls far short of the $\sqrt{2}$ ring ratio characteristic of lunar multiring basins (Hartmann and Wood 1971). In any event, the rings of Caloris are very unlike those of the lunar Orientale basin, to which they have often been compared (McCauley 1977; McCauley et al. 1981).

The difficulty in identifying rings in the youngest, best preserved basins renders claims of their recognition in ancient, degraded, poorly perceived basins suspect (see Chapter by Spudis and Guest). Radar altimetric coverage of three of these ancient "basins" by Harmon et al. (1985) provides no unambiguous support for their multiringed nature.

Even if Caloris, the 625-km diameter Beethoven, and the 400-km diameter Tolstoj crater (McKinnon 1981) are accepted as multiring basins, such basins are much rarer on Mercury than on the Moon. On the basis of ring tectonic theory, this implies that Mercury's lithosphere was relatively thick compared to the Moon's lithosphere at the end of heavy bombardment when

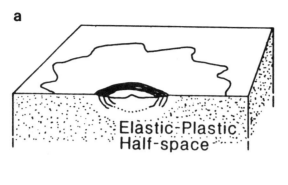

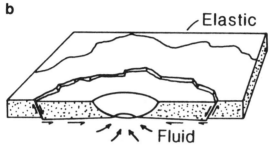

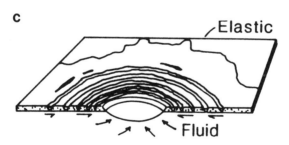

Fig. 6. Ring tectonics. (a) Craters forming in a lithosphere much thicker than the crater's depth do not initiate rings. (b) When the crater's depth is greater than the lithosphere thickness, flow of the underlying asthenosphere toward the crater cavity produces one or more ring fractures. (c) A large number of rings form when the lithosphere is much thinner than the crater's depth. This is the form of Valhalla basin on Callisto. (Figure after McKinnon and Melosh 1980.)

the largest craters formed. Rough estimates indicate a lithospheric thickness in excess of about 100 km for Mercury at this time (McKinnon 1981), compared with thicknesses of 25 to > 75 km for the Moon at the end of the mare basalt flooding, depending upon location (Solomon and Head 1980).

McKinnon (1981) offered three additional possible explanations for the paucity of identifiable multiring basins on Mercury: viscous relaxation of topography (as noted above), extensive intercrater plains formation, and subsidence of lithospheric ring blocks. Intercrater plains formation was especially abundant in pre-Caloris times and preferentially filled the topographic lows created by impacts (Leake 1982). The complete burial of interior ring structures, if any, in young basins such as Beethoven and Caloris, by thick deposits of smooth plains (see, e.g., the Chapter by Spudis and Guest), is a younger geological manifestation of the same process. McKinnon (1981) noted the subsidence of lithospheric ring blocks of the Caloris Montes formation, and argued that *if* the Mercurian crust were thin or mafic (as opposed to felsic), then the potential existed for widespread subsidence of fractured ring segments and subsequent burial by plains (presumably volcanic). This hypothesis was not offered as proof that the Mercurian lithosphere was negatively buoyant compared to the interior, but in the spirit that our knowledge of crustal composition and thickness is quite poor and that some plausible accretion scenarios do not lead to a lunar-like crust.

B. Far-field Tectonic Stresses

Fleitout and Thomas (1982) and Thomas et al. (see their Chapter) propose that the tectonic effects of the Caloris impact extend far beyond the basin's main scarp. Employing a model rather similar to ring tectonics, they envision that the formation of a basin which penetrates a lithosphere already compressed by planetary contraction would relieve the deviatoric stresses within the basin and allow the entire lithosphere to shift slightly basinward. Stresses parallel to great circles passing through the basin center thus become extensional relative to stresses acting parallel to small circles centered on the basin. This stress difference produces thrust faults radial to the basin, initiated by the circumferential compression. Stress differences are largest at the basin rim and die away rapidly as the antipode is approached, where all horizontal stresses are equal.

This model was discussed in Sec. II.A. in the context of lobate scarp orientation. Although Thomas et al. (1982; Chapter by Thomas et al.) do find evidence favoring a preferred orientation of lobate scarps radial to Caloris, the data are from a part of Mercury within 90° of the antipode of the basin where stress differences are not expected to be large. The importance of far-field tectonic effects is thus unclear at the present time. It is especially unclear if the likely state of stress in the Mercurian lithosphere is considered. Turcotte (1983) has shown that secular cooling of a planet induces compression in the upper lithosphere and *extension* in the lower lithosphere. The latter arises because, in the lower lithosphere, the thermoelastic stresses due to temperature decreases dominate those caused by the cooling and shrinking of the "fluid" interior. Thus, in the model of Fleitout and Thomas (1982), the overall response of the lithosphere is weighted by the vertical integral of compression and extension prior to the formation of Caloris; the lithosphere may or may not shift basinward. In order to guarantee that compression dominates, it is probably necessary to impose additional global contraction due to inner core freezing. It is not likely, however, that the lithosphere was completely in (non-hydrostatic) compression, for the voluminous eruptions of the exterior smooth plains some time after the impact (discussed in the next section) probably required the prior ascent of magma into the lower lithosphere. This ascent would be markedly aided by any extensional stress (Solomon 1978).

C. Tectonics of the Interior Plains

Caloris' main ring scarp surrounds an interior circular plain. This plain, which is composed either of impact melt or is veneered with extrusive lavas, is extensively ridged and fractured (Fig. 7). This type of plain is unique to the Caloris basin on Mercury; no similar plains occur on the Moon or Mars. The principal tectonic elements on this plain are ridges and troughs that appear to be grabens.

The ridges are irregular to sinuous in plan, forming a crude polygonal network on the floor of the basin (Fig. 8). They range from 50 to 300 km long, 1 to 12 km wide, and 100 to 500 m high, with rounded crests and gently sloping (5° to 10°) flanks that are convex upward in cross section (Dzurisin 1978). The ridges inside Caloris outline two crude interior arcs with diameters of about 700 and 1000 km. Other arc identifications are possible (Maxwell and Gifford 1980). The arcs may be the traces of interior rings within Caloris, telegraphed through the plains by some kind of structural inheritance. These ridges are morphologically similar to lunar and Martian mare ridges and probably reflect a compressional tectonic environment.

The ridges are cut by irregular troughs hundreds of kilometers long, 1 to 10 km wide, and 100 to 500 m deep with flat floors and abrupt, steep walls (Fig. 8). The troughs are oriented with a strong concentric trend and a weaker radial trend that produces a broadly polygonal pattern (Dzurisin 1978). Since the troughs cut the ridges, the troughs are younger. They are best interpreted as graben that formed in an extensional tectonic environment.

Dzurisin (1976) suggested that the compressional stresses which formed the ridges on the interior Caloris plains were due to subsidence of the basin interior associated with the extrusion of the smooth plains units surrounding Caloris. After the emplacement of these plains, the depressed basin's interior rises isostatically, changing its stress state to extension and forming the network of grabens.

This scenario predicts that the interior of Caloris cannot contain a mascon similar to those within lunar basins, because isostatic rebound cannot overshoot from undercompensation to overcompensation. However, Caloris is located improbably close to Mercury's hot pole (the intermediate axis of the planet's dynamic figure) suggesting that it is associated with a large positive mass anomaly (Goldreich and Peale 1968). Since the tectonic interpretation rules out a mascon within the basin, Melosh and Dzurisin (1978b) suggested that the smooth plains outside the basin are only partially isostatically compensated and that they form an annular positive mass anomaly surrounding the basin. This interpretation has been strengthened by the recent radar results (Chapter by Harmon and Campbell; Chapter by Clark et al.) which show that

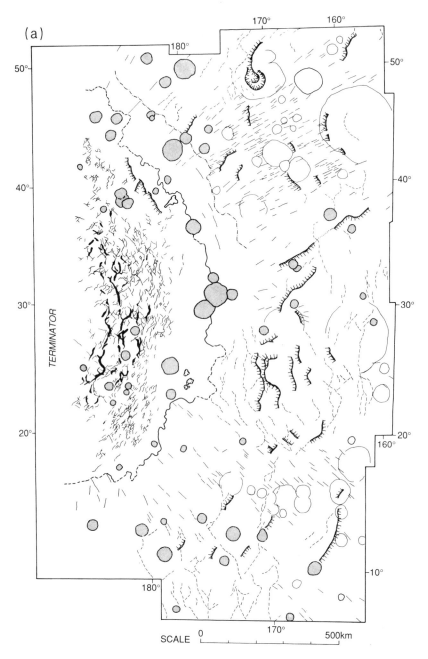

Fig. 7. Troughs and scarps in and around Caloris basin. (a) illustrates troughs and fractures, (b) is a map of ridges. The irregular continuous line follows the rim crest of Caloris' main ring scarp. A dash-dot line to the northeast of the main rim is the faint secondary ring. (Figure from Strom et al. 1975.)

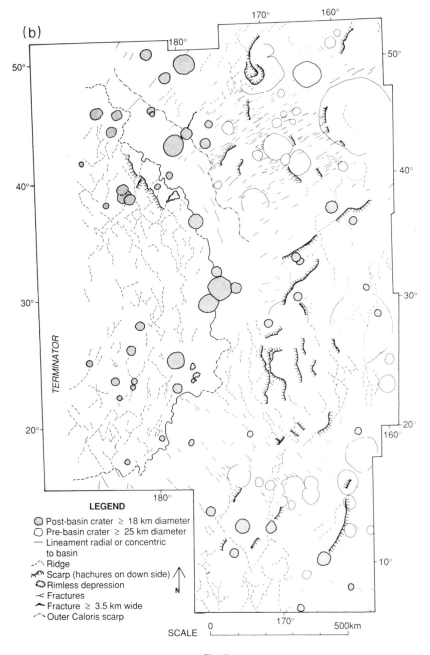

(b)

TERMINATOR

LEGEND

- ⬭ Post-basin crater ≥ 18 km diameter
- ○ Pre-basin crater ≥ 25 km diameter
- — Lineament radial or concentric to basin
- ⌁ Ridge
- ⌁ Scarp (hachures on down side)
- ⬡ Rimless depression
- ⌁ Fractures
- ⬛ Fracture ≥ 3.5 km wide
- ⌁ Outer Caloris scarp

N

SCALE ├────────┤ 500km
0

Fig. 7-b.

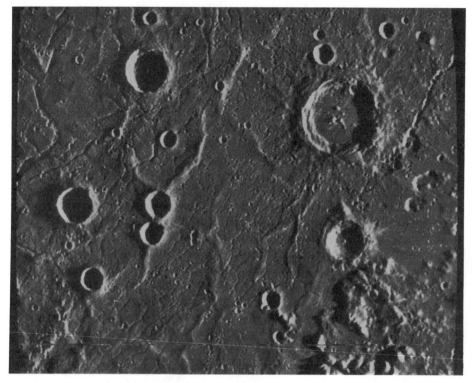

Fig. 8. Part of the Caloris basin floor showing ridges and troughs on the interior plains. The largest crater is about 60 km in diameter (Image FDS 126).

the smooth plains extend into the hemisphere unimaged by Mariner 10 and that the surfaces of the smooth plains do not follow the geoid, but instead sag downward in their central areas, some by as much as 2.5 km with respect to the surrounding terrain. The crust of Mercury has evidently been depressed by the weight of the lavas that form these plains. Harmon et al. (1985), however, also noted the presence of topographic highs at the "hot" poles. These may be important in determining Mercury's dynamic figure as well.

The presence of a large annular load associated with the smooth plains outside Caloris has its own tectonic consequences. McKinnon (1980) showed that such a load can induce concentric graben formation within the basin, independent of any hypothetical isostatic uplift. The pattern and extent of the observed grabens is consistent with a lithosphere between 75 and 125 km thick (McKinnon 1986), in agreement with the lithosphere thickness deduced from the scarcity of multiring basins.

The annular, or ring, load was modeled using the thick-plate flexure theory of Melosh (1978), in which shell curvature is parameterized. The vertical load, σ^l_{zz}, was assumed to be axially symmetric about the center of the basin with a radial profile given by

$$\sigma^l_{zz} = L \left(\frac{r}{2a}\right)^4 \exp\left[-\frac{1}{2}\left(\frac{r}{a}\right)^2 + 2\right] \tag{11}$$

where r is the radial distance, a is basin radius, and L is the load maximum at $r = 2a$. Stresses, strains and displacements were calculated as a function of lithospheric thickness H. Some results are shown in Fig. 9 for Caloris. Surface tectonic style using the faulting criteria of Anderson (1951) is shown in the upper frame, and shearing strength intensity, the square root of the second invariant of the stress deviator, at the surface is shown in the lower frame. This latter quantity is, at yield, roughly equal to half the yield stress for a variety of failure criteria, and is a good measure of the stress available to cause

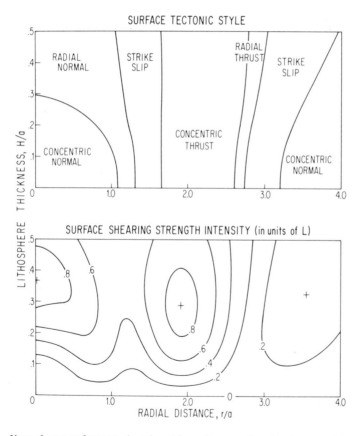

Fig. 9. *Upper frame:* surface tectonic style and *lower frame:* surface shearing strength intensity for Caloris due to a ring load, as a function of radial distance and lithosphere thickness normalized to the basin radius.

faulting. The main concentrations of surface stress are under the ring load and in the basin. The predicted thrust faulting beneath the load and normal faulting within the basin are both observed. More importantly, the dominant concentric trend of the basin normal faults is consistent with the ring load hypothesis, provided the lithosphere of Mercury was ≲ 125 km thick at the time of faulting. Simple updoming within the basin would produce normal faults of predominantly radial orientation.

The lower limit to the lithosphere thickness of ~75 km is set by requiring sufficient deviatoric stress to initiate faulting for reasonable load magnitudes. This can be estimated by a variety of means. Strom et al. (1975) gave a lower thickness for the circum-Caloris plains of 10 km, although this probably includes basin ejecta. Estimates by de Hon (1979) are much thinner. The amounts of sagging deduced from the radar measurements of Harmon et al. (1985) are also good minimum estimates, because even such a large-scale structure as Caloris is not completely compensated. Several kilometers of nearly compensated basalt are also in accord with the Caloris position, nearly on the meridian of the long axis of Mercury's dynamical figure but at 30° north (Melosh and Dzurisin 1978b; Willeman 1984).

In an extreme version of McKinnon's model, the ridges inside Caloris formed as normal mare ridges due to subsidence of the interior plains under their own weight. The interior plains nearly reached isostatic equilibrium, creating a near-zero, but positive mass anomaly within the basin (in contrast to Melosh and Dzurisin's model which predicts a small negative mass anomaly at this time). Subsequent emplacement of the exterior smooth plains and formation of the annular load then flexed the basin center upward and initiated the predominantly concentric grabens.

D. Mare Ridges on the Smooth Plains

The origin of the smooth plains surrounding Caloris has been a hotly debated topic. Murray et al. (1974) and Strom et al. (1975) proposed a volcanic origin based on the plain's similarity to the lunar maria. On the other hand, Wilhelms (1976) and Oberbeck et al. (1977) proposed that the plains are basin ejecta similar to the lunar Cayley formation. Recent study of post-Caloris crater densities on the plains units and ejecta facies (Strom 1984; Kiefer and Murray 1987; Chapter by Spudis and Guest), however, has shown that the smooth plains significantly postdate the ejecta blanket and were emplaced over an extended period. Thus, an extrusive volcanic origin for the plains appears to be the most reasonable alternative.

The smooth plains are traversed by numerous sinuous ridges (Fig. 10) that are similar in both form and size to mare ridges observed on both lunar and Martian lava plains. These ridges form in a compressional environment due to the sagging of the crust under the superimposed lavas. Their occurrence only enhances the basic similarity of Mercurian, lunar and Martian tectonic processes.

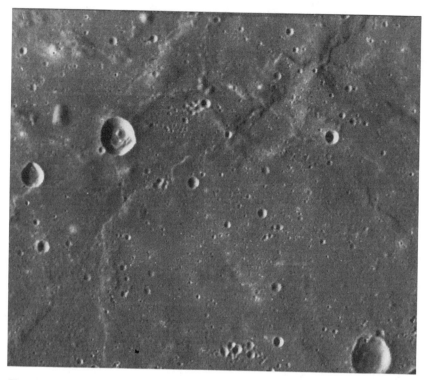

Fig. 10. Mare ridges on the smooth plains outside Caloris. The picture is 111 × 148 km (Image FDS 70).

E. Hilly and Lineated Terrain

An unusual terrain unit (originally called "weird terrain" by the Mariner 10 imaging team) consists of hills and depressions which disrupt pre-existing landforms (Fig. 11). This terrain occurs in an area which is directly antipodal to Caloris. The hills are 5 to 10 km wide and 100 to 1800 m high (Strom 1984). The hills and troughs are preferentially oriented along Mercurian grid directions.

Schultz and Gault (1975) proposed that this terrain is due to strong seismic shaking produced by focusing of seismic waves from the Caloris impact. They found similar terrain antipodal to Imbrium and Orientale on the Moon. Numerical simulations by Hughes et al. (1977) showed that the Caloris impact could have caused vertical ground movement at the antipode on the scale of about one kilometer and accelerations approaching one lunar gravity. (However, we note that these calculations may not apply in detail to Caloris, because the transient crater formed by Hughes et al.'s [1977] initial energy is as big as Caloris itself and thus would collapse to a much larger structure.) Such displacements could have induced profound modification of the pre-existing ter-

Fig. 11. Hilly and lineated terrain near the antipode of Caloris basin. The picture is 543 km across (Image FDS 27370).

rain. The ancient tectonic fabric of the Mercurian grid presented numerous planes of weakness that were reactivated by the shaking, which accounts for the predominance of grid directions in the modified terrain.

IV. SUMMARY

Mercury's tectonic activity was confined to its early history as a planet. Its endogenic tectonic activity was principally due to a small change in the shape of its lithosphere (by tidal despinning) and a small change in area (by shrinkage due to cooling). These events produced the ancient tectonic grid and the lobate scarps, respectively. This low degree of activity was ultimately due to Mercury's small size and, perhaps, deficiency of radioactive elements with respect to the other terrestrial planets. Both factors conspired to give Mercury a low surface heat flow and, therefore, a thick lithosphere that was difficult to disrupt.

Exogenic processes, specifically impact, have produced more abundant tectonic features, but even here Mercury's thick lithosphere has suppressed the formation of the spectacular multiring basins so familiar on the Moon and

Callisto. Many features associated with the Caloris basin are due to loading of Mercury's thick lithosphere by extrusive lavas or subsidence due to magma withdrawal. Seismic shaking by the impact that created Caloris produced a terrain type that is best developed on Mercury, but which is not unique even here.

Figure 12 summarizes the probable tectonic history of Mercury and shows the relative sequence of events. The earliest known tectonic activity is the pervasive fracturing that is recorded in the Mercurian global tectonic grid and which was probably produced by tidal despinning. Cooling subsequently brought the lithosphere into compression and produced the lobate scarps, incidentally shutting off extrusion of the lavas that formed the intercrater plains, according to Solomon (1977). A few enigmatic features such as Mirni ridge also formed at about this time.

The long, drawn-out interval of lobate scarp formation was sharply punctuated by a major impact that created Caloris and its associated tectonic features. For a short time, Mercury enjoyed a localized episode of vigorous tectonic and volcanic activity that ended with the emplacement and subsidence of the smooth plains surrounding Caloris. A few lobate scarps appear to have formed after the smooth plains, and then the tectonic evolution of the planet, as far as we know it, came to an end.

Mercury is a planet which is tectonically stillborn. The few tectonic features which are present, however, yield insight into the first processes that

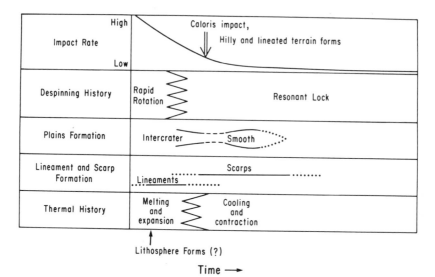

Fig. 12. Schematic illustration of the major tectonic and related events in Mercury's history. The time scale is intentionally not specified, because this is a relative time plot. (Figure after Strom 1984.)

affected its lithosphere. This insight may be applied to the earliest tectonic events on planets like Mars and perhaps the Earth, where subsequent tectonic events have either obscured or erased the most ancient tectonic record. This ability to peer back to Mercury's very early history is especially valuable considering Mercury's unique place in the solar system. The innermost planet possesses a very high iron/silicate ratio compared to solar abundances, one that may require a catastrophic explanation. Mercurian tectonics provide a testable link to the poorly understood accretionary epoch.

Further advances in our knowledge of Mercurian tectonics depend heavily on the acquisition of additional data on Mercury's surface. Although radar observations from the Earth are now yielding important insights, it would be better to obtain spacecraft images covering both the hitherto unimaged hemisphere and the areas imaged by Mariner 10 at higher resolution than is currently available.

In the meantime, the Mariner 10 data set is not exhausted and further interpretation, coupled with modeling, can be expected to produce a slow harvest of results. Questions such as the true nature of the Mercurian grid, the reason for the paucity of Orientale-type multiringed basins on Mercury, and the relationship between the Caloris impact and planetary tectonics may receive fresh insights from the study of analogous problems on other planets.

Acknowledgments. These studies of Mercurian tectonics have been supported over the years by NASA's Planetary Geology and Geophysics Program. This chapter was substantially improved by the attention of reviewers J. A. Burns and M. Parmentier.

TECTONIC HISTORY OF MERCURY

PIERRE G. THOMAS, PHILIPPE MASSON, AND LUCE FLEITOUT
Université Paris XI

Mercury exhibits four important tectonic features: (1) a global grid system, (2) lobate scarps, (3) structures associated with the Caloris basin and (4) local extensional features. The global lineament or grid pattern is characterized by four major directional trends in the equatorial areas: N 20° E, N 50° E, N 45° W and N 20° W. This pattern was already present when the Caloris impact occurred, and could partially be a result of tidal despinning. Lobate or arcuate scarps, which are randomly distributed on the surface, were created before and after the Caloris event. They are thought to be compressive features resulting from cooling and contraction of the core. The third tectonic element is characterized by features occurring outside and inside the Caloris basin, the largest known impact basin on the planet. The fourth consists of features due to local extensional or downlift motions. Lobate scarps are not randomly oriented, but are preferentially oriented radial to the Caloris basin. The tectonic features in the interior of the Caloris basin (peripheral ridges crosscut by central fractures) are quite different from their lunar counterparts. Therefore, we propose and formulate an interaction between lithospheric compression due to core cooling and the Caloris basin. This interaction temporarily oriented the compression over the entire planet, and induced a shortening of the Caloris radius that has ridged and fractured the basin inside. We describe two areas which show evidence of more local tectonic movements. The first, near the Tolstoj basin, is an elongated bulge with outer arc extensional features. The second, near the Phidias crater, exhibits an area of subsidence appearing volcano-tectonic in origin. These two examples show that tectonic movements may occur independent of global compression or large basin formation.

The tectonic evolution of small planetary bodies such as Mercury or the Moon is usually interpreted as the result of the planets' thermal evolution, sometimes in conjunction with external perturbations such as large impacts or tidal effects. Thermal contraction or expansion theoretically could produce global crustal compressional or extensional stresses; these stresses would generate randomly oriented compressional or extensional features. Tidal despinning, in which an ellipsoidal planetary body could become spherical, could produce nonrandomly oriented tectonic features. This chapter summarizes the tectonic features observed on Mercury, and relates these observations to the accepted models of planetary evolution. New models are also proposed to explain the observed tectonic features.

I. STRATIGRAPHY OF MERCURY: A SUMMARY

The stratigraphic history of Mercury was first described by Murray et al. (1974) and by Trask and Guest (1975). More recent studies concern only details and do not change the general history. The major part of the known surface of Mercury consists of old terrains, heavily cratered during the end of post-accretional heavy bombardment, which appear similar to the cratered terrains of the Moon or Mars. This widespread terrain unit of cratered plains is named the intercrater plains (designated "Pi" on USGS geological maps). After the end of post-accretional bombardment, a major impact created the Caloris basin (1300 km in diameter), and modified the landscape surrounding the basin. The seismic energy from the Caloris impact was focused at its antipode creating the hilly and lineated terrains (Schultz and Gault 1975). This major impact was followed by volcanic activity inside and around the margins of Caloris basin that created the smooth plains. A few additional impacts modified Mercury's surface after the end of this volcanic activity.

II. THE GLOBAL LINEAMENT PATTERN

Since the work of Fielder (1963b) and Strom (1964), the word "lineament" is commonly used for all rectilinear features of unclear origin on planetary surfaces including valleys, ridges, scarps, rilles, craters chains, linear portions of central peaks or crater rims, albedo contrasts and other linear alignments of unknown origin. On the Moon, the lineaments are not randomly oriented. Fielder and Strom both have shown that by deleting radial lineament systems associated with the major lunar impact basins, three prominent systems remain, trending NW-SE, NE-SW and N-S. This is interpreted in terms of tectonic stresses, producing a planetary-wide fault system termed the lunar grid, at least partially due to tidal spindown caused by the synchronization of the Moon's rotational period and revolution around the Earth. The grid pattern was created very early in the Moon's history, as implied by the subdued morphology and degraded form of the features associated with it. Local stresses due to impacts or other causes could reactivate movements along these pre-existing directions, resulting in the actual valley, ridges, and other features observed today.

Several authors (see, e.g., Fielder 1974; Trask and Guest 1975; Masson and Thomas 1977; Dzurisin 1978; Thomas 1978; Schaber and McCauley 1980) have described local or global Mercurian lineament patterns, but an exhaustive and detailed compilation of lineaments has not been previously undertaken. In order to characterize further the tectonic patterns proposed earlier, we have examined 2500 high- and medium-resolution pictures of Mercury for all rectilinear features (Thomas and Masson 1983), and produced two types of maps. One was based on the observation of high- or medium-resolution pictures and another was the result of direct interpretation of small-scale computer USGS photomosaics of each of Mercury's quadrangles. This has resulted in lineaments maps for the 9 quadrangles of Mercury (see example in Fig. 1).

In order to determine the origin of these lineaments, we assumed that the lineaments resulting from secondary craters associated with an impact are randomly oriented in azimuth. However, tectonic lineaments are not randomly oriented in azimuth: strike-slip faults are theoretically oriented 45° with respect to the main horizontal stress; reverse faults are perpendicular; and normal faults parallel to this stress. Consequently, it is assumed that major trends in azimuth distribution of lineaments may indicate the influence of tectonic processes. A planetary-wide homogeneity of this distribution would probably indicate the existence of a grid pattern. The origin of each lineament is unclear, and the actual morphology of each lineament is often subdued and probably does not represent its initial shape. Therefore, all rectilinear features were mapped, except those which were obviously not due to tectonic effects such as secondaries. The different types or ages of the lineaments were not taken in account; only their orientations were mapped.

Fourier transforms of the mapped lineaments were generated for each quarter of a quadrangle map of Mercury, producing an azimuthal rose diagram showing the relative importance of the lineament directions. The resulting Mercurian grid is illustrated by plotting the obtained Fourier transforms on a map of Mercury (Fig. 2). On the basis of this map, the following conclusions may be drawn:

1. The lineaments are not randomly oriented for each quarter of quadrangle, but exhibit well-defined trends.
2. The lineament trends show a gap in the azimuthal direction of solar illumination, suggesting that the illumination angle greatly influences the detectability of lineaments. This effect, which emphasizes lineaments oriented perpendicularly to the sunlight direction and minimizes the parallel ones, would produce a unimodal distribution, however, and not the multiple maxima which are observed. The combination of these two results

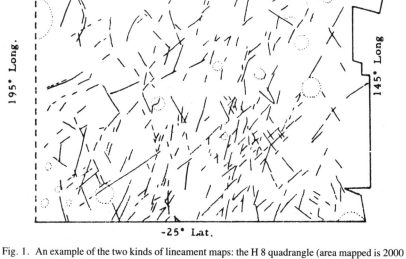

Fig. 1. An example of the two kinds of lineament maps: the H 8 quadrangle (area mapped is 2000 km across). *Upper:* Type one map (I) based on high- or medium-resolution pictures. Lower: Type two map (II) is based on the global photomosaics of the USGS.

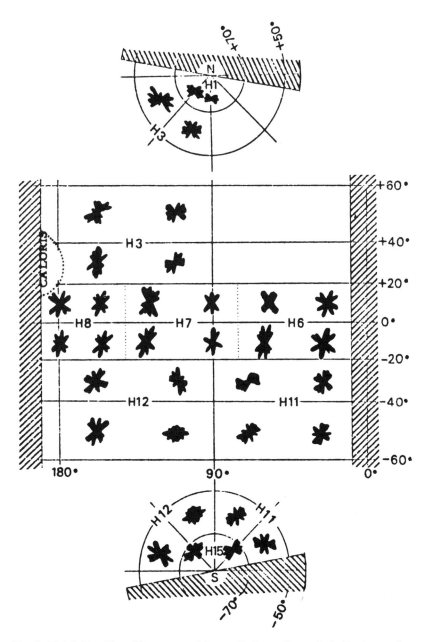

Fig. 2. Global disposition of lineaments on Mercury. Each rose diagram is the Fourier transform of all mapped lineaments in each quarter of each quadrangle map (type II). This map clearly exhibits that the different maxima are not randomly oriented, especially in the equatorial areas, where 4 directions are well represented: N 20° E, N 50° E, N 45° W and N 20° W.

indicates the existence of one or more tectonic processes which are locally responsible for the lineament directions.

3. The global map (Fig. 2) shows that the different maxima are not randomly oriented, but are more or less constant in azimuth for the entire planet, especially in the low-latitude areas (between $+60°$ and $-60°$ lat), where four directions are well represented: N 20°E (the most significant), N 50°E, N 45°W and N 20°W (the least significant). In the polar areas, the different maxima and the azimuthal distribution of these maxima are not as well defined although an E-W direction is common.

Discussion

Most of the lineaments seem to correspond to a pre-existing old grid pattern. The abundance of lineaments is higher in the intercrater plains than in the smooth plains, indicating the great age of the majority of the lineaments. On all of the terrestrial planets, large impact craters can cause tectonic stresses beyond their rims, which generally reactivate the local pre-existing tectonic trends (Strom 1964; Eppler et al. 1983). Features of this origin may exist aound the Caloris basin. Moreover, the impact which produced Caloris basin also caused tectonic movements at its antipodal point (Schultz and Gault 1975). These movements around and antipodal to the basin both show the same orientation as those in the intercrater plains. Thus, the lineament grid seems to have existed before the end of the major bombardment, and certainly before the Caloris impact.

This global pattern of Mercurian lineaments may reflect a planetary-wide change in the shape of the early Mercurian lithosphere. Two models have been proposed to explain the alteration of an anisotropic global shape of a planet: despinning and reorientation.

Theoretically, despinning induces NW-SE, NE-SW and N-S trends in the low latitudes, with also an E-W trend in the polar areas (Burns 1976; Melosh 1977b). Despinning associated with a global contraction or expansion will induce alterations in the relative intensity and in the latitudinal distribution of lineaments, but will not change their directions (Pechmann and Melosh 1979). The despinning model explains partially the N-E and N-W trending lineaments on Mercury, but fails to explain the lack of a N-S lineament trend and the presence of the important N 20°E and N 20°W lineament trends.

A possible reason why the Caloris basin lies on the axis of minimum moment of inertia is that a large positive free-air anomaly, or mascon, is associated with Caloris (Murray et al. 1974). Thus, it is possible that this basin altered the orientation of the planetary surface with respect to the rotational pole. Theoretically, such a reorientation could induce a global lineament grid pattern (Melosh 1980). The Mercurian grid, which already existed before the Caloris impact, exhibits lineament trends different from those predicted by Melosh's theory, and it is unlikely that it is due to such a reorientation. Moreover, the present orientation of the grid is symmetrical with regard to the actual equator, and could not be the result of the reorientation of a pre-existing pattern.

No single model explains the observed grid completely. The tidal despinning partially explains this grid, but cannot be the only mechanism that changed the planet's shape. This may reflect deficiencies in the analytical forms of this model, or unmodeled properties of the Mercurian lithosphere. Further theoretical studies, comparison between other planetary lineament grids, and mainly the study of the unknown side of Mercury are required to provide significant understanding of this intriguing Mercurian grid.

III. THE GLOBAL COMPRESSIVE PATTERN

The surface of Mercury exhibits morphologic features unique in the known solar system (Fig. 3). These features, named lobate scarps by Strom et al. (1975b), and arcuate scarps by Dzurisin (1978), represent the most significant geologic characteristic of Mercury. Strom et al. (1975b) described these features as follows (Fig. 3): "Lobate scarps are relatively steep and long escarpments which usually show a broadly lobate outline on a scale varying from a few to tens of kilometers. They vary in length from about 20 km to over 500 km, and shadow measurements indicate that their heights range from a few hundred meters to about 3000 m. Often, individual scarps transect several terrain types, including craters." Their morphology is generally sharp and they can be easily distinguished from the lineaments previously described.

Murray et al. (1974), Strom et al. (1975b), Cordell and Strom (1977) and Dzurisin (1978) have interpreted these scarps as compressive features, because of their thrust fault morphology. Moreover, in a few cases, lobate scarps show a horizontal offset of a large center rim (shortening perpendicular to the scarp elongation). Lobate scarps are apparently present at all latitudes and longitudes. Despite these numerous compressive structures, however, there is a lack of graben or other tensional features (except in local regions) which would compensate for the shortening shown by the lobate scarps.

The scarp's morphology is often relatively well preserved. Sometimes these scarps transect the smooth plains. These observations indicate that most of the scarps (at least their final development) postdate the end of the major bombardment episode, and, consequently, the formation of the grid.

Scarp statistics show that there is no significant dominant trend in azimuth; i.e., scarps are seen trending in all azimuths and randomly oriented (Cordell and Strom 1977). Because of their thrust fault-like morphology, randomly oriented azimuth, and the lack of extensional features on the entire known surface of Mercury, these scarps are usually interpreted as the result of planetary contraction by cooling and shrinkage of the presumed large iron core, resulting in a 1 to 2 km decrease in the radius of Mercury (Strom et al. 1975b).

We undertook directional studies using the tectonic map of the Mercurian

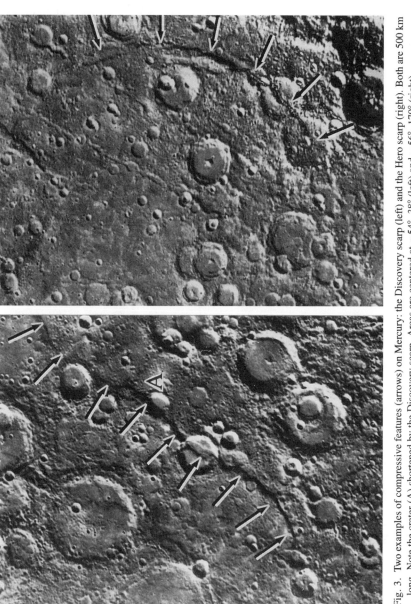

Fig. 3. Two examples of compressive features (arrows) on Mercury: the Discovery scarp (left) and the Hero scarp (right). Both are 500 km long. Note the crater (A) shortened by the Discovery scarp. Areas are centered at −54°, 38° (left) and −56°, 170° (right).

southern hemisphere drawn by O'Donnell (1979; Fig. 4a), a data base which did not exist at the time of the previous studies. This map was prepared by stereoscopic analysis of all high- or medium-resolution stereopairs available after the Mariner 10 mission. Stereopairs exist only for the areas covered by the first and second Mercury encounters, which consist of areas south of the equator and between +10° and +190° lat. Despite the limited coverage of this map (a quarter of the planet only), we used it for our directional studies for the following reasons: first, the sample of lobate scarps mapped stereoscopically is more complete; only the largest and the youngest scarps are observed in nonstereoscopic mode. Second, there is an important illumination effect: scarps parallel to the direction of sunlight are only visible as faint arcuate lineaments, and not as arcuate scarps. The stereoscopic analysis that determines differences in the heights of scarps avoids these inconveniences theoretically. We then transferred all the compressive features mapped by O'Donnell (1979) onto the orthographic maps of the three stereographically covered quadrangles H 11, H 12 and H 15 (USGS 1976, 1977, 1984), without adding or taking out any scarps.

Following the same procedure used for the lineament studies, directional analysis was attempted for each map by the use of Fourier transform processing. This analysis shows that for each quadrangle, the scarps are not randomly oriented, but exhibit significant preferred azimuths (Fig. 4b): NE-SE for the H 11 quadrangle, NW-SE for H 12, and E-W for H 15. Thus, it seems that one or more geologic processes locally influenced the orientation of the main stress (σ_1) and the formation of scarps. This result differs significantly from the previous studies (showing randomly oriented scarps), but is not contradictory: the dominant trends are different for each quadrangle, and, taken together, all azimuths are present. We find, however, that the azimuthal random distribution of the scarps observed for the whole mapped area is caused by the reference system used in the mapping process (the N-S system), and not due to their orientation. By plotting the Fourier transform diagrams onto a sphere or onto a stereographic projection of the southern hemisphere (Fig. 4b), one can observe that the dominant directions are more or less parallel for the H 12 and H 15 quadrangles, and that they show only a minor angular difference for the H 11 quadrangle. Therefore, assuming that the majority of the scarps mapped by O'Donnell are compressive features, we cannot explain such an arrangement only through global contraction by core cooling; an anisotropic stress field homogeneous in direction for a quarter of Mercury must have existed. The direction of the main horizontal stress, perpendicular to the scarps' trends, is different from the direction of the main stress induced by the despinning model, and cannot be due to such a process.

It should be noted that the compressive pattern is not randomly oriented with respect to the Caloris basin: the elongation of the scarps is mainly radial to the Caloris basin, which implies that the main horizontal stress σ_1 is tangential to the basin (Fig. 4c).

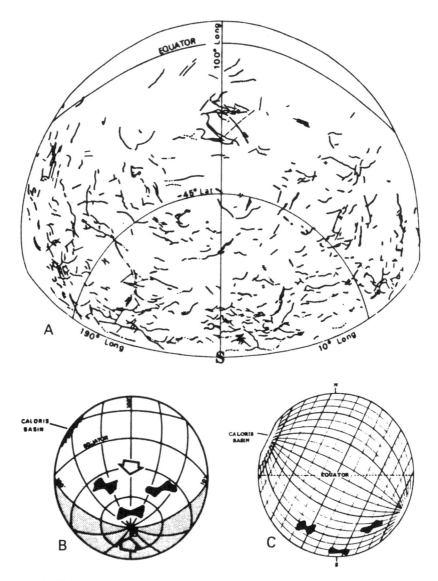

Fig. 4. (A) Distribution of compressive features on the south polar projection of Mercury, deduced from the stereoscopic analysis of the area by O'Donnell (1979). (B) Global distribution
of the Fourier transforms of the compressive features of Mercury. In this frame of reference,
they are more or less parallel, and seem to be due to more or less parallel compressive stresses
(arrows). (C) Global distribution of the Fourier transforms of the compressive features on a grid
using Caloris to define the frame of reference. Note that the scarps extend mainly radially from
Caloris, suggesting that the dominant compressive stress was tangential to the basin.

The observed scarp directions suggest the existence of two possible compressive mechanisms:

1. The orientation of the stress in the southern part of the known Mercurian surface would be due only to local causes; the geometric relationships between the orientation of the scarps and Caloris would be fortuitous. A stereoscopic analysis of the entire surface of Mercury in the future will confirm or disprove this hypothesis.

2. The oriented stress field could be due partially to the Caloris basin, despite the long distance between the scarps and the basin. This hypothesis of a "Caloris effect" is reinforced by the presence of a graben field farther away from Imbrium (the largest proven impact basin on the Moon) containing graben which are mainly radial and tangential to Imbrium (Mason et al. 1976). This "Imbrium effect" shows the possible existence of a far-field tectonic influence around a large impact basin. One should note that, even if it exists, we do not assume that the "Caloris effect" is the only cause of all of the compressive stresses which have existed before and after the Caloris impact. It only explains the statistical importance of some directions that have been temporarily emphasized.

So, awaiting a new Mercury mission in order to confirm or negate this hypothesis with complete stereoscopic coverage, one can assume, and try to formulate, this large-basin effect.

Discussion and Theoretical Formulation

The model developed here is schematically described in Fig. 5. Assuming that the rim of the crater acts as a free boundary and that Mercury was submitted to a uniform compression before the Caloris impact due to the cooling of the planet, the relaxation of the stress on the crater's rim induces an elongation and a displacement of the lithosphere towards Caloris. The least compressive stress is in the direction radial to Caloris and the greatest compressive stress is in the concentric direction. This explains why the scarps form preferentially parallel to the radial direction.

The amplitude of the radial and tangential membrane stresses as a function of the distance to Caloris is discussed in the following. In the calculations presented here, we take into account the effect of the load of ejecta on the lithosphere after the impact. As shown previously (Fleitout and Thomas 1982), this loading has small tectonic consequences but it can be important for the geoid, the moment of inertia and the orbital dynamics of the planet. We consider a reference grid in which the crater is at the north pole. The co-latitude used to indicate the distance to the crater is such that $\delta = 0$ on the point antipodal to the crater and $\delta = \pi$ at the center of the crater. The lithosphere of Mercury is considered to be an elastic shell resting on a viscous fluid. We calculate how a circular free boundary corresponding to the crater's rim can modify an initially uniform stress field. We neglect flexural stresses.

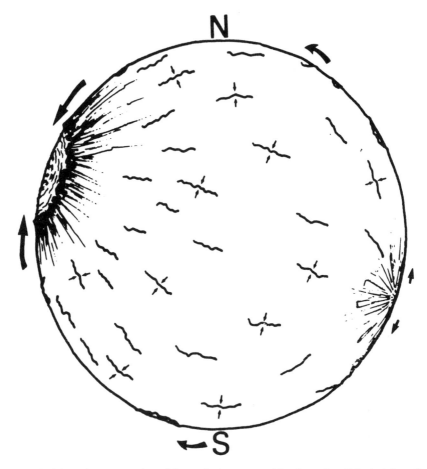

Fig. 5. Schematic representation of the mechanism proposed for the preferential orientation of the lobate scarps on Mercury. The exterior arrows show the direction of the displacement of the lithosphere. The interior small arrows indicate the main compressive stress.

The local equilibrium equations in the vertical and the radial direction may be written, respectively (Veining Meinesz 1947),

$$s_r \rho g + \frac{T}{R}(\sigma_\alpha + \sigma_\delta) = -h\rho g \tag{1}$$

and

$$\frac{d\sigma_\delta}{d\delta} = (\sigma_\alpha - \sigma_\delta)\cot \delta. \tag{2}$$

In Eqs. (1) and (2), s_r is the vertical displacement of the lithosphere, T the lithospheric thickness, R the radius of Mercury, σ_α and σ_δ the stresses in the concentric and radial direction with $h\rho g$ representing the weight of the overburden (here, h is the ejecta thickness, ρ the ejecta density and g the Mercurian gravity force).

We also use the two stress-strain relations in an elastic spherical shell,

$$\sigma_\delta - v\sigma_\alpha + vh\rho g = \frac{E}{R}\left(\frac{ds_\delta}{d\delta} + s_r\right) \tag{3}$$

and

$$\sigma_\alpha - v\sigma_\delta + vh\rho g = \frac{E}{R}(s_\delta\cot\delta + s_r). \tag{4}$$

where s_δ is the displacement in the direction radial to Caloris, and E and v are Young's modulus and the Poisson ratio (1000 kbar and 0.25, respectively). By combining these equations (Fleitout and Thomas 1982), one obtains a system of three equations:

$$(1 + v)(\sigma_\alpha - \sigma_\delta) = \frac{E}{R}\left(s_\delta\cot\delta - \frac{ds_\delta}{d\delta}\right) \tag{5}$$

$$\sigma_\delta[(1 + u)^2 - (v - u)^2] =$$
$$\frac{E}{R}\left[(1 + u)\frac{ds_\delta}{d\delta} + (v - u)s_\delta\cot\delta - h(1 + w)(1 + v)\right] \tag{6}$$

$$(1 + u)\frac{d^2s_\delta}{d\delta^2} - (1 + u)\frac{ds_\delta}{d\delta}\cot\delta -$$
$$[(1 + u)\cot^2\delta + (v - u)]\,s_\delta - (1 + w)(1 + v)\frac{dh}{d\delta} = 0 \tag{7}$$

where $u = TE/R^2\rho g$ and $w = v\rho gR/E$.

The displacement in the radial direction s_δ as a function of δ can be easily found if one solves Eq. (7) using a standard two-point boundary value-solving program. When $\delta = 0$, s_δ is set equal to zero. On the crater rim, the stress σ_α induced by the elongation must compensate the prestress. Using Eq. (6), this yields the second boundary condition.

Once s_δ is obtained, the values of σ_α and σ_δ follow from Eqs. (6) and (5). The height of ejecta has been taken equal to

$$h = 0.14R_0^{0.74}\,\frac{\theta_0^2\sin\theta_0}{\theta^2\sin\theta} \tag{8}$$

where $\theta = \pi - \delta$, R is the radius of Caloris basin and $\theta = 16°$, its angular radius. In order to avoid divergence between $0°$ and $10°$, h is taken equal to its value for $\theta = 10°$ in this interval.

Figure 6 presents the values of the radial displacement for various values of the prestress and of the elastic lithosphere thickness. Except for the case without prestress, the lithosphere is pulled toward Caloris. The total displacement can be of the order of 1 km. The stresses corresponding to the case where the prestress is 2 kbar and the lithospheric thickness is 50 km are drawn in Fig. 7. A large difference between σ_α and σ_δ persists at large distances from Caloris. The stress σ_δ becomes less compressive near Caloris but the compression in the tangential direction is increased: the difference between σ_α and σ_δ is 200 bar $45°$ away from the crater's rim. This difference between σ_α and σ_δ is, of

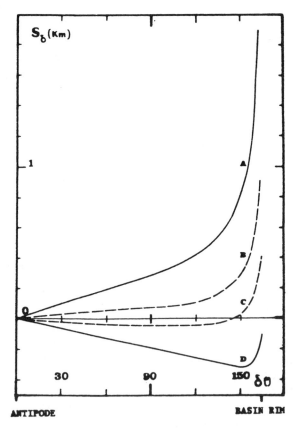

Fig. 6. Azimuthal displacement s_δ towards Caloris as a function of the angular distance to Caloris antipode δ. Curve A corresponds to the case of a 50-km thick lithosphere with a 2 kbar prestress; curve B corresponds to a 100 km thick lithosphere with a 1 kbar prestress; curve C to a 250 km lithosphere with a 0.5 kbar prestress; and curve D to a 50-km thick lithosphere with no prestress. The basin rim is at $\delta = 164°$.

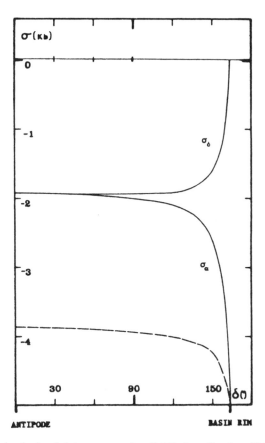

Fig. 7. Tangential and azimuthal stresses σ_α and σ_δ (full line) as a function of δ for a 50-km thick lithosphere with a 2 kbar prestress. The sum $\sigma_\alpha + \sigma_\delta$ (dashed line) is proportional to the height of uncompensated material. The proportionality constant is approximately 200 m of topography for 1 kbar.

course, not responsible for the compression of Mercury due to a global cooling of the planet. However, we think that the preferential orientation of the compressive scarps is due to this differential stress.

The sum $\sigma_\alpha + \sigma_\delta$ has been plotted because it represents a quantity proportional to the noncompensated topography (Turcotte et al. 1981). One can see in Fig. 7 that an excess of noncompensated masses exists around Caloris. This is due to the strong support of loads by membrane stresses. This excess of mass may very well be responsible for the anomaly in inertia moment that has induced the capture of Mercury's rotation with Caloris oriented towards the Sun (Goldreich and Peale 1966; Melosh and Dzurisin 1978b).

The difference in the elastic stresses σ_α and σ_δ calculated here will remain captured as residual stresses. Its tectonic effect can appear a long time after the formation of Caloris. On the other hand, movements similar to those described here could have occurred just after the impact under the influence of the large extensional stress due to the depression. However, the opposite movements and deformations should have occurred as the basin filled up.

The displacement of the lithosphere toward Caloris might have taken place over a relatively long period of time: as the viscoelastic lower lithosphere relaxed, the stresses concentrated in the upper layers and s_8 increased slowly. As shown below, the tectonic consequences of this slow relaxation can be seen in the tectonic features inside Caloris.

IV. THE CALORIS BASIN

The Caloris basin on Mercury is the largest structural feature observed by Mariner 10 (Fig. 8a). This basin is similar in appearance and size to the largest lunar basins and was undoubtedly caused by an impact (Strom et al. 1975b). Its external system of valleys and ridges, and its highly ridged inner plain will be described here first with regard to the tectonic history. We will then discuss the possible origins of these tectonic features.

External Structures

The perimeter of the Caloris basin is defined by a ring of irregular mountains about 1300 km in diameter called the Caloris Montes formation (McCauley et al. 1981). Outside this ring lies an extensive system of ridges and valleys, particularly well developed to the NE and SE where it starts at the edge of the mountains and extends outward to a distance of about one basin diameter. The results of previous studies (Trask and Guest 1975; Strom et al. 1975b; McCauley 1977; McCauley et al. 1981) suggest that these features originated in the same manner as the lunar Imbrium sculpture (Gilbert 1893), i.e., faulting, or, more probably, ejecta and sculptures associated with secondary impacts from Caloris basin.

There are three stratigraphic units in this area:

1. The old pre-Caloris terrains (the intercrater plains) that lie at a distance of about 1.5 radii from the basin edge;
2. Secondary impacts and Caloris ejecta, named "lineated facies of the Van Eyck formation" and "Odin formation" (McCauley et al. 1981) that cover intercrater plains. This formation consists of numerous, long, more or less radial hills and intervening grooves.
3. Smooth plains material, which may be post-Caloris volcanic flows (Trask and Guest 1975) that excessively embay the intercrater plains and Caloris-related terrains.

The "Van Eyck formation" (Fig. 9) consists of high blocks separated by depressions embayed by smooth plains material. The tops of these high blocks

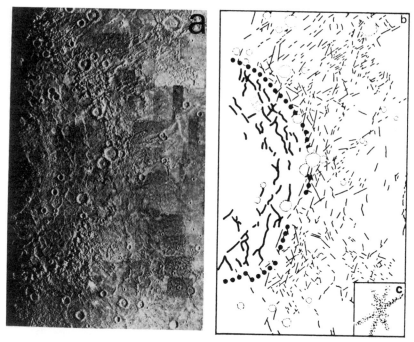

Fig. 8. (a) Photomosaic of the 1300 km diameter Caloris basin, showing the different geologic units: the ridged and fractured inner plains, the mountain ring and the ejecta blanket. (b) Tectonic map of the Caloris area with the same scale and orientation as for (a). / : rectilinear lineaments (Thomas 1978); ⌐ : compressive ridges inside the basin (Maxwell and Gifford 1980); ... : approximate limit of the basin. Insert (c) is a schematic representation of the Fourier transform of the tectonic map (b) analyzing only the outer lineaments. Note that these four directions are also the directions of the Mercurian grid (see Fig. 2).

exhibit the classic morphology of ejecta. However, these blocks are often sharply bounded by rectilinear scarps, which separate the ejecta from the plains. These limits do not show the curvilinear-to-swirly morphology typical of the ejecta blanket of younger basins, such as the Hevelius formation around Orientale basin on the Moon, but look like faults. Many, but not all, of these scarps are radially oriented toward Caloris basin. The smooth plains generally lie in the depressions between the blocks of ejecta. However, some high blocks are also covered by smooth plains. This disposition extends outward to a distance of more than one basin radius from the Caloris Montes formation.

The rectilinear scarps and other lineaments of the Caloris surroundings were mapped (Thomas 1978, 1980) (Fig. 8b), and directional analysis was then attempted by using coherent optics for generating Fourier transforms of the mapped lineaments. This analysis shows that these scarps and lineaments are

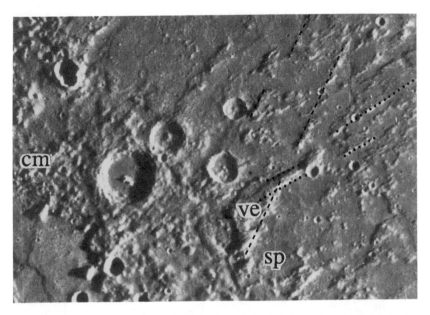

Fig. 9. Northeastern part of the Caloris basin, showing the Caloris Montes formation (cm), the Van Eyck formation (ve), the smooth plains (sp), and the radial and nonradial lineaments. Scene is 400 km across.

not randomly oriented, but have significant preferred azimuthal trends: N 20° E, N 50° E, N 45° W and N 20° W (Fig. 8c). These preferential directions are the most important orientations of the Mercurian grid defined in Sec. II.

On the basis of these morphological and azimuthal observations, we propose a tectonic origin for these scarps and lineaments. During, or shortly after the Caloris impact and deposition of ejecta, but before the smooth plains flooding, tectonic stresses created uplifted blocks of ejecta and downwarped depressions, now embayed by smooth plains material. The few high blocks now covered by smooth plains indicate that the last tectonic episodes occurred immediately after the flooding. Because the smooth plains are slightly younger but close to the age of Caloris (Trask and Guest 1975), one can assume that this tectonic episode took place during or somewhat after the impact, and that it was probably related to it. Because this scarp and lineament pattern shows the same dominant orientation as the Mercurian grid, we can assume that these motions were due to the rejuvenation of the old grid pattern by the Caloris impact. This indicates that the Mercurian grid is older than the Caloris impact.

On the Moon, the large basins are not surrounded by such tectonic dislocations. The graben and ridge systems frequently associated with the lunar basins are contemporaneous to the basin filling by the mare materials, i.e., considerably younger than the impact itself, and occurred only in the immediate vicinity of the basins.

One of the main differences between the lunar basins and Caloris exterior is the absence of well-expressed outer scarps outside Caloris. Large impact basins generally show multiring structures, both inside and outside the limits of the transient crater excavation. The inner ones are generally interpreted as rebounded central peaks which grew up into rings (Stuart Alexander and Howard 1970; Hartmann and Wood 1971; Murray 1980). Such structures, if they exist inside the basin, are now completely covered by the interior smooth plains. The outer ones are now classically interpreted as the result of asthenospheric flows toward the basin center in the case of a basin deeper than the lithospheric thickness (Melosh and McKinnon 1978; McKinnon and Melosh 1980; Melosh 1982b). Such well-expressed outer scarps exist around lunar, Martian and Galilean satellite basins (McCauley 1977; Wilhelms 1973; Thomas and Masson 1984a; Melosh 1982b). As noted by Strom et al. (1975b), however, such a well-expressed outer scarp does not exist around Caloris. There is only a weak, faint and subdued inward facing scarp, at a distance of 100 to 160 km from the main scarp. This outer ring exists only in the NE part of the basin, and it is about 500 km long (a complete outer scarp at such a distance would be 5000 km long).

One could envision that outer rings are completely buried by the smooth plains material. Nevertheless the annulus of smooth plains surrounding Caloris is not complete, and such rings do not exist in the areas without smooth plains flooding. Thus, this lack of a complete outer ring seems to have geologic significance and may be explained by a very thick lithosphere, or by some other process.

Interior Structure

Like many lunar mare-filled basins, the Caloris interior is intensely ridged with broad arch-like features (Fig. 10) oriented in concentric and radial directions (Fig. 8b). This tectonic pattern is predominantly located in the outer part of the basin interior. Several interpretations of the ridge system have been proposed for Caloris, such as motions inside the basin during the later stages of the impact itself (McCauley 1977) or a loss of support caused by the withdrawal of magma (Dzurisin 1978). However, both the lunar and Mercurian basin ridges have been generally interpreted as being the result of concentric compressive stresses produced by regional subsidence due to the weight of the volcanic material filling the basin (Strom et al. 1975b; Solomon and Head 1976; Lucchitta 1976).

In spite of these similarities, there are four substantial differences between processes forming the Caloris and the lunar basin ridges.

(1) The lunar ridges affect infilling mare materials which are considerably younger than the impact itself. The Caloris ridges, on the other hand, affect plains materials which are slightly younger but close to the same age as the impact: the ejecta blanket and the interior plains show about the same crater density, and there are no "Archimedes type" craters, i.e. craters which oc-

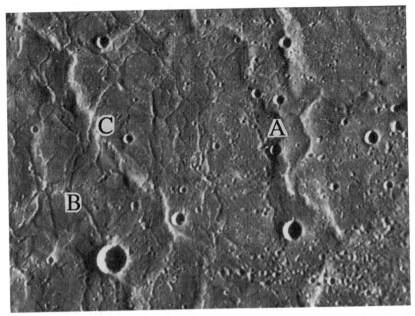

Fig. 10. Part of the floor of the Caloris basin showing the ridges (A), the fractures (B) and their intersection (C). Scene is 250 km across.

curred after the basin formation but before its flooding (Trask and Guest 1975). The lack of young craters modified by the ridges indicates that this compressive tectonic event probably occurred after plains formation. These age relations are not contradictory to subsidence by loading, but they suggest that the compressive stresses may be directly related to the impact, which cannot be the case on the Moon, because of the difference in timing.

(2) As shown by Maxwell and Gifford (1980), the ridge orientations in the Caloris basin are more strongly concentric than those in the flooded lunar basins. This strong circular symmetry suggests that the Caloris ridges are caused by basin-controlled stresses. To the contrary, the ridge orientations in the lunar basins only indicate an approximately concentrically symmetrical state of stress. This difference in stress-field geometry suggests a difference in origin.

(3) The tectonic effects associated with the basin subsidence due to the volcanic loading theoretically produce extensional stresses and graben outside the area of load application (Solomon and Head 1979, 1980). Such circumferential graben are present outside the lunar basins but not around Caloris (see, e.g., Maxwell and Gifford 1980). It is possible to explain this lack of extension features by the global crustal compression which would prevent the formation of extensional graben. This reasoning does not, however, explain the fracture pattern in the center of Caloris basin.

(4) The major difference between Caloris basin and lunar basins is the fractured central floor of Caloris, which shows a pattern that appears to be unique in the solar system (Fig. 10). The morphology of these fractures is similar to tensional fractures or graben. They only occur in the central area of the basin, and display both radial and concentric patterns. In the intermediate areas of the basin's floor, located between the fractured center and the ridged periphery, some fractures are observed on the top of the ridges and are clearly due to the same motions which created the fractured central floor (outer arc fractures). Most of the fractures transect the ridges and are not modified by them, so that this fracture pattern appears to postdate the ridges slightly (Strom et al. 1975*b*; Melosh and Dzurisin 1978*b*; Dzurisin 1978*b*). This extension has been interpreted (Strom et al. 1975*b*; Dzurisin 1978*b*) as the result of an isostatic uplift due to the excavation of an incompletely compensated basin.

Discussion

The lack of a well-defined ring outside the Caloris basin limits, and the occurrence of extensional features after the compressive ridging inside the basin are two of the most intriguing characteristics of the Mercurian surface. In fact, it is difficult to envision how a subsidence due to loading could be followed by an uplift due to incompletely compensated excavation. Melosh and Dzurisin (1978*b*) noticed that "inertial forces are negligible for isostatic rebound, so that no overshoot of the isostatic state is possible. Equilibrium in case of a filled depression would therefore be asymptotically approached, with a decreasing positive free air anomaly. Under no circumstances will uplift occur in response to a positive anomaly." Therefore, either the compression, or the extension, and perhaps both are due to nonisostatic phenomenon.

In response to that problem, Melosh and Dzurisin (1978*b*) and Dzurisin (1978) proposed another explanation. They proposed that the late extensional features are due to isostatic uplift caused by uncompensated basin excavation, while the anterior compressive features are not due to subsidence by loading, but to subsidence owing to magma withdrawal from Caloris basin to form smooth exterior plains. In that model, the Caloris basin corresponds to a negative free air anomaly, and cannot be the origin of the positive anomaly which would be present in this area, as shown by the position of Caloris on the Mercury's axis of minimum moment of inertia. Melosh and Dzurisin have then calculated that an average thickness of 400 m of uncompensated smooth plains around Caloris could explain the astronomical position of the basin (i.e., the Caloris basin is located on Mercury's axis of minimum moment of inertia), even if Caloris corresponds slightly to a negative free air anomaly.

This explanation does not take into account two points. (1) If there was a magma withdrawal from Caloris to its surrounding plains, then the top of the Caloris inner plain would be at least at the same level as that outside, and probably slightly above, but not below. Preliminary photometric measurements (Hapke et al. 1975) suggest, however, that the basin floor may be several kilometers below its surroundings. (2) This hypothesis does not explain the absence of outer basin rings.

It is possible to envision another hypothesis to explain this complex geometry, as a magma intrusion beneath the basin which could have induced a bulging in a previously isostatically downlifted basin. However, this hypothesis does not explain the lack of outer rings. As an alternative view, we have suggested a global interpretation of the Caloris tectonics that may explain all of the described observations and differences with the Moon (Thomas and Masson 1984b).

The areal distribution of the lobate scarps on Mercury (Thomas et al. 1982; Sec. III) and the graben location on the Moon (Mason et al. 1976) suggest the existence of far-field tectonics associated with a large impact basin (Fleitout and Thomas 1982; Sec. IV). At the time of the Caloris impact, we assume that the Mercurian lithosphere was in a compressive state because of the core cooling, but that the basin interior was a zero stress area with a completely broken and molten lithosphere just after the impact. Then, the boundary between this zero stress area and the exterior prestressed lithosphere induced lithospheric motions directed towards the basin center. This motion temporarily oriented the compressive stress on the entire lithosphere and induced a decrease of the Caloris radius; the expected decrease of the Caloris radius is about 1.5 km (Fleitout and Thomas 1982; Sec. IV).

This radius decrease, inferred from observations of Mercurian surface structures distant from Caloris, may explain many features in the Caloris area as listed below.

1. The compressive state of the Mercurian lithosphere near Caloris basin prevented outer scarps from forming, because they compensated the extensional stress due to asthenospheric motions induced by the Caloris cavity.
2. The motion of the Caloris surroundings induced dislocations around the basin, which occurred preferentially along the pre-existing grid pattern.
3. The radius decrease induced the concentric folding of the internal plains near the basin rim. This folding occurred just after the filling, when the basement under the basin was still hot and soft enough, covered only by a very thin superficial rigid layer.
4. After the cooling and the thickening of the internal rigid surficial layer, the radius decrease could only induce a central flexure and a bulge. The fracture pattern would be the result of the "outer arc" extension due to this buckling of the lithosphere inside Caloris.
5. The motion stopped when the internal rigid surficial layer was thick enough to support the compressive stress transmitted by the surrounding lithosphere. The tectonic behavior of Caloris was then the same as that in other areas of the planet.

The global radius decrease could be geologically contemporaneous with the impact, but it cannot be an instantaneous phenomenon such as a central peak formation, and must have continued for a long time. We do not assume that the possible radius decrease is the unique origin of the tectonics in this area. It is compatible with isostatic motions: either subsidence by loading or uplift by incompletely compensated excavation may be possible, but not both.

Such motions did not occur around the other young Mercurian basins, because they were not large enough to have induced global mechanical perturbations of the whole lithosphere just after the impact. On the Moon, the late mare basalts buried the possible tectonic features inside Imbrium, a unique, young sufficiently large basin. The absence of tectonic dislocations outside this basin and the presence of outer concentric scarps indicate that membrane stresses did not play the same role for Imbrium as they did for Caloris. This is consistent with the fact that the Moon was not in a state of contraction at the time of the Imbrium impact.

V. LOCAL TECTONIC ACTIVITIES

Mercury is not considered a tectonically active planet. Discussion of the tectonic features of Mercury generally describe only the lobate scarps, the Caloris related events, and the lineaments pattern. Nevertheless, some local areas exhibit morphological features which may be due to tectonic motions of internal origin, but which do not seem to be related to a global model. In order to characterize such areas, we describe here the region between the Tolstoj, Zeami and Phidias craters (H 8 quadrangle).

The Tolstoj-Zeami Area

The distinct nature of the NE part of the terrain surrounding Tolstoj basin (Fig. 11) was first described by Trask and Guest (1975) who referred to it as lineated terrains. These terrains consist of lines of hills, scarps and valleys that extend as far as 200 to 300 km NE from the Tolstoy basin. Trask and Guest related lineated terrains to the ejecta of an unknown basin located on the hemisphere unobserved by Mariner 10.

Schaber and McCauley (1980) interpreted this area as the radially lineated and grooved rim ejecta from Tolstoj basin. However, these authors noticed a very unusual attribute of the ejecta. Despite Tolstoj's great age and its partial embayment by the very old intercrater plains, the ejecta blanket appears to be remarkably well preserved. They noticed also that the valleys and grooves are not seen all around the basin, but only to the SW and more predominantly to the NE of the basin. From the stereophoto interpretation, they also suggested that the Tolstoj ejecta have been upwarped to an altitude higher than the surrounding plains.

Problems with interpreting lineated terrains as the ejecta of Tolstoj include the following points (see Fig. 11b).

(1) The hills and grooves are not radial to the center of Tolstoj, but are oriented in a pattern of parallel straight lines. The grooves located in the

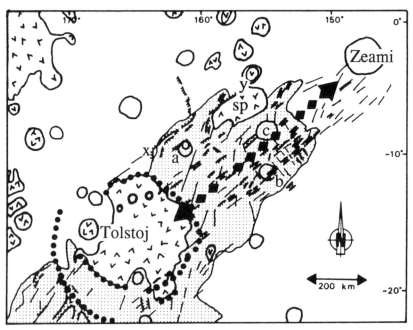

Fig. 11. (a) Photomosaic of the Tolstoj-Zeami area. (b) Simplified sketch map of the Tolstoj-Zeami area, with the same scale and orientation as in (a). .... : the rings of Tolstoj; 〰 : valleys and furrows; / : other lineaments; ♦ ♦ ♦ : axis of the bulge; ○ : young craters; ▽ : smooth plains; ▦ : terrains mapped as Tolstoj ejecta by Schaber and McCauley (1980); □ : intercrater plains.

center of the pattern are effectively radial to the basin center, but the grooves located near the limits of the lineated terrains show a tangential orientation to the basin.

(2) Detailed stereoscopic analysis shows that the uplift of this area is an elongated bulge between Tolstoj and Zeami. The long axis of the bulge is oriented N 45°E. It exhibits the same azimuthal direction as the superposed grooves and hills. The NW limit of the bulge is a subdued scarp 450 km long, and corresponds approximately to the NW limit of the hills and the grooves (X-Y on Fig. 11b).

(3) The bulge and its grooves and hills pattern are parallel to one of the main trends of the Mercurian grid (NE-SW).

(4) The bulge and its grooves and hills are parallel to the long dimension of an elongated patch of smooth plains located on the flank of the bulge (SP on Fig. 11b).

(5) The boundary between smooth plains and intercrater plains is remarkably linear at small scale throughout the whole H 8 quadrangle. The bulge is parallel to the general direction of this main boundary between the smooth plains and the intercrater plains (Fig. 12).

(6) The hills and grooves affect some craters which clearly postdate Tolstoj's ejecta. The NW rim of the crater located at −10° lat, 161° long (*a* on

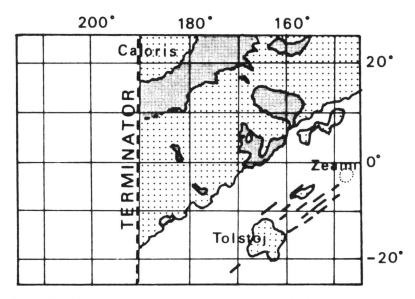

Fig. 12. Simplified sketch map of the H 8 quadrangle. Note that the Tolstoj-Zeami valleys and furrows are parallel with the smooth plains-intercrater plains limit. Shown are (heavily shaded) Caloris ejecta, smooth plains (dotted areas), and intercrater plains (unshaded). Hatch marks indicate the Tolstoj-Zeami valleys and furrows.

Fig. 11b) is obliterated by a trending ridge at N 45°. This crater is classified as a C_3 crater by Schaber and McCauley (1980) and is about the same age as the smooth plains. The bottom of the C_5 crater (younger than the smooth plains) located at −11° lat, 155° long (b on Fig. 11b) is crosscut by a prominent NE trending groove. Elsewhere (−8° lat, 155° long, c on Fig. 11b), a valley seems superposed by a C_4 crater.

(7) On the highest part of the bulge (near −10° lat, 153° long), the main system of NE trending valleys crosscuts a slightly older but clearly visible system of SE trending grooves that exhibit about the same morphology, but which cannot be Tolstoj ejecta.

Interpretation

Schaber and McCauley (1980) propose three possible explanations for this unusual area: (1) control of the ejecta pattern by prebasin structures; (2) preferential burial along structural trends of an originally symmetrical ejecta blanket by the intercrater plains material; or (3) formation of Tolstoj by an oblique impact from the NW that produced an ejecta blanket with bilateral symmetry. None of these explanations completely accounts for the morphological and chronological characteristics of this area. Thus, we propose a tectonic hypothesis for the Tolstoj-Zeami region: the area could consist of the "extrados" extensional tectonic features (horst and graben) of a tectonically uplifted area. Therefore, this area would show examples of extensional features on Mercury.

Bulging is unlikely to have occurred in a single tectonic phase for the following reasons: (1) its direction is controlled by the Mercurian grid; (2) the smooth plains material limits are quite rectilinear in the NW of the bulge, and parallel to the bulge (Fig. 12) (this rectilinear limit indicates that this region probably would have been slightly upwarped at the time of the smooth plains flooding); and (3) some valleys are crosscut by C_4 craters, while others transect C_4 craters; the motions could have partially stopped at the C_4 times, but locally could have continued elsewhere.

Thus, it appears that the tectonic development of this area occurred over a long period of time, and is probably due to a deep and long-lived internal source, similar in some respects to Tharsis Regio on Mars, despite the large difference in the timing and scale. The existence of such large-scaled internal activity and inhomogeneities must be taken into account in further models of Mercurian internal structure and history.

The Phidias Area

The crater Phidias (+11° lat, 150° long) has unusual characteristics (Fig. 13). It is filled by smooth plains material, and classified as a very old C_2 crater by Schaber and McCauley (1980). Despite the absence of secondaries, central peak and wall terraces, it exhibits a very sharp break between wall and floor materials, and walls and surrounding plains. This wall is sharp in the northern

 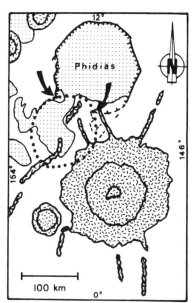

Fig. 13. Photomosaic (left) and simplified sketch map (right) of the Phidias area. Shown are the intercrater plains (unshaded), the smooth plains (dotted areas) and the young crater material and secondaries (mottled areas). ⚡ indicates rectilinear limits of the Phidias depression, and dotted line the boundary of the "ghost crater." Note that the very old "ghost crater" seems to transect the Phidias depression (see arrows).

part of its rim where it is circular in plan. It is also sharp on the eastern rim, where it is rectilinear in plan with a N 20° E trend. In the western part, the wall consists of en echelon rectilinear segments trending N 20° E. The SSE part of the rim is subdued, and the floor-wall boundary is indistinct. The SSW part of the rim is crosscut by a very old "ghost" crater (older than C_1 according to Schaber and McCauley).

Outside the Phidias crater, the terrains between Tyagaraja and Phidias seem to be affected by relatively subdued horst and graben.

Interpretation

It is difficult to envision how impact-related phenomena could be the only origin of many of the characteristics described above, especially with regard to the apparent differences in ages for the Phidias walls, and the geometric relationship between Phidias and the ghost crater. We thus propose that the Phidias depression is not due to an impact, but rather due to a tectonic (or volcano-tectonic) subsidence of a nearly circular area. The subsidence resulted in fault walls around three-quarters of the depression, controlled by the N 20° E trend, the most important direction of the Mercurian grid. This subsidence may be described as a caldera-like motion.

These two examples, both located in the H 8 quadrangle, show that tectonic motions may occur independently of global compression or of large basin formation. Such motions could be local ones and they might be due only to "volcano-tectonic" events such as inside Phidias. But motions which affected large areas during long periods of time certainly indicate large-scale internal activity, as between Tolstoj and Zeami.

VI. GENERAL CONCLUSION

In this section, we summarize the tectonic history of Mercury. After formation of a solid lithosphere, but before the end of heavy meteoritic bombardment and the Caloris impact, a global pattern of linear lithospheric fractures appeared. This grid seems partially due to tidal spindown (that produced the NE-SW and NW-SE lineaments), but also to an unknown process(es) that produced an important N 20° E trend. During the same time and continuing afterward, global contraction due to core cooling induced a compressive state in the lithosphere. This planet-wide compressive stress was responsible for the development of arcuate and lobate scarps interpreted to be thrust faults.

During this compressive state, the impact which created Caloris basin occurred, completely breaking and melting the lithosphere beneath the basin. The lithosphere, already under compression, tended to relax its stresses by stretching towards the cavity produced by the Caloris impact. This induced a global displacement of the lithosphere towards the basin center. These motions temporarily oriented the pre-existing compressive stress on the entire planet, and induced tectonic motions around and inside the basin, i.e., the concentric ridges and graben.

Independently of these global motions, some areas exhibited unobtrusive vertical motions and bulging, and associated tectonic features. The long duration of such uplifting indicates probable deep and long-lived internal sources. Furthermore, local volcano-tectonic motions could occur during the smooth plains flooding.

A number of puzzling questions persist in this outline of Mercurian tectonic history. The main problem is the complete ignorance of 60% of the surface. New data about the unknown part of Mercury may or may not be compatible with the proposed scenarios. Hopefully, a future mission will answer these questions.

Acknowledgments.   The investigations for this study were carried out under CNES and INAG-INSU grants. The numerical calculations were performed in the computing center CIRCE (Orsay, France).

MERCURY'S THERMAL HISTORY AND THE GENERATION OF ITS MAGNETIC FIELD

G. SCHUBERT, M. N. ROSS
University of California at Los Angeles

D. J. STEVENSON
California Institute of Technology
and

T. SPOHN
Westfaelische Wilhelms Universität

Mercury probably formed hot with early differentiation of an iron core. If this core is pure iron, then it would have frozen very quickly. However, volatile-bearing planetesimals most likely contributed to the accretion of the planet Mercury, adding a small amount of sulfur. Under these conditions, the strongly depressed Fe-S eutectic has prevented the Mercurian core from freezing completely. Detailed models incorporating subsolidus convection of the Mercurian mantle indicate that, to date, it may be possible to maintain convection in the outer fluid core of Mercury, perhaps allowing for dynamo generation of the observed magnetic field. A total sulfur abundance of around 2 to 3% by mass relative to iron allows for both the rapid growth of an inner core prior to cessation of early bombardment (to satisfy the geological constraint of little planetary contraction over geologic time), and the possibility of ongoing outer core convection. The role of tidal heating in the inner solid core is also assessed and found to be a potentially significant contribution to the core energy budget in favorable circumstances, enhancing or prolonging outer core thermal convection. Permanent magnetism is at best a marginal explanation of the magnetic field. A fluid outer core is a desirable feature of the present structure. However, it is not certain whether the criteria for dynamo generation are currently satis-

fied; estimates of the field expected for a dynamo are one to two orders of magnitude in excess of the observed field. Alternative explanations, including the possibility of a weak thermoelectric dynamo, are discussed.

If we are to understand the origin of Mercury's magnetic field, then it is essential that we determine the structure and temperature of the planet's interior. Thermal history modeling must play a central role in this endeavor because of the paucity of observational constraints. The construction of a thermal evolution model necessarily begins with assumptions about Mercury's initial temperature and structure. In the first two sections of this chapter, we discuss the initial state and conclude that Mercury probably formed hot and differentiated into an inner core and silicate mantle contemporaneous with or shortly after its accumulation. Planets, which start their evolution vigorously convecting at high temperature, have the property of retaining no memory of the details of their initial state. Having set the stage for the thermal history model, we review briefly previous attempts at simulating Mercury's thermal evolution. The thermal history model of Stevenson et al. (1983) is discussed at some length and extended to include the effects of tidal heating in Mercury's solid inner core. Finally, we discuss the implications of thermal history modeling for the source of the magnetic field.

I. INTERNAL STRUCTURE

The cosmic abundance of heavy elements leaves little doubt that iron must constitute 60 to 70% of Mercury's mass in order to account for the planet's high density (Urey 1951,1952; Reynolds and Summers 1969; Kozlovskaya 1969; Siegfried and Solomon 1974). Though the mass percentage of iron is insensitive to how the iron is distributed, the nature of this distribution, i.e., whether the iron is in a central core or homogeneously distributed in oxide form or some combination of both extremes, is the major unknown about Mercury's internal structure. The existence of a planetary magnetic field does not unambiguously distinguish among the possible iron distributions as long as neither permanent magnetism nor the hydromagnetic dynamo can be unequivocally eliminated as the source of the field. However, in Sec. VIII we argue against permanent magnetism and accordingly offer the magnetic field as observational evidence that Mercury has a large, at least partially liquid iron core. Support for an iron core in Mercury is provided by the lunar-like appearance and spectral characteristics of its surface (McCord and Adams 1972a,b; Murray et al. 1974,1975; Hapke et al. 1975; McCord and Clark 1979; Vilas et al. 1984; Vilas 1985). Mercury "looks" as if it is differentiated into a silicate mantle and an iron core. The planet's high albedo (somewhat larger than the albedo of the lunar highlands) does not allow for a large amount of metallic iron in its surface. A weak or nonexistent (Vilas 1985; also Chapter by Vilas) Fe^{2+} absorption feature limits the amount of iron in orthopyroxenes at the surface. Mercury's high density suggests that it is iron rich while its surface seems iron poor, a circumstance favoring concentration of iron into a core.

The question of whether Mercury has an iron core will only be settled with the acquisition of new data by orbiting or lander spacecraft. If Mercury is seismically active we could determine if a core exists by placing seismometers on its surface; however, a seismic experiment might not be straightforward in the unlikely event that Mercury's core is solid. A determination of Mercury's axial moment of inertia C would suffice to distinguish an iron core from a more homogeneous distribution of iron; with M standing for Mercury's mass and R its radius, $C/(MR)^2 \simeq 0.34$ for models of Mercury with an iron core (Basaltic Volcanism Study Project 1981; Chapter by Peale, while a Mercury model with a homogeneous distribution of iron would give $C/(MR^2) \simeq 0.4$. Peale (1976,1981) has discussed how measurements of the gravitational coefficients J_2 and C_{22}, and the dynamical quantities ϕ (amplitude of the 88 day physical libration of the principal axis of minimum moment of inertia) and θ (obliquity of the spin axis to the orbit normal) could be combined to determine $C/(MR^2)$. Radio tracking of a Mercury orbiter would give J_2 and C_{22}. The

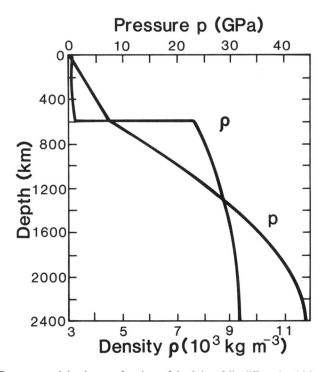

Fig. 1. Pressure p and density ρ as functions of depth in a fully differentiated Mercury model (figure after Siegfried and Solomon 1974).

quantities θ and ϕ would be most readily determined from the positions of a network of transponders. A sufficiently detailed survey of the spatial structure of the magnetic field, and electromagnetic induction observations, might provide still other means of detecting Mercury's core.

In the following, unless stated otherwise, we assume that Mercury has an iron-nickel core and a silicate mantle. Various modeling assumptions about the composition and equations of state all yield a similar core radius of about ¾ of the planetary radius. Depth profiles of pressure and density for a representative model of a differentiated Mercury (Siegfried and Solomon 1974) are shown in Fig. 1.

II. TIMING OF CORE FORMATION AND THE INITIAL STATE FOR THERMAL HISTORY MODELING

There are strong theoretical arguments to support a hot initial state for Mercury with early core formation. Indeed, if Mercury is a differentiated planet, then its surface geology provides observational confirmation of the early core separation event. Core formation must have occurred prior to the end of the period of heavy bombardment (about 4 Gyr ago) because Mercury would have undergone a large expansion (an increase in radius of about 15 km) upon core separation and there are no large-scale extensional features on Mercury's heavily cratered terrain to attest to such an occurrence (Solomon 1977). In principle, accretional heating alone can account for the energy requirement of early core differentiation. The gravitational potential energy per unit mass made available for heating upon homogeneous accumulation of a planet is $3GM/5R$ (with G equal to the universal gravitational constant, M the planet mass, and R the planet radius). This energy is enough to raise the temperature of Mercury by nearly 5500 K (assuming a specific heat of 1 kJ $kg^{-1}K^{-1}$). Clearly, there is a surfeit of energy for melting and differentiation, since the silicate melting temperature in Mercury rises to only about 2500 K at great depth and the accumulating planetesimals may have initial temperatures of about 1000 K (the equilibrium condensation temperature for Fe and $MgSiO_3$ at Mercury's orbit is about 1400 K [Lewis 1972]). The major uncertainty, of course, is the fraction of the available accretional energy that is retained as heat inside the planet instead of being radiated away. Models of accretional heating of the Earth and Moon (Kaula 1979,1980) suggest that an energy retention factor of 10% is a minimum with values as large as 50% being plausible. A retention efficiency h of only 20% would suffice to insure accretional melting of nearly all of Mercury.

Figure 2 shows accretional temperature profiles for Mercury based on the model equation (Kaula 1980; Schubert et al. 1986)

$$T(r) = \frac{hGM(r)}{cr}\left\{ 1 + \frac{ru^2}{2GM(r)} \right\} + T_e \qquad (1)$$

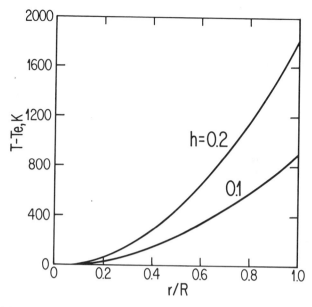

Fig. 2. Model excess accretional temperature $(T - T_e)(K)$ vs normalized radius r/R. The temperature T_e is the background temperature of planetesimals accumulating to form Mercury. The parameter h is the fraction of gravitational potential energy retained within the growing planet as internal thermal energy.

where c is the specific heat, T_e is the ambient temperature during accretion, $u^2/2$ is the approach kinetic energy per unit mass of planetesimals forming Mercury, and $M(r)$ is the mass of the planet internal to the radial distance r. The calculations for Fig. 2 assume $ru^2/[2GM(r)] \ll 1$, c has a value of 1 kJ $kg^{-1}K^{-1}$, and a constant density planet. The temperature equation is based on the assumption that when the planet's radius is r, the retained fraction of impact heating is distributed uniformly over the surface and that no radial redistribution of heat takes place over the accumulation time of the planet. The high near-surface temperatures of the model profiles would certainly be mitigated by radial heat transfer in the actual accreting planet and by radiative heat loss from its surface.

Other sources of energy might contribute to the early differentiation of Mercury though they must be considered speculative, given our present knowledge of them. They include: (1) short-lived radioactivity (Lee et al. 1976; Runcorn et al. 1977); (2) electromagnetic heating (Sonett et al. 1975); (3) tidal heating upon despinning (though Burns [1976] has concluded that this is unimportant); and (4) heating accompanying the high temperature, late-stage dissipation of the solar nebula (Cameron 1985a). If Mercury contains a full solar complement of U and Th, then long-term radioactive heating is sufficient to differentiate Mercury into an inner core and silicate mantle within about 1 Gyr after accretion even if Mercury was cold following its accumulation (Siegfried and Solomon 1974; Solomon 1976; Toksöz et al. 1978).

Given that core separation occurred contemporaneously with or shortly after accretion, then the initial temperature of Mercury is certain to have coincided with the silicate solidus in the mantle and a near-liquidus adiabat in the iron core. The likelihood of this high-temperature initial state is enhanced by the large gravitational energy release that accompanies core formation. This energy is unquestionably retained and will raise Mercury's temperature by about 700 K (Solomon 1976).

The high initial temperature of Mercury insures that there was vigorous convection in the planet's mantle and core at the start of its thermal evolution. Because of the thermostat effect provided by the strong temperature dependence of mantle viscosity (Tozer 1967), early vigorous convection cools Mercury rapidly at first. Later, when the planet is not as hot, mantle convection is less vigorous and the planet cools at a slower rate. The early vigorous convection and rapid cooling precludes any retention of the details of the initial state (Schubert et al. 1979).

III. THERMAL HISTORY MODELING: PRELIMINARY REMARKS AND LITERATURE REVIEW

The previous two sections set the stage for the thermal history modeling we are about to present. The models will assume a fully differentiated initial state with the mantle at the solidus temperature and the core at a near liquidus adiabat. However, before proceeding with the discussion of these models, we will briefly review previous attempts to simulate Mercury's thermal evolution because many of these earlier papers assumed different initial conditions. Additionally, most of these papers neglected one or more of the physical-chemical processes important in Mercury's thermal history.

Several papers (Siegfried and Solomon 1974; Fricker et al. 1976; Sharpe and Strangway 1976; Solomon 1976,1977,1979; Solomon and Chaiken 1976) have allowed for mantle heat transport only by the process of conduction. However, conduction is not the principal means of heat transfer in the mantle. Heat transfer by subsolidus convection is more efficient than conduction and it is generally agreed that mantle convection regulates the temperature in terrestrial planetary interiors (Runcorn 1962; Tozer 1967; Turcotte and Oxburgh 1969; Schubert et al. 1969).

A number of the early papers on Mercury's thermal history (see, e.g., Siegfried and Solomon 1974; Solomon 1976) predicted rapid solidification of the core, thereby creating a problem for a core dynamo explanation of the present-day magnetic field. Fricker et al. (1976) suggested a solution to the dilemma in terms of the melting temperature difference between iron and silicates at the core-mantle boundary. Their models yielded molten cores at the present day, but they did not allow for the more efficient cooling of the core by mantle convection. The studies of Cassen et al. (1976), Gubbins (1977*b*) and Toksöz et al. (1978) emphasized that subsolidus mantle convection exacerbated the problem of early core solidification. Cassen et al. (1976) suggested that retention of radiogenic heat sources in the lower mantle could keep the core from freezing. Toksöz et al. (1978) prevented core solidification by adding radioactive heat sources to the core, a solution also suggested by Siegfried and Solomon (1974), Toksöz and Johnston (1977), Sharpe and Strangway (1976) and Solomon (1977). These authors also considered late core formation as a possible way to arrive at a present-day molten core, a suggestion at variance with the surface geologic record.

The most plausible way for Mercury to have retained a partially molten core at the present time, against the efforts of mantle convection to freeze the core, is for Mercury to have incorporated a light alloying element into its core. This solution to the core freezing problem was first mentioned by Cassen et al. (1976) and Solomon (1976); it has been explored in considerable quantitative detail by Stevenson et al. (1983) who emphasized that radial mixing of planetesimals within the primordial solar nebula would cause Mercury to accrete bodies containing S and other volatiles even though equilibrium condensation models (Lewis 1972; Grossman 1972) predict negligible S in Mercury. Sulfur is the only reasonable candidate; oxygen has been advocated for the Earth (Ringwood 1977; McCammon et al. 1983), but O is not sufficiently soluble in iron at the low internal pressures of Mercury. The reduction in melting temperature with the addition of S is crucial to the maintenance of a liquid outer core in Mercury, for the calculations of Stevenson et al. (1983) show that Mercury's core should be largely solidified at the present time. The increase with time in the S content of the liquid outer core, as the solid Fe inner core grows larger, progressively lowers the melting point of the molten outer core and retards the complete solidification of the core.

The amount of radial contraction that Mercury could have sustained since the end of the early period of heavy impact cratering is constrained by the surface geologic record. Strom et al. (1975*b*; see also Strom 1979) measured the inclinations, heights and lengths of lobate scarps and deduced that these compressional features limited the amount of surface area reduction to less than about 10^5 km^2, corresponding to a decrease in radius of about 2 km. Solomon (1976,1977,1979) has stressed that this limited post heavy-bombardment contraction of Mercury places an upper bound of about 1100 km on the radial extent of core solidification for a core entirely fluid 4 Gyr ago. Complete core solidification would shrink the planet about 15 km in radius (Solomon 1976), far in excess of the geologically inferred contraction. The large degree of core solidification in some of the models of Stevenson et al. (1983) and in some of the models presented later in this chapter might be at variance with this constraint. However, much depends on how early core freezing begins and on how rapidly it proceeds; if a substantial amount of core freezing can occur prior to the end of the heavy bombardment and preservation of terrain (at about 4 Gyr ago), then the constraint on extent of core freezing imposed by the observed 2 km contraction is weakened. Even the association of the scarp-inferred 2 km contraction with global cooling of the planet is in doubt if tidal despinning contributes to scarp formation (Burns 1976; Melosh and Dzurisin 1978a; Dzurisin 1978; Pechmann and Melosh 1979). The argument for scarp formation by tidal despinning rests on the claim of a preferred north-south orientation of equatorial scarps (Melosh and Dzurisin 1978a; Dzurisin 1978) that is disputed by Cordell and Strom (1977). An argument against the influence of tidal spindown on surface tectonics is the absence of the polar extensional features on Mercury that are predicted by despinning (Melosh 1977b). We will return to this issue in the next section when model results are discussed.

A few papers on Mercury's thermal evolution have considered very different scenarios to the initially hot, fully differentiated model preferred by us. Majeva (1969) studied the conductive thermal history of undifferentiated Mercury models. Similar models were also presented by Siegfried and Solomon (1974) and Sharpe and Strangway (1976). They prevented core differentiation by assuming a low-temperature accretion of the planet and a low concentration of the radioactive heat sources U and Th. Solomon (1976) showed that heating of Mercury by its full solar complement of U and Th would lead to core differentiation for almost any initial temperature distribution. In the cold, undifferentiated Mercury models of Sharpe and Strangway (1976), Mercury's present magnetic field is attributed to a fossil remanent magnetization of the planet's deep interior.

Thermal history models that include mantle convection allow studies of other aspects of Mercury's structure, e.g., its lithosphere thickness. Mercury is a one-plate planet (Solomon 1977) whose lithosphere has thickened with time as the planet has cooled (Schubert et al. 1979). Model calculations presented in this chapter give lithosphere thicknesses of about 180 km; previous models (Schubert et al. 1979) that did not account for internal heat sources give present-day lithosphere thicknesses of about 300 km. Some part of the 2 km contraction of Mercury inferred from the geologic record could be attributed to the cooling and growth of Mercury's lithosphere (Solomon 1977); a portion of lithospheric growth occurs prior to the end of the period of early heavy bombardment. The results of Stevenson et al. (1983) indicate that, at present, Mercury's mantle is convecting beneath its lithosphere though the convection is not very vigorous. In the thermal models of Toksöz et al. (1978), mantle convection ceases after about 2.5 Gyr of evolution. Pertinent to the question of whether Mercury's mantle is presently convecting is Runcorn's (1977) suggestion that Mercury's resonant state of rotation is maintained by the solar torque on the second degree nonhydrostatic component of a mantle convection system. However, because only a small difference in equatorial moments of inertia is required to establish and maintain the resonant rotation (Goldreich and Peale 1966), this difference could be maintained by stresses in a cold lithosphere.

In the next section, we present detailed results of a Mercurian thermal history model that includes:

1. Thermal coupling of a fully differentiated mantle and core;
2. Whole mantle convection parameterized by a simple Nusselt-Rayleigh number relation that controls the rate at which heat escapes from the core;
3. Thermally and/or chemically driven convection in the core;
4. Radiogenic heat production confined to the mantle;
5. Hot initial states with the mantle at the solidus and the core at a near-liquidus adiabat;
6. A light-alloying constituent in the core;
7. A growing inner core (if needed) that excludes the light-alloying material which then mixes uniformly upward through the outer core;
8. Pressure and composition-dependent freezing curves for the core;
9. Latent heat and gravitational energy release upon inner-core growth;
10. A temperature-dependent mantle viscosity;
11. Tidal heating in an inner core.

Except for the addition of inner-core tidal dissipation, the model is essentially identical to the one of Stevenson et al. (1983).

IV. THERMAL HISTORY MODEL

The model shown in Fig. 3 is defined by concentric shells that represent the conductive lithosphere $R_P > r > R_\ell$, the convecting mantle $R_\ell > r > R_c$, the liquid outer core $R_c > r > R_i$, and the solid inner core $r < R_i$ (r is radial distance from the center of the Mercury model). Also shown in Fig. 3 are the melting curves for pure iron in the inner core $T_{mi}(r)$ and a binary iron alloy in the outer core $T_{ma}(r)$. A light element (e.g., S) in the outer core is required to reduce the outer-core melting temperature and prevent complete core freezing. The likelihood that Mercury was fully differentiated at the start of its thermal evolution has been discussed in Sec. II. Other details of the internal structure are suggested by general principles thought relevant to a single-plate planet with a convecting mantle (Schubert et al. 1979,1980; Stevenson et al. 1983). The mantle is subsolidus throughout, consistent with the lack of volcanic activity in the last 3 Gyr (Gault et al. 1977; Strom 1984). Temperatures and physical properties are spherically averaged within each shell. The model includes a thickening lithosphere and a freezing inner core. The temperature profile in the lithosphere is approximated by a conductive steady state. The temperature varies linearly in the thermal boundary layers at the top and bottom of the mantle (and, possibly, in the boundary layer at the top of the inner core if significant heat sources exist in the inner core). Temperatures follow an adiabat between boundary layers in the mantle; in the liquid core, thermal

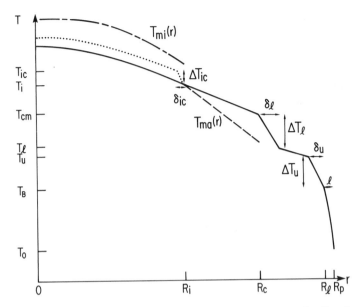

Fig. 3. Sketch of model temperature T profile in Mercury. The temperature rises linearly across all thermal boundary layers and adiabatically across the mantle, and inner and outer core regions. The inner core may contain a thermal boundary layer (dotted line) or it may not (solid line). Also shown are the melting curves of pure iron T_{mi} and the iron alloy in the outer core T_{ma}. R_p: planet radius; R_c: outer core radius; R_i: radius of inner-outer core boundary; T_B: temperature at the base of the lithosphere; T_u: temperature at the base of the upper thermal boundary layer in the mantle; T_ℓ: temperature at the top of the lower thermal boundary layer in the mantle; T_{cm}: temperature at the core-mantle interface; T_i: temperature at the inner-outer core boundary; T_{ic}: temperature at the base of the upper thermal boundary layer within the inner core.

boundary layers are assumed negligible. Heat sources include uniformly distributed radioactive heating in the mantle and lithosphere and tidal heating in the inner core. The time t evolution of the system is found by writing energy balances for each layer. For the lithosphere with thickness ℓ,

$$\rho c (T_u - T_B) \frac{d\ell}{dt} = F_\ell - F_s \qquad (2)$$

where ρ and c are the density and specific heat of the lithosphere (assumed equal to their respective mantle values), T_u and T_B are the temperatures at the top of the convecting mantle and the bottom of the lithosphere (assumed constant in time), and F_ℓ and F_s are the heat fluxes into and out of the lithosphere. The energy balance for the mantle is

$$\frac{4\pi}{3} (R_\ell^3 - R_c^3)\left(Q - \rho c \eta \frac{dT_u}{dt} \right) = 4\pi(R_\ell^2 F_\ell - R_c^2 F_c) \qquad (3)$$

where Q is the uniform volumetric rate of radioactive heat production, R_ℓ, R_c and T_u are defined in Fig. 3, η is a factor that accounts for the adiabatic temperature rise across the mantle, and F_c is the heat flux from the core. We assume that the chemical and gravitational energy released upon inner-core freezing is instantly available at the core-mantle boundary for transfer into the mantle. The core energy balance, without internal heat sources in the core, is

$$(L + E_G)4\pi R_i^2 \rho_c \frac{dR_i}{dt} - 4\pi R_c^2 F_c = \frac{4\pi}{3} R_c^3 \rho_c c_c \eta_c \frac{dT_{cm}}{dt} \qquad (4a)$$

where L is the latent heat of Fe solidification, E_G is the gravitational energy made available upon concentration of the light element into the outer core, T_{cm} is the temperature at the core-mantle boundary, R_i is the radius of the inner core, ρ_c and c_c are the density and specific heat of the core, and η_c is a factor that accounts for the adiabatic rise of temperature in the core. If tidal dissipative heating in the inner core is accounted for, then separate energy balances must be written for the outer and inner cores

$$4\pi R_i^2 F_i + (L + E_G)4\pi R_i^2 \rho_c \frac{dR_i}{dt} - 4\pi R_c^2 F_c = \frac{4\pi}{3} R_c^3 \rho_c c_c \eta_c \frac{dT_{cm}}{dt} \qquad (4b)$$

$$W_i - 4\pi R_i^2 F_i = \frac{4\pi}{3} \rho_c c_c R_i^3 \eta_i \frac{dT_i}{dt} \qquad (4c)$$

where W_i is the total power generated by the inner core tidal heat source, T_i is the temperature at the inner-outer core boundary, F_i is the heat flux from the inner core, and η_i accounts for the adiabatic temperature rise in the inner core.

The core is assumed to be initially completely molten consistent with models of primordial differentiation and core formation (Safronov 1978; Kaula 1979; Schubert et al. 1980; Basaltic Volcanism Study Project 1981). At later times the inner-outer core boundary is located by determining the pressure within the core at which the core alloy melting curve intersects the outer-core adiabat. The alloy melting curve as a function of pressure p is given by

$$T_{ma} = T_{mo} (1 - \alpha\chi)(1 + T_{m1}p + T_{m2}p^2) \qquad (5)$$

where T_{m1} and T_{m2} are constants, α is a constant determined by the high-pressure phase diagram of the iron binary system, and χ is the weight fraction of the light element. The outer-core adiabat is given as a function of pressure by

$$T_a = T_{cm} \left[\frac{1 + T_{a1}p + T_{a2}p^2}{1 + T_{a1}p(r = R_c) + T_{a2}p^2(r = R_c)} \right] \qquad (6)$$

where T_{a1} and T_{a2} are constants. The simultaneous solution of Eqs. (5) and (6) provides the pressure at the inner-outer core boundary. The radius of the inner core R_i is found by assuming that gravity within the core equals rg/R_c, where g is the gravity at the outer-core boundary assumed equal to the surface gravity; R_i is given by

$$R_i = \{2[p(r = 0) - p(r = R_i)]R_c/\rho_c g\}^{1/2} \qquad (7)$$

The thicknesses of the thermal boundary layers in Fig. 3, δ_u, δ_ℓ, and δ_{ic} determine the heat fluxes F_u, F_ℓ, and F_i. For a boundary layer with a linear temperature profile, $F = k\Delta T/\delta$, where k is thermal conductivity. In a vigorously convecting fluid, the horizontally averaged boundary layer thickness δ is given by (Turcotte and Oxburgh 1967)

$$\delta = d(\mathrm{Ra}_{cr}/\mathrm{Ra})^\beta \qquad (8)$$

where d is the depth of the convecting region, Ra_{cr} is the critical Rayleigh number for the onset of convection, β is a constant, and the Rayleigh number (Ra) is given by $g\alpha_m(\Delta T_u + \Delta T_\ell)(R_\ell - R_c)^3/\nu_m\kappa_m$ for the mantle and $g\alpha_c(\Delta T_{ic})R_i^3/\nu_c\kappa_c$ for the inner core. Here α is the coefficient of thermal expansion, κ is thermal diffusivity, ν is the kinematic viscosity and ΔT_u, ΔT_ℓ, and ΔT_{ic} are the temperature differences across the upper mantle, lower mantle and inner core thermal boundary layers, respectively (subscripts m and c refer to mantle and inner core). Equation (8) is valid for constant viscosity fluids. However, because of the strongly temperature-dependent viscosity, the lower boundary layer may be thinner than the upper boundary layer. Booker and Stengel (1978) suggest that the local critical Rayleigh number for thinning of the lower boundary layer is $g\alpha\Delta T_c\delta_c^3/\nu_c\kappa = 2 \times 10^3$ where ν_c is the average temperature within the boundary layer. We have used a thinned boundary layer thickness instead of Eq. (8) for the lower boundary layer whenever it is smaller than that given by Eq. (8). The parameterization (Eq. 8) has been widely accepted for use in planetary thermal evolution simulations (Schubert et al. 1979,1980); the limitations and approximations involved with the use of Eq. (8) are discussed in these references.

The model heat sources we have so far discussed include radioactive heating in the mantle by the decay of U and Th and release of latent heat and gravitational energy upon inner-core freezing. Because Mercury's dynamical state is dominated by the tidal interaction with the Sun, it is reasonable to suppose that tidal heating might be a significant heat source. Burns (1976) examined the thermal contribution from the tidal slowing of a presumed initial noncommensurate rotation and found it insignificant. Present-day tidal heating in Mercury is a consequence of rotation with respect to the Sun (period = 175.94 days) and the eccentricity of the orbit (mean eccentricity = 0.175)

(Cohen et al. 1973). Tidal heating in the anelastic mantle can be shown to be unimportant at the present time.

A large inner solid Fe core, however, might be capable of generating an amount of tidal dissipation comparable to the heat loss along the adiabat in the outer core. Following Ross and Schubert (1986) and Sabadini et al. (1982), we have calculated inner-core tidal heating for a three-layer model of Mercury with elastic mantle, inviscid outer core, and viscoelastic inner core of variable radius as in Fig. 3. The inner core is represented by a Maxwell body with characteristic temperature-dependent shear modulus μ_i and viscosity ν_i. Inner-core tidal heating might be significant because the solid iron Maxwell time ($\tau = \rho_c \nu_i / \mu_i$) could be near the tidal forcing period.

Rotation with respect to the Sun and orbital eccentricity result in a varying tidal potential at the planet's surface given to first order by (Kaula 1964)

$$
W = V_0 \left\{ \left[\left(\frac{1}{4} + \frac{e}{8} \right) P_2^2(x)\cos(2\phi) - j\left(\frac{1}{4} + \frac{7e}{8} \right) P_2^2(x)\sin(2\phi) \right] e^{jn/2t} \\
- j\left[\frac{7e}{8} P_2^2(x)\sin(2\phi) \right] e^{j3/2nt} - \frac{3e}{2} P_2^0(x)e^{jnt} \right\}
\tag{9}
$$

where $n = 2\pi/(87.97 \text{ days})$, $V_0 = n^2 R_p^2$, e is orbital eccentricity, θ and ϕ are the Mercurian colatitude and longitude, respectively, $x = \cos\theta$, and P_2^0 and P_2^2 are Legendre functions. Mercury's complex love number k^* is calculated for the three frequencies appearing in the potential (Eq. 9). The time averaged integration over the surface of the product of the applied potential and the lagged response Wk^* yields the tidal dissipation (Zschau 1978).

V. MODEL PARAMETER VALUES

Many of the physical parameters in Eqs. (1)–(7) are reasonably well known and do not require extensive discussion; the values for model parameters not discussed below can be found in Table I. The core radius is 1840 km

TABLE I
Model Parameter Values

R_p	$= 2.44 \times 10^6$ km	η	$= 1.1$
R_c	$= 1.84 \times 10^6$ km	k	$= 4$ W m^{-1} K^{-1}
ρ_m	$= 3300$ kg m^{-3}	g_s	$= 3.8$ m s^{-2}
ρ_c	$= 8600$ kg m^{-3}	Ra_{cr}	$= 1000$
T_o	$= 440$ K	α	$= 3 \times 10^{-5}$ K^{-1}
T_{mo}	$= 1880$ K	β	$= 0.3$ (in Eq. 8)
T_{m1}	$= 1.36 \times 10^{-11}$ K Pa^{-1}	κ	$= 10^{-6}$ m^2 s^{-1}
T_{m2}	$= -6.2 \times 10^{-23}$ K Pa^{-2}		
T_{a1}	$= 8.0 \times 10^{-12}$ K Pa^{-1}		
T_{a2}	$= -3.9 \times 10^{-23}$ K Pa^{-2}		

and pressures at the core-mantle boundary and planet center are 10 GPa and 40 GPa, respectively. The coefficients T_{m1} and T_{m2} reproduce the data of Liu and Basset (1975). The coefficients for the core adiabat T_{a1} and T_{a2} provide a zero pressure Grüneisen parameter γ of 1.6 (Stacey 1977a). The coefficients for the core liquidus were chosen according to the Lindeman law of melting (Stacey 1977a). Because $\gamma > 2/3$, the core alloy liquidus is steeper than the adiabat (Stevenson 1980).

Volumetric radioactive heating in the mantle and lithosphere decays with time according to $Q = Q_o \exp(-\lambda t)$. The correct values of Q_o and λ for Mercury are unknown and depend on the geochemical model chosen. Since we assume sulfur to be the light alloying element in the core, a strict equilibrium condensation model of Mercury (Lewis 1972; Grossman 1972) is ruled out. A small amount of sulfur can be explained by a variety of processes such as inner solar system planetesimal mixing or nonequilibrium condensation (Ringwood 1979; Basaltic Volcanism Study Project 1981; Wetherill 1985a). Geochemical models that allow about 1 wt.% sulfur in Mercury's core assume that Mercury has an Earth-like uranium and thorium content yet is potassium depleted (Basaltic Volcanism Study Project 1981). Accordingly, we assume values for Q_o and λ corresponding to K depleted chondritic material, $Q_o = 1.7 \times 10^{-7}$ W m^{-3} and $\lambda = 1.4 \times 10^{-17}$ s^{-1}. The model is not very sensitive to the exact choice of Q_o and λ since the most uncertain parameter, K concentration, is most important shortly after formation when primordial heat dominates the thermal evolution (Schubert et al. 1979,1980).

The latent heat release upon inner core freezing is assumed equal to the zero pressure value $L = 250$ kJ kg^{-1}. The precise value of gravitational energy available as a result of mass redistribution within the core depends on the radius of the inner core and the alloy chemistry. The power available from gravitational energy is given by (Merrill and McElhinny 1985)

$$P_G = \int_{V_c} \Phi_c \dot{\rho}_c dV \qquad (10)$$

where $\Phi_c = 2/3\pi G\rho_{ic}(3R_c^2 - r^2)$ is the approximate gravitational potential in the core, $\dot{\rho}_c$ is the time rate of change of density in the core, and the integral is over the total core volume V_c. Effects of net core volume changes are small (Gubbins 1977a) and have been neglected in Eq. (10). There are two competing contributions to the integral in Eq. (10), one from the increase in density with time at the inner core-outer core boundary and one from the gradual decrease in density throughout the outer core. The expression (10) for P_G can be rewritten to take these two contributions explicitly into account

$$P_G = \int_{V_{oc}} \Phi_c \dot{\rho}_{oc} dV + \int_{S_{ic}} \Phi_c (\rho_{ic} - \rho_{oc}) \dot{R}_i \, dS \qquad (11)$$

where $\dot{\rho}_{oc}$ is the time rate of change of density in the outer core, and the integrations are over the volume of the outer core V_{oc} and surface of the inner core S_{ic}, respectively. By assuming $\dot{\rho}_{oc} = \left(\dfrac{\rho_{ic}}{\rho_S}\right) \dot{\chi} \Delta\rho =$ constant in the outer core, where $\Delta\rho$ is the difference in density between the light element (ρ_S) and iron, and using conservation of light constituent mass in the form $\chi = \chi_o R_c^3/(R_c^3 - R_i^3)$, where χ is the light element mass concentration in the outer core with χ_o its initial value ($\chi \leq \chi_e$, the eutectic value of about 0.25 at the pressures of the Mercurian core [Usselman 1975]), we obtain from Eq. (11)

$$P_G = 2\pi g_c R_c^4 \dot{\chi} \Delta\rho \left(\frac{\rho_{ic}}{\rho_S}\right) \left[\frac{1}{5}(1 - \xi^5) - \frac{\xi^2}{3}(1 - \xi^3)\right] \qquad (12)$$

where $\xi = R_i/R_c$ and $g_c = \frac{4}{3}\pi G \rho_c R_c$ is the local gravitational acceleration at the outer core boundary. In the limit $\xi \to 1$, $P_G \propto (1 - \xi)^2 \dot{\chi}$, a physically reasonable result since both the amount of material undergoing redistribution per unit time, and the distance through which it is redistributed, scale as $1 - \xi$, the nondimensional thickness of the outer core.

Equation (12) for P_G can be rewritten using $P_G = E_G \dot{m} \, (\dot{m} = 4\pi R_i^2 \dot{R}_i \rho_c)$ as

$$E_G = \frac{2\pi G R_c^2 \chi_o \Delta\rho}{(1 - \xi^3)^2} \left(\frac{\rho_{ic}}{\rho_S}\right) \left[\frac{1}{5}(1 - \xi^5) - \frac{\xi^2}{3}(1 - \xi^3)\right] . \qquad (13)$$

For the Earth, $\xi = 0.35$ and $E_g = 3.0$ MJ kg^{-1} for $\chi_o = 0.1$, comparable to previous estimates (Loper 1978). The limiting value of $\chi = \chi_e$ sets an upper limit on inner core radius of $\xi \leq (1 - \chi_o/\chi_e)^{1/3}$. Figure 4 shows E_G as a function of ξ for different values of χ_c.

The value of α in Eq. (5) depends on the identity of the light alloying element in the core. Essentially any binary Fe system can be described by Eq. (5) as long as $\chi << 1$, but as argued above, only sulfur is suitable for Mercury because of the low pressures in its core (Fig. 1). It should be noted that $\alpha\chi$ in Eq. (5) is a universal variable in that our results with sulfur are equally valid for silicon, for instance, by scaling χ according to $\alpha_S \chi_S = \alpha_{Si} \chi_{Si}$. For pressures in Mercury's core, $\alpha = 2$ (Usselman 1975). The initial wt.% sulfur χ_o is a variable parameter in our model.

The kinematic viscosities of the mantle ν_m and inner core ν_{ic} are assumed to be functions of the temperatures just beneath the upper boundary layers in the respective regions (see Fig. 3)

$$\nu_m = A_m \exp(E_m/T_u) \qquad (14a)$$

$$\nu_{ic} = A_{ic} \exp(E_{ic}/T_{ic}) \qquad (14b)$$

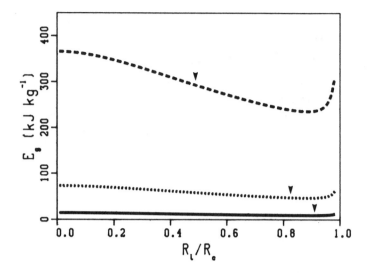

Fig. 4. Gravitational energy release E_G as a function of $\xi = R_i/R_c$. The solid, dotted, and dashed lines are for $\chi_o = 0.002$, 0.01 and 0.05, respectively. Arrows indicate current value of ξ for each value of χ_o.

where $A_{m,ic}$ and $E_{m,ic}$ are constants. In order to match the Earth's present-day mantle viscosity of about 10^{21} Pa s (Cathles 1977), Stevenson et al. (1983) chose $A_m = 4000$ m^2 s^{-1} and $E_m = 52,000$ K in their thermal models. This value of E_m is consistent with laboratory measurements of deformation in minerals and rocks thought to comprise the Earth's upper mantle (Weertmann and Weertmann 1975; Post 1977). Because of Mercury's undeniably refractory nature, its mantle is likely "stiffer" in the sense that the material exhibits higher viscosity than Earth mantle material at the same temperature and pressure. We assume that the melting temperature in Mercury's mantle is about 25% higher than in the Earth's mantle (at the same pressure) and use $A_m = 4000$ m^2 s^{-1} and $E_m = 57,000$ K. The bottom of the lithosphere is defined as the depth at which mantle material no longer takes part in the convective flow, assumed here to be the depth at which $\rho_m \nu_m = 10^{24}$ Pa s; accordingly, T_B is 1467 K.

The relative importance of inner core tidal heating depends strongly on the rheology of iron which is unknown at high temperature and pressure. Low pressure creep data summarized by Frost and Ashby (1982) indicate deformation of γ Fe by grain boundary diffusion at the high temperatures ($T/T_m > 0.95$) suggested by thermal history simulations. Grain boundary diffusion viscosity is linear in stress and directly proportional to the cube of the mean iron grain size h (Frost and Ashby 1982). Extrapolation of the data summarized in Frost and Ashby (1982) to the stress appropriate for Mercurian solar tides gives $A_{ic} = 3.6 \times 10^6 h^3$ m^2 s^{-1} (h in millimeters). The actual grain size in

Mercury's inner core is unknown. We show later that only a small range of grain sizes results in significant inner core tidal heating. The grain boundary activation energy is relatively well known and we use $E_{ic} = 19,000$ K for the high temperature γ Fe. Frost and Ashby (1982) also give the iron shear modulus as a function of temperature as $\mu = 81(1 - 0.91T/T_m)$ GPa.

The required initial and boundary conditions are $T(R_P) = T_s = 440$ K, T_u $(t = 0) = 2000$ K, T_{cm} $(t = 0) = 2600$ K, and R_i $(t = 0) = 0$. T_{cm} is chosen so that the core-mantle boundary is just subsolidus; results are not sensitive to the exact choice.

VI. RESULTS OF THERMAL HISTORY MODELING: NO INNER CORE TIDAL DISSIPATION

In order to identify better those aspects of Mercury's thermal evolution that might be influenced by inner core tidal heating, we first present results of integrating the thermal history equations without an internal heat source in the inner core. In the following section, we will highlight the changes to the thermal history that are specifically attributable to inner core tidal dissipation. We focus the presentation of model results mainly on quantities that are observable; these include surface heat flow (this might eventually be measured), changes in planetary radius (a record of which may be preserved in the surface geology), and the inferred magnetic dipole strength. Other quantities of particular interest include the onset time and evolution of inner core growth and the heat flow at the core-mantle boundary. The latter quantity essentially determines whether convection in the outer core is thermally or chemically driven.

The decrease in Mercury's radius with time is due principally to the growth of the inner core and the cooling of the mantle. As the inner core grows, the total volume of the core changes by an amount (Solomon 1977)

$$\Delta V_c = \int_0^{R_i} (\Delta V_{\gamma \to \ell}/V_\gamma) \, 4\pi r^2 \, dr \qquad (15)$$

where $\Delta V_{\gamma \to \ell}$ is the specific volume change upon solidification of liquid Fe to the γ phase and V_γ is the specific volume of the γ Fe at relevant pressures. The secular cooling of the mantle results in a further volume change given approximately by $\Delta V_m = -V_m \alpha [T_u(t = 0) - T_u(t)]$, where V_m is the volume of the mantle. It follows that the radius change for Mercury is approximately

$$\Delta R_p = (\Delta V_c + \Delta V_m)/4\pi R_p^2 \ . \qquad (16)$$

Although imperfect knowledge of both the dynamo process (Stevenson 1983; Merrill and McElhinny 1985) and Mercury's magnetic field (see Chapter by Connerney and Ness) preclude a definitive assessment of magnetic field generation in a Mercury thermal model, we estimate the strength of the field that a model might be capable of producing by following the approach of Stevenson et al. (1983). The power available for magnetic field generation can be expressed as

$$P_d = E_g \dot{m} + \bar{\epsilon}(L\dot{m} - P_s - P_{ad}) \tag{17}$$

where $\dot{m} = 4\pi\rho_c R_i^2 \dot{R}_i$, $\bar{\epsilon}$ is an average Carnot efficiency, P_s is the time rate of change of core heat content, and P_{ad} is the adiabatic heat flow from the outer core. Gubbins (1977a) has shown that $E_g\dot{m}$ is entirely available for magnetic dissipation. If the energy supply was provided at the inner core-outer core boundary, then $\bar{\epsilon} \simeq \Delta T/T$, where ΔT is the total temperature drop across the assumed adiabatic outer core. Plausible parameter choices for the evaluation of ΔT are a Gruneisen γ of 1.6 and a bulk modulus of $(80 + 6p)$ GPa, where p is the local pressure in GPa (Stevenson 1981). It then follows that $\Delta T/T$ is $\sim 0.24(1 - \xi^2)$, but any value in the range 0.2 to 0.3 is plausible. The value of $\bar{\epsilon}$ is reduced if a large fraction of the energy supply is distributed throughout the outer core. In the limiting case of a uniform heat source, $\bar{\epsilon} \simeq 0.24[1 - 0.6(1 - \xi^5)/(1 - \xi^3)]$. Intermediate cases may be appropriate weighted averages, but if the persistence of convection is in doubt, then one may need to consider the possibility that only part of the outer core convects.

The quantity P_d is the power available to overcome the ohmic dissipation in the current system that generates the magnetic field. Having no detailed model of how organized fluid motion might generate a magnetic field in Mercury, we can make no direct calculation of magnetic field strength. It is useful nonetheless to apply a simple model that relates P_d to dipole moment strength in order to gain some insight into the history of Mercury's magnetic field. Stacey (1977b) assigned the field-generating current to a core-sized toroid with radius $r_1 = (R_c + R_i)/2$ and cross-sectional radius $r_2 = (R_c - R_i)/2$. The magnetic dipole moment μ generated by azimuthal currents with ohmic dissipation P_d in a toroid with resistivity Ω is

$$\mu = 10^2 [P_d \pi^2 r_1^3 r_2^2/2\Omega]^{1/2} \, A \, m^2 \tag{18}$$

where the numerical constant is empirical and is based on application of the model to the Earth. Even though Mercury's magnetic field generating mechanism may not be similar to the Earth's, it is convenient to normalize μ with respect to the Earth's present-day dipole moment. For R_c and R_i in km and P_d in TW, we obtain

$$\mu_M/\mu_E = 11(R_c - R_i)[P_d(R_c + R_i)]^{1/2}. \tag{19}$$

We present results of thermal evolution simulations for three different values of χ_o, 0.002, 0.01 and 0.05 weight fraction. Cosmochemical arguments give plausible values for χ_o ranging from 0 in a strict equilibrium condensation model (Lewis 1972; Grossman 1972) up to about 0.05 if substantial mixing occurred among planetesimals in the inner solar system (Basaltic Volcanism Study Project 1981; see Chapters by Goettel and by Lewis). Figure 5 shows the evolution of the inner core radius. Inner core growth begins earlier than about 1 Gyr for $\chi_o = 0.002$ and 0.01 but it is substantially delayed to 2.5 Gyr for $\chi_o = 0.05$ due to significant lowering of the alloy solidus temperature. Inner core radius increases rapidly during the early phase of core growth and a core of substantial radius exists by about 1 Gyr for $\chi_o = 0.002$ and 0.01. Inner core growth rate decreases monotonically with time, as the rate of planetary cooling decreases with the decreasing vigor of mantle convection and with the increasing concentration of light constituent in the outer core. If χ_o is greater than about 0.07, inner core growth does not begin until after 4.6 Gyr.

Figure 6 shows the evolution of lithosphere thickness ℓ which is not very dependent upon χ_o; present-day ℓ is about 180 km. This is smaller than previous estimates of ℓ (Schubert et al. 1979) in which lithospheric cooling was not moderated by the internal heat sources (e.g., mantle radioactivity) of the present model. The vigor of mantle convection is reflected in the Rayleigh number (Ra) and the relative thicknesses of the thermal boundary layers. The models begin with $Ra = 7 \times 10^5\ Ra_{cr}$; the boundary layers are 10 km thick. By 0.5 Gyr, $Ra = 10^4\ Ra_{cr}$ and the upper and lower boundary layers are 33 km and 31 km thick, respectively. At 4.6 Gyr, $Ra \approx 80\ Ra_{cr}$ and the boundary layers are about 100 km thick (only slightly dependent upon χ_o). While Mercury's mantle is probably convecting today, the flow is not vigorous.

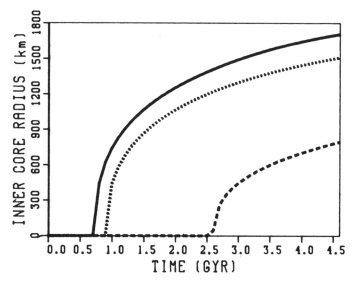

Fig. 5. Inner core radius as a function of time for three values of initial sulfur weight fraction. The solid, dotted and dashed lines are for $\chi_o = 0.002$, 0.01 and 0.05, respectively.

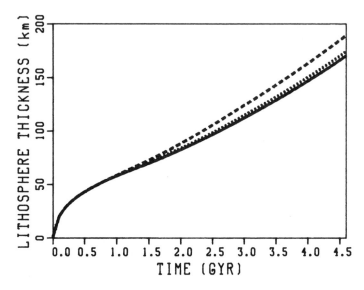

Fig. 6. Lithosphere thickness as a function of time for three values of initial sulfur concentration. Notation as in Fig. 5.

The calculated history of ΔR_p is shown in Fig. 7. Until inner core growth begins, ΔR_p primarily results from mantle cooling; this remains true for $\chi_o = 0.05$ up to 4.6 Gyr. For $\chi_o = 0.05$, the inner core comprises only about 10% of the total core volume at the present day. Smaller values of χ_o result in large inner cores and therefore large planetary contraction. Some unknown portion of Mercury's net ΔR_p is recorded in the formation of lobate scarps on the surface (Murray et al. 1974; Strom et al. 1975b; Gault et al. 1977; Strom 1979). Accordingly, it is of interest to relate Fig. 7 to the 1 to 2 km of planetary contraction associated with the scarps. We assume that the stabilization of Mercury's surface geology occurred after about 1 Gyr (subsequent to heavy bombardment, and perhaps the Caloris event, and extensive volcanism). Figure 8 shows ΔR_p since 1 Gyr as a function of χ_o. Based on a 1 to 2 km ΔR_p constraint, our model favors χ_o values greater than about 0.02. Models with less initial sulfur produce large inner cores and contraction greater than 2 km, in agreement with Solomon (1977).

Surface and core-mantle boundary heat flow are shown in Figs. 9 and 10, respectively. As with other terrestrial planets including the Moon, a portion of the surface heat flow is due to secular cooling (Schubert et al. 1979,1980). In the model represented by Fig. 9, about 15% of the present-day surface heat flow is due to secular cooling. Figure 10, which includes the conductive heat flow along the outer core adiabat, shows that the present-day heat flow out of the core is subadiabatic for all choices of χ_o. Therefore, convective motions in the core at the present time must be driven by chemical stirring due to inner

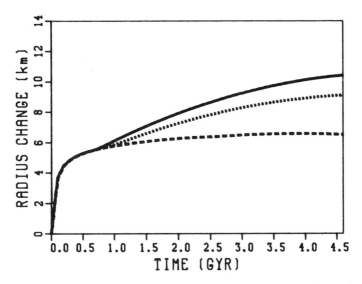

Fig. 7. Change in Mercury's radius due to inner core growth and mantle cooling vs time with initial sulfur concentration as a parameter. Notation as in Fig. 5.

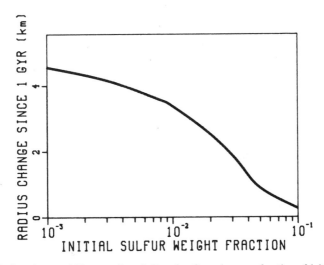

Fig. 8. Radius change of Mercury since 1 Gyr after formation as a function of initial sulfur concentration.

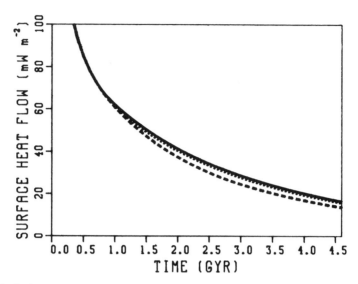

Fig. 9. Surface heat flow as a function of time for three values of initial sulfur concentration. Notation as in Fig. 5.

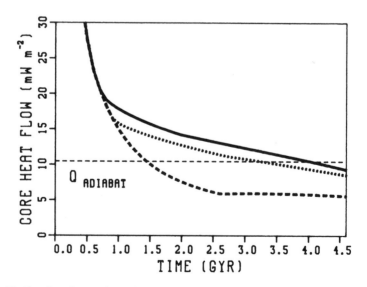

Fig. 10. Core heat flow vs time with initial sulfur concentration as a parameter. Notation as in Fig. 5. The conductive heat flow along the adiabat at the top of the outer core is indicated as $Q_{adiabat}$.

core freezing. Core heat flow is greater for smaller χ_o values as a result of larger $\dot{m} = 4\pi\rho_c R_i^2 \dot{R}_i$.

The results of this model differ in several respects from that of Stevenson et al. (1983): (1) we use a significantly "stiffer" mantle rheology; (2) a conducting lithosphere has been included; and (3) E_G has been explicitly calculated. The most important of these is the choice of rheology. While Stevenson et al. (1983) report a relatively thinner outer core and cooler mantle, our results do not differ qualitatively from theirs.

Figure 11 illustrates the evolution of the magnetic dipole moment μ (normalized to Earth's present value μ_E) based on Eq. (19). Early ($t < 1$ Gyr) in the evolution of each of these models (distinguished by different values of χ_o), before the start of inner core growth, the dynamo is generated by secular cooling, and core convection is thermally driven. Later, after the inner core begins to grow, a combination of gravitational and thermal energy drives the dynamo. For $0.03 < \chi_o < 0.07$, however, the heat flow out of the liquid core becomes insufficient to drive thermal convection before inner core growth begins, and dynamo action ceases for a time until inner core growth provides sufficient energy release (illustrated by the $\chi_o = 0.05$ case). Similar behavior takes place in the Venus models of Stevenson et al. (1983), and this may explain the lack of a present-day magnetic field for that planet. Dynamo action also ceases before 4.6 Gyr for $\chi_o < 0.001$. In this case, dynamo action shuts off because of insufficient gravitational energy release ($E_G \simeq 6$ kJ kg^{-1}) upon inner core freezing.

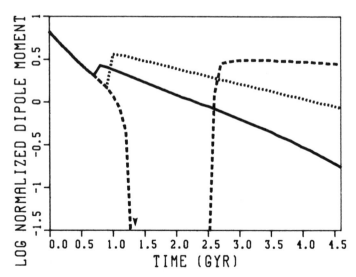

Fig. 11. Model magnetic dipole moment (normalized to the Earth's present-day model moment) as a function of time. Initial sulfur weight fraction is a parameter. The notation is the same as in Fig. 5. Arrow indicates cessation of dynamo action.

For the particular thermal model discussed here, the persistence of a magnetic field to the present day requires initial core sulfur weight fraction in the range $0.001 \leq \chi_o \leq 0.07$. This range is consistent with the initial sulfur concentration expected on the cosmochemical grounds previously discussed. The exact range of χ_o that results in a present-day dynamo is model dependent; most important is the mantle rheology. For example, a more Earth-like mantle rheology moves the limits on χ_o to $0.01 \leq \chi_o \leq 0.15$; a slightly stiffer mantle ($E_m = 60,000$ K) precludes present-day dynamo action for any χ_o.

The observed value of μ_M/μ_E is about 4×10^{-4} (Ness et al. 1974a,1975). Our results show that Mercury's magnetic field is considerably smaller than can be inferred from thermal modeling and the scaling discussed above, an observation used by Stevenson (1987) to propose thermoelectric magnetic field generation for Mercury.

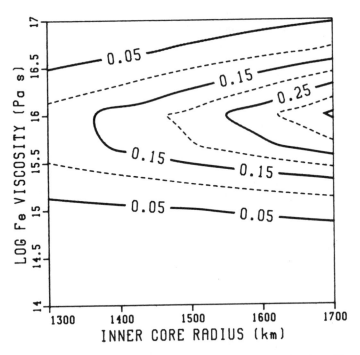

Fig. 12. Present-day tidal heating as a function of both current inner core viscosity and inner core radius. Contour values represent the heat flux in mW m^{-2} at the core-mantle boundary due to tidal dissipation in the solid inner core. The upper limit on core radius corresponds to $\chi_o = 0.002$, the lower limit to $\chi_o = 0.03$. The lower limit on viscosity corresponds to $A_{ic} = 4.5 \times 10^5$ m^2 s^{-1} (grain size = 0.5 mm) in Eq. (14b); the upper limit corresponds to $A_{ic} = 4.5 \times 10^8$ m^2 s^{-1} (grain size = 5 mm) in Eq. (14b).

VII. RESULTS OF THERMAL HISTORY MODELING: EFFECTS OF INNER CORE TIDAL DISSIPATION

Results of thermal models that include tidal heating in the inner core are shown in Fig. 12. The range of present-day R_i corresponds to the range of initial χ_o ($0.002 \leq \chi_o \leq 0.03$) and the range of present-day inner core viscosity corresponds to the range of Fe grain size (0.5 mm $\leq h \leq 5$ mm). The contours of present-day tidal heating are expressed in terms of heat flow at the core-mantle boundary. The tidal heat flow now existing is comparable to core heat flow with no inner core heat sources, only for a small range of inner core viscosity. In addition, only relatively large inner cores (and therefore small values of χ_o) generate significant heating. Inner core growth rate $\dot{R}_i$ and temperatures in the planet are not significantly altered from the previously discussed model; the inner core is slightly hotter as the result of a nonadiabatic temperature increase of less than 10 K across the inner core boundary layer.

The value of A_{ic} of about 6×10^7 m^2 s^{-1} (inner core grain size of 2.5 mm) results in the maximum amount of tidal heating. We have no way of knowing the grain size in Mercury's inner core but 2.5 mm is a plausible value. Figure 13 shows the percentage of current heat flow at the core-mantle boundary and surface that originates from tidal heating for $A_c = 6 \times 10^7$ m^2 s^{-1}. For $\chi_o = 0.002$, about 4% of the heat currently leaving the core is the result of tidal heating, less than the amount contributed by secular cooling. At the planet's surface, 2% of the surface heat flow might be the result of tides.

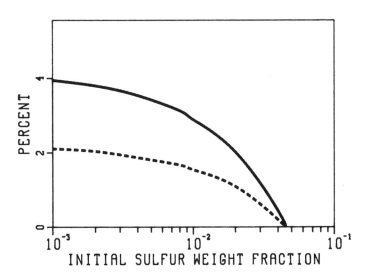

Fig. 13. Percent of present-day core heat flow (solid curve) and surface heat flow (dotted curve) contributed by inner core tidal heating as a function of initial core sulfur weight fraction. $A_{ic} = 6 \times 10^7$ m^2 s^{-1}.

These percentages would be larger if the actual gravitational energy release mechanism is less efficient than assumed.

Tidal heating, though relatively small, might affect the convection mechanism in the outer core. In Fig. 10, with no inner core heat source, the heat flow out of the core for $\chi_o = 0.01$ falls below the rate of heat loss conducted along the outer core adiabat after about 3 Gyr, and thermal convection in the outer core ceases. However, with tidal heating, thermal convection could be maintained for an additional 225 Myr.

VIII. FURTHER DISCUSSION OF THE IMPLICATIONS FOR MERCURY'S MAGNETIC FIELD

In the absence of any data directly relevant to Mercury's present thermal state, the observed magnetic field assumes particular importance as an indirect indicator and possible test of evolution models. This is especially true for any future mission to Mercury since it will then be possible to characterize the field geometry more fully, and to look for dynamical behavior (i.e., any change in the field since the Mariner 10 encounter). Unfortunately, we shall see that the interpretation of the existing data is sufficiently uncertain that no convincing test of thermal models is yet possible. Nevertheless, the thermal models suggest the preferred class of possible explanations of the field: those explanations which involve the existence of a liquid outer core. We justify this view below.

The magnetic field data are described at length in the Chapter by Connerney and Ness. Here it is sufficient to recall that the surface field appears to be dipolar with a typical strength ~300 nT and a dipole tilt ~10° relative to the rotational axis. The data are grossly insufficient for a complete description of the field geometry, however, because of the nature of the flyby trajectories. It is agreed that the field has an internal source, but beyond that lies little consensus. The possible explanations of the field are of three kinds: permanent magnetism, a hydrodynamic dynamo or "something else". We discuss each possibility in turn.

Permanent Magnetism

The Earth's magnetic field is predominantly produced in the core, but it includes a small crustal component ~10 nT, mostly associated with short-wavelength ($\lesssim 10^3$ km) variations in magnetization (Lowes 1974). The issue is whether one could imagine a contribution thirty times larger in the cold outer layers of Mercury. This is difficult for the following reasons:

1. It requires a thick, cold outer layer of Mercury at the time the magnetization was acquired. If we make the plausible assumption that metallic iron is the dominant magnetic mineral, then, for example, the shell would need to be over 100 km thick, contain ~5% free iron and be subjected to an

imposed field of 10^{-4} T (1 Gauss) in order for the present inferred magnetization to be explained (Stephenson 1976). This constraint is partially relaxed if the iron is very fine grained. Since the Curie temperature is $\sim$1040 K, a shell of this thickness is plausible for the present Mercury which has a diurnally averaged (subsurface) temperature $\sim$450 K and a conductive temperature gradient $\sim$5 K km^{-1} (see also Fig. 6 and the accompanying discussion of lithosphere thickness). However, the temperature gradient was larger in the past, and the shell may have been thin at the time when magnetization was possible.

2. If the remanence were acquired from an internal dynamo (a possibility despite the magnetostatic theorem of Runcorn [1975; see also Stephenson 1976; Srnka 1976]), then the dynamo must have had few or no reversals. These would reduce the net magnetization (Srnka and Mendenhall 1979).

3. If, instead, the field is imposed by an external source, then it must have been acquired very early in the history of the solar system, either from the primordial Sun or from the solar nebular field (presumably the field responsible for the observed remanence of many primitive meteorites). But this is precisely the epoch at which the temperatures within Mercury are least favorable for magnetization. No plausible accretion model exists in which Mercury forms cold; some of the recent ideas for Mercury formation involve complete melting and partial vaporization (Cameron 1985a; see also origins chapters). Most importantly, the time scale for decay of a primordial solar or nebular field would be $\leq$ 1 Myr, governed by the time necessary for the Sun to reach the main sequence, but the time scale for growth of a cold lithosphere on Mercury, governed by thermal diffusion, is $\gtrsim$ 100 Myr. Other (nonthermal) ways of acquiring remanence would not give a global field.

In conclusion, an explanation of Mercury's field using solely permanent magnetization is marginal at best, primarily because of the expectation that Mercury was never substantially colder than it is now, and was probably much hotter.

A Hydromagnetic Dynamo

In the years immediately after the Mariner 10 encounter, a dynamo explanation of the Mercurian field was often discussed but seemed difficult to support because of the belief that the iron core would freeze completely in $\leq$ 1 Gyr (see, e.g., Siegfried and Solomon 1974). Subsolidus convection would efficiently extract heat from the mantle and only exacerbate this problem (Cassen et al. 1976). However, the Earth's core has only frozen to a small extent because of the presence of light alloying constituents in the outer core which depress the freezing point. Sulfur, even in small amounts, can achieve the same effect inside Mercury, as the detailed models based on Stevenson et al. (1983) and discussed earlier in this chapter have demonstrated. The crucial point is that the self-regulated mantle temperature, determined by the physics of subsolidus convection, is higher than ~1300 K, the eutectic of the Fe-S system. The detailed models indicate that the outer liquid core of Mercury is currently about $3.1 \times 10^3 \chi_o^{1/2}$ km thick, provided $\chi_o \ll 1$. The cosmic abundance is $\chi_o \approx 0.3$; the terrestrial value could be as large as 0.1 (Ahrens 1979; Morgan and Anders 1979) but could also be ~0.03 (Ringwood 1977). A value of $\chi_o \gtrsim 0.01$ could plausibly result from the mixing of planetesimals across the terrestrial planetary zone during the accretion of Mercury (Wetherill 1985a). The question is: What outer core thickness is sufficient for dynamo action?

Unfortunately, dynamo theory lacks the predictive (quantitative) power to determine this. The following estimates are only suggestive. If the core is nearly adiabatic, as expected if the outer core is convective, then the conductive heat flow F_{cond} is about 11 mW m^{-2}. The detailed evolution models show that the actual present-day core heat flow is less (Fig. 10). This is possible if the temperature gradient is just slightly subadiabatic, since then the convective heat flux F_{conv} is negative. However, the second law of thermodynamics must be satisfied:

$$\psi \simeq \frac{\Delta T}{T} F_{conv} + F_{grav} \gtrsim 0 \qquad (20)$$

where ψ is the total dissipation per unit area and ΔT is the temperature drop across the outer core (Gubbins 1977a; Stevenson 1983). In order for $\Psi > 0$ and convection to be occurring at the present time, the detailed models of the previous section require 800 km $\gtrsim d \gtrsim$ 150 km. Moreover, application of Kolmogorov scaling (Golitsyn 1979) or simple mixing-length theory (Stevenson 1979) suggests that the convective velocity v_{conv} is roughly

$$v_{conv} \simeq 0.1 \ (\Psi/\rho)^{1/3} \qquad (21)$$

where ρ is the fluid density. If the relevant dimensionless number characterizing dynamo activity is $R_m \equiv v_{conv} \ d/\lambda$, where $\lambda \approx 1$ m^2 s^{-1} is the magnetic diffusivity of liquid iron (Stacey 1977b), then the plausible dynamo criterion $R_m \gtrsim 10$ (see, e.g., Moffatt 1978) is currently satisfied for an outer core thickness just slightly less than 800 km and slightly larger than 150 km. In other words, a dynamo is "easy" provided convection occurs at all (Stevenson 1983).

It seems possible that Mercury has a dynamo provided $0.001 \lesssim \chi_o \lesssim 0.07$. But what field strength is expected for this dynamo? The problem is that the field should be larger than its observed value as discussed earlier. This can be appreciated by the following simple argument (Gubbins 1977a; Stevenson 1983,1984). Since the magnetic diffusivity is the largest diffusivity in the fluid (i.e., much larger than the thermal diffusivity or the kinematic viscosity), ohmic dissipation should be the dominant contribution to Ψ. From Eq. (20) and the detailed models previously presented, the total present-day ohmic dissipation is of order 3200 d^3 (km) W. For a current density j, this should be of order Vj^2/σ where V is the volume enclosing the current and σ is the electrical conductivity. Assuming the current arises predominantly from the toroidal field B_T, as seems likely for most planetary dynamos (Moffatt 1978), it follows that

$$\Psi \ (W \ m^{-2}) \sim 10^3 \ B_T^2 \ R_c^2/(\mu_o^2 \ \sigma d) \tag{22}$$

where $\mu_o = 4\pi \times 10^{-7} NA^{-2}$ and the numerical constant should actually be $4\pi^3$ but is mostly empirical (it yields about the right answer for the Earth). As long as the outer core is convecting, it follows that for Mercury

$$B_T \ (T) \sim 6 \times 10^{-8} \ d^2 \ (km) \tag{23}$$

about three orders of magnitude larger than the observed poloidal field B_p for reasonable values of d. The problem is that B_T/B_p is only of order ten for Earth and it is difficult to understand why it would be enormously larger for Mercury. One possibility is that the part of the dynamo process responsible for B_T, presumably the action of differential rotation on B_p, is much stronger than the process generating B_p, presumably the action of small scale convection on B_T. The possibility that this might be related to Mercury's slow rotational rate does not seem likely. The differential rotation responsible for generating B_T should not increase greatly as planetary rotation decreases. The α-effect, which in simple models generates B_p, is insensitive to rotation until the rotation period drops below the convective overturn time, a circumstance that cannot be made self-consistent for Mercury. Another possibility is that Mercury is a different kind of dynamo than the other planets. Stevenson (1984) proposed that Mercury's dynamo is energy limited, meaning that it is unable to achieve the magnetostrophic state that is believed to be dynamically preferable, because the energy source is small. However, this argument still has difficulty avoiding the estimate of B_T given above. Perhaps the currents within Mercury are very complicated. Since the current is the curl of the field, the field magnitude could be accordingly small. None of these arguments is particularly satisfying and the dynamo explanation of Mercury's field is accordingly in some doubt. The most attractive aspects of the dynamo explanation are that it seems possible in principle, and (unlike permanent magnetism) it has no difficulty producing a large enough field.

Alternative Explanations

The history of planetary magnetism is replete with alternate hypotheses to the well established pair discussed above (Stevenson 1974, 1983). Many of these involve intrinsically weak phenomena (e.g., thermomagnetic or thermolectric effects) which offer no prospect of explaining large fields such as the Earth's field, but deserve more careful assessment for Mercury. One recent conjecture (Stevenson 1987) is a compromise between a true homogeneous dynamo and a purely thermoelectric effect. This theory, termed a thermoelectric dynamo, makes use of the thermoelectric electromotive force (emf) set up at Mercury's core-mantle boundary because this boundary is distorted by the mantle convection above. The resulting currents provide a purely toroidal field which can be distorted by outer core turbulence to produce a poloidal field. In this model, the current density is given as

$$j \sim \sigma_m Q \Delta T_{\ell cm} / R_c \qquad (24)$$

where σ_m is the mantle conductivity (since most of the voltage drop is in the low-conductivity part of the circuit), Q is the thermopower and $\Delta T_{\ell cm}$ is a typical lateral temperature variation on the core-mantle boundary. The associated toroidal magnetic field is

$$B_T \sim \mu_o \sigma_m Q \Delta T_{\ell cm} \qquad (25)$$

and the poloidal field is $B_p \sim R_m B_T$, where R_m is the magnetic Reynolds number for small scale turbulence. For the plausible choices of $Q \sim 10^{-3}$ V K^{-1}, $\sigma_m \sim 10^3 \, \Omega^{-1} m^{-1}$, $\Delta T_{\ell cm} \sim 1$ K and $R_m \sim 10$, one obtains $B_p \sim 10^{-5}$ T, about equal to the observed value. The estimate for $\Delta T_{\ell cm}$ is given by $h\beta$, where h is the typical amplitude of topography at the core-mantle boundary and β is the magnitude of the temperature gradient in the core. For the Earth, h is approximately a few km (Richards and Hager 1984; Gudmundsson et al. 1986) and a similar but somewhat smaller value should apply to Mercury. The value of β should be close to adiabatic, about 1 K km^{-1}. More detailed models (Stevenson 1987) reveal that the above estimates for B_T and B_p may be optimistic by an order of magnitude. However, it may still be possible to explain the observed field, provided the conductivity of the deep mantle is $\sim 10^3 \, \Omega^{-1} m^{-1}$, a value larger than that at the same pressure level in the Earth, but possibly comparable to that for the deepest mantle of the Moon (Hood and Sonnett 1982). This model may also predict a complicated geometry for the magnetic field; it is not clear whether the data preclude this. We need to return.

IX. SUMMARY AND DISCUSSION

Thermal evolution models suggest that permanent magnetism is a problematic explanation for Mercury's magnetic field. The dynamo mechanism remains an attractive possibility, but it seems capable of producing a field that is embarrassingly larger than that observed. Perhaps this merely indicates that we have an insufficient appreciation of the diversity of planetary dynamos. An alternative, speculative possibility is a thermoelectric dynamo. Either dynamo explanation requires a thin outer liquid core and provides support for that aspect of the thermal model described in this chapter. The existence of a liquid outer core means that the core must contain a small amount of sulfur. As little as a few tenths of a percent sulfur results in a fluid outer core that may be capable of dynamo action. On the other hand, more than about 7% sulfur precludes any inner core growth, and there is insufficient thermal energy to drive a dynamo in such a model at the present time. The amount of planetary contraction is also strongly affected by the amount of sulfur in the core. Less than about 2% initial sulfur results in planetary contraction greater than the value inferred from geologic considerations. An initial sulfur content $\geq 0.1\%$ facilitates dynamo action. A sulfur concentration of about 2 to 3% accommodates both the geologic and magnetic constraints.

Whether tidal dissipation in Mercury's inner core has played any role in the planet's thermal evolution depends sensitively on the unknown rheology of Fe in the inner core. In models which adopt Fe rheological parameter values that maximize tidal dissipation effects, it is found that inner core tidal heating can prolong the operation of a thermally driven outer core dynamo perhaps to the present day. Inner core tidal heating in such models can contribute up to 4% to the core heat flux and 2% to the surface heat flux at the present time, depending on the sulfur concentration in the core. Even under the most favorable of circumstances, inner core tidal dissipation has a negligible effect on the growth of the inner core.

The rheology of Mercury's mantle is also uncertain. We assumed a relatively stiff rheology compared to the Earth; other rheologies, even Earth-like, are possible. A softer mantle would have removed heat more effectively, resulting in a cooler planet and a larger present-day inner core. Conversely, Mercury could have stopped convecting at some point in its evolution if its rheology is stiff enough. In this case, inner core growth would have slowed, resulting in a smaller currently existing inner core.

The amount and distribution of radioactive elements in Mercury's mantle may be different from the state assumed in our model. Mercury may have received a larger concentration of uranium and thorium than the Earth (Basaltic Volcanism Study Project 1981) as assumed here, though this is at odds with the sulfur requirement. In addition, we did not address the issue of upward concentration of radioactive elements through differentiation. A model with 60% less mantle radioactive heat production has a cooler interior, less vigorous mantle convection and a slightly larger inner core than the model with a full complement of mantle radioactivity.

Although there are many uncertainties in the details of our model, we believe that its main features properly characterize Mercury's internal structure, thermal history and magnetic field generation: a hot start with early core differentiation followed by gradual cooling via mantle convection; the early formation of an inner core with growth continuing to date; a large present-day solid inner core and a thin liquid outer core containing a light element such as sulfur; early magnetic field generation driven by thermal convection in a largely molten core; present day magnetic field generation in the fluid outer core driven by release of gravitational energy and latent heat upon inner core growth.

THE ROTATIONAL DYNAMICS OF MERCURY AND THE STATE OF ITS CORE

S. J. PEALE

University of California at Santa Barbara

The dynamical evolution of Mercury's spin angular momentum controlled by the dissipative processes of tidal friction and relative motion between a solid mantle and liquid core is shown to lead naturally to the current state of rotation. This is a very ordered state where Mercury's spin angular velocity is exactly 1.5 times its mean orbital angular velocity (a spin-orbit resonance) and with the spin vector fixed in a frame precessing with the orbit and locked in a direction which deviates a small but significant amount from the direction of the normal to the orbit plane (a so-called Cassini state). If Mercury had a large molten core over much of its history, the initial spin rate may be constrained to a value less than twice the mean orbital angular velocity and the initial angular separation of the spin and orbital angular momenta (obliquity) to less than 90°. Theoretical arguments are not conclusive about the current existence of a molten core in spite of the detection of an intrinsic magnetic field. The need for a more precise knowledge of Mercury's interior that would constrain theories of planetary magnetic field generation and thermal history and that might constrain the initial rotation state strongly motivates direct observational determination of core properties. It is technically feasible to make precise measurements of the amplitude of the physical libration about the resonance spin rate, the obliquity, and the lowest degee gravitational harmonic coefficients to determine the existence and extent of a Mercurian molten core.

I. INTRODUCTION

The discovery that Mercury's rotation is not synchronous with its orbital motion by Pettengill and Dyce (1965) using the then new planetary radar at

Arecibo, Puerto Rico stimulated a flurry of activity among dynamicists in a rush to explain a phenomenon so contrary to the dogma of the day of how planetary spins should tidally evolve. Within a year of the discovery of the 59 day period, it had become well understood how Mercury's tidal evolution was modified by spin-orbit coupling to a nonsynchronous but commensurate spin (Colombo 1965; Goldreich and Peale 1966; Colombo and Shapiro 1966). Possible histories of Mercury's obliquity (the angle between the spin and orbital angular momenta) as determined by tidal dissipation were then described (Peale 1973; Ward 1975). These histories were based on a generalization of the theory of rotating planetary bodies, which treated dissipations from tidal and rotational distortion as perturbations of a conservative Hamiltonian analysis (Goldreich and Peale 1970; Peale 1972).

Apparently an initially rapidly spinning Mercury was slowed by tidal dissipation until it was captured into the observed rotation state, where the spin angular velocity is locked to 1.5 times the mean orbital angular velocity. Counselman (1969) pointed out that the probability of capture would be greatly enhanced if Mercury had a molten core. This idea received little attention until the Mariner 10 spacecraft revealed Mercury's intrinsic magnetic field (Ness et al. 1974a), which seemed to imply the existence of a conducting molten core (Ness et al. 1975a). The calculated enhancement of capture probabilities by such a molten core is very large. The fact that Mercury could have passed through the spin-orbit resonance with the spin locked to twice the orbital mean motion (44 days) places rather severe constraints on the core viscosity and the tidal dissipation function Q (Peale and Boss 1977a,b). The liquid core-solid mantle interaction also modifies the history of the obliquity (Goldreich and Peale 1970; Peale and Boss 1977b), but does not significantly change the final state. In this state the spin axis is fixed at a constant obliquity in the orbit frame of reference as the latter precesses about the normal to the plane of the solar system (Peale 1969,1973).

Although the existence of a liquid core has been inferred from the magnetic field (Ness et al. 1975a), early thermal history calculations had to be somewhat contrived in order to keep an initially molten core from cooling and solidifying before the present time (Siegfried and Solomon 1974; Cassen et al. 1976). This led Stephenson (1976) to propose remanent magnetism in a cool crust as the source of the measured field. The remanence could have been induced by a primordial dynamo which ceased to function as the core solidified, and an external dipole would persist if the permeability of the planet were nonuniform (Runcorn 1975; Srnka 1976). However, Stevenson (1987) has pointed out that frequent reversals of the dynamo-induced dipole and the likely thinness of a crust cool enough to establish the remanence may preclude ordered remanence sufficient for the observed field (see also the Chapter by Schubert et al.). It is possible for the outer regions of the core to remain molten until the present time if a small amount of sulfur is present in the initial melt (Stevenson et al. 1983; Chapter by Schubert et al.). Such sulfur would be concentrated in the outer liquid phase as iron crystallized onto the growing inner core, which would thereby depress the freezing temperature of the remaining liquid. The existence of such a liquid layer offers at least the possibility of a dynamo-generated field or a field due to currents generated thermoelectrically from temperature differences along the core-mantle boundary (Stevenson 1987).

Knowledge of whether or not Mercury has a liquid core is crucial to theories of the generation of magnetic fields in terrestrial-type planets and to theories of planetary thermal histories; it makes an experimental determination of the existence and extent of the Mercurian molten core of utmost importance. The fixed nonzero obliquity and the spin angular velocity fixed at 1.5 times the orbital mean motion comprise an ordered state where the obliquity depends on several planetary parameters. Measurement of the obliquity, the lowest degree gravitational harmonic coefficients, and the physical librations about the commensurate spin rate would allow the determination of the extent of any liquid core (Peale 1976). The importance of this determination to planetary science makes it a primary motivation for renewing our spacecraft investigations of this unique planet.

The purpose of this chapter is twofold: first, we review the descriptions of the dynamical evolution leading to Mercury's current rotation state as the natural outcome of known physical processes; second, we describe the experiment to measure the core properties and show that it is technically feasible. This experiment depends on Mercury's current occupancy of a so-called Cassini spin state (Peale 1969) where the spin axis is fixed in a frame precessing with the orbit at a constant obliquity, and on the spin rate which is locked to the value of 1.5 times the orbital mean motion except for physical libration. Section II outlines the development of the variational equations used to describe Mercury's rotational evolution. These equations are used in Sec. III to describe the possible evolution of Mercury's obliquity to its current value if the evolution is dominated by tidal dissipation. The tendency of the tides to bring Mercury toward smaller obliquities as the spin slows, which is established in Sec. III, implies that the obliquity was probably modest as the stable spin-orbit commensurabilities were approached. In Sec. IV it is shown that the original analysis for the capture of Mercury into the spin-orbit resonances, which assumed zero obliquity, is only slightly modified for the modest obliquities now expected on approach. Section V describes the modifications to the rotational history if Mercury has had a large molten core over much of its history. Here, severe constraints on initial conditions and/or core properties are indicated, where constraints on initial conditions are almost absent if tides alone controlled the evolution. Next, in Sec. VI we describe the experiment to measure the extent of a liquid core, establish the necessary precision of the measurements for the four parameters, and discuss a few schemes which have been proposed actually to acquire the data. The conclusions are summarized in Sec. VII, where we end by pointing out that Mercury is a unique planet in the solar system and should not be neglected in continuing spacecraft investigations.

II. VARIATIONAL EQUATIONS

A complete specification of the rotation state of a rigid planet requires six variables which together give the magnitude and orientation of the spin angular momentum relative to a coordinate system fixed in the orbit and relative to a system fixed in the planet. If the spin vector does not coincide with the axis of maximum moment of inertia, it will precess in a cone about this axis in the body system of coordinates. The changing position of the equator results in a periodic flexing of the body which dissipates energy and drives the spin toward coincidence with the axis of maximum moment, which is the minimum energy configuration for the given angular momentum. The time constant τ for damping this wobble for Mercury is given by (Peale 1973; Burns and Safronov 1973)

$$\tau = \frac{38}{5} \frac{\mu Q}{p a_e^2 \dot\psi^3} = 2 \times 10^7 \text{ yr} \tag{1}$$

where $\mu = 5 \times 10^{11}$ dyne cm^{-2} is the rigidity, $\rho = 5.4$ g cm^{-3} is the mean density, Q is the specific dissipation function for which we adopt a value of 100 which is typical for rocks on Earth, $\dot\psi = 2\pi/(58.65 \text{ day})$ is the current spin angular velocity and $a_e = 2439$ km is the equatorial radius. Since this time constant is short compared with the age of the solar system, we may assume Mercury has rotated about its axis of maximum moment of inertia throughout history and reduce the number of variables needed to describe the rotation state to four.

A coordinate system with origin at the center of mass is chosen and the Sun is viewed as being in a relative orbit about this point. Unit vectors $\mathbf{I},\mathbf{J},\mathbf{K}$, define an XYZ system of coordinates fixed in the orbit with $\mathbf{K}$ being parallel to the orbital angular momentum and $\mathbf{I}$ being directed toward the ascending node of the orbit on the plane of the solar system. Unit vectors $\mathbf{i},\mathbf{j},\mathbf{k}$ define an xyz principal axis coordinate system fixed in an undeformed Mercury with the z axis being that of maximum moment of inertia coincident with the spin axis and the x axis being that of minimum moment of inertia. The variational equations can then be written (Peale 1973)

$$\frac{d\alpha}{dt} = -\frac{\partial H}{\partial \psi}$$

$$\frac{d\psi}{dt} = \frac{\partial H}{\partial \alpha} - \frac{Z}{\alpha} \frac{\partial H}{\partial Z}$$

$$\tag{2}$$

$$\frac{dZ}{dt} = -\frac{1}{\alpha}\frac{\partial H}{\partial \Omega} + \frac{Z}{\alpha}\frac{\partial H}{\partial \psi}$$

$$\frac{d\Omega}{dt} = \frac{1}{\alpha}\frac{\partial H}{\partial Z}$$

where α is the magnitude of the spin angular momentum ($= \alpha \mathbf{k}$), ψ is the angle between $\mathbf{i}$ and $\mathbf{K} \times \mathbf{k}$ with the latter vector being along the ascending node of the equator plane on the XY plane, $Z = \mathbf{k} \cdot \mathbf{K}$ is the direction cosine of the spin angular momentum along the Z axis, and Ω is the angle between $\mathbf{K} \times \mathbf{k}$ and $\mathbf{I}$. (The angles Ω, $\cos^{-1}(Z)$, and ψ are just the Euler angles orienting the xyz system relative to the XYZ system.) The Hamiltonian H is written

$$H = \frac{1}{2}\,\boldsymbol{\alpha} \cdot [I]^{-1}\,\boldsymbol{\alpha} - \boldsymbol{\alpha} \cdot \boldsymbol{\mu} + V + V_T \tag{3}$$

where $[I]^{-1}$ is the inverse of the inertia tensor and $\boldsymbol{\mu}$ is the precessional angular velocity of the orbit.

$$V = -\frac{GM_{\odot}M_M}{a}\sum_{l=2}^{\infty}\left(\frac{a_e}{a}\right)^l\sum_{m=0}^{l}\sum_{p=0}^{l}\sum_{q=-\infty}^{\infty} G_{lpq}(e)\,F_{lmp}(Z)$$

$$\times\left[C_{lm}\binom{\cos v_{lmpq}}{\sin v_{lmpq}}\begin{matrix}l\text{-}m\ \text{even}\\ l\text{-}m\ \text{odd}\end{matrix} + S_{lm}\binom{\sin v_{lmpq}}{-\cos v_{lmpq}}\begin{matrix}l\text{-}m\ \text{even}\\ l\text{-}m\ \text{odd}\end{matrix}\right] \tag{4}$$

is the gravitational potential energy (less the central terms which have no effect on the rotation) of a point Sun of mass $M_{\odot}$ in Mercury's field where M_M is Mercury's mass, G is the gravitational constant, a is the semimajor axis of the orbit, $G_{lpq}(e)$ are expansions in the orbital eccentricity e, $F_{lmp}(Z)$ are functions of the obliquity $\cos^{-1} Z$ which have been defined and tabulated by Kaula (1966, pp.34,38), and C_{lm} and S_{lm} are harmonic coefficients of Mercury's gravitational field which are moments of the mass distribution.

Also,

$$v_{lmpq} = (2 - 2p)(\gamma - \Omega) + (2 - 2p + q)M + m(\pi - \psi) \tag{5}$$

where $M = n(t - t_0)$ is the mean anomaly with n being the mean orbital angular velocity, t the time and t_0 the time of perihelion passage, and γ is the angle from the X axis to the perihelion of Mercury's orbit. (See Peale [1973] for a more complete derivation of the above Eqs. 2 and 3.)

The lowest order secular term of the potential energy due to the distribution of mass caused by the tidal distortion are given by

$$V_T = - \frac{k_2 GM_{\odot}^2 a_e^2}{a^{*3} a^3} \sum_{m=0}^{2} \frac{(2-m)!}{(2+m)!} (2 - \delta_{0m})$$

(6)

$$\times \sum_{p,q} F_{2mp}(Z) F_{2mp}(Z^*) G_{2pq}(e) G_{2pq}(e^*) \cos(v_{2mpq}^* - \epsilon_{2mpq} - v_{2mpq})$$

where $k_2 = 1.5/[1 + 19\mu/(2\rho g a_e)] \approx 0.05$ is the potential Love number of degree 2 (see, e.g., Munk and MacDonald 1960) with ρ being the mean density, μ here the rigidity, g the acceleration of gravity at the surface and a_e the equatorial radius of Mercury. δ_{0m} is the Kronecker delta. The starred variables refer to $M_{\odot}$ treated as the mass causing the tide, and the unstarred variables refer to $M_{\odot}$ treated as the mass reacting to the tidal potential. The variables are equated in the arguments only after the partial differentiations with respect to the unstarred variables are performed in the variational Eqs. (2). The secular part of the tidal potential retained in Eq. (3) as V_T results solely from the dissipation of the energy stored in the tidal bulge. This dissipation is modeled by assuming a phase lag in the response of Mercury to the tide-raising potential due to the Sun, where Mercury being distorted by a periodic tide is analogous to a dissipative harmonic oscillator being forced at a frequency which is small compared to the natural frequency of the oscillator. This phase lag for a particular term in the expansion is represented by $\epsilon_{2mpq} = 1/Q_{2mpq}$ in Eq. (6), where Q_{2mpq} is the specific dissipation function (see, e.g., Munk and Mac-Donald 1960) appropriate to the particular term. The phase lag has the same sign as $\dot{v}_{lmpq}$. The high tide at a particular subsolar point on Mercury thus occurs slightly after the Sun passes over that point. The displacement of the double-ended tidal bulge from the Mercury-Sun line leads to the secular torque which tends to slow Mercury to a rotation state which is synchronous with its orbital motion. If Mercury's obliquity is not zero, the tidal bulge is displaced out of the orbit plane and there is a component of the torque perpendicular to the spin angular momentum. Both the obliquity and the spin magnitude thus have a history of tidal change.

The XYZ system is fixed in the precessing orbit (250,000 yr period), which leads to the term proportional to the precessional angular velocity μ in Eq. (3). With $\alpha = \alpha k$ and

$$\mu = -\mu[J \sin \iota + K \cos \iota]$$

(7)

where ι is the inclination of the orbit plane to the invariable plane and where, again, the X axis is along the ascending node of the orbit plane on the invariable plane, we can write

$$-\alpha \cdot \mu = \alpha\mu[-\sin \iota (1 - Z^2)^{1/2} \cos \Omega + Z \cos \iota].$$

(8)

Equations (2) with Hamiltonian (3) without the V_T term describe the rotation of Mercury considered as a rigid body. Generally, there will be rotational and tidal increments to the components of the inertia tensor and small changes in V due to the elastic distortion, some of which are contained in V_T. These effects are treated as perturbations to the rigid body motion, where any functional dependence of the increments in the inertia tensor are ignored when the partial derivatives are taken in Eqs. (2). When rotation is not about a principal axis, the rotational increments in the inertia tensor components lead to the dissipation which damps the nonprincipal axis rotation (free wobble). The additional potential due to the tidal distribution of mass leads to a secular change in the rotational angular momentum. For the persistent principal axis rotation assumed here, the rotational increments in the inertia tensor and in V are constant and are absorbed in the principal moments of inertia. The tidal increments in the inertia tensor are periodic and are neglected, so we can assume the inertia tensor remains diagonal with constant principal moments of inertia $A < B < C$.

For a fixed, circular orbit, the end point of tidal evolution is a rotational angular velocity which is synchronous with the orbital mean motion at an obliquity of zero. Both Mercury and the Moon are tidally evolved, yet the Moon has a nonzero obliquity (6°41′), and Mercury is not rotating synchronously with its orbital mean motion (Fig. 1). Both orbits are inclined with respect to a nearly invariable plane, and both orbits are eccentric. Solar torques on the Moon's orbit cause it to precess in a retrograde sense with a period of 18.6 yr, and planetary torques on Mercury's orbit cause such a regression with a 250,000 yr period. The orbital precession shifts the final tidally evolved state away from a zero obliquity. Tides try to bring the spin into coincidence with the orbit normal which is itself changing direction. The result for both bodies is the occupancy of a Cassini state ([Peale 1969], where the spin axis is fixed in a frame precessing with the orbit at a constant obliquity). For the Moon, the obliquity is the easily measurable 6°41′, whereas the obliquity of Mercury is less than 0°.5 and is currently bounded only from above (Klaasen 1976).

The large eccentricity of Mercury's orbit (0.206) allows stabilization of spin angular velocities which are nonsynchronous but commensurate with the orbital mean motion (Goldreich and Peale 1966; Colombo and Shapiro 1966). As we noted at the beginning of this chapter, Mercury currently occupies the spin-orbit commensurability for which the spin angular velocity is 1.5 times the orbital mean motion (Fig. 1). The stabilization of a nonsynchronous spin state follows from the fact that the solar torque on the permanently deformed planet varies as the inverse cube of the separation and is therefore maximal when Mercury is at the perihelion of its orbit. This maximal torque attempts to align the axis of minimum moment of inertia (long axis) with the direction to the Sun at perihelion. The commensurability insures that Mercury has nearly the equivalent orientation on subsequent passes of the perihelion to allow

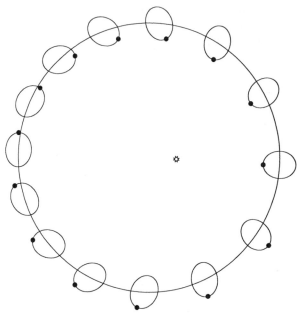

Fig. 1. Rotation of the planet Mercury during one orbital period beginning at the perihelion and
proceeding counterclockwise around the orbit. The spacing of the small ellipses representing
Mercury is at equal time intervals.

reinforcement of the alignment at each pass. Mercury's long axis thus librates
about the direction to the Sun at perihelion. In Secs. III and IV we describe
how Mercury's current rotation state follows from Eqs. (2) and (3) from almost any initial conditions and thereby establish the basis for using the current
dynamics to probe Mercury's interior, as discussed in subsequent sections.

III. EVOLUTION OF THE OBLIQUITY

In this section we concentrate on the possible historical changes in Mercury's obliquity under the assumption that tidal torques are the only outside
influence. It was pointed out in Sec. II that both the Moon and Mercury occupy Cassini spin states where the spin axis is fixed with a constant obliquity
in the XYZ frame precessing with the orbit. Necessary conditions for this
configuration are

$$dZ/dt = d\Omega/dt = 0. \tag{9}$$

In the absence of tides, or if Mercury is locked in a stable spin-orbit commensurability, α is conserved *on the average*, so $\partial H/\partial\psi = 0$ and the conditions in
Eqs. (9) are equivalent to

$$\partial H/\partial \Omega = \partial H/\partial Z = 0 \tag{10}$$

by the third and fourth of Eqs. (2). But if we average over short-period terms, H depends only on Z, Ω, α and ψ. The orbital parameters a and e are constant and the mean anomaly M disappears in the averaging process or is eliminated in resonant terms, whereas the perihelion position γ must be retained for phase relationships. Equations (10) then show that the Cassini states correspond to extremes in the averaged Hamiltonian under the condition of conserved angular momentum (Peale 1969). As such, we expect dissipative processes to drive the spin vector to one of these extremes. But there are either two or four extremes (Colombo 1966; Peale 1969; Beletski 1972) so what conditions determine the choice?

The Cassini states are determined for Mercury by keeping only the constant and long-period terms in H for $l = 2$ under the assumption that $\alpha/Cn = \dot{\psi}/n = 1.5$, where $n = \dot{M}$ is the orbital mean motion. The form of the averaged Hamiltonian $\langle H \rangle$ for other (nonresonant) spins will follow simply by setting the coefficients of the resonant terms to zero. The Hamiltonian so averaged is (Peale 1974)

$$\langle H \rangle = \frac{\alpha^2}{2C} + \alpha\mu[Z \cos \iota - (1 - Z^2)^{1/2} \cos \Omega \sin \iota]$$

$$+ \frac{GM_\odot}{a^3} \{(C - A/2 - B/2)F_{201}G_{210}$$

$$- \frac{B - A}{4} [F_{220}G_{201}\cos(2\gamma - 2\Omega - 2\psi_0) \tag{11}$$

$$+ F_{221}G_{213} \cos(2\psi_0)$$

$$+ F_{222}G_{225} \cos(-2\gamma + 2\Omega - 2\psi_0)]\}$$

where $C_{20} = -(C - A/2 - B/2)/M_M a_e^2$ and $C_{22} = (B - A)/4M_M a_e^2$ are the only two nonzero second degree harmonics because of our choice of principal axes and where (Kaula 1966)

$$F_{210} = \frac{1}{2} - \frac{3}{4} Z^2 \qquad G_{210} = (1 - e^2)^{-3/2}$$

$$F_{220} = \frac{3}{4} (1 + Z)^2 \qquad G_{201} = \frac{7}{2}e - \frac{123}{16}e^3 + \frac{489}{128}e^5 + \cdots$$

$$\tag{12}$$

$$F_{221} = \frac{3}{2} (1 - Z^2) \qquad G_{213} = \frac{53}{16}e^3 + \frac{393}{256}e^5 + \cdots$$

$$F_{222} = \frac{3}{4} (1 - Z)^2 \qquad G_{225} = \frac{81}{1280}e^5 + \cdots \quad.$$

We have replaced ψ by $1.5M + \psi_0$ in Eq. (11), where ψ_0 is the value of ψ when Mercury is at perihelion and is only slowly varying. Of the terms with $B - A$ as a factor, the first is far dominant for all obliquities less than about 100° and we can neglect the other two terms (Peale 1974). The conditions of the extremes in $\langle H \rangle$, which define the values of the variables which allow occupancy of the Cassini states, now become

$$\frac{\partial \langle H \rangle}{\partial \psi} = 0 = \frac{-GM_\odot}{a^3} \frac{(B - A)}{2} F_{220} G_{201} \sin(2\gamma - 2\Omega - 2\psi_0) \quad (13)$$

$$\frac{\partial \langle H \rangle}{\partial \Omega} = 0 = \alpha\mu(1 - Z^2)^{1/2} \sin \iota \sin \Omega$$

$$\qquad\qquad - \frac{GM_\odot}{a^3} \frac{(B - A)}{2} F_{220} G_{201} \sin (2\gamma - 2\Omega - 2\psi_0) \quad (14)$$

$$\frac{\partial \langle H \rangle}{\partial Z} = 0 = \alpha\mu \left[\cos \iota + \frac{Z}{(1 - Z^2)^{1/2}} \cos \Omega \sin \iota \right]$$

$$\qquad + \frac{GM_\odot}{a^3} \left[-\frac{3}{2} Z (C - A/2 - B/2) G_{2 i 0} \right. \quad (15)$$

$$\qquad \left. - \frac{3(B - A)}{8} (1 + Z) G_{201} \cos(2\gamma - 2\Omega - 2\psi_0) \right] .$$

Equation (13) requires $\gamma - \Omega - \psi_0 = 0$ or π, which means that the long axis of Mercury must be nearly aligned with the Sun when Mercury is at perihelion. This is just the condition for completely damped librations about the stable resonant angular velocity (see, e.g., Colombo 1965; Goldreich and Peale 1966). Equation (14) requires $\Omega = 0$ or π, which requires the spin angular momentum to be in the plane defined by the orbit normal **K** and the precessional angular velocity **μ**. These conditions together define the three Cassini laws for the Moon (see, e.g., Kaula 1968); Eq. (15) then determines the values of Z (i.e., the obliquity) corresponding to the Cassini states. The corresponding conditions for a nonresonant spin are obtained by omitting all the resonant terms, i.e., those containing the factor $B - A$, which would average to zero.

$\langle H \rangle$ and α are constants of the motion when $\dot\psi = 1.5n$ and, in the absence of tides, for all nonresonant spins. We can then construct the following function from the averaged Hamiltonian:

$$D = Z \cos \iota + Y \sin \iota - RZ^2 - S(1 + Z)^2 \quad (16)$$

where

$$D = \frac{\langle H \rangle}{\alpha\mu} - \frac{1}{2} \frac{\alpha}{Cn} \frac{n}{\mu} - \frac{R}{3} \quad (17)$$

is a constant of the motion with $Y = -(1 - Z^2)^{1/2} \cos \Omega = \mathbf{k} \cdot \mathbf{J}$ being the component of the normalized spin along the Y axis and

$$R = \frac{3}{4} \frac{Cn^2}{\alpha\mu} \frac{C - A/2 - B/2}{C} G_{210}(e)$$

$$S = \frac{3}{16} \frac{Cn^2}{\alpha\mu} \frac{B - A}{C} G_{201}(e) \quad \text{for} \quad \frac{\alpha}{Cn} = 1.5, 0 \leq \theta \leqslant 100°$$

$$S = 0 \qquad\qquad \text{for} \quad \frac{\alpha}{Cn} \neq 1.5, 0 \leq \theta \leq 180° \tag{18}$$

with $\theta = \cos^{-1}(\mathbf{k} \cdot \mathbf{K})$ being the obliquity. Since $X^2 + Y^2 + Z^2 = 1$, allowed trajectories of the spin vector in the frame precessing with the orbit are defined by the intersection of the parabolic cylinder defined by Eq. (16) with the unit sphere. For a given value of α, the position of the parabolic cylinder along the Y axis is determined by the value of $\langle H \rangle$. The limited range of positions which allow intersection of the parabolic cylinder with the unit sphere corresponds to the limited change in the energy effected by changing the orientation of Mercury in the gravitational field of the Sun. The spin is fixed in the orbit frame at points where the parabolic cylinder is just tangent with the unit sphere, so the tangent points determine the position of the Cassini states. (See Peale [1974] for a more complete discussion of the variation of parameters.) Figure 2

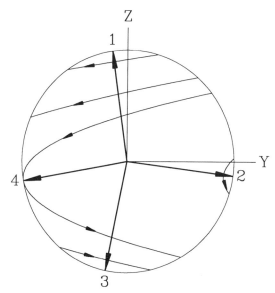

Fig. 2. Cassini states and representative precession trajectories of Mercury's spin angular momentum.

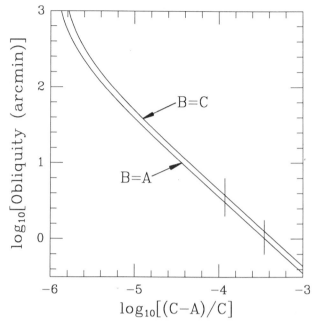

Fig. 3. Magnitude of Mercury's obliquity in Cassini state 1 as a function of $(C - A)/C$.

shows the positions of the Cassini states for parameters appropriate to Mercury but with projections on the XZ plane of representative precession trajectories having an exaggerated parabolic opening rate for clearer definition. (Figure 4 below shows the true parabola shapes.) The critical precession trajectory
of infinite period through state 4 is also shown. States 1, 2 and 3 are stable and
state 4 is unstable, as can be seen from the nature of the spin precession
trajectories near each state. Mercury apparently occupies state 1 with an obliquity very near zero degrees. Figure 3 shows the dependence of the obliquity
on the value of $(C - A)/C$ for Mercury in state 1 with the indicated range
marked by vertical lines estimated from perturbations of the Mariner 10 spacecraft (Anderson et al. 1987). (Figure 3 corrects about a factor 2 error in the
Fig. 6 of Peale [1969].)

From Fig. 2 we see that the spin axis generally traces out precession
trajectories which surround a stable Cassini state if the spin does not occupy
one of the states. Since the obliquity is no longer constant around a spin
precession trajectory, we must determine the effect of the tides on the obliquity by numerical integration. D in Eq. (16) is a function of the obliquity θ
through Y and Z. In principle, the time derivative of D could be integrated
using the general Hamiltonian by changing variables from the set used in Eqs.
(2) to the direction cosines of α, X,Y and Z which are used to define D. The
change in θ is then determined from the change in D. In practice, the family of

trajectories is sampled by determining the change in D for one complete circuit of a trajectory, beginning at a point in the YZ plane and ending when the trajectory has returned to the YZ plane going in the same direction. Then

$$\Delta D = \oint_0^\tau \frac{dD}{dt} \, dt \tag{19}$$

where τ is the period of the trajectory which is determined by first integrating the variations in the direction cosines around the trajectories with $V_T = 0$. Then

$$\left\langle \frac{d\theta}{dt} \right\rangle = \frac{\Delta\theta}{\tau} = \frac{\Delta D}{\tau(dD_i/d\theta_i)} \tag{20}$$

where D_i is the expression for the initial value of D in terms of the initial obliquity in the YZ plane, θ_i. This procedure is valid for both resonant and nonresonant rotations, as the effect of the change in α for the nonresonant rotation vanishes to first order (Peale 1974, Eq. 29). Since D is constant in the absence of the tides, the total variation in D is expressed in terms of the partial derivatives of the tidal potential alone:

$$\frac{dD}{dt} = \{[2RZ + 2S(1 + Z) - \cos \iota](1 - Z^2) + YZ \sin \iota\}$$

$$\times \frac{1}{\alpha(1 - Z^2)} \left. \begin{matrix} \dfrac{\partial V_T}{\partial \Omega} - Z\dfrac{\partial V_T}{\partial \psi} & \alpha/Cn \neq 1.5 \\[2mm] \dfrac{\partial V_T}{\partial \Omega} - \dfrac{\partial V_T}{\partial \psi} & \alpha/Cn = 1.5 \end{matrix} \right\} \tag{21}$$

which is substituted into Eq. (19) to evaluate ΔD and subsequently $\langle d\theta/dt \rangle$.

The drift of α from tidal dissipation is shown for four magnitudes of the spin in Fig. 4. For α/Cn approaching the current value of 1.5, the tides drive the spin toward either state 1 or state 2. However, for Mercury's parameters the spin must be inside the rather small region defined by the critical trajectory through state 4 before the tides will select state 2 which lies nearly in the orbit plane. For values of $\alpha/Cn > 2$ there is an intermediate trajectory passing between states 1 and 2 toward which the tides tend to drive the spin vector. This behavior is observed for the obliquity evolution of a planet with a fixed orbit (Goldreich and Peale 1970), and it results from the fact that the tide tends to reduce the component of the angular momentum that is perpendicular to the orbit plane faster than it reduces the component in the orbit plane when the spin is much greater than $2n$. This intermediate trajectory approaches Cassini state 1 as α/Cn approaches 2, and virtually all initial conditions drift toward state 1 for slower spins. State 3 with obliquity near $180°$ is always unstable to

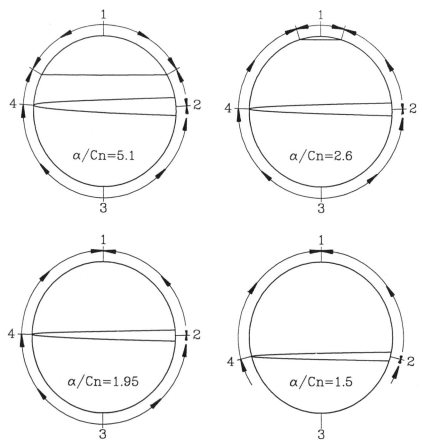

Fig. 4. Drift of Mercury's spin angular momentum vector under the action of tidal dissipation with $(C - A/2 - B/2)/C = (B - A)/C = 10^{-4}$.

the effects of tidal dissipation. Even if the obliquity were to start out near 180°, the spin would reach state 1 with almost unit probability. As the spin drifts away from state 3, it eventually traverses the lower part of the critical trajectory through state 4, which can be pictured as a very thin ellipse folded around the unit sphere with the tips just meeting at state 4. As the spin traverses the top part of the critical trajectory and approaches state 4, it can either reverse its direction, retrace a trajectory just inside the lower half, and drift toward state 2, or it can continue just above the critical trajectory in the same direction and drift toward state 1. This trajectory is noisy from the high-frequency terms which we have averaged away, so there may be many passes before the final choice is made. However, near the critical trajectory the rate of drift toward state 1 is about ten times faster than the rate of drift toward state 2. It is thus very unlikely for the spin to end up in state 2 near the orbit plane unless somehow placed very close initially. For almost all initial conditions, tidal dissipation would drive Mercury's spin to state 1 provided that the spin magnitude was not trapped into a spin-orbit resonance with $\alpha/Cn > 2$. From Fig. 4, we see that a stable spin resonance at, say, $\alpha/Cn = 2.5$ might have forced Mercury to precess forever around state 1 with a substantial average obliquity.

We turn now to the reason why Mercury, in fact, did not stop in any of the higher-order spin-orbit resonances even though they would have been stable. The explanation is reasonable if we maintain our assumption that tidal interaction is the only dissipative process, but perhaps not so reasonable when an alterative scheme is considered in the following sections.

IV. SPIN-ORBIT COUPLING

In the first treatments of spin-orbit coupling in the solar system (Goldreich and Peale 1966; Colombo and Shapiro 1966), it was assumed that the obliquity remained zero as tidal dissipation brought the spin rate toward its currently observed value ($\alpha/Cn = 1.5$). We have seen in the last section that the obliquity must eventually be driven to Cassini state 1 with a very small obliquity as α/Cn decreases below 2, so it is likely that Mercury approached at least the 1.5 resonance with a small obliquity and it probably passed through a few higher order resonances with θ not too far from zero in spite of the fact that the tides cause a drift toward nonzero obliquities for $\alpha/Cn > 2$ (see Sec. III). The formalism of Eqs. (2) includes arbitrary obliquities and we shall see below that the original treatments are not substantially altered by a modest obliquity.

The spin-orbit coupling analysis follows from the first of Eqs. (2), where again only $l = 2$ terms are kept, $C_{21} = S_{21} = S_{22} = 0$ from the choice of principal axes and only terms with coefficient C_{22} remain after differentiation of V with respect to ψ. We write $\psi = p'M + \psi_0$, where p' is a half integer, and keep only those terms coming from V with arguments which can be slowly varying to yield

$$\frac{d\alpha}{dt} = -\frac{2GM_\odot M_M a_e^2}{a^3} C_{22} \sum_q F_{220} G_{20q} \sin[2(\gamma - \Omega)$$
$$+ (2 + q - 2p')M - 2\psi_0]$$
$$+ F_{221} G_{21q} \sin[(q - 2p')M - 2\psi_0]$$
$$+ F_{222} G_{22q} \sin[-2(\gamma - \Omega)$$
$$+ (-2 + q - 2p')M - 2\psi_0]$$
$$+ \frac{\partial V_T}{\partial \psi} . \tag{22}$$

For a zero obliquity ($Z = 1$), $F_{221} = F_{222} = 0$ (Eqs. 12), and only the first term in the sum survives. On the other hand, for the current resonance, $p' =$

1.5 which requires $q = 1$ to eliminate M from the first argument, $q = 3$ in the second argument, and $q = 5$ in the third (to select the slowly varying terms). But $G_{lpq} = O(e^q)$ in Eq. (22), such that the last two terms have very small coefficients for $p' = 1.5$ and can be neglected. Similarly, for $p' = 2$, the slowly varying arguments have $q = 2$ in the first term, $q = 4$ in the second, and $q = 6$ in the third. Thus, for all the spin orbit resonances with $p' > 1.5$, the first term in the sum over q in Eq. (22) is far dominant except for obliquities very close to 180°. ($\theta > 90°$ for positive p' actually corresponds to retrograde rotation.) As Mercury would have been driven toward obliquities approaching Cassini state 1 over its entire history, reasonably small obliquities should have prevailed during the encounters of all important spin orbit resonances, and we can keep only the first term in the sum over q in Eq. (22). Now $\dot{\Omega}$ and $\dot{\theta} << \dot{\psi}$ and have negligible accelerations. Hence, $\alpha \approx C(\dot{\Omega} + \dot{\psi})$ for small obliquities and $d\alpha/dt$ is approximately C times the second derivative of the slowly varying argument of the sine. If we write

$$\beta = \psi_0 + \Omega - \gamma \tag{23}$$

which is nearly the angle between the axis of minimum moment of inertia (x axis) and the direction to the Sun when Mercury is at perihelion, then Eq. (22) is well approximated for small obliquities by

$$C\ddot{\beta} + \frac{(B - A)F_{220}G_{20|2p' - 2|}n^2}{2} \sin 2\beta = \frac{\partial V_T}{\partial \psi} \tag{24}$$

which is just the pendulum equation appropriate to a given resonance used by Goldreich and Peale (1966), except for the slight reduction in the restoring torque from $F_{220}(Z < 1)$. We have replaced C_{22} by $(B - A)/4M_M a_e^2$ and $GM_\odot/a^3$ by n^2 in Eq. (24). This equation is applicable for both libration within a resonance and circulation near the resonance.

As described by Eq. (24), a given resonance is stable if the maximum restoring torque on the equivalent pendulum derived from V (coefficient of sin 2β in Eq. 24) exceeds the tidal torque $\partial V_T/\partial \psi$. For a zero obliquity and a circular orbit

$$\frac{\partial V_T}{\partial \psi} = \langle T \rangle = -\frac{3k_2 GM_\odot^2 a_e^5}{2a^6 Q} \tag{25}$$

where $1/Q = \epsilon_{2200}$ has been used. With $k_2 = 0.05$, $C_{22} = (B - A)/4M_M a_e^2$ and $C \approx 0.33 M_M a_e^2$, a resonance is stable against disruption by the tides if (Goldreich and Peale 1966)

$$\frac{B - A}{C} > \frac{7 \times 10^{-8}}{Q|G_{20|2p' - 2|}(e)|}. \tag{26}$$

Since $(B - A)/C = O(10^{-4})$ (Anderson 1987) and with $Q = 100$, an eccentricity of 0.206 implies that all spin-orbit resonances with $p' \leqslant 4.5$ should be stable. As it is likely that Mercury passed through some of these resonances, we can understand its escape from these resonances and its ultimate capture in the resonance for which $p' = 1.5$ in terms of a probability of capture as it passes through a resonance.

The first integral of Eq. (26) is

$$\frac{1}{2}\dot{\beta}^2 - \frac{1}{4}\frac{(B-A)}{C}F_{220}G_{20|2p'-2|}n^2 \cos 2\beta = \int \frac{\langle T \rangle}{C}d\beta + E_0 = E \quad (27)$$

where $\langle T \rangle = \partial V_T / \partial \psi$ is the averaged tidal torque, E_0 is a constant of integration and E is referred to as the energy. A value of E equal to the coefficient of the cosine in Eq. (27) separates rotation from libration. As $\langle T \rangle < 0$, the approach to the resonance is characterized by E approaching this critical E from above. In Fig. 5 the approach to a resonance is shown in a schematic plot of $\dot{\beta}^2/2$ versus β. The maximum value of β is reached after the equivalent pendulum goes over the top of its support for the last time before reversing direction in a libration. If the magnitude of the mean tidal torque is decreasing with β, the plot of $\dot{\beta}^2/2$ after $\dot{\beta}$ has changed sign will fall below that before the change in sign as shown in Fig. 5. There is then a chance for a second zero in

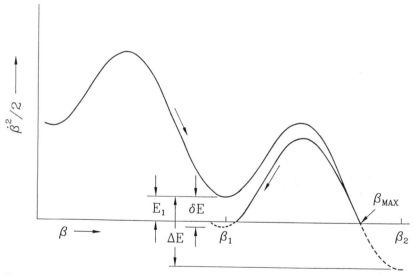

Fig. 5. Schematic diagram of the relative specific kinetic energy of Mercury's rotation as a function of phase as a spin-orbit resonance is encountered. The diagram illustrates capture into a resonance. In the actual case, reversals of $\dot{\beta}$ will initially occur very near the minima marked by β_1 and β_2. (Figure after Goldreich and Peale 1966.)

$\dot{\beta}$ and a subsequent damping of the librations of the spin about the resonant value. Whether or not there will be a second zero in $\dot{\beta}$ depends on the value of E_1, which is the value of $\dot{\beta}^2/2$ as the equivalent pendulum goes over its support for the last time before reversing direction. We cannot know the value of E_1 *a priori*, but if we assume that the probability of finding E_1 anywhere in the range from 0 to its maximum value of ΔE in Fig. 5 is uniform, then the probability of capture into the resonance is given by

$$P = \frac{\delta E}{\Delta E} \qquad (28)$$

where ΔE is the magnitude of the decrease in $\dot{\beta}^2/2$ in a single rotation of the equivalent pendulum, and δE is the difference in the magnitudes of $\dot{\beta}^2/2$ as the equivalent pendulum passes over the top the last time before $\dot{\beta}$ reaches 0 and as it would pass over the first time in the reverse direction. Figure 5 illustrates a capture.

The magnitude of the mean tidal torque decreases with $\dot{\beta}$ for all models which have been used extensively (see, e.g., Goldreich and Peale 1966), so a capture probability can always be calculated. The specific dissipation function Q strictly applies to simply periodic oscillations. The periodic tidal torque is expanded into a series of simply periodic terms whose arguments have unique frequencies, and a Q is associated with each term. This expansion is the derivative of Eq. (6) with respect to ψ, after which differentiation the starred and unstarred variables are set equal. With $1/Q \propto$ frequency, the magnitudes of all the terms in the sum decrease continuously as the spin angular velocity is reduced. For Q constant, the magnitude of a term does not change as the frequency is reduced, but the term changes sign with its frequency. The total mean tidal torque is reduced in steps as the frequencies of successive terms change sign as the spin is reduced. (See Goldreich and Peale [1966] for a discussion of these and other tidal models.) The capture probabilities calculated for the above two tidal models are shown in Table I.

The actual values of the capture probabilities are seen to be strong functions of the particular tidal model assumed, but they have the common property of getting larger as the order of the resonance is reduced. For each tidal model, the resonance for $p' = 1.5$ has the largest capture probability of all those resonances presumed to have been encountered previously. The zero capture probability for $p' = 1$ (in the $1/Q \propto$ frequency case) arises because the average tidal torque vanishes *before* a rotation rate synchronous with the orbital mean motion is reached in the highly eccentric orbit (see, e.g., Goldreich and Peale 1966). Therefore, the occupancy of the 3/2 spin-orbit resonance by Mercury in spite of the stability of higher order resonances can be understood in terms of the increasing capture probability as the order of the resonance is reduced.

We have arrived at Mercury's current rotation state without requiring any special circumstances and with essentially no restrictions on the initial condi-

TABLE I
Capture Probabilities as a Function of the Resonance for $e = 0.2$[a]

p'	$1/Q$ = Constant	$1/Q \propto$ Frequency
2.5	0.03	0.0066
2.0	0.15	0.016
1.5	0.73	0.067
1.0	1.0	0.0

[a]Table after Goldreich and Peale 1968.

tions. Its current spin rate, which is commensurate but not synchronous with its orbital mean motion with an obliquity near zero, and consistent with occupancy of Cassini state 1, is the natural outcome of tidal evolution in an eccentric, precessing orbit. This highly ordered rotation may provide a means of making a unique geophysical measurement, the determination of the existence and extent of a Mercurian liquid core. At the same time, the possible existence of a liquid core requires a re-examination of the rotation history. We turn to these topics in the next section.

V. WHAT IF THERE IS A LIQUID CORE?

The Mariner 10 flyby of Mercury revealed a distinct bow shock wave and magnetopause characteristic of the interaction of the solar wind with an intrinsic magnetic field of the planet (Ness et al. 1974a; see also Chapters by Connerney and Ness, and Russell et al.). Except for a very slight chance that the intrinsic field is the result of remanent magnetism in a cool crust (Stephenson 1976), the field is thought to be most likely due to motions in a conducting fluid core. These motions could generate the field either by dynamo action (Ness et al. 1975a) or by thermoelectric currents at the core-mantle boundary (Stevenson 1987). Although the existence of such a molten core has not been ascertained otherwise (see Sec. VI), and some thermal history calculations have questioned whether it could have been preserved over geologic time scales (Plagemann 1965; Majeva 1969; Siegfried and Solomon 1974; Fricker et al. 1976; Cassen et al. 1976), the magnetic field evidence is sufficiently strong that it is imperative that we investigate the dynamical consequences of a molten core on the rotation history of the planet.

If there is slippage between the liquid core and the mantle, a second source of dissipation of rotational energy results which has an effect added to that of the tide. There would always be some relative motion between the core and mantle during the periodic variation in the spin angular velocity of the mantle from the gravitational torque on the planet due to the axial asymmetry. The added dissipation of the libration about resonant angular velocities leads to drastically enhanced probabilities of capture into the spin-orbit resonances

(Goldreich and Peale 1967; Counselman 1969; Peale and Boss 1977a). If Mercury is not in a Cassini state, the resulting precession of the spin angular momentum about the Cassini state in the frame precessing with the orbit can also lead to relative motion between core and mantle. The same gravitational torque on a rotationally or otherwise distorted planet causing the mantle to precess also acts on the molten core, which is at least rotationally distorted. However, the torque on the core is insufficient to cause it to precess at the same rate as the mantle (Toomre 1966), and a first approximation is to neglect the torque on the core altogether. This assumption does not guarantee a significant relative core-mantle motion, since pressure coupling forces nearly coprecession unless the core ellipticity $\epsilon < \dot{\Omega}/\dot{\psi}$ (Toomre 1966), where the ellipticity arises from the rotation. This condition reduces to

$$\frac{\dot{\psi}^2}{\pi G \rho} < \frac{|\dot{\Omega}|}{\dot{\psi}} \approx \frac{3}{2} \frac{n^2}{\dot{\psi}^2} \frac{(C - A/2 - B/2)}{C} |\cos \theta| \qquad (29)$$

where ρ is the core density. With $1.2 \times 10^{-4} \lesssim (C - A/2 - B/2)/C \lesssim 3.4 \times 10^{-4}$ (Anderson et al. 1987), this condition is satisfied for obliquities far from 90° if the rotation period is greater than 16 to 21 days, which includes most of the spin-orbit resonances that are likely to be stable (Eq. 26). Equation (29) is written for a fixed orbit, but Mercury's orbit precesses so slowly that the precession rate around Cassini states 1 and 3 is nearly what it would be about the orbit normals if the orbit were fixed. For a motion near Cassini state 2 (near $\theta = 90°$), there is little relative motion between the core and mantle due to precession, as the precession rate is very slow (Eq. 29).

For obliquities slightly displaced from ~90°, any differential motion between the core and mantle due to precession will tend to drive the spin angular momentum to an orbit normal for a fixed orbit, and to Cassini states 1 or 3 for the precessing orbit of Mercury. The torque on the planet causing the precession lies in the orbit plane, such that the normal component of the angular momentum is conserved. The rotational energy dissipated by the differential core-mantle motion must then come entirely from the component of spin angular momentum lying in the orbit plane. The obliquity is driven toward Cassini state 1 near $\theta = 0°$ if $\theta < 90°$ and toward Cassini state 3 near $\theta = 180°$ if $\theta > 90°$. The high mean density of Mercury leads in the models of the interiors to a rather large core of iron alloy with a radius about three-fourths the radius a_e. This means that the core-mantle interaction should easily dominate the tides and drive Mercury to Cassini state 3 if the obliquity were ever greater than 90°. This mechanism has been invoked as a way to stabilize the obliquity of Venus at the observed value near 180° (Goldreich and Peale 1970). The fact that Mercury's obliquity is now very small implies that it was never greater than 90° *if* there were a primordial liquid core. For obliquities less than 90°, a liquid core-solid mantle interaction would accelerate the drift to Cassini state 1. For likely core properties, the time constant for the decrease of $\tan \theta$ is only about 10^5 yr (Peale and Boss 1977b). The presence of such a core would thus force the tidal evolution of the spin magnitude through the series of spin-orbit resonances to occur at very small obliquities, and we are justified in making this assumption in the investigation of the effect of the molten core on the capture probabilities, or as we shall see, to establish the constraints on the core properties which allow Mercury to reach its present spin from substantially higher values.

A core-mantle interaction was invoked by Goldreich and Peale (1967) to allow capture of Venus into a possible spin resonance with the orbital motion of the Earth (a resonance now known not to be occupied). A substantial probability of capture in this resonance was obtained for a range of core viscosities, whereas the capture probability due to tidal effects alone was negligible. The results of this application to Venus imply that the relatively much larger core of Mercury should lead to unit capture probability into some resonances for some range of core viscosities. Counselman (1969) has verified this hypothesis by showing that unit capture probabilities can result from a core-mantle interaction for the resonances corresponding to $p' = 1.5$, 2 and 2.5. If capture of Mercury into a resonance above that currently occupied, say $p' = 2$, is certain for a range of kinematic viscosity ν, then ν must lie outside that range or Mercury's initial spin angular velocity $\dot{\psi}$ was less than $2n$. Otherwise, it could never have reached the $p' = 1.5$ resonance where it is currently found. Since capture into a spin-orbit resonance is more difficult as the order of the resonance is increased, we expect certain capture into the $\dot{\psi} = 2n$ resonance for a wider range of viscosities than for any higher-order resonance. Escape from the $\dot{\psi} = 2n$ resonance therefore puts the most severe constraints on the viscosity of a Mercurian molten core.

The equations describing the rotations of the mantle and core follow from Eq. (23):

$$C_m \ddot{\beta}_m = -C_m \omega_0^2 \sin 2\beta_m + \langle T \rangle - \kappa(\dot{\beta}_m - \dot{\beta}_c) \tag{30}$$

$$C_c \ddot{\beta}_c = \kappa(\dot{\beta}_m - \dot{\beta}_c) \tag{31}$$

where

$$\omega_0^2 = \frac{3}{2} \frac{B_m - A_m}{C_m} n^2 G_{20|2p' - 2|}(e) \tag{32}$$

and where we have added a core-mantle interaction proportional to the difference in angular velocities. The subscripts m and c refer to mantle and core, respectively. The torque on the core due to the external field is neglected, which is justified because the slow rotation of Mercury near $\dot{\psi} = 2n$ ensures that a hydrostatic core is nearly spherical. The form of the core-mantle coupling with constant κ is consistent with a viscous coupling between a sphere and a shell.

Equations (30) and (31) (with a different form of ω_0^2) have been used to determine expressions for the probability of capture into a synodic spin orbit resonance between Venus and Earth under the assumption of strong ($\kappa/C_c\omega_0 \gg 1$) and weak ($\kappa/C_c\omega_0 \ll 1$) coupling. Capture probabilities are small for very strong coupling since there is relatively little motion between the core and mantle with subsequently little energy dissipation. In the limit, the core is solid with no relative motion and no dissipation. Capture probabilities are also small for very weak coupling since the mantle moves almost independently of the core with little energy dissipation. The limit is a zero viscosity core and no dissipation. The capture probability is maximal for intermediate viscosities with maximum energy dissipation. Capture probabilities for the Venus synodic resonance at intermediate values of ν were determined by Monte Carlo numerical techniques. The relatively large core of Mercury, together with the much greater stability of the resonances, means that the capture probabilities reach unity well within the approximations of strong or weak coupling, and numerical calculations are unnecessary. We can also keep $\langle T \rangle$ constant in Eq. (30) since the tidal contribution to the determination of $\kappa/C_m\omega_0$ for which the capture probability just reaches unity is negligible (Peale and Boss 1977a).

The capture probability expressions derived for Venus can be applied directly to Mercury if we use the restoring torque appropriate to Mercury (Eq. 30) in place of that for Venus (see Goldreich and Peale 1967, Eqs. 21 and 27):

$$P = 2\left[1 - \frac{3\sqrt{2}\pi\langle T \rangle \kappa(1 + C_c/C_m)^{3/2}}{8\omega_0^3 C_c^2} \right]^{-1} \quad , \quad \frac{\kappa}{C_m\omega_0} \gg 1 \quad (33)$$

$$P = 2\left[1 - \frac{\sqrt{2}\pi\langle T \rangle}{4\omega_0\kappa(1 + C_c/C_m)} \right]^{-1} \quad , \quad \frac{\kappa}{C_m\omega_0} \ll 1 \quad (34)$$

where the definition from Fig. 5 and Eq. (28) is used. We can relate κ to the kinematic viscosity of the core material in the case for which the boundary layer at the core-mantle interface is laminar by the following argument. If we ignore the first two terms on the right-hand side of Eq. (30), Eqs. (30) and (31) can be solved to yield an exponential decay of the differential angular velocity in terms of κ. Since the time constant for the equilibration of the angular velocities of a spherical container and the fluid inside are known in terms of the kinematic viscosity, we can equate the time constant derived from Eqs. (30) and (31) with the latter time constant to yield κ as a function of ν. There results for a laminar boundary layer,

$$\frac{C_c C_m}{\kappa(C_c + C_m)} = \frac{r_c}{(\nu\dot{\psi}_m)^{1/2}} \quad , \quad \frac{\dot{\psi}_m r_c^2}{\nu} \gg 1$$

$$\frac{C_c C_m}{\kappa(C_c + C_m)} = \frac{r_c^2}{\nu} \quad , \quad \frac{\dot{\psi}_m r_c^2}{\nu} \lesssim 1 \quad (35)$$

where the time constant for the first of Eqs. (35) (Greenspan and Howard 1963) is derived for no density stratification in the core and depends on circulation of core fluid through the Ekman boundary layer whose thickness is small compared to the core radius r_c. The second of Eqs. (35) is the ordinary viscous diffusion time constant, and it is appropriate when v is so large that the boundary layer is no longer small compared to r_c.

The right-hand side of the first of Eqs. (35) is appropriate for a completely fluid core, whereas the left-hand side assumes the core behaves as a rigid body. Differential velocities in the core make the equality in Eq. (35) only approximate. The relation between v and κ in Eq. (35) should define any constraints on v to within a factor of a few, but the approximation should be noted by the reader. If the solid inner core was large at the time of passage through the $2n$ spin-orbit resonance, the mass of the fluid shell could have been a relatively small fraction of the total. The model of core coupling in Eqs. (30) and (31) is then a better approximation, and we can calculate κ in terms of v directly from the stress across the Ekman boundary layer of thickness $(v/\dot{\psi}_m)^{1/2}$ (Greenspan and Howard 1963). The liquid shell is generally much thicker than the Ekman layer (chapter by Schubert et al.). To determine κ, we assume the bulk of the fluid layer to be rotating as a rigid body at angular velocity $(\dot{\psi}_m - \dot{\psi}_c)/2$, write the increment of torque due to the viscous stress at a point on the spherical boundary of radius r_c, and integrate the torque increments over the sphere. There results

$$\kappa = \frac{4\pi}{3} \rho_f r_c^4 (v\dot{\psi}_m)^{1/2} \tag{36}$$

where ρ_f is the density of the fluid in the shell. The dependence of κ on v and $\dot{\psi}_m$ in Eq. (36) is identical to that in the first of Eqs. (35), as it should be since the viscous stress is exerted across the same Ekman layer in both cases. If we assume $r_c = 0.75a_e$, $C_c/C_m = 1$ and $\rho_f = 7$ g cm^{-3}, the coupling for a given value of v is stronger in Eq. (36) by a factor of about 1.8 over that in the first of Eqs. (35).

The shaded region of Fig. 6 gives the ranges of v and Q, for which the probability of capture into the resonance with $\dot{\psi} = 2n$ is unity when $(B - A)/C = 10^{-4}$ (Peale and Boss 1977a), where the value of $(B - A)/C$ is approximately that estimated by Anderson et al. (1987), the average value of $e = 0.17$ (Cohen et al. 1973), and $C_c/C_m = 1$ are assumed. The left boundary of the region corresponds to the weak coupling limit and the right boundary to the strong coupling limit of Eqs. (35). The left boundary of the shaded region is moved to the left by a factor of 1.8^2 if the coupling constant from Eq. (36) is used instead of the first of Eqs. (35). The expansion of the range of viscosity leading to $P = 1$ as Q increases is due to the fact that δE in Eq. (30) is due entirely to the core-mantle interaction in Eqs. (30) and (31), since we assumed $\langle T \rangle$ is constant, whereas ΔE is due entirely to the decrease in the average value

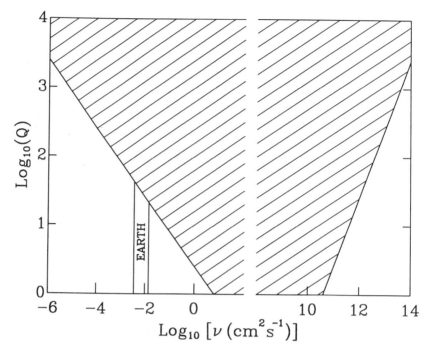

Fig. 6. Values of the tidal Q and the core viscosity ν for certain capture of Mercury into the $2n$ spin-orbit resonance. (Figure after Peale and Boss 1977a.)

of $\dot{\beta}_m$ from $\langle T \rangle$. Since ΔE is reduced as Q is increased, a smaller δE suffices for the same capture probability (Eq. 28), which allows a smaller value of ν in the weak coupling limit and a larger value of ν in the strong coupling limit. The range of ν for $P = 1$ also expands as $(B - A)/C$ or e is increased (Peale and Boss 1977a).

The range of kinematic viscosity ν for which $P = 1$ for any other spin-orbit resonance, including 1.5n, is easily inferred from Fig. 6. From Eqs. (6), (32) and (35), we see that the changes in the expressions for the capture probabilities in Eqs. (33) and (34) for a change in the resonance involve changes in ω_0, $\langle T \rangle$ and, for weak coupling only, κ. Since P maintains the same functional dependence on ν for any resonance, changing the resonance for which P is being determined just involves changing the factor by which ν^ζ is multiplied in the denominators of Eqs. (33) and (34), where $\zeta = -1/2$ for weak coupling and $\zeta = 1$ for strong coupling. The probability of capture into another resonance is obtained from the probability of capture into the $2n$ resonance simply by multiplying ν by a constant factor. The boundaries on the shaded region of Fig. 6 are simply expanded or contracted parallel to themselves for lower- and higher-order resonances, respectively. For the 1.5n resonance, the boundaries are expanded by factors of between 5 and 10 depending on the tidal model (Peale and Boss 1977a). Although it is not necessary to appeal to a core-mantle interaction to account for the occupancy of the $1.5n$ spin-orbit state (Sec. IV), the inclusion of nearly all reasonable kinematic viscosities in the region where $P = 1$ for all values of the tidal Q would make this resonance almost impossible to avoid if the model above is appropriate to the actual core-mantle interaction.

In Fig. (6) escape from the $2n$ resonance is possible with a laminar boundary layer if (v,Q) lies in the triangular region in the lower left corner. (We ignore the high-viscosity escape zone, since such a high viscosity for the core material is little different from no molten core at all.) The vertical lines in Fig. 6 spanning $v = 0.01$ correspond to bounds on the viscosity of the Earth's core estimated by Gans (1972). If the viscosity of a Mercurian core falls within these limits, $Q \lesssim 25$ would allow escape from the $2n$ spin-orbit resonance. This Q is somewhat smaller than the value of 100 typically found for the Earth's mantle material, and it is reduced even further if the stronger coupling implied by Eq. (36) is used to move the left boundary of the shaded region a factor 3 farther to the left. We might well be led to suspect that Mercury's primordial spin rate was less than $2n$.

This suspicion is greatly reinforced when one considers the fact that the boundary layer may be turbulent. Peale and Boss (1977a) discuss the consequences of such turbulence and find that even the limited region of the v,Q plane which allows escape from the $2n$ resonance with a laminar boundary layer is essentially eliminated. However, this result is based on the crudest of models of turbulent flow. Inferences were made about the critical Reynolds number based on experiments with somewhat inappropriate geometries. Also the standard, although unjustified, substitution of the turbulent viscosity wherever v appears in the above equation was used. Given the uncertainty of the effects of turbulence and the variation in the coupling between a solid inner core and the mantle through a fluid shell as a function of the shell thickness, there is a real need for laboratory experiments dealing with the proper geometries for a more secure understanding of the core-mantle coupling. For these reasons, this discussion of the quantitative effects of turbulence should not be taken too seriously, and above constraints on the core for escape from the $2n$ resonance should therefore be considered only initial estimates. Still such boundary layer turbulence may exist and it would expand the region of unit capture probability.

Thus, it is entirely possible that Mercury never rotated much faster than its current rate after its accretion was essentially complete. That Mercury could have rotated with such a slow initial spin is thought to be exceedingly unlikely by many planetary scientists, since most solar system bodies which have not obviously been affected by dissipative processes have reasonably fast spins. However, we observe Uranus and Venus rotating in a retrograde sense. Uranus must have essentially its primordial angular velocity, and *if* the retrograde spin of Venus is stabilized by a liquid core-solid mantle interaction, it, too, must have started retrograde (Goldreich and Peale 1970). Given that primordial planetary spins are observed to range from retrograde to prograde relative to their orbital motion, there can be no reason that the statistical increments in spin angular momentum during the accretion process could not lead to any value of the spin in between, including a slow prograde spin. In any case, it is clear that capture into the $1.5n$ resonance would be almost certain if Mercury approached the resonance with a large molten core, and if the above time constant for the relaxation of a differential angular velocity between the core and mantle in the weak coupling limit is appropriate.

It would be important to know whether or not a molten core actually exists. Perhaps one could then infer if the rotation history was dominated completely by the effects of tidal dissipation, or whether that history was severely constrained and modified by a liquid core-solid mantle interaction. More fundamental constraints on theories of thermal histories and planetary magnetic field generation would also ensue from a knowledge of Mercury's interior. If the core is still molten, perhaps a sulfur-iron eutectic is responsible for its persistence over the age of the solar system. If the core has solidified, what is the source of the observed magnetic field and why was there insufficient sulfur available to prevent its complete solidification? In the next section we outline an experiment, which is based only on the ordered dynamical state of the current rotation, that has been proposed to determine the existence and estimate the radius of a Mercurian molten core.

VI. IS THERE A MOLTEN CORE?

The discovery of an intrinsic magnetic field for Mercury by the Mariner 10 flyby (Ness et al. 1974a) and the dominant belief that an internal dynamo is its source (Ness et al. 1975a) are the primary motivations for believing that Mercury currently has a molten core. However, modelers of the thermal history of the planet are not unanimous in their predictions of the current state of the core. In a model in which Mercury is differentiated, Siegfried and Solomon (1974) imply that an initially molten core would tend to cool and solidify over geologic time scales. Whether or not a given model with differentiation has a molten core which persists to the present depends critically on assumptions about initial conditions (the time of core formation), the time varying distribution of radioactive elements within the planet, thermal conductivities, and the efficiency of convective heat transport. Fricker et al. (1976) find that a liquid outer shell 500 km thick may persist to the present if one accounts for a thermal barrier at a liquid iron-solid silicate core-mantle boundary. The thermal barrier results from the higher melting point and lower thermal conductivity of the silicates. The liquid iron core temperature rises above the melting curve near the core mantle boundary thereby decreasing the core temperature gradient. If the solid mantle is convecting, however, it must contain a density of heat sources comparable to that of the Earth's mantle-wide average if the core is to remain molten (Cassen et al. 1976).

Since we cannot be certain about the convective heat transport or the distribution of heat sources, theoretical predictions about the nature of Mercury's core must remain inconclusive. The observational experiment that has been proposed (Peale 1976) and which we describe here is nonseismic and appears to be accessible with currently available technology. It should unambiguously determine the existence and extent of a liquid core. Whether or not the core is found to be molten, the result will have profound effects on inferences about the rotational history discussed in Sec. V, on theories of planetary magnetic field generation, and on theories of the thermal history.

There are two necessary conditions on the core-mantle interaction for the experiment to work: (1) the core must *not* follow the 88 day physical librations of the mantle; (2) the core must follow the mantle on the time scale of the 250,000 yr precession of the spin in Cassini state 1. We shall assume these two conditions are satisfied to develop the method and later establish the constraints on the core viscosity for which they are satisfied.

The physical libration of the mantle about the mean resonant angular velocity arises from the periodically reversing torque on the permanent deformation as Mercury rotates relative to the Sun. The amplitude of this libration is given by (Peale 1972)

$$\phi_0 = \frac{3}{2} \left(\frac{B-A}{C_m} \right) \left(1 - 11e^2 + \frac{959}{48} e^4 + \cdots \right) \simeq \frac{B-A}{C_m} \qquad (38)$$

where the moment of inertia in the denominator is that of the mantle alone since the core does not follow the librations. The core is assumed axially symmetric so it does not contribute to $B - A$. We have seen in Secs. IV and V that dissipative processes will carry Mercury to Cassini state 1 with an obliquity θ close to $0°$. Equation (15), which relates the obliquity in the state to the differences in the moments of inertia and other orbital parameters, can be written in the form

$$K_1(\theta) \left(\frac{C-A}{C} \right) + K_2(\theta) \left(\frac{B-A}{C} \right) = K_3(\theta) \qquad (39)$$

where the moment of inertia in the denominator is now that of the total planet since the core is assumed to follow the precession. Note that the precession here is not the relatively rapid precession of the spin about the Cassini state in the frame rotating with the orbit, which the core is not likely to follow (Sec. V), but rather it is the precession of the orbit (with the much longer period) in which frame the spin is locked if Mercury occupies the exact Cassini state. In Eq. (39), $K_2(\theta) < K_1(\theta)$ and can sometimes be neglected as it has been in the classical determination of $C - A/C$ for the Moon (Peale 1969).

In Sec. III it was pointed out that

$$C_{20} = -\frac{C - A}{M_M a_e^2} + \frac{1}{2}\frac{B - A}{M_M a_e^2}$$

(40)

$$C_{22} = \frac{B - A}{4M_M a_e^2} .$$

Equations (40) can be solved for $(C - A)/C$ and $(B - A)/C$ in terms of C_{20} and C_{22}, where the latter two quantities can be determined by tracking one or more artificial satellites. Substitution of the solutions of Eqs. (40) into Eq. (39) yields a numerical value for $C/M_M a_e^2$ since the K_i are known once the obliquity θ is measured.

Measurement of the amplitude ϕ_0 of the physical libration determines $(B - A)/C_m$ (Eq. 38) from which three known factors give

$$\left(\frac{C_m}{B - A}\right) \left(\frac{B - A}{M_M a_e^2}\right) \left(\frac{M_M a_e^2}{C}\right) = \frac{C_m}{C} \le 1.$$

(41)

A value of C_m/C of 1 would indicate a core firmly coupled to the mantle and most likely solid. If the entire core or the outer part is fluid, $C_m/C \simeq 0.5$ for the large core size ($r_c \simeq 0.75$) in current models of the interior (Cassen et al. 1976).

We have made the implicit assumption that Mercury occupies Cassini state 1. But the obliquity corresponding to this state depends on the orbital inclination to the invariable plane and on the orbital eccentricity, both of which are variable (see, e.g., Cohen et al. 1973).

$0.11 \le e \le 0.24$	10^6 yr period
0.006 amplitude	10^5 yr period
$5° \le \iota \le 10°$	10^6 yr period
$0°.25$ amplitude	10^5 yr period.

(42)

If the precession period of the angular momentum around Cassini state 1 is much less than the above periods for the variations in the orbital parameters, the solid angle traced out by the spin angular momentum is an adiabatic invariant (Goldreich and Toomre 1969). This precession period is about 1800 yr for the Anderson et al. (1987) values of $(C - A/2 - B/2)/C \simeq (B - A)/C \simeq 10^{-4}$, and we expect Mercury therefore to be very close to the instantaneous location of the Cassini state (Peale 1974). Determination of the four parameters, C_{20}, C_{22}, θ and ϕ_0, thus allows the above measurement of Mercury's core.

Are the necessary conditions on the core-mantle interaction likely to be satisfied? In Eqs. (37) two time constants for the decay of a differential rotation of a spherical cavity and its contained fluid were used to relate the coupling constant between the core and mantle to the kinematic viscosity:

$$\tau = \frac{r_c}{(\nu\dot\psi)^{1/2}} \qquad \tau = \frac{r_c^2}{\nu} \tag{43}$$

where the first applies to small viscosities and the latter to large viscosities. If $\tau \gg 88$ day, the core will not follow the mantle, and if $\tau \ll 250{,}000$ yr, the core will follow the precession of the mantle angular momentum. These conditions correspond to

$$4 \times 10^{-4} < \nu < 5 \times 10^8 \text{ to } 4 \times 10^9 \text{ cm}^2 \text{ s}^{-1}. \tag{44}$$

Since this range includes all possible values for the viscosity of likely core material (see, e.g., Gans 1972), the experiment should work if the core-mantle coupling is primarily of a viscous nature.

What are representative values of C_{20}, C_{22}, θ and ϕ_0, and what are tolerable errors in their measurement? The first estimates of C_{20} and C_{22} have been obtained from tracking the Mariner 10 spacecraft (Anderson et al. 1986):

$$C_{20} = -(6.0 \pm 2.0) \times 10^{-5}$$
$$C_{22} = (1.0 \pm 0.5) \times 10^{-5}. \tag{45}$$

This value of C_{20} and a slightly smaller value of $C_{22}(= -C_{20}/8)$ were used by Peale (1981) in estimating tolerable errors of measurement. Representative values of the obliquity and of the libration amplitude used in the 1981 estimates are

$$\theta \approx 7 \text{ arcmin}$$
$$\phi_0 \approx 30 \text{ arcsec} \tag{46}$$

where θ follows from Fig. (3) and ϕ_0 from Eq. (38) with $\phi_0 = 0.854(B - A)/C_m$ for $e = 0.206$, $C_m/C = 0.5$ and $C/M_M a_e^2 = 0.34$.

If we designate the four parameters, with nominal values as given in Eqs. (45) and (46), by η_i then we can write

$$\Delta\left(\frac{C_m}{C}\right) = \sum_i \frac{\partial}{\partial\eta_i}\left(\frac{C_m}{C}\right)\Delta\eta_i \tag{47}$$

which gives

$$\frac{\Delta(C_m/C)}{(C_M/C)_0} = f_1\frac{\Delta C_{20}}{C_{20}^0} + f_2\frac{\Delta C_{22}}{C_{22}^0} + f_3\frac{\Delta\theta}{\theta_0} + f_4\frac{\Delta\phi}{\phi_0}$$
$$= -0.87\frac{\Delta C_{20}}{C_{20}^0} + 0.87\frac{\Delta C_{22}}{C_{22}^0} - 0.98\frac{\Delta\theta}{\theta_0} - \frac{\Delta\phi}{\phi_0} \tag{48}$$

where the form of the coefficients f_i can be found in Peale (1981), with the numerical values corresponding to the nominal values of the η_i in Eqs. (45) and (46). These nominal values are represented by zero sub- and superscripts in Eq. (48). If all four parameters are known to 10%, the maximum error in C_m/C is

$$\Delta \left(\frac{C_m}{C} \right) = \sum_i \left| f_i \frac{\Delta \eta_i}{\eta_i^0} \right| = 37\% \qquad (49)$$

and the probable uncertainty is

$$\Delta \left(\frac{C_m}{C} \right) = \left[\sum_i \left(f_i \frac{\Delta \eta_i}{\eta_i^0} \right)^2 \right]^{1/2} = 19\%. \qquad (50)$$

Even for maximum uncertainty, we could have $C_m/C = 0.5 \pm 0.18$ which would distinguish the molten core. Of course this conclusion depends on our assumptions of the nominal values of the measured parameters. (Figure 3 indicates that θ may be closer to 3 arcmin than to 7.) To allow for significant deviations from these assumptions, we should obtain C_{20} and C_{22} to two significant figures and require $\Delta\theta$ and $\Delta\phi$ both to be 1 arcsec or less to be assured that meaningful bounds on C_m/C are obtained.

From experience with the Moon, tracking of one or two artificial satellites placed at optimal distances should provide C_{20} and C_{22} to the desired accuracy. The measurement of θ and ϕ would be easy for an observer on Mercury's surface, but the necessity of making the measurements remotely offers some special challenges. There are many compromises between confidence in the ultimate accuracy obtainable for a given scheme, and the expense of carrying out that scheme. All methods of determining the two angles to the desired accuracy will require one or more landed transponders of some kind on the surface. Three landed transponders in a triangular array, which are capable of transmitting to the Earth, would yield the most secure and most rapidly interpreted data (C. C. Counselman and A. B. Whitehead, personal communications, 1977). (One arcsec measured from the center of Mercury corresponds to almost 12 meters on the surface.) This technique has already been used to determine Mars' pole position to 4 arcsec (Reasenberg et al. 1977). It is probably prohibitively expensive to place three rather sophisticated landers on Mercury's surface. Ranging to the surface from a satellite has the advantage of lower power requirements and it offers the possibility of using passive reflectors on the surface; but it has the added disadvantage of introducing the uncertainty in the satellite position into the data reduction. A single surface transponder is probably the most that can be expected if any mission to Mercury is undertaken, during the next decade, and preliminary calculations by P. Bender (personal communication, 1986) indicate that meaningful measurements of the angles with satellite ranging to a single transponder may be possible. This scheme should be carefully evaluated because of its economy.

Another proposal involves the hard landing of a single optical instrument which would track star trajectories across its field (M. C. Malin, personal communication, 1977). This is a much more complex instrument than a transponder, but it has the advantage of determining both θ and ϕ in a single sidereal day.

It is clear that it is technically feasible to measure the four necessary parameters to the required accuracy to determine the extent of a Mercurian liquid core. The importance of this measurement as a constraint on thermal histories and magnetic field generation in terrestrial planets justifies a serious effort to make the measurements.

VII. SUMMARY

Mercury's current rotation state could be a natural outcome of tidal evolution alone from almost any initial conditions. The obliquity migrates toward small values on the same time scale as that for the retardation of the spin angular momentum. The precession of the orbit causes the final obliquity not to be zero, but slightly displaced into Cassini state 1, where the value of the obliquity can be used to determine a relation among the differences in the moments of inertia. In this state, the spin vector remains coplanar with the orbit normal and the normal to the plane of the solar system about which the orbit precesses. The spin magnitude could have reached its current value of exactly 1.5 times the orbital mean motion, stabilized by resonant torques on the permanent deformation, by cascading down through several other stable spin-orbit resonances as tides retarded the spin. The selection of the current state over the other stable states can be understood in terms of an increasing probability of capture as the order of the state is reduced. Capture into the state where $\dot{\psi} = 1.5n$ is more probable than that for any higher-order state, although the actual values of the capture probabilities are strongly dependent on the tidal model chosen, and none of the tidal models used leads to a certain capture into the existing state.

The freedom of choice of initial conditions which would lead to the current state over the history of the solar system vanishes if Mercury has had a molten core over much of its history. First, obliquities greater than 90° would not have been possible when $\dot{\psi} \lesssim 4n$, since a liquid core would have lost its pressure coupling to the mantle and would have lagged behind the latter's precession. The dissipation resulting from the relative core-mantle motion would have driven the spin vector to Cassini state 3 (a retrograde rotation) against the tide's attempts to drive it toward smaller obliquities. Then the retrograde rotation would have been maintained as long as the core remained molten. If the obliquity was somewhat less than 90° at the time the core became only viscously coupled with the mantle, the core-mantle dissipation would have greatly accelerated the slow tidal evolution toward Cassini state 1 with a time constant for the decay of tan θ of only about 10^5 yr. Although the tidal friction alone would maintain substantial nonzero obliquities for $\dot{\psi} > 2n$, the core-mantle interaction would force small obliquities for $\dot{\psi} \lesssim 4n$ for which the core is only viscously coupled to the mantle.

In the presence of a large molten core, nearly all of the stable spin orbit resonances would have been approached with Mercury occupying Cassini state 1 at a small obliquity. The core-mantle interaction then leads to drastically increased capture probabilities for all of the resonances over that obtained with a tidal torque which decreases with $\dot{\psi}$. In fact, capture is certain for several of the lower-order resonances for a range of values of the viscosity ν, where the range broadens as the order of the resonance is reduced. Capture into the currently occupied spin-orbit resonance is probabilistic ($P < 1$) if tides are the only dissipative process. But the range of core material ν for which $P = 1$ for this resonance for a core-mantle interaction spans all reasonable values so that capture into the $1.5n$ resonance would seem almost impossible to avoid, provided that capture by earlier resonances has been avoided. Escape from the $\dot{\psi} = 2n$ resonance as tides drag the spin through is possible but only if the core is constrained to have a kinematic viscosity at least as small as that estimated for the Earth, if the specific dissipation function for the tides is 25 or less, and if the boundary layer is laminar. A turbulent boundary layer may eliminate even this small window through the $2n$ resonance, and Mercury would have had to get by the $2n$ resonance before the molten core formed. As differentiation and core formation are thought to occur early in the history of terrestrial planets, Mercury's primordial spin may have had to be less than $2n$.

Although the effects of a liquid core-solid mantle interaction on the rotation history are dramatic, the reader should note that an oversimplified treatment of fluid mechanics was used in the analysis. If the core were stably stratified, for example, circulation of fluid through the Ekman boundary layer would be frustrated, and the time constant which was used for decay of a differential velocity between core and mantle in the low viscosity limit would not be appropriate. It is not clear, however, that a stratified core is consistent with a dynamo generation of the magnetic field. The time constants used also refer to more or less steady-state flows. Do these flows become established on a libration time scale? If not, the capture probabilities due to the core-mantle interaction may be drastically reduced and the constraints on the core properties for escape from the $2n$ resonance would be relaxed. The core fluid mechanics should be done more carefully to better constrain a liquid core's influence on the rotation history.

Although a liquid core could drastically constrain Mercury's rotation history from that which might have occurred if tides were the only dissipative process, the observed occupancy of the $1.5n$ spin-orbit resonance in Cassini state 1 does not present any theoretical or philosophical problems. We can use the ordered rotation state to determine the extent of such a liquid core; the proposed experiment depends essentially on the fact that the amplitude of the physical libration of the mantle about the resonant spin angular velocity is much larger if the core does not follow the libration (liquid core) than if the core does follow it (solid core). The obliquity in the Cassini state and the lowest degree gravitational harmonics complete the set of parameters needed to measure C_m/C. Whether or not Mercury's core is currently molten would provide vital constraints both on the theories of planetary magnetic field generation and on the theories of the thermal history of terrestrial planets. The measurements of the four required parameters to a sufficiently high precision for meaningful interpretation is technically feasible. Also, the determination of the extent of a liquid core by this technique can be done on no other planet, and the resulting constraints on the theories would be unique.

Acknowledgments. It is a pleasure to thank J. Lissauer for reading the manuscript and thereby improving the presentation. The author is supported in part by the Geology and Geophysics Program of the National Aeronautics and Space Administration.

MERCURY'S MAGNETIC FIELD AND INTERIOR

J. E. P. CONNERNEY
Goddard Space Flight Center

and

N. F. NESS
University of Delaware

The discovery and all available in situ observations of the magnetic field of Mercury were obtained by a single NASA spacecraft launched on 3 November 1973. During the first and third of three close encounters with the planet Mercury, Mariner 10 passed briefly through a small but Earth-like magnetosphere. Quantitative analyses of these observations established the presence of an internal field, with a dipole moment of ~300 nT-R_M^3 (1 R_M = 2439 km). However, estimates of the dipole moment vary considerably (factor of 2) among different authors, largely as a result of different model assumptions and techniques employed. Analyses are particularly difficult because of the relatively large magnetic field due to magnetopause and magnetotail currents, large temporal field fluctuations, and a relatively small magnetospheric cavity. We review the observations and analyses of Mercury's magnetic field with emphasis on model completeness and nonuniqueness. We demonstrate that the lack of agreement among models is due to fundamental limitations imposed by the spatial distribution of available observations. The mere existence of an internal magnetic field generated by an active dynamo is, however, a hard constraint on the composition and thermal evolution of the interior of Mercury. We review models of Mercury's interior consistent with the dynamo requirement and explore the cosmochemical implications of such models.

The Mariner 10 spacecraft, known prelaunch as the Mariner Venus-Mercury spacecraft, was launched on 3 November 1973. Its primary objective was to encounter and conduct the first close observations of the planet Mercury on 29 March 1974. Due to a lack of sufficient launch vehicle capability, it was necessary to use a gravity assist (the first such in spaceflight) from a near-flyby of Venus on 5 February 1974 in order to achieve an encounter with Mercury at the heliocentric distance of 0.46 AU. The post-Venus trajectory was very serendipitous as its heliocentric orbital period (176 days) was nearly exactly commensurate with the orbital period of Mercury (88 days). As a result, two additional encounters with Mercury were achieved on 21 September 1974 (Mercury II) and 16 March 1975 (Mercury III). Due to the exhaustion of attitude control and thruster gas, no further encounters were accomplished.

The celestial mechanics of the encounters are such that the trajectory of the spacecraft projected onto the Mercury orbital plane crosses the planet-sun line at an angle of ~85°. In other words, the spacecraft approaches the planet from a local time of 1900 hr and departs near a local time of 0700 hr. Relative to the orbital plane, the spacecraft approaches the planet from below by about 20° and departs ~20° above. The one variable, critical in intersecting the Hermean magnetosphere, is the altitude and subspacecraft latitude of the point of closest approach. For Mercury I this was selected to be 705 km on the night side at $-2°$ while the Mercury III encounter occurred towards the north pole at 327 km with a latitude of ~68°. Table I summarizes the spacecraft trajectories and interplanetary field conditions during the two close encounters with Mercury.

TABLE I
Magnetometer Observations

Feature	Mercury I (29 March 1974)	Mercury III (16 March 1975)
Interplanetary Field	~ 18 nT	~ 20 nT
Bow shock		
inbound	~ perpendicular	~ parallel (waves upstream)
outbound	~ parallel (waves upstream)	~ perpendicular
Deduced magnetosphere subsolar distance	1.6 ± 0.2 R_M	1.4 ± 0.2 R_M
Closest approach (dark side)		
altitude above surface	705 km $= 0.29$ R_M	327 km $= 0.13$ R_M
latitude	2°S	68°N
maximum field	98 nT	400 nT

In order to study the gross characteristics of the magnetic field, plasma and charged particle environment of the planet, the spacecraft was instrumented to measure magnetic fields, low-energy solar wind and magnetospheric plasma and energetic charged particles. The magnetometer instrumentation on Mariner 10 was the first flight of dual, triaxial magnetometers (Ness et al. 1971). This dual system of magnetometers was developed in order to permit analytical separation of the ambient magnetic field in space from the significant and variable spacecraft magnetic field which would contaminate observations by a single magnetometer. Two identical triaxial fluxgate magnetometer sensors were located on a boom of 5.8 meters length, with one sensor at the outermost end and with the second sensor positioned 2.3 meters inboard from the outer end.

For the first encounter, the multiple range magnetometer system was preset in the ± 128 nT range with a corresponding quantization step size of 0.26 nT. For the second encounter, the instrument was set in the ± 512 nT range with a corresponding step size of 1 nT. The magnetic field was measured once every 40 msec and was subsequently processed and averaged following transmission to ground stations. A preliminary discussion of flight data processing and techniques employed in calibrating the data were described by Lepping et al. (1975). The magnetometer instrumentation was described by Seek et al. (1977) and a summary of the Mariner 10 magnetic field and trajectory data for both encounters was given by Lepping et al. (1979). These data have been deposited in the National Space Science Data Center since 1977.

Mariner 10 discovered that Mercury possesses an intrinsic magnetic field sufficient to deflect the solar wind flow. These results were first described by Ness et al. (1974a) and Ogilvie et al. (1974). Additional observations regarding the existence of an intrinsic planetary magnetic field and magnetosphere were obtained on the third encounter and reported by Ness et al. (1975b) and Hartle et al. (1975a). No results regarding Hermean magnetosphere and magnetic field were obtained from the second encounter with Mercury because that trajectory was designed to optimize imaging coverage of the daylit planetary surface. The closest approach distance of 50,000 km was approximately an order of magnitude larger than the estimated distance from the planetary center to Mercury's bow shock at the subsolar point, under nominal solar wind conditions.

It is the purpose of this chapter to review critically the analyses of magnetic field data obtained by Mariner 10 with respect to quantitative spherical harmonic representations of the planetary magnetic field. An alternative method for determination of the equivalent magnetic dipole field intensity based upon the observed positions of the magnetopause and bow shock boundaries in comparison with those at Earth has been given elsewhere (Ness 1977; Slavin and Holzer 1979b). A more detailed description of Mariner 10 observations can be found in the reviews of Ness (1978,1979a,b).

I. MERCURY I ENCOUNTER OBSERVATIONS

The Mercury encounter observations are conveniently displayed in a body-centered, inertial coordinate system, due to the planet's slow rotation (with a period of 58.44 days) and the relatively short duration of the Mariner 10 encounters. A brief 33 minutes elapsed between observations of the entrance and exit to the bow shock during the first encounter. The spacecraft spent about 17 of those minutes within Mercury's magnetosphere, the region in which the planet's magnetic field excludes the solar wind.

Figure 1 shows the Mercury I spacecraft encounter trajectory and magnetic field observations within the magnetosphere in the Mercury ecliptic (ME) coordinate system. This is a right-handed, body-centered cartesian coordinate system in which the $X-Y$ plane is parallel to the ecliptic plane, X is positive in the direction of the Sun and Z is positive northward. Mercury rotates (slowly) about an axis nearly aligned with the ecliptic normal. In Fig. 1, 6 s averaged magnetic field vectors and the trajectory of Mariner 10 are shown, viewed along the ecliptic normal (bottom panel) and viewed from the

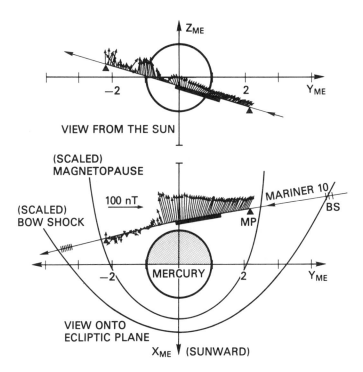

Fig. 1. Mercury I encounter (29 March 1974) observations. Projection of the observed magnetic field onto the Mercury ecliptic $Y-Z$ (top) and $X-Y$ planes.

Sun (top panel). Appropriately scaled terrestrial bow shock and magnetopause boundaries are also shown (Ness 1979b).

From the time of Mariner 10's entry into the magnetosphere until just after closest approach at a radial distance of 1.29 R_M, near the midnight meridian, the magnetic field increased steadily in magnitude from ~45 nT to ~100 nT. During this time the field maintained the anti-sunward orientation characteristic of a planetary magnetotail. Observations obtained during the post-midnight half of the Mariner 10 encounter are, in comparison, chaotic, evidencing dramatic spatial and time variations of Mercury's magnetosphere which are interpreted with more difficulty. During this time substantial magnetic field variations occurred in association with the arrival of intense energetic electron fluxes at the spacecraft (Simpson et al. 1974; Eraker and Simpson 1986). Siscoe et al. (1975) attributed these events to brief magnetospheric substorms by analogy with their terrestrial counterparts. Two such events, each persisting for 2 to 3 minutes, are readily identified in Fig. 1, top panel, by the marked decrease in the magnetic field component normal to the ecliptic. Also occurring in this interval of time was the apparent Mariner 10 traversal of the magnetotail current sheet, evidenced in the bottom panel of Fig. 1 as an abrupt decline of the field component oriented parallel to the ecliptic. Again by analogy with the terrestrial magnetosphere, these observations suggest the entry of Mariner 10 into the cross tail current sheet just after closest approach (Ness et al. 1974a; Whang 1977) as the spacecraft crossed the equatorial plane. However, the observations do not clearly show Mariner 10's emergence from the current sheet, which would appear in the figure as a sunward-directed field on the outbound leg of the trajectory. The appearance of nearly coincident time variations (substorms) and spatial variations (passage of the spacecraft through the tail current sheet) complicated considerably attempts to describe the average configuration of Mercury's magnetosphere.

Much if not most of the magnetic field observed by Mariner 10 along the Mercury I encounter trajectory is due to currents flowing external to Mercury, on the magnetopause boundary and in the cross tail current sheet. The very gradual increase in field magnitude with decreasing radial distance from the planet, particularly evident in the normal Z field component, is much less than the radial variation expected ($1/r^3$) of sources confined to the planet's interior. However, the magnitude of the observed field, and the Earth-like magnetospheric structure suggests immediately the presence of a dipole internal field with the same polarity as the Earth's and oriented approximately along the ecliptic normal.

II. MERCURY III ENCOUNTER OBSERVATIONS

The Mercury III encounter trajectory was a dusk to dawn, nightside encounter with the closest approach of only 1.14 R_M, much like that of Mercury I. However, in order to maximize the opportunity to study the internal field,

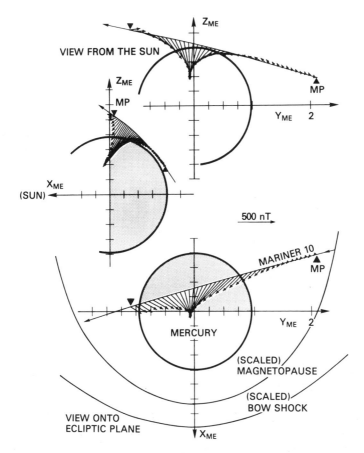

Fig. 2. Mercury III encounter (16 March 1975) observations.

Mariner 10 was targeted to pass very nearly over the pole at an altitude of 327 km. The Mariner 10 orbital period was exactly twice the 88 day orbital period of Mercury. Because Mercury is locked into a 3:2 spin-orbit resonance, completing 3 rotations for every 2 orbital periods, the position and orientation of Mercury during the third encounter was just as it was during the first encounter. Thus, the solar-oriented ME coordinate system described earlier is also appropriate for the Mercury III encounter.

Figure 2 shows the Mercury III encounter trajectory and magnetic field observations obtained within Mercury's magnetosphere in the ME coordinate system. In addition to the two projections shown previously for the Mercury I encounter, this figure also shows a projection onto a plane perpendicular to the ecliptic and containing the planet-sun vector. The scaled magnetopause and bow shock boundaries are the same as in Fig. 1.

Mariner 10 passed through the small Hermean magnetosphere in a brief 14 minutes. The magnetic field increased in magnitude from ~20 nT observed upon entry into the magnetosphere to a maximum of 402 nT observed just after closest approach at a distance of 1.14 R_M. In contrast to the first encounter, the internal origin of the observed field is clearly evidenced by the approximate $1/r^3$ increase in field strength as the spacecraft approaches the planet. The geometry of the field is approximately that of a dipole oriented nearly along the ecliptic normal. In this presentation, emphasizing the dominant presence of the internally generated planetary magnetic field, the field contribution of magnetopause and magnetotail currents is less apparent. In addition, the effects of magnetotail and magnetopause currents are less pronounced along the over-the-pole trajectory of the third encounter, compared with the low-latitude trajectory of the first encounter. However, the confinement of the Hermean magnetosphere inferred from the magnetopause observations is consistent with that of the first encounter. Near the outbound magnetopause crossing, the magnetotail geometry can be seen in the progressive sunward shift in field orientation.

III. CONCEPTUAL MODEL MAGNETOSPHERE

The Mercury III encounter observations firmly established the presence of an intrinsic magnetic field of internal origin, a conclusion that could only tentatively be arrived at on the basis of the observations made during the first encounter (Ness et al. 1974a). Early interpretations of the Mercury observations recognized the qualitative similarities between the extensively studied magnetosphere of the Earth and the observations obtained at Mercury. The Hermean bow shock and magnetopause observations are consistent with a magnetosphere size a factor of 7.5 smaller than the terrestrial magnetosphere, when normalized by the planetary radius. Thus, the planet Mercury occupies a very large volume fraction of its magnetosphere, as is illustrated in Fig. 3, based on the magnetosphere model of Whang and Ness (1975). In this figure, the magnetopause and magnetotail currents are schematically indicated as they intersect the noon-midnight meridian plane; the model of Whang (1977) utilized an image dipole to confine the planetary field to the appropriate volume. Also shown in Fig. 3 are the 6 s averaged encounter observations of Mercury I projected onto the noon-midnight meridian plane. The third encounter trajectory (only) is indicated by the dashed line.

The field orientation observed during the first half of the Mercury I encounter, as Mariner 10 approached the $Z = 0$ plane, is qualitatively consistent with the magnetotail geometry in Fig. 3. The disturbed magnetic field observed during the second half of the encounter illustrates the extreme time variability of Mercury's magnetosphere, and the difficulty of distinguishing between spatial and temporal variations.

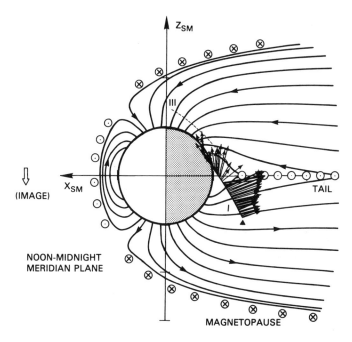

Fig. 3. Noon-midnight meridian plane projection of the Hermean magnetosphere. Magnetopause and tail currents are schematically indicated crossing (at 90° to) the meridian plane. Field lines (after Whang and Ness 1975) assume a dipole internal field, tail current sheet and image dipole. Mercury I encounter observations and trajectory are shown projected onto the meridian plane.

Two very large variations in the magnetic field, which occur in association with the arrival of energetic electrons at the spacecraft, appear to be magnetospheric substorms (Siscoe et al. 1975; Eraker and Simpson 1986). The magnetic signature of a Hermean substorm appears predominantly in the Z component as a diminution of the field by approximately 10 to 30 nT, lasting 2 to 3 minutes. The encounter observations of Mercury I suggest that the diminution of the field occurs before the energetic electron fluxes were detected at the spacecraft, as is illustrated in Fig. 4. This figure shows the components of the observed field in ME coordinates throughout encounter. The variation of the magnetic field attributed to the substorm events, which immediately preceeds the energetic electron fluxes detected at the spacecraft, is indicated by the stippled area in the top panel. The component of the field in the ecliptic plane is relatively unchanged by the events, as is also apparent in high time resolution data (Eraker and Simpson 1986). This behavior is qualitatively consistent with a brief intensification of the cross tail current as field lines are drawn farther tailward, followed by a rapid relaxation which restores the original field configuration, terminating with the energetic elec-

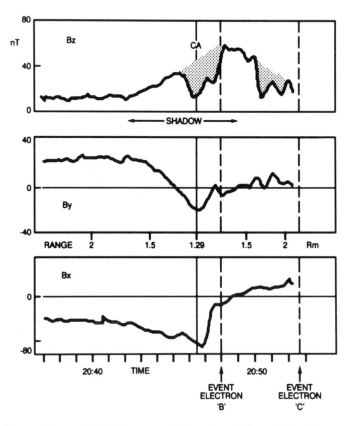

Fig. 4. Mercury I magnetic field 6 s averaged observations in the cartesian Mercury ecliptic
coordinate system. Stippled area indicates inferred field diminution prior to energetic electron
events.

tron events at the spacecraft. A very similar substorm event occurred during
the Mercury III encounter. During this event, the field decreased by ~10 nT
for a period of 2 to 3 minutes. again terminating with the arrival of energetic
electrons at the spacecraft. A detailed analysis of the substorms is given by
Eraker and Simpson (1986) and Baker et al. (1986c). See also the Chapter by
Russell et al.

The dynamic variations of the Hermean magnetosphere are quite large
relative to the total field, complicating attempts to obtain an accurate magnetic
field model from the encounter observations. Due to the brevity of the flyby
encounters, a 2 or 3 minute temporal variation of the magnetic field due to a
substorm can appear similar to the spatial variation of an internal, planetary
field. It is therefore necessary to restrict analyses to periods representative of
the "quiet" magnetosphere. It is also advisable to remain aware of the rela-

tively large naturally occurring variations of the magnetic field in assessing how well a model field fits the observations.

IV. MODELS AND METHODS

Spacecraft observations of a planetary magnetic field are interpreted utilizing some form of the spherical harmonic analysis traditionally used in studies of the Earth's magnetic field. In the usual spherical harmonic approach, it is assumed that the observations were obtained in a source-free region of space. The magnetic field B can then be expressed as the gradient of a scalar potential function V

$$B = -\nabla V \tag{1}$$

which can be written as the sum of two potentials, representing sources internal and external to the set of observations. The potential is expanded in terms of spherical harmonics

$$V = V^e + V^i = a \sum_{n=1}^{\infty} (r/a)^n T_n^e + (a/r)^{n+1} T_n^i \tag{2}$$

where r is the distance to the planet's center, a, is the planet radius, and the T_n^e and T_n^i are given by

$$T_n^i = \sum_{m=0}^{n} P_n^m (\cos\theta) [g_n^m \cos(m\phi) + h_n^m \sin(m\phi)] \tag{3}$$

$$T_n^e = \sum_{m=0}^{n} P_n^m (\cos\theta) [G_n^m \cos(m\phi) + H_n^m \sin(m\phi)]. \tag{4}$$

The P_n^m are the associated legendre functions with Schmidt normalization, and the g_n^m, h_n^m, G_n^m, H_n^m are the internal and external Schmidt coefficients; θ and ϕ are the polar and azimuthal angles of a spherical coordinate system. In the traditional approach, the series (Eq. 2) is truncated at some order $n = N_{max}$ according to the complexity of the field and the availability of observations, and the Schmidt coefficients are chosen to minimize, in a least squares sense, the difference between the model field and the observations.

The magnetic field due to local sources (e.g., magnetotail or other currents distributed within the region of measurement) cannot be accommodated by the scalar potential representation (Eq. 2). Two techniques are frequently used in coping with appreciable fields due to local sources. The first is to restrict the analysis to those observations obtained in source-free space. In the case of the Mercury I encounter observations, this can be accomplished by modeling only those observations obtained between the inbound magnetopause crossing and the tail current sheet penetration just prior to closest approach (refer to Fig. 1). The high-latitude Mercury III observations are obtained in source-free space essentially from magnetopause to magnetopause. Alternately, one can attempt to model explicitly the field due to local currents, which requires additional knowledge of and/or assumptions regarding the geometry and distribution of currents in space. Present knowledge of magnetospheric current systems allows for, at best, models only approximating those encountered by spacecraft in a planetary magnetosphere.

Models of Mercury's magnetic field all use some form of a spherical harmonic expansion for the internal field, but differ in the treatment of fields of external origin. All models of Mercury's magnetic field which use a spherical harmonic expansion for the external field as well as the internal field are summarized in Table II. Those which use an explicit model in the representation of external fields are summarized in Table III. The models of Whang (1977) employ an image dipole approximation for the field of magnetopause surface currents and an infinite half-plane current sheet for the crosstail current system. The other models in Table III use a (scaled) terrestrial magnetospheric field for the field due to magnetopause and magnetotail currents.

All of these analyses assume that the same external field was present throughout both encounters. In each table the extent of data used by each author in obtaining a field model is indicated in abbreviated form by the Roman numerals I and III corresponding to the Mercury I and III encounters. The parenthetical fraction following the numeral I indicates the fraction of the first encounter observations used in model fitting. All authors with the exception of Whang (1977) have elected to discard the (relatively disturbed) observations obtained during the latter half of the first encounter.

The salient parameters of the internal field models obtained by each author are listed in the Tables in a form which facilitates a direct comparison

TABLE II
Magnetic Field of Mercury Spherical Harmonic Models

Model[a]	$|g_1^0|(nT\text{-}R_M{}^3)$	Data	References
I1, offset, no E	227	I(1/2)	Ness et al. 1974
I1E2	350	I(1/2)	Ness et al. 1975
I1E1	342 ± 15	III	Ness et al. 1976
I1E1,I1E2	330 ± 18	III $S_1, S_2 \cdots$	Ness 1979

[a]InEm ≡ Internal spherical harmonic of order n (=1, dipole; 2, quadrupole;...) + external spherical harmonic of order m (=1, uniform; =2, $f(r, \theta, \phi)$).

TABLE III
Magnetic Field of Mercury Magnetosphere Models

Model[a]	$|g_1{}^0|/|g_2{}^0|$	Data	Reference
I1 + ID + tail	266	I + III	Whang 1977
I1 + $g_2{}^0$ + ID + tail	165/117		
I1 + $g_2{}^0$ + $g_3{}^0$ + ID + tail	166/75/48		
I1 + ⊕ MP, tail	154	I(1/2)	Jackson and Beard 1977
I1 + $g_2{}^0$ + ⊕ MP, tail	136/88		
as above	298	III	
	200/95		
as above	271	I(1/2) + III	
	177/113		
as above, $g_2{}^0$	179/196	I(1/2) + III	Ng and Beard 1979
replaced by (X_0, Y_0, Z_0)	$(Z_0 \sim 0.19)$		
as above, ⊕ tail	207	I(1/2) + III	Bergan and Engle 1981
improvements	$(Z_0 \sim 0.17 \pm 0.02)$		

[a]ID = Image dipole; ⊕MP = Earth analog magnetopause; ⊕ tail = Earth analog magnetotail.

(rotated to dipole coordinates) but ignores details necessary for application of the models as intended by the authors. For that purpose reference to the original publication is recommended. The internal field parameters listed are the dipole coefficient ($g_1{}^0$) axisymmetric quadrupole ($g_2{}^0$) and axisymmetric octupole ($g_3{}^0$) coefficients, where appropriate, or a spatial offset of the dipole. A small displacement of the dipole along the dipole axis is essentially equivalent to an axisymmetric quadrupole ($g_2{}^0$) term with magnitude

$$g_2{}^0 \simeq 2\,\Delta z\,g_1{}^0 \tag{5}$$

where Δz is the displacement in units of planet radius. This formula has been used to facilitate comparisons between models parameterized in terms of spherical harmonics and those presented as displaced dipoles.

A lack of agreement among estimates of the dipole moment is apparent upon examination of Tables II and III. Evidently, models which include an axisymmetric quadrupole ($g_2{}^0$) term are characterized by a smaller dipole ($g_1{}^0$) term, relative to dipole-only models, as are models based on encounter I observations (with the exception of the Ness et al. [1975a] model). The correlation between estimates of the dipole ($g_1{}^0$) and axisymmetric quadrupole ($g_2{}^0$) terms is illustrated in Fig. 5. Each field model has been entered in Fig. 5 on the $g_1{}^0$–$g_2{}^0$ plane, identified with symbols coded to the appropriate reference, without distinction as regards to modeling methodology. Models with

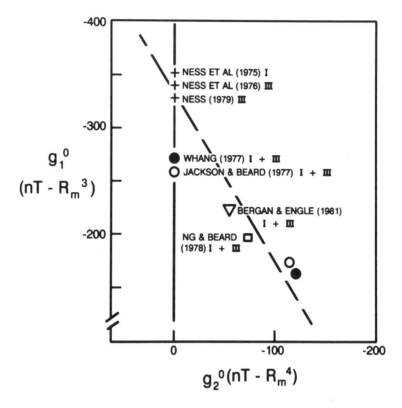

Fig. 5. Estimates of Mercury's magnetic field organized by dipole (g_1^0) and axisymmetric quadrupole (g_2^0) terms. Symbols are coded to appropriate author(s).

large axisymmetric quadrupole terms, or equivalently a large displacement along the dipole axis, lie close to the dashed line with slope $\Delta g_1^0 \simeq -1.5$ Δg_2^0 indicated in Fig. 5. This peculiar relationship between parameters of the various internal field models has previously been cited (Ness 1978,1979*b*) as evidence of "spatial aliasing" of the model coefficients, i.e., correlated errors among the model parameters as a result of the limited extent of the observations.

V. CRITICAL ANALYSIS

It is not necessary to reproduce in detail each model in Tables II and III to understand the lack of agreement among them. The differences in modeling techniques and in the chosen representation of the field of external origin are of lesser importance than the one element each has in common: observations limited to the Mariner 10 encounter trajectory. We demonstrate in this section that the apparent lack of agreement among these models is a manifestation of model nonuniqueness, a natural consequence of the limited spatial extent of the observations. Model nonuniqueness can be effectively studied using generalized inverse methods (Connerney 1981). These methods are designed to reveal parameters or combinations of parameters that are not well determined by the observations. These poorly constrained parameter vectors are unique to a particular spacecraft trajectory and limit the usefulness of any magnetic field model obtained from the observations.

For the following illustrative analysis we assume the internal field can be parameterized by the spherical harmonic coefficients g_1^0, g_1^1, h_1^1, g_2^0 representative of the models in Table III. The field of external currents is represented as a spherical harmonic expansion (Eq. 4) to order 1, equivalent to a uniform field. In this example, the external field is treated independently for the two encounters. This allows for changes in the average configuration of the magnetosphere in response to different solar wind conditions at the time of each encounter. More importantly, it allows for different uniform external fields in two different regions of the Hermean magnetosphere: just under the magnetotail current sheet (Mercury I observations) and at high latitude near the ecliptic north pole (Mercury III observations). This is a practical means of accommodating the spatial variation of the field due to magnetopause and magnetotail currents without admitting an excessive number of terms in the spherical harmonic expansion (Eq. 4). Observations throughout the first half of the Mercury I encounter and the entire Mercury III encounter are assumed.

Using a generalized inverse methodology, the least squares solution to the linear system described above can be written as a summation over independent, orthonormalized parameter vectors which are linear combinations of the original parameters (Connerney 1981, Eq. 9). These new parameter vectors are the independent eigenvectors of parameter space and depend only on the trajectory and the model. More to the point, these parameter vectors are naturally ordered by the singular value decomposition according to how sensitive the observations are to them. To study the nonuniqueness of *any* solution to this linear system, we need only consider the most poorly determined parameter vector. For the linear system described above, representative of the models listed in Table II, the combination of parameters that the data are *least* sensitive to is characterized by the (orthonormalized) parameter vector:

$$
\begin{aligned}
g_1^0 &= 0.76 \\
g_1^1 &= 0.23 \\
& \qquad\qquad \text{internal field} \qquad\qquad (6) \\
h_1^1 &= 0.03 \\
g_2^0 &= -0.50
\end{aligned}
$$

$$G_1^0 = -0.24$$
$$G_1^1 = 0.13 \quad \text{external field}$$
$$H_1^1 = 0.00 \quad \text{during Mercury I} \tag{7}$$

$$G_1^0 = 0.17 \quad \text{external field}$$
$$G_1^1 = -0.12 \quad \text{during Mercury III.} \tag{8}$$
$$H_1^1 = 0.00$$

This vector can be regarded as the direction in 10-dimensional parameter space along which the model is most poorly constrained by the available observations. Given any least squares solution (model), it is possible to construct another solution which results in a minimal change to the fit between model and observations, by adding a scalar multiple of the parameter vector given above to the model parameters.

A small amount of observation noise will result in a relatively large excursion of the model solution in the direction (of parameter space) of this vector (see, e.g., Connerney 1981, Eq. 13). Errors in the original model parameters (g_n^m, h_n^m, . . .) will tend to be correlated accordingly. The dashed line in Fig. 5, along which all of the published magnetic field models lie, is the projection of the most poorly constrained parameter vector onto the g_1^0–g_2^0 plane, with the slope

$$\Delta g_1^0 = -\frac{0.76}{0.5} \Delta g_2^0 = -1.52 \, \Delta g_2^0 \,. \tag{9}$$

Since this direction is uniquely determined by the Mariner 10 trajectory, it appears that the lack of agreement among models is a natural consequence of the limited observations. The distribution of these models in the g_1^0–g_2^0 plane is simply a manifestation of the model nonuniqueness that is inherent in the spatial distribution of the available observations. This unsatisfactory situation can only be resolved by the acquisition of additional observations (not along the Mariner 10 trajectory).

The results obtained above and conclusions drawn are rather insensitive to the details of the model representation chosen. We illustrate this by briefly considering different models and different subsets of the Mariner 10 observations. For example, when only encounter observations from Mercury III are used, discarding all Mercury I observations, the slope of the dashed line in Fig. 5 is altered only slightly ($\Delta g_1^0/\Delta g_2^0 = -0.76/0.46 = -1.65$). Figure 6 shows graphically a set of four models differing by a multiple of the least well-determined parameter vector. The root mean square (rms) difference between model and observations for these models ranges from ~5 to ~8 nT, compared with the maximum field of ~400 nT measured by Mariner 10. The main dipole (g_1^0) term can be varied from ~100 nT-R_M^3 to well over 300 nT-R_M^3 with little change in the rms of the fit, as long as the other model parameters are adjusted accordingly. This figure also illustrates well the correlation be-

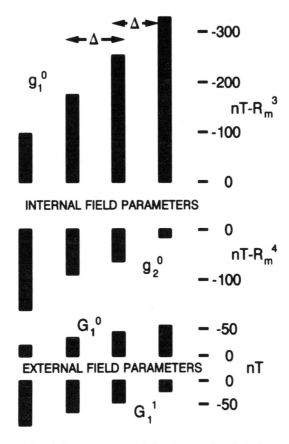

Fig. 6. Representation of the parameter correlations inherent in spherical harmonic models of Mercury's magnetic field fitted to Mariner 10 observations.

tween the internal field parameters and those of the external field. In another example, using all of the observations and a more complex field model containing terms to order 2 in both the internal (e.g., all quadrupole) and external spherical harmonic expansion, the slope of the dashed line in Fig. 5 is unchanged ($\Delta g_1{}^0/\Delta g_2{}^0 = -0.48/0.32 = -1.50$).

Perhaps a more vivid demonstration of the $g_1{}^0 - g_2{}^0$ parameter correlation is shown in Fig. 7, which shows contours of rms residuals for quadrupole models fitted to Mercury III observations only. This figure shows the results of a Monte Carlo inversion in which all possible combinations of the parameters $g_1{}^0$ and $g_2{}^0$ have been tested against the data. In each case, the parameters $g_1{}^0$ and $g_2{}^0$ have been held constant and all other parameters adjusted to minimize the rms residuals. The contoured rms residuals so obtained reveal precisely the same model nonuniqueness found using the singular value decomposition

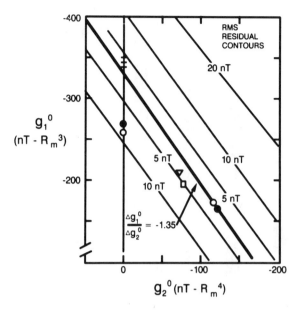

Fig. 7. Contours of the root mean square residual (observed-model) for quadrupole models of Mercury's magnetic field (I 2E1) fitted to Mercury III observations.

and generalized inverse analysis. With the latter, model nonuniqueness is explored by examining the properties of the linear system. With the Monte Carlo method illustrated here, it was necessary to try all possible $(g_1^0 - g_2^0)$ models to find those that were indistinguishable. Clearly, independent of detailed modeling considerations, all of the published models lie along the direction of ambiguity in parameter space. They are, on the basis of these magnetic field observations only, practically indistinguishable.

Additional information may help to resolve the ambiguity or nonuniqueness problem, in lieu of additional magnetic field observations. Several authors have used the observed magnetopause and bow shock positions, along with an estimate of the solar wind pressure at the time of the Mariner 10 encounter, to deduce the magnitude of the dipole moment. Use of the solar wind pressure balance condition requires knowledge of the radial distance to the magnetopause along the Sun-planet line, the magnetopause stand-off distance. Since Mariner 10 observed the magnetopause over the darkened hemisphere, it is necessary to infer the stand-off position from observations obtained well downstream. This is difficult at best, because several observations are required to infer the stand-off distance, and in all likelihood the magnetopause boundary moved in response to fluctuations in the solar wind ram pressure during the encounter. Estimates of the dipole obtained from bow shock and/or magnetopause positions (only) range from approximately

200 nT-R_M^3 (Russell 1977) to approximately 400 nT-R_M^3 (Slavin and Holzer 1979b).

A consideration of the probable magnetic field due to solar wind interaction may help discriminate between the internal field models illustrated in Fig. 6. We evaluate this possibility by considering the following simple model. Assume, for simplicity, that the magnetic field of Mercury is confined by the solar wind to a spherical cavity of radius r_o (planet radii). Requiring the normal component of B to vanish at the boundary of the cavity results in a relationship between the dipole and the uniform external field (G_1^0 term): $G_1^0 = 2/(r_o^3) g_1^0$. Analysis of Mercury III encounter observations (Fig. 6) illustrates that the external G_1^0 term is highly correlated with the dipole; thus, if G_1^0 were exactly known, one of the models with the appropriate G_1^0 would be selected. (For this over-the-pole encounter, the uniform external field approximation may be reasonably accurate. The lack of symmetry of the magnetosphere cavity about the dawn-dusk meridian remains a concern in this simple analysis. However, the Mercury III encounter was nearly confined to the dusk-dawn meridian plane, minimizing problems associated with the spherical cavity assumption.) Taking $r_o = 2 R_M$ as an estimate of the radial distance to the magnetopause over the pole, in the dawn-dusk meridian (Lepping et al. 1979), we obtain $G_1^0 \simeq 0.25 g_1^0$. The actual G_1^0 may be rather less than this, because the neglected magnetic field due to magnetotail currents will partially cancel the magnetopause field over the (north ecliptic) pole. Unfortunately, it seems that this condition is largely already satisfied by each of the models in Fig. 6. In essence, the relationship $G_1^0 \simeq 0.25 g_1^0$ is not useful new information. The most poorly determined parameter vector for the Mercury III encounter observations, projected on the g_1^0–G_1^0 plane, leads to the relationship $G_1^0 = (0.17/0.76) g_1^0 \simeq 0.22 g_1^0$. The two analyses offer nearly the same constraint on the ratio of g_1^0/G_1^0. Thus, one would need a very accurate determination of G_1^0 to discriminate among the models in Fig. 5.

VI. SUMMARY DISCUSSION OF MODELS

A very conservative conclusion, based on the discussion thus far, is that Mercury possesses a substantial dipole field of internal origin, with an axis of symmetry nearly aligned with the ecliptic normal (at encounter epochs). The magnitude of the dipole obtained assuming no higher order contributions to the internal field is $g_1^0 \sim 330$ nT-R_M^3. Other solutions with characteristically smaller dipole moments can and have been obtained (Whang 1977; Jackson and Beard 1977; Ng and Beard 1979; Bergan and Engle 1981). However, all solutions are subject to the inevitable limitations imposed by the Mariner 10 encounter trajectories. What is worse, all such solutions, considered together, delineate rather effectively the model nonuniqueness inherent in the Mariner trajectory. There is no persuasive evidence for the existence of a large g_2^0 term; just as there can be no conclusive proof that there is no large g_2^0 term.

Inclusion of the axisymmetric g_2^0 term only in quadrupole models of Mercury's magnetic field is not justified on physical grounds (other dynamos have nonaxisymmetric quadrupole terms) or on the basis of other observations (e.g., charged particle data).

VII. CONSIDERATIONS REGARDING ORIGIN OF THE INTERNAL FIELD

Following the observations at first encounter, there was some speculation that the observed magnetic field might be due to a solar wind induction effect (Herbert et al. 1976). This possibility was promptly disposed of with the observations from the Mercury III encounter when a maximum field intensity of 400 nT was observed. No plausible induction mechanism could explain a field intensity so far in excess of (factor of 20) the interplanetary (inducing) field. The characteristics of the planetary magnetic field, deduced from spherical harmonic analyses, yielded consistent results with respect to dipole moment and orientation. In spite of the fact that the interplanetary conditions were substantially different, this held with respect to both momentum flux and interplanetary field orientation.

With a clear identification of the global magnetic field of Mercury being due to an internal source, the remaining possibilities are either an active dynamo or a remanent magnetic field due to permanently magnetized sub-Curie point temperature layers in the planet. Note that even with a perfect description of the global magnetic field of any planet, it is not possible from data taken at a single epoch to distinguish unambiguously between the contributions due to an active dynamo from those due to remanent magnetization.

In the case of the Earth, it is well known that the nondipole magnetic field, defined as the residual field obtained by subtracting the dipole component from the total observed field, drifts westward. In addition, the location of the dipole axis at the Earth has varied considerably such that during the last three centuries it has moved mainly equatorward by approximately 8°. Also, the magnitude of the dipole term is steadily decreasing. These three secular variations of the geomagnetic field provide compelling evidence that the Earth's magnetic field is primarily that associated with an active internal dynamo (Gubbins 1977b).

Unfortunately, there are no adequate dynamo theories which predict either the existence or the characteristics of a planetary magnetic field (Gubbins 1974; Stevenson 1983; Stevenson et al. 1983). A number of studies regarding the possibility of remanent magnetization as the explanation for the Hermean field have been conducted (Ness 1978,1979a; Srnka 1976; Stephenson 1976). However, no plausible proposal for the origin of such a field has been forthcoming.

The existence of an active dynamo places certain constraints on the present-day internal structure and temperature. The most basic constraint is that there must now exist an electrically conducting region in the interior in which electrical currents flow, driven by some energy source. The most traditional mechanism is one where convective stirring of the electrically conducting material is driven by either thermal energy due to radiogenic isotope decay, gravitational settling, and/or core formation.

In the case of Mercury, this mechanism is less than certain due to the several studies on thermal evolution which show that the inner core region of Mercury would solidify or freeze out early in its history (Solomon 1976; Cassen et al. 1976). This leaves only a thin shell-like region which, at present, is not considered by most dynamo theorists as a probable source region for traditional planetary dynamo processes to be active. In order to prevent the core from freezing out, various "fixes" have been proposed to either provide more internal heat, retain the heat longer, or lower the melting point of the core material. These include increasing the amount of additional internal heat with an enriched uranium-thorium fraction, providing thermal blanketing by reducing the thermal diffusivity of the mantle, or lowering the melting point by adding an alloying element such as sulfur. The latter is effective even in small initial concentrations because the weight percent of sulfur in the remaining melt increases as the solid iron core forms. A recent proposal utilizing metallic silicon as an alloying element is more attractive from a cosmochemical point of view.

Numerous attempts to deal with this problem of core solidification have been marginally successful. A recent effort by Stevenson (1987) attempts to explain the Hermean magnetic field as due to a thermo-electric dynamo in which the current flow is driven by temperature differences at an irregular core mantle boundary. Stevenson's mechanism assumes helical convective motions in a thin outer layer to produce the modest poloidal field which is observed externally.

The facts remain clear that Mariner 10 discovered substantial evidence for an intrinsic global magnetic field. However, a plausible explanation in the framework of traditional planetary dynamos has been frustrated by thermal evolution models in which a large inner core freezes out prematurely, thus inhibiting the formation of a traditional dynamo. Certainly, additional data are required to describe quantitatively the global magnetic field of the planet and perhaps to identify, in the characteristics of its multipolar expansion, any possible feature or secular variation to elaborate on the nature of the dynamo. Since no future spacecraft missions to explore Mercury seem realizable in the remaining years of this century, this issue will remain for subsequent generations to consider, contemplate and vigorously address.

THE MAGNETOSPHERE OF MERCURY

C. T. RUSSELL
University of California at Los Angeles

D. N. BAKER and J. A. SLAVIN
Goddard Space Flight Center

Mercury's intrinsic magnetic field is strong enough to stand off the solar wind well above the surface of the planet under usual solar wind conditions. Thus, it has a magnetopause and a bow shock which are quite similar to their terrestrial counterparts, albeit much smaller in linear dimension. However, the absence of any significant atmosphere or ionosphere alters the flow of current in the Mercurian magnetosphere from the patterns at the Earth and may affect the transfer of energy from the solar wind to the magnetospheric plasma. Thus, the magnetosphere of Mercury is unique in the solar system. In the limited data provided by Mariner 10, Mercury's magnetosphere strongly resembles a miniature terrestrial magnetosphere in which everything simply happens more quickly and repeats more often than in the terrestrial magnetosphere. Nonetheless, estimates of the power requirements for the observed particle acceleration event are somewhat higher than expected. Many of the ambiguities inherent in the interpretation of the Mariner 10 data can be resolved only by the acquisition of new in-situ observations.

I. INTRODUCTION

Mercury is of great interest to those studying planetary magnetospheres. Mercury has an intrinsic magnetic field, one which is strong enough to stand off the flowing solar wind plasma well above the surface of the planet. Thus, its interaction with the solar wind has similarities to the interactions of Earth,

Jupiter, Saturn and Uranus with the solar wind. However, Mercury is also different from these other planets in ways that are important for determining the properties of, and processes occurring in, a planetary magnetosphere. First, Mercury has the most rarefied atmosphere of any of the planets with known magnetospheres. Atmospheres are important to planetary magnetospheres because they provide the lower boundary of most magnetospheric systems. When neutral atmospheric atoms become ionized they are affected by the magnetic and electric fields in the planetary magnetosphere and they then, in turn, affect those electric and magnetic fields. In particular, the electrons and ions which result from atmospheric photoionization can carry electric current. At the Earth, Jupiter, Saturn and possibly Uranus, the electrical conductivity of the near surface region of ionization, known as the ionosphere, is high enough that significant currents flow parallel to the planetary surface. These currents are dynamically important in that they affect how ionized matter is transported throughout the planetary magnetosphere. At Mercury there is a dynamically insignificant ionosphere and the surface is expected to be highly insulating. Thus, the dynamics of the Mercurian magnetosphere should be different from the dynamics of the Earth's magnetosphere. We cannot simply scale terrestrial processes to Mercury if the terrestrial process involves a strong ionospheric current.

Another important difference in the Mercurian magnetosphere is its size. The magnetic moment of Mercury is over 1000 times smaller than that of the Earth and the solar wind at Mercury is stronger than at 1 AU. The size of the magnetosphere is determined by the balance between the magnetic pressure of the planetary magnetic field and the dynamic pressure of the flowing solar wind plasma. Both pressures combine to give a magnetosphere whose linear dimensions are only 5% of those of the Earth. Since the solar wind moves at the same average speed at Mercury as at the Earth and other natural velocities of the plasma at the Earth and Mercury are similar, within a factor of two or three, events in the solar wind pass by the Mercurian magnetosphere in much shorter time than by the terrestrial magnetosphere. If some wave in the plasma flowing past Mercury is unstable and grows with time, it has a shorter time in which to grow at Mercury than it would near the Earth. As a result, the amplitudes of disturbances in the plasma at Mercury could be substantially different than at the Earth. Also, the high velocity of Alfvén waves at Mercury has led Slavin and Holzer (1979a) to propose that reconnection at Mercury may be more efficient than at the Earth.

Another important difference about Mercury is that the planet occupies a much larger fractional volume of its magnetosphere than the Earth, Jupiter, Saturn or Uranus occupy of their magnetospheres. The inner regions of the magnetospheres of these other four planets contain quite stable and, in some cases, quite intense radiation belts. However, the equivalent region in the Mercury magnetosphere is below the surface of the planet and so the stably trapped charged particle environment of Mercury is probably quite benign.

The above three differences do not exhaust the distinctive features found at Mercury. The fact that the interior of Mercury should become highly electrically conducting close to the surface may affect the compressibility of the magnetosphere. The solar wind density and temperature and the solar and galactic cosmic ray particle populations are somewhat different than at the Earth. This too may affect the nature of the interaction. Other differences may exist at Mercury that we do not yet appreciate, but, even in the absence of any other differences, we can be sure that Mercury presents us with sufficient contrasts that we can test our models of how magnetospheres work. Thus, those who are concerned with planetary magnetospheres are very interested in the magnetosphere of Mercury. Unfortunately, there is very little data with which to work.

In this review we will examine the available data and the possible implications of these data. We divide the review into two major sections, the solar wind interaction and the processes in the magnetosphere. Before treating these two major topics, we review the history of the investigation of the Mercury magnetosphere, the plasma and energetic particle environment in which Mercury is situated, and briefly the planetological properties of relevance to the magnetosphere.

History

The only spacecraft to have visited Mercury is Mariner 10, which was launched on 2 November 1973, and flew by Mercury on 29 March 1974, after a Venus swing by on 5 February 1974. Mariner 10 re-encountered Mercury on 21 September 1974 and 16 March 1975 after which it was no longer tracked. The first encounter (M I) passed the night side of the planet coming within 707 km of the surface at 2046:38 on 29 March 1974. The second encounter (M II) was a distant dayside flyby, coming within 50,000 km of Mercury at 2059:01 UT on 21 September. The third encounter (M III) was another close night-time flyby, coming within 327 km of the surface at 2239:23 on 16 March 1975. (These trajectories are discussed in more detail below; see Figs. 6 and 11.)

Mariner 10 was a three-axis stabilized spacecraft carrying three instruments which could contribute to the understanding of the magnetosphere. The first was a fluxgate magnetometer that had two ranges of $\pm$ 16 nT and $\pm$ 128 nT, each quantized to 10-bit accuracy for a digital window of 0.030 or 0.26 nT depending on range. The magnetometer could measure field strengths above 128 nT to a maximum of 3188 nT through the application of bias fields. During the first encounter the magnetic field stayed below 128 nT on each sensor (Ness et al. 1974a). On the third Mercury encounter the magnetic field reached 400 nT (Ness et al. 1975b). The fluxgate magnetometer included two redundant sets of triaxial sensors spaced 2.3 m apart on a 5.8 m boom. The difference in the readings of these two sets of sensors was used to put limits on the contribution to the measured field from spacecraft sources (Ness et al. 1974*b*). The contribution to the measured field from the spacecraft was estimated to be < 4 nT (Ness et al. 1974*a*). The magnetic field was sampled at a rate of 25 vectors per second.

The plasma science instrument consisted of a sophisticated ion and electron analyzer observing the sunward direction and a less elaborate electron instrument observing in the antisunward direction (Ogilvie et al. 1977). These instruments were mounted on a motor-driven scan platform which could operate at a scan rate of either 1 or 4 deg s^{-1}. The former rate was used at Mercury. The sunward-facing detector did not function properly and never detected counts above the cosmic ray background. Thus, all conclusions about the nature of the low-energy plasma environment of Mercury come from the rear facing electron detector. The solar wind bulk speed could be determined from these data but only to about $\pm$ 50 km s^{-1}. The instrument was a hemispherical electrostatic analyzer with 15 energy channels logarithmically spaced in energy between 13 and 715 eV. The instrument was stepped continuously through the 15 energy steps dwelling at each energy for 0.4 s and obtaining a complete energy spectrum in 6 s.

The third instrument was an energetic particle detector consisting of two telescopes, the Main Telescope (MT), and the Low Energy Telescope (LET) (Simpson et al. 1974; Eraker and Simpson 1986). The MT consisted of six detectors inside a plastic scintillator which was in anticoincidence with the telescope elements. The energy losses in 3 of the detectors (1, 2 and 5) were determined by 256-channel pulse-height analyzers. The LET was designed to measure low-energy protons and helium nuclei in the presence of a high intensity of low-energy electrons. For both telescopes the events accumulated for all particle range intervals were read out every 0.6 s together with the pulse height information. The telescopes were intended to measure electrons from 175 keV to 30 MeV and protons from 500 keV to 68 MeV. However, the detectors had some sensitivity to lower-energy particles when simultaneous deposition of energy by more than one particle piled up in the detector. Had instrumentation been carried by Mariner 10 to measure energies between the kilovolt range and the hundreds of kilovolt range, compensation could have been made for such pile-up effects. Since such instrumentation was not carried, there remains substantial ambiguity about the exact energy spectrum of the observed particles.

Interplanetary Plasma Environment

Mercury is unique in being both the closest planet to the Sun and, except for Pluto, having the most eccentric orbit (i.e., perihelion = 0.31 AU; aphelion = 0.47 AU). These two factors result in its magnetosphere being exposed to much denser and hotter solar wind plasma and more intense in-

TABLE I
Interplanetary Conditions[a]

Planet	R (AU)	V_{SW} (km s^{-1})	N_P (cm^{-3})	B (nT)	T_P (10^4K)	T_E (10^4K)
Mercury	0.31	430	73.	46	17	22
	0.47	430	32.	21	13	19
Venus	0.72	430	14.	10	10	17
Earth	1.0	430	7.0	6.0	8.0	15
Mars	1.5	430	3.1	3.4	6.1	13
Jupiter	5.2	430	0.26	0.83	2.7	8.7
Saturn	9.6	430	0.076	0.44	1.8	7.1
Uranus	19.1	430	0.019	0.22	1.1	5.6
Neptune	30.2	430	0.0077	0.14	0.82	4.8
Scaling	—	R^0	R^{-2}	$(2R^{-2}+2)^{1/2}/2R$	$R^{-2/3}$	$R^{-1/3}$

[a]Table adapted from Slavin and Holzer (1981).

terplanetary magnetic fields than any of the other planetary magnetospheres. Table I lists the principal interplanetary plasma and magnetic field parameters, their radial scaling laws, and extrapolated values based on typical 1 AU measurements. It is at once apparent that not only are all of the parameters higher at Mercury, but the eccentricity of Mercury's orbit produces very significant variations between perihelion and aphelion. One possible implication is that magnetospheric activity at Mercury may experience a strong semiannual variation.

Fortunately, the interplanetary environment between 0.31 and 0.47 AU has been well observed by the 1974-1980 Helios 1 and 2 missions (Musmann et al. 1977; Marsch et al. 1982). In Fig. 1 histograms of hourly averaged Helios 1 and 2 magnetic field magnitude, solar wind density, and velocity over the radial distances appropriate to Mercury have been compiled. The distributions are all quite broad, possibly as a result of both the 0.31 to 0.47 AU radial gradients and the solar cycle variations in these quantities over the years 1974 to 1980. The mean values all agree well with Table I and the factor of 4 to 9 increases in interplanetary magnetic field (IMF) magnitude and solar wind density over the 1 AU values are very evident.

The Helios observations may also be used to calculate derived parameters that play important roles in determining the nature of the solar wind-planetary interaction. Collisionless shocks, for example, are classified in terms of the Mach number (the ratio of the velocity of the solar wind to the velocity of compressional waves), the ratio of plasma thermal to magnetic field pressure (usually referred to as the beta of the plasma), and the angle that the magnetic field makes to the local normal to the shock surface, θ_{BN} (Tid-

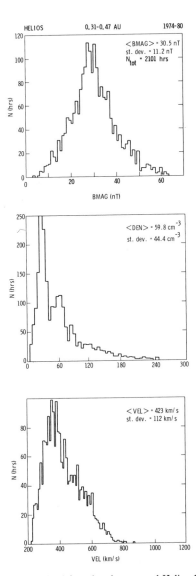

Fig. 1. Solar wind statistics derived from hourly averaged Helios 1 and Helios 2 observations obtained in the years 1974 to 1980 for the range of heliocentric distances corresponding to the orbit of Mercury (0.31 AU to 0.47 AU). Top panel shows histograms of field magnitude, middle panel shows density and bottom panel shows solar wind velocity. Means and standard deviations are indicated. (Data courtesy of NSSDC and Helios investigators.)

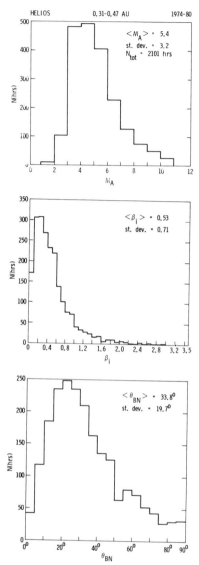

Fig. 2. Solar wind statistics derived from hourly averaged Helios 1 and Helios 2 observations for the orbit of Mercury for parameters expected to influence solar wind interaction with Mercury. Top panel shows Alfvén Mach number, middle panel shows ion beta and bottom panel shows angle between interplanetary magnetic field and shock normal at subsolar point.

man and Krall 1971; Greenstadt 1985). Furthermore, the nature of the bow shock determines the conditions in the magnetosheath including its width and shape, the amount of compression, the levels of turbulence, and the draping pattern of the magnetic field line (Russell 1985). Figure 2 shows that the Alfvénic Mach number and ion beta are much lower than is typical at the Earth and lower still than observed for the outer planets. So, too, should the magnetosonic, or compressional, Mach number and the total (electron plus ion) beta of the plasma be lower. These lower values for Mach number and beta will affect the nature of several of the processes occurring in the solar wind, but the median values at Mercury are within the range of values observed on occasion at 1 AU.

Energetic Charged Particle Environment

In addition to the highly variable solar wind plasma environment near Mercury, there are a number of other important sources of exogenic particle fluxes. These other particle populations include solar cosmic rays (SCR), galactic cosmic rays (GCR), Jovian electrons, and low-energy particles (~ 1 MeV) produced within the interplanetary medium. The energetic nuclei that pervade the inner solar system have a broad range of energies and compositions (see Table II). The nuclei of both the GCR and SCR components are primarily protons with a few percent alpha particles and about 1% heavy nuclei.

TABLE II

Energies, Mean Fluxes, and Interaction Depths of the Two Types of Cosmic Ray Particles[a]

Radiation	Energies ($MeV\,nucleon^{-1}$)	Mean Flux (particles $cm^{-2}\,s^{-1}$)	Effective Depth (cm)
Solar cosmic rays			
Protons and helium nuclei	5–100	~ 100	0–2
Iron group and heavier nuclei	1–50	~ 1	0–0.1
Galactic cosmic rays			
Protons and helium nuclei	100–3000	3	0–100
Iron group and heavier nuclei	~ 100	0.03	0–10

[a]Table taken from Reedy et al. 1983.

Galactic Cosmic Rays. The detailed spatial and temporal variations of the cosmic ray components at any point in the heliosphere are strongly controlled by the Sun's activity and the IMF. GCR particles probably originate in supernova remnants and/or the interstellar medium (Lingenfelter 1979) and subsequently are transported by various processes to the solar system. Observations near the Earth have clearly established that < 1 GeV n^{-1} GCR particles are modulated by about a factor of 10 during the solar cycle, while for $E \geq 10$ GeV n^{-1} the GCR ion population is not very strongly affected by solar variations (cf. Reedy et al. 1983 and references therein). As is evident from Table II, the GCR particles are highly penetrating, and they can represent a significant sputtering source for an airless planet like Mercury (McGrath et al. 1986). Clearly, owing to Mercury's weak intrinsic field, virtually all galactic cosmic rays will have essentially direct access to the entire planetary surface. Because of the stronger IMF and the greater proximity to the Sun, however, we would expect greater modulation of the GCR flux at Mercury's orbit during a given solar cycle. The surface layers of Mercury should have embedded within them a fascinating record of solar variability over the eons due to this GCR modulation effect.

Solar Cosmic Rays. Table II shows that SCR particle intensities are much higher than the GCR intensities, but the characteristic energies are much lower. Figure 3 (Castagnoli and Lal 1980; Reedy et al. 1983) shows the average solar proton spectrum as compared with the GCR proton flux for different degrees of solar-cycle modulation. As is clear from the figure, SCR proton fluxes tend, on average, to dominate the GCR fluxes below ~500 MeV. However, it should be kept in mind that just a few large solar flares produce most of the solar cosmic rays during any given solar cycle (Lal 1972). Calculations of the transport of solar particles predict that the integrated fluence of particles should obey an inverse square (R^{-2}) relationship, where R is the distance from the Sun (Zwickl and Webber 1977). Thus, the average spectral character of SCR particles should be similar at Mercury and at 1 AU, but the absolute intensities would be factors of 4 to 9 higher at Mercury than at Earth. As with GCR particles, the SCRs will leave an important and distinctive historical record of solar activity in Mercury's surface.

Neugebauer et al. (1978) have tried to estimate the highest likely solar flare proton and electron fluxes as a function of heliocentric distance. The SCR component would likely be dominant under most circumstances. As shown by Neugebauer et al. (1978), one can estimate the average SCR electron flux fairly effectively based upon a knowledge of the SCR ion fluxes. Unlike the GCR and SCR nucleonic component, the solar electron fluxes will not produce much of a record in the Mercury surface layers (Reedy 1977), but solar electrons could be an important source population for the Mercurian magnetosphere as is the case for Jovian electrons discussed below.

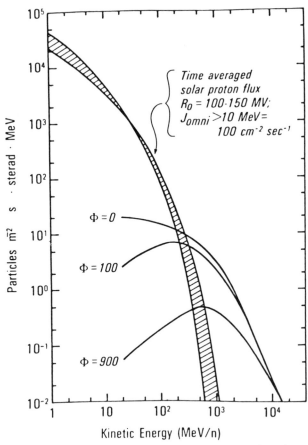

Fig. 3. The long-term average fluxes of solar protons determined from lunar data as a function of energy. The omni-directional flux, J_{omni} of solar proton flux with rigidity, R_o in the range 100 to 150 MV is approximately 100 cm^2 s^{-1}. These fluxes are much larger at low energies than the galactic cosmic ray (GCR) proton fluxes for different modulation levels. The curve $\phi = 0$ refers to no modulation while $\phi = 900$ is typical of GCR fluxes during solar maximum. (Figure after Reedy et al. 1983.)

Solar Neutrons.   The estimation of the solar flare neutron flux and fluence near Mercury is difficult because a neutron flux of solar origin has not yet been unambiguously detected at 1 AU. This failure can be understood since all experimental attempts to identify a solar neutron flux have been made in Earth satellites or balloons, where two physical constraints make detection of solar neutrons very difficult (Neugebauer et al. 1978). First, the approximately 12 minute half-life of neutrons reduces greatly the flux of low-energy neutrons which could survive to 1 AU. Second, local neutron background generated in the spacecraft, mainly the cosmic radiation, has placed severe limits on the

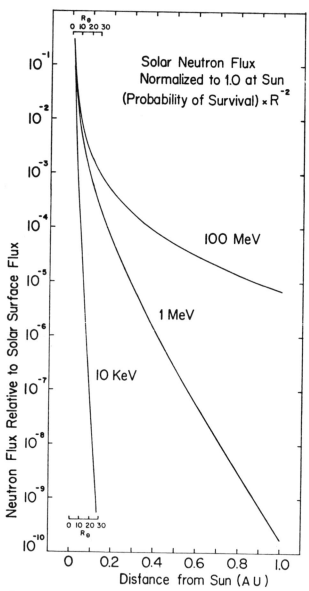

Fig. 4. The dependence of the normalized solar neutral flux as a function of radial distance from the Sun for three neutron energies, 10 keV, 1 MeV, and 100 MeV. (Figure after Neugebauer et al. 1978.)

sensitivity for detection. Indeed, only upper limits on neutron fluxes have been reported (Lockwood et al. 1973; Kirsch 1973). Figure 4 (Neugebauer et al. 1978) illustrates the effect of the decay of neutrons of various energies as a function of radial distance. The measurement of this dependence has been proposed as a means to identify solar neutrons and to separate the solar from background neutron fluxes (Anglin et al. 1972). Two components of solar neutron fluxes are expected to be present in the heliosphere. First, a quasi-steady solar neutron production primarily from extended areas of solar activity (mainly 10 keV to 1 MeV neutrons). Second, an impulsive neutron production from solar flares associated with collisions between accelerated charged particles and the chromospheric and coronal material (mainly 1-100 MeV neutrons).

During large solar flares, the region near Mercury may be strongly illuminated with solar neutrons. A spacecraft near the planet could probably readily detect and characterize the solar neutron spectrum thereby learning a great deal about the mechanisms operative in solar flares. Moreover, the interaction of solar neutrons with the Mercurian surface could be detected remotely using neutron albedo fluxes and gamma ray sensor systems (Simpson et al. 1974), thereby giving a remote diagnostic technique for probing the Mercurian regolith from satellite altitudes.

Jovian Electrons.   Jovian electrons, both at Jupiter and in the interplanetary medium near Earth, have a very hard spectrum that varies as a power law with energy (see, e.g., Mewaldt et al. 1976). This spectral character is sufficiently distinct from the much softer solar and magnetospheric electron spectra that it has been used as a spectral filter to separate Jovian electrons from other sources (Krimigis et al. 1975; Mewaldt et al. 1976; Chenette et al. 1977). A second Jovian electron characteristic is that such electrons in the interplanetary medium tend to consist of flux increases of several days duration which recur with 27 day periodicities (Teegarden et al. 1974; Mewalt et al. 1976). The Jovian electrons at 1 AU typically occur during the declining phase of solar wind streams and the 27 day periodicity of the Jovian electron increases was attributed by Conlon (1978) to the effects of recurrent, high-speed solar wind streams overtaking slower solar wind plasma. These form corotating interaction regions (CIR) in the heliosphere (see Fig. 5). A third feature of Jovian electrons at 1 AU is that the flux increases exhibit a long-term modulation of 13 months which is the synodic period of Jupiter as viewed from Earth (Chenette et al. 1977). Every 13 months, Earth and Jupiter are directly connected along the interplanetary magnetic field. Eraker and Simpson (1979) reported a broad range of synoptic data from Mariner 10 which showed that Jovian electron features persist in interplanetary space as close as 0.46 AU from the Sun. Their observations show the characteristically hard spectrum of the Jovian source; they show the modulation of the relativistic electron fluxes by CIR at the solar synodic period; the data also show

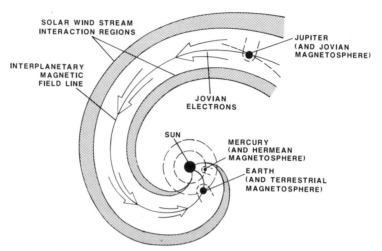

Fig. 5. Propagation of Jovian electrons to Earth and Mercury along the spiral magnetic field of the interplanetary medium, as seen from north of the ecliptic. (Figure after Baker 1986.)

the long-term intensity modulation (7 month period) due to the synodic period of Jupiter as seen by Mariner 10. Figure 5 illustrates the orbit of Mercury and the likely relationship of Mercury's magnetosphere to the Jovian electron transport. As revealed by the data of Eraker and Simpson (1979,1986), the Mercury encounter (M I) by Mariner 10 on 29 March 1974 occurred during the height of a Jovian electron increase in the interplanetary medium. Thus, it is very likely that Jovian electrons were enveloping the Mercury magnetosphere during this encounter. In analogy with the model developed for the Earth (Baker et al. 1979,1986), it has been suggested that Jovian electrons could supply a spectrally hard electron population which would have a significant influence inside the magnetosphere of Mercury (Baker 1986).

Planet

Mercury is the smallest of the terrestrial planets with a radius of 2440 km, intermediate between the Earth's Moon and Mars in size. It rotates more slowly than the Moon, rotating with a period of 59 days compared with the Moon's 28 day period. It differs from the Moon in the sense that its axial rotational period is not synchronous with its orbital period. Rather it orbits the Sun every 88 days so that every two Mercurian years the same side of the planet faces the Sun. This slow rotation and close proximity to the Sun implies a very high dayside surface temperature ($\gtrsim$ 630 K). The night side, in long periods of darkness and with no appreciable atmosphere, is cold ($95 \leq T \leq 130$ K).

Our knowledge of the surface of Mercury comes primarily from the nearly 3000 photos taken by Mariner 10, having resolutions ranging from 1 to 3 km and covering about 40% of the surface of the planet. Consideration of all of the optical, thermal and radar-albedo information has led to the conclusion that many of the properties of the surface of Mercury are similar to that of the Moon (Chase et al. 1976; Hapke 1977). This suggests that the surface of Mercury consists of a fragmented layer similar in grain-size distribution to the lunar regolith. As we will discuss further below, there remain important questions about the electrical and thermal conductivities of such a regolith material.

Mercury is observed to have the highest mean (uncompressed) density of any planet (5.3 g cm^{-3} versus 4.1 g cm^{-3} for the Earth (at 10 kbar) (Ringwood 1979). This high average density is taken to imply a composition of 70% metallic-phase material and 30% silicate-phase material. As noted above, Mercury possesses an internal magnetic field. If this is due to a planetary dynamo, this requires a molten core (see, e.g. Ness 1979). Estimates place the core-mantle boundary temperatures below the melting point of FeNi, suggesting an admixture of lighter material such as O or S (Gault et al. 1977). Assuming a dynamo origin for the magnetic field (see, e.g., Stevenson et al. 1983), the dynamo region could be very close to the planetary surface. The internal magnetic field probably consists of both dipolar and higher-order components but the results of the Mariner 10 flyby and its very limited coverage of the planet leave an intrinsic ambiguity between the contribution of the dipole and higher-order moments (cf. the chapter by Connerney and Ness). The observed low field magnitude compared with the magnitude that could be reproduced by a dynamo, operating in a core of the size of that of Mercury, implies that the Mercury dynamo may not be self-sustaining and may involve thermoelectric currents (Stevenson 1987). Any remote observation of temporal variations of the magnetic field close to Mercury could also provide information about the electrical conductivity of the mantle.

The atmosphere of Mercury was found by Mariner 10 to consist of an exosphere (Broadfoot et al. 1974) of neutral He (600 cm^{-3} surface density) and atomic hydrogen (8 cm^{-3}). Two principal candidates for producing the exospheric population are direct solar wind accretion on the surface and decay of thorium and uranium in the planetary crust. Recent, high-resolution spectral measurements of Mercury show emission in sodium D lines (Potter and Morgan 1985a). This suggests a substantial sodium population in Mercury's atmosphere (Ip 1986), possibly due to photo-sputtering of the planetary surface (cf. Cheng et al. 1987). Potassium has also been discovered (Potter and Morgan 1986). Mariner results indicated only a modest ionosphere at Mercury with an upper-limit electron density of 10^3 cm^{-3}. Direct observations in the polar cap reported by Ogilvie et al. (1977) showed electron densities of 0.1 cm^{-3}. At higher altitudes in Mercury's magnetotail, the observed electron densities were commonly 1 cm^{-3}. All of the present evidence, therefore, suggests at most a very modest atmosphere and ionosphere at Mercury. This, in turn, would indicate that solar wind plasmas, and any solar energetic or magnetospheric particle bursts, should be able to impact rather directly onto the planetary surface, except as they are stood off by the planetary magnetic field. Furthermore, magnetospheric processes should not be strongly affected by effects of the ionospheric conductivity at Mercury as they are at the Earth (see, e.g., Ogilvie et al. 1977; Hill et al. 1976). A major difference between the terrestrial and Mercurian magnetospheres should be in the sources for their plasma. At the Earth the solar wind is thought to be the dominant source (see, e.g., Hill 1974) but with a significant, and possibly very important ionospheric contribution (see, e.g., Baker et al. 1982). A major objective of magnetospheric studies at the Earth over the past two decades has been to separate quantitatively the two sources (see, e.g., Johnson 1983). The lack of a strong ionospheric source at Mercury may provide an opportunity for the study of how solar wind plasma enters a magnetosphere and is eventually precipitated onto the surface or lost down the tail. The plasma densities reported by Mariner 10 at high altitudes in the magnetosphere were higher than those observed at the Earth by a factor of about four, the ratio of the solar wind flux at 1 AU to 0.5 AU. This result was interpreted as confirming that the solar wind was the principal magnetospheric plasma source at Mercury as is observed at the Earth (Ogilvie et al. 1977).

II. SOLAR WIND INTERACTION

When the flowing solar wind plasma encounters a planetary magnetic field, it is deflected to either side of the planetary field and forms a magnetic cavity. The magnetic field exerts a pressure and the gradient in that pressure at the leading edge of the magnetic cavity exerts a force on the plasma. This force in turn stops the forward motion of the solar wind plasma and deflects it. This basic understanding of the formation of a magnetic cavity, or magnetosphere as it is called, allows us to calculate where the boundary of the magnetosphere lies and to first order the configuration of the magnetic field within the magnetosphere. We cannot calculate many of the properties of the thickness of this boundary, or magnetopause, nor the dynamics of the magnetosphere solely from the pressure balance since these other properties of the magnetosphere depend on other aspects of the interaction. However, knowledge of where the magnetopause is located for a particular solar wind dynamic pressure provides a good first order estimate of the strength of the planetary magnetic field, i.e., its intrinsic magnetic moment.

The magnetopause is also important because it is the site of energy transfer from the solar wind plasma to the magnetospheric plasma. What happens here determines how much energy eventually gets deposited on the planetary surface by the solar wind. We have no experience with a magnetopause not connected to an ionosphere except that gained at the four Mariner crossings of the Mercury magnetopause. We note that this energy deposition has been proposed to be a candidate for remote sensing with the VLA (Baker et al. 1986b).

In order that the solar wind be deflected by the planetary magnetic field, the solar wind must somehow sense the presence of the planetary obstacle. This information is transmitted upstream by a compressional wave, called a magnetosonic wave by plasma physicists. This magnetosonic wave is the analog of the usual sound wave in a collisional gas. The solar wind flows much faster than the speed of propagation of magnetosonic waves throughout all of the solar system except for a small region near the Sun. As a result, a standing bow shock wave is found upstream of each of the planets. This shock wave heats, slows and deflects the solar wind around the planetary obstacle. The Earth's Moon does not have such a shock wave because it absorbs rather than deflects the solar wind. That Mercury does have a bow shock was the first indication of the presence of a planetary magnetic field. The location of a bow shock is an indication of how large an obstacle to the solar wind is present, but the location of the shock also depends on several properties of the plasma, such as Mach number, which are themselves variable. Hence, it is not as straightforward to deduce the strength of the planetary magnetic field from the shock position as it is from the magnetopause position.

Shocks themselves are interesting phenomena in space plasmas because the processes which act to heat and deflect the solar plasma can also accelerate particles to extremely high energies. The region upstream of a collisionless shock which is connected to the shock by interplanetary magnetic field lines is called the foreshock and is replete with many interesting energetic charged particle and plasma wave phenomena (cf. Russell and Hoppe 1983).

The three Mariner 10 passes each showed evidence for the presence of a planetary magnetosphere. The Mercury I and III passes detected the bow shock and magnetopause and directly sampled the nightside magnetosphere. Mercury II, despite being a dayside encounter 50,000 km upstream, also may have observed effects of the upstream particles and waves. In this section, we examine the observations of the upstream waves, the bow shock, the magnetopause and the structure of the nightside magnetosphere.

Boundary Locations

Both the plasma instrument and the magnetometer detected clear signs of a bow shock and a magnetopause on both the inbound and outbound portions of the Mercury I and III passes (Ogilvie et al. 1974; Ness et al. 1974a). The locations of these boundaries and the Mercury I and III trajectories are shown in Fig. 6. The magnetic field data from the Mercury I pass are shown in Fig. 7. The inbound bow shock is clearly defined by the sharp rise in field magnitude. There are multiple crossings of the boundary as it moves back and forth at velocities greater than that of this spacecraft. The behavior is not unlike that seen at the bow shock of other planets with magnetospheres. The diffuse nature of the outbound bow shock is associated with the fact that the interplanetary magnetic field was oriented at a small angle to the shock normal. This same behavior is observed at the Earth (cf. Russell 1985).

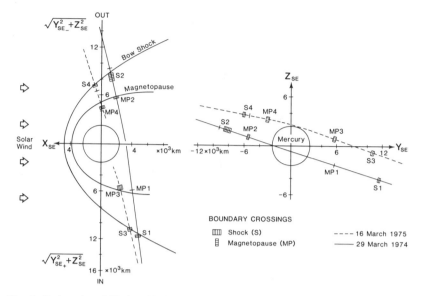

Fig. 6. Trajectories of Mariner 10 during the first and third Mercury flybys. Left-hand panel shows the projection on the ecliptic plane. Right-hand panel shows the view from the Sun.

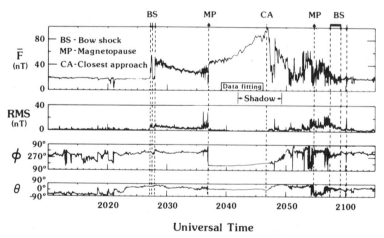

Fig. 7. Mariner 10 magnetic field measurements during the 29 March 1974 Mercury I flyby. Top panel shows the magnetic field magnitude, the next panel down shows the standard deviation of the magnetic field overall the components, the next panel shows the ecliptic longitude of the field and the bottom panel shows the ecliptic latitude. BS: bow shock; MP: magnetopause; CA: closest approach.

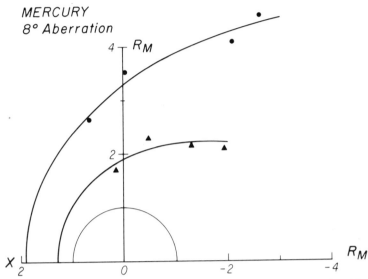

Fig. 8. Location of the bow shock and magnetopause crossings observed by Mariner 10 rotated by 8° to account for aberration of the solar wind velocity by planetary motion. Cylindrical symmetry of the surfaces has been assumed. (Figure after Russell 1977.)

The motion of Mercury in its orbit around the Sun causes an aberration of the otherwise nearly radial flow of plasma from the Sun. Correcting for this aberration in Fig. 8, we obtain the boundary crossing locations shown (Russell 1977). Extrapolating from these crossings to the subsolar region has large uncertainty because of the uncertainty in the shape of the boundary, the precise direction of the solar wind and the effects of flapping of the boundary. Nevertheless, if we do extrapolate, we obtain a subsolar magnetopause distance of 1.35 ± 0.2 R_M and a subsolar shock standoff distance of 1.9 ± 0.2 R_M. These distances give a ratio of 1.41 ± 0.3 compared to 1.3 for the Earth. The value for this ratio should depend on the shape of the magnetosphere and the Mach number of the solar wind (Spreiter et al. 1966). The Mach number is lower on average at Mercury, and the larger ratio at Mercury than at the Earth is in accord with our expectations. However, the value is very uncertain. We can make no specific deductions about the shape of the magnetopause based on this ratio (in contrast to what was possible for Venus [Russell et al. 1985]) or from the shape of the bow shocks as has been done for other planets by Slavin and Holzer (1981).

The position of the subsolar magnetopause can be used to determine the magnitude of the intrinsic magnetic moment of the planet if the dynamic pressure of the solar wind is known and if the interior of the magnetosphere contains a low-beta plasma. If the thermal energy density of the plasma approaches that of the magnetic field, i.e., the beta = 1 condition, then some of the pressure that holds off the solar wind is supplied by the plasma and the simple formula below does not apply.

If we take a magnetospheric shape factor appropriate for a gas dynamic interaction with a ratio of specific heats of 5/3 and ignore plasma effects and tail current effects (Schield 1969), we obtain in SI units,

$$M = 6.1 \times 10^{-4} R_{sp}^3 (\rho V^2)^{1/2}. \tag{1}$$

Here R_{sp} is the subsolar magnetopause radius in m; ρ is the density of the solar wind in kg m^{-3}; V is the velocity of the solar wind in m s^{-1} and M is the magnetic moment in T m^3.

If we substitute the measured solar wind values from the electron instrument as reported by Slavin and Holzer (1979a), we obtain a moment of 1.5×10^{12} T m^3. If we use their higher estimates of the pressure obtained by estimating the field strength at the subsolar point based on the observed strength at the terminator, we obtain a value of 2.7×10^{12} T m^3. We note that if we use all the corrections proposed by Slavin and Holzer (1979a), we obtain a value of $6 \pm 2 \times 10^{12}$ T m^3. Estimating the moment in this way is thus uncertain by at least a factor of two. It has an advantage over inverting the direct observations of the magnetic field in that the position of the magnetopause is globally determined rather than due to local effects. Direct observations are available only from a limited portion of the planet and may not be truly representative of the entire planet (cf. Chapter by Connerney and Ness). On the other hand, our limited knowledge of the exact dynamic pressure of the solar wind during the encounters and the necessity of extrapolating from the terminator region makes this global method also uncertain.

The magnetosphere shields the planet from the effects of the solar wind both as a supplier of hydrogen and helium and as a scavenger of any outgassing and sputtering products. Siscoe and Christopher (1975) and later Goldstein et al. (1981) examined the statistics of location of the subsolar magnetopause of Mercury to determine how often the solar wind would strike the surface of Mercury. While both used slightly different assumptions about the strength of the magnetic moment, the properties of the solar wind and how much tangential stress was applied to the magnetosphere, they both concluded that the solar wind seldom strikes the surface of Mercury. Figure 9 shows probability curves for the standoff distance of the magnetopause for two different assumptions about flux transfer to the magnetotail by the tangential stress applied by the solar wind. The probability of the solar wind striking the surfaces of Mercury for a magnetic moment of 2.5×10^{12} T m^3 varies from 6.1×10^{-5} (aphelion with no tangential stress) to 6.6×10^{-2} for perihelion with tangential stress.

Recently, D. Beard (personal communication, 1985) has advocated the use of the formula,

$$R/R_{sp} = 1 + 0.0851 \; \phi^2 + 0.0251 \; \phi^4 \tag{2}$$

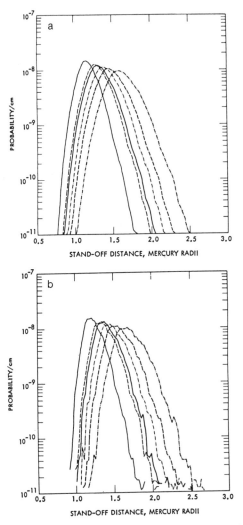

Fig. 9. Panel (a) shows the distribution of the solar wind standoff distance for the case of Earth-like flux transfer. The heavy lines are for aphelion solar wind conditions; the thin lines are for perihelion conditions. The solid curves correspond to a magnetic moment of $2.5 \times 10^{12}\text{T m}^3$, the dashed curve to $3.5 \times 10^{12}\text{T m}^3$ and the dot-dash curve to $4.5 \times 10^{12}\text{T m}^3$. Panel (b) shows the same as panel (a) but with no flux transfer. (Figure after Goldstein et al. 1981.)

to extrapolate to the subsolar point where ϕ is the angle of the observations from the subsolar point measured in radians. When applied to the Mercury I and III magnetopause observations, this formula gives a smaller subsolar radius than used above, $1.2\ R_M$. This value in turn is equivalent to a somewhat smaller planetary moment and solar wind would impact the planetary surface more often.

Finally, two other more esoteric effects may affect the location of the magnetopause boundary. First, if the relative amount of magnetic flux in the tail to that in the magnetosphere is larger at Mercury than at the Earth, then the shape of the magnetopause will be different and the subsolar point will be closer to the planet for the same magnetic moment (Slavin and Holzer 1979a). The tangential drag on the magnetopause exerted by viscosity and by the phenomenon known as reconnection governs the amount of flux in the tail (cf. Russell and McPherron 1973). However, since there is little or no ionosphere at Mercury and since the ionosphere at the Earth plays a major role in modulating magnetic flux transport to and from the tail, it is difficult to assess by how much Mercury's tail differs from that of the Earth.

The second effect is the stiffening of magnetic field lines by the high electrical conductivity of the planetary core and mantle (Hood and Schubert 1979; Suess and Goldstein 1979). These highly conducting regions will oppose any forces that attempt to bend the field lines embedded in them until the currents set up in them decay. Thus, the core and mantle stiffen the field lines most against rapid solar wind variations and least against long-term variations. All these effects add uncertainty to estimates of how often the solar wind can directly impact the surface of Mercury.

Bow Shock and Upstream Waves

When the interplanetary magnetic field lies nearly parallel to the normal to the bow shock, the bow shock is locally very turbulent. Simultaneously, elsewhere on the bow shock where the interplanetary magnetic field lies at right angles to the bow shock normal, the shock is much more well defined. This situation was exemplified by both Mariner 10 shock encounters. Figure 7 shows a quiet magnetic field at the inbound shock crossings where the magnetic field was nearly orthogonal to the shock normal, and a very disturbed magnetic field at the outbound shock where the magnetic field lies almost parallel to the shock normal. On Mercury III, as illustrated in Fig. 10, the situation was reversed. The inbound shock was the quasi-parallel shock and the outbound was the quasi-perpendicular shock.

As mentioned in the introduction, the solar wind Mach number and beta of the solar wind at Mercury's orbit are on average lower than at the Earth. These parameters affect the structure of the shock, so that there will in general be less ion heating at Mercury than at the Earth (cf. Gosling and Robson 1985). The magnetic profile will at least for quasi-perpendicular shocks be more regular because the bow shock is weaker. On the other hand, the interplanetary magnetic field is oriented more radially out from the Sun at Mercury's orbit. Thus, there will be a more frequent occurrence of quasi-parallel bow shocks at Mercury than at the Earth.

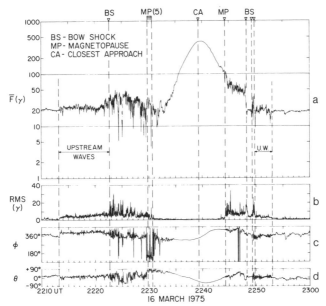

Fig. 10. Mariner 10 magnetic field measurements during the Mercury III flyby. Panel (a) shows the magnetic field magnitude, panel (b) shows the standard deviation of the magnetic field over all three components. Panel (c) shows the ecliptic longitude of the magnetic field and panel (d) shows the ecliptic latitude. BS: bow shock; MP: magnetopause; CA: closest approach; UW: upstream wave. (Figure after Ness et al, 1975a.)

Figure 11 shows the magnetic field profile across the quasi-perpendicular inbound shock on Mercury I (Ness et al. 1974a). The structure and waves present are not unlike those seen at the Earth's bow shock (Fairfield and Behannon 1976). In front of the outbound quasi-parallel shock, waves were observed in the 5 to 10 s period range that were very similar to those seen at periods of 20 to 60 s at the Earth. Hoppe and Russell (1982) have shown that, in fact, these same waves seem to be present in front of the bow shocks of all the planets and that their wave frequencies are proportional to the interplanetary magnetic field at the planet.

Mariner 10 passed Mercury well upstream of the bow shock during the Mercury II encounter. Figure 12 shows the geometry of this passage and the measured magnetic field and plasma parameters. Panel (c) shows a number that is representative of the high-energy (~600 eV) flux. Panel (f) shows the impact parameter of the field lines passing through the spacecraft. This parameter tells how close to the planet the field line comes. Some, but not all, of the strong peaks occur when the impact parameter is small, suggesting that Mariner 10 may have detected upstream electrons and associated magnetic disturbance 50,000 km upstream of Mercury.

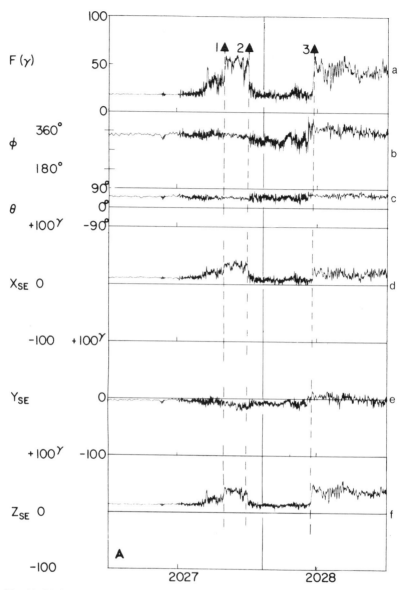

Fig. 11. Mariner 10 magnetic field measurements at 0.04 s resolution across the bow shock on the inbound leg of the I pass. Panel (a) shows the field magnitude. Panels (b) and (c) show the ecliptic longitude and latitude and panels (d), (e) and (f) show the three solar ecliptic coordinates. (Figure after Ness et al. 1974a.)

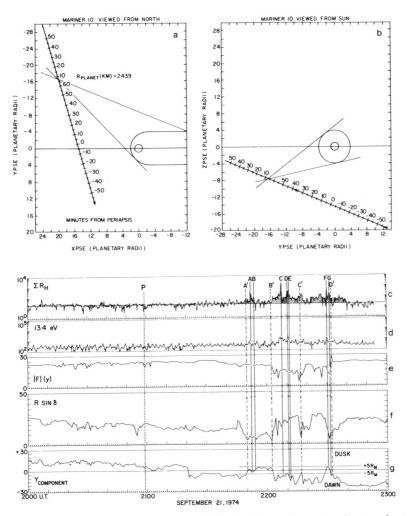

Fig. 12. Panels (a) and (b) show the trajectory during the Mercury II encounter. An approximate bow shock is shown to illustrate the requirements for field line intersection with the bow shock. The panels below show low energy electron data obtained during this pass. In (c) is the sum of the highest-energy channel counting rates and in (d) the counting rate of the 13.4 eV channel. Below that is the magnetic field magnitude (e), the impact parameter (f) and its component on the Y solar ecliptic direction (g). (Figure after Ogilvie et al. 1977.)

Energetic electrons ($E > 60$ keV) and ions ($E > 80$ keV) almost 10^7 km from Mercury have been reported as probably coming from Mercury (Kirsch and Richter 1985). The enhancements in the direction coming from Mercury are small and close to the level of statistical significance. Kirsch and Richter (1985) attribute the putative energetic particles to substorms on Mercury rather than bow shock acceleration. Their argument regarding the source of these particles is not convincing because they minimize the possible energization of the ions at the bow shock by using a 20 nT IMF in their calculation whereas, when discussing reconnection two paragraphs later, they maximize the energization by using an IMF that is twice as large. Since it is quite likely that, in fact, no particles were observed coming from Mercury, it seems moot to discuss the efficiency of the possible acceleration mechanisms.

In the same paper that discusses the upstream waves and bow shock, Fairfield and Behannon (1976) also examined the waves in the magnetosheath. Nothing unusual or particular about the Mercurian magnetosheath was observed. The low-energy electrons detected by the plasma analyzer were also quite typical of a planetary magnetosheath (Ogilvie et al. 1977), i.e., they were hotter than the solar wind and cooler than in the magnetosphere.

Structure of the Magnetopause

The magnetopause is the boundary between the flowing, shocked, solar wind plasma in the magnetosheath and the magnetic field and plasma of the magnetosphere. At the Earth, the plasma outside the magnetosphere is usually much more dense and cooler than that inside and the field strength is lower outside than inside. Since there is a change in the magnitude and the direction of the magnetic field at the magnetopause, a current flows at this boundary. The smallest thickness this boundary could have, under the assumption that no strong electric fields are present, is one proton gyroradius. This would occur if the current layer consisted only of particles that started in the magnetosheath and were turned around by the magnetospheric magnetic field and returned to the magnetosheath (cf. Willis 1971). However, this is not how the terrestrial magnetopause behaves (Berchem and Russell 1982). The magnetopause is many gyroradii thick because there is a population of so-called trapped ions that circulate in the boundary and do not return to the magnetosheath or the magnetosphere each orbit. It is of interest to determine if a magnetopause that is not connected to an ionosphere behaves in a similar way. Although it is somewhat difficult to make the measurement from a single spacecraft, the rough estimates that can be made from the Mariner 10 data indicate that the Mercury magnetopause is similar in thickness to that of the Earth (Russell and Walker 1985).

At the Earth, magnetic field lines are transported from the dayside magnetopause to the magnetotail by the process of reconnection (cf. Russell and McPherron 1973). This process may take place in a steady-state manner (cf. Sonnerup et al. 1981) or it may take place in a non-time stationary manner (Russell and Elphic 1979). It is very difficult to show that reconnection is taking place in a steady manner without 3-dimensional high-resolution plasma measurements. Mariner 10 was not so instrumented. On the other hand, the signature of non-steady or patchy reconnection at the Earth's magnetopause has a quite characteristic pattern at the Earth's magnetopause in the magnetic

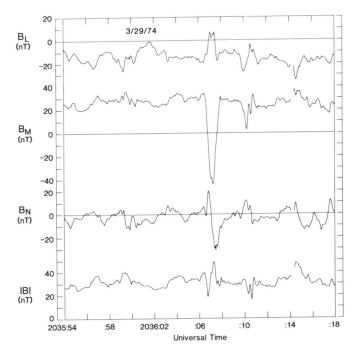

Fig. 13. The three components of the magnetic field in boundary normal coordinates near the magnetopause on the first inbound pass of Mariner 10. The time resolution is 0.04 s. The N-component points outward along the normal to the magnetopause. (Figure after Russell and Walker 1985.)

field. To see this signature, it is best to display the measurements in a coordinate system that is oriented in the plane of the magnetopause. Figure 13 shows high-resolution magnetic field records taken just outside the first inbound magnetopause on the Mercury I pass and is displayed in "boundary-normal" coordinates (Russell and Walker 1985). The feature at 2036:06 which is strongest in the B_M component looks very similar to the structures associated with terrestrial patchy reconnection and which have been called flux transfer events (FTEs). The strength of the magnetic signatures is similar to the terrestrial events but their durations are much shorter. The Mercurian flux transfer events are about 400 km across or about 6% of the size of a terrestrial event. Since the Mercurian magnetosphere is about 5% of the size of the terrestrial magnetosphere, the Mercurian events have the same relative scale as the terrestrial events. They also occurred more frequently, about once a minute at Mercury, compared with about once every 8 minutes on the Earth. Figure 14 shows an artist's conception of a series of flux transfer events connecting a planetary magnetosphere to the solar wind (Kuznetsova and Zeleny 1986).

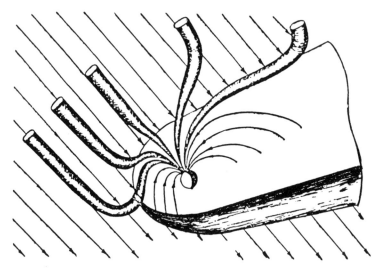

Fig. 14. Artist's conception of a series of flux transfer events connecting a planetary magne-tosphere to the solar wind. (Figure after Kuznetsova and Zeleny 1986.)

The small scale sizes of these FTEs at Mercury have been interpreted to be due to the limited time for growth of the structures as they are blown back by the solar wind from the subsolar region (Kuznetsova and Zeleny 1986). One, therefore, might expect that at Jupiter the FTEs would grow to a much larger dimension. However, observations show that at Jupiter the FTEs are of similar size to those at the Earth. Kuznetsova and Zeleny (1986) hypothesize that there is an upper limit to the size of FTEs, and that they cannot grow to a size that exceeds by too many times the thickness of the magnetopause layer. At Mercury, then, the short transit time from the subsolar point limits the growth, and at Jupiter the magnetopause thickness limits the ultimate size.

We note that no evidence has been reported at the terrestrial magne-topause for smaller FTEs near the subsolar point. An alternate possibility is that FTEs grow at the subsolar point until they reach a certain size limited either by the size of the magnetosphere or by the thickness of the magne-topause, whichever provides the smaller upper limit.

Magnetosphere and Magnetotail

The crucial nightside and high-latitude portions of the magnetosphere were well sampled by Mariner 10, but the dayside magnetosphere was not probed directly. The dimensions of the dayside magnetosphere, however, were inferred from modeling and extrapolation of the magnetopause and bow shock surfaces (see, e.g., Ness et al. 1974a; Ogilvie et al. 1977; Russell 1977; Slavin and Holzer 1979a). Slavin and Holzer (1979a) obtained solar wind standoff distances of 0.3 to 1.1 R_M above the surface of the planet which were

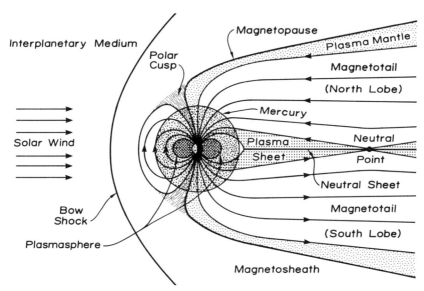

Fig. 15. The planet Mercury scaled so that its magnetosphere occupies the same volume as the terrestrial magnetosphere. In this noon-midnight meridian, it is seen that the planet occupies a large fraction of the inner magnetosphere including all of the plasmasphere as well as the region of the most intense Van Allen belts.

dynamic-pressure-corrected. The variability was attributed to the effects of dayside magnetic merging and magnetospheric dynamics as will be considered in later sections.

The fact that Mercury occupies a large fraction of its magnetosphere significantly alters the expected structure of the magnetosphere. Figure 15 illustrates this in the noon-midnight meridian plane by superimposing a scaled Mercury on the Earth's magnetosphere. Simply because of size alone without any other effects included, the features occurring in the inner magnetosphere, the plasmasphere and the energetic radiation belts would all be absent. The plasma sheet would almost touch the surface of the planet near midnight. The polar cap field, consisting of field lines that entered the magnetotail, would extend to much lower planetary latitudes on the night side. Figure 16 shows the orthogonal view from the north. The solid lines with arrows now show the motion of low-energy, or cold plasma, which drifts across magnetic field lines that are all perpendicular to the page, because of the electric field applied to the magnetosphere by the solar wind. The strength of this electric field is proportional to the strength of the coupling between the solar wind and the magnetosphere. There is also a component of the electric field due to the rotation of the Earth and its highly conducting ionosphere. This rotation produces a set of closed streamlines in the inner magnetosphere of the Earth. On these streamlines, plasma can circle the Earth essentially forever and build up

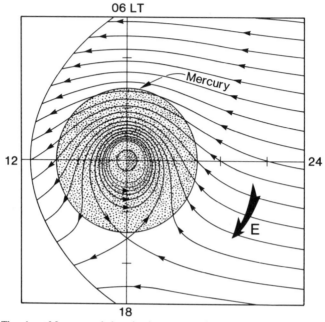

Fig. 16. The planet Mercury scaled so that its magnetosphere occupies the same volume as the terrestial magnetosphere. In this equatorial plane projection, it is seen that the planet blocks the entire region of corotating plasma. However, on Mercury we would not expect such a region, even if the planet were smaller relative to the magnetosphere, because of the lack of a dense atmosphere and ionosphere and because of the slower rotation. In the Mercurian magnetosphere the streamlines of the flow should be much straighter.

in density from sources in the ionosphere at low altitudes. This leads to the formation of the high-density, cold plasma region called the plasmasphere. Outside that region the plasma will be lost, convected out of the magnetosphere, through the magnetopause, at least whenever the coupling between the solar wind and the magnetosphere is strong.

Mercury rotates more slowly than the Earth; it does not have a substantial ionosphere and has very little ionospheric plasma with which to supply the magnetosphere. Thus, even if Mercury did not occupy much of its magnetosphere, we might expect it not to have a plasmasphere. Furthermore, the flow lines should be different from those sketched in Fig. 16; they should be much straighter than the terrestrial ones. Any cold or low-energy plasma that gets into the equatorial region should be swept out through the magnetopause. This should be an important loss process for the Mercurian atmosphere. Any ionized component will find itself swept out of the magnetosphere into the solar wind.

A simple way mentally to compare Mariner 10 observations with the same measurements at the Earth is to use a scaling where $1 R_M$ equals $8 R_\oplus$

(Siscoe et al. 1975). Using such conversions, the M I encounter, had it taken place at Earth, would have corresponded to an entry into the magnetosphere at $X = -13$ $R_\oplus$ and an exit at $X = -5$ $R_\oplus$ with closest approach at $X = -8$ $R_\oplus$, $Z = 3$ $R_\oplus$, and the spacecraft would leave the magnetosphere at $X = 0.5$ $R_\oplus$, $Z = 10$ $R_\oplus$. While these scalings are only approximate and ignore the very different boundary conditions on these two magnetospheres (e.g., the conducting ionosphere at the Earth vs a poorly conducting regolith at Mercury), they do indicate that the Mariner 10 data set consists of primarily near-tail and high-latitude polar cap observations.

Magnetic Field Observations. Figure 17 (Ness et al. 1975*b*) displays an overview of the Mercury I (29 March 1974) magnetic field vectors projected on the plane perpendicular to the solar direction and on the ecliptic plane. This display complements the time series shown in Figs. 7 and 10. The corre-

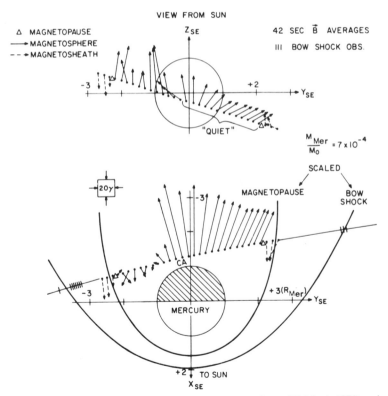

Fig. 17. Observed magnetic field vectors from the Mercury I pass (29 March 1974) projected onto the ecliptic plane (bottom) and on the plane perpendicular to the Sun line (top). (Figure after Ness et al. 1975*b*.)

sponding diagram for Mercury III (Ness et al. 1976) is not as instructive because the field is less tail-like at the low altitudes of the Mercury III pass. As expected on the basis of the scalings for a terrestrial-type magnetosphere, the region adjacent to the inbound M I magnetopause is the south lobe of the magnetotail. There follows a smooth monotonic increase in field magnitude to a peak field strength of 98 nT at closest approach (Ness et al. 1974b,1975b). Shortly after closest approach, the field magnitude dropped and a strong current sheet was crossed which reversed the polarity of the X component of the magnetic field. The magnetic field decrease is believed to be diamagnetic and to correspond to the entry into a high-beta central plasma sheet analogous to the region separating the two lobes of the magnetotail at the Earth (Hartle et al. 1975; Ogilvie et al. 1977a). The structure of the current sheet is further investigated in Fig. 18 where the high-resolution (i.e., 25 vectors per second) magnetic field measurements have been subjected to a minimum variance

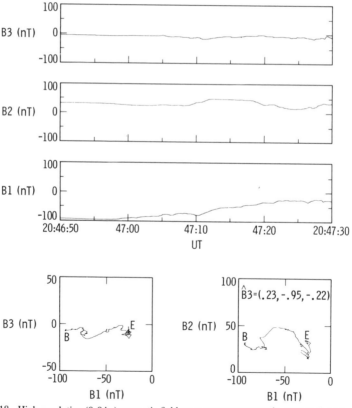

Fig. 18. High-resolution (0.04 s) magnetic field measurements across the current sheet during the M I pass in the minimum variance coordinate system. The minimum variance is along the B3 direction.

analysis (cf. Sonnerup and Cahill 1967). In the top panels the field is plotted in an orthogonal coordinate system where the maximum variation takes place along the B1 axis and the minimum is in the B3 direction. For a weak near-planet extension of the cross-tail current sheet, the minimum variance direction should be predominantly out of the magnetic meridian containing the dipole field. The analysis in Fig. 18 indicates that this is indeed the case with the minimum variance direction essentially antiparallel to the Mercury Solar Orbital (MSO) Y-axis. These results are consistent with the contention that it was the cross-tail current sheet which Mariner 10 crossed shortly after closest approach as opposed to a temporal variation.

The outbound M I magnetic field observations were very disturbed and characterized by large-amplitude fluctuations. Based on the measurements from all of the Mariner 10 instruments during this period and theoretical calculations, Siscoe et al. (1975) argued that Mercury's magnetosphere is subject to frequent, short-duration, intense substorms which are directly analogous to the substorm phenomena observed in the Earth's magnetosphere. The Mercury III high-latitude magnetic field observations in Fig. 10 are far quieter, especially quieter than those seen after closest approach in Fig. 7, with a very smooth rise and decline from the peak field magnitude of 400 nT at closest approach. This second encounter, with its very strong, well-ordered magnetic fields, proved definitively that Mercury possessed a significant intrinsic field which interacted with the solar wind to generate a magnetosphere. In terms of the modeling of the Mercury magnetic field, the two Mariner 10 encounters were highly complementary. The magnetic fields measured during Mercury I, because of its greater altitude, were predominantly due to the magnetospheric current systems driven by solar wind interaction (Ness et al. 1975a). While not as useful for the derivation of the intrinsic planetary magnetic field, the Mercury I observations did aid in the modeling of the magnetosphere (Whang 1977; Jackson and Beard 1977) and the subtraction of these large external fields from both the Mercury I and Mercurian III data sets. As treated in the Chapter by Connerney and Ness, the Mercury dipole moments inferred from the Mariner 10 observations vary from 6 to 2×10^{22} G cm^{-3} depending principally upon the magnitude of the higher-order moments included in the harmonic analyses (Ness 1979; Slavin and Holzer 1979b).

Figure 19 shows the magnetospheric field lines in the noon-midnight meridian. It is difficult to determine how long a planetary magnetotail extends even on a well-studied planet. Theoretically this depends on the viscosity of the solar wind and the merging efficiency. One recent estimate of the tail length gives a length of 15 to 60 R_M and a polar cap radius of $22° \pm 4°$, assuming a merging efficiency of 0.1 to 0.2 (Macek and Grzedzielski 1986).

Plasma Observations. Counting rates for selected energy channels from the plasma electron experiment (Ogilvie et al. 1974,1977) on the Mercury I and III passes through the magnetosphere are displayed in Fig. 20. At the

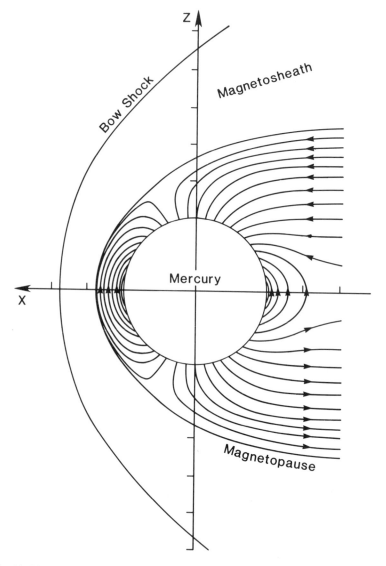

Fig. 19. Magnetic field lines and a realistic magnetopause and bow shock location for a model
Mercurian field. (Figure after Jackson and Beard 1977.)

bow shock inbound and outbound for both passes, the electrons are hotter in
the magnetosheath than in the solar wind. This can be seen as an increase in
the flux at the lowest energy. The entrance into the magnetosphere for both
encounters is marked by a rise in the high-energy flux and a drop of the low-
energy fluxes. These electrons are similar to those seen in the plasma sheet

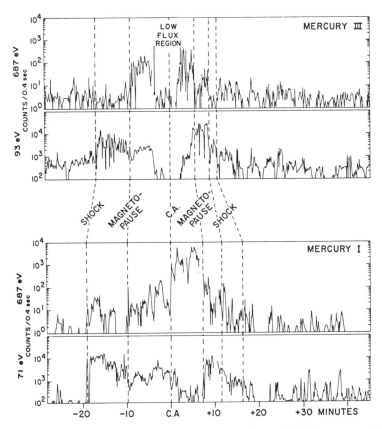

Fig. 20. Counting rates of the electron spectrometer on the Mercury I and III passes at selected energies to illustrate the similarities and differences of the two passes. (Figure after Ness 1979.)

boundary layer at the Earth. No cold, dense plasmas indicative of ionospheric or plasmaspheric particles were encountered. This result is consistent with the $1 R_M$ to $8 R_\oplus$ scaling which would scale these terrestrial features to a location well below the surface of the planet. Most of the Mercury I and Mercury III trajectories placed the spacecraft in the vicinity of the midplane plasma sheet and its horn-like extension to lower altitudes. Figure 21 shows energy spectra obtained in the plasma sheet on the two passes. The plasma sheet densities at Mercury are higher than those measured at Earth by a factor of five which is close to the ratio of the solar wind density at 0.5 to 1 AU. This finding supports the hypothesis that the solar wind is their primary source (Ogilvie et al. 1977) in the absence of a significant ionosphere. The magnetospheric implications of the recent discovery of exospheric Na at Mercury (Potter and Morgan 1985a) will be considered in a later section.

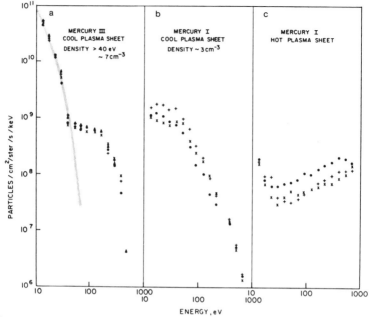

Fig. 21. Three electron spectra obtained by the Mariner 10 electron spectrometer. Panel (a) shows a cool plasma sheet spectrum from M III. The high flux at low energies is due to photoelectons. The shaded line is a scaled photoelectron spectrum from IMP-6. Panel (b) shows a cool plasma sheet spectrum from the M I encounter; panel (c) a hot plasma sheet spectrum from M I. (Figure after Ogilvie et al. 1977.)

In the case of the Mercury III passage at higher latitude, Fig. 20 indicates that the spacecraft penetrated a very low-flux region near closest approach. On the basis of the magnetic field observations (Ness et al. 1975a), this region should correspond to the low-altitude extension of the lobes of the magnetotail. These so-called polar cap regions at the Earth are populated by very low fluxes of solar wind electrons which enter the tail along open field lines which connect to the solar wind (see, e.g., Baker et al. 1986c and references therein). At Mercury the large size of the planet relative to the magnetosphere requires that the lobe magnetic fields intersect the surface over a larger area than at the Earth. Conservation of magnetic flux arguments by Ness et al. (1975a) indicate that the colatitude of the polar cap at Mercury is ~22°, or nearly twice that observed at the Earth.

Figure 22 (Ogilvie et al. 1977) displays the principal plasma regions detected by Mariner 10. Straight lines mark the Mariner 10 trajectories and the magnetic equator based on the magnetic field modeling. The fact that the two passes were nearly parallel to the magnetic equator is consistent with the

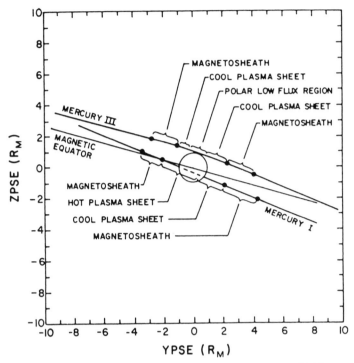

Fig. 22. The trajectories of Mariner 10 on the first and third encounters with Mercury as seen from the Sun. The plane shown is the Y-Z planetary solar ecliptic projection. The various regimes found in the plasma electrons are indicated as well as the location of the magnetic equator or current sheet. (Figure after Ogilvie et al. 1977.)

near-continuous contact of the spacecraft with portions of the plasma sheet which lies near the magnetic equator at these altitudes. The one region not yet discussed is the hot plasma sheet segment of Mercury I which began around closest approach and extended through the rest of outbound passage. It is this region that was marked by the large magnetic fluctuations that Siscoe et al. (1975) interpreted as being temporal in nature and associated with substorm activity. During substorms at the Earth there is a large scale conversion of magnetic energy into charged particle heating and acceleration in the plasma sheet (Bame et al. 1967; Russell and McPherron 1973). This phenomenon is reproduced in Fig. 20 where the electron population shifts rapidly to higher energies. The peak electron energy was beyond the 690 eV upper threshold of the electron plasma detector, but it was estimated to be around 1 keV (Ogilvie et al 1977) as is typical for terrestrial substorms. This event will be discussed in more detail in the following section.

III. ENERGETIC PARTICLES AND MAGNETOSPHERIC DYNAMICS

As discussed in the introduction, Mariner 10 carried no instrumentation to measure energetic particles from about 1 to 100 keV. This is the energy range in which much of the energy flux in the terrestrial magnetosphere is observed. Furthermore, there were no Mariner 10 ion observations and most of the terrestrial energy flux is contained in the ions. Thus, it is difficult to make direct comparisons of the processes which energize the Mercurian and terrestrial plasmas. The major evidence for a substorm-like phenomenon at Mercury comes from the sudden entry into a region of hot electrons after closest approach on Mercury I, as discussed above, and the observations of energetic particle bursts with the charged particle telescopes. As mentioned earlier, the possibility of pileup of lower-energy particles to exceed the detectors' energy threshold, and our inability to use independent measurements to identify when this has occurred and make allowances, colors somewhat the interpretation of these energetic particle bursts.

Energetic Particle Bursts

The principal results on energetic particle bursts in Mercury's magnetosphere came from flyby M I on 29 March 1974. The upper half of Fig. 23 shows data from Eraker and Simpson (1986) for the period 2030 to 2100 UT. Panel (a) shows counting rates from a sensor designed to measure electrons with $E \geq 170$ keV, while panels (b) and (c) show concurrently measured plasma number density ρ and electron temperature T. Panels (d), (e) and (f) show magnetic field data including field magnitude B, azimuth γ, and inclination Θ. The field data are presented in a right-handed coordinate system in which X is toward the Sun and Z is perpendicular to Mercury's orbital plane.

Of note in Fig. 23 (upper half) is the fact that Mariner crossed the magnetopause (MP) at 2037 UT inbound, reached closest approach at 2046:40 UT, and subsequently crossed the magnetopause again at ~2054:30 UT outbound. Several bow shock (BS) crossings were seen between 2057 and 2059 UT. Panel (a) shows several enhancements of the energetic electrons both in the magnetotail (events A, B, B' and C) and in the magnetosheath (events D and D').

Siscoe et al. (1975) and Ogilvie et al. (1977) have shown that the energetic particle burst periods in Fig. 23 tended to be times of substorm-like behavior. According to these authors, Mariner 10 entered the near-tail below the plasma sheet, on the dusk side. The sheath field was northward on tail entry while the field inside the magnetopause was very tail-like and relatively quiet. As seen in Fig. 23, the magnetic field strength increased with time as the planet was approached and, as discussed above, the higher-energy plasma electrons decreased in intensity (see also Christon et al. 1986). Shortly after closest approach, the field strength decreased rapidly and the field inclination

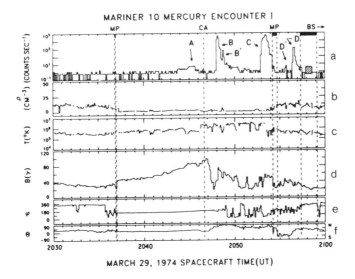

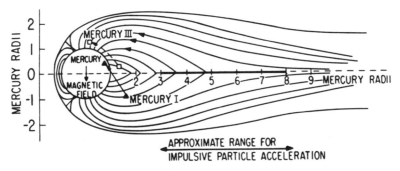

Fig. 23. Upper half of the figure: Energetic particle measurements together with the plasma and magnetic field data from the M I pass showing the relationship of the four particle bursts to the activity in the plasma and the magnetic field. The lower half of the figure shows the magnetospheric model used by Eraker and Simpson (1986) to explain these bursts.

increased markedly. This indicated a transition from a tail-like to a dipole-like field orientation. In the terrestrial case (Baker et al. 1984) this would be a clear signature of the onset of a substorm expansive phase. Between 2047 and 2054 UT there were several large changes in the magnetic field strength and these occurred in the same time period as did the large energetic particle bursts *B*, *B'* and *C*. Even the particle burst *A* prior to closest approach appears to have some (weak) magnetic field changes associated with it. The magnetic field was strongly southward upon Mariner's outbound exit from the magnetotail.

Siscoe et al. (1975) thus took special note of the fact that the IMF switched from northward to southward while Mariner was in the Mercurian magnetosphere. They suggest, in analogy with the Earth's case, that this initiated strong sunward plasma sheet convection and, presumably, enhanced magnetotail energy storage. In this picture, when Mariner was about halfway through the tail, the southward IMF initiated a series of substorms. Siscoe et al. showed by scaling arguments that substorm time scales should be of order 1 to 2 minutes in Mercury's case compared with 30 to 60 minutes in the terrestrial case (see, also, Slavin and Holzer 1979a). Hence, several substorms in a 20 minute period are not unreasonable for Mercury.

Eraker and Simpson (1986) developed this scenario further and suggested that the substorms in Mercury's magnetotail resulted from magnetic reconnection (neutral line formation) in the range of 3 to 6 R_M on the night side. They suggested, in close analogy with the Earth, that this substorm reconnection resulted in the impulsive acceleration of energetic particles and that this mechanism was the source of the bursts seen in panel (a) of Fig. 23. The magnetotail model envisaged by Eraker and Simpson and the likely region of particle acceleration is shown in the lower half of Fig. 23.

An important discovery of Simpson et al. (1974) and of Eraker and Simpson (1986) was that the energetic electron bursts detected by Mariner 10 were highly modulated with a period of 5 to 10 s. This is illustrated by the detailed field and energetic particle plot shown in Fig. 24. This exhibits high-resolution data for event C which commenced just prior to 2053 UT on 29 March 1974. Particularly striking features of this event were the rapid rise of flux by several orders of magnitude at the onset and the subsequent strong flux modulation at 6 to 8 s periods. The flux modulation was particularly strong during the decay phase of the event (i.e., from 2053:30 to 2054:15 UT).

Eraker and Simpson (1986) discussed an interpretation of the modulated energetic electron bursts as seen in Fig. 24 in terms of individual particle acceleration events of several seconds duration. Thus, each periodic increase in the events B, B', C and D, as shown in Fig. 23, upper half—see Eraker and Simpson 1986—was taken as a new episode of magnetic reconnection in Mercury's magnetotail. Assuming the nominal response of the instrument to electrons and taking a minimum area for the size of the region of enhanced particle flux, they obtain an energy of 10^{10} J per 6 s burst. With a 1% efficiency for acceleration, the magnetosphere would have to supply energy at a rate of 10^{11} W to power the peak particle intensities. Assuming a maximum area for the region of enhanced particles, they obtain a peak power of 10^{13} W. Thus, of order 10^{11} to 10^{13} W of power would have to be extracted by Mercury's magnetosphere from the solar wind to power each 6 s burst. This solar wind energy input rate is comparable to the solar wind energy transfer rate estimated for the terrestrial magnetosphere (Baker et al. 1984). Such a large

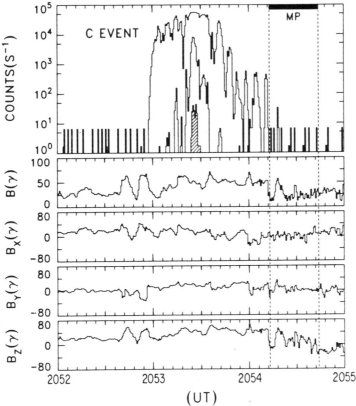

Fig. 24. High-resolution plot of the *C* event (M I pass) showing the modulation in both the magnetic field and the count rate of energetic particles. (Figure after Eraker and Simpson 1986.)

energy extraction rate for Mercury's magnetosphere would be surprising given that its cross-sectional area to the solar wind flow is ~700 times smaller than that of the Earth. It is also surprising that the tearing-mode instability would repeat that rapidly. Considerations that would lower the required energy would be if there were very significant electron pile up or a contribution of protons to the observed count rate or if the bursts observed were quite rare requiring a long period of energy accumulation before the energy could be released in these bursts.

In analogy with terrestrial substorm models, Baker et al. (1986*a*) assumed that, during a substorm on Mercury, a neutral line forms across a large fraction of the width of the tail at 3 to 6 R_M from the planet. A major effect of this neutral line would enhance greatly the sunward convection of plasma and magnetic flux in the plasma sheet (Siscoe et al. 1975), thereby rapidly returning flux to the dayside magnetosphere. In conjunction with neutral line formation, there would be a strong planetward collapse of magnetic field lines which produces a much more dipole-like field region in the midnight sector, and a compression wave (Russell and McPherron 1973; Moore et al. 1981) which moves toward the planet at substorm onset, heating and accelerating plasma as it moves. Baker et al. (1986a) hypothesized that energetic particles are impulsively accelerated in the magnetotail of Mercury and are then transported closer to the planet and injected on closed field lines. They then drift around the planet several times as a relatively coherent bunch (drift echoes) as is seen in the terrestrial case. Baker et al. calculated adiabatic drift times in reasonable accord with the 6 to 8 s periodicities seen in the recurrent bursts of the Mariner data. It is expected that, in each successive passage of the particle bunch through the dayside region, particles would be lost through the magnetopause; one would also expect that particles would be lost into the tail each time the bunch passed through the nightside region.

The event D (Fig. 23) seen in the magnetosheath would presumably correspond (in this model) to bursts of particles on draped sheath field lines lost through the magnetopause as the particle bunch passes around on the day side of the system. The broad overall decrease in the pulse peaks (decay of peak intensity) would be due to expected losses of particles from the system.

Baker et al. (1986c) concluded that the available magnetic field, plasma, and energetic particle data support a model in which energetic particles are produced primarily near the tail reconnection region some 3 to 6 R_M from the planet. It is argued that for a brief period (a few seconds) there exist large ($\sim$500 kV) electric potentials near the tail neutral line due to an induced electric field. Quantitative estimates suggest that a region some 2 to 3 R_M wide must be involved in the inductive ($\partial B / \partial t$) process.

A variant on this model has been proposed by Christon et al. (1986). They note that as illustrated in Fig. 25 for event B there is a close correlation between the waves in the magnetic field and the fluctuations in the count rate during the decay phase of the burst. They propose that there are only two bursts during the B event, both of which have injection fronts at which the field and particles suddenly increase and following which the particles gradually decay. The analogous process on the Earth is called the multiple onset substorm. Christon et al. (1986) also attribute the bulk of the count rate of the energetic particle instrument to particles with energies above but in the neighborhood of 35 keV. Thus, their model and interpretation of the data requires lower overall voltage drops in the magnetosphere and a lower energy deposit in particle acceleration, placing fewer demands on the efficiency of energy extraction by the Mercurian magnetosphere from the solar wind than either of the other two models.

Relativistic Electron Events

The data taken by Mariner 10 in 1974 suggest two quite different kinds of energetic particle increases (see Fig. 23). The most evident difference between

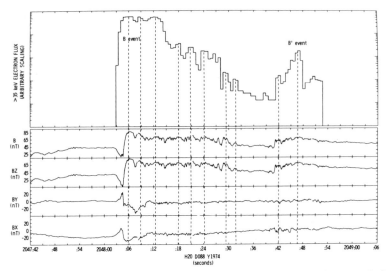

Fig. 25. High-resolution plot of the *B* event showing the modulation in both the magnetic field and the count rate of energetic particles. (Figure after Christon et al, 1986.)

the *A* and *B* events, for example, is the much higher peak counting rate in the *B* event ($\sim 10^5$ c s^{-1}) compared to the peak of the *A* event (< 10 c s^{-1}). In fact, the true *B* event count rate was really much higher than shown due to electronics saturation effects (see, e.g., Christon et al. 1979). The *A* event increased gradually toward a peak intensity and the entire event lasted for a period of over two minutes. In contrast, the *B* event lasted for only about 30 s at which time the *B'* increased occurred (Fig. 25). Furthermore, the *B* event exhibited a very fast rise time (~ 1 to 2 s) followed by a more gradual decay. This is reminiscent of fast rise-slow decay substorm events in the terrestrial magnetosphere (Baker et al. 1984) and is completely different from the *A* event.

A further distinction between the *A* and *B* events is in the magnetic field variations which occur in concert with the particle bursts. For event *A* there was a smooth, gradual increase (for the most part) of the field strength with no major fluctuations. Even in the detailed plots of the magnetic field components (Ness 1979; Christon et al. 1986), there were only modest breaks in the field traces associated with the *A* injection event. In contrast, the *B* event showed major changes in the field strength (Fig. 25) and overall field component configuration at the time of the event onset (Eraker and Simpson 1986; Christon et al. 1986). The field direction signatures at ~ 2048 UT are consistent with the kind of field reconfiguration (return toward a dipolar state) which occurs at terrestrial substorm onset.

A final point of contrast between the A and B events is in the inferred electron energy spectrum. As shown by Eraker and Simpson (1986), the A event is well described by a power law spectrum ($\partial J/\partial E = k\, E^{-\gamma}$) with the spectral index $1.5 \lesssim \gamma \lesssim 2.0$. On the other hand, the B event electron spectrum was found to be much softer and if a power law spectrum is again assumed, the B event requires $\gamma \gtrsim 7$.

In the case of the terrestrial magnetosphere, the source of relativistic electron populations at geostationary orbit (6.6 R_E) has been a continuing puzzle. The occurrence of distinctive relativistic electron bursts in the Mercurian system constitutes a surprising result within the framework of the generation mechanism of an Earth-like substorm. Baker et al. (1979,1986a) have argued that Jovian electrons make a dominant contribution at higher energies in the Earth's outer magnetosphere. They suggested that the Jovian electron population enters the distant magnetotail and plasma sheet (where the fields are relatively weak) essentially unattenuated. The spectrally hard Jovian electron population then becomes part of the pre-existing plasma sheet population and begins to participate in the magnetospheric dynamical processes. This model suggests that Jovian electrons would be swept toward the Earth as part of the overall plasma sheet convection.

The model assumes that the electron's first adiabatic invariant ($\mu \sim E/B$) is conserved during the convection and injection process. Thus, as the electrons are transported from interplanetary space and the deep tail (where B is small) to the region of synchronous orbit (where B is large), there will be an increase of B by a factor of 20 to 50. Preservation of μ and a spectrum extending down to of the order of 100 keV as $E^{-1.5}$ suggests an increase of flux at a given energy by a factor of 100 or more due to this "adiabatic acceleration" effect. Even larger flux enhancements in the magnetosphere are possible if fluxes of Jovian electrons can be trapped and stored in the outer magnetosphere for many convective time scales. Throughout the magnetosphere, the Jovian electrons would dominate the magnetospheric population (at energies above hundreds of keV) since the Jovian spectrum is so much harder than the substorm-generated magnetospheric spectrum.

Baker (1986) has also advocated this model for Mercury to explain the A event. During normal, quiescent conditions, the Jovian population would probably dominate the Mercurian tail fluxes at most energies. During an episode of greatly enhanced tail convection (as when the IMF suddenly turns southward, for example), it would be expected that much of the tail electron flux would be swept strongly toward Mercury. Under such circumstances, there would be a large injection of Jovian electrons into the inner magnetosphere of Mercury. Thus, one would expect a burst-like increase of Jovian electrons for 1 to 2 minutes during periods of enhanced solar wind-magnetosphere coupling. As in the terrestrial analog, the Jovian electrons at Mercury are suggested to be a high-energy tracer population which has a distinctively hard-energy spectrum. This feature distinguishes the Jovian electrons in a clear way from solar or magnetospheric sources. Such a mechanism obviates the need for the magnetosphere of Mercury to produce highly relativistic, spectrally hard electron bursts by internal generation mechanisms.

In the case of Mercury, one has a very dynamic magnetosphere where it is highly unlikely that persistent, long-term particle trapping occurs (see, e.g., Ogilvie et al. 1977). Similarly, it seems unlikely that the magnetosphere of Mercury is stable on long enough time scales for substantial particle radial diffusion to occur (Hill et al. 1976). Furthermore, the planet fills a large fraction of the magnetospheric phase space so that trans-L diffusion at low altitude and high latitudes is impossible. All of these facts about Mercury argue against the kind of recirculation mechanism required by the internal generation model at the Earth. Yet, very high-energy electron bursts occur at Mercury, and this therefore suggests that the Jovian (external) source model is operative there.

Energy Input and Dissipation Rates

As a result of Mercury magnetosphere-solar wind interaction and the concomitant substorm-like processes, several possible remotely observable effects are expected (Baker et al. 1986b). First, because of magnetic interaction between the solar wind and Mercury's outer magnetosphere, there may be "open" polar cusp and polar cap field lines onto which solar wind plasma may have relatively direct access. This inflow of solar wind plasma could potentially constitute a large incident particle kinetic energy flux if one considers the entire magnetospheric cross section as a collection area. Assuming a magnetospheric obstacle size of radius 3 R_M (1 R_M = 2440 km) to the solar wind flow (see, e.g., Ness 1979), one has $A_M = \pi R^2 \sim 2 \times 10^{18}$ cm^2. Taking a solar wind speed of 300 to 800 km s^{-1} and a density of 20 to 50 cm^{-3} (Slavin and Holzer 1979b), the kinetic energy flux (ρV^3) is estimated between 1 and 40 erg cm^{-2} s^{-1}. Using the above area estimate, this gives 10^{10} to 10^{12} W of incident power, and even a small fraction of this intercepted power dumped onto the darkside polar caps could produce a substantial surface heating effect.

A more likely remotely observable effect at Mercury is the result of magnetospheric substorm energy dissipation. From examination of the available particle and field data, it is possible to estimate the solar wind electromagnetic energy extraction rate of the system and also the peak energy dissipation rates during explosive substorm onsets. Such estimates can be obtained based on scaling from the terrestrial case (Siscoe et al. 1975; Baker et al. 1986c) or they can be obtained directly from hot plasma and energetic particle observations made by Mariner 10 (see Eraker and Simpson 1986). Magnetospheric scaling arguments (Siscoe et al. 1975) suggest a solar wind energy input rate at Mercury of order 10^9 to 10^{10} W. Even higher input rates ($\sim 10^{11}$ W) might be required if certain substorm dissipation models apply (see, e.g., Eraker and Simpson 1986).

The amount of energy stored in the tail of Mercury and its rate of conversion into hot plasma has been estimated by several means. Siscoe et al. (1975) estimate substorm dissipation in the range 10^{11} to 10^{12} J; if this is dissipated on a time scale of ~ 10 s, then as much as 10^{11} W power is available. Alternative estimates of the substorm energy dissipation suggest 10^{12} to 10^{14} J of energy at a minimum during a substorm burst (Eraker and Simpson 1987, in preparation). Again using ~10 s dissipation times, this suggests peak substorm powers of 10^{11} to 10^{13} W or more may be available. Since the body of Mercury fills so much of the magnetosphere, much of the hot plasma and energetic particle power (perhaps $\geq 10^{12}$ W, during a major substorm burst) would be dumped directly onto the planetary surface.

A majority of the substorm-dissipated power should be confined to the band of magnetic flux tubes that map into the Mercury tail. This set of field lines which define the Mercury auroral zones is illustrated in Fig. 26. It is suggested that the precipitated hot plasma and energetic particles from substorms in this (primarily) nightside latitudinal band (from $\sim45°$ to $\sim70°$ lat) would constitute a large energy influx onto the cold regolith surface (Baker et al. 1986b). The total surface area of Mercury is $\sim8 \times 10^{17}$ cm^2. If one assumes an auroral precipitation region of $\sim 1/10$ the planetary surface and if one assumes a peak power of 10^{14} W, on occasion, as much as $\gtrsim 1$ mW cm^{-2} may sometimes be incident on the ~110 K surface.

The calculations of Baker et al. (1986b) suggest that appropriate groundbased infrared telescopes could provide useful information about the solar wind-Mercury interaction. They point out that the infrared data could allow nearly instantaneous assessment of magnetospheric reconfigurations and sequences of energy deposition. Radio observations, on the other hand, would provide time-averaged heating information relating to deeper layers

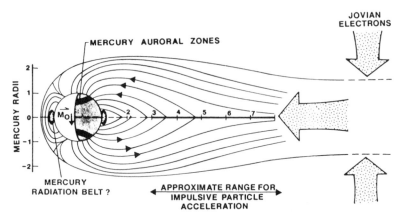

Fig. 26. Schematic illustration of acceleration of Jovian electrons and their eventual precipitation in a Mercurian auroral oval.

beneath the Mercury surface. Furthermore, by mapping the brightness temperature of the planet across the day-night terminator in two frequencies, one should be able to calibrate the electrical and thermal properties of the Mercurian regolith.

Baker et al. (1986b) therefore conclude that solar wind-magnetosphere interactions at Mercury should produce several effects which would in principle be observable from groundbased or near-Earth facilities: (1) there may be direct interaction of the solar wind with the surface of Mercury which could produce measurable heating effects; (2) substorm-like magnetotail processes within Mercury's magnetosphere may precipitate particles carrying in excess of 10^{12} W power into narrow latitudinal bands on the cold dark side of Mercury, producing surface-heated auroral zones; and (3) the presence of Jovian electrons within the inner Mercury magnetosphere may produce transient, very low-frequency synchrotron-emitting radiation belts. Thus, an observational program based on these ideas would hold great promise for providing remote information about the intrinsic properties of Mercury's magnetic field, about the planetary surface, about the dynamics of the Mercury magnetosphere, and about the interaction of this magnetosphere with the solar wind at ~0.4 AU. Such information has the possibility of greatly illuminating issues of general magnetospheric physics through comparative studies with the Earth, Jupiter, Saturn and other such systems. It would also permit a significant clarification of physical conditions near Mercury which would be crucial for proper definition of future Mercury-orbiting spacecraft that may be launched by NASA (Yen 1985) or other agencies.

OUTSTANDING PROBLEMS AND RECOMMENDATIONS

Based upon two magnetospheric passes by the Mariner 10 spacecraft, we have gained a superficial glimpse of the workings of Mercury's magnetosphere. We recognize many apparent similarities to the terrestrial system, but many questions remain. In future satellite missions to Mercury, it would be crucial to probe the dayside magnetosphere and magnetopause. Similarly, a Mercury probe should explore the deep magnetotail region ($r \gtrsim 10$ R$_M$). Particularly valuable data could result from a highly elliptical, polar orbiting spacecraft.

A number of questions come to mind, based upon analogies with terrestrial magnetospheric analogs:

1. Are there permanent, or quasi-permanent, radiation belts near Mercury just above the planetary equator?
2. Is there anything equivalent to a geomagnetic storm on Mercury, including a major ring current development?
3. How does Mercury produce high fluxes of energetic particles during substorms and what implications does that have for the terrestrial magnetosphere?

4. How large (in latitude and longitude) are the auroral regions on Mercury, and what is the auroral energy dissipation rate?

5. What is the energy transfer rate into Mercury's magnetosphere from the solar wind? Is this process more efficient under the solar wind conditions prevalent at 0.4 AU?

6. How is the dayside reconnection process affected by the lower Alfvénic Mach number in the solar wind at Mercury?

7. How much magnetic flux is contained in the magnetotail? What is the size and shape of the magnetosphere?

8. How is global magnetospheric convection effected on Mercury without a substantial polar ionosphere?

9. Are plasmoids produced in the magnetotail during substorms as they are in the terrestrial case?

10. What are the true ion-to-electron flux ratios throughout the magnetosphere, both at quiet and at disturbed times? This question applies both at plasma and suprathermal energies.

11. What are the electron and ion energy spectra and angular distributions at all points in the magnetosphere, especially during substorms?

12. What plasma instabilities are associated with the larger loss cones present in the magnetosphere of Mercury?

13. How do the magnetospheric electric fields, both transverse and parallel to the magnetic field, affect the loss of the sodium atmosphere upon photoionization?

14. What is the plasma and energetic ion composition within the magnetosphere of Mercury, i.e., what are the ultimate plasma sources?

15. How does the magnetosphere of Mercury behave during intense solar flare bombardment? Where does this energetic flux hit the surface of the planet?

16. What is the role of particle and photon sputtering in producing an atmosphere and exosphere at Mercury and does such sputtering affect substorm dynamics?

17. How often and where does the solar wind hit the planetary surface?

18. What is the source of the observed MHD fluctuations in the magnetosphere? Are they caused by processes in the magnetosphere or in the solar wind?

19. What is the neutron flux from the Sun at quiet and disturbed times?

20. What are the dipole and higher-order moments of the planetary field? If rich in harmonic content, how does this affect the solar wind interaction and magnetospheric processes?

These and many other questions strongly suggest that a planetary orbiter mission for Mercury would be highly desirable. In fact, a pair of simple spacecraft near Mercury, one at low-altitude (polar) and the other in a highly elliptical orbit to monitor the deep tail or the upstream solar wind, would be well suited to address these questions. Because of the high "metabolism rate" of Mercury's magnetosphere, a one-year mission at Mercury could provide a wealth of data on the magnetospheric dynamics, including a full range of substorm and (possible) storm activity.

Many questions about the intrinsic magnetic field properties and surface conductivities (thermal and electrical) remain crucial to understanding the workings of the inner magnetosphere. As discussed above, the remote radio and infrared observations suggested by Baker et al. (1986b) hold some potential for addressing several magnetospheric and surface questions until a spacecraft mission can be undertaken. Only an orbiter program, however, can fully resolve the magnetospheric structure and dynamics issues.

Acknowledgments. The preparation of this review chapter was supported at the University of California at Los Angeles by a contract from National Aeronautics and Space Administration. Work at Los Alamos National Laboratory was performed under the aegis of the U. S. Department of Energy and at the Jet Propulsion Laboratory, California Institute of Technology under contract with NASA.

THE MERCURY ATMOSPHERE

D. M. HUNTEN
University of Arizona

T. H. MORGAN
NASA Johnson Space Center

and

D. E. SHEMANSKY
University of Arizona

The known gases in Mercury's atmosphere are H, He and O, discovered by the ultraviolet spectrometer on Mariner 10, and Na and K, discovered from the ground by Potter and Morgan. Dayside number densities at the planet's surface are estimated to be between 100 and 4×10^4 cm^{-3}, with probable large night-side enhancements for hydrogen and helium. Other possible gases are considered; the only ones likely to be present in comparable abundance are H_2 and H_2O. The atmosphere is technically an exosphere, but the gas-surface interaction is very different from the interaction of a normal exosphere with the atmosphere below. Quantum-mechanical effects alter the velocity distribution and the rate of migration across the surface, and accommodation to the local temperature is inefficient. Probable sources are the solar wind for hydrogen and helium, and evaporation of meteoroidal material for the alkalis and water, with a possible contribution to the former by sputtering and photosputtering. Solar-wind ions generally do not enter the atmosphere directly, but rather via Mercury's magnetosphere. They tend to be implanted in surface materials, and later be displaced to the atmosphere by the impact of subsequent ions; most hydrogen is probably released as H_2. The dominant sink for all of the atoms seems to be photoionization. Following photoionization, most ions are recycled to the sur-

face and neutralized, although a significant fraction are swept up in the flows of the magnetosphere and solar wind. For H_2O photodissociation dominates, and may be a substantial source of H, H_2 and O. Solar radiation pressure is a large effect, especially for Na and K, but its role as a sink is probably small except for unusually fast atoms.

I. INTRODUCTION AND OVERVIEW

The atmosphere of Mercury is tenuous; the gas particle density is low enough that the planet's surface forms the exobase boundary. In other words, atoms collide with the surface more often than with each other. Five elements are known to be present in the atmosphere: oxygen, sodium, helium, potassium, and hydrogen, in order of decreasing abundance. Helium, hydrogen, and oxygen were discovered by the Mariner 10 airglow spectrometer (Broadfoot et al. 1974,1976), while sodium and potassium were discovered with groundbased instrumentation (Potter and Morgan 1985,1986a). In all cases, the basis of the identification was the observation of emission by resonant scattering of sunlight. Table I contains wavelengths, flux, subsolar point density, scale height and other observational details for each known constituent, as well as similar information for the Moon. We cannot say that these five elements constitute a complete list of the major species in the atmosphere of Mercury, because the total pressure of the known species is almost two orders of magnitude less than the upper limit of the atmospheric pressure, 10^{-12} bar, set by the Mariner 10 occultation experiment. Table II contains the measured upper limits of abundances and densities for other possible species at the location of the planet's terminator. The relationship between the data shown in Tables I and II is uncertain because little is known about the distribution of the gases over the surface. Limited observational evidence and uncertain atmospheric theory both suggest a large concentration of gases on the night side; the transition between night and day side in the terminator region is even less understood.

The physical quantities listed in Table I provide only a very schematic description of the atmosphere of Mercury, and much is unknown. The mean free path of an atom in Mercury's atmosphere is greater than the scale height of any of the components so that the atoms of the gas move on ballistic trajectories. The first collision experienced by a typical atom in the atmosphere is with the surface. The velocity distribution and extent of the atmosphere are controlled by poorly understood gas-surface interactions and velocity-dependent loss processes.

Early work on the lunar and Mercurian atmospheres, discussed in Sec. V, assumed that the gas-surface interaction could be represented in the same way as a true exobase, the level in an atmosphere above which collisions between atoms can be considered negligible. Although this assumption is invalid, it still serves as a useful point of departure. An exobase is not a true boundary;

TABLE I
Known Gases on Mercury and Moon

Species	Wavelength (Å)	Mercury[a]			Moon[b]	
		g^c (ph atoms^{-1} s^{-1})	Brightness (R)	N_0 (cm^{-3})	N_0 (Day) (cm^{-3})	N_0 (Night) (cm^{-3})
H	1216	5.3×10^{-3}	70, 720	23, 230[d]	<10	—
He	584	5.1×10^{-5}	70	6.0×10^3	2×10^3	4×10^4
O	1304	2.1×10^{-5}	63	4.4×10^4	—	—
Na	5890, 5896	2.45, 1.22	$1-10\times10^5$	$1.7-3.8\times10^4$	—	—
K	7664, 7699	3.24, 1.67	5×10^3	5×10^2	—	—
Ar[e]	869	5.5×10^{-8}	—	$<6.6\times10^6$	1.6×10^3	4×10^4

[a] Densities or upper limits based on resonance scattering.
[b] Density or upper limits based on *in situ* measurements.
[c] Scattering coefficient (photons atoms^{-1} s^{-1}) at Mercury aphelion and no Doppler shift. For Na and K they vary (see Sec. IV).
[d] Hot and cold components.
[e] Mercurian values inferred from lunar values.

TABLE II
Number Densities at Mercury Terminator: Upper Limits from Mariner 10
Occultation Experiment

	Max. Abundance[a] (10^{15}cm^{-2})	Cross Section (10^{-17}cm^2)	Number Density (10^7cm^{-3})
He	3.7	0.8	2.6
Na	5.0	0.6[b]	11
K	5.0	0.6[c]	14
O	2.5	1.2	4.2
Ar	0.9	3.5	3.1
H_2	2.9	1.0	1.4
O_2	0.9	3.5	2.5
N_2	0.9	3.3	2.3
CO_2	0.4	7.4	1.6
H_2O	0.8	3.6	1.5

[a]Observed upper limit to slant integrated density; value from Broadfoot et al. (1976) except as shown in footnotes b and c.
[b]From cross sections of Samson (1982).
[c]Assumed equal to Na.

atoms and molecules are continually crossing it from below and above. In the exosphere, they are maintained close to a Maxwellian velocity distribution by collisions that occur mainly *below* the exobase.

An important part of our information on Mercury's atmosphere is the distribution of helium over the day side and twilight regions. Lack of agreement with conventional exospheric theory was a major clue that this theory is not applicable. The temperature of the Earth's exobase typically varies by a factor of about 1.3 between day and night, and there is a corresponding excess of H atoms on the night side. This excess can be understood as a balance between a lateral flow of atoms that is more rapid from warm to cold than the reverse. To first order, the density is expected to vary as $T^{-5/2}$, where T is the exospheric temperature (see Sec. V). The observed noon and midnight temperatures on Mercury at aphelion are 575 and 110 K; the ratio of helium densities would therefore be 80. A Monte Carlo simulation of this transport by Smith et al. (1978) gives a ratio of helium densities of 150. This difference is probably due to the inclusion of Jeans escape from the dayside atmosphere in the Monte Carlo calculation. The loss rate by escape is comparable to the flow rate to the night side, and the dayside density is reduced to half what it would be under steady state conditions. The value inferred from Mariner 10 airglow data was 50; the fact that it is lower than both theoretical estimates is probably due to two comparable effects:

1. Helium atoms do not efficiently accommodate their velocity to the temperature of the local surface; indeed, they retain more than 90% of their incident energy after a collision;

2. Interaction of the helium atoms with the surface may not produce velocities sufficient for escape of the atoms.

The physical interaction of the atmosphere with the surface has been treated only in a most rudimentary way in model calculations, and only for sodium and helium. For other elemental species, the only modeling of exospheric distributions has been done with conventional exospheric theory, at least partly because of a lack of physical parameters describing the gas-surface interaction. These interactions are not understood in the required detail (Shemansky and Broadfoot 1977). Relevant properties such as the surface composition and microstructure on Mercury are essentially unknown. The discussion of the surface interactions of atmospheric species on Mercury is limited to known properties of lunar surface material, which is justified only by similarities in the albedo and the general geological appearance. Use of the lunar analogy for Mercury suffers from obvious differences such as proximity to the Sun (temperature) and the presence of an organized magnetic field on Mercury. Perhaps even worse, it is necessary to simplify grossly the surface composition to pure quartz, crystalline SiO_2.

None of the species known to be present in the atmosphere of Mercury can remain bound to the planet for a time comparable to the planet's age, so that there must be a source for each of the elements. Several likely sources of hydrogen and helium have been discussed, but their relative importance is not well defined. The sources of sodium and potassium are even less understood. Several processes may act to remove atoms from the atmosphere, but their relative importance has not been established in all cases. The magnetic field of Mercury may control either the removal or supply of some species to the atmosphere, but the extent and nature of this role have not been established. Examination of Table I reveals remarkable similarities between Mercury and the Moon, for the few comparisons that can be made. We shall examine each of these issues after a discussion of the observations.

II. PRE-MARINER STUDIES

The planning of the Mariner 10 airglow and solar-occultation experiments (UVS, ultraviolet spectrometers) was based on numerous unsuccessful groundbased attempts to detect an atmosphere on Mercury, either spectroscopically or by its scattering of light. This work is now summarized.

The chapter by Dollfus (1961) describes his visual polarimetric measurements at various wavelengths and positions on Mercury's image. Variations with both variables were interpreted as giving evidence for an atmosphere with a surface pressure of about 1 mbar (one millibar is equal to 1000 dynes cm^{-2} or 100 Pa). As he pointed out, it was necessary to assume that the effects were not due to the properties of the surface. This assumption was examined in detail by O'Leary and Rea (1967) who concluded that known surface effects could indeed explain the results, and suggested that the 1 mbar should be regarded as an upper limit. Further measurements by Ingersoll (1971) extended into the ultraviolet; his photoelectric technique gave no spatial resolution. He found no evidence for gas and reduced the upper limit to a value that depends on composition, but is 0.28 mbar for CO_2.

In a survey of Mercury's atmosphere written for a symposium, Field (1964) considered, along with Dollfus' suggestion, the possibility that an atmosphere might be convecting heat to the night side (then believed to be in permanent darkness). At this time, the 3:2 relationship between orbital and rotational periods and 176-day diurnal period (Pettengill and Dyce 1965) had not yet been established, and there was some evidence that the dark side of Mercury was warmer than expected. Although Field suggested a thin atmosphere, he did not suggest a pressure. In addition to their spectroscopic study, mentioned below, Belton et al. (1967) adapted Field's suggestion to the case of an object with a diurnal cycle. In this version, heat moves vertically through the soil, downwards during the day and upwards at night. The presence of gas can greatly increase the thermal conductivity of a fine powder, and might be revealed by a warmer nighttime temperature. Earth-based measurements available then could not decide the issue. Morrison (1970) presented a thorough review of computations of the thermophysics of the surface, taking into account the varying distance of Mercury from the Sun. The Mariner 10 radiometer (Chase et al. 1976) showed low nightside temperatures that are consistent with a gas-free powder, and fully consistent with the much lower bounds on the pressure set by the UVS.

A number of groups attempted to detect the absorption spectra of common gases, particularly CO_2 and CO; this work is summarized in Table III. Moroz (1965) suggested a marginal detection, which was adopted in a review by Rasool et al. (1965), along with Dollfus' polarimetric argument, as evi-

TABLE III
Spectroscopic Upper Limits for Mercury's Atmosphere

Gas	Wavelength (μm)	Upper Limit		Authors
		(cm-Å)	(μbar)	
CO_2	0.87	not seen		Adams and Dunham (1932)
CO_2	1.6	not seen		Kuiper (1952, p. 352)
CO_2	0.87	5700	4000	Spinard et al. (1965)
CO_2	1.6	~2000[a]	1500	Moroz (1965)
CO	2.35	10	5	Moroz (1965)
CO_2	1.6	200	150	Binder and Cruikshank (1967)
CO_2	1.2	58	40	Bergstralh et al. (1967)
CO_2	1.05	500	360	Belton et al. (1967)
CO_2	2.04	0.2	0.15	Fink et al. (1974)

[a]According to the re-analysis by Belton et al. (1967).

dence for a surface pressure of a few millibars. However, a re-analysis by Belton et al. (1967), with an improved curve of growth consistent with their own new data, showed only an upper limit. With this correction, Table III shows only a sequence of increasingly tight upper limits, the state of knowledge when the Mariner 10 experiments were planned and selected. For a nominal dayside temperature of 500 K, the bounds correspond to number densities of 10^{14} cm^{-3} for CO and 2×10^{12} cm^{-3} for CO_2, far above the Mariner 10 limits in Table II.

The last two columns of Table I give a summary of measurements of the Moon's atmosphere (reviewed by Hodges et al. 1974). Most of the results are from the landed or orbiting mass spectrometers (Hodges et al. 1972; Hodges 1973b; Hodges et al. 1973; Hoffman et al. 1973; Hodges and Hoffman 1974), which suffered severe interference during the day from gases emitted by the large quantities of hardware remaining in the vicinity. Attempts to detect H Lyman α were frustrated by the same large interplanetary background that is responsible for the noise level at high altitudes in Fig. 3 below (Sec.III.C) (Fastie et al. 1973).

III. MARINER 10 OBSERVATIONS OF THE ATMOSPHERE

The encounters of Mariner 10 with Mercury took place on 29 March 1974, 21 September 1974 and 16 March 1975. All occurred at aphelion, with the identical hemisphere in sunlight. A good description of encounter geometry, details of events and spacecraft experimental configuration can be found in the book by Dunne and Burgess (1978), and a scientific review is given by Gault et al. (1977).

The Mariner 10 spacecraft included two atmospheric science experiments, the airglow spectrometer and the occultation spectrometer (Broadfoot 1976; Broadfoot et al. 1977a,b). The first of these was designed to observe airglow at preselected wavelengths corresponding to the resonance transitions of ionized and neutral helium and of neutral neon, argon, xenon, hydrogen, oxygen and carbon. A separate detector was provided for each line, giving a small spectral window (~20 Å) centered on the resonance wavelength. Two background channels were also provided. Observations during each encounter included scans obtained as the slit was stepped across the planet's visible disk or allowed to drift across it. The second instrument, the occultation spectrometer, used a grating at grazing incidence to obtain simultaneous observations in each of four bands (~75 Å) centered on 470, 740, 810 and 890 Å. These bands lie in the strong ionization continuum of any likely gas, and the four channels were chosen to give rough information on ionization thresholds and therefore identities of any gas detected. The instrument observed the Sun as it was occulted by the limb of the planet. The radiometer experiment (Chase et al. 1976) provided important auxiliary information on the local surface temperature.

A. Resonance Scattering

The intensity I of an airglow emission is conventionally expressed in Rayleighs (R), the photon intensity per steradian multiplied by $4\pi\ 10^{-6}$. Following Chamberlain and Hunten (1987), we write for an optically thin medium

$$4\pi I = 10^{-6}p(\theta)gN \text{ (in Rayleighs)} \tag{1}$$

where $p(\theta)$ is the phase function for single scattering (1 for isotropic scattering, a common approximation), g is an emission rate factor (Table I) in photons atom^{-1} s^{-1}, and N is the line-of-sight abundance in atoms cm^{-2}. The value of g depends on the solar flux at the relevant wavelength and on the oscillator strength of the transition; tables exist for 1 AU (Chamberlain and Hunten 1987; Hunten 1967; Table I). The abundance can be written $N = \eta N_0$, where N_0 is the zenith abundance and η is the airmass factor, equal to the secant of the zenith angle except near the horizontal. For a horizontal line of view tangent to the surface, right through a spherical atmosphere of radius R and scale height H, $\eta = (2\pi R/H)^{1/2}$.

The Mercury-Sun Doppler shift can become large enough to affect the incident solar flux by a large factor. For the alkali metals, this effect is discussed below (Sec. IV). In principle, the influence can be even larger for the ultraviolet resonance lines, but the Mariner 10 encounters all took place near Mercury's aphelion, with a negligible Doppler shift.

B. Helium

Figure 1 shows the brightness observed in the Helium I 584 Å airglow channel during a slow drift across the planet at encounter III, together with the projection of the slit on the disk of the planet during encounter. Also contained in the figure is the result of a model calculation to be discussed below. Figure 2 shows the intensity distribution observed in the same channel during a drift above the bright limb of the planet at encounter I. The observations of helium constitute the best data set for any species observed during the Mariner 10 encounters with Mercury, and their analysis deserves some discussion. The photon flux distribution across the visible disk of the planet can provide directly only column densities averaged over the field of view. To determine the global distribution of helium atoms, one must construct a 3-dimensional model of the atmosphere, combine it with the observational geometry, and predict the observed emission rate. This analysis is necessarily iterative, and must take account of the complications of gas-surface interactions introduced in Sec. I and discussed in Sec. V.

Shown in Fig. 1 as a solid line and in Fig. 2 as a dashed line are the predicted fluxes from Smith et al. (1978) whose work represents the most thorough attempt to treat the helium-surface interaction. The same model was used for both figures. Differences can be seen near the bright limb (Fig. 2), in

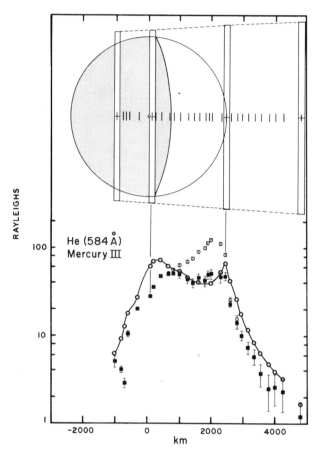

Fig. 1. Mariner 10 ultraviolet spectrometer (UVS) scan of Mercury's disk and subsolar atmosphere, at the helium resonance wavelength. The upper diagram illustrates the projected size of the field of view at four different positions in the scan. Filled symbols have been corrected for planetary albedo, and the open ones are the observed intensities. Circles joined by a curve represent a model described by Smith et al. (1978), which starts with a Maxwell-Boltzmann flux distribution at the surface and includes losses by thermal escape and photoionization. The temperature field is based on Chase et al. (1976) for the evening side of the planet.

the 3000–4000 km region above the bright limb (right side of Fig. 1), and above the dark side of the planet (left side of Fig. 1). Although the model shown represents the best fit that could be obtained, there are significant differences between the model and the observations. The ratio of the number densities at the antisolar and subsolar points is inferred to be ~50. All of these discrepancies make it clear that even the best available model of the gas-surface interaction is still inadequate. More details appear in Sec. V.

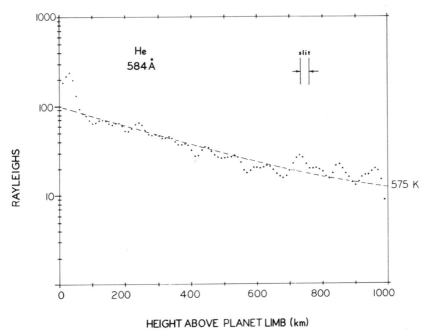

Fig. 2. Close-up scan of the helium distribution above the subsolar limb by the Mariner 10 ultraviolet spectrometer (UVS) (Broadfoot et al. 1976). The projected slit width is also shown. The dashed line is a barometric distribution for 575 K. The bump below 50 km is from planetary albedo.

C. Hydrogen

Figure 3 shows the altitude distribution of the hydrogen 1216 Å emission above the subsolar point determined from encounters I and III. This can be compared with the helium data in Fig. 2. The feature near 200 km is not well understood but may be an artifact unrelated to atomic hydrogen emission. It is not possible to regard this data set as a single population characterized by a single kinetic temperature. Shemansky and Broadfoot (1977) argue that the hydrogen above the subsolar point is in essence a 2-component system consisting of a cold component whose scale height is characteristic of that associated with darkside temperatures, and a second smaller component apparently thermally coupled with the surface near the sub-solar point. The presence of the "cold" population has never been quantitatively explained. Drift scans across the terminator did not lead to a measurable scale height.

D. Oxygen and Other Constituents

The 1304 Å channel detected radiation from oxygen atoms during encounter III. While the signal-to-noise level precluded determination of a scale

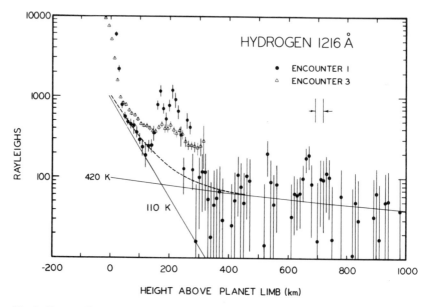

Fig. 3. Same as Fig. 2, but for H in its resonance line, Lyman α. Data from Encounters I and III are included, and a fit to an arbitrary 2-temperature distribution is shown. Information from another scan to higher altitudes contributed to this fit (Shemansky and Broadfoot 1977). Again, planetary albedo appears at the lowest altitudes.

height from the data, it was possible to determine the subsolar point density shown in Table I.

Absorption cross sections are large in the extreme ultraviolet (<1000 Å) for the most likely atmospheric gases, and the Mariner 10 occultation experiment was designed to use these large cross sections to search for the presence of an atmosphere. No measurable absorption was detected in any of the channels. Broadfoot et al. (1976) placed upper limits on common species based on both the occultation and airglow experiments. These are contained in Table II (which includes gases actually detected by the more sensitive resonance scattering method and shown in Table I).

The radio occultation experiment on Mariner 10 (Fjeldbo et al. 1976) also placed an upper limit on the ionospheric electron density, which led the authors to suggest a limit of 10^6 cm^{-3} for the total atmospheric density. This value, which is lower than the results obtained in the EUV occultation, was obtained by analogy with Venus' ionosphere at a similar level. It is not obvious that it would be sustained by a quantitative model of the actual situation at Mercury. Until such a model is worked out, we recommend use of the EUV limits.

Of the undetected gases, by far the most likely to be present is H_2, formed by reaction between implanted H atoms and subsequently released as further atoms overload the surface grains. (This assertion is controversial; even the present authors cannot agree on it.) As discussed in Sec. VI, a simple scaling from the helium density suggests a dayside value $\sim 10^4$ cm^{-3}, which could be lowered if thermal escape is a more important sink than photoionization. If the maximum estimate is correct, H_2 is almost as abundant as Na. Other hydrides, such as CH_4 and H_2O, should be several orders of magnitude less important. Apart from helium, the most important radiogenic gas is argon, observed on the lunar day side. If the source strengths are similar, and the sinks scale with the solar flux, the dayside density on Mercury might be 300 cm^{-3}.

There is a strong possibility that meteorites are a major source of water vapor, estimated in Sec. VI as having a number density of 2×10^4 cm^{-3}. The lifetime of water vapor is controlled by photodissociation, which in turn is a substantial source of H, H_2 and O.

Hoffman and Hodges (1975) cite evidence for very small amounts of CH_4, NH_3, CO_2, Ne and Ar on the Moon in addition to the established He. CO has not been detected. The rarity or absence of such gases may reflect the absence of a substantial source in the face of the ever-present photoionization sink.

IV. NEW ATMOSPHERIC SPECIES

Sodium D-line emission in the spectrum of Mercury was observed in 1985, and potassium resonance line emission was reported a year later (Potter and Morgan 1985a, 1986a). In each case, the emissions coincided approximately with the bright disk, and the Doppler shifts of the emission lines and the solar spectrum were consistent with the radial velocities of Mercury relative to the Earth and the Sun. Day-to-day variation in the brightness of the sodium emission was noticeable.

Although the discovery is recent, the frequent appearances of Mercury have made it possible to study the spatial extent, to make abundance estimates, and to discuss sources and sinks of sodium and potassium.

The large eccentricity of Mercury's orbit and its rapid orbital motion produce large Doppler shifts, different for reflected solar radiation and resonance emission. The Sun-Mercury component also changes the effective intensity of the exciting radiation, because of the broad, deep Fraunhofer lines due to solar sodium atoms. These and other topics in radiative transfer are discussed next, followed by a description and interpretation of the data.

A. The Motion of Mercury and the Strength of Emission

The solar spectrum incident on an atom of sodium approximately at rest in the Mercurian atmosphere appears Doppler-shifted due to the instantaneous radial velocity of the planet relative to the Sun, v_{rs} in Fig. 4. The effective flux for the resonance transition is therefore modulated. For a shift of zero, the

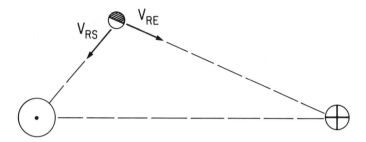

Fig. 4. Illustration of the relevant velocities in the Sun-Mercury-Earth system.

fluxes at D_1 and D_2 are only 5.55 and 4.90% of the nearby continuum. At maximum shift, ~ 10 km s^{-1}, this fraction rises to 45%. Emission in the rest frame of the planet is observed from the Earth with a Doppler shift v_{re} in Fig. 4. The solar spectrum reflected from the surface is shifted by the sum, v_{rs} + v_{re}. Actual data, Figs. 5 and 6, illustrate this. Fig. 5 shows the observed spectrum of Mercury near 5900 Å when v_{rs} was 9.6 km s^{-1} and v_{re} was 34.8 km s^{-1}.

Mercury is a difficult object to observe. In twilight, it appears behind a relatively dark sky, but the time interval in which the planet is visible is short and, because of the large airmass, the seeing is poor and water-vapor lines are strong. In the daytime, the seeing is poor because the Sun is visible, and the sky foreground has a greater surface brightness than Mercury. Spatially re-solved spectroscopy has considerable potential for providing information, but is very difficult to achieve in practice. When Mercury is at greatest elongation

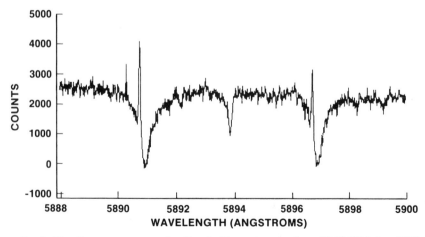

Fig. 5. The discovery spectrum of sodium in Mercury's atmosphere (20:20 UT 3 Jan. 1985) (Potter and Morgan 1985).

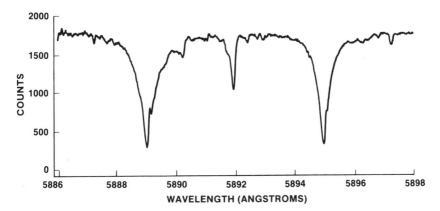

Fig. 6. Another spectrum of the sodium region, showing the Mercurian lines almost hidden in the solar absorptions (16:40 UT 22 Feb. 1986).

from the Sun, its illuminated dimensions are 4–5 by 6–8 arcsec, comparable to a typical seeing disk for the restricted observing conditions.

The observed spectrum is a composite, containing emission from the atmosphere of Mercury, reflected light from the surface, and terrestrial absorption lines due in most cases to H_2O or O_2. Daytime spectra also contain scattered sunlight from the Earth's atmosphere. Terrestrial emission may occur for both sodium and potassium, particularly in twilight, but is not Doppler shifted and thus could not be mistaken for emission from Mercury. The scattered sky component, if present, must be subtracted from the data. The region near the sodium D-lines is relatively free of terrestrial lines, though an H_2O line occurs at 5889.5 Å on the short wavelength side of D_2 and there are many weaker lines. Terrestrial oxygen absorptions partially mask the stronger member of the potassium doublet, but there are frequent occasions when favorable Doppler shifts make observations possible.

B. Physics of Emission in the Sodium D-lines

A large amount of literature exists concerning the sodium layer in the Earth's atmosphere from whose airglow can be obtained temperature, column density and height (Chamberlain 1961; Hunten 1971). More recently, the layer has been studied by lidar, a method that is not applicable to other planets at present. The sodium emissions associated with Io require similar calculations (Chamberlain and Hunten 1987; Brown and Yung 1976). The phenomena are further discussed in Sec. VII. The following brief discussion can be supplemented by the above references and the literature cited in them.

Figure 7 shows an energy-level diagram of the levels that give rise to the sodium D-lines at 5895.92 and 5889.95 Å. The major hyperfine splitting is 21 mÅ. The oscillator strengths are 0.655 for D_2 and 0.327 for D_1. The reported

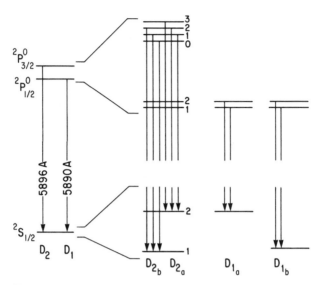

Fig. 7. Energy levels and hyperfine transitions for the sodium D-lines.

measurements of the column density of sodium in the atmosphere of Mercury fall near 10^{11} atoms cm^{-2} and the derived temperature of the gas is at least ~500 K. The optical depth at the center of the D_2 line is ~1: the atmosphere of Mercury is not optically thin and radiative-transfer corrections are required to derive the conditions of the atmosphere from observations. This is particularly true if the data refer to regions near the terminator or limb, for which the incident or emergent radiation has traversed a long slant path. From a knowledge of the wavelengths and oscillator strengths of each hyperfine component, and assumed temperature and abundance, a line profile for each D line is obtained. The path lengths for extinction and scattering are found from the geometry of a particular observation, and the sum gives the monochromatic optical depths for each. For some of the available observations, it has been necessary to average over variable geometry for the areas imaged on the slit. In the case of the sodium resonance doublet, the abundance is usually derived from the ratio of the integrated intensities, which does not require absolute intensity calibration.

At any apparition, a wide range of observing geometries is present across the apparent disk of the planet, and must be considered in interpretation of the results, particularly in estimating abundances. Figure 8 shows the variation of the D_2/D_1 ratio with zenith optical depth for three different geometries. This calculation, which follows Brandt and Chamberlain (1958), assumed a temperature of 500 K, high enough so that hyperfine structure can be ignored to first order. For appreciably lower temperatures, this approximation is not adequate (Chamberlain et al. 1958).

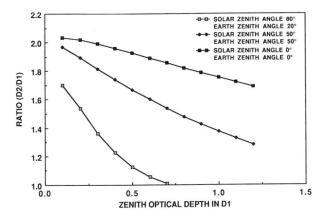

Fig. 8. Sodium D_2/D_1 ratio as a function of zenith abundance, represented by the D_1 optical depth. The three curves represent three different geometries, as indicated.

C. The Density, Physical Extent and Temperature of Sodium

Column densities inferred from the D_2/D_1 ratio range from 1 to 2×10^{11} atoms cm^{-2} (Potter and Morgan 1986b). Corrections for hyperfine structure were not included. On average, small values of sodium column density correlate with large solar radial velocities v_{rs}, and large values occur near perihelion and aphelion when the radial velocity is small. Figure 9 is a plot of column density against a measure of the radiation force (Potter and Morgan 1987). Because of the deep Fraunhofer line, radiation pressure effects increase rapidly with radial velocity. Thus, the measurements may reflect increased loss of sodium from the dayside atmosphere when the radiation pressure is large, or perhaps the sodium is being pushed away to the night side, even if it is not escaping. Although this effect is reminiscent of the predictions of Smyth

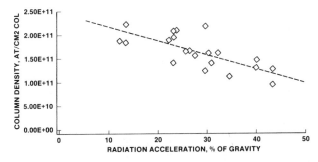

Fig. 9. Sodium abundance as a function of solar radiation pressure (Potter and Morgan 1987).

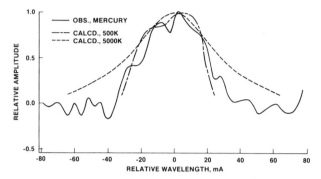

Fig. 10. A partially resolved spectrum of sodium D_2, along with a calculated line for a temperature of 500 K (8 August 1986) (Potter and Morgan 1987).

(1986) and Ip (1986), it probably involves appreciably different conditions. Their models require velocities corresponding to much higher temperatures than are observed, as discussed next (see also Sec. VI).

A high-resolution spectrum of the D_2 line, obtained by Potter and Morgan (1985b), is shown in Fig. 10. The 20 mÅ hyperfine splitting is partially resolved, and the FWHM is 55 mÅ, corresponding to a temperature of 500 K, close to that of the surface of the planet included in the slit. At this temperature, the most probable speed is 600 m s^{-1}. In order for radiation pressure to remove sodium effectively, the sodium atoms must have a velocity of at least 2 km s^{-1} (see Smyth [1986] for an in depth discussion). A small population of high-velocity sodium atoms could still be present in the wings of the emission feature, but clearly such atoms are not prevalent. If they are in fact present, but just below the observational limit, losses due to radiation pressure could become significant.

Despite the difficulty of obtaining good spatial resolution, some interesting preliminary observations of sodium are available. Schneider et al. (1985) reported an equator-to-pole variation, and a sunward shift of approximately 1 arcsec in the centroid of the emission relative to the neighboring reflectance spectrum from the planet. They concluded that a genuine separation *might* be present, although different scattering laws for the two spatially distributed sources could also produce the observed shift. Using the same instrumentation and under similar conditions, Tyler et al. (1986) saw no sunward spatial shift, but did confirm the equator-to-pole variation. They also found a north to south ratio greater than unity.

Although Potter and Morgan (1987) were unable to confirm the observations of Schneider et al. or Tyler et al., such differences might be due to differences in viewing geometry. Neither group reports an antisolar tail or any extension of the sodium emission above the dark side of the planet.

D. Physics of Emission in the Potassium Resonance Lines

Potassium and sodium are in the same column of the periodic table, and the spectroscopic terms of the allowed transitions of the two elements are quite similar. However, the potassium lines are optically thin so that a detailed radiative-transfer calculation is not required in order to determine the column densities. The two lines are at 7665 and 7699 Å with oscillator strengths again in the ratio 2:1; because of the large spacing, simultaneous high-resolution observations are difficult to obtain. The stronger line, at 7665 Å, is not usable for Earth airglow studies because it is absorbed by O_2 (see Fig. 11). For Mercury, the Doppler shifts can move it clear of the O_2 absorption where it can be observed.

E. Observations of Potassium

Figure 11 shows the 7665 Å line in the spectrum of Mercury. The average column density on 16 November 1985 was 1×10^9 atoms cm^{-2}, estimated by calibrating the intensity against an assumed intensity in the nearby continuum reflected by Mercury's surface. The sodium abundance was estimated

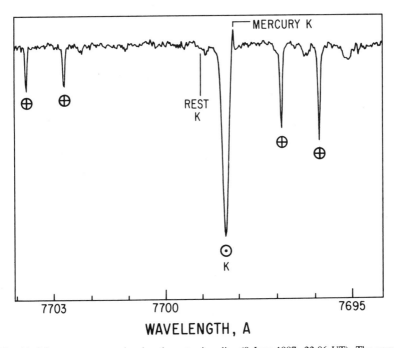

Fig. 11. Mercury spectrum showing the potassium line (8 June 1987, 22:06 UT). The corresponding solar absorption is just to its left, and nearly all the remaining absorptions are by atmospheric O_2. Sky subtraction is almost perfect (figure courtesy of A. Tyler and R. Kozlowski).

from the D_2/D_1 ratio; the sodium-to-potassium ratio, Na/K, is 100. As a comparison, we note that the Na/K ratio for lunar materials ranges from 2 to 7 (Kopal 1974, p. 161), and for the CI chondrites, the Na/K ratio is 13 (Wasson 1985). However, large Na/K ratios can be found in some cometary material. For example, the Na/K ratio observed in Comet Ikeya-Seki was 1000 (Preston 1967), and the VEGA and Giotto measurements of the composition of Comet P-Halley dust grains lead to a Na/K ratio of 188 (Jessberger and Kissel 1987). This probably indicates that similar sink processes are acting to remove potassium and sodium in both the atmosphere of Mercury and in comets that closely approach the Sun, and that the potassium sink is stronger. However, the Na/K ratio on Mercury would also be as it is observed if much of the meteoritic material inside the orbit of Mercury were cometary in origin and deficient in potassium, and if most of the alkalis present in the regolith of Mercury were meteoritic in origin.

V. MODELS OF THE MERCURY ATMOSPHERE

A. Prior Work on He and H and a Comparison of Mercury vs Earth

Previous sections have discussed the available observations of H and He and the conversion of intensities to column densities along the line of sight. Some information about the atmosphere (scale heights, for example) can be retrieved directly from such data, but the iterative process of modeling and comparison must be adopted to extract the full content. If the physical processes that control the atmosphere are well enough understood, this iterated comparison leads to detailed understanding of the atmosphere. Among the most important products which can be derived are inventories for each gas, essential for discussion of sources and sinks.

The atmospheres of the Moon and Mercury are collisionless to a good approximation and thus it was natural to apply the techniques used to study the exospheres of the Earth and other planets. The physics of the terrestrial exosphere is well studied; for example, the distribution of source energies and directions to be used in model calculations is described by the definitive paper of Brinkmann (1971) (cf. Chamberlain and Campbell 1967; Smith et al. 1978). An ideal exosphere, defined by a lower boundary in detailed balance (thermodynamic equilibrium) and a loss-free diffuse upper boundary, contains particles in a Maxwell-Boltzmann kinetic energy distribution (see Feynman et al. 1963; Chamberlain 1963); the ideal exosphere is barometric. Observed exospheres always show deviation to varying degrees from the ideal: escape of light atoms and diurnal variations of exospheric temperature perturb the velocity distribution.

Early research on the exospheres of the Moon and Mercury was thus a direct extension of calculations in which atmospheric particles formed the lower boundary. However, the solid interface at the Moon and Mercury pre-

sented an entirely different problem in the sense that the kinetics of gas particles are controlled by heterogeneous collisions. This fundamental difference
was not recognized in much of the early work on the subject. Even the most
recent models for the Moon and Mercury contain only rudimentary accounting
of the physical interaction at the surface. The physics of gas-surface interaction is crucial for understanding atmospheric evolution on these bodies and
must be understood before atmospheric observations can be interpreted in
terms of sources, sinks and evolutionary processes. Everything pertaining to
the planetary atmosphere depends on the details of the interaction between the
gas atom and the surface, along with the competing loss processes.

The following assumptions are implicit in the published models of helium in the atmosphere of Mercury by analogy with a standard exosphere:
though there is no net trapping of atoms on the surface, an individual atom
that collides with the surface may subsequently have a long (undefined) residence time (chemisorption); and the impact of a helium atom with the surface
and the subsequent return of an atom to the atmosphere are independent.
Further, it is assumed that the velocity distribution of the atoms returned from
the surface is a Maxwell-Boltzmann flux distribution appropriate to the local
surface temperature. For a given distribution of surface temperatures, it is then
possible to produce a model atmosphere using Monte-Carlo techniques, and
from this to calculate a brightness distribution to compare with the data. This
class of model was formulated by Hartle (1971) and subsequently used by
Hartle et al. (1973,1975b), Hodges (1974), Curtis and Hartle (1978) and
Hodges (1980a). Smith et al. (1978) noted, however, that the interaction of
helium, and to a lesser extent any light atom, with the surface under Mercurian conditions is typically a scattering event in which the interaction between the surface lattice and the helium atom is brief and, on average, results
in only a small exchange of energy. In short, the interaction of helium with the
surface of Mercury violates all of the assumptions given above.

Shemansky and Broadfoot (1977) extended these considerations. The
collision of a helium atom with the surface is typically free-free; the consequences include (a) a departure from Maxwell-Boltzmann statistics in the
sense that, for a given kinetic temperature, the number of high-velocity atoms
is decreased, and (b) a much smaller variation of the surface densities from the
subsolar point to the antisolar point than that calculated by Smith et al. Unfortunately, work on the topic was suspended at that point; these ideas have not
been pursued to the construction of a model atmosphere of the planet and
comparison with the observations. There was, however, some further discussion of the philosophy and methods of computation (Hodges 1980a,b;
Shemansky 1980).

B. Components of the Atmosphere

The view of the atmosphere taken in this chapter is summarized in Fig.
12. Confusion can easily arise if the distinctions to be discussed are not kept

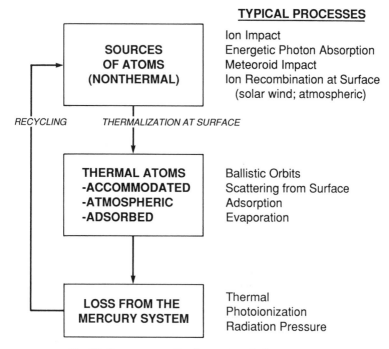

TYPICAL PROCESSES

SOURCES OF ATOMS (NONTHERMAL):
Ion Impact
Energetic Photon Absorption
Meteoroid Impact
Ion Recombination at Surface
(solar wind; atmospheric)

RECYCLING   THERMALIZATION AT SURFACE

THERMAL ATOMS -ACCOMMODATED -ATMOSPHERIC -ADSORBED:
Ballistic Orbits
Scattering from Surface
Adsorption
Evaporation

LOSS FROM THE MERCURY SYSTEM:
Thermal
Photoionization
Radiation Pressure

Fig. 12. Processes operating for an atmospheric component.

in mind (Appendix A). *Ambient atoms* consist of atmospheric atoms having energies accommodated to the surface, and their counterparts physically adsorbed on the surface. These two components are in equilibrium with each other; we shall argue that the adsorbed component is by far the smaller, except perhaps for some elements on the night side. *Source atoms* are nearly always more energetic; they are defined as the immediate products of energetic processes such as those shown in the right-hand column. Some may be lost immediately from the atmosphere; the rest lose their energy in impacts with the surface and become ambient atoms.

In this discussion, *sinks* are defined as processes that remove atoms from the planetary environment. Loss processes operate on both source and ambient atoms, but in different proportions. We make the approximation that the sinks operate primarily on the ambient atoms, not on the source atoms, although the proportion undoubtedly depends on the species. Because of the extremely low atmospheric density, any chemical reactions occur on the surface. The mere impact of an atmospheric atom with the surface is not normally considered a sink, because the atom will almost certainly scatter or be thermally evaporated in a very short time. Only if the atom becomes chemically bound to the surface is the impact counted as a loss event for ambient atoms.

A consequence of this view is the existence of two distinct time scales, a longer one for the sources and sinks and a shorter one for interaction of the atmospheric particles with the surface. The latter is the ballistic or free-fall time, which is $\sqrt{2}$ v/g for projection at 45° from the horizontal at speed v. Setting v equal to the modal velocity $U = (2kT/m)^{1/2}$, we obtain

$$t_b = \frac{2}{g}\left(\frac{kT}{m}\right)^{1/2} = 2\left(\frac{H}{g}\right)^{1/2} = 52\left(\frac{T}{M}\right)^{1/2} \tag{2}$$

where M is the atomic mass in amu and the temperature T is in Kelvins. Typical dayside values are 250 and 1300 s for thermalized Na and H; a nightside value for Na is 100 s. As a reference for comparison, we may take the shortest well-established loss time, 10,000 s for photoionization of Na; the ratio to the ballistic time t_b is around 40.

Thus, the typical life history of a sodium atom would be: ejection from the surface with energy of a few thousand degrees; several hops with duration decreasing from 1000 to 250 s as the atom thermalizes; ionization after 40 hops. The ion may be lost from the planet, re-implanted in the subsurface, or neutralized at the surface (Sec. VI). Much of the justification for the above statements is given for the atmospheric atoms in the rest of this section, and in the next section for the sources and sinks.

As mentioned in Sec. I, there is a large excess of helium on the cold night side. This excess can be understood as a balance between a lateral flow that is more rapid from warm to cold than the reverse (Hanson and Patterson 1962; McAfee 1967; Hodges and Johnson 1968; Chamberlain and Hunten 1987). To first order, the flow rate scales as $T^{5/2}$, where T is the exobase temperature. To justify this, we consider exospheric atoms to be "hopping" in a random walk from place to place on the exobase and note that the average length of each hop is approximately equal to the scale height. The distance between hemispheres is of the order of the planetary radius r. Since the motion is a random walk, the number of hops is of the order $(r/H)^2 = \lambda^2$ (Eq. 4, Sec. VI), and the migration time is

$$t_m \simeq t_b\lambda^2 = 2gr^2\left(\frac{m}{kT}\right)^{3/2}. \tag{3}$$

For helium at 500 K, this time is 4×10^4 s. The flow rate between hemispheres is proportional to $N/t_m = nH/t_m$, where N is the integrated density. Thus, if a steady state is attained, the product $nT^{5/2}$ should be constant. The H density n on the Earth does vary approximately as $T^{-5/2}$, as expected for "zero net ballistic flow," or ZNBF (Quessette 1972). The surface of Mercury is not a true exobase, but the same tendency should apply there, with other parameters also significant. Quantitatively, the global atmospheric distribution depends on both the kinetic energy distribution imparted by the free-free collision process and the accommodation efficiency of the velocity of H atoms to the local surface temperature. Nonuniform distributions of sources and sinks are also likely to be significant, especially for the shorter-lived species.

The Monte-Carlo computation for helium by Smith et al. (1978) included the migration process and loss by thermal escape, under the assumption of an accommodated Maxwellian distribution. Without this escape, the result would be expected to follow the $T^{-5/2}$ rule, which gives a night/day ratio of 80. Most of the thermal escape is from the hot day side, and the effect should be to reduce the density there below the steady-state value; the night/day ratio was found to be 150, corresponding to a dayside reduction factor of 0.53. Although it was not explored at the time, this explanation seems reasonable. The $1/e$ escape time (Eq. 7, Sec. VI) would be $\sim 10^5$ s under the assumed conditions; the time to 0.53 is only slightly longer than the migration time of 4 $\times$ 10^4 s obtained above.

As discussed in Sec. V.F below, the migration time for the alkali metals is considerably longer than the ionization time, and the $T^{-5/2}$ rule is unlikely to apply at all.

C. The Gas-Surface Interaction on Mercury

The subject of gas-surface interactions is interdisciplinary, and terms are not always used consistently. Definitions as used here are therefore presented in Appendix A of this chapter. Once again we call attention to the definitions of ambient atoms and source atoms, discussed above and in Fig. 12.

Standard exospheric theory assumes that the downcoming atom is fully accommodated to the exobase temperature and, if ejected back into the exosphere, the initial and final energies are independent. This theory has been applied to Mercury by assuming that the atoms stick to the surface on impact, and the surface simply acts as a saturated source of atoms in a Maxwell-Boltzmann distribution. In reality, most atmospheric atoms are scattered on collision with the surface, and in general there is a relationship between the pre- and post-collision energies of the atom. Physical-scale adsorption collisions on the sunlit surface, when they occur, have durations of only 10^{-13} to 10^{-11} s. The theory of gas-surface interactions is outlined in Appendix B, to supplement the semi-qualitative summary given here. Figure 13 schematically illustrates the potential energy of an atom near a surface. If bound, it vibrates around the potential minimum with average energy kT. For convenience, we express energies E as equivalent temperatures E/k; 1 eV corresponds to 11605 K. Chemical bonds tend to have a strength of a few eV, whereas physical bonds are much smaller, as shown in the second column of Table IV. The observed energy distributions of the atmospheric atoms point to a dominance of physical-scale interaction with the surface even for the chemically active species.

Figure 13 includes a dashed portion representing an *activation energy* E_a. Atomic electron configurations virtually always cause the development of a barrier at intermediate internuclear separations in the collision process. In

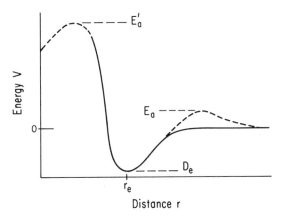

Fig. 13. Schematic potential curve for an atom near a surface. D_e is the well depth, E_a is a possible activation energy, exaggerated for visibility, and E_a' represents an activation energy for diffusion through the lattice.

general, the activation energy is an important factor in limiting the rate of chemical reaction, and sets a lower limit on the kinetic energy of dissociation. For collisions of ground state hydrogen atoms this quantity is very small. However, in collisions of more complex systems including molecular hydrogen, activation energy is a significant rate-determining quantity, and defines the properties of the mutual diffusion coefficient. The interaction terms causing a positive potential at intermediate separations vary in origin from system to system. Finite total electron spin in the interaction (atomic or multiatomic) introduces positive terms in the dispersion energy (see Hirschfelder et al. 1964). Exchange or valence energy terms can also produce positive interaction energies (Buckingham and Dalgarno 1952). It is therefore likely that gas-surface physical interaction also shows positive dispersion energies, but the question is one of magnitude. Hydrogen bimolecular reactions show E_a $\sim D_e/20$, while more complex atomic reactions such as those involving Na show $E_a \sim D_e/10$ (Hirschfelder et al. 1964). For the physical-scale interactions considered here, the value of E_a is expected to be much less than kT, and therefore negligible.

A trapped atom of radiogenic or solar-wind origin is subject to a much larger barrier, indicated in Fig. 13 as E_a'. For lunar material the activation energy for outgassing of helium is ~ 2 eV (Ducati et al. 1973). Helium diffusing out of the subsurface may then have initial energies of this order.

An atmospheric atom colliding with the surface enters Fig. 13 from the right with energy above zero, corresponding to atmospheric temperature. The atoms will gain or lose energy near the potential barrier in the inner turning-point region, and will be adsorbed only if energy is lost to the lattice structure in sufficient amount to fall below zero. Otherwise the collision will simply scatter the atom.

TABLE IV

Gas-Surface Physical Interaction Parameters for α Quartz[a]

Species	H	He	O	Na	K
Well depth[b] D_e(K)	254	102	920	3000	2800
Well distance[b] r_e (Å)	3.39	2.57	3.11	3.48	4.04
Depth to v = 0[b] D_0(K)	156	54	837	2900	2720
Vibration period $\tau(10^{-13}$ s)	2.0	4.7	2.7	1.8	2.5
600 K adsorption time τ_a(s)	2.6×10^{-13}	5.1×10^{-13}	1.3×10^{-12}	2.3×10^{-11}	2.3×10^{-11}
100 K adsorption time τ_a(s)	9.5×10^{-13}	8.1×10^{-13}	1.8×10^{-9}	0.71	0.16
600 K accommodation coefficient α	0.08	0.05	0.11	0.62	0.26

[a]Sources: Kunc and Shemansky (1985) for helium; Shemansky and Kunc (in preparation) for all others.
[b]D_e, r_e, are illustrated in Fig. 13. D_0 is the dissociation energy to the separated-atom limit from the lowest vibrational level.

An adsorbed atom has energy less than zero in Fig. 13, and is oscillating with the vibrational period of order 10^{-13} s. It exchanges vibrational energy with the solid, and after a number of vibrations of order exp (D_0/kT) is likely to have energy D_0 or greater, enough to escape the well (cf. Eq. B1 in the Appendix). If we take Na in Table IV as a specific example, an atom at 600 K in a well with a depth of 0.25 eV, or 2900 K, requires less than 130 vibrations, or 2×10^{-11} s. At 100 K, these values rise to 4×10^{12} and 0.7 s. Since the ballistic times are 100 s or greater, fewer than 1% of sodium atoms are expected to be adsorbed to the surface, even on the night side; for other observed atoms this fraction is much smaller still. For the much hotter subsolar surface, it appears safe to conclude that the majority of collisions do not involve adsorption for any of the known atmospheric gases, and that the few that do adsorb are rapidly re-emitted.

Along with these very short interaction times goes a strong possibility that the speed distribution of the atmospheric atoms is non-Maxwellian. There are extra drains on the faster atoms: they may be able to escape, and even the gravitationally bound ones are exposed for longer times to loss processes such as photoionization. Restoration of the high-energy tail must therefore occur during interactions with the surface, which appear to be very inefficient. The only detailed information available is for helium, but the same ideas can be expected to apply to the other light atoms and especially to H.

Shemansky and Broadfoot (1977) report the result of a calculation of the He 1-dimensional energy distribution in equilibrium with a surface of SiO_2 (Fig. 14). A Maxwellian distribution is shown for comparison, and the deviations are obvious, especially at high energies. This calculation, described in Appendix C, is intended to give a statistical equilibrium between the gas atoms and the surface. It appears that the energy transfer in a single collision is limited to the maximum phonon energy. (A phonon is a quantum of the lattice vibrations in a solid; the typical energy corresponds to the Debye characteristic temperature.) If the system were a closed, isothermal cavity in thermodynamic equilibrium, the velocity distribution would approach a Maxwellian after a large number of collisions (including, of course, rare collisions among atoms). However, Mercury's atmosphere is very different: the surface itself is not in thermodynamic equilibrium, and the approach of the atmosphere to a steady state must take into account competing processes. The need for replenishment of high-speed atoms has already been mentioned; moreover, the planet is very far from isothermal. Further work is required.

The same computations give an *accommodation coefficient* α for helium, shown in Table IV as 5% (Shemansky and Broadfoot 1977; Kunc and Shemansky 1985). As defined in Appendix A, α is an efficiency for energy transfer in a collision with the surface. However, this quantity alone, determined by a classical calculation, cannot define the energy distribution function in the gas because it contains no information about energy dependence.

The principal characteristics of surface reactions with light atoms have

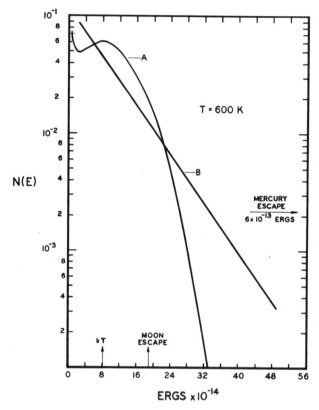

Fig. 14. Calculated one-dimensional energy distribution for helium atoms interacting with a
surface of SiO_2 at 600 K (curve A). Curve B is a thermal distribution for the same temperature
(Shemansky and Broadfoot 1977).

been summarized by Shemansky (1980). The two effects mentioned above as
controlling factors in the global atmospheric distribution of gas are closely
coupled physically: a low accommodation coefficient is accompanied by a
truncated energy distribution. The reason is that energy exchange reactions
are limited to single phonons, defined by the Debye characteristic temperature
of the solid. A low accommodation coefficient implies rapid transport of gas,
because successive hops along the surface are strongly coupled in direction as
well as speed. Thus, the effective mean free path along the surface may be
greatly increased. This high mobility must, however, be reduced by microscopic surface roughness and by the inherent angular scattering distribution.
Another effect, which decreases mobility, is a small reduction in the mean
energy of the gas below that calculated from the surface temperature because
of the truncated energy distribution. Although these effects are in opposition,
it seems likely that the net effect is an increased mobility. The magnitudes

probably depend on the parameters of the particular gas. Heavier atoms are expected to approach a Maxwellian distribution in statistical equilibrium more closely because multiple-phonon exchanges are more probable (see Table IV). Moreover, the processes (such as escape of the faster atoms) forcing the distribution away from being thermal are weaker.

Gas-surface interaction theory generally assumes that the dominant energy exchange involves the surface normal component of the velocity vector of the gas atom. The quality of the surface consequently plays a role in determining both the effective accommodation coefficient and the angular scattering distribution, and hence the transport properties.

D. Helium

The enhanced transport rate discussed in the previous section presumably explains the discrepancy between observation and earlier model calculations that implicitly assumed an accommodation coefficient $\alpha = 1.0$. The truncation of the tail of the energy distribution is predicted to reduce greatly the thermal escape rate of helium from the subsolar atmosphere in comparison with that expected for a gas in thermodynamic equilibrium at the surface temperature. The theory of gas-surface interaction thus directly affects ideas on atmospheric evolution. Similar considerations have yet to be applied to the other atmospheric gases, and will have to recognize that almost every physical aspect depends on the species.

All atmospheric model calculations published to date are limited by the assumption of thermodynamic equilibrium at the surface. Calculations by Hartle et al. (1973,1975b), Hodges (1974) and Curtis and Hartle (1978), all follow essentially the same computational method, whose properties and limitations have been discussed above. All of these papers generate their velocity distributions by a method intended to simulate an atmospheric gas in thermodynamic equilibrium at the altitude of the surface, and at the local surface temperature. However, Smith et al. (1978) show that the previous calculations contain an inadvertent mix of two different source distributions, giving a nonbarometric atmosphere under idealized conditions. They presented results of a Monte Carlo calculation using a source flux correctly formulated for a gas in thermodynamic equilibrium. The temperature distribution used was that measured by Chase et al. (1976). The night/day density ratio for helium at the antisolar and subsolar point was found to be ~150 (see also Sec.V.B above). A separate calculation based on the analytic formulation of Hodges (1973a) produces essentially the same result (Smith et al. 1978).

The comparison of model and data (Fig. 1; Broadfoot et al. 1976; Smith et al. 1978) indicates a night/day ratio ~50. This discrepancy is the basis for the statement made repeatedly above that a theory based on a fully accommodated gas-surface interaction does not account for the observations. The required large mobility across the surface may be consistent with the He accommodation coefficient for a single scattering event. Table IV shows $\alpha = 0.05$

for alpha-quartz and it is expected to be similar for many other inorganic surfaces (Shemansky and Broadfoot 1977); this value may be small enough to explain the data. However, the translation of a mean accommodation coefficient into an atmospheric distribution is not a simple matter. The classically calculated accommodation coefficient is a single number, and does not provide the necessary probability distribution for energy exchange. Without this, it is not possible to calculate distribution functions for kinetic energy and direction, which affect mobility, height distribution and the rate of thermal escape. The angular distribution of the scattered atom depends both on the detail of the energy exchange process and the quality of the surface (Shemansky and Broadfoot 1977).

The night/day ratio is also affected by asymmetry in the loss process. In the calculations of Smith et al. (1978), Jeans escape was the major sink, whereas, in fact, photoionization probably dominates, with a different distribution and a considerably smaller rate. However, it too acts primarily on the day side and should also act to somewhat increase the night/day ratio.

Table V shows the estimated global content of helium based on Mariner 10 observations, and with the assumption of a nightside density enhancement of 50; it corresponds to the number quoted by Goldstein et al. (1981).

E. Hydrogen

Figure 3 illustrates the distribution of H Lyman α radiation above the subsolar surface, already described in Sec.III.B. There are unexplained features, specifically the presence of two components and their small scale heights (Shemansky and Broadfoot 1977) (see Sec.III.C). (The bump in encounters I and III near 200 km is a complete mystery.) One component appears to have a temperature (420 K) 40% below the local surface temperature. This component is best seen in high-altitude data, not reproduced here, obtained from a greater distance (Broadfoot et al. 1976). Even if the disagreement is outside the errors, it may be that the appropriate surface temperature is a dayside average rather than the local value. The dominant population in the subsolar atmosphere is very cold (100–150 K; Fig. 3). The H abundance of the cold component is ~50% larger than that of the warm component. The obvious source for cold atoms is the dark side, but such atoms would have to survive a large number of collisions with warmer surfaces without gaining energy. Hydrogen probably does have a higher mobility than helium, even though the accommodation coefficient ($\alpha = 0.08$, Table IV) may be larger. Perhaps further study will show that the nightside source is viable.

Broadfoot et al. (1976) suggested that the small scale height reflects that of a source gas, namely H_2O, the H atoms being produced by photolysis. As discussed below in Sec. VI, this process seems to require an unreasonably large amount of water vapor.

Alternately, one could consider interactions with the local surface along with selective thermal escape of the faster atoms. Mercury's escape velocity is

TABLE V
Estimated Atmospheric Physical Properties on Mercury

	H	He	O	Na	K	Notes
SCALE HEIGHTS (km)						
Subsolar H_s	1330	330	83	58	33	a
Antisolar H_a	230	57	14	10	6	b
Dayside content (10^{27} atoms) N_d	4	100	140	60	0.7	c
Nightside content (10^{27} atoms) N_n	4	670	2200	60	0.7	c
PHOTOIONIZATION TIME (s)	5.5×10^6	3.8×10^6	8.8×10^5	12,800	7500	d
PRODUCTION RATES, (10^{22} s^{-1})						
Radiogenic, P_r	—	0.7–4.6	—	—	—	e
Sputtering, P_{sp}	—	—	30–300	3–30	<1	f
Photosputtering, $P_{\phi s}$	—	—	?	10;<2000	—	f,g
Meteoritic, P_m	—	—	—	2–14	0.3	h
Interplanetary medium, P_{ipm}	1.4	0.2	—	—	—	e,o
Solar wind + magnetosphere, P_{sw}	230–850	10–37	—	—	—	e,i
LOSS RATES, (10^{22} s^{-1})						
Jeans	<<80	—	—	—	—	j
Photoionization	<0.14	<13	<50	≤600	≤10	k
Solar-wind, magnetosphere ionization	0.09	0.06	1.3	0.04	.0008	l
Radiation pressure, nonthermal atoms	—	—	—	?	—	m
ATMOSPHERIC LIFETIME (hr)	$(8–70)\times10^3$	$(7–70)\times10^3$	>7000.	>5.	>3.	n

a: $T = 575$ K; b: $T = 100$ K; c: From Table I and above scale heights. Area of a hemisphere $= 3.75 \times 10^{17}$ cm^2. Night/day density ratio taken as 1 for H, 50 for He, 100 for O, and 1 for Na, K; d: Kumar (1976) and Sec. VI; e: Goldstein et al. (1981); f: Ip (1986); g: McGrath et al. (1986); h: Sec. VI; Morgan et al. (1987); i: H value assumed to be 23 times He; j: Sec. VI; k: Upper limits assume no recycling of ions from magnetosphere to surface; l: He value from (e); others are probably comparable if scaled by abundances; m: Smyth (1986); Ip (1986); n: Ratio of estimated content to loss rate. It does not include possible losses of source atoms. o: Assumed deposition efficiency 0.3.

4.25 km s^{-1}, and the modal velocity U, defined above, see Eq. (2), is 3 km s^{-1} at 600 K. Almost the entire upper half of the Maxwellian distribution must be essentially missing, and surface collisions are very inefficient at repopulating it. It seems unlikely that the resulting distribution would resemble that for 100 K, but in the absence of any computations there is no point in speculation. The global content of atomic hydrogen is poorly defined because the strange distribution observed above the subsolar surface is not well understood, and measurements are not available elsewhere on either hemisphere.

The calculated subsolar scale height H_s = 1330 km for atomic hydrogen shown in Table V is ~40% larger than the measured value, as mentioned above. The global content given in Table V (M = 8 × 10^{27} atoms) is estimated by assuming the same amount on night and day sides. This assumption is justified by the presence of the cold component in the subsolar H distribution and the further assumption that this cold gas has been transported from the dark hemisphere.

F. Sodium and Potassium

The physical interaction potential D_0 for the Na-quartz system calculated by Shemansky and Kunc (1987) is estimated to be ~3000 K or ~0.26 eV in depth (Table IV). The accommodation coefficient is large (~0.6); thermalization at the surface is efficient, in accord with the observed line width. Model calculations attempting to define source and atmospheric distribution have been published by Ip (1986), Smyth (1986) and McGrath et al. (1986), and are also discussed in the next section. There has been a tendency to assume that the sodium is hot and occupies a very extended height region, but the evidence does not support this idea. The velocity distribution appears to correspond to the surface temperature (Fig. 10), and only a few bounces from the surface should be sufficient to approach this condition. The scale height at 600 K is only 60 km. Nevertheless, Fig. 10 could not be used to refute (or support) the presence of a sodium component with a temperature of several thousand degrees and comparable area, but 2 to 3 times the width and 1/2 to 1/3 the height.

Radiation pressure exerts a strong antisunward force, especially when the Mercury-Sun Doppler shift is large, and is integral to the models of Ip and of Smyth. Such transport may be effectively dampened by efficient thermalization at the surface and the short photoionization time. Observations of the behavior of sodium throughout the Mercurian year, such as those in Fig. 9, promise to shed light on this issue.

The discussion by McGrath et al. (1986) does not clearly distinguish between source and ambient atoms, but the lifetime they adopt pertains to the latter. They conclude that the principal source (of atmospheric atoms) is "thermal desorption," or evaporation. To this we would add that only a very small fraction of the free sodium is adsorbed to the surface, and that the majority of the impacting atoms are most simply regarded as scattering, rather than condensing and then evaporating. In this connection, it is interesting to look at the volatility of sodium metal, simply as a rough indication of what to expect at Mercury. A typical dayside density of 3×10^4 cm^{-3} would be attained at a temperature of around 290 K, a rather cool room temperature. Much smaller densities would, however, be expected in the 100 K range, to the extent that such an analogy is applicable.

Under steady state conditions, an ambient, thermal atmosphere with a relatively large accommodation coefficient should have most of its atmospheric content in the unobserved dark hemisphere. This situation probably does not exist for sodium and potassium. The migration time for Na, estimated from Eq. (3), is $\sim 10^6$ s, far longer than the photoionization time, and only an order of magnitude shorter than the 88-day alternation of day and night. Radiation pressure, suggested as the explanation of Fig. 9, may produce some nightside enhancement, but it is probably small. The entry for the nightside content in Table V therefore does not include any enhancement. If the principal source is global, there could be a large nightside enhancement.

Although this discussion is phrased in terms of sodium, much of it pertains equally to potassium, given how much less is known about its atomic physics and its behavior on Mercury. Relevant figures are given in Tables IV and V.

VI. SOURCES AND SINKS

The "ambient atoms" in the central box of Fig. 12 have been discussed in Sec. V; this section concentrates on the upper and lower boxes, the generation of source atoms and the loss of ambient atoms. The possible sources discussed in the literature are capture and neutralization of solar-wind ions, impact of atoms from the interstellar medium, sputtering of surface material by impact of energetic ions or photons and vaporization by meteoroid impact. For helium (and argon, detected only on the Moon), degassing of radiogenic atoms is another possibility, although Ducati et al. (1973) suggest that this source is small compared with the solar wind. Although some source atoms probably have energies in the thermal range, it seems likely that most have higher energies, and extend to correspondingly greater heights. Ejection of chemically bound atoms, with binding energy of a few eV, is likely to give velocities higher than thermal, especially in photo-ejection or sputtering by a photon or an ion with excess energy. Moreover, if an atom must overcome an activation energy E_a to be released, it should have kinetic energy equal to at least E_a. The absence of such energies in the data of Fig. 10 does not rule out the presence of energetic atoms, but suggests that they are less than half the total population, at least for sodium.

The rate at which any of the atmospheric species is replenished to make up for loss is difficult to define with a satisfactory degree of accuracy. Observationally, the source particles appear to be masked by an ambient pool of gas; observed mean kinetic energies are at or below the surface temperature. As argued in the previous section, the residence time of an atom must be several times longer than the characteristic time for accommodation with the surface.

Thermal escape is probably the most important loss process for H, and perhaps also He, but these rates are not readily calculated because the velocity distributions are highly non-Maxwellian. (Escape from the Moon is likely to be thermal for both H and He.) For the other gases, as well as He, the lifetimes are probably limited by photoionization. Radiation pressure affects the escape rate of source atoms of Na and K, but the ambient atoms are much less affected. The rates of photoionization are reasonably well defined for all of the species, but the efficiencies with which the ions are lost from the system are very uncertain. Table V gives rough estimated quantities for the known atmospheric species.

The next two subsections discuss general characteristics of sinks and sources; sinks are treated first because the rates are easier to estimate. These concepts are then applied to the individual groups of atoms.

A. Sinks and Lifetimes

Thermal Escape. As mentioned above, light gases should escape rapidly from a planet as small and as hot as Mercury if their energy distribution is Maxwellian. Thermal escape is conveniently discussed in terms of the dimensionless parameter

$$\lambda = r/H(r) \tag{4}$$

where the scale height is

$$H(r) = \frac{kT}{mg(r)} = \frac{kTr^2}{mg_0} \tag{5}$$

where r is the planet-centered radial distance, $H(r)$ is the atmospheric scale height, and g_0 is gravitational acceleration at r_0, the radius of the solid planet. Small values of λ_0 imply very rapid escape from the planet's surface; the time scale can be obtained by taking the ratio of the integrated density $n_0 H(r_0)$ to the Jeans escape flux:

$$\tau_J = \frac{2\sqrt{\pi}H(r_0)\exp(\lambda_0)}{BU(1+\lambda_0)} \ . \tag{6}$$

As before, U is the most probable thermal velocity. The quantity B multiplies the flux obtained for a pure Maxwell-Boltzmann distribution to correct for a reduced population at high energies (Brinkmann 1971; Shizgal and Blackmore 1986). It is usually 0.5 or greater for a standard exosphere, but is expected to be much smaller for Mercury and the Moon. To obtain a crude idea of the lifetime, we can take values (except in the exponential) appropriate for helium on the day side of Mercury: $H \sim 300$ km, $\lambda_0 \sim 10$, $U \sim 1$ km s^{-1}; the lifetime in seconds (Eq. 6) becomes roughly

$$\tau_J \sim \frac{100}{B} \exp \lambda_0 \qquad (7)$$

With $\lambda_0 = 5$ to 10 and B not much smaller than unity, we obtain extremely short times of 15000 and 2×10^6 s. Since B is almost certainly very small for H and He, Eq. (7) actually gives only an extreme lower bound. For atomic oxygen, λ_0 is around 40, for which this lifetime approaches the age of the solar system.

As Fig. 14 shows, collisions of helium atoms with a quartz surface are very inefficient at populating the higher energies, and a similar statement can be surmised to hold for H. Simultaneously, the escape process is strongly draining the same energies. As the escape velocity is around 1.4 U for H at subsolar temperatures, and 2.8 U for He, the effect on the escape rate and the velocity distribution is certainly large. Just how large will not be known until a self-consistent numerical calculation has been carried out.

Radiation Pressure. The acceleration of an atom of mass m due to resonance scattering of solar radiation is

$$a_n = (\pi h e^2 v_i / m_e m c^2)\, (\pi F_{v_i})\, (f_i / R^2) \qquad (8)$$

where e, m_e are the charge and mass of the electron, v_i and f_i are the frequency and oscillator strength of the resonance transition, πF_v is the solar flux at 1 AU (in photons cm^{-2}s^{-1}) in the rest frame of the atom, and R is the distance from the Sun in AU.

The solar flux at the resonance wavelengths of Na and K not only excites the observed scattered light, but is also strong enough to produce significant accelerations of the atoms. (For H, the acceleration is roughly equal to that of solar gravity, ~ 4 cm s^{-2}, but still much smaller than that of a planet.) The presence of a deep Fraunhofer line affects the radiation pressure in exactly the same way as the resonance scattering; Doppler shifts relative to the Sun can increase the acceleration by an order of magnitude or more (Smyth 1986*b*). For sodium, taking into account both fine-structure components, the acceleration can reach 200 cm s^{-2}. In a single dayside hop of 250 s, the increment of velocity is as great as 500 m s^{-1}. The value for potassium is always larger, because the Fraunhofer lines are neither as deep nor as wide, even though the continuum flux is less by a factor of 2/3.

The relevant radial velocity includes the motion of the atom relative to that of the planet, which changes as the atom is accelerated; it is therefore necessary to follow the motion with a computer. Such computations have been carried out by Ip (1986) and Smyth (1986) for Na. The effect on the escape rate is minor unless the ejection velocity from the surface is at least 2 km s^{-1} (3 km s^{-1} if the Mercury-Sun radial velocity is small). To obtain appreciable populations at these velocities from a thermal distribution would require temperatures of 1000 to 2000 K for sodium. Compared with the example in the previous paragraph, such velocities allow a longer time for the acceleration to operate, and also a reduced planetary gravity. Under such conditions, a comet-like tail is predicted and the loss rate from the planet is enhanced. The effect is larger if the planet is already moving away from the Sun. If the planet's motion is towards the Sun, the atom moving in the opposite direction sees the weaker radiation near Fraunhofer line center and is accelerated less.

The importance of radiation pressure as a loss mechanism for the alkalis therefore depends crucially on whether there is a substantial population of nonthermal atoms in the atmosphere, that is, source atoms that have not been thermalized by collision with the surface. According to Table IV, the accommodation coefficients are large for the alkali metals, and most of the atmosphere should be thermalized. This conclusion accords with the observed D_2 line profile (Fig. 10), although a substantial component at a few thousand degrees cannot be ruled out in the absence of data with a higher signal-to-noise ratio.

Since radiation pressure requires a substantial time to build up a significant velocity, the competition by photoionization as a loss mechanism must be considered. For illustration, with the maximum acceleration of 200 cm s^{-2}, the time required to build up a speed equal to the escape velocity (4.25 km s^{-1}) is 2100 s, not much less than the ionization time of 6000 to 12,000 s.

Ionization. Photoionization times are short, especially for the alkali metals (Table V) where the lifetime in daylight is around 3 hr (half as great at perihelion). Whether the ion is lost from the planet depends entirely on the characteristics of the magnetosphere and solar wind environment. If the magnetic field were an undisturbed dipole, the ion would rise or fall at the same rate as its parent atom, being simply constrained to spiral along a field line until it would hit the surface and be neutralized. However, the actual field is rapidly convecting, its outer parts driven downstream by the solar wind; an equivalent picture is that there are large electric forces on the ions, much larger than gravity. Although many of the ions are therefore swept away in the solar wind, some fraction are aimed back to the surface, particularly at high latitudes or on the night side, where they can be neutralized and re-emitted. Goldstein et al. (1981) discussed these processes and settled on an escape fraction β of 0.5 for He, with a range from 0.25 to 1; Ip (1986) prefers 0.8 or 0.9. On grounds similar to those discussed below, McGrath et al. (1986) suggest $\beta \sim 0.1$ for Na, which is low enough to have a significant effect on sodium budget assessments.

Thermal alkali atoms, in particular, are confined within 100 to 200 km of the surface, and ions produced there may be at least partially protected from magnetospheric sweeping. Such protection could be very important if the surface were a conductor, which it almost certainly is not. Possible effects are the presence of a weak ionosphere and a small inward component of the electric field which would carry ions into the surface. Such effects need quantitative assessment; at present we can only guess that they will lower the escape fraction and adopt the 0.1 of McGrath et al. Thus, for sodium near aphelion, an average atom would be ionized every 10^4 s and would be recycled 10 times before escaping from the planet, for a total lifetime of 10^5 s, just over an Earth day. The sodium budget, discussed below, suggests that a still smaller escape fraction may be required.

B. Sources

Atmospheric H and He would seem likely to arise from neutralized solar-wind ions, and O, Na and K from meteorites or from Mercury's crust and interior. However, helium could be degassed from the interior, as on the Earth, although Ducati et al. (1973) conclude that atoms from this source are negligible in the gases trapped in lunar grains. The elements discovered by Mariner 10 were discussed by Kumar (1976), and capture and loss processes involving the solar wind by Banks et al. (1970) and Goldstein et al. (1981). Since the discovery of the sodium atmosphere, papers have been published by Ip (1986), Smyth (1986) and McGrath et al. (1986).

Capture of Solar Wind. Mercury's magnetic field is just strong enough to deflect the solar wind away from the surface under average conditions, but not all of the time (Chapter by Russell et al.; Goldstein et al. 1981). Rapid magnetospheric convection is probably accompanied by entrainment of solar-wind ions, many of which are precipitated to the planet in the same way as auroral ions on Earth. The ions impinging on the surface are probably implanted initially. They must be released in some separate process, which may be thermal or involve displacement by another newly arrived ion. These processes have been surveyed in detail by Goldstein et al. (1981), who give many references to earlier literature. They conclude that a fraction between half and all of the implanted gas is eventually released. Formation of H_2 seems likely to be the fate of much of the hydrogen, but theoretical arguments to the contrary have been suggested. Such arguments tend to be based on the notion of small quantities of H trapped at isolated sites in a nearly ideal crystal lattice, hardly applicable to the lunar surface nor that of Mercury. In addition, the probable rarity of H atoms trapped in soil grains (cf. Ducati et al 1973) may suggest that they are unlikely to encounter one another. We find below that there is a severe deficiency of atomic H on both bodies, and formation of molecules is by far the most likely explanation.

The results of the study by Goldstein et al. (1981) are summarized in Table V. Further discussion appears below in Sec. VI.C. They find that the solar wind has direct access to the surface of Mercury only ~6% of the time;

this access is supplemented by indirect magnetospheric processes. It is possible that this estimate may be too low: the UV albedo of Mercury is lower than that of the Moon (Wu and Broadfoot 1977), and a decreased albedo is a characteristic of proton irradiation (as opposed to α particles: Zeller et al. 1966). According to the latter authors, chemical reactions in the solid are most efficiently produced by protons having energies of 5 to 20 eV.

Retention and diffusion of gases in lunar glasses (the major component of the regolith) have been thoroughly discussed by Ducati et al. (1973). Helium is retained much more tightly than in synthetic glass of similar composition; they suggest that the structure is altered by the irradiation received on the Moon. Both He and Ne are released in the range between 900 and 1300 K, corresponding to an activation energy for diffusion of around 2 eV. The incident solar wind should be able to saturate the surface, to levels of 10^{17} and 10^{16} cm^{-2} for H and He, in 10 yr. The amounts observed are lower by 2 orders of magnitude. As another indication, the He/Ne ratio is ~50 instead of 600 in the solar wind, and the neon itself is reduced by a factor ~50 relative to nitrogen. It is clear that, for atmospheric purposes, hydrogen and helium are not retained in the solid. The H density may even remain so low that H_2 formation is inefficient. If H atoms are released with nonthermal energies, they would be rapidly lost from the planet. In addition their very large scale height would make them difficult to detect, because they could not easily be distinguished from the background of the interplanetary medium.

Meteoroids. The supply of Na and K by meteoroid bombardment has been discussed by Morgan et al. (1987). The vapor production depends on the mass and velocity of the object and the material properties of object and target surface. At velocities of 2 to 3 km s^{-1} or very small masses, the mass of vapor may be only a few percent of the projectile mass (Eichhorn 1976). Above 30 km s^{-1}, the mass of vapor may be many times the impactor mass (O'Keefe and Ahrens 1976). Vapor temperatures range from 2500 to 5000 K (Eichhorn 1978), and most of the sodium in the vapor should be atomic.

According to Lienert et al. (1981), the density of material varies with distance r from the Sun as $r^{-1.3}$. Fluxes at the Earth (Zook 1975), scaled to Mercury, predict a mean impact velocity near 30 km s^{-1} and a rate of 400 to 800 g s^{-1} for the whole planet. For a sodium content of 0.13% by mass, the source ranges from 1.4 to 5.6 $\times$ 10^{22} atoms s^{-1}; the highest figure assumes equal amounts of sodium from meteoroid and surface. If the Na/K ratio is 7 in the meteoroids and much greater in the crust, the potassium source ranges from 0.2 to 0.4 $\times$ 10^{22} s^{-1}. Because of the high temperatures, some of the sodium, and even potassium, atoms produced may be able to escape from the planet.

Meteoroids may be a much more potent source of water vapor than of alkalis. The meteoroid flux at the Earth could consist primarily of carbonaceous objects containing 10% water by mass, giving a source strength of 240×10^{22} s^{-1}. The photodissociation time at aphelion is 2×10^4 s (Kumar 1976), and the global average is twice as great; the resulting column density is 10^{11} cm^{-2} and the number density 2×10^4 cm^{-3}. The primary products of dissociation are H and OH, but some 13% of the events give H_2 and O (Gombosi et al. 1986). OH photolyzes rapidly, but may also adsorb to the surface and react with H or OH to recycle H_2O; if so, the effective lifetime and the abundance of the latter may be increased by a factor of ~8. Even without this effect, the sources of H and O would be 240 and 30, in units of 10^{22} s^{-1}. All of these estimates could be considered upper bounds, though we feel that they may actually be realistic.

Sputtering and Photodesorption. The most thorough discussion of these processes is that of McGrath et al. (1986); earlier work has been published by Potter and Morgan (1985a) and Ip (1986). In our terminology, McGrath et al. concluded that thermal evaporation is the most likely source of atmospheric atoms, and found that neither sputtering nor photodesorption is adequate. Here we look at their potential for generating source atoms. It should be noted that the second line of their Table 1 is mislabeled: it shows the incident flux, not the sputtered flux. The latter can be obtained by dividing the column densities N in the last line of their table by the assumed (ballistic) lifetime, 1000 s. Multiplication by the appropriate surface area gives the global production rates in Table V. For the ionic processes, we used the total surface area of the planet, 7.5×10^{17} cm^2; for photosputtering, this is divided by 4 to give the projected area.

A strong limit on the total rate of sputtering by all processes can be obtained from studies of lunar samples (McDonnell 1977; Carey and McDonnell 1978). This rate, 0.031 Å/yr (±30%) for the Moon translates to 0.19 Å/yr at Mercury's mean solar distance. If the total number density of Mercurian surface material is 8.9×10^{22} atoms cm^{-3} and the number fraction of Na is 0.002, the rate of sodium production is limited to 1.1×10^4 atoms cm^{-2} s^{-1}, or 0.8×10^{22} atoms s^{-1} for the entire surface. A fresh surface sputters much more quickly, but the rate settles down to the value given. Material richer in Na could produce a proportionally larger rate.

Processes that derive the sodium from the surface of Mercury are, in the end, limited by the rate of production of new regolith. On the Moon, the amount of new regolith typically deposited is 300 to 400 g cm^{-2} per Gyr (Langevin and Maurette 1978), corresponding to 1.9 to 2.2 m Gyr^{-1}. The process is not continuous and much of the material may be emplaced in a few episodes, particularly crater-forming events. Thus, the cratering record should be a good guide to the regolith turnover rate, and these records appear to be very similar for the Moon and Mercury (Murray et al. 1975; Wetherill 1975). Much of the material included in the estimates above is reworked surface material, so that a supply rate of "new sodium" calculated from the depositional rate is an upper limit. With 400 g cm^{-2} Gyr^{-1} and a sodium atomic fraction of 0.002, we find a maximum supply rate of 6.1×10^5 atoms cm^{-2} s^{-1} or, for the whole planet, 140×10^{22} atoms s^{-1}.

The supply rate estimated above for meteoroid influx is about 2 orders of magnitude smaller and therefore is not significantly affected by the upper limit.

C. Hydrogen and Helium

The global contents of hydrogen and helium were estimated in the previous section and appear in Table V, along with the source and sink rates discussed next. The deposition rates of solar-wind H and He P_{sw} follow the study of Goldstein et al. (1981), who suggest an uncertainty of about an order of magnitude. For their adopted solar-wind helium flux of 7.9×10^7 cm^{-2} s^{-1}, the quantity of He incident on a disk the size of Mercury is 1500×10^{22} cm^{-2} s^{-1}. Thus, the amount actually collected is around 1% of what it would be without the magnetic field. Goldstein et al. did not explicitly give numbers for H; those in Table V follow from their He/H ratio of 0.045.

Shemansky and Broadfoot (1977) suggest that the dominant loss process for helium is photoionization, because the high-energy tail of the thermal distribution is depleted. If so, the mean lifetime is at least the ionization time, ~10^7 s or 100 days, and the global loss rate is 11×10^{22} s^{-1}. However, ions may still return to the surface and be neutralized; Goldstein et al. (1981) suggest a loss efficiency β of 0.5, within a factor of 2 either way, and thus the minimum loss rate shown in Table V is 3×10^{22} s^{-1}. The estimated source and sink strengths are in reasonable balance, especially if β is fairly near 1. Semiannual or diurnal effects should be observable for helium, but the Mariner 10 observations were confined to aphelion.

The solar-wind source strength for H is several hundred $\times$ 10^{22} s^{-1}, and the likely source from photolysis of H_2O is comparable. Because the observed abundance is less than that of helium, and the ionization lifetimes are similar, the ionization loss rate is an order of magnitude less, and is 3 orders of magnitude smaller than the source. Some loss of atomic hydrogen may be driven by solar radiation pressure (Smyth 1986), but again the small densities argue against a large effect. Only two alternatives remain: thermal escape and chemical combination into an unobservable molecule such as H_2, unless there is a large loss of source atoms.

Another possible source of H is photolysis of H_2O, suggested by Broadfoot et al. (1976) as an explanation of the cold component of Fig. 3. The required number density of H_2O would be 2×10^6 cm^{-3} for a loss rate assumed to be 1000×10^{22} s^{-1}. The H_2O amount suggested above is 2 orders of magnitude less. If the lifetime of the cold H is increased by this factor, the mechanism could be viable.

Direct use of the Jeans equation with the 420 K component of Fig. 3 gives a lifetime of 10^4 s and a loss rate of 80×10^{22} s^{-1}. On the unlikely assumption that a similar lifetime would apply to the cold component, which has twice the abundance (Table I), the thermal sink still does not seem adequate. Any *ab initio* estimate of thermal loss must include a computation of the velocity distribution, which has not been done. The distribution must be even more distorted than that for He (Fig. 14), which is itself uncertain. Of course, freshly released ("source") atoms may escape immediately, if they are directed upwards.

Even with the maximum likely thermal loss rate, the supply of H seems to exceed the loss by a factor of 3, and a much larger factor seems possible. Similarly, Hodges (1973b), comparing the lunar observational upper limit for H with a value based on his model (which is severely criticized above), concluded that there is a severe depletion. He suggested formation of molecules such as H_2. Thomas (1974) suggested H_2O, as explaining the deficiency. The molecules H_2, He, CH_4 and H_2O have been identified in the lunar subsurface as produced from the solar wind (Gibson 1977). Surface grains contain a large quantity of implanted H, and although the details are obscure, the formation of H_2 seems inevitable.

A rough upper bound to the expected H_2 density can be estimated by scaling from He, and by assuming that in steady state nearly all of the incident H is released from the surface grains as H_2. The ratio of the photoionization times is 0.21 (Kumar 1976), and photodissociation is much slower. The dayside density should therefore be $\sim 10^4$ cm^{-3}, and the nightside enhancement similar to that for helium. If thermal or nonthermal escape is significant, as it could well be, this estimate would be reduced. For CH_4 and H_2O, much smaller abundances would be expected for two reasons: small solar-wind fluxes of C and O (about 10^{-2} that of He), and short lifetimes against photodissociation. In principle, the surface silicates offer an additional source of O, but it seems likely that the available atoms have long since been used up. Densities of 1 cm^{-3} or less seem reasonable.

D. Sodium, Potassium and Oxygen

Although it is likely that all of the constituents vary with time, available observations are limited to sodium (Potter and Morgan 1987). Significant night-to-night variations appear to occur, and there is little doubt that variations occur on a 100-day scale. These are consistent with the time scale of a few days associated with photoionization.

Sources.   As discussed above (Sec. VI.B) and summarized in Table V, there are three major candidates for the principal sodium source: ionic sputtering, photosputtering, and meteoroid impact. Unless the limit from the lunar erosion rate derived above in Sec. VI.B (0.8×10^{22} s^{-1}) can somehow be refuted, the sum of the first two sources cannot exceed it, and meteoroids, at around 8×10^{22} s^{-1}, seem to dominate. However, the uncertainties are large enough that this conclusion could be upset.

The magnitude of the sodium production due to either photodesorption or impact vaporization is proportional to the elemental abundance of sodium in the regolith, and calculations made to date have used very small values for the elemental abundance of sodium. The elemental abundance of sodium in the regolith of Mercury is often based on that of the lunar regolith which is quite small, and it is often argued that the temperature of the protoplanetary nebula was too hot in the region in which Mercury was formed to include any significant accretion of sodic silicates. Yet the Moon is clearly a less differentiated body than Mercury. Even if the global elemental abundance of sodium in a planetary body were small, differentiation may still concentrate that small amount into the crust. Thus, sodium is much more abundant in the crust of the Earth (over 2%) than in the planet as a whole. This is not to say that the abundance of sodium in the crust of Mercury is that high, but does say that 0.2% is a conservative number, and that an order of magnitude variation is possible due to the elemental abundance alone.

Sinks. With an escape fraction $\beta = 0.1$ adopted for the alkalis in Sec. VI.A, the loss rate for sodium due to photoionization is 60×10^{22} s^{-1}, larger by a factor ranging from 4 to 30 than the estimated meteoritic source. Several ways can be visualized to close this gap: a still smaller value of β; relaxation of the constraint on sputtering set by the lunar erosion rate; a larger meteoroid influx; or meteoroids richer in sodium.

Again, the discussion of sodium can be adapted to potassium, with the somewhat different numbers shown in Table V.

The presence of *atomic oxygen* has received almost no attention, perhaps because its presence was not announced until two years after the first Mariner 10 encounter (Broadfoot et al. 1976). Possible sources include sputtering by ions and atoms from the solar wind and photolysis of water vapor brought in by meteoroids.

E. Summary

The solar wind is the major source of hydrogen and helium in the atmosphere of Mercury, although a portion of the helium observed must be of radiogenic origin. Sodium, potassium and oxygen are derived from the surface or from meteoritic material. The sodium, in particular, may be converted from a cation in the regolith silicate assemblage by impact vaporization, or by photodesorption. Impact vaporization following meteoritic impact must be present, and provides a source of sodium and potassium in the event that the regolith is deficient in alkalis. Photodesorption will be a significant source if sodium can be supplied to the planet's extreme upper surface (to which photodesorption is limited). Charged-particle sputtering of surface materials may be the principal source of oxygen, but photolysis of water vapor is another possibility.

Loss following photoionization plays a major and possibly dominant role in the loss of helium, sodium, potassium and oxygen. The dominant loss mechanism for hydrogen is probably formation of molecules, but there are no published rate calculations to show that formation of H_2 or H_2O in or on surface grains can account for the required rate of removal of hydrogen to attain a steady state. Loss of sodium and hydrogen due to radiation pressure may be important over portions of the orbit of Mercury.

The practical demonstration that the sources and sinks for any elements have been identified is that the calculated rates for addition and removal are in approximate balance. Our understanding of the following topics is still very incomplete:

1. The magnetosphere of Mercury;
2. The elemental and mineralogical composition of the regolith;
3. The meteoritic complex inside 0.5 AU;
4. Surface physics and surface chemistry.

That the desired balance has not been achieved for sodium, hydrogen and potassium may simply reflect this ignorance, rather than neglect of important sources or sinks. Carefully conceived observing programs may be able to test the importance of candidate sources or sinks for particular elements in the atmosphere. A better understanding of the atmosphere may provide valuable insights about many other aspects of the planet Mercury.

VII. INTERCOMPARISONS

A. The Moon's Atmosphere

Inspection of Table I does not reveal any large discrepancies between the atmospheric contents of the Moon and Mercury, given their vastly different distances from the Sun and different escape velocities. Although H was not detected on the Moon, the upper limit is similar to the density of the hot component on Mercury (cf. Fig. 3); it is a factor of 8 less than the cold component. Such a factor could be due to the greater solar-wind flux at Mercury, to absence of the cold component on the Moon, or to unsuitable geometry for detecting it. The helium contents of the two bodies are fairly similar. The landed experiments did not yield much information on diurnal variations, because they were swamped by gas from the payload and spacecraft on the day side.

The Moon has been observed with the same equipment that discovered sodium on Mercury, but no sign of emission has ever been found. It must be remembered that the intensity for the same amount of vapor tends to be two orders of magnitude smaller. Two effects each contribute about equally: the Moon-Sun Doppler shifts are always small, and the distance from the Sun is large.

All of the physical processes that might work to add or remove the constituents of the atmosphere of Mercury are also at work on the Moon. While the rates are different, they can, in principle, be calculated with much more certainty for the Moon than for Mercury. The Moon's surface composition is known, and the particle and field environment has been determined. The meteoroid flux, the velocity distribution of the meteoroids, and the elemental composition of the meteoritic material are also known at 1 AU. In short, a model of the atmosphere of Mercury can also be adapted to make predictions about the atmosphere of the Moon.

B. The Earth's Atmosphere

A very large amount of literature exists on observations and interpretations of alkali metals, especially sodium, which appear as a layer in the mesopause region, at an altitude of about 90 km. The layer thickness at half maximum is 10 km or less. The discovery of this layer (Bernard 1938) and subsequent study up to 1969 were carried out under twilight conditions, where the brightness of the sky is greatly reduced relative to the daytime, and height information can be obtained as the shadow of the Earth's limb scans through the layer (Chamberlain 1961; Hunten 1967). The corresponding dayglow was observed by rockets (Hunten and Wallace 1967; Meier and Donahue 1967). Lidar observations have been carried out more recently (see, e.g., Gardner et al. 1986), with similar results but a much better definition of the height profiles. A typical abundance is 3×10^9 atoms cm^{-2}, perhaps an order of magnitude smaller than is seen at Mercury. The sodium source is commonly believed to be meteoroid ablation, which reasonably explains the observed densities if the sodium content is a few percent (Gadsden 1968; Hunten 1984).

The presence of free sodium is believed to stem from the large abundance of oxygen atoms in the upper atmosphere, through reactions of the type

$$NaO + O \rightarrow Na + O_2. \tag{9}$$

The cutoff at the bottom of the layer coincides with a large decrease in the density of O atoms; however, at and above the peak of the layer, the entire inventory of available sodium is expected to be in atomic form. Sodium ions can be detected by rocket-borne instruments, but are much less abundant than neutral atoms. The ions are certainly being produced rapidly, but are evidently recycled in some unknown way. The lifetime is probably limited by other processes such as deposition on dust particles, agglomeration into large aggregates and subsequent fallout. The rates of such processes are not well determined, but they are probably much slower than photoionization at Mercury.

A typical ratio of Li:Na:K is 1:8000:160, but it is variable with latitude, time, and season and its relationship to the sources is complex and obscure. This matter is further complicated by the presence of lithium injected by rocket experiments and, in the past, thermonuclear explosions. It is likely that the potassium source is the same as that of sodium.

Perhaps some day we will understand terrestrial sodium well enough to illuminate its Mercurian counterpart, but that day has not arrived.

C. Io's Atmosphere

Bright sodium emission from Io was unexpectedly discovered by Brown in 1973 and shown to be in an extended cloud to tens of Io radii by Trafton. Brown and Yung (1976) have provided a thorough discussion of the physics and of the first few years of observational results. There are many other reviews, the most recent that of Nash et al. (1986). Io, like Mercury, exhibits large changes in radial velocity which modulate the intensity of solar excitation by about an order of magnitude. Potassium is also present (Münch et al. 1976; Trafton 1977), and the Na/K ratio is 8 to 20.

Emission from ionized sulfur was discovered by Kupo et al. (1976). By the time of the Voyager encounter, it had been established that Io's orbit is surrounded by a sulfur torus with ion density of a few thousand per cm^3 and an electron temperature of ~50 eV. These results were confirmed and greatly extended by Voyager 1 (Broadfoot et al. 1979; Bridge et al. 1979); in particular, oxygen ions are also present, and the ratio of oxygen to sulfur is 2. Since SO_2 is a prominent constituent of Io's surface and atmosphere, it is impossible to avoid the suspicion that it, or its dissociation products, provide the major source of matter for the torus. The torus plasma, in turn, bombards Io and its atmosphere, producing collisions which are thought to create the major source of ions. Sinks are even harder to specify, although it is clear that ionization by electron impact is very important for neutral Na. In a second process, charge exchange converts corotating ions to fast neutrals that can escape the Jupiter system, but is not a net sink of charge. The third major process is magnetospheric radial diffusion.

Conditions at Io and Mercury are quantitatively very different. Io is bombarded by plasma ions at 55 km s^{-1}; Mercury is bombarded part of the time by protons and α-particles at hundreds of km s^{-1} and a much lower density. Io is immersed in Jupiter's strong magnetic field; Mercury has its own weak field and a very active regime of magnetospheric convection. It remains to be seen whether each has much to tell us about the other.

VIII. SUMMARY AND OUTLOOK

The persistent reader may have reached this point with an unsettled feeling. There are many open issues and many opportunities for further modeling and data interpretation. Work on the Mariner 10 results was broken off after 1978, with the encounters of Pioneer Venus and Voyager. The discoveries of sodium and potassium are so recent that they have not been fully assimilated, and there is plenty of scope for fresh work.

Nevertheless, a few statements can be made about the atmosphere of Mercury. The known gases H, He, O, Na and K are summarized in Table I. The surface number density of molecular hydrogen may approach 10^4 cm^{-3}, comparable to He and O. If argon can be scaled from the Moon, its density would be a few hundred per cm³. Nothing else is likely to exceed 1 cm⁻³. The longer-lived and highly mobile gases He and perhaps H_2 and H should show nightside enhancements by a factor of around 50.

Capture of the solar wind is the probable source of hydrogen and helium, with an intermediate step of implantation, followed by displacement as further ions arrive. Ion sputtering or photosputtering probably eject oxygen from the surface, and must still be considered for sodium and potassium. For these latter two elements, however, impact evaporation of meteoroids seems more likely. Meteoroids may also be a large source of water vapor, which in turn supplies H, O and H_2. An important sink for all gases is photoionization and entrapment in the solar wind and magnetospheric flows, although it is likely that 90% or more of these ions are recycled to the surface. Thermal loss cannot be quantified at present, because the velocity distributions of the light elements are strongly non-Maxwellian, but may still be important for atomic and molecular hydrogen.

Transport across the surface, and speed and height distributions in the atmosphere, are strongly influenced by the nature of the atom surface interaction. Any adsorption to the surface lasts for times less than a second, especially on the day side; the typical collision is a scattering, not an adsorption. For helium, these collisions give an energy distribution that is highly deficient at high energies relative to a Maxwellian. This tendency is reinforced by escape of faster atoms. Similar effects are likely for hydrogen.

In the near future, several kinds of work could be pursued. Observation of the sodium and potassium lines can give information about temporal changes and spatial distributions. The Mariner data have much better spatial information, especially vertical distribution, than can be obtained from the Earth, and address lighter elements whose behavior is very different. Analysis of all of these data can benefit from further calculations of gas-surface interactions and migration over the surface; the results in Fig. 14 and Table IV represent only a start.

APPENDIX A: DEFINITION OF TERMS

The interdisciplinary nature of the study of the atmosphere of Mercury has led to some confusion in terminology. We therefore define terms used in connection with aspects of the subject encountered in this chapter.

Surface: The outermost layer of atoms of the solid component of the planet. This definition differs from that normally used in studies of solid geochemistry (see, e.g., Gibson 1977).

Sub-Surface: The region within ~500 μm of the surface.

Adsorption; physisorption: Physical-scale bonding of a gas atom to a surface with a duration of at least one vibrational period. Bonding energies are <0.5 eV. In Fig. 13, the atom is oscillating in the potential well at r_e.

Adsorption; chemisorption: Chemical-scale bonding of a gas atom to a surface. Bonding energies are >0.5 eV.

Absorption or entrapment: Chemical-scale bonding of a gas atom in the subsurface.

Energy accommodation coefficient (α): Coefficient of fractional energy exchange of a gas atom with a surface, including both adsorption collisions and free-free collisions. The equation for α at the macroscopic level is defined by

$$\alpha = (E_2 - E_0)/(E_1 - E_0) \qquad \text{(A1)}$$

where E_0 is the mean energy per atom of the impacting particle, E_2 is the mean energy per atom leaving the surface, and E_1 is the mean energy per atom in the limiting case of thermal equilibrium with the surface. Similar coefficients can be constructed to specify the fractional momentum exchange of gas atoms with the surface.

Sticking coefficient (S): Coefficient referring to chemical-scale bonding to the surface. The quantity S is the rate of adsorption per incident particle. The adsorption lifetime is undefined in the usage of this parameter, and it is therefore a useful quantity only for chemical-scale bonding in which residence time is long compared to typical experimental time scales. In the context of the present subject the sticking coefficient is generally not a useful quantity, and usage of the term can be unphysical and misleading.

Heat of adsorption (D_0): The dissociation energy measured from the zero-point energy level, which is somewhat higher than the minimum shown in Fig. 13.

Activation energy (E_a): Peak of the potential barrier above the separated atom asymptote, occurring at physical-scale internuclear distances; the minimum energy required for chemical reaction (see Fig. 13).

Exosphere, exobase: The outermost part of an atmosphere, in which collisions can be neglected to first order and the atoms can be regarded as executing ballistic orbits. Its bottom, the exobase, is taken as the level where the local mean free path is equal to the scale height. As emphasized in the text, there are important differences between a normal exosphere and an atmosphere like that of Mercury.

Ambient atoms: Atoms defined here as those atoms occupying the central box in Fig. 12; they include atmospheric atoms (defined next) and those adsorbed briefly to the surface.

Atmospheric atoms: Atoms actually in the ballistic atmosphere, with near-thermal energies.

Source atoms: Atoms freshly ejected from the surface or freshly generated in some other way, such as neutralization of an ion; in practice, distinguished from atmospheric atoms by their higher energies.

APPENDIX B: GAS-SURFACE INTERACTIONS

This Appendix gives some of the technical details behind the material in Sec. V.C. The residence time for adsorption is approximated by

$$\tau_{ad} = \tau_0 \exp (D_0/kT) \tag{B1}$$

where τ_0 is a vibration time for the van der Waals potential, around 10^{-13} s, and D_0/k is the heat of adsorption expressed as an equivalent temperature (see Fig. 13). With the values for helium shown in Table IV, the residence time at 200 K is found to be $\sim 10^{-12}$ s. The collision frequency, or flux to the surface, for an atmospheric number density n is

$$\nu = \left(\frac{kT}{2\pi m} \right)^{1/2} n \tag{B2}$$

which is $\sim 2 \times 10^{11}$ cm^{-2} s^{-1} at a nightside number density of 10^7 cm^{-3}. If the sticking coefficient is 1, the coverage is 0.2 atom cm^{-2}. Any approach to full coverage requires conditions not found on Mercury: namely pressure approaching one bar, temperature well below 100 K, or chemical bonding to give a much larger value of D_0. Observations of the day side imply mean energies corresponding approximately to the surface temperature, and there is no reason to believe that the "bonding" is other than physical.

Estimates of the heats of adsorption and accommodation coefficients depend on the composition of the surface material. Unfortunately we have no direct determination of surface composition, and it is necessary to resort to analogy with the Moon. The spectral reflectivities of Mercury and the Moon are very similar in shape and magnitude from infrared wavelengths to the EUV (McCord and Adams 1972*b*; Wu and Broadfoot 1977), and we assume similar surface compositions. The dominant surface materials on the Moon are calcium aluminum silicates and iron-rich silicates, although SiO$_2$ has been mentioned (Cadenhead et al. 1972). Calculations of gas-surface interaction characteristics have concentrated on the surface structure of quartz (Shemansky and Broadfoot 1977; Kunc and Shemansky 1981). This is hardly an ideal situation, but is forced by the availability of the necessary laboratory information. There is some justification in that one important quantity is the Debye characteristic temperature, which depends mainly on the average atomic mass of the material.

The heats of adsorption of H, He, O, Na and K on α-quartz have been calculated by Kunc and Shemansky (1985) and Shemansky and Kunc (in preparation). These quantities are given in Table IV, along with crude estimates of accommodation coefficients. Table IV includes other basic physical properties required for a qualitative description of atmospheric characteristics, as well as for quantitative calculations. However, the accommodation coefficients (α) are based on classical theory and are not truly applicable to detailed calculation (Shemansky and Broadfoot 1977).

The kinetic energy distributions of the atmospheric gases on Mercury appear to be controlled by physical collisions with the surface. This argument is based on the observation that the mean energies of the particles are at most characteristic of the temperature of the solid surface. Source particles may however be involved in chemical-scale interaction, which may involve activation energies. If so, the dissociation process would typically provide energies in excess of 1000 K/atom (see Shemansky and Broadfoot 1977). The presence of gas essentially confined to the surface temperature then implies that the majority of interactions with the surface are on a physical scale.

Another approach to estimating the rate of adsorption on the surface of Mercury assumes an activation energy E_a which forms a small barrier. The flux (Eq. B2) is multiplied by the Boltzmann factor $\exp(-E_a/kT)$ (Glasstone et al. 1941,p.351). No information is available on the size of such an activation energy, or even its presence; in this situation it is common practice to use the Hirschfelder semi-empirical rule

$$E_a \sim \frac{D_0}{20} \, . \tag{B3}$$

The Boltzmann factor does not differ much from unity unless $E_a > kT$, or $D_0 > 20\,kT$; even at $T = 100$ K, D_0/k would have to exceed 2000 K. For sodium it is about 3000 K, and the Boltzmann factor is ~ 0.2 on the night side. If we assume that the gas density at the surface on the antisolar side is ~ 100 times greater than observed on the day side, the number density $n(\text{Na}) = 2 \times 10^6$ cm^{-3}. At $T = 100$ K, the downward flux at the surface is then $\nu(\text{Na}) \sim 3 \times 10^{10}$ cm^{-2} s^{-1}. The adsorption lifetime from Eq. (B1) is 0.7 s, and the areal density on the surface is 4×10^9 cm^{-2}, less than 10^{-5} of a monolayer ($\sim 6 \times 10^{14}$ cm^{-2}). Thus, it is almost certain that no significant part of the surface of Mercury is saturated in adsorbed gas.

This treatment assumes that the atoms are mobile in the adsorbed state, which is only partially true for sodium with its relatively large value of D_0. The adsorption rate for immobile adsorbed atoms involves a concentration of activated sites denoted by [CS] (Glasstone et al. 1941,p.349). For sodium [CS] $= 6 \times 10^{14}$ cm^{-2}; the adsorption rate is given by

$$N[CS] \frac{kT}{h} \frac{\omega_s}{F} \, exp \, \frac{-E_a}{kT} \qquad (B4)$$

where

$$F = (2\pi mkT)^{3/2} \left(\frac{\omega_g}{h^3} \right) \qquad (B5)$$

is the partition function per unit volume for the gas phase, h is the Planck constant, and $\omega_s = \omega_g = 1$ are the vibrational and rotational degeneracies of adsorbed and gas-phase atoms, respectively. With (B4) the adsorption rate is found to be reduced by about 2 orders of magnitude. The concentration of adsorbed Na given above is therefore an upper limit.

The rather long lifetime for Na on the antisolar surface does present an opportunity for formation of molecules. This possibility requires further research, and can only be discussed in a general way here. In the case of sodium and potassium on the dark surface of Mercury, the rate-limiting factors for combination reactions are the mobility and activation energy on the surface, rather than the adsorption lifetime. It is then possible to have reactions on the surface forming van der Waals Na$_2$ and K$_2$, and with a much lower probability, chemically bound versions of the same molecules. Two cases can be distinguished: reaction of two adsorbed atoms, and reaction of an incoming atom with an adsorbed one. If we assume a high mobility, a crude estimate of the rate of formation of van der Waals molecules from adsorbed atoms can be obtained (Glasstone et al. 1941,p.373):

$$N^2 \left(\frac{\sigma}{m} \right) \left(\frac{\pi}{kT} \right)^{1/2} exp \left(\frac{-D_0}{kT} \right) \qquad (B6)$$

where σ is the atomic diameter. The result for the dark surface of Mercury is $\sim$300 cm^{-2} s^{-1}, a negligibly slow rate. The interaction of the downward flux with adsorbed atoms (of surface density N_{ad}) yields a similar rate:

$$NN_{ad} \left(\frac{kT}{h} \right) \left(\frac{\omega_s}{F} \right) exp \left(\frac{-E_0}{kT} \right) \qquad (B7)$$

where E_0 is the activation energy for formation of van der Waals or chemically bound molecules. A suitable value for E_0/k is 100 K, which gives a formation rate of $\sim$200 molecules cm^{-2} s^{-1}. The effect of the formation of molecules on the atmospheric distribution and loss rates is probably negligible, because the dissociation of chemically bound Na_2 or K_2 yields kinetic energy of ~ 0.2 eV (2200 K) per atom.

APPENDIX C: HELIUM ENERGY DISTRIBUTION

The energy distribution shown in Fig. 14 was obtained in the following way. The controlling factor for the helium atoms is the heterogeneous collisions with the surface. The collisionally equilibrated system can be described by the general expressions, Eqs. (C.1) and (C.2)

$$P_{ij} = v\sigma_{ij}n(E)dE \qquad (C1)$$

where P_{ij} represents the probability per solid population element s_i in level i for excitation to level j by collision with the helium population in the energy range E to $E + dE$, $n(E)$ is the differential population distribution of helium atoms, v is the collision velocity, and σ_{ij} is the excitation cross section.

$$L_{ji} = v\sigma_{ji}n(E)dE \qquad (C2)$$

is the deactivation probability per solid population element s_j in level j for a reciprocal transition to level i, and σ_{ji} is the cross section for the deactivation process in the solid population. As a matter of convenience, the cross sections σ are related to the collision strengths Ω_{ij} defined by

$$\sigma_{ij} = \frac{h^2}{4\pi m v^2} \frac{\Omega_{ij}}{\omega_i} \qquad (C3)$$

where m is the atomic mass, and ω_i is the degeneracy of level i. The principle of detailed balance, which defines a thermodynamic equilibrium, requires each microscopic process to be balanced by its inverse. A necessary condition for this principle is that

$$\Omega_{ij} = \Omega_{ji}. \qquad (C4)$$

As a consequence, in thermodynamic equilibrium, the collision strengths, which contain the energy-dependent physical properties of the reactions, cancel out in the equations of equilibrium (such as the Saha equation). In thermodynamic equilibrium, no knowledge of the physical properties is required to define the populations of the excited states.

In the case of interest here, the ratios represented by Eqs. (C1) and (C2) are exceedingly small relative to reactions in the solid and the rates in the equilibrium established by radiative balance with solar input and conduction between the surface and the subsurface. The vibrational populations s_j in the solid therefore have no dependence on the atmospheric atoms, whereas the differential energy distribution $n(E)$ depends on collisional coupling to the solid. The much larger rates for gas-solid collisions relative to homogeneous rates then place control of $n(E)$ on the population distributions s_j and the physical properties of the energy-dependent collision strengths Ω_{ij}. The distribution of populations in the solid cannot adjust to produce detailed balance with the gas, in analogy to the collisionally excited ion populations in a low-density plasma. Here the populations are controlled by radiative loss rather than by detailed balance with the electron population (see, e.g., Osterbrock 1974).

The calculation illustrated in Fig. 14 is based on collision strengths $\Omega_{ij}(E)$ derived from the one-dimensional theory of Devonshire (1937) which describes the collision process for light gas atoms as being dominated by single-phonon events. The general characteristics of the interaction are summarized by Shemansky (1980).

Note added in proof: Sodium and potassium in the lunar atmosphere have been observed by Potter and Morgan (1988) and the sodium also found by Tyler et al. (1988). Near the subsolar point, the two groups find essentially identical number densities for sodium, 50 to 60 cm^{-3}, and the Na/K ratio is between 3 and 9. The sodium scale height is consistent with the surface temperature. The ratio of sodium densities on Mercury and Moon is therefore around 400, much greater than any likely ratio of meteoroid impact fluxes. Potter and Morgan suggest a reduced loss rate for Mercury, probably due to its magnetic field, while Tyler et al. favor an increased source strength, suggesting that Mercury's interior is warm enough to permit diffusion of sodium atoms to the surface.

References

Potter, A.E. and Morgan, T.H. 1988. Discovery of sodium and potassium in the atmosphere of the Moon. *Science,* in press.

Tyler, A.L., Kozlowski, R.W.H., and Hunten, D.M. 1988. Observations of sodium in the tenuous lunar atmosphere. *Geophys. Res. Lett.* 15, in press.

PRESENT BOUNDS ON THE BULK COMPOSITION OF MERCURY: IMPLICATIONS FOR PLANETARY FORMATION PROCESSES

KENNETH A. GOETTEL

Harvard University

The bulk composition of Mercury is virtually unconstrained by the present mea-
ger data set. Compositions ranging from extremely refractory-rich to volatile-
rich may be consistent with the present data. The extreme, end-member models,
however, are judged implausible because the conditions under which such com-
positions could be produced are very restrictive and thus improbable. An inter-
mediate, moderately refractory model for the composition of Mercury is present-
ed. Additional data are essential for a better understanding of Mercury's com-
position. Because of Mercury's end-member position among the terrestrial
planets, a better knowledge of Mercury would contribute substantially to a bet-
ter understanding of the origin and composition of all of the terrestrial planets.

I. INTRODUCTION

The planet Mercury is an end member among the terrestrial planets in several fundamental characteristics, including density, mass and heliocentric distance. Because of its end-member position, Mercury provides a key test of competing models for the origin and composition of all of the terrestrial planets.

The mean density of Mercury, 5.43 ± 0.01 g cm^{-3}, is well determined from the observed mass and mean radius of the planet (Anderson et al. 1987). Because of its small mass, the correction of mean density to zero-pressure density is subject to little uncertainty. This zero-pressure density is about 5.3 g cm^{-3}, which is much higher than the zero-pressure densities of Earth and Venus (about 4.0 g cm^{-3}) or Mars (about 3.75 g cm^{-3}). The principal conclusion that has been drawn from Mercury's extremely high zero-pressure density is that the planet is substantially enriched in metallic iron relative to the other terrestrial planets (i.e., Mercury's Fe/Si ratio is higher than that of the other terrestrial planets or solar composition material). Thermal evolution models (see, e.g., Solomon 1976) indicate that separation of metal and silicate components should have occurred in Mercury and it is thus generally assumed to have a massive Fe-rich core which constitutes about ⅔ of the mass of the planet.

The conclusion that Mercury is enriched in Fe is based, however, entirely on indirect evidence (e.g., cosmochemical abundances). Based solely on the mean density of Mercury, there are numerous combinations of heavy and light elements that would match the mean density of the planet, and numerous core compositions which would satisfy the density constraints (e.g., Na-U-Ti, Co-W-S, Lu-V, Be-Er). The above caveats notwithstanding, the high zero-pressure density of Mercury is almost certainly due to a marked enrichment of Fe because Fe is the only heavy element sufficiently cosmochemically abundant to constitute a major fraction of a terrestrial planet. It is with respect to iron enrichment that Mercury is most clearly distinct in composition from the other terrestrial planets.

If the responsible mechanism(s) were understood, Mercury's iron enrichment could provide important insight into the relative efficacy of several competing processes governing the formation of all of the terrestrial planets. Understanding the processes responsible for the Fe content of Mercury requires knowledge of the bulk composition of the planet but the bulk composition of Mercury is extraordinarily poorly constrained by present data. Compositional extremes ranging from an extremely refractory-rich (volatile-poor) planet to a volatile-rich (even water-rich) planet cannot be excluded on the basis of currently available, rigorous constraints on the bulk composition of Mercury. Stated bluntly, we really know very little about the composition of Mercury and additional data are urgently needed.

II. CONSTRAINTS ON COMPOSITION

The bulk composition of Mercury can be discussed in terms of three first-order parameters: core composition, mantle (plus crust) composition, and core/mantle ratio. Core composition and mantle composition are probably coupled to a considerable extent because most processes governing composition affect both the core and mantle compositions. The core/mantle ratio, however, could be largely or completely decoupled from the compositions of the core and mantle because the processes governing the core/mantle ratio (i.e., Fe/Si fractionation) may be largely separate from the processes governing composition. For example, aerodynamic fractionation (Weidenschilling 1978) or giant impacts (Chapter by Wetherill) are mechanisms which could produce the Fe enrichment in Mercury. Fe/Si fractionation by these mechanisms, however, would proceed virtually independent of the compositions of the mantle and core of Mercury.

The mean density of a planet is a direct constraint on bulk composition: rock-metal, ice-rich, and gas-rich planetary bodies are readily distinguished by mean density alone. However, in the absence of further geophysical constraints (e.g., moment of inertia factor or core radius), the mean density of Mercury provides few constraints on the chemical details of composition which are necessary in order to choose between competing models. Major differences in silicate compositions (e.g., abundances of FeO, alkalis, water) which profoundly affect interpretations of planetary composition and origin, are essentially indistinguishable on the basis of mean density. Specifically, mantle compositions ranging from ultra-refractory CaO- and Al_2O_3-rich, to moderately refractory magnesium silicate (low FeO), to extremely volatile-rich (high alkalis, FeO and water) can all be accommodated by the mean density of Mercury, with suitable differences in core/mantle ratio or core density or both.

The mean density of Mercury also barely constrains the composition of the core. For example, the density of pure FeS is only about 10% less than the zero-pressure density of Mercury. Therefore, Mercury models in which the core contains substantial amounts of sulfur are completely consistent with the mean density of the planet; likewise, cores which contain substantial amounts of other possible light elements (e.g., O or Si) cannot be excluded on the basis of mean density.

The FeO content of Mercurian silicates would be one indicator of the refractory-rich vs volatile-rich character of the silicate portion of the planet because FeO abundance increases monotonically with decreasing condensation temperature in the solar nebula, and thus FeO content may be correlated with the abundances of other nonrefractory species including alkalis, FeS and water. The FeO content of surface materials on Mercury is constrained principally by Earth-based reflectance spectrophotometry data. McCord and Clark (1979) interpreted a shallow absorption feature at about 0.9 μm as due to Fe^{2+} in orthopyroxenes and estimated the FeO content of surface materials to be about 5.5%. Vilas (1985) reported that the FeO band was absent from more recent spectra obtained with a high resolution CCD detector system. It appears that there is at most a few percent FeO in Mercurian surface materials and perhaps much less. Even these meager results, however, are subject to model-dependent interpretation before they can be applied to the bulk (mantle plus crust) silicate fraction of Mercury; the surface abundance of FeO could be modified by addition of FeO-rich material or by loss of FeO during an episode of volatilization of surface materials.

The bulk composition of Mercury is virtually unconstrained by the present data set. Attempting to infer the bulk composition of Mercury is thus largely a matter of building a self-consistent model for planetary origin, testing the model against planets for which more data exist, and then deducing the composition of Mercury from the postulates of the model.

III. MODELS FOR THE COMPOSITION OF MERCURY

The broad spectrum of models for the composition of Mercury which are allowed by the present constraints is outlined in the following three sections. The intent of this discussion is not to argue in detail the merits of one model vs another, but rather to summarize the implications of the various models for the composition of Mercury and to emphasize the wide range of possible models.

Extreme End-Member Models

The assumption that Mercury is a refractory-rich, volatile-poor planet has been widely held for at least the past decade (see, e.g., Kaula 1976). This assumption appears to be based in large part on the pioneering work of Lewis (1972,1974) in which he derived models for planetary compositions which were based on calculations of condensation in the solar nebula. The basic premise in Lewis' work is that heliocentric gradients in temperature and pressure in the solar nebula produced heliocentric gradients in the composition of condensed material in the nebula, and that such compositional gradients are preserved in the present bulk compositions of the planets. In the extreme, end-member case of this model, Mercury is an extraordinarily refractory planet because the silicate component must be only partially condensed and thus only extremely refractory components are incorporated into Mercury. The Mercury composition predicted by this end-member model is shown in Table I. The silicate component is very high in Al_2O_3, CaO and TiO_2; high in MgO

TABLE I
COMPOSITION OF MERCURY, wt.%

	Extreme Refractory-Rich Model		Preferred Model	Extreme Volatile-Rich Model
Mantle[a]				
Al_2O_3	16.62		3.5–7	3.26
CaO	15.19		3.5–7	3.03
TiO_2	0.72		0.15–0.3	0.14
MgO	34.58		32–38	32.06
SiO_2	32.58		38–48	45.04
FeO	0		0.5–5	15.07
Na_2O	0		0.2–1	1.40
H_2O	0		a little	a lot
Core				
Fe	92.48		88–91	76.22
Ni	7.52		6.5–7.5	6.20
S	0		0.5–5	17.58

[a]Plus crust (i.e., bulk silicate fraction of Mercury).

relative to SiO_2 because of incomplete condensation of silicates; and essentially devoid of FeO, alkalis, water and other volatile species. The core is essentially pure Fe-Ni alloy without S, Si or O.

The antithesis of the refractory-rich model for the composition of Mercury is that Mercury is a refractory-poor, volatile-rich planet. The consensus that Mercury is a refractory-rich planet has been so widespread that the possibility that Mercury could be volatile-rich has not been considered seriously, even though such a model is entirely consistent with the observed mean density of the planet. A volatile-rich model for Mercury is presented in Table I. This model composition for Mercury assumes that the silicate portion of Mercury has the same composition as the Mars model computed by Goettel (1983). This Mars model assumed that Mars was composed of the solar proportions of the major rock-forming elements, with the oxidation state of Fe (i.e., the FeO to total Fe ratio) adjusted to produce mantle and core densities and a mantle/core ratio consistent with the rigorous bounds derived by Goettel (1981). Wänke and Dreibus (1985) presented a similar model for the composition of Mars which was based on the interpretation of Mars as the SNC meteorite parent body. Thus, the volatile-rich model composition given in Table I appears to be a reasonable approximation to current understanding of the composition of moderately volatile-rich material in the inner solar system (i.e., Mars-like material). In this model, the core of Mercury would be high in S and larger than pure Fe-Ni cores; the mantle would be high in FeO and alkalis and contain substantial amounts of water and other volatiles.

Mercury could possibly have a Mars-like composition (albeit with a separate mechanism for Fe/Si enrichment) if mixing were sufficiently vigorous in the solar nebula during the accretion process to homogenize the formation region of the terrestrial planets compositionally. This possibility is admittedly extreme (i.e., there is some evidence supporting a compositional gradient within the terrestrial planet group); however, the homogeneous terrestrial planets hypothesis cannot be excluded if one allows a few additional assumptions. For example, the difference in FeO content between the Earth's mantle (about 8%) and the Martian mantle (about 15%), one of the benchmarks of a compositional gradient among the terrestrial planets, can be eliminated if the Earth's mantle originally had about 15% FeO with the excess above the present upper mantle abundance either sequestered in the lower mantle or incorporated into the Earth's core. Likewise, the spectrophotometric evidence suggesting low FeO content on the surface of Mercury is somewhat ambiguous. Even if the surface of Mercury were proven to have virtually no FeO, however, it is still possible that the interior could be FeO-rich if the surface were depleted by a volatilization episode. If the volatile-rich model is carried to its logical extreme, then Mercury could potentially be the most volatile-rich terrestrial planet (e.g., if stochastic fluctuations in the accretion process resulted in Mercury's accretion of a large protoplanet formed originally in the outer fringes of the terrestrial planets' accretion zone).

Moderately Refractory-Rich Models

There are three major groups of models which predict a moderately refractory-rich Mercury:

1. Variations of Lewis' (1972,1974) condensation temperature model;
2. Models requiring the post-accretion loss of a significant fraction of Mercury's silicate component by volatilization (Ringwood 1966; Cameron 1985; Fegley and Cameron 1987) or by giant impact (see, e.g., Chapter by Wetherill 1987);
3. Models based on incorporation of highly reduced, refractory components (see, e.g., Morgan and Anders 1980; Wänke and Dreibus 1986; Chapter by Wasson 1987).

These groups of models are discussed briefly below.

The original condensation temperature model for the composition of Mercury (Lewis 1972) attempted to explain both the composition of the silicate portion of Mercury *and* the Fe/Si ratio by a single (very narrow) condensation temperature and pressure for the material constituting Mercury. In its end-member form, this model is not plausible because of the inevitability of temporal and spatial gradients in the composition of condensed material, and because of the finite (perhaps large) width of the accretion zone feeding the growing planet Mercury. Developments in the understanding of the accretion process (see, e.g., Hartmann 1976; Cox and Lewis 1980; Chapter by Wetherill) indicate that material accreted by Mercury cannot originate only in immediate proximity to Mercury, but must also include material (in model-dependent proportions) from the entire region of the terrestrial planets. Barshay (1979) examined the compositional implications of a finite feeding zone for Mercury and concluded that the extreme Fe content of Mercury cannot be explained simply by accretion of ultra-refractory, silicate-poor material; rather, some physical process of Fe/Si separation is required to account for the high density of the planet.

There are three principal effects of relaxing the end-member assumptions in the condensation-temperature model for Mercury. First, the proportion of Fe in the condensate drops sharply as silicates are fully condensed. Second, the Mg/Si ratio drops rapidly towards the solar system value as Si condenses fully (enstatite increases at the expense of forsterite). Third, the abundances of nonrefractory species (alkalis, FeO, FeS, water and other volatiles) increase gradually, with a corresponding gradual decrease in the proportions of the more refractory species. Lewis (see his Chapter) has extensively explored the relationships between accretion sampling algorithms and the predicted composition of Mercury by assuming Gaussian sampling functions (centered near Mercury's heliocentric distance) and varying the half width of the sampling distribution. Lewis' quantitative results map out a spectrum of models in multidimensional composition space ranging from the end-member, extremely refractory model to moderately refractory models; this spectrum of composition models varies principally in the three ways listed above.

The second group of models which predict a refractory-rich Mercury includes models invoking post-accretion loss of a substantial fraction of Mercury's silicate component either by volatilization from the surface or by giant impact. Surface-volatilization models (see, e.g., Ringwood 1966; Cameron 1985a; Fegley and Cameron 1987) drive whatever the initial composition of Mercury was towards the refractory direction; that is, the volatilization processes, whatever the detailed mechanism(s), deplete Mercury in volatile components such as alkalis, FeS, FeO and water. In chemical detail, the resulting composition varies in a model-dependent manner; qualitatively, however, the range of compositions predicted by volatilization models maps out a spectrum quite similar to the spectrum of models presented in the Chapter by Lewis.

The chemical effects of removing a substantial fraction of Mercury's silicate component by giant impacts are somewhat less clear. Lewis (his Chapter) suggested that abundances of refractories would not be greatly enhanced by a giant impact event because ejection of a feldspar-rich crust would deplete CaO and Al_2O_3 along with alkalis and that the FeO content after impact would reflect the primordial oxidation state of the planet at the time of accretion. However, recent analyses of the physics of giant impacts suggest that extensive melting and vaporization of material occur during the event (Benz et al. 1987). Thus, it appears probable that substantial loss of volatile and moderately volatile species would occur during a giant impact event. The chemical effects of giant impacts, while differing in details from the effects of volatilization, will also drive Mercury in the refractory-rich direction.

The third group of refractory-rich models for Mercury are those in which the refractory nature of Mercury is a postulate of the model. Morgan and Anders (1980) computed a detailed model for Mercury, based on their seven-component model for planetary compositions. In this model, Mercury is moderately refractory, with a mantle FeO content of 5.5% based on the spectrophotometric results of McCord and Clark (1979). Wänke and Dreibus (see their Chapter) suggested that Mercury could be highly reduced with Si incorporated into the core; their model assumes a composition similar to that of the enstatite chondrites. Wasson (see his Chapter) also suggested that enstatite chondrites may be a good model for the composition of the silicate component of Mercury. This group of models is more eclectic than the groups discussed previously; nevertheless, despite some differences (e.g., the presence of reduced Si in some models), the range of compositions predicted for Mercury by this group of models follows the general trend of the spectrum of compositions mapped out in the Chapter by Lewis.

Preferred Model

The extreme, end-member models for the composition of Mercury are implausible because the circumstances under which such compositions would be produced during the formation of the terrestrial planets are very restrictive. More plausible, and thus preferred, models lie between the extremes (i.e., that Mercury is somewhat enriched in refractories, but not nearly as enriched as in the end-member, refractory-rich model). The preferred model that is presented in Table I is based on the following assumptions.

1. The three most refractory components (Al_2O_3, CaO, TiO_2) are not fractionated with respect to each other because they were fully condensed in all regions from which Mercury accreted material.
2. The three most refractory species are moderately enriched above solar proportions (i.e., the Al/Si, Ca/Si and Ti/Si ratios are greater than the solar ratios) because Mercury accreted some material in which magnesium silicates were only partially condensed.
3. The Mg/Si ratio in Mercury is somewhat higher than the solar value because the refractory condensates are enriched in Mg and correspondingly depleted in Si. The Mg/Si ratio, however, is near the solar value because Mg and Si were both fully condensed in the regions from which Mercury accreted most of its material.
4. Mercury contains significant amounts of moderately volatile components, including alkalis, FeO, FeS and water. The absolute abundances of these components, as well as their relative proportions, are strongly dependent on the extent of mixing of materials from different heliocentric distances and thus are poorly determined.
5. The Fe/Si ratio of Mercury was increased by a physical mechanism, perhaps aerodynamic fractionation (Weidenschilling 1978) and/or giant impact (Wetherill 1987), which operated largely independently of the composition of the silicate component of Mercury.

The preferred model for Mercury is a moderately refractory composition, with refractory components (e.g., Al_2O_3, CaO, TiO_2) enriched above solar proportions and with moderately volatile components (e.g., alkalis, FeO, FeS, water) significantly depleted below solar proportions. The core of Mercury contains much less than the solar proportion of S relative to Fe, but enough S to affect the thermal evolution of the core (in particular, enough S to prevent complete freezing of the core).

This preferred model was framed in the context of the condensation-temperature model for composition gradients in the solar system. The intent in presenting it was not, however, to exclude other models from consideration. The ranges of compositions presented in this model can also be produced by models in which surface volatilization and/or giant impacts have modified the original composition of Mercury and by models incorporating moderately refractory components into Mercury.

IV. CONCLUSIONS

The present data base for Mercury is so limited that virtually no rigorous bounds can be placed on the bulk composition of the planet: compositions ranging from extremely refractory-rich to volatile-rich may be consistent with the limited data. It is important to note that many of the models discussed are not mutually exclusive. For example, Mercury could have had an initial composition governed by accretion sampling of materials from various heliocentric distances, with the Fe/Si ratio determined by aerodynamic fractionation, and the original composition modified by episodes of surface volatilization and giant impacts.

Mercury, because it is an end-member terrestrial planet with respect to heliocentric distance and density, has the potential to be a key indicator of the processes governing the formation and composition of all of the terrestrial planets. Acquisition of additional data is absolutely essential for a better understanding of the composition of Mercury. Quantitative data from a Mercury mission including orbiter and lander chemical analyses, determination of the moment of inertia factor and (ideally) seismic profiling of the interior would vastly increase our understanding. The direct relevance of a better understanding of Mercury to a better understanding of the processes governing the origin, composition and evolution of all of the terrestrial planets suggests that a Mercury mission should have a very high priority.

THE BUILDING STONES OF THE PLANETS

JOHN T. WASSON
University of California at Los Angeles

Chondritic meteorites are the only materials known to have formed in the inner portion of the solar nebula, and thus, they almost certainly are samples of the nebular material that accreted to form the terrestrial planets. Unfortunately, we have relatively little compositional information about the planets other than the Earth, and we know least about Mercury. Cosmochemical arguments favor the view that the most reduced chondrites (EH and EL) formed nearest, the most oxidized (CM and CI) farthest from the Sun. New evidence favoring the formation of enstatite chondrites near the Sun includes rare-gas interelement abundances of Venus similar to those in EH chondrites, and a reflection spectrum of Mercury that shows no evidence of FeO in the surficial silicates. Projections of group O-isotope compositions onto the terrestrial fractionation line allow the nebular equilibration temperatures to be inferred for the FeO-rich fractions; in order of decreasing temperature, chondritic meteorites are grouped as: EH,EL-IAB-SNC-Euc-H,L,LL-CV,CO-CI. This sequence is essentially identical to that based on degree-of-oxidation and refractory-lithophile abundance trends. There are two main classes of models for the origin of Mercury: Mercury either consists of (1) chondritic materials that have experienced a mechanical 4–7 fold-enrichment in metal, or of (2) high-temperature, volatile-free materials formed by the distillation or incomplete condensation of silicates. Spacecraft determination of the Na and K content of Mercurian basalt will indicate which class of model is correct.

I. INTRODUCTION: PLANETARY ACCUMULATION PROCESSES

In order to model the terrestrial planets successfully we need to know their bulk compositions. Unfortunately, for planets other than the Earth and the Moon, the only direct evidence consists of well-determined whole-planet densities and semiquantitative analyses of rocks and soils at two sites on Mars and about six sites on Venus. We therefore must use indirect information to infer plausible compositions. In this chapter, I present the case that the best approach to understanding the bulk composition of the terrestrial planets is to use available evidence to infer the formation locations of the various chondrite groups, since the chondrites are the "building blocks of the planets."

This approach is justified provided there is reason to believe that the chondrites formed over a wide range of distances from the Sun, and that their compositional differences mainly result from differing formational conditions at these different locations. Wasson and Wetherill (1979) summarize the arguments indicating that, although most meteorite parent bodies are now "stored" in orbits having semimajor axes ranging from 1.8 to 4.0 AU, some of these objects could have formed at much more distant locations, at least as near the Sun as the inner edge of Venus' orbit at ∼0.5 AU.

Current models for the formation of planets from the solar nebula call for a gradual growth in the size of bodies, and a coupled increase in the relative velocities among the bodies (Safronov 1969). Starting with grains and gas, the first step in the sequence appears to have been a "settling" of grains to the nebular midplane followed by gravitational collapse to form 10^{13} to 10^{16} g planetesimals. Further growth occurred by collisional accretion among bodies. When the largest bodies reached ∼10^{18} g, gravitational perturbations during near misses resulted in increased orbital eccentricities and increased relative velocities among the growing bodies. By the time these bodies reached the 10^{26} to 10^{27} g masses of the terrestrial planets, eccentricities of subplanetary bodies were large enough to allow them eventually to be swept up or out by the present-day planets.

In classic planet accumulation models (see, e.g., Safronov 1969), an overwhelming fraction of each planet consists of materials that underwent planetesimal formation within that planet's zone, a band in the nebular midplane within which the local planet's gravitational field was stronger than that of any other planet. A minor fraction of the accumulated materials formed in other zones.

A recent Monte Carlo study by Wetherill (1985a) yielded the contrary result that the materials originally formed within the inner solar system were mixed to a high degree, and thus that bulk compositions of the terrestrial planets should be the same. However, the starting conditions (radial, number and size distribution of bodies) and the physics (degree of fragmentation during collisions) were not necessarily more plausible than other sets of initial conditions that would yield planets consisting mainly of local materials. To avoid interzonal mixing, most unaccreted matter must remain in relatively circular orbits. Collisions tend to reduce the eccentricity of orbits, and the orbits of very small bodies become more circular as a result of drag from the nebular gas (if still present). Conditions that produce runaway accretional growth of a few large bodies tend to minimize the amount of interzonal mixing.

The chondritic meteorites are rocks formed from nebular grains. This is attested to by numerous facts, especially those recorded in the most unequilibrated chondrites: the generally solar compositions, the lack of evidence of melting, the juxtaposition of grains having grossly different compositions, the correspondence of phase compositions with those expected to form under nebular conditions.

A long-standing problem is the definition of the process that led to the agglomeration of nebular grains. Weidenschilling (1980) has suggested that Van der Waals forces would lead to grain sticking, but these forces may be too weak. The generally accepted model is that, following grain sedimentation to the nebular midplane, intergrain velocities damped down to the point where gravitational collapse could form planetesimals having masses of 10^{13} to 10^{16} g (Safronov 1969; Goldreich and Ward 1973). Following planetesimal formation, there was no longer exchange between solids and gas, and the composition of the chondritic matter was fixed. There was some minor interzonal mixing of different kinds of chondritic matter as the planetesimals accreted together to form larger bodies.

Because of random grain motions prior to gravitational collapse, these km-size planetesimals must have been uniform in composition throughout. This is consistent with the remarkable constancy (in samples as small as 400 mg) in the abundance of all but the most volatile elements among members of each of the chondrite groups. Some key compositional characteristics of the 10 chondrite groups are listed in Table I.

A fundamental question is whether all the planetesimals which formed within one region (e.g., within 1 or 2% of a given orbital radius) had the same composition. Planetesimal formation occurred over a period of time, and particle settling into the midplane region must have occurred throughout this period. If the compositions of the later-settling particles were different from those of the earlier-settling particles, and if the amount of the later particles was sufficient to make them a resolvable component in the later-forming planetesimals, then differences in planetesimal composition may be recognizable in the set of chondrites in our museums. It seems likely that the period of planetesimal formation was short compared to that required for the accretion of asteroid-size (20 km $< r <$ 500 km) bodies, in which case we should find some breccias consisting of mixtures of the two kinds of closely related materials.

Among known chondrites the only serious candidates for such sequential formation are the CM and CO chondrites. These do form mixed breccias, and some of their components (e.g., chondrules, metal) are virtually identical in texture and composition. Kallemeyn and Wasson (1981) infer that, because of their larger mean particle size, the CO chondrules would represent the earlier generation. In contrast, the ordinary chondrites show no resolvable evidence of sequential formation at a single location. Intergroup-mixed breccias are very rare, and in the best-studied case (St. Mesmin) the mixing event seems to

TABLE I
Properties of Chondrite Groups[a]

Clan	Group	$\frac{refr^b}{Si}$	$\frac{Fe^b}{Si}$	$\frac{FeO^c}{FeO+MgO}$	$\frac{Fe\ met^c}{Si}$	$\delta^{18}O$ (‰)	$\Delta^{17}O$ (‰)	Chondrule size[d]	Chondrule freq[e]	Fall freq[f]
Refract. rich	CV	1.35	0.87	35	0.6–19	1.5	−3.6	0.9	46[g]	0.72
Minichondrule	CO	1.10	0.90	35	2.3–15	−0.9	−4.5	0.3	18	0.60
	CM	1.13	0.93	43	0.1–0.5	7	−3	0.3	12	2.17
Volatile rich	CI	1.00	1.00	45	<0.1	17	+1	—	<0.1	0.60
	LL	0.76	0.62	27	2.7–11	4.9	1.3	0.8	70	7.95
Ordinary	L	0.77	0.66	22	17–22	4.6	1.1	0.6	70	38.43
	H	0.79	0.93	17	46–52	4.2	0.8	0.5	70	33.25
IAB inclusion	IAB	~0.7	~0.70	6	~50	5.0	5.0	—	—[e]	0.95
Enstatite	EL	0.60	0.76	0.05	47–57	5.6	0.0	≤1.0	≥12	0.84
	EH[h]	0.59	1.13	0.05	68–72	5.3	0.0	0.5	20	0.72

[a]Table after Wasson 1985.
[b]Refractory lithophile and bulk Fe abundances are CI-normalized.
[c]Atom or mole ratios × 100.
[d]Median diameter in mm.
[e]Frequency in %; metamorphism has destroyed the record in IAB and EL.
[f]Percent of observed falls.
[g]41% chondrules in "oxidized" CV; 50% chondrules in "reduced" CV.
[h]EH chondrites are unequilibrated; although their mean FeO/(FeO+MgO) ratios range up to 0.2 mol%, they have more metallic Si in the metal than EL chondrites. I have arbitrarily set their FeO/(FeO+MgO) ratio equal to that in the equilibrated EL chondrites.

have occurred only 1.3 Gyr ago (Schultz and Signer 1977; Chou et al. 1981), not during the main period of planet formation.

During the past 1.5 decades, reflection spectra have been measured for many asteroids. These show that most of the asteroids in the main belt 2.1 to 3.5 AU from the Sun belong to one of two categories: low-albedo, neutral-colored C asteroids and moderate-albedo, reddish S asteroids. The S asteroids are more common inside 2.5 AU, the C asteroids beyond 2.5 AU (Gradie and Tedesco 1982). Comparison of laboratory spectra for meteorites and separated minerals with asteroid spectra (see, e.g., Gaffey and McCord 1979) commonly shows significant differences. Perhaps the most serious problem is that, despite numerous indications that the three groups of ordinary chondrites originate in the main belt, there is no common asteroid spectral class that matches laboratory spectra of these objects. As a result, Gaffey (1984) states that no large belt asteroids can be the parent bodies of the ordinary chondrites. However, there are numerous poorly understood factors that can alter the spectral properties of the outermost mm of an asteroid, and asteroid spectra may not be identical to those obtained from freshly ground samples of subregolithic materials. As a result, Wasson and Wetherill (1979) and Chapman and Wetherill (1987) consider it likely that ordinary chondrite bodies are present but not recognized in the asteroid belt, probably among members of the S-class. It does not seem possible to use asteroid spectra to constrain the chondrite formation sequence described below.

Here I present evidence and arguments indicating that the CM and CO chondrites formed beyond 2.5 AU, farther from the Sun than the ordinary chondrites. It seems likely that the single generation of accumulation processes recorded in the ordinary chondrites are more representative of those occurring during the formation of the terrestrial planets, and that, inside 2 AU, only one kind of planetesimal formed at each distance from the Sun.

II. PROPERTIES OF THE CHONDRITES

Chondritic meteorites of any one group are uniform in composition even down to small sample sizes; Kallemeyn and Wasson (1981) report relative sample standard deviations of $\leq 6\%$ for most nonvolatile lithophile (oxide-forming) elements in 300-mg samples for the four groups of carbonaceous chondrites. Similarly low scatter is observed for samples of the other major groups. Some fraction of this is analytical uncertainty, thus the sampling variability must be smaller.

The compositional parameters listed in Table I are those that show relatively large differences between groups, differences that are generally much greater than sampling and analytical uncertainties. The ten groups are divided into 6 clans consisting of 1 to 3 closely related groups. The 3 ordinary-chondrite groups comprise the type clan. These meteorites are very similar in terms of their refractory lithophile abundances and textures (e.g., chondrule sizes and structures). The chief properties that resolve the three groups are their abundance of siderophiles (represented by the Fe/Si ratio), their fractions of oxidized Fe (represented by increasing $FeO_x/(FeO_x + MgO)$ and decreasing abundance of metallic Fe, (Fe_{met}/Si)). [FeO_x represents Fe in oxidation states $+2$ or $+3$ bound to O.] In Table I the clans are ordered in terms of decreasing refractory lithophile abundance.

There are simple arguments indicating that the degree of iron oxidation should increase with increasing distance from the Sun. The degree of oxidation of nebular phases is controlled by the pH_2O/pH_2 ratio, which is near 4×10^{-4} (see, e.g., Wai and Wasson 1977). Calculations for the equilibrium relationship:

$$H_2(g) + FeO(s) = Fe(s) + H_2O(g) \tag{1}$$

show that Fe is the stable phase at high nebular temperatures (≥ 900 K), FeO at low nebular temperatures. It seems certain that there was always a radially decreasing temperature gradient within the nebula. Thus, had all solids separated from the nebula at the same moment, the fraction of oxidized iron should increase radially outwards.[a] In fact, in a nonturbulent nebula, planetesimal formation by gravitational collapse should occur at successively later times with increasing distance from the Sun, and thus produce a still steeper gradient on a plot of iron oxidation against semimajor axis.

On the basis of the ordinary chondrite evidence, I concluded that there was little temporal variation in the composition of planetesimals at each inner-solar-system location, and thus that it is reasonable to associate the increase in the mean degree of Fe oxidation of the chondrite groups with formation at increasing distances from the Sun.

As shown in Fig. 1, the chondritic abundances of refractory lithophiles tend to increase with increasing degree of oxidation. There seems to be no way to explain refractory abundance variation in terms of differences in equilibration temperatures, since all chondrite groups show evidence of volatile-element equilibration temperatures well below the condensation temperatures of the refractory elements. Kallemeyn and Wasson (1981) suggested that the difference results from increasing mean grain size of refractory lithophile components relative to those of the common silicates with increasing distance from the Sun. As a result, the fraction of the refractory lithophile component that settled to the midplane prior to planetesimal formation increased with increasing distance. Petrographic data indicate that the mean size of the major

---

[a]A key assumption in this treatment is that the H/O and C/O ratios in the nebular gas were essentially the same at all times and locations. The only grounds to question this assumption are that continuing nebula accretion may involve parcels of interstellar material having different compositions, and these may differ in angular momentum and thus accrete to the nebula at different mean heliocentric distances. Condensation of metal oxides alone could not significantly alter these ratios. In the inner solar system, temperatures never dropped low enough (~ 160 K) to allow H_2O to *condense*.

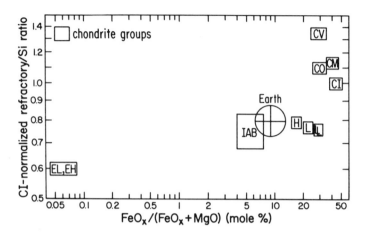

Fig. 1. Plot illustrating the mean abundance of refractory oxide-forming metals (Al, Ca, Ti, etc.) that generally increase with increasing $FeO_x/(FeO_x + MgO)$. The fraction of oxidized Fe increases with decreasing nebular temperature, and should increase with increasing distance from the Sun. The refractory abundance depends on the accretion efficiency of carriers such as the white, refractory-rich inclusions in CV chondrites. The mean size of unevaporated refractory residues (and thus their accretion efficiency) probably increased with increasing distance from the Sun.

refractory lithophile carrier (i.e., the refractory inclusions) is larger than that of the common lithophiles in the CV, CM and CO chondrites. For example, Kornacki and Wood (1984) report that more than half the refractory inclusions in CV Allende have dimensions >1 cm, and a mass-weighted mean size of refractory inclusions can be estimated to be ~600 μm. In contrast, Housley and Cirlin (1983) state that CV matrix consists predominantly of 2 to 10 μm-size olivine grains. In the ordinary and enstatite chondrites, the refractory lithophile carrier grains are too small to be generally recognizable, although a few have been discovered (Bischoff and Keil 1984).

The oxygen-isotope ratios are also key classificational parameters. All groups are well resolved except the two enstatite groups (EH and EL) and two of the carbonaceous groups (CV and CO), and these two pairs are partially resolved.

It seems likely that the mean composition of solar-system oxygen was on or near the terrestrial fractionation (*TF*) line, a line with slope of 0.52 that includes all terrestrial O data. We have used the *TF* line as a reference and tabulated $\Delta^{17}O$, the difference between the measured $\delta^{17}O$ value and 0.52 $\delta^{18}O$. As noted by Wasson (1977), it seems likely that the fraction of presolar solids that failed to exchange with the nebular gas (the main O reservoir) increased with increasing distance from the Sun, and thus that the absolute value of $\Delta^{17}O$ should tend to increase with increasing distance from the Sun. We will return to this point in Sec. V.

The fall frequency of meteorite groups provides information about their relative abundances in Earth-crossing orbits. This is important because these abundances are related to the abundances in the (mainly asteroidal) source regions and, less directly, to the original abundances in the formation regions. Unfortunately, just to calculate the number in Earth-crossing orbits we require corrections for numerous biases including the strength of the meteoroid (its resistance to fragmentation in the atmosphere), the velocity with which the meteoroid strikes the atmosphere (those with velocities ≥ 25 km s^{-1} are essentially never recovered), and possible differences in size distributions. And if we were able to correct for these, we would only know the relative abundance of the different classes in the metastable (~ 10 Myr epoch) Earth-crossing orbits, at least one major perturbation removed from the long-term storage orbits and, for some classes, one or more additional major perturbations away from the solar-system location where the object formed.

Because of the short dynamic lifetime and the stochastic nature of major collision fragmentations and perturbations, the meteorites falling during any 10-Myr epoch may be quite different from those falling in earlier or later epochs. As a result, a high observed flux of a particular group (e.g., the present-day high flux of L chondrites) could simply reflect the fact that an asteroid-size L chondrite body was either collisionally fragmented while near an asteroid-belt resonance (see, e.g., Wisdom 1985) or was perturbed into an Earth-crossing orbit, with meteoroids subsequently removed by cratering.

Despite this caveat, it cannot be an accident that the three chondrite groups having the highest fall frequencies are the very closely related H, L and LL ordinary chondrite groups. As summarized by Wasson (1972,1985), evidence such as cosmic-ray ages, ^{39}Ar-^{40}Ar ages and O-isotope compositions indicate that each of these groups originated in separate bodies at nearly the same distance from the Sun. Clearly, these bodies are now stored in a location very favorable to injection in Earth-crossing orbit. Formation at a location distant from the asteroid belt requiring random walk through numerous orbits to arrive at the present-day storage location would tend to mix these bodies together with those formed at other locations. Therefore, it is nearly certain that the ordinary chondrites formed in the asteroid belt, perhaps near an important resonance such as the 3/1 period resonance with Jupiter (Wisdom 1985), 2.5 AU from the Sun.

Most, if not all, chondrites are breccias. In most, the admixed materials are genetically related (genomict), often differing chiefly in terms of their degrees of metamorphic reheating. As discussed by Wasson (1972) and Scott and Rajan (1981), it seems that the metamorphic heating occurred in small, (10 km-size) bodies before these accreted to form the final, genomict parent body. Since there is little evidence for admixture of other closely related (e.g., H or LL into L) materials, the smaller bodies must have been in nearly circular orbits before accretion to the final body.

Some of these breccias contain exotic clasts; in most cases, the foreign material closely resembles CM chondrites (Wasson and Wetherill 1979). A simple explanation involves the exotic material being in an eccentric orbit that crossed the more circular orbits of each of the ordinary chondrite hosts (and howardite hosts as well). As summarized by Wasson and Wetherill (1979), breccias consisting of clasts of one kind of ordinary chondrite in a host consisting of another kind (e.g., an H clast in an LL chondrite) are rare. The implication is that, during most of the periods when these ordinary chondritic (also howarditic) regoliths were active, CM chondrite-like materials accounted for most of the objects in eccentric orbits that crossed the orbits of these other classes. If the host chondrites or howardites formed in the inner solar system (say, inside 3 AU), the CM materials formed at a greater distance from the Sun, and were later perturbed into orbits having perihelia in the inner solar system.

Meteorites from the Moon have been discovered in Antarctica. Their absence from observed falls is surprising, but a plausible stochastic explanation is that impacts that eject them from the lunar gravity field occur infrequently, on a time scale comparable to or greater than their relatively short ($\sim$100 ka according to Arnold [1965]) mean orbital lifetimes prior to accretion to the Earth. Also, the Antarctic lunar meteorites are so small ($<$30 g) that they would probably have remained undiscovered had they fallen during historical times.

More important for understanding where the meteorites formed is the hypothesis that the SNC differentiated meteorites formed on Mars (Wasson and Wetherill 1979; Bogard et al., 1984; Wänke 1987). The chief arguments favoring this view are the high, Earth-like volatile contents of the SNC meteorites (Stolper 1979), and the close isotopic resemblance between the rare gases and N trapped in the glass of the SNC Antarctic meteorite EETA79001 and the data on these elements determined by instruments on the Viking spacecraft that landed on the Martian surface (Bogard et al. 1984; Becker and Pepin 1984; Wiens et al. 1986; Swindle et al. 1986). If the SNC meteorites are from Mars, studies of these meteorites can be used to constrain the chemical and isotopic composition of Mars and of its chondritic building stones. My assessment of the evidence is that it is about 50% probable that the SNC meteorites are from Mars. With this caveat, I will assume that they are from Mars in the remainder of the chapter.

III. PROPERTIES OF THE PLANETS

The surface density of the nebula can be estimated from the masses and compositions of the planets, and simple assumptions about the dimensions of the "feeding zones" within which the planets accreted. Wasson (1985, Fig. VIII-1) shows estimates of nebular surface densities and pressures based on the masses of the inner planets, and their Fe contents. The Fe contents of Mercury, Venus, Earth and Mars are roughly 560, 280, 280 and 250 mg/g, respectively; these estimates are discussed below. The mass in the Jupiter region is based on estimates that Jupiter contains about 20 solar masses of matter heavier than H and He (Hubbard 1984), and the assumption that, like comets, this heavy matter is roughly half ice and half chondritic, an estimate that is not inconsistent with spacecraft studies of Comet Halley (Geiss 1987). A smooth curve drawn through the surface densities of the Venus, Earth and Jupiter regions shows that a large amount of matter is missing from the regions where Mercury, Mars and the asteroids formed.

Figure 2 shows the densities of the inner planets, their best defined composition-dependent property. The actual densities have been adjusted to laboratory conditions of 1 atm pressure, 298 K temperature using simple planetary models and equation-of-state relationships (Reynolds and Summers 1969). These densities are based on models calling for unrealistically large amounts of Fe in the Earth, but no models based on chondritic Fe contents have been developed. Uncertainty estimates on these densities are based on my own assessment of the uncertainties in equation-of-state data and in the bulk mineralogy of the interiors of the planet. The shaded band shows the density range observed in the anhydrous chondrites (all groups except CM and CI).

The chief driver of these density variations is the weight fraction of Fe in the planet. The other key variable is the fraction of compounds of the light

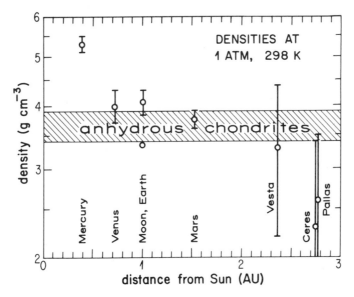

Fig. 2. Densities of the inner planets calculated at a temperature of 298 K and a pressure of 100 kPa (Reynolds and Summers 1969). Relatively large uncertainties are associated with the 100 kPa density estimates of Venus and Earth. Excepting that of Mercury, densities of the inner planets and asteroids are consistent with anhydrous chondritic compositions. The high density of Mercury requires an enhancement in Fe by a factor ≥ 4.

elements, particularly H_2O, but the weight fraction of such substances in the terrestrial planets is so minor as to be negligible.

One explanation is that Mercury's nonchondritic composition reflects a silicate depletion that occurred as a result of one or more high-energy impacts that removed 75 to 80% of Mercury's mantle, or about 60% of the original mass of a reduced chondritic body. The escape velocity of Mercury is 4.3 km s^{-1}; that of the hypothetical proto-Mercury would have been ~5 km s^{-1}. According to Wetherill (1976a) the mean interparticle velocity at 1 AU was about 8 km s^{-1} during the final stages of planetary accretion. Scaling the velocity to Mercury's location nearer the Sun yields an interparticle velocity of about 13 km s^{-1}. The attraction of Mercury's gravitational field would only increase such a velocity to ~14 km s^{-1}. By definition the object that struck Mercury was smaller than proto-Mercury. It seems highly unlikely that such an object with a velocity of ~14 km s^{-1} could remove its own mass and 60% of proto-Mercury's mass from the gravitational potential well. Cameron et al. (see their Chapter) computed results supporting this conclusion; they found that a (low-probability) on-axis collision requires a minimum velocity of 20 km s^{-1}, and a (more probable) off-axis collision requires a velocity of 35 km s^{-1}. For this reason I conclude that this model is implausible and will not discuss it further.

Variations in the weight fraction of Fe in chondritic matter can result from two nebular processes: (1) the fractionation during accretion/agglomeration of siderophiles (probably mainly as metal) from silicates, and (2) the oxidation of metallic Fe to FeO or $FeO_{1.5}$. Contrary to the commonly held view, the "partial density" of O in metallic oxides is relatively high (~3.1 g cm^{-3}), thus the oxidation of Fe yields a relatively minor decrease in the bulk density of a planet. For example, in chondritic material with a 100 kPa density of 3.70 g cm^{-3}, the oxidation of 200 mg/g Fe to oxide in the form of FeO dissolved in Fe_2SiO_4 would only decrease the density to 3.66 g cm^{-3}. Conversion of 200 mg/g Fe to Fe_3O_4 yields a density of 3.64 g cm^{-3}. Differences in the fraction of oxidized Fe are not able to account for the apparent density difference (0.3 g cm^{-3}) between the Earth and Mars; this difference, which may not be real, would require that the Earth contain more bulk Fe (i.e., have a higher Fe/Si ratio).

Although the 1 atm densities calculated for Venus and the Earth both lie outside the range of anhydrous chondrites (Fig. 2), the uncertainty limits overlap the chondrite range. These densities are based on Fe/Si ratios ~1.4 times the ratio in CI chondrites, but there are no chondritic meteorites with Fe/Si ratios > 1.1 times this ratio. No effort has been made to use equation-of-state models to constrain the range of possible Fe contents of the Earth, but it seems certain that the CI-chondrite Fe/Si ratio is not ruled out. It therefore seems reasonable to assume that the Fe contents of Venus and the Earth are ~280 mg/g, similar to that expected in moderately reduced and anhydrous CI chondrites. A key cosmochemical question that high-pressure geophysics should address is whether accurate phase equilibrium and equation-of-state data and plausible assumptions about mantle and core compositions do allow the Earth to have such a chondritic composition, and thus a 100 kPa density nearer that calculated for Mars.

The much larger difference in density between the Earth and Mercury can only be understood in terms of a metallic Fe content in Mercury far higher than that observed in chondrites. The degree of fractionation required to account for the high 100 kPa, 298 K density of Mercury can be estimated from a simple model based on the highly reduced EH chondrites (the basis for choosing the EH chondrites is discussed below). The EH condrites consist of about 240 mg/g metal, 13 mg/g schreibersite ($(Fe,Ni)_3P$), 120 mg/g troilite (FeS) and 627 mg/g other oxides and sulfides. The mean density of EH chondrites is close to 3.70 g cm^{-3} (Mason 1966); the combined density of metal, troilite and schreibersite ("opaques") is 6.39 g cm^{-3}, and that of the remaining sulfides and oxides ("oxides") is 2.96 g cm^{-3}. If we assume that opaques fractionated coherently, we find that an increase in the opaque/oxide ratio of a factor of 7 (an Fe content of 640 mg/g) is required to achieve the estimated 100 kPa density of 5.2 g cm^{-3}. If one hypothesizes that Mercury is somewhat more reduced than the EH chondrites, the increase in the opaque/oxide ratio can be lower, perhaps as low as a factor of 4, and the estimated Fe content is 560 mg/g.

Note that the conclusion that most Fe in Mercury is present as metal is independent of whether the mechanism that produced the mechanical fractionation was nebular or impact removal of mantle silicates. If appreciable FeO was in the (chondritic or planetary) silicates, the fraction of the silicates that need to be removed would be even higher than the fractions calculated in the previous paragraph. An additional argument favoring a nebular fractionation is the observation that the nebula has shown us that it can provide a factor of 2 fractionation in Fe/Si; this is the mean difference between the EL and EH chondrite groups. To produce a Mercury-type Fe/Si ratio we require only an additional factor of 4. Weidenschilling (1978) has pointed out that radial drag forces may be effective at separating metal grains from silicates at the innermost edge of the nebula.

As discussed in the previous section, there is good reason to expect that the mean oxidation state of chondritic planetesimals increased with increasing distance from the Sun. Thus it is highly desirable to be able to determine or infer the relative degrees of oxidation of the terrestrial planets. As noted above, the 1-atm densities of the planets are only slightly affected by the degree of oxidation of Fe. The only remaining evidence is the composition of surface materials; such evidence must be considered even though, for various reasons (alteration by hydrosphere or atmosphere, contamination by late-accreted materials), it may not be representative of the bulk planet. The two compositional ratios determined by spacecraft experiments that measure degree of oxidation are $FeO_x/(FeO_x + MgO)$ and FeO/MnO. Wasson

(1985,p.191) summarized the data available through 1984. Surficial material on Venus and Earth have similar ratios, consistent with similar degrees of oxidation; Mars is distinctly more oxidized.

The older reflection spectra of Mercury were interpreted to show a small absorption near 900 nm, indicative of the presence of some FeO in pyroxene. McCord and Clark (1979) attempted to maximize the precision in their determination of the center of the absorption band. Their result, 890 nm, was surprising because curves given by Adams (1974) show that this value implies nearly no FeO. But if there was no FeO, there should be no absorption. Figure 3 shows the latest development in this saga, a recent reflection spectrum of Mercury by Vilas (1985); she concludes that indeed no FeO absorption is present. This result, if it stands the test of time, is highly significant with regard to the composition of bulk planet: because the FeO/MgO ratio is higher in melts than in coexisting solids, planetary differentiation and lava extrusion can only increase the surface FeO/(FeO + MgO) ratio relative to that of the bulk planet. If the surface of Mercury is highly reduced, the bulk planet must also be highly reduced. A caveat must be added; solar wind irradiation will gradually reduce surficial FeO. If there is much FeO in the Mercurian surface rocks, it seems plausible that the fraction of the surface covered by fresh craters might be great enough to make FeO resolvable from Earth. However, complete confidence that the degree of oxidation is very low requires spacecraft-based spectral studies of fresh craters.

The K/U ratio has been used to show that the Earth and Moon are variably depleted in volatiles (K is volatile, U is refractory) relative to all groups of

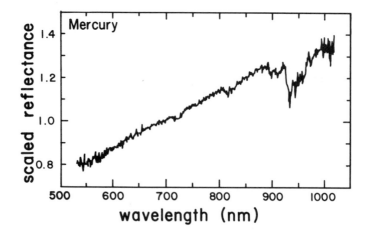

Fig. 3. The Mercury reflection spectrum measured by Vilas (1985). In this reflection spectrum, the absorption minima are all attributed to H_2O in the Earth's atmosphere. There is no absorption in the region near 900 nm attributable to FeO, and the data are consistent with there being no FeO in the lavas that cover the face of Mercury.

chondrites. The method relies on a shared geochemical property; both elements are incompatible with mantle minerals, and thus strongly (and subequally) enriched in crustal materials. These two elements also share a property that enhances their detectability by remote sensors: they (or their decay products) emit gamma rays. The currently available data, summarized by Wasson (1985), show similar K/U ratios in the Earth and Venus (0.15, CI-chondrite normalized). No data are available for Mercury, and K data for Mars do not appear to be reliable. If SNC meteorites are from Mars, then the Martian CI-normalized ratio is ~0.17, similar to that on Earth and Venus. The K/U data are important for what they can tell us about volatile loss during planetary formation. I am skeptical regarding the alternative view, that primitive nebular materials may have had such low values; this seems implausible precisely because there is so little variation in the ratio among the chondrite groups. As noted by Wasson and Warren (1979), the loss of volatiles may have been produced by outgassing associated with igneous processes occurring in asteroid-size bodies prior to their accretion to the planet.

As discussed in Sec. VII, the concentration of alkali elements in endogenous Mercurian materials (e.g., lavas) offers a key test for choosing between different formation models. Potter and Morgan (1985a) discovered Na in the tenuous atmosphere of Mercury, but for several reasons it does not appear possible to use such observation to determine the Na content of the lavas: (1) the physics of the sputtering process is too poorly understood to allow the Na concentration on the surfaces of solids to be calculated; (2) recycling may lead to a surficial enhancement of Na; and (3) appreciable amounts of the labile Na may have originated as accreted interplanetary dust.

IV. RARE-GAS INTERELEMENT RATIOS

The Viking mission to Mars and the Pioneer Venus mission gathered data on rare gases in the atmospheres of these planets. Abundances of Ne, Ar, Kr and Xe are illustrated in Fig. 4. Rare gases are especially important because they are inert; with the possible exception of Xe, they undergo no chemical evolution following the formation of the planet. The missing terrestrial Xe is not in the crust, the ocean or Antarctic ice (Bernatowicz et al. 1985). It is possible that it is in the lower mantle; the fraction retained is not well known. Ozima and Podosek (1983) estimate it to be ~25% of that in the atmosphere, but cannot rule out a much larger fraction. Under certain conditions, however, rare gases can escape from a planet. Mercury has lost its entire atmosphere, and He has been lost from all the terrestrial planets. The low content of the rare gases in the Martian atmosphere is not understood; appreciable escape must have occurred, but it is not known whether this was mainly during accretion, or during later loss "events" or "episodes." Episodic, hydrodynamic loss leads to relatively small interelement and mass fractionations (Hunten et al. 1987).

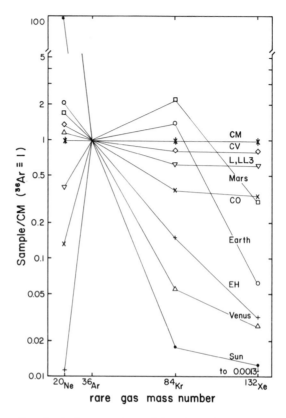

Fig. 4. Rare-gas/Ar ratios in the atmospheres of three inner planets and in several groups of chondrites normalized to element/Ar ratios in CM chondrites. The rare gases in planetary atmospheres probably originated as gases trapped in the chondritic solids accreting to the planet. Low Xe in the planets may indicate trapping in the interiors; high Kr/Ar ratios probably indicate selective loss of Ar. A close relationship between Venus and the EH chondrite is suggested by the similarities in their Kr/Ar and Xe/Ar ratios. Venus, Earth and Mars data are from Donohue and Pollack (1983); solar data from Ozima and Podosek (1983); EH (South Oman) data from Crabb and Anders (1981); CM, CO and CV data from Mazor et al. (1970); L and LL3 data from Heymann and Mazor (1968). 70% confidence limits on the chondritic isotope/^{36}Ar ratios are typically 25 to 30%.

All rare-gas patterns in meteorites and planetary atmospheres are fractionated relative to that in the Sun. The fractionations are always in the direction of heavier gases being enriched relative to lighter gases. The trend in the fractionations is consistent with volatility decreasing and/or reactivity with solids (including physical adsorption) increasing with increasing atomic number. An unanticipated result of the spacecraft missions was the discovery that the elemental abundance ratios of the heavy rare gases varied in systematic fashion from Mars to Earth to Venus; the ^{84}Kr/^{36}Ar and ^{132}Xe/^{36}Ar ratios decrease by a factor of 20 to 50 through this sequence (Fig. 4). There is less than a factor of 2 variation in the $^{20}Ne/^{36}Ar$ sequence, and it is not systematic with orbital radius; the Earth's ratio is highest, that of Venus is lowest. We will ignore the Ne/Ar ratios in the following discussion.

For the most part the Kr/Ar and Xe/Ar ratios are expected to reflect the ratios in nebular materials accreted to the planets. Under current conditions there is no appreciable escape of rare gases other than He from Venus, Earth and Mars. With the possible exception of Ne on Mars, it is likely that thermal escape was never significant even under the high-temperature conditions occurring during planetary accretion. Of course, explosions resulting from the accretion of large objects occasionally stripped off portions of the atmosphere, but such loss events should have been an inefficient means of fractionating interelement ratios. If there were a large number of these stripping events, the current ratios in the atmosphere might reflect mainly those in the last materials accreted.

Given this picture, the high Kr/Ar ratios in Mars and the Earth are mysterious. Since the ratio in CM chondrites is higher than those in other chondrite groups, it seems possible that this is the highest ratio generated in nebular processes. Since any trapping of rare gases in present-day planetary solids would remove more Kr than Ar, the higher ratios on Mars and the Earth may indicate that some atmospheric Ar escaped during accretion.

The low Xe/Ar and Xe/Kr ratios on Mars and the Earth are generally attributed to the sequestering of $\geq 90\%$ of the planetary inventory of Xe in the interiors of these bodies, even though there is no direct evidence that the amount trapped is $>25\%$ (Ozima and Podosek 1983). Because of these uncertainties in Ar and Xe loss mechanisms, it is impossible to try to establish a specific link of the Martian or terrestrial rare-gas patterns to that in one of the common groups of chondrites (e.g., to those shown for CM, CV, CO, or L, LL3 chondrites). The isotopic composition of terrestrial Xe is also different from that in any group of chondrites.

The rare-gas pattern shown for Venus (Fig. 4) is that obtained by the neutral mass spectrometer on the lander from the Pioneer Venus mission (Donahue and Pollack 1983). This data set is consistent with sets obtained from other instruments with the exception of the Kr values determined by the Venera 11 and 12 neutral mass spectrometers (Istomin et al. 1980). The Venus pattern is very different from those of Mars and the Earth, and rather similar to that of the Sun.

Only one group of chondrites has a rare-gas pattern similar to Venus' patterns—the EH chondrites, and especially the South Oman EH chondrite studied by Crabb and Anders (1981), and shown as the EH curve in Fig. 4. The close resemblance of the South Oman and Venus patterns is circumstantial evidence that EH chondrite-like materials were important building stones of Venus, and thus that EH chondrites formed near 0.7 AU. The solar-like patterns are also generally consistent with expectations that chondritic materials forming nearer the Sun would have higher solar-wind components in their primordial rare gases (Wetherill 1981). Since Mercury has not retained its atmosphere, we have no basis to fit it into the rare-gas interelement trend.

V. THE O-ISOTOPE COSMOTHERMOMETER REVISITED

Onuma et al. (1972) suggested that the $^{18}O/^{16}O$ ratio in chondritic meteorites could be used to estimate the temperature at which their oxide minerals last equilibrated with nebular gases. They derived equilibrium partition relationships for the oxygen isotopes between the main nebular gases and solid phases, and an inferred mean isotopic composition of the nebula. However, Onuma et al. (1974) withdrew the proposed cosmothermometer after Clayton et al. (1973) showed that oxygen isotopes had not equilibrated among the petrologic components of carbonaceous chondrites. Onuma et al. (1974) also revised their estimated O-isotopic composition of the nebula to bring it into better agreement with the carbonaceous chondrite data. The new composition resulted in discordancy between ordinary-chondrite O-isotope temperatures and so-called "accretion temperatures" based on volatile metal abundances (Keays et al. 1971).

The data of Clayton and coworkers (see, e.g., Clayton et al. 1976; Clayton and Mayeda 1984) show that chondrite groups occupy well-defined fields in Fig. 5, a $\delta^{17}O-\delta^{18}O$ diagram. [The terms $\delta^{17}O$ and $\delta^{18}O$ are in effect $^{17}O/^{16}O$ and $^{18}O/^{16}O$ ratios: $\delta^{17}O = ((^{17}O/^{16}O)_{sample}/(^{17}O/^{16}O)_{standard} - 1)$ expressed in ‰, i.e., in parts per thousand; $\delta^{18}O$ is analogously defined.] Two important reference lines are also shown: (1) the terrestrial fractionation (*TF*) line shows the loci of all terrestrial samples, and has a slope of 0.52 (Matsuhisa et al. 1978); (2) the carbonaceous-chondrite anhydrous minerals *CCAM* line shows the loci of high-temperature oxides separated from CV, CO and CM chondrites; it has a slope of about 0.94 (see, e.g., Clayton and Mayeda 1984). The *CCAM* line is generally interpreted to be a mixing line. Some recent experiments by Heidenreich and Thiemens (1985) have shown that a similar slope can be generated in kinetically controlled chemical systems. Because there is doubt that such systems played significant roles in the solar nebula, I choose to follow the traditional interpretation that the *CCAM* line results from the mixing of two components, one having values at least as low as the lowest measured in chondrite inclusions ($\delta^{17}O = -42$, $\delta^{18}O = -40$‰; henceforth two numbers in parentheses are $\delta^{17}O$ followed by $\delta^{18}O$, both in ‰), the other at least as high as the intersection of the *CCAM* line with the *TF* line (+5, +10).

Clayton et al. (1985) argued that two gaseous reservoirs, one "terrestrial," one ^{16}O rich, are required to explain the O-isotope variations in meteorites. They recognized that it is difficult to isolate gaseous reservoirs, but suggested that they "existed either at different times or in different regions."

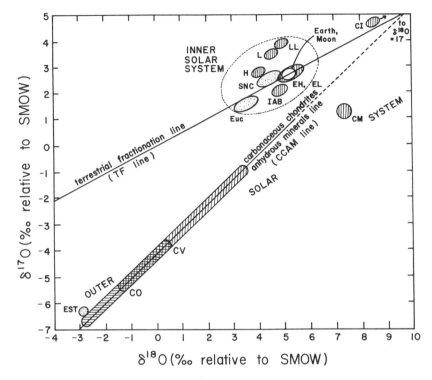

Fig. 5. O-isotopic compositions of meteorite groups that generally occupy relatively small fields on a plot of $^{17}O/^{16}O$ vs $^{18}O/^{16}O$, here shown in the normalized format (see text). The more common and more reduced groups of meteorites cluster near the composition of the Earth and Moon, and are believed to form in the inner solar system. The more rare and more oxidized meteorites occupy a field stretching across the diagram from the lower left to a region beyond the upper right corner. Terrestrial samples describe a line with slope 0.52 designated the terrestrial fractionation line, *TF*.

Wood's (1981) scenario calls for a solid reservoir having O at $\delta^{17}O = -42$, $\delta^{18}O = -40‰$, and a gaseous reservoir with $\Delta^{17}O \geq 0‰$ ($\Delta^{17}O = \delta^{17}O - 0.52\,\delta^{18}O$). Wood argues that the absence of appreciable isotopic variations in elements such as Mg and Ca argues against there being more than one solid reservoir, though he envisions his solid reservoir to be 95% normal solids having O near the *TF* line, and 5% anomalous solids whose O was pure ^{16}O. Relatively little of this pure ^{16}O entered the gaseous reservoir.

These authors envision their models to be simple, and thus able to stand the test of Occam's razor. I question the value of this simplicity. The O and major metals (Si, Mg, Ca and Al) are injected into interstellar space from a wide spectrum of stars. I find it implausible that the O that became gases (CO, H_2O) should have mainly come from one set of stars, whereas the O that became bound to metals mainly from another set of stars.

I find it more plausible that stars produced a wide range of O isotopic compositions, and, following ejection into interstellar space, that the O formed metal oxides and gaseous oxides (CO, H_2O) having more or less the same O-isotopic ranges. I suggest that some parcels of interstellar matter differed from others in O-isotope composition, and that the solid particle clumps differed in size from one presolar parcel to the next.

Since there is evidence for the processing of ordinary chondrite materials at temperatures ≥ 1300 K, and strong evidence that ordinary chondrites originated beyond 2 AU, it is very likely that nebular solids which formed at 1 AU underwent complete exchange with the gas, and therefore that the mean O-isotope composition of solar matter accreted to that region lay near the *TF* line. It is a plausible assumption that the main reservoir, the gas, lay near the *TF* line at all locations and at all times during the accretion of the nebula.

But, if O-isotope compositions of the solids spanned a wide range, why are the high-temperature oxides in carbonaceous chondrites so ^{16}O rich? I think the answer has to do with the kinetics of evaporation during the infall heating of the solar nebula (Chou et al. 1976). If the highest temperatures were reached during the infall of each parcel of presolar material, the particles that would evaporate most rapidly are those rich in the more volatile oxides (such as ferroan silicates and hydrated minerals), and those that had small sizes and thus high surface-to-volume ratios, because infall heating is by the collision of high-velocity ions or molecules with particle surfaces. According to this scenario the ^{16}O-rich materials avoided complete vaporization because they were relatively refractory and/or had large presolar sizes (of course, any surviving particle preserved its O-isotope composition independent of the *final* size). Large presolar sizes are also indicated by the relatively large mean sizes of the refractory inclusions (i.e., refractory residues) in CV, CO and CM chondrites that best preserve the record of ^{16}O-rich oxygen.

If the reason the ordinary chondrites lie above the *TF* line of Fig. 5 relates to the fact that unevaporated solids at this location were ^{16}O poor, this difference could reflect a temporal change in the O-isotopic composition of infalling protosolar matter. There is general agreement that, relative to carbonaceous chondrites, ordinary chondrites formed closer to the Sun in a region where temperatures were higher and the duration of the high-temperature period longer. Perhaps the ^{16}O-rich materials preserved in carbonaceous chondrites fell into the nebula during an early period when temperatures were highest; at the ordinary chondrite (OC) location, such materials evaporated completely. The ^{16}O-poor presolar materials that pulled the solids above the *TF* line fell in fairly late, and (since large refractory materials in ordinary chondrites are rare and generally contaminated by common or volatile elments) were smaller in size than those earlier refractory materials that survived at the CV, CO and CM locations. The absence of a correlation between O anomalies and isotopic variations in refractory metals such as Ca that correlate with the O-isotope variations probably has one of the following simple explanations: (1) the bulk of each metal became bound to well-mixed O in the interstellar medium rather than in stellar ejecta; or (2) the bulk of the metals did bond to O in cooling stellar ejecta but most of the local oxygen available in regions having isotopically anomalous metals was "normal" rather than extreme in composition.

It is not appropriate in this chapter to discuss critically all aspects of these three attempts to generate scenarios that can account for the meteoritic O-isotope variations. I have presented my scenario because it is the one which I will use in the following revision of the O-isotope cosmothermometer.

Unequilibrated ordinary chondrites are mixtures of materials that formed over a wide range of temperatures. However, the high-temperature condensates or unevaporated residues were altered to a greater-or-lesser degree as the nebula cooled, and thus only preserve a ghost of the high-temperature record. In fact, in most classes of chondrites the high-temperature minerals are largely found in chondrules, and compositional evidence indicates that chondrules formed at relatively low (≤ 600 K) nebular temperatures by the flash melting of mixtures of solids formed over a range of temperatures (Grossman and Wasson 1983).

The low-temperature record is better preserved. However, the major and minor oxide-forming elements condensed from the nebula at temperatures ≥ 1000 K (Wasson 1985), and low-temperature materials such as the fine-grained, FeO-rich matrix of ordinary chondrites (Huss et al. 1981) probably formed by the alteration of oxides and metal formed at high temperatures. Because oxygen exchange was probably incomplete during this alteration process, the O isotopes in matrix can be expected to retain some record of processes occurring at higher temperatures. Despite this caveat, low-temperature nebular materials (including bulk samples of CI chondrites) often have O-isotope compositions near the *TF* line, an indication that most O was originally in the gas, and entered the solid either during condensation or exchange with nebular gas (or in CI or CM chondrites, possibly with low-temperature nebular condensates such as H_2O).

My basic approach in constructing a cosmothermometer is the same as that of Onuma et al. (1972). I assume that the nebular gas had the same O-isotope composition at all times and places. Small displacements of ± 0.5‰ in either $\delta^{17}O$ or $\delta^{18}O$ in this composition have no significant effect on the results. Because temperatures were highest and gas-solid equilibration most complete in the innermost part of the nebula, it is highly probable that the composition of the nebular gas was on or near the *TF* line. We handle the problem that most chondrite groups do not lie on the terrestrial fractionation line by observing that studies of separated fractions of unequilibrated CV and ordinary (H, L, LL) chondrites form arrays that intersect the *TF* line, and that fine-grained, FeO_x-rich materials tend to be nearer the *TF* line than is the bulk meteorite.

Reproduced on Fig. 6 is a diagram from Clayton et al. (1985) showing

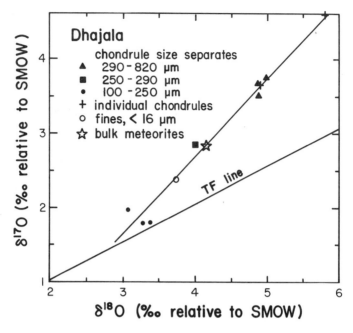

Fig. 6. O-isotope compositions of separates from the Dhajala H3.8 chondrule forming an array with a slope of 0.93 intersecting the terrestrial fractionation *TF* line at ^{18}O = 2.9. The finest chondrule separates and the matrix sample fall nearer than the whole-rock sample to the *TF* line, as expected if the finest materials most nearly equilibrated with the nebular gas. The diagram is a slightly revised version in Clayton et al. (1985, Fig. 3).

the results of an O-isotope study of size fractions of the Dhajala H3.8 chondrite. Dhajala has experienced mild metamorphism, but since Fe and Mg have only experienced limited diffusion, and O is more strongly bound to the silicate lattice, the nebular O-isotope record should be well preserved. The Dhajala samples consist mainly of chondrule-size separates, but also include two large (>820 μm) chondrules, a fine (<20 μm) matrix sample and a whole-rock sample. The matrix is significantly nearer the *TF* line than is the whole rock, and the finest chondrule-size separates are nearer still. It is too early to assess the significance of the relative locations of the latter two samples. In particular, we need to know the $FeO_x/(FeO_x + MgO)$ of the small chondrules, and O-isotope and petrographic data for a few additional matrix samples; matrix samples are known to vary in their characteristics. In a recent study, Grossman et al. (1987) determined the O-isotope composition of 5 matrix-rich samples from LL Semarkona. The two purest samples showed $\Delta^{17}O$ values 0.3 to 0.4‰ below whole-rock Semarkona, and ~0.6‰ below whole compositions of LL chondrites. Some of these samples showed very high $\delta^{18}O$ values interpreted to be the effects of aqueous alteration, and as a result, the data differ from the Dhajala data in not forming a sharply defined trend.

The Dhajala chondrules, augmented by chondrules from L3 Hallingeberg and LL3 Semarkona (Gooding et al. 1983), are plotted again in the upper part of Fig. 7. The same line with slope 0.93 is shown. The Gooding et al. chondrule O-isotope data show a significant correlation with compositional data on the same chondrules (Gooding et al. 1980; Gooding 1979): $\Delta^{17}O$ correlates positively with the refractory lithophile Al and negatively with the FeO/(FeO + MgO) ratio in the olivine (Shirley 1983). Cosmochemical models predict that the last solids to evaporate or the first solids to condense in the solar nebula would have high Al contents and that the last materials to equilibrate with the solar nebula will have high FeO/(FeO + MgO) ratios; thus the trends are precisely those expected if the last materials to equilibrate with the nebular gas prior to agglomeration had $\Delta^{17}O \sim 0$, and high-temperature mate-

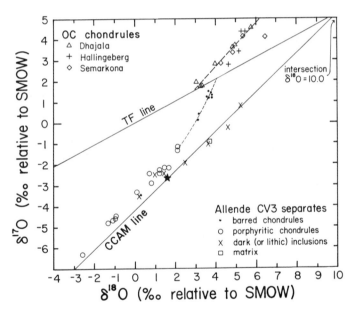

Fig. 7. O-isotope diagram showing the points of the Dhajala chondrules from Fig. 6 plotted above the *TF* line, along with additional L3 Hallingeberg and LL3 Semarkona chondrules from Gooding et al. (1983). The regression from Fig. 6 fits the data providing one Semarkona chondrule is ignored. The points below the *TF* line represent separates from the Allende CV3 chondrite (Clayton et al. 1977, 1985). Samples other than chondrules form a linear array (the *CCAM* line) with *TF* line intersection at $^{18}O = 10.0$. Fine, FeO-rich samples (matrix and dark inclusions) generally plot between the bulk composition (the star) and the *TF* line. Allende chondrules form a curved array to the left of the *CCAM* line; the slope increases with increasing ^{18}O. We suggest that the *TF* line intersection of the *CCAM* array reflects the composition of materials equilibrating at low nebular temperatures at the CV location, and the much lower chondrule intersection is mainly produced by volatile loss during chondrule formation.

rials had $\Delta^{17}O$ ~2 and $\delta^{18}O$ ~6‰. Intersection of the OC chondrule line with the *TF* line gives us a $\delta^{18}O$ value of 2.9‰ for our cosmothermometer.

Two different arrays are found in separates from the Allende CV3 chondrite (Clayton et al. 1985). In Fig. 7 we show the portions of these arrays nearest the *TF* line; many additional separates plot along the *CCAM* line at values going down to −42, −40‰ (Clayton et al. 1977). The chondrule points were not included in calculating the *CCAM* regression line shown here; they are treated separately below. This line has a slope of 0.94 and intersects the *TF* line at ^{18}O = 10.0‰. The dark inclusions and matrix samples are all fine grained and FeO_x rich. The matrix and four of the dark inclusions are nearer the *TF* line than is the whole rock, shown by the star, as predicted by our model. The *CCAM-TF* intersection at $\delta^{18}O$ ≃10.0‰ is another point for our cosmothermometer. That this point differs from the OC point by 7‰ suggests that the CV and (since they also fall along the *CCAM* line) CO chondrites formed relatively far in time and/or space from the OC formation region.

The loci of the Allende chondrule data are curious (Fig. 7). All chondrule points are left of the *CCAM* line, and the degree of divergence increases with increasing $\delta^{18}O$. The slope increases from about 1.0 at $\delta^{18}O$ ~0 to 1.5 at $\delta^{18}O$ ~2.5 to ~1.9 at $\delta^{18}O$ ~3.5‰. Clayton et al. (1985) suggest that both the OC- and Allende-chondrule arrays were generated by exchange of the molten chondrule with the nebular gas. However, Shirley (1983) carried out a dimensional analytical study showing that the survival of alkalis and other volatiles in chondrules is inconsistent with there being sufficient time for such an oxygen exchange process. Shirley's (1983) Na loss rates are too low, since Na loss decreases with increasing pO_2, and the Tsuchiyama et al. (1981) data he used were taken at pO_2 values much higher than those in the solar nebula. Based on their exchange model, Clayton et al. (1985) state that "the two groups appear to converge toward a common O-isotope composition in the most extensively exchanged chondrules." The data plotted in Fig. 7 call such a convergence into question, since both the slopes near the *TF* line and the intersection of the curves with the *TF* line are distinctly different (although additional OC chondrule [Clayton et al. 1985] data show more scatter and thus more uncertainty in slope and intersection).

McSween (1985) reports that the barred Allende chondrules plotting nearest the *TF* line are also highest in their FeO/(FeO + MgO) ratios. McSween explains the relationship between FeO and O-isotope composition in terms of the Clayton et al. (1985) model: the FeO-rich chondrules remained molten longer and thus underwent more exchange with the nebular gas. However, in a solar mixture of the elements, FeO is unstable at melting temperatures. Because there is abundant evidence that chondrules solidified within ~1 s, too little time to allow appreciable exchange of O existed. Because of the simplicity of the cosmochemical model, it seems more plausible that the high-FeO content and an O-isotopic composition near the *TF* line were both already present in the precursor materials. Both were the result of an approach to gas-solid equilibrium at low nebular temperatures.

But can an approach to equilibrium at the CV location lead to two data arrays that intersect the *TF* line at distinctly different points? I think not. I suggest that the *CCAM* intersection is the one produced by gas-solid interactions, and the chondrule intersection is an artifact that either resulted from nonequilibrium flashing-off of a minor volatile component during the chondrule-forming event, or by a late contamination of the region by $\delta^{17}O$-rich solids of the sort that dominate the OC region. If the precursor materials lay on the *CCAM* line but contained a fine component with the composition of CM matrix (4, 12), flashing off ~40% of the oxygen in the form of this material would displace the O-isotope composition of the residuum to the position of the upper cluster of barred olivine points. Forty percent seems implausibly large; if this model is correct, I suspect that the lost material was much farther from the *CCAM* line than is CM matrix. An alternative might be that, in addition to the loss of matrix-like materials, the chondrule precursors were slightly contaminated by the ^{17}O-rich refractory solids we postulated to be responsible for the OC fields lying above the *TF* line. This is consistent with my earlier suggestions that these materials fell into the nebula fairly late, and that chondrule formation by flash heating was also a late process that occurred after the local nebular temperature had fallen below 600 K (Grossman and Wasson 1983).

For most groups, we do not have data arrays to use to determine the intersection points along the *TF* line. However, the EH, EL and IAB chondrite groups lie near the *TF* line, and extrapolation from their mean position to the *TF* line along a line having a slope of 1 should not introduce much error. We tabulate the intersection values in Table II.

TABLE II
Projections of O-Isotope Compositions onto the Terrestrial Fractionation Line and Inferred Equilibration Temperatures of the Fine Fraction.

Group[b]	$\delta^{18}O$-TF (‰)	$\Delta^{18}O_{12}$ (‰)	T[a] (K)		$\Delta^{18}O_{13}$ (‰)	T[a] (K)	
EH, EL	5.5	−6.5	1070	(650)	−7.5	930	(660)
IAB	5.4	−6.6	1060	(650)	−7.6	920	(660)
SNC	4.2	−7.8	930	(660)	−8.8	840	(670)
Euc, etc.	3.8	−8.2	870	(670)	−9.2	800	(670)
H, L, LL	2.9	−9.1	800	(680)	−10.1	740	(690)
CV, CO	10.0	−2.0	—	620	−3.0	—	620
CI	17.	+5	—	460	+4	—	490

[a]Temperatures based on the difference $\Delta^{18}O$ between $\delta^{18}O$-*TF* and mean nebular O compositions of $\delta^{18}O_{neb} = 12.0‰$ and $13.0‰$ [$\Delta^{18}O = (\delta^{18}O$-*TF*$) - \delta^{18}O_{neb}$] are read off the dashed curve in Fig. 8. Parentheses indicate the less plausible temperature (see text).
[b]Groups are listed in order of decreasing equilibration temperature.

Also listed in Table II are intersections estimated for two differentiated meteorite parent bodies—the body that produced the eucrites, howardites, diogenites, mesosiderites and probably the pallasites, and the body (possibly Mars) that produced the SNC clan. In both cases the O-isotope compositions (Fig. 5) are very near the *TF* line, so the extrapolation should introduce little error. It is also justifiable to use the composition of the shergottite basalts to estimate the whole-planet composition. These are relatively primitive basalts produced by high degrees of melting (indicated by flat rare-Earth patterns, and modest concentration enhancements over chondritic levels). Kyser et al. (1982) showed that O-isotope compositions of primitive terrestrial basalts such as flood tholeiites and mid-ocean ridge basalts are essentially the same as those in upper-mantle rocks.

In Fig. 8 are curves calculated by Onuma et al. (1972) showing the difference of $\delta^{18}O$ values for minerals and nebular gas at an H_2 pressure of 10 Pa. The dashed curve shows how the partitioning should change as the dominant silicate goes from olivine at high to pyroxene at intermediate to a layer-lattice silicate at low nebular temperatures, and is the one used to calculate equilibration temperatures for all groups except the CI group, where the layer-lattice curve was required. The olivine and pyroxene curves pass through a

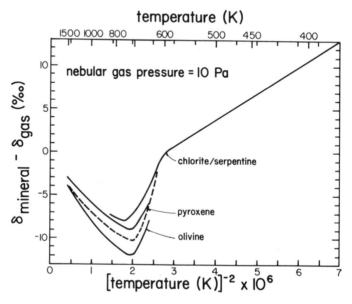

Fig. 8. A plot showing the differences in the $\delta^{18}O$ values of nebular minerals and gas under equilibrium conditions at 10 Pa total pressure that depend on temperature and on the mineral involved. A dashed line approximates the path of mean solids that are dominantly olivine at high, pyroxene at intermediate, and layer-lattice silicates at low nebular temperatures (curves calculated by Onuma et al. 1972).

minimum when CO converts to CH_4 and H_2O becomes the major nebular gaseous species. Because of the minimum, most $\Delta^{18}O$ values yield two possible equilibration temperatures.

To calculate nebular equilibration temperatures from the *TF* intersection $\delta^{18}O$-*TF* values and the dashed curve in Fig. 7 requires the $\delta^{18}O_{neb}$ value of the nebular gas. This is not known, but values in the range 11.5 to 13.5‰ yield reasonably plausible equilibration temperatures. I assumed values of 12 and 13‰ for $\delta^{18}O_{neb}$ gas, and calculated $\Delta^{18}O = \delta^{18}O$-*TF*-$\delta^{18}O_{neb}$. The inferred equilibration temperatures are listed in Table II. For those cases where two temperatures can be inferred, I have chosen one as more reasonable, based on the textural evidence in the meteorites; in each of these cases, I chose the higher temperature based on the evidence that these groups formed in portions of the nebula where extensive volatilization of presolar solids had occurred. The less plausible temperature is in parentheses.

My equilibration temperatures are systematically higher than those of Keays et al. (1971) based on the condensation of volatile metals. This is not unexpected. Keays et al. estimated the lowest temperature at which volatile metals equilibrated with the gas, whereas I have estimated the mean temperature at which matrix O condensed from the gas.

The meteorite groups are listed in order of decreasing nebular equilibration temperature in Table II. The chondrite list is in the same order as that based on the $FeO_x/(FeO_x + MgO)$ and refractory lithophile abundance trends (Fig. 1). The two groups of differentiated meteorites fall between the IAB and ordinary chondrites. Because of the possibility of exchange with H_2O inside a parent body, the CI-chondrite temperature should be taken *cum grano salis*.

With the exception of IAB, the rank of the first 5 groups or clans in Table II is the same as one would infer on the basis of the mean $\Delta^{17}O$ values (Table I), i.e., the mean vertical distance from the *TF* line. This is consistent with the view that the nearer to the Sun that groups formed, the smaller the fraction of anomalous presolar solids that survived and became incorporated during agglomeration.

VI. SUMMARY OF FORMATION LOCATIONS

My best assessments of where the different meteorite groups formed is shown in Fig. 9. The placement is the same as that discussed by Wasson and Wetherill (1979), augmented by the differentiated meteorite groups: SNC and the igneous clan (Euc, etc.). There are insufficient data to determine which group of the enstatite-chondrite and ordinary-chondrite clans formed nearest the Sun. My current view is that the least fractionated (EH,H) formed farthest from the Sun.

My preferred model calls for forming the carbonaceous chondrites at the outer edge of the asteroid belt or beyond. I favor this because the evidence for forming the ordinary chondrites in the inner portion of the belt (at roughly 2.5

Proposed Formation Locations of Meteorites

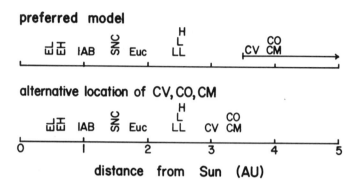

Fig. 9. Sketches summarizing the inferred formation locations of the chondrite groups and two clans of differentiated meteorites. Enstatite chondrites probably formed inside 0.7 AU, ordinary chondrites in the inner part of the asteroid belt, and carbonaceous chondrites in the outer reaches of the asteroid belt or beyond. Equilibration temperatures based on O isotopes are consistent with SNC meteorites forming at Mars' distance from the Sun.

AU near the 3/1 resonance) is reasonably strong and self consistent. The differences between ordinary chondrites and carbonaceous chondrites in abundance of refractory lithophiles and, especially, in the composition and degree of equilibration of O isotopes are so great that a reasonably large gap must exist between their respective formation locations.

The evidence favoring the view that the highly reduced enstatite chondrites formed inside 1 AU continues to mount: the Venus rare-gas abundances are similar to those in EH South Oman; the inferred O-isotope-based nebular equilibration temperatures are higher than those for any other group; and Vilas' (1985) reflection spectrum of Mercury shows no evidence of FeO in the silicates.

VII. APROPOS MERCURY

This interpretation of the chondritic evidence leads to the conclusion that the planet Mercury could have formed from building blocks that were closely related to the enstatite chondrites, though mechanically enhanced in metal and possibly other phases that tend to associate with metal in planetary or nebular settings. The model requires no processes different from those that formed the chondrites or the asteroids. The high density of Mercury requires an Fe/Si ratio 4 to 7 times greater than that in EH chondrites. It is possible that this metal/silicate fractionation was produced by selective accretion of the cores of differentiated asteroids; this process also requires a reduced composition for the original composition of the asteroid—a composition at least as reduced as that of H chondrites—and the more reduced, the easier to explain the Fe/Si fractionation. It seems equally plausible that the mechanical separation occurred in the solar nebula, and was precisely the same process (selective sticking, fragmentation, settling or radial drag) that produced EH chondrites with an Fe/Si ratio 2 times higher than that for EL chondrites.

The above "reduced-chondrite" models strongly contrast with some other models regarding the origin of Mercury. Lewis (1972) proposed that the Fe/Si fractionation occurred because planetesimal formation occurred at very high nebular temperatures ($\sim$1400 K at a plausible nebular pressure of 10^{-3} atm at 0.4 AU). This model is inherently implausible because it calls for planetesimal formation only within a very narrow temperature range, after condensation of Fe-Ni metal at $\sim$1400 K but before condensation of Mg_2SiO_4 at $\sim$1385 K. Such a high, narrowly constrained cutoff temperature seems implausible because (1) such high temperatures would imply turbulence that prevented planetesimal formation; (2) even if planetesimal formation were able to occur at such high temperatures, there seems to be no plausible reason why, at lower temperatures, the more volatile materials would not have formed their own planetesimals, and thus have participated in the planet-building process at 0.4 AU; and (3) some inmixing of materials from adjacent zones having lower equilibration temperatures must have occurred. No low-mass nebular model yields temperatures of 1400 K that last for the $\sim$10-Myr period necessary for Mercury to accrete. The Lewis (1972) model leads to compositional predictions that contrast with those of the reduced-chondrite model regarding the composition of Mercury. Like the reduced-chondrite model, it predicts that silicates on Mercury have very low FeO contents, but it also predicts very low contents of all volatile elements including Na, K and S.

In the Cameron (1985a; Fegley and Cameron 1987; also see Cameron et al.'s Chapter) model, the silicate depletion on Mercury is attributed to vaporization following planet formation. Temperatures are assumed to be in the 2500 to 3500 K range for > 10^4 yr. Such high temperatures for such a long period after the formation of the nebula and the subsequent formation of proto-Mercury involve energy sources far greater than those present in most nebular models, even high-mass nebula models such as Cameron's. There are other potential problems, such as the ability of the nebular wind to remove the hot atmosphere (Cameron 1985a). In any case, the compositional predictions of this model are qualitatively similar to those of the Lewis (1972) model: all volatiles would have been lost.

Thus, there are two basic cosmochemical classes of Mercury formation models: (1) mechanical fractionation of reduced-chondritic material, and (2) high-temperature loss of most silicates and all volatiles as gases. The former allows the nebula to cool to low temperatures and requires no processes other than those that formed the enstatite chondrites, although mechanical fractionation after asteroid formation may have also played a role. The latter either requires accretion at very high and sharply defined nebular temperatures, or requires high-temperature distillation of proto-Mercury after it had formed as a Mars-size planet. I find the reduced-chondrite models much more plausible. In any case, spacecraft can be used to choose between these models. A spacecraft landed in a freshly cratered basalt flow should readily determine whether alkali contents are chondrite-like (as expected if the metal/silicate fractionation occurred in the nebula), asteroid-like (~0.1 chondritic, as expected if the metal/silicate fractionation was produced by the loss of mantle materials from asteroid-size bodies), or much lower (<0.01 chondritic, as predicted by the Lewis or Cameron-Fegley models).

Acknowledgments. I am indebted to numerous colleagues for informative discussions: C. R. Chapman, R. N. Clayton, K. A. Goettel, R. Jeanloz, A. E. Rubin, D. N. Shirley, D. J. Stevenson, F. Vilas and G. W. Wetherill. I thank R. Demonteverde, L. Mikami, A. Pang and H. Qian for technical assistance. This research was mainly supported by a grant from the National Science Foundation.

ORIGIN AND COMPOSITION OF MERCURY

JOHN S. LEWIS

University of Arizona

A suite of condensation-accretion models for Mercury have been calculated for nebular models with steep (adiabatic) radial temperature gradients and for a range of dynamically plausible accretion scenarios. The present work confirms that the accretional sampling of local metal-rich condensates cannot account for the large density difference between Mercury and the other terrestrial planets. Also, such accretion scenarios sample the available preplanetary solids so widely that significant amounts of FeO (on the order of 1% of the planetary mass) and alkali metal oxides (a few tenths of a percent) would be expected in the planet. The present composition of Mercury is discussed in the context of the three leading models for metal enrichment in Mercury: (1) preferential comminution and loss of silicate grains during accretion; (2) partial volatilization of the planet by the early Sun; and (3) stripping of the crust and upper mantle by giant impacts. Modification of regolith composition by late infall of meteoritic material is also considered.

I. INTRODUCTION

Theoretical models for the composition of solar nebula condensates as a function of distance from the Sun (Lewis 1972, 1974) predict that the preplanetary solids formed at Mercury's distance were extremely reduced, virtually devoid of FeO and volatiles. Iron sulfides and oxides would be absent; refractory minerals rich in calcium, aluminum, titanium and rare earths would be fully condensed; metallic iron would be partially condensed; and magnesium silicates would be only slightly condensed. The density of the local solids would be significantly higher than that of materials present near the orbits of Venus and Earth, but would not be as high as the deduced uncompressed density of Mercury.

Plausible statistical sampling algorithms for assembling a planet out of small bodies (Safronov 1969; Hartmann 1976; Cox and Lewis 1980; Chapter by Wetherill) show that the material actually accreted by a planet growing at Mercury's orbit should not be exclusively derived from a local "feeding zone", but should rather be drawn from a very wide region spanning the formation positions of all the terrestrial planets. The eccentricities of the orbits of the accreting bodies will naturally and unavoidably be pumped up by the cumulative effects of gravitational interactions with the largest bodies in the swarm, and mean eccentricities of 0.15 or higher are produced. Smaller mean eccentricities are not only incompatible with the perturbing effects of the growing planetary bodies, but also give rise to a planetary system strikingly different from the solar system as we see it: planets would be far less massive and far more numerous, with 10 to 20 planets inside the asteroid belt. Any distribution of orbital elements that leads to a small enough number (3 to 5) of terrestrial planets also leads to high encounter velocities and violent, disruptive collisions between very massive bodies.

Further, Wetherill finds one case in which the proto-Mercury wanders extensively during accretion, reaching out at one point almost to the orbit of Mars (see Chapter by Wetherill). While such an extreme excursion may be a pathological case, broad sampling of interplanetary material is surely plausible. However, the sampling cannot be too large, or it would, contrary to observation, erase the density differences between the terrestrial planets.

The only prior attempt to model both the condensation and accretion processes contributing to the terrestrial planets (Goettel and Barshay 1978; Barshay 1981) showed clearly that the high density of Mercury cannot be attributed solely to accretion of the planet out of iron-rich local condensates. That work indicated that some other major fractionation effect such as collisional comminution of silicates with consequent preferential accretion of metal (Weidenschilling 1978), post-accretion vaporization of silicates (Bullen 1952; Ringwood 1966; Cameron 1985a; Fegley and Cameron 1987), or ejection of silicates by giant impacts (Chapter by Wetherill) has strongly influenced the present bulk composition of Mercury.

It is always desirable to test theoretical models against observational data. In the case of Mercury, however, the paucity of reliable evidence is daunting. The only surface species that has ever been claimed by observers is FeO, and the upper limit on its concentration is a few percent (McCord and Clark 1979). Vilas (1985) shows that FeO is in fact not detectable in some of the best spectra available. We have no knowledge of the distribution of FeO with depth in the regolith, and concentrations on the order of 1% could plausibly be present "dusted on the surface" in some regions as a result of the impact of FeO-rich planetesimals.

The uncompressed density of Mercury is near 5.3 g cm^{-3}. The planet has so little mass that the correction to the observed density is very small (about 3%), and hence the final result is very reliable. The corresponding values for the uncompressed densities of Venus and Earth are close to 4.0 (after a correction of about 30% to their observed densities), and Mars is near 3.75.

The recent observation of a tenuous envelope of sodium and potassium vapor about Mercury (Potter and Morgan 1985a, 1986a) places no useful constraints on the abundance of these elements in the bulk Mercurian regolith, since even a small admixture of late-arriving asteroidal material would serve as an adequate source for solar-wind sputtering of these elements.

II. CONDENSATION-ACCRETION MODELS

A model for the composition of rock-forming solids vs heliocentric distance in the primitive solar nebula (Lewis 1972, 1974; Barshay and Lewis 1975; Lewis et al. 1979; Fegley and Lewis 1980), which is calibrated against such benchmarks as the FeO content of the Earth, the density of Mars, and the condensation of water ice at Jupiter, has been used to define the equilibrium behavior of the elements Si, Mg, Fe, Ca, Al, Na, Ni, K, S, H, O, C, N, P, F and Cl (Fig. 1a,b). The treatment of the disequilibrating effects of nebular turbulent mixing on the condensation chemistry of carbon and nitrogen by Lewis and Prinn (1980), and the detailed consideration of the chemistry of nitrogen by Fegley (1983) have also been incorporated. This model, which was originally developed at a time when the only compositional datum on the asteroid belt was the identification of 4 Vesta as a basaltic achondrite (McCord et al. 1970), made the prediction that primitive bodies in the asteroid belt would be found to be dominated by carbonaceous chondritic material, and that ordinary chondrites would be rare or absent in the belt. Recent reviews of asteroid taxonomy, such as that by Gradie and Tedesco (1982), clearly show that this is the case. This simple model of the chemical behavior of solar material successfully predicts some of the most general trends in the composition of solid solar system materials, including their overall density trend vs heliocentric distance. However, as emphasized by Lewis (1974), "It would be wholly unwarranted to conclude that, for example, 100% of the mass of Venus originally condensed at 900 ± 150 K. . . . Rather, it can only be claimed that the *bulk* composition of Venus is dominated by material formed in this temperature range. The relative contributions of higher- and lower-temperature condensates can only be assessed from elemental abundance data for Venus." Similarly, the abundances of low-temperature condensates, such as FeO, in Mercury are reflective of the details of the accretion process, not just the local nebular chemistry.

The accretion of the terrestrial planets has been treated under a wide variety of assumptions by several authors. Safronov (1969) pioneered the

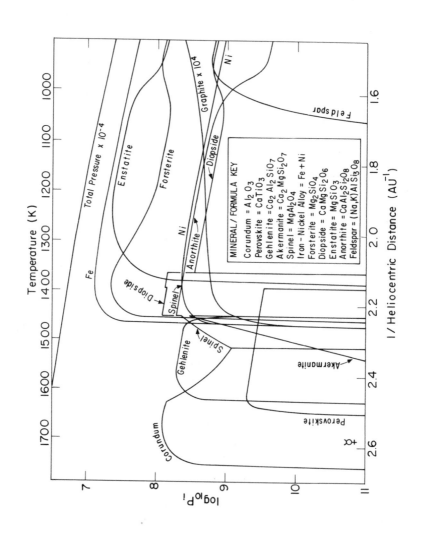

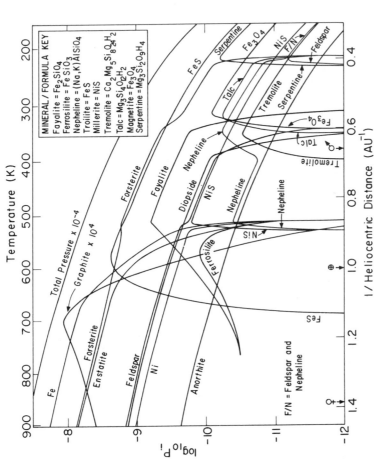

Fig. 1. Radial distribution of condensates in the solar nebula, after Barshay and Lewis (1980). The molar abundances of the principal condensates are displayed here as equivalent pressures: 10^{-9} bars of solid Ni, for example, means that the amount of solid nickel per unit volume is the same as in a gas containing 10^{-9} bars partial pressure of Ni vapor. These units facilitate comparison of gas and condensate abundances. This graph is the source of the Ca, Al, Mg, Si, Fe, Ni, C and S chemistry used in this chapter. Data on the chemistry of Na, K, P, Cl, H and F are taken from Fegley and Lewis (1980). The chemistry of N is from Fegley (1983).

study of the celestial mechanics of accretion without the use of computerized numerical simulations, and was the first to treat the "stirring" of planetesimal populations by gravitational interactions with small numbers of more massive bodies embedded in the swarm. Hartmann (1976) considered the terminal phase of accretion by treating the accretionary fate of the numerous small bodies remaining after most of the masses of the terrestrial planets were already accreted. He found that the small-body population was widely sampled by the planets during the last phase of their accretion.

Cox and Lewis (1980), applying a new and efficient two-dimensional algorithm for calculating the orbits of hundreds of interacting bodies, verified that such wide sampling occurred throughout much of the accretion process, not just in its terminal phases. Wetherill (see his Chapter), having extended this computation scheme to three dimensions, finds that the mean orbital eccentricities of accreting planetesimals are typically near or above 0.15, and that growing planets can migrate extensively over a range of distances from the Sun. The latter authors, in attempting to define the limiting envelope of initial conditions required for the formation of a planetary system grossly similar in overall properties to our own solar system, agree that wide sampling of the planetesimal swarm by each accreting planet is unavoidable. Each planet is dominantly (60 to 90%) composed of material that originated closer to it than to any other planet, but 10 to 40% of the mass of each planet is derived from what would formerly have been regarded as the feeding zones of other planets. Mercury may readily accrete several percent of its mass from the vicinity of Mars.

This chapter employs a simple accretion model that convolves the predicted chemical composition of preplanetary solids vs heliocentric distance with an accretion probability distribution. The temperature profile in the inner solar system is taken to be adiabatic with central-plane temperature T_c varying according to

$$T_c = 550 \, r^{-1.1} \tag{1}$$

where T is in Kelvins and the heliocentric distance r is in AU (see Lewis 1974). The nebula is adiabatic with $Cp/R \simeq 7/2$, so that the central-plane pressure P_c (bar) is

$$P_c \propto T_c^{3.5} = 10^{-4} r^{-3.85}. \tag{2}$$

The surface density of the nebula σ is then

$$\sigma \propto \frac{P_c r^{3/2}}{T_c^{1/2}} \propto r^{-1.8}. \tag{3}$$

For convenience, the accretion sampling function is taken to be Gaussian over heliocentric distance (semimajor axis a). Any realistic accretion history involves stochastic processes, and will result in "lumpy" multimodal accretion sampling distributions for each planet. The use of smooth accretion sampling functions herein is not an attempt to deny the complexity of the real accretion process, but rather to model the most likely outcome averaged over the accretion of a large number of Mercuries. I see no clear theoretical or practical advantage to postulating 2-component or 3-component models of Mercury, in which the number of undetermined adjustable parameters (semimajor axis, Gaussian half width, and integrated mass of each component) quickly escalates to high numbers, requiring the generation of astronomical numbers of models, most of very low (but unknown) probability. Besides this mathematical obstacle, there is the fact that the physical basis for 2-, 3-, or multi-component models is not at all understood. I, therefore, present no such models.

The present models are constrained to have their sampling peaks near the present semimajor axis of Mercury and have as their principal adjustable parameter the half width HW of the Gaussian. The Gaussian half widths are chosen to cover the entire range of plausible values suggested by the numerical accretion experiments described above, from 0.05 AU to 0.45 AU. The mass of each accreted material m_i is then given by

$$m_i - 2\pi \int_{r_{in}}^{r_{out}} G(r,HW) \; \sigma_i(r) r dr \qquad (4)$$

where the inner and outer limits of integration are set to span the terrestrial planet region and G is the Gaussian function.

The abundances of condensates are digitized in 60 heliocentric distance bins with widths of 0.025 AU, extending from 0.3 AU to 1.8 AU from the Sun. The nebular pressure and temperature at the midplane range from (10.3 mbar, 2068 K) at the inner boundary to (0.011 mbar, 288 K) at the outer boundary. Temperatures at the inner boundary are so high that no condensates are present there. The location of the outer boundary is not of great importance for Mercury models because plausible sampling probabilities are very small at that distance. The chemical data are used to calculate the molar concentration of each condensate for each radial distance bin. These molar abundances are converted into condensate mass per radial distance interval, and then convolved with the sampling probability function. The composition for each of a number of locations and half widths of the sampling function is calculated both as wt. % of the basic condensate species (Fe metal, FeO, FeS, CaO, etc.) and as wt. % of the constituent elements. The abundances of the elements used in these calculations (Table I) are generally those of Cameron (1973), which are strongly influenced by analyses of chondritic meteorites. Other abundance compilations generally differ from Cameron's by no more than a few percent for the major rock-forming elements. The abundances of iron and nickel have been raised above the chondritic value because the densi-

TABLE I
Abundances of the Elements

Element	Normalized Abundance (Si = 10^6 atoms)
H	31.8×10^9
O	21.5×10^6
C	11.8×10^6
N	3.74×10^6
Mg	1.05×10^6
Fe	1.05×10^6
Si	1.00×10^6
S	0.50×10^6
Al	85.00×10^3
Ca	72.00×10^3
Na	60.00×10^3
Ni	60.00×10^3
P	9600
Cl	5700
K	4200
F	2450

ties of Venus, Earth and Mars all suggest an Fe/Si ratio slightly higher than chondritic (Lewis 1972).

The time dependence of nebular conditions is not treated explicitly. Cooling, which increases the local abundances of low-temperature condensates, would be indistinguishable from an offset in the temperature of the nebular adiabat or a change in the Gaussian half width.

III. RESULTS

In general, the results confirm the principle (Goettel and Barshay 1978) that no plausible sampling of the products of nebular condensation can provide a planet with the density and metallic iron content of Mercury. All plausible accretionary models of Mercury ($HW \geq 0.2$ AU) have metallic iron contents $< 33.4\%$ by weight. The maximum core mass (defined as the sum of the weight fractions of metallic iron, nickel, troilite (FeS) and schreibersite (Fe_3P) is close to 36%. The highest metallic iron contents are associated with narrow sampling functions located close to the condensation point of iron. The metal mass declines slowly with increasing half width as more oxidized iron (mantle FeO and core FeS) is accreted.

Figure 2 displays the results for metallic iron content in wt. %. The low value for planets accreted at small heliocentric distances is due to incomplete condensation of iron in the region being sampled, and the slow fall off of metallic iron content noticed for very wide sampling functions is due to the conversion of metallic iron into FeS and FeO at greater heliocentric distances.

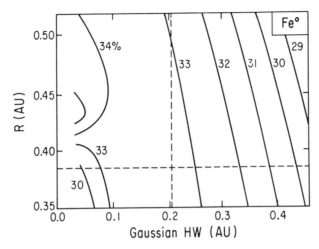

Fig. 2. Abundance of metallic iron in Mercury for a suite of accretion models. The accretion models are characterized by the location of the center of the Gaussian sampling function R and its half width HW, both given in AU. The dashed lines indicate the present orbital semimajor axis and eccentricity of Mercury. The most plausible accretion scenarios sample at least as wide a range of distances as are presently swept by Mercury, and hence, lie in the right-hand half of the diagram.

Figure 3 gives the wt. % FeS in total planetary material. The sulfur content of the planet is extremely sensitive to the width of the sampling function: a planet accreted at Mercury's present heliocentric distance with a Gaussian half width of 0.1 AU would contain only about 100 ppt (parts per trillion: 1 part in 10^{12}) of FeS, whereas accretion at the same location with a half width of 0.2 AU would provide 0.02% FeS. The most plausible models have FeS contents of 0.1 to 3% by weight. Such FeS abundances are sufficient to play a major role in the facilitation of planetary melting and differentiation by enabling eutectic melting in the Fe-Ni-FeS system and, may be responsible for maintaining Mercury's magnetic field by keeping the core at least partly melted. Thermal history models that assume Fe-FeS eutectic melting in Mercury therefore seem reasonable from both chemical and dynamical perspectives.

FeO differs significantly from FeS in condensation behavior. FeS condenses quite abruptly beginning at 680 K, but FeO is present as a solute in olivine and pyroxene at all temperatures at which they are condensed. The oxidation of Fe metal to FeO is favored by lower temperatures, and the FeO content of the solids is accordingly at a minimum close to the Sun (Fig. 4). FeO contents on the order of 1% are expected for plausible accretion models, although concentrations as high as 2 to 3% are not ruled out.

The abundance of total core-forming material (Fe, FeS, Ni and Fe_3P) is shown in Fig. 5. Note the general slow decrease of core mass with the width

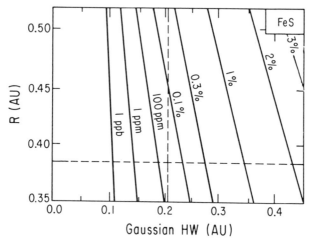

Fig. 3. Abundance of FeS in Mercury for a suite of accretion models. See caption of Fig. 1 for explanations.

of the sampling function, due to the inclusion of FeS and loss of FeO formed at greater distances from the Sun. This weak variation of core mass and core density preserves the overall uncompressed density of the planet as ~4.2 g cm^{-3} over the entire range of reasonable formation conditions (compared to ~5.3 g cm^{-3} for Mercury).

The concentrations of refractory components are of course high because of their stability at high condensation temperatures: their abundances drop off

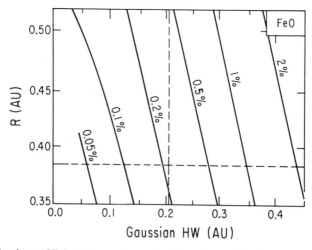

Fig. 4. Abundance of FeO in Mercury for a suite of accretion models. See caption of Fig. 1 for explanations.

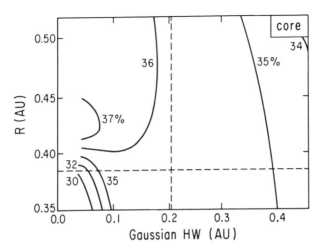

Fig. 5. Abundance of core material in Mercury for a suite of accretion models. Core material is defined as metallic iron, nickel, FeS and Fe₃P. See caption of Fig. 1 for explanations.

steadily because of dilution by much more massive low-temperature condensates as the heliocentric distance or the width of the sampling function is increased. The behavior of the refractories is illustrated in Fig. 6 by alumina. Lime (CaO) and other refractory oxides, such as titanium dioxide, generally behave similarly, except that small differences in their condensation temperatures cause discernible differences in behavior close to the Sun when very narrow sampling functions are employed.

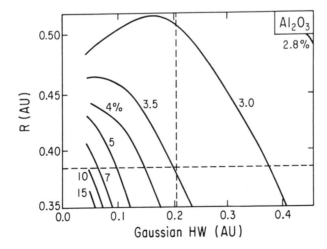

Fig. 6. Abundance of alumina (Al₂O₃) in Mercury for a suite of accretion models. See caption of Fig. 1 for explanations.

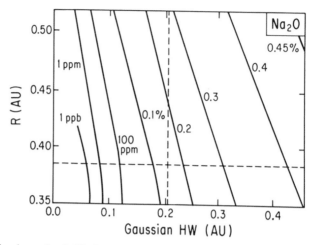

Fig. 7. Abundance of soda (Na_2O) in Mercury for a suite of accretion models. See caption of Fig. 1 for explanations.

Sodium and potassium are moderately volatile elements with condensation temperatures near 1000 K (Fegley and Lewis 1980), well above the formation temperature of FeS. The detection of a tenuous envelope of sodium and potassium vapor about Mercury raises the question of whether primordial retention of alkalis or late infall of alkali-bearing asteroidal and cometary materials has provided these elements to Mercury. We find that all plausible sampling functions for Mercury give bulk sodium oxide contents of at least 0.15% (Fig. 7). Potassium (not graphed) behaves very similarly: all of the broader sampling functions ($HW > 0.15$ AU) give potash abundances that are 15 to 30% of the soda abundance.

Effects of Fractionation Processes

Since direct condensation-accretion modeling of Mercury uniformly gives too low a density for the planet, we must explicitly consider the compositional effects of the major fractionation processes that have been proposed. These three processes, respectively, attribute the high density to (a) preferential accretion of metal (Weidenschilling 1978); (b) the ejection of most of the silicate portion of the planet by giant collisions (Chapter by Wetherill); or (c) vaporization of a similar mass of material by heating during the formation of the Sun (Bullen 1952).

If the impact hypothesis is correct, then the residual silicate material on Mercury would be of mantle composition. The FeO content would reflect the primordial oxidation state of the material of the planet at the time of accretion. The concentrations of refractories would not be greatly enhanced, since the ejection of a feldspar-rich crust would severely deplete both calcium and aluminum along with the volatile alkali metals. The evaporation model, on the other hand, leaves Mercury very strongly enriched in refractories and severely depleted in alkalis and FeO. Finally, if Weidenschilling's qualitatively appealing scenario for comminution and loss of silicate grains during accretion is correct, there would be little fractionation between the various major silicate minerals. The bulk composition of Mercury would be well described by the core and noncore components of the present chemical models, except that the ratio of core to noncore material would be increased from about 0.32 to about 1.5.

It is of interest to determine whether late infall of cometary and asteroidal materials onto Mercury may have had a significant effect on the abundances of any of these important materials. By comparison, the regolith of the Moon contains about 0.2% by weight of asteroidal metal with mean composition similar to that of the metal grains that make up 10 to 25% by weight of the L and H chondrites. This indicates the presence of about 0.8 to 2% weight of asteroidal material in the lunar regolith. If an oxidized chondritic impactor containing 10% FeO were to be mixed in with 100 times its own mass of FeO-free silicates, the FeO content of the mixture would be 0.1%. This suggests that regional variations in the FeO content of the Mercurian regolith may be found to be associated with particular impact structures. In any event, even if Mercury were formed devoid of FeO, late infall could provide up to a few tenths of a percent of FeO in the present regolith. This is about the same as the concentration of primordial FeO expected in the narrowest, most FeO-poor accretion scenarios presented above.

Assessing the late-infall mechanism for sodium and potassium is rather more difficult because of the very uncertain behavior of these volatile elements during an energetic impact. The maximum regolith alkali oxide content from a late infall contribution of 1% of the regolith mass would, neglecting volatilization, be about 0.01%. All of the plausible alkali accretion models presented herein provide at least 20 times this amount of soda and potash.

Typical compositions for the selective accretion, giant impact and volatilization models are displayed somewhat schematically in Fig. 8. The effects of adding up to 5% of average chondritic matter to the regolith by late impacts are also shown. It should be borne in mind that regolith contents of impact debris are probably much less than 5%. The figure clearly indicates that late infall cannot much influence the composition of the regolith for giant impact models, but may readily dominate the alkali and FeO content in the volatilization model.

The abundances of the atmophiles H, Cl, F and N are, as expected, very sensitive to the width of the sampling function. Hydrogen abundances found for Gaussian half widths of 0.2 are less than one ppb (part per billion: 10^{-9}), reaching as high as 1 to 10 ppm (part per million: 10^{-6}) for half widths of 0.35 to 0.40 AU. Because of the high solubility of elemental carbon in metallic iron, however, the predicted carbon abundance is much less sensitive to the

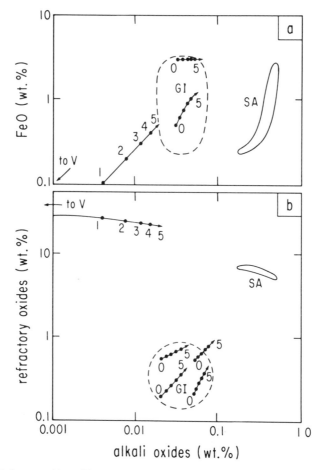

Fig. 8. Bulk composition of Mercury for several different histories. Composition is here parameterized in terms of three factors, the FeO content, alkali content (soda plus potash) and
refractory oxide content (calcium plus aluminum plus titanium oxides) of bulk silicate material.
SA denotes plausible compositions for Weidenschilling's selective accretion model, GI indicates crudely the compositional range for the giant impact model, and V is the residual Mercurian mantle from the volatilization model (< 0.1% FeO, ~40% refractories and ~0% alkalis). Part (a) gives the FeO-alkali plane and part (b) is the refractory oxide-alkali plane. The
modifying effects of late infall of 0% to 5% of average chondritic material on several representative regolith compositions are indicated by arrows labeled 0 to 5. Mixing lines on a log-log
plot are curved.

width of the sampling function. For half widths of 0.2 to 0.4 AU, carbon
abundances of 0.2 to 1 ppm are expected. This prediction takes on added
significance when we realize that the coexistence of carbon with FeO during
melting and differentiation allows outgassing of significant masses of carbon
monoxide or carbon dioxide: 1 ppm of carbon, fully oxidized to carbon diox-

ide and outgassed, provides over 10^{21} grams of atmosphere. This is 100 times the maximum plausible amount of radiogenic argon.

IV. CONCLUSIONS

A suite of condensation-accretion models for Mercury spanning a range of condensation temperatures and accretion sampling functions appropriate to Mercury provides specific predictions of the expected range of composition of Mercury at its time of formation. The principal conclusions to be drawn from these models are:

1. The observed density of Mercury is confirmed to be too high to be explainable by straightforward condensation and accretion models. Some additional important mechanism is required to provide core masses in excess of about 37% of the planetary mass.
2. Dynamically plausible accretion scenarios for Mercury provide, at the time of formation, a few tenths of a percent of alkali oxides, 2.8 to 3.5% alumina, 30 to 33% metallic iron, roughly 0.1 to 3% FeS, 0.2 to 3% FeO, and core mass fractions of about 34-36%.
3. These compositional models can, if modified simply to take into account the nonselective loss of most of the silicate component of the planet during accretion, provide compositional predictions for Weidenschilling's mechanism for the accretion of a metal-rich Mercury. Mercury's silicate portion would, in this case, contain 3.6 to 4.5% alumina, roughly 1% of alkali oxides and between 0.5 and 6% FeO.
4. If giant collisions have stripped the crust and upper mantle of Mercury late in accretion, then the present surviving silicate portion of the planet need not be enriched in refractories: the material lost would contain the calcium- and aluminum-rich crust. For the same reason, alkalis could be severely depleted by such an impact because of loss of the feldspar-rich crust. The FeO content of the planet, however, might be the same as in the Weidenschilling model.
5. If post-accretion heating and vaporization of the crust and mantle of Mercury were important, then very severe alkali and FeO depletion would accompany extreme enrichment of refractories (Fegley and Cameron 1987). The very low FeO content would be diagnostic of this particular mechanism.
6. Modification of the surface composition of Mercury by late impact events may be important. If the volatilization scenario of Fegley and Cameron is correct, then the present surface inventories of FeO and alkali metals would be dominated by late infalling asteroidal and cometary material. If the giant-impact model is correct, then the surviving primordial FeO content is probably larger than the late-infall contribution, but infall could enhance the alkali abundances. If Weidenschilling's preferential-accretion

model is correct, then infall is probably not important as a source of either alkalis or FeO. The free-metal content of the regolith would then be the most sensitive measure of the amount of infall.

Acknowledgment. This work was supported in part by a grant from the Planetary Atmospheres program office of the National Aeronautics and Space Administration. I am grateful to W. Farrand for his valuable assistance in developing the computer software used in this project.

FORMATION OF MERCURY AND REMOVAL
OF ITS SILICATE SHELL

A. V. VITYAZEV, G. V. PECHERNIKOVA AND V. S. SAFRONOV
USSR Academy of Sciences

The model for formation of Mercury by gas-free accumulation of planetesimals provides a natural way for understanding Mercury's low silicate-to-iron ratio. Impact heating would rapidly lead to differentiation of Mercury: a metallic core surrounded by a silicate mantle. Subsequent high-velocity impacts by late-arriving planetesimals would erode away much of the silicate crust and mantle; such silicates would be accumulated by Venus and fall into the Sun.

It is very likely that Mercury, like the other terrestrial planets, formed by accumulation of solid particles and bodies. Its small mass, high density, large orbital eccentricity, unique spin and other rather unusual physical and dynamical properties are closely related to its position at the inner edge of the solar system.

Several ways of explaining the specific properties of Mercury have been suggested. We shall not discuss these various suggestions here but give only our hypothesis about the origin of Mercury that can be deduced from the model of planetary accumulation, according to our views of this process.

The model of gas-free accumulation explains adequately the late stage of evolution of the inner region of the preplanetary disk and the formation of terrestrial planets. Such characteristics as the number of planets (3–5), their masses, average eccentricities and inclinations of their orbits, spins and inclinations of rotational axes have all been evaluated in analytical form and obtained by numerical calculations (see, e.g., the review by Safronov and Vityazev 1985). It seems to us that the hypothesis of giant gaseous protoplanets (Cameron 1978,1985b; see Chapter by Cameron) cannot give such results; while his scenario may be unacceptable for the solar system, perhaps it can be used for explaining some other (less regular) planetary systems in our Galaxy.

Obviously not all the features of the solar planetary system are satisfactorily explained today. An intriguing problem of planetary cosmogony and comparative planetology is the composition of the terrestrial planets. Many scientists think that the difference in densities between Mars, the Earth and Venus is not so large that we should assume significant differences in the iron/silicate content. For Mercury this is not the case and a mechanism of effective separation of iron from silicates and some way of removing silicates from this zone should be found.

According to Pechernikova and Vityazev (1979,1980; Vityaz and Pechernikova 1981), the embryos of the terrestrial planets had a chance to experience close encounters and collisions with bodies of a comparable size. Wetherill (1985a; see his Chapter) came to the same conclusions on the basis of his numerical simulations. Large bodies very effectively heated the growing planets (Safronov 1978). Heating due to the collision of comparable bodies m and m' with relative velocity v is approximately

$$\Delta T = \frac{v^2 m m'}{C_p (m + m')^2} \tag{1}$$

where the specific heat $C_p = 10^7$ erg cm^{-3}K^{-1}—for simplicity, we assume here uniform distribution of heat. It is easy to see that after reaching v $\gtrsim$ 3 km s^{-1}, conditions for melting and differentiation of material in the interior of the embryos arise. Moreover, when v becomes higher than the escape velocity v_e from embryos, after collisions among embryos, part of material can leave an embryo and go into planetocentric and even heliocentric orbits. Such considerations lead us to the model for removing the silicate shell from proto-Mercury (Vityazev and Pechernikova 1984,1985; Vityazev et al. 1984). In this model, we can explain the planetary variations of Fe/Si even assuming quasi-homogeneous initial distribution of this quantity in the preplanetary disk; however, one cannot exclude the possibility of an aerodynamic sorting mechanism pointed out by Weidenschilling (1978).

Our calculations show that at a later stage of accumulation, the largest bodies in the inner zone including Mercury increased their eccentricities during close encounters. When eccentricities (including that of Mercury) increased to $\gtrsim$ 0.2, the velocities of bodies relative to Mercury reached on the average about $10 \cdot \sqrt{2}$ km s^{-1}, i.e., they became much higher than the escape velocity from proto-Mercury (current v_e from Mercury's surface is 4.3 km s^{-1}). The situation was apparently the same as in the asteroid zone—the impact energy became high enough to stop the accumulation of Mercury and even to reverse it to the loss of mass from the surface of the planet. One can expect that at macroimpacts about 20% or more of the impact energy is partitioned into kinetic energy. Thus the mass ejected into heliocentric orbits could be about twice the mass of the impacting body.

Due to the high-velocity impacts, major heating, melting and silicate-iron differentiation of the proto-Mercury interior should have occurred. Although thermal convection and impact mixing of matter in the subsurface layer opposed such differentiation, there are reasons to expect that iron sank to the center and the upper shell was considerably enriched in silicates. Impact erosion of the upper layer removed mainly silicates from the planet, leading to a considerable Fe/Si fractionation. The matter ejected from Mercury's sphere of influence was partly swept out by growing Venus; some fraction of ejected material fell on the Sun due to diffusion and Poynting-Robertson drag. Calculations cited above have shown that the position of proto-Mercury could have been closer to Venus than it is now.

It is interesting to note that this model can explain not only Mercury's enrichment in iron but also a small silicate excess (1–2%) for Venus compared with that of the Earth.

Estimates show that a similar but less significant erosion process could have happened on the surfaces of other planets. In the case of the Earth, some fraction of its primitive silicate mantle which was ejected from the surface could have led to a more efficient growth of a prelunar disk (Pechernikova et al. 1984). It seems that such a process may be a major cause of iron-silicate fractionation in the inner region of the solar system. The inclusion of macroimpacts into the model of gas-free accumulation gives a better explanation of peculiarities in the system of the terrestrial planets.

ACCUMULATION OF MERCURY FROM PLANETESIMALS

G. W. WETHERILL
Department of Terrestrial Magnetism

In order to understand better the accumulation history of Mercury-size bodies, 19 new Monte Carlo simulations of terrestrial planet growth have been calculated. Three cases are presented, involving different assumptions regarding the initial state of the final stage of planetary accumulation and the degree and ease with which planets can be collisionally disrupted. It is found that the same conditions that lead to Mars-size giant impacts on Earth and Venus imply a more catastrophic fragmentation history for Mercury-size bodies and the fragments from which they accumulated. In accordance with these results, one can speculate that the large iron core of Mercury may be a consequence of a giant impact that removed the silicate mantle from a previously differentiated body. Less extreme fragmentation events may also contribute to chemical fractionation processes for which bodies in this mass range may be especially susceptible. It is also found that the terrestrial planets, including Mercury, will accumulate material originating over the entire terrestrial planet range of heliocentric distances. Although there is significant correlation between the final heliocentric distance of the planets and that of the planetesimals from which they were formed, this tends to reduce the relative importance of more primordial chemical fractionation processes.

I. INTRODUCTION

It is not hard to remember from our childhood the best part of building with blocks a mighty tower was knocking it all down when it was done. As a nursery rhyme goes: Brick upon brick, Stick upon stick, This is the way Our castle will grow; It shall soon rise Up to the skies, Then Crumble, Tumble All with one blow! This chapter examines the way Nature may have played this same game during the formation of the planets, particularly the smaller bodies, such as Mercury. Numerical simulation of planetary accumulation plays an important part in this discussion. But it should be emphasized at the outset that the goal of this work is not the arrogant and hopeless task of trying to build within a computer a detailed model of the early solar system. Rather, an attempt is being made to evaluate in a disciplined way the conceptual framework for considering the simultaneous formation of several planets; and, if it turns out to be appropriate, replace concepts such as individual feeding zones for each planet, strict variation of chemical composition with heliocentric distance, and accumulation of planets from much smaller bodies, with the newer concepts of widespread radial mixing, giant impacts, and major mass loss from even the largest bodies. In accordance with the subject of this book, the origin of Mercury will be discussed in this context.

Probably the most puzzling and important feature of the planet Mercury is its high uncompressed density, indicating an Fe/Si ratio of about twice that of the other terrestrial planets. There are a number of processes by which Fe/Si fractionation may occur. These include differential condensation from a hot solar nebula (Lewis 1972), aerodynamic sorting of small bodies at the outset of planetesimal accumulation (Weidenschilling 1978) and volatilization of silicates from a small planet at high temperatures (Cameron 1985a; Fegley and Cameron 1987; see Cameron's Chapter). This chapter considers a possible additional mechanism for Fe/Si fractionation during the formation of Mercury-size bodies, namely catastrophic fragmentation following high-velocity collisions with bodies of only slightly (0.1 to 0.2 times) smaller mass. A giant impact of this kind may be responsible for the formation of the Earth's Moon (Hartmann and Davis 1975; Cameron and Ward 1976), including the associated Fe/Si fractionation.

A result of both 2-dimensional (Cox and Lewis 1980; Lecar and Aarseth 1986; Wetherill 1980a) and 3-dimensional (Wetherill 1980a,b,1985a,1986) studies of the later stages of planet formation, is that the largest impacts on a growing terrestrial planet are often in this range of 0.1 to 0.2 times the mass of the larger colliding body. In the earlier of these calculations, in which only 100 initial bodies were included, it appeared possible that this result was primarily an artifact of the necessarily large initial mass of the initial bodies (1.2×10^{26} g). Extension of these calculations to 500 initial bodies led to an identical result, however. In this case, considerable accumulation (coagulation of 25 initial bodies) is required before a Mars-size impactor becomes available. Starting assemblages for simulation of the final stages of accumulation approximately overlap particle-in-a-box investigations of the initial stages of accumulation (Wetherill and Stewart 1986). It therefore seems likely that calculations of the kind reported in this chapter are capable of elucidating the general features of the growth of a terrestrial planetary swarm, assuming that it began with most of its mass in an assemblage of small (1 to 10 km diameter) planetesimals spanning a limited mass range (a factor of ~50) distributed throughout the terrestrial planet region. Because of the simplistic characteristics of these models, as well as the actual stochastic nature of the accumulation process, it would be foolish to suggest that modeling of this kind could predict the exact number, positions, and masses of the observed terrestrial planets. Rather, as mentioned earlier, the principal value of this work is to examine the relevance of older qualitative concepts regarding the simultaneous growth of several planets, and, when appropriate, suggest new concepts. For example, the conventional idea that each individual planet had its own unique feeding zone (see, e.g., Schmidt 1957) now seems unlikely. Instead, accumulation is likely to be associated with radial migration of bodies over a range comparable to the entire range of heliocentric distances observed in the final, observed, terrestrial planet system.

II. METHOD OF CALCULATION

In this chapter, the method of calculation follows that described in earlier papers (Wetherill 1985a, 1986). In the new calculations reported here, attention was directed to identification of those events in which impacts were catastrophic in the sense that more than half the energy required for collisional disruption, or more than half the angular momentum required for rotational instability was involved in the impact. Furthermore, the discussion is directed toward those cases where, as a result of the stochastic nature of the accumulation process, a small planet is found at the inner edge of the final swarm. The evolution of this body is then examined in more detail than in earlier work, in the hope that this simulated Mercury will provide insights into the possible evolutionary history of that particular planet.

The criteria for catastrophic events were obtained in the following way. The impact energy available for breakup, in the center of mass frame is

$$U_{\text{imp}} = \frac{K_a}{2} (V_{\text{esc}}^2 + V_{\text{rel}}^2) \frac{M_1 M_2}{(M_1 + M_2)} \qquad (1)$$

where M_1 and M_2 are the masses of the two colliding bodies, V_{rel} is their relative velocity at infinity, V_{esc} is their mutual escape velocity $\sqrt{[2G(M_1 + M_2)/(R_1 + R_2)]}$ when their surfaces are in contact, and K_a is the fraction of the impact energy assumed to be available for breakup, the remainder being dissipated as heat.

The total energy required to break up a body with the combined mass of the projectile and target is

$$U_{\text{break}} = U_{\text{grav}} + Q$$
$$= 0.3 V_{\text{esc}}^2 (M_1 + M_2)^{2/3} (M_1^{1/3} + M_2^{1/3}) + Q(M_1 + M_2) \qquad (2)$$

where U_{grav} is the gravitational binding energy, and Q (taken to be 3×10^{10} erg g^{-1}) is the energy assumed to be converted into heat during disruption, primarily by melting. Breakup occurs when $U_{imp} > U_{break}$. In all collisions not leading to breakup, it is assumed that a fraction $K_b/(1 - K_a)$ of U_{imp} went into heating the final body, up to a maximum of 3×10^{10} erg g^{-1}. When the accumulated heat reached this value, it was designated as melted. If the target was melted, the criterion for breakup is simply

$$U_{imp} > U_{grav}. \tag{3}$$

In those cases where $0.5\ U_{grav} < U_{imp} < U_{grav}$, the target was designated as cracked, indicating that significant, but not total, mass loss might be expected.

The angular momentum L_{imp} supplied by the impact was calculated from

$$L_{imp} = \frac{M_1 M_2}{(M_1 + M_2)}\ V_{rel}\ d \tag{4}$$

where d is the unperturbed distance of closest approach of the two bodies. The angular momentum at the threshold of rotational instability of the combined body is

$$L_{rot} = \frac{\sqrt{2}}{5}\ V_{esc}(M_1 + M_2)(R_1 + R_2)\ \sqrt{\frac{(M_1 + M_2)^{1/3}}{M_1^{1/3} + M_2^{1/3}}} \tag{5}$$

when rotational instability is considered to exist if the rotational angular velocity of the body exceeds the Keplerian planetocentric orbital angular velocity at the surface of the planet. As a point of reference, it may be pointed out that the angular momentum of the Earth-Moon system is about 1/4 that required to satisfy this criterion for rotational instability. In the present calculations, occurrence of rotational instability is simply noted, but otherwise no account of it is taken. Although the physical effects associated with impacts of this magnitude are obviously very complex, it seems likely that the processes by which angular momentum stability is restored to such a system are highly relevant to the physical and chemical evolution of the planet, and its possibly temporary satellites. When $L_{imp} > L_{rot}$, it was noted that the impact was sufficient to cause rotational instability. Those cases where $L_{imp} > 0.5$ L_{rot} were also noted.

In principle, catastrophic breakup can occur on bodies of any size. When the relative velocities and sizes of projectiles and target actually found in these calculations are used, however, it is noted that the larger Earth- or Venus-size bodies are much less vulnerable to breakup than smaller Mars- or Mercury-size bodies. This may be seen from Fig. 1, taken from an accumulation calculation published earlier (Wetherill 1985a). The larger bodies (black squares)

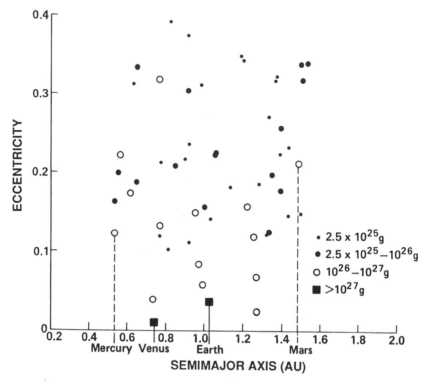

Fig. 1. Snapshot of simulated accumulation of five hundred 2.5×10^{25} g bodies after 9 Myr. The largest ($> 10^{27}$ g) bodies have eccentricities < 0.05, whereas smaller bodies have, on average, higher eccentricities, reaching values as high as 0.3 to 0.4. (Figure taken from Wetherill 1985a.)

have eccentricities of < 0.05, and their largest collisions occur with intermediate mass bodies (open circles) with eccentricities of ~ 0.15 leading to relative velocities ~ 5 km s^{-1}. The largest impacts of the intermediate mass bodies will be with smaller bodies (large solid circles), some of which have eccentricities ~ 0.35, leading to relative velocities of ~ 13 km s^{-1}. The implications of this are given in Table I, where these relative velocities are expressed in terms of a partially grown "Earth" with an escape velocity of 10 km s^{-1} and collisions with a ratio of projectile-to-target mass of 0.2 are considered. Assuming both the target bodies are melted, impacts with "Earth" will at most provide an impact energy of only 1/5 that required for breakup, whereas impacts with energies exceeding that required for breakup can occur for a body smaller than "Earth" by a factor of 10. Thus, it may be expected that the consequence of major impacts on the Earth are likely to be limited to events comparable to the formation of the Moon, whereas those on Mercury can be far more catastrophic.

TABLE I
Impacts onto Targets[a]

Target Mass	0.8 Earth Mass	0.08 Earth Mass
$(V_{esc})_{target}$	10 km s^{-1}	4.6 km s^{-1}
V_{rel}	5 km s^{-1}	13 km s^{-1}
V_{rel}/V_{esc} (mutual)	0.575	3.25
U_{imp}/U_{grav}	0.21	1.79

[a]Projectile mass/target mass = 0.2 and $U_{imp}/U_{grav} = 0.153 \, (1 + V_{rel}^2/V_{esc}^2)$ from Eqs. (1) and (2) in the text.

It is also possible that bodies of the size of Mercury may experience a comparable net loss of mass, both to escape or planetocentric orbits, as a consequence of subcatastrophic cratering impacts, e.g., cratering by ~100 km planetesimals impacting at velocities at ~10 to 15 km s^{-1} (see Chapter by Vityazev et al.). These phenomena have not been included in the calculations, but deserve future study.

Three new groups of accumulation simulations were carried out:

A. 500 initial bodies of mass equal to 2.52 × 10^{25} g (~1/3 lunar mass) distributed randomly with equal mass per unit area between 0.6 and 1.2 AU. Initial eccentricities had random values between zero and 0.02, initial inclinations between zero and 0.01. Prior to melting, it was assumed that 25% of the impacting energy was available for internal heating, the remainder available for disruption. After melting occurred, all of the impacting energy was available for overcoming the gravitational binding energy and disrupting the body. When the criteria defined by Eqs. (1) and (2) for breakup or rotational instability were met, this fact was noted, but no loss of material was assumed, and the body was allowed to continue its growth and orbital evolution as if nothing had happened.

B. 500 bodies with masses, semimajor axes, eccentricities, and inclinations chosen as in group A. When the criterion for breakup as defined for the previous group, was met, the combined body was assumed to break into four fragments of equal mass, and the independent subsequent evolution of each fragment was followed. In one case, it was assumed that 75% of the impacting energy was available for internal heating, rather than the usual 25%. This had no obvious effect on the final distribution of bodies.

C. 151 bodies distributed in mass and semimajor axis in accordance with a possible outcome of an initial stage of accumulation in which local runaway growth occurred. It is found (Wetherill and Stewart 1986) that such early runaway growth may be expected with considerable confidence if the primordial swarm of 1 to 10 km diameter bodies contained a single seed body with a mass at least ~50 times greater than the mid-point mass of the swarm m_p, defined by

$$\int_0^{m_p} \frac{dn(m)}{dm}\, m\,dm \; = \; 0.5 M_{\text{tot}} \tag{6}$$

where $dn(m)$ is the number of bodies in an interval of width dm, and M_{tot} is the total mass of the swarm (1.26×10^{28} g). Inclusion of other physically realistic phenomena (Watherill and Stewart 1987) are found to facilitate runaways even in the absence of a seed. It was assumed that the swarm was initially confined primarily to the interval between 0.7 and 1.1 AU. Rather than cutting off the swarm abruptly at these distances, a tail at both ends of the mass distribution was assumed (Fig. 2).

It was assumed that the local runaway ended when the largest body in each annular zone became isolated from its nearest neighbors by 4 Hill sphere radii, in accordance with earlier results (Birn 1973; Wetherill and Cox 1984,1985) for the range of influence of bodies in nearly circular orbits. This implies a mass distribution that varies with semimajor axis a and surface density $\sigma = 20$ g cm^{-2} in accordance with

$$M(a) \; = \; (8\pi\sigma)^{3/2}\, a^3/(3 M_\odot)^{1/2} \tag{7}$$

and semimajor axis spacing Δa,

$$\Delta a \; = \; 4[M(a)/3 M_\odot]^{1/3} a \tag{8}$$

where $M_\odot$ is the solar mass.

In the range of heliocentric distance between 0.7 and 1.1 AU, in which most of the mass is concentrated, strict observance of this procedure leads to 30 bodies of mass ranging from 6.4×10^{26} g at 1.094 AU to 1.7×10^{26} g at 0.702 AU. The tails, where the original surface density was assumed to be less, contain smaller and more closely spaced bodies, including some as small as 4×10^{23} g.

It was found, however, that strict observance of this mass distribution was not useful to our purpose here, because it did not lead to bodies of mass and position analogous to the observed planet Mercury. Although this matter has not been studied in detail, it appears that the root of the problem lies in the concentration of the largest initial bodies in the region between 0.9 and 1.1 AU. The mass distribution was therefore modified to be more uniform by adding to each body a mass

$$M_u \; = \; 3 \times 10^{27}\, (0.9\text{AU} - a)\, R_x \tag{9}$$

where R_x is a random number between 0 and 1.

The initial masses of bodies actually used in the range 0.7 to 1.1 AU are shown in Fig. 2 by the vertical bars. It would be premature to say whether this result implies that an initial mass distribution of the kind described by Eqs. (7) and (8) could or would not actually lead to the present distribution of planets.

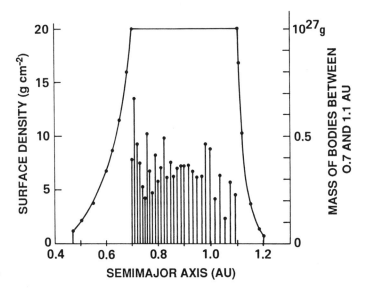

Fig. 2. Mass distribution in the terrestrial planet region for group C of 151 bodies distributed in mass and semimajor axis in accordance with a possible outcome of early runaway growth. The solid curve represents the surface density originally assumed (left-hand scale). As discussed in the text, this uniform surface density was modified into a more random distribution between 0.7 and 1.1 AU. Each vertical bar represents an individual body in this mass range, of mass given by the right-hand scale.

As far as is known, this initial distribution of 151 bodies represents a dynamically stable system, but quite unlike the observed system of terrestrial planets. In order to explore its potential for a more realistic outcome, it was assumed that the eccentricities of the circular orbits were increased (e.g., by secular perturbations following the formation of Jupiter, or by more local resonances) to values sufficiently high to permit their orbits barely to overlap, after which they were allowed to evolve as a result of their mutual gravitational perturbations.

For this group of calculations, it was assumed that, prior to melting, 75% of the impacting energy in the center-of-mass reference frame was available for heating. This had the effect of facilitating melting, which in turn facilitated breakup, because after melting, the full impact energy was available for breakup. As a consequence, the full development of an accretion-fragmentation mass distribution was found most clearly in this group of calculations.

III. ·RESULTS OF CALCULATIONS

The calculations reported in this chapter, together with those in several earlier papers (Wetherill 1980a,b,1985a,1986) represent to my knowledge the only 3-dimensional "simulations" of the orbital evolution of planetesimals during their growth into the planets. The number of bodies used and variety of phenomena investigated (e.g., range of initial conditions, fragmentation, tidal disruption, low-velocity enhancement of the gravitational cross section) exceeds considerably those included in 2-dimensional calculations (Cox and Lewis 1980; Lecar and Aarseth 1986). Nevertheless, the results presented here should be considered preliminary, not in the sense that the author is very likely to contribute much more in the foreseeable future, but in the sense that these results are still very early and must wait further development of quantitative and internally consistent models for the concurrent formation of the Sun, the giant planets, the terrestrial planets, and the smaller bodies of the solar system. Although they will play a role, simple things like future availability of super super computers are not likely to suddenly make everything clear. It seems inevitable that as future workers labor to understand how all these events fit together and reconcile them with present and future observations, serious problems will arise, some things will have to give, and despite my best intentions, it could well be the things that are said in this chapter.

Assuming that some of this theory survives destruction of the sort described in the opening words of this chapter, one could speculate as to what this may be. This "surviving remnant" could be identification of the linkage that connects various phenomena with one another, and thereby provides tests of the internal consistency of models of solar system formation, and facilitates comparison of theory with observation.

In the present work it is possible to identify the essential factors that connect a number of phenomena—giant impacts, catastrophic disruption, wide-spread radial migration, impact origin of satellites, tendency toward chemical homogeneity, implantation of terrestrial material into the asteroid belt. This is the presence in the terrestrial planet region at the outset, i.e., within $\sim 10^6$ yr of the formation of the Sun of a number of large planetesimals or planetary embryos ($\gtrsim 10^{26}$ g) that exceed the number of observed present planets. If they were once there, somehow nature has reduced the number of bodies to four. The only known natural process capable of accomplishing this reduction is collision of these large bodies with one another. These physical collisions must be accompanied by more frequent close encounters and these will increase the relative velocities of the bodies to the vicinity of the escape velocity of the larger bodies. The corresponding eccentricities will span a major portion of the present terrestrial region.

Our admittedly limited understanding of the earlier stages of planetesimal growth implies that an intermediate stage that includes excess embryos or "failed planets" should be considered at least to be plausible, even if runaway growth occurs during the initial stage of planet formation. No quantitative theory leading to an alternative, i.e. only four embryos, has been developed. If this proves to be possible, it is by no means clear that the alternative system of bodies can be expected to evolve into an assemblage of terrestrial planets similar to those observed today, e.g., in number, size, chemical composition, absence of stranded small bodies, orbital eccentricities, inclinations and spin states, existence and properties of the Moon, etc. On the other hand, it is entirely conceivable that by a combination of creative ingenuity and serious theoretical effort, (not just assertion of "my scenario") one or more such alternatives will be found. Ultimately, selection between such alternatives will depend on how well they can be accommodated into a future general theory of solar system formation that agrees with observation of our solar system and others as well. Only time will tell.

The three sets of calculations reported above share this property of having excess embryos. They differ somewhat in the choice of parameters controlling when breakup and rotational instability will occur. These differences lead to some variation in the frequency of occurrence of catastrophes of various kinds (Table II). Despite these differences, catastrophic events of both kinds are predicted under this range of assumptions.

The final outcome of six accumulation calculations in accordance with the starting conditions of group A are shown in Fig. 3. In every case the initial swarm evolves into four or five final planets, with only a small loss of mass (2 to 5%) by ejection from the system. In more detail, the results vary considerably. It is not known to what extent this variability represents a defect of the modeling or, on the other hand, outlines the range of probable terrestrial planet systems, of which we have as yet observed only one example. The innermost planet tends to be larger, and at greater semimajor axis than the observed planet Mercury. It is possible that this outcome could be changed by "massaging" the initial distribution and the nature of the fragmentation process, but this will not be attempted here.

Of the six calculations, case 1 is most similar to the observed distribution of terrestrial planets. In the subsequent discussion, details of the evolution of its "Mercury" will be described, not in the belief that this outcome can be claimed to be typical, but because it is the one most likely to provide insight into events associated with our solar system.

The growth of the innermost planet of case 1 is depicted in Fig. 4. Over a period of 26 Myr, 28 initial bodies of about 1/3 lunar mass coagulated to form a

TABLE II
Average Number of Catastrophic Events per Simulation
Between Bodies with Combined Mass Between
1.0×10^{26} g and 1.0×10^{27} g

Group	Breakup	Crack	Rotational Instability	1/2 Rotational Instability
A	2.0	5.7	7.2	21.8
B	2.1	4.4	5.6	20.2
C	4.2	1.8	3.0	7.8

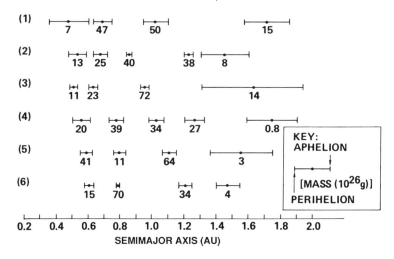

Fig. 3. Final outcome of 6 accumulation calculations in group A (five hundred 2×10^{25} g bodies, no explicit fragmentation). The semimajor axes of the final planets are indicated by points; the line through each point extends from the perihelion to the aphelion of the planet. The numbers under each point indicate the final mass of the body in units of 10^{26} g. Case 1 is discussed in detail in the text.

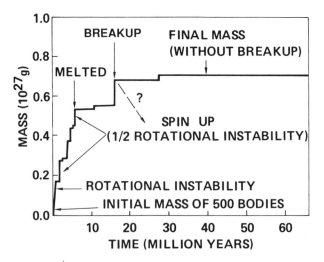

Fig. 4. Growth of the innermost inner planet "Mercury" for case 1 of group A (see Fig. 3). The growth is punctuated by a number of giant impacts, one of which is sufficiently energetic to cause major disruption of the planet.

final body of mass 7.1×10^{26} g, about twice the mass of the present planet Mercury. If during the course of the growth, one identifies the largest member of each colliding pair with the final planet, the growth began with the collision of two bodies each with the uniform initial mass (2.52×10^{25} g), and with semimajor axes of 0.63 and 0.68 AU within $< 6 \times 10^4$ yr. The mass increased at an approximately uniform rate for the next 6 Myr. This growth was punctuated by three events that imparted large amounts of angular momentum to the system, 110%, 67% and 70% of that required to result in rotational instability. Although the resulting systems remained gravitationally bound, it is likely that some mass shedding would accompany these impacts (Boss 1986), possibly leading to satellite systems, and to some mass loss from the system. The last of these impacts provided the remaining internal heat required to melt the body. The rate of growth then decreased, being primarily represented by a very large impact at 16 Myr when the semimajor axis was 1.4 AU, for which the impact energy of 3×10^{11} erg g^{-1} exceeded the gravitational binding energy by a factor of about 3. The mass of the impactor was 0.19 that of the final planet and the relative impact velocity was 20 km s^{-1}.

It seems inevitable that an impact of this kind would lead to major fragmentation of the body. More detailed discussion of this matter will require further work of the kind reported by Benz et al. (1986a,b). In their work (see Chapter by Cameron), computer simulation of an impact of comparable energy resulted in total loss of a silicate mantle with preservation of a molten iron core. One may therefore hypothesize that an event of this kind could result in a planet with a large iron core similar to that of Mercury. In the case under discussion, most of the smaller-size silicate material would be swept up by "Earth" and "Venus", with which the small body "Mercury" could offer little competition.

The semimajor axis of simulated "Mercury" varied considerably during its growth (Fig. 5). Over a period of 16 Myr, the initial semimajor axis of 0.6 AU grew in a rather erratic way up to 1.4 AU. Following the major impact at that time, during the next 18 Myr, the semimajor axis was perturbed into its final value of 0.48 AU. This particular history of radial migration is not an essential feature of the formation of planets in the position of Mercury, but neither is it rare (see, e.g., Wetherill 1986,Fig. 3). The more general feature is that bodies with masses $\lesssim 10^{27}$ g are usually found to experience radial excursions comparable to the dimensions of the swarm. Most of these collide with $\gtrsim 10^{27}$ g bodies ("Earth" and "Venus") while crossing the orbits of those bodies and constitute the giant impacts of the kind proposed to be responsible for the formation of the Moon. Those that escape from the domain of these large planets constitute smaller terrestrial planets that may be identifiable with Mercury and Mars.

Too much focus on the radial excursions of "Mercury" should not obscure the fact that the last bodies to collide with that planet deserve to be designated "Mercury" just as much or as little as the first two. As accumulat-

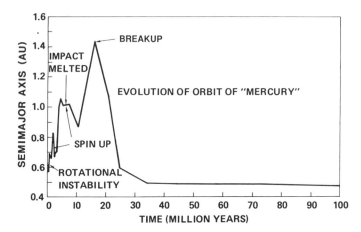

Fig. 5. Evolution of the semimajor axis of "Mercury" for case 1 of group A. During the course of its growth, the heliocentric distance of the body spans the entire terrestrial planet region.

ing bodies migrate through the terrestrial planet region, they accumulate bodies at various heliocentric distances, most of which have also experienced a history of radial migration. The end result of this process is illustrated in Fig. 6, where the "provenance" of the terrestrial planets is given. All of the four final planets have collected some material from all portions of the original swarm. About half of "Mercury" consists of material originally formed beyond 0.8 AU. For this reason, preservation of distinct chemical or isotopic signatures characteristic of a specific heliocentric distance would not be expected if planets formed in the manner considered here. On the other hand, Fig. 6 exhibits clear trends; "Mars" and "Earth" accumulate more material from the larger heliocentric distances than do "Venus" and "Mercury". Because of the stochastic nature of the accumulation process, histograms analogous to that of Fig. 6 vary considerably from one calculation to another. They characteristically exhibit widespread mixing. Nevertheless, preservation of correlation between final heliocentric distance and provenance is a common feature. This result should not be thought of as a surprising outcome of numerical calculations; a very similar result was found simply from consideration of conservation of energy and angular momentum (Wetherill 1978).

Discussion of the origin of Mars is peripheral to the scope of this chapter. There are considerations peculiar to that planet that would have to be included in an adequate discussion of Mars. These would include the role of resonances analogous to the v_6 secular resonance, of major importance to the dynamics of this region of the solar system at present (Wetherill 1979,1985b), as well as the matter of the relative ease with which a small planet in Mars' position would be contaminated by material removed from the asteroidal region. Nevertheless, it should be mentioned that in the framework of planetary ac-

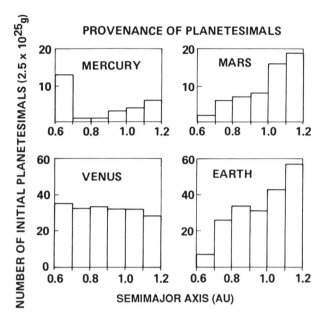

Fig. 6. Distribution of initial semimajor axes of the planetesimals that formed the final planets for case 1, group A.

cumulation models of the kind considered here, Mars-size bodies are also vulnerable to catastrophic impacts.

In the particular accumulation under discussion, the 1.5×10^{27} g body in the position of "Mars" experienced two impacts exceeding half that required for breakup and rotational instability. The second of these occurred after the body had been melted, and presumably formed a core. There are some factors, barely studied as yet, that seem to introduce an asymmetry between the inner and outer edges of the accumulation region, with regard to catastrophic events. For example, there is a tendency for residual bodies to be trapped in the deep solar potential well near the position of Mercury, until they strike that planet. In contrast, residual bodies crossing the orbit of Mars are more loosely bound and are likely to be ejected from the solar system, following a major perturbation by Earth. These appear to be illustrated by the present work. Whether these, together with a more refined treatment of the origin of Mars, will significantly reduce the probability of Mars' undergoing an impact sufficiently large to have observable consequences requires further study.

In the set of accumulation calculations made using the starting conditions of group A following a disruptive encounter, no explicit loss of mass from the merged bodies was assumed. Readers are thus free to imagine for themselves the outcome of the impact. For example, a result, such as that found by Benz et al. (1986*b*), of extensive fragmentation and loss of the mantle but retention of an iron core, would leave a 3×10^{27} g body, resembling Mercury, after the impact. The effect of the orbital perturbations of this body by the larger "Earth" and "Venus" are not strongly dependent upon the mass of this body. It is therefore entirely plausible to suppose that this fragment may have experienced the same migration into 0.48 AU shown in Fig. 5.

Probably the most serious objection to this failure of explicitly recognizing the effects of fragmentation is the possibility that fragmentation may also eliminate the small (~Moon-size) bodies required to fragment $\sim 7 \times 10^{26}$ g "pre-Mercurys". This question is addressed in a simplified way in the calculations made corresponding to group B of Sec. II.

Here it is assumed that when the conditions for breakup are met, the body fragments into four equal-size pieces moving in nearly the same orbit. Of course, it is not claimed that this is a physically realistic mode of breakup. It would be more realistic to break the bodies into some kind of a power-law size spectrum, or in the case of superfragmentation of the sort studied by Benz et al., a single residual body of higher density. A simplified breakup model was chosen in order to avoid the suggestion that the model was intended to be physically correct, and rather simply to introduce into the calculation an accretion-fragmentation hierarchy at the lower end of the mass distribution, without introducing any more bodies than necessary. Even with four fragments, their vulnerability to further fragmentation is large enough to set into motion the generation of enough smaller fragments to increase problems of computational time and memory size considerably. A significantly larger number of fragments would lead to a divergence in the number of fragments.

In order to prevent the number of residual bodies from exceeding 500, collision fragments smaller than 5×10^{23} g were either removed from the swarm or prevented from further fragmentation. No difference was found between these two alternative ways of dealing with this problem. Such bodies were always melted. This, together with their low gravitational binding energy and their usual high-velocity orbits precludes further growth. Their fate is to be ground down to fine dust unless sweep-up by planets occurs first. The amount of mass involved (~2% of the total mass of the swarm) is less than that lost by ejection into escape orbits, and has a negligible effect on the overall growth of the planets.

The approximate validity of this simplified approach of breaking into four fragments is related to the insensitivity of a fragmentation hierarchy to the details of the individual fragmentation process, as shown by Dohnanyi (1969). In his work, the resulting steady-state fragmentation integral radius spectrum tends toward a power law with an exponent of -2.5. This exponent is not a consequence of the way rocks break, but instead is a geometric consequence of area varying with r^2 and volume with r^3, which in the absence of a significant effect of gravity leads to a steady-state self-similar radius spectrum with fractal dimension 2.5.

The results of the 9 accumulation calculations are shown in Fig. 7. In calculation case 1 (group B), a small planet in the position of Mercury was found and this will be discussed in more detail. The growth and radial migration of "Mercury" are shown in Fig. 8 and 9, respectively.

Although the accumulation history of this body is not as dramatic as that discussed for case 1 of group A, it is far from tranquil. In this case, no total disruption occurred. On three occasions, however, an impact energy exceeding half that required for breakup occurred (82%, 55%, 59%). For the first two of these impacts, the increase in angular momentum was more than half that required for rotational instability. The differences between these two accumulation histories is not principally a consequence of explicit introduction of fragmentation in group B, but simply the result of the essentially stochastic, even chaotic, nature of planetary accumulation of the kind under consideration.

The initial semimajor axis of "Mercury" was 0.71 AU. During the next two million years, during which about half its growth occurred, the semimajor axis migrated in an erratic manner between 0.65 and 0.85 AU. During this time it accumulated smaller bodies present in that region. The original posi-

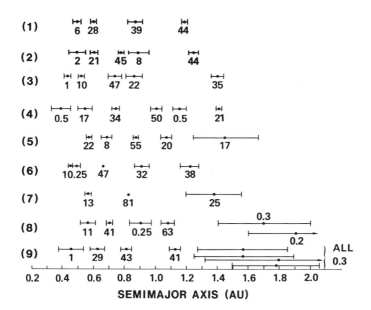

Fig. 7. Final outcome of the 9 accumulation calculations of group B. Following an impact energetic enough to fragment the body, it is assumed to break into four equal mass fragments, all of which are melted. When fragmentation is permitted, there is a tendency for a number of very small bodies sometimes to remain in orbits with semimajor axes > 1.5 AU (e.g., case 9) after times > 10⁹ yr. This effect may not be real, because of the neglect of secular resonance effects that affect bodies in orbits of this kind. See Fig. 3 for explanation of points, bars and numbers.

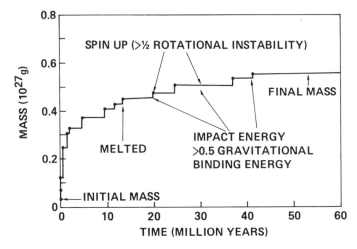

Fig. 8. Growth of "Mercury" for case 1 of group B. In this case no breakup actually occurred. At three times, however, impacts with energy exceeding 1/2 the gravitational binding energy occurred, and major mass loss may be expected.

tion of these smaller bodies tended to be concentrated in this range of heliocentric distance, but also included a number of bodies from outside this region, including some with original positions between 1.0 and 1.2 AU. By the time the semimajor axis migrated into ~0.5 AU, the original swarm was quite well mixed and impactors originated from a wide range of semimajor axes. The impacting bodies that caused the very energetic impacts at ~30 and ~40 Myr, were themselves fragments of disrupted bodies that had formed and grew beyond 1.0 AU. The large impact at ~20 Myr was caused by an unfrag-

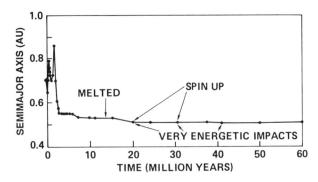

Fig. 9. Evolution of semimajor axis of "Mercury" for case 1 of group B. In this case less, but still considerable, radial migration occurred. The difference between this case and that of Fig. 5 is not primarily a consequence of the explicit inclusion of fragmentation, but rather represents stochastic fluctuations in these accumulation histories.

mented body that formed at 0.61 AU, but spent almost all of its life beyond 1.0 AU. Again, although some memory of possible primordial compositional gradients in the solar nebula may be expected to be reflected as trends in the composition of the final planets, the wide-range mixing precludes preservation of pure compositions characteristic of original heliocentric position. Any chemical or isotopic components formed at a specific heliocentric distance may be expected to contaminate all of the final planets.

The results of four accumulation calculations for case C, possibly representing the outcome of a runaway initial stage, with explicit breakup, is shown in Fig. 10. Again, small planets are found at the inner edge of the swarm. Detailed analysis of the first calculation, reveals, however, a mode of formation of these bodies that is qualitatively different from that found in the other calculations discussed above. In the foregoing calculations, a single body of $\sim 5 \times 10^{26}$ g grew from 2.5×10^{25} g initial bodies, and then experienced impacts sufficiently energetic to result in major mass loss. In contrast, the small body in calculation number 1 of Fig. 10, was never larger than its final mass of 2.0×10^{26} g.

Nevertheless, its growth involved fragmentation in a major way. It was formed by the accumulation of 41 bodies, all of which were previously melted collision fragments. In this case the fragmentation accretion hierarchy is characterized by a "neutral growth mass region" in the vicinity of the mass of Mercury. Bodies significantly larger are unlikely to be fragmented. They simply grow until they are swept up by a larger body or become a final planet. Significantly smaller bodies are fragmented to form a quasi-steady state fragmentation hierarchy. These bodies are either captured by bodies at least as

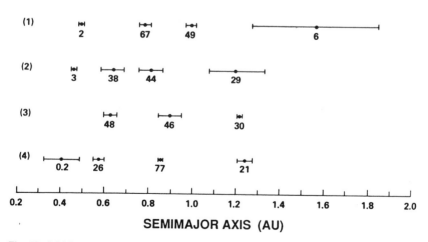

Fig. 10. Initial swarm with a population of small bodies inside 0.6 AU (3% of total mass). The figure shows the final state of accumulation calculations for group C, in which fragmentation was explicitly introduced. The initial semimajor axis and mass distribution is shown in Fig. 2. See Fig. 3 for explanation of points, bars and numbers.

large as those in the neutral growth mass region, or are fragmented further, ultimately to fine dust, removable by solar emanations. In the region of neutral growth, the effects of growth and fragmentation are comparable, and whether or not the one or the other ultimately is dominant is determined stochastically.

The complex growth history of this body is illustrated by Fig. 11. In this diagram, time increases from left to right, but an exact time scale is not given, because this leads to drafting problems that obscure the central concept. The final growth of "Mercury" began with the formation of an embryo with a mass of 3×10^{25} g at 23 Myr by collisional disruption of two bodies (designated 9B and 12 on Fig. 11). This body then grew with little change in heliocentric distance, by accumulation of 39 smaller bodies until its final mass of 2.0×10^{26} g was reached at 74 Myr. Although none of these impacts were sufficiently energetic to disrupt this melted body, in three cases the energy of impact exceeded half the gravitational binding energy. All of these impactors were themselves melted fragments of collisions between other bodies.

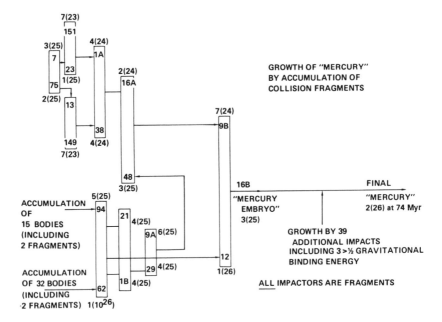

Fig. 11. Complex growth history of case 1 of group C. Each rectangular box represents a disruptive collision. The numbers within the boxes are the internal identification number of the bodies in the computer output. Adjacent to these numbers, outside the boxes, the mass (in grams) of the body is indicated at the time of impact. For example, 7(23) means 7×10^{23} g. The collision fragments that belong to the "family tree" of final "Mercury" are identified by the arrows between the boxes. Finally, after 23 Myr a 3×10^{25} g "Mercury" embryo was formed (body 16B) that grew without further total fragmentation to 2×10^{26} g after 74 Myr.

This final stage of accumulation was preceded by an intermediate stage characterized by repeated fragmentation and collision between prior bodies. Altogether 57 bodies, from a range of heliocentric distances collided, grew, fragmented, and recombined, sometimes with one another. The final product of this stage of growth, the Mercury embryo, contained 15% of the mass of the final planet. Detailed tracing of the "family tree" of the 39 previously fragmented impactors was not attempted. A rough look at their orbital evolution shows that most of this material was probably derived from the tail of the initial distribution (Fig. 1) sunward of 0.7 AU. Nevertheless, a comparable mass was delivered to this region from the principal part of the mass distribution between 0.7 and 1.1 AU, primarily as collision fragments. This material became well mixed in the collision hierarchy of bodies within 0.6 AU and must have contributed significantly to the final "Mercury". This mode of formation of "Mercury" is quite different from that found in the calculations described from group A and group B, and does not correspond to the simple picture of a previously melted Mars-size Mercury embryo stripped of its core by a single energetic impact. This process could also lead to metal-silicate fragmentation if the physical properties of the silicate fragments favored their dispersal, following continuing fragmentation, by emanations from an early active Sun.

Although this matter has not been studied in detail, the most likely cause of the difference between the results found for group B and group C is not the different initial size distribution, but the higher fraction of impact energy available for melting in the calculations of group C, as described in Sec. II. Case 2 of group B also assumed 75% of the impacting energy was available for melting, and led to a well-developed accretion-collision hierarchy similar to those of group C. In this case, a body 3.0×10^{26} g grew in the region between 0.95 and 1.3 AU and was struck at 21 Myr by a body of 1/4 its mass, the impact energy being barely (13%) above the energy required for breakup. One of the fragments of mass 1.3×10^{26} g migrated out to a semimajor axis of 1.74 AU at 30 Myr before being perturbed by $> 10^{27}$ g bodies into 0.5 AU. At this distance it continued to grow to its final mass of 1.8×10^{26} g entirely by accumulation of previously melted 10^{23} to 10^{24} g collision fragments that had become trapped in the potential well at ≤ 0.55 AU.

The projectiles responsible for this final bombardment of "Mercury" could be considered as "Vulcanoids" constituting a peculiarly Mercurian population of the sort considered earlier (Wetherill 1975), and could result in the cratering record of Mercury being different from that of the other terrestrial planets (Chapman 1976; Wetherill 1976b; Hartmann 1977; Davis et al. 1986). The population of Vulcanoids found in these simulations differs from those proposed by Davis et al. in that they constitute the large-body end of a fragmentation hierarchy, the smaller mass component of which is therefore less vulnerable to rapid collisional destruction than the small primordial planetesimal population proposed by these authors. For some time, the collisional loss

of small 10 to 100 km basin-forming bodies would be compensated by replacement following collision of large bodies, until the larger bodies were
themselves removed by collision with Mercury or by mutual destruction. It
would be of interest to model this problem quantitatively.

The essential difference between the results of the group A and all but
one of the group B calculations, and those of group C is the ease with which
fragmentation can occur. This, as well as the outcome of these massive fragmentation events, is poorly constrained at present. When fragmentation becomes relatively easy, a full collision hierarchy develops up to the $\sim 10^{26}$ g
mass range, reducing somewhat the number of bodies available for disruption
of 6×10^{26} g bodies. At the same time, a continuously renewed supply of
collision fragments becomes available for accumulation onto bodies in the 1 to
6×10^{26} g range. In this case, accumulation of previously fragmented bodies
in the mass range of 10^{24} to 10^{25} g becomes a major mechanism for growth of
planets with the mass and position of Mercury, and also leads to the formation
of Vulcanoids. When fragmentation is more difficult, any original population
of $\sim 10^{25}$ g objects becomes fully available for destruction of proto-Mercury.
The transition between these two regimes is continuous, and spans the range
of plausible possibilities for the nature of the fragmentation process.

IV. CONCLUSIONS

It was found earlier (Wetherill 1985,1986) that for a range of initial conditions, the final stage of growth of Earth-size planets was accompanied by
giant impacts, equal to or greater than the present mass of Mars. Under these
circumstances, the impacts proposed to explain the origin of the Moon (Cameron and Ward 1976; Hartmann and Davis 1975; Benz et al. 1986a,b; Thompson and Stevenson 1986) arise in a natural way.

In this chapter, it is found that these same, as well as similar initial
conditions, imply that planets of the size of Mercury would be expected to
have experienced even more catastrophic impact events. In some cases, these
are sufficiently energetic to disrupt the planet or to cause it to become rotationally unstable. It is likely that planets of this size and the Moon are in the
vicinity of a regime of neutral growth. Larger bodies, e.g., Venus and Earth,
can grow without major mass loss. Smaller bodies will be fragmented until
they are swept up by a planet-size body or nongravitationally removed from
the solar system in the form of dust by solar emanations.

After reaching this neutral growth region, bodies tend to remain at more
or less a constant mass, as a result of a balance between relatively low-velocity impacts of relatively small bodies (themselves collision fragments), and
more energetic impacts that result in major mass loss. This same impact history will often lead to differentiation and core formation in bodies in the Mercury size range. For this reason, one may speculate that the relatively large iron
core of Mercury may have been the consequence of the loss of its silicate

mantle following a giant impact. In addition, iron-silicate fractionation by a less drastic winnowing effect might result from a hierarchical fragmentation cascade involving differentiated cores and mantles, as a consequence of their differing mechanical properties.

It is also found that under these same conditions, Mercury-size bodies experience wide migration of their semimajor axes during their growth. As a result, they may be expected to accumulate bodies originally formed over the entire terrestrial planet range of heliocentric distances. This will have the same effect of partially smoothing chemical differences associated with primordial fractionation processes.

For these reasons, chemical fractionation processes peculiar to planets (e.g., as opposed to asteroid-size planetesimals) deserve consideration similar to that accorded the chemical processes associated with earlier stages of solar nebula and planetesimal evolution.

Acknowledgments.   I wish to thank K. Ford and A. Boss for their collegial contribution toward the existence and operation of our computing system, and to B. Pandit for its maintenance. The help of J. Dunlap in manuscript preparation is appreciated. This work was supported by a grant from the National Aeronautics and Space Administration and is part of a departmental program also partially supported by NASA.

THE STRANGE DENSITY OF MERCURY: THEORETICAL CONSIDERATIONS

A. G. W. CAMERON
Harvard-Smithsonian Center for Astrophysics

BRUCE FEGLEY, JR.
Massachusetts Institute of Technology

WILLY BENZ
Harvard-Smithsonian Center for Astrophysics
and

WAYNE L. SLATTERY
Los Alamos National Laboratory

Two classes of models which have been advanced to explain the high density of Mercury are reviewed and contrasted. These models invoke either the differing volatilities of iron and silicates or disruptive collisions to fractionate the two phases. We also contrast equilibrium condensation and planetary vaporization models, both of which fall within the first broad class considered. Our review indicates that equilibrium condensation models are unable to account for the observed high density of Mercury without invoking special mechanisms such as unrealistically narrow planetary accretion zones. However, we find that distinctive chemical differences, which are potentially testable by spacecraft experiments, provide means for distinguishing between planetary vaporization and large impact scenarios.

I. INTRODUCTION

Urey (1951,1952) first noted that the planet Mercury must have an iron to silicate ratio much larger than that for any other terrestrial planet because of its anomalously high density. The observed mean density of 5.44 g cm^{-3} (uncompressed ~5.3 g cm^{-3}) implies an iron to silicate mass ratio in the range from 66 : 34 to 70 : 30, about twice that of any of the other terrestrial planets, the Moon or the Eucrite Parent Body (Basaltic Volcanism Study Project 1981). The mean density of the Earth is 5.52 g cm^{-3}, corresponding to an uncompressed density of ~4.45 g cm^{-3} (Lewis 1972).

Three broad classes of models have been suggested to account for the anomalous composition of Mercury. One class of models employs scenarios which invoke differences in the physical properties (e.g., density, ferromagnetism, mechanical strength) of iron and silicates to achieve the required iron/silicate fractionation (Harris and Tozer 1967; Orowan 1969; Weidenschilling 1978; Smith 1979). A second class of models invokes the differing volatilities of iron and silicates to fractionate the two phases. Both equilibrium condensation (Lewis 1972) and vaporization (Bullen 1952; Ringwood 1966; Cameron 1985a) models are included within this class. A third class of models, relatively little explored so far, involves a scenario including major planetary collisions which are suggested to blow off the bulk of the silicate mantle from the original Mercury protoplanet (Smith 1979; Wetherill 1985c; Chapter by Wetherill).

In this chapter we shall ignore the first class of models and discuss the second and third classes. We shall have relatively little to say about the equilibrium condensation model because we find little evidence in its favor; indeed, our principal purpose here is to point out its shortcomings within the second class of models. We shall review at somewhat greater length the recent work by Cameron (1985a) and by Fegley and Cameron (1987) on the planetary vaporization model. We shall also report on the third class of planetary collision calculations carried out at Los Alamos by Benz, et al. (1986a,1987b).

There is an important distinction between the second and third classes of models. In the third class of models, physical fractionations of iron and silicates occur, and we would expect that Mercury would be composed of iron plus essentially chondritic silicates, with a diminution of the total mass; there would be little chemical fractionation, except to the degree that differentiation of the silicates (into crust and mantle) would remain apparent after physical fractionation. In the second class of vaporization models, volatility effects would lead to extensive chemical fractionations, and we would expect that the silicate phase in Mercury would undergo large compositional changes as a function of the extent of the postulated vaporization. This evolved composition would later be diluted by infalling planetesimals composed essentially of chondritic silicates. These compositional differences are an important means of distinguishing between the proposed models and are potentially testable by a future Mercury mission.

II. THE EQUILIBRIUM CONDENSATION MODEL

In this early model (Lewis 1972), a very simple scenario was envisioned. The primitive solar nebula is supposed to be hot, isothermal at a given radius, and the accretion of the grains and the cooling of the nebula are assumed to be slow relative to gas-condensate reaction rates so that complete gas-solid equilibrium is attained. At pressures which conventional wisdom postulates for the nebula near the region of formation of Mercury (10^{-5} to 10^{-3} bar), metallic iron condenses at a slightly higher temperature than do the magnesium silicates. The theory seizes on this distinction between iron and silicates to postulate that this temperature gap can be responsible for the formation of an iron-rich planet.

Several authors have noted that this qualitatively appealing but unrealistically simple model is actually unable to account for the high uncompressed mean density of Mercury without recourse to such special mechanisms as aerodynamic sorting or to an unrealistically narrow accretion zone (Basaltic Volcanism Study Project 1981; Weidenschilling 1978; Lewis and Prinn 1984; Barshay 1981). The basic problem is that the condensation temperatures of Fe metal and of the magnesium silicates ($MgSiO_3$ and Mg_2SiO_4) are really so close together that it is extremely difficult to separate the two phases during accretion. This has stimulated the invention of special mechanisms for removal of silicate grains or for concentration of metal grains.

If condensation occurred under conditions more oxidizing than solar, as suggested by Mo and W depletions in many Ca-, Al-rich inclusions (Fegley and Palme 1985), then this problem is exacerbated because silicate condensation temperatures increase with increases in oxygen fugacity while that of Fe metal remains constant (Bartholomay and Larimer 1982; Palme and Fegley 1987).

On the other hand, if condensation occurred under conditions more reducing than solar which appears required for the enstatite chondrites (Larimer and Bartholomay 1979; Wasson's Chapter), then the silicate condensation temperatures decrease with a decrease in oxygen fugacity while that of Fe metal remains constant (Larimer and Bartholomay 1979). However, there are a number of problems with such a scenario. Large separations in the Fe metal and Mg_2SiO_4 condensation temperatures do not occur until the carbon/oxygen ratio is increased to about unity. No astrophysically reasonable mechanisms for changing the nebular C/O ratio from the solar value of 0.6 to the required value of unity have been proposed.

A further difficulty arises from the condensation of another suite of minerals including elemental carbon (graphite), SiC, CaS, MgS, AlN and TiN instead of ordinary silicates at C/O ratios above one. The large amount of graphite ($\rho = 2.25$ g cm^{-3}) makes the production of a high density, iron-rich planet dependent upon having special mechanisms for separating the graphite from iron in the planet; there would be 2.5 times as much graphite (by mass) as Fe metal in the condensate. This problem might be avoided by assuming that condensation occurs at C/O ratios just below one where graphite will not form (Larimer and Bartholomay 1979). But in this case, the other unusual minerals still form in preference to ordinary silicates. There is no evidence from remote sensing for any such exotic mineralogy on the surface of Mercury, which resembles the lunar surface (Lewis and Prinn 1984, and references therein) with some differences (Chapter by Vilas). An additional point is that the mineral cohenite (Fe$_3$C) replaces Fe at a C/O ratio of unity. Cohenite is 2% less dense than Fe metal, so that it cannot be ruled out solely on this basis. Si-bearing Fe alloys may also form under reducing conditions (Lewis et al. 1979). However, these will lead to decreases in the density of the metal phase (e.g., the density of FeSi is 6.1 gm cm^{-3}, 22% less dense than Fe itself). It should be noted that equilibrium condensation under conditions sufficiently reducing to decrease significantly the silicate condensation temperatures (but not to produce graphite) predicts a volatile-rich Mercury containing about 12% of the solar carbon abundance, 4% of the solar nitrogen abundance, and 100% of the solar sulfur abundance. This is why we do not pursue this equilibrium model further.

III. PLANETARY VAPORIZATION

Cameron (1985*b*) has estimated that, during the evolution of the solar nebula, temperatures at the position of formation of Mercury were in the range of 2500 to 3500 K. Current studies of star formation in dense interstellar molecular clouds limit the time interval involved between the beginning of collapse and the turn-on of stellar luminosity to about 10^5 yr, in order that calculated evolutionary tracks in the HR diagram can fit observations of young clusters (Mercer-Smith et al. 1984). The high temperature region of the solar nebula near the radius of formation of Mercury may thus have been maintained for a time of the order of a few times 10^4 yr. The high temperatures indicated here are not subject to too much uncertainty, because that is the temperature range needed to radiate away the gravitational potential energy released when about one solar mass of gas is moved inside the orbit of Mercury (Cameron 1985*b*).

A protoplanetary form of Mercury can only have formed within the primitive solar nebula early enough to be exposed to these high temperatures in the nebula if gravitational instabilities occurred within the gas phase (forming giant gaseous protoplanets), and if small solid particles (clumps of interstellar grains from which ices have evaporated) were able to settle within the gaseous protoplanets to form very refractory bodies. Such gaseous protoplanets can only form in the nebula very early, before much mass has had a chance to collect at the center of the disk. They are very large and quite cool, and the gas is expected to evaporate quite soon after formation as the temperature in the nebula increases (Cameron et al. 1982). This would leave a condensed protoplanet. Cameron has assumed that the original Mercury protoplanet, if close to a chondritic silicate/iron ratio, would be about 2.25 times the mass of the present Mercury.

The nebula surrounding any such early proto-Mercury would be optically thick (Cameron 1985b). Within the relevant time scale, the blackbody radiative energy input to the planet is much more than sufficient to vaporize the entire mantle (Cameron 1985a). The protoplanet would also be situated in a strong solar nebula wind, composed of both a fixed and a fluctuating component, with a velocity relative to the protoplanet of the order of 1 km s^{-1}. The fixed component arises from the fact that the gas is partially supported in the radial direction by a pressure gradient, whereas the protoplanet is not; thus, there are two different orbital velocities needed for the centripetal support of the gas and the protoplanet. The fluctuating component comes from the turbulent velocity expected to be present within the disk and responsible for the turbulent viscosity that dissipates the disk on such a short time scale. It is this same turbulence that prevents gravitational instabilities from occurring either in the gaseous or the solid components of the nebula during the strongest part of the dissipation. Cameron has suggested that this wind may be able to carry away most of the vaporized mantle of the planet, although the process only marginally satisfies the energy requirements.

The physical scenario is as follows. Inflowing radiant energy heats the surface rocks of the protoplanet and increases the rate of vaporization of the rocky constituents. The resulting vapors accumulate until the surface pressure of the resulting atmosphere reaches the value that corresponds to chemical equilibrium; actually this phase would be quickly achieved so that the condition for chemical equilibrium can be assumed for the entire period. It is this atmosphere which forms the interface between proto-Mercury and the nebular wind. In Cameron's models the surface pressure of the rock vapor products is some two orders of magnitude greater than the pressure in the surrounding solar nebula; thus, solar nebula hydrogen can reach the planetary surface only by forced downward turbulent mixing, which is very inefficient in the face of continued removal of the rock vapor atmosphere by the wind and the continued renewal of the atmosphere by vaporization at its base.

The energy required for removal of the rock vapor atmosphere can reside only in the kinetic energy of the nebula wind. The wind will force turbulent mixing of the outer layers of the rock vapor atmosphere and the solar nebula. The effective scale height of the planetary atmosphere will be increased as the mean molecular weight decreases by such mixing. This process will continue to increase the height of the planetary atmosphere until the turbulent flow of the solar nebula wind past the protoplanet is able to entrain parcels of the atmosphere and carry them away. Cameron has estimated that this interface will exist at an effective planetary atmosphere radius of about three times the radius of the planet, where the kinetic energy of the nebula wind impacting the protoplanet is an order of magnitude greater than the energy required to remove the vaporized mantle to infinity in about 3×10^4 yr. Thus, the operation of this scheme requires that the kinetic energy of the wind be used for this purpose with about 10% efficiency. It is not known whether this is possible.

The temperature range 2500 to 3500 K is above the liquidus temperatures (at 1 bar) of chondritic silicates and refractory mineral assemblages in Ca,Al-rich inclusions, so at least the surface layers would be molten. Furthermore, the postulated scenario requires proto-Mercury to be formed very hot, and hence we make the reasonable assumption that the entire silicate phase of proto-Mercury would be a silicate magma. The calculations in Cameron (1985a) were based on the vaporization of a silicate magma with the composition $MgSiO_3$ and did not attempt to examine the vaporization chemistry of more complex silicate materials. That task was carried out by Fegley and Cameron (1987).

Fegley and Cameron (1987) report the results of a series of calculations to study the compositional evolution of a silicate magma (with initial composition being approximately chondritic) as a function of the extent of vaporization. The composition of a totally molten silicate magma and of the vapor phase in equilibrium with it was calculated using a multicomponent gas-melt chemical equilibrium code based on the silicate melt solution models developed by Hastie and coworkers (1982a,b,1984,1985). The calculations were done as a function of the degree of vaporization at 2500 to 3500 K. Both ideal and nonideal magma solution models were studied.

The initial composition of the magma was assumed to be a mixture of the oxides SiO_2-MgO-CaO-Al_2O_3-TiO_2-Na_2O-K_2O-FeO-UO_2-ThO_2-PuO_2 in relative solar proportions (Anders and Ebihara 1982) except for FeO which was set to 0.1 times the MgO abundance. The remainder of the Fe solar abundance was assumed to form the core of proto-Mercury. The actinides and K were included in the calculations to study the effect of vaporization on the amounts of heat-producing radionuclides left in the planet after partial vaporization. Inclusion of the alkalis also provides a comparison with Hastie's work. The use of a more refractory starting composition would not significantly affect the results of these calculations. In fact, such compositions appear as intermediate stages during the vaporization process.

The calculated compositions were for the equilibrium between the gas and melt phases. Thermodynamic data were obtained from standard sources (Stull and Prophet 1971; Chase et al. 1971–1975, Glushko et al. 1978–1982) as well as elsewhere in the literature (see Fegley and Cameron 1986 for detailed references). An ideal magma was considered to be one in which the activity coefficients of the constituent oxides were taken to be unity. In a nonideal magma these coefficients were usually orders of magnitude smaller; the procedure of Hastie and his coworkers assumes that the magma is composed of a large number of pseudospecies, so that the abundances of the basic oxides, which are included among the pseudospecies, are greatly reduced, along with the corresponding activities. The vaporization calculations agreed well with experimental results of Hashimoto (1983), giving some confidence in the method.

Results of the vaporization calculations for the composition of the proto-Mercury mantle in the ideal solution case at 3000 K are shown in Fig. 1. The first elements lost during vaporization are the alkalis, Na and K, with K being more volatile. The next element lost is Si, followed by Fe and Mg. At the highest degree of vaporization (~4.6% residual mantle mass fraction), ~95% of the Al, ~40% of the Ca and ~15% of the Ti remain in the mantle.

Figure 2 shows the results for the vaporization of the mantle in the non-ideal solution case, also at 3000 K. Several interesting features are displayed. The alkalis are still the most volatile elements but Fe is the next element lost, followed by Si and Mg. However, Mg loss from the magma accelerates after about 80% of the silicate has been vaporized. At the highest degree of vaporization considered (96% of original silicate lost), more Si remains in the magma than Mg. Loss of Ca and Ti is also occurring at this point. More Ti (53%) but less Ca (33%) remain in the magma relative to the ideal magma case.

The calculated mantle density for these cases is shown in Fig. 3. The ideal solution case is shown as the dashed line; it has a complicated structure. The density increases to a maximum at 3.71 g cm^{-3} when approximately 28% of the original silicate remains. The mantle density then decreases with

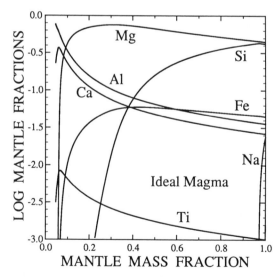

Fig. 1. The composition of the mantle as a function of its remaining mass fraction as vaporization proceeds, for an ideal magma. The elements shown are balanced by oxygen in the mantle. K is not shown because it coincides with the right frame starting close to the bottom.

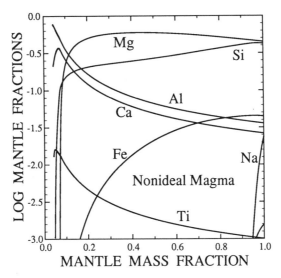

Fig. 2. The composition of the mantle as a function of its remaining mass fraction as vaporization proceeds, for a nonideal magma. The elements shown are balanced by oxygen in the mantle. The unlabeled line in the lower right corner is potassium.

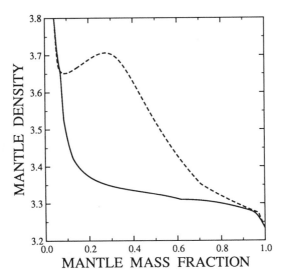

Fig. 3. The density of the mantle as a function of the remaining mantle mass fraction as vaporization proceeds. The dashed line is for an ideal magma; the solid line is for a nonideal magma, both at 3000 K.

increasing vaporization until about 10% of the silicate remains and the density is about 3.65 g cm^{-3}. The density then starts increasing again with further vaporization and is virtually indistinguishable from the density profile for the nonideal magma case, shown as the solid line. However, in the nonideal case there is a marked difference in the behavior of the density in the earlier stages of the vaporization. After quickly rising to about 3.3 g cm^{-3} as the alkalis are lost, the density then stays at nearly the same value until almost 80% of the mass has been lost. Further vaporization then causes a rapid rise in the density.

The calculated mean uncompressed density of the planet, shown in Fig. 4 for the ideal (dashed-line) and nonideal (solid-line) cases, monotonically increases with the degree of vaporization in each case. The uncompressed mean density of 5.3 g cm^{-3} for Mercury is reached after about 72% (ideal) to 79% (nonideal) of the mantle has been vaporized. The calculated composition of the silicate phase at this point is given in Table I, which also includes the results of nonideal calculations for temperatures of 2500 K and 3500 K. The (possible) later accretion of ordinary silicate (MgO-, SiO$_2$-rich) planetesimals would change these compositions by adding more SiO$_2$, FeO and alkalis. However, the observed mean density of the planet could then be obtained only if more than 72% to 79% of the mantle had originally been vaporized and lost, leaving the remaining fraction of the original mantle even more refractory.

The heat-producing radionuclides behave in two quite different fashions during vaporization. Loss of K occurs early during the vaporization process. U is also lost during vaporization and is depleted much more than Pu (approx-

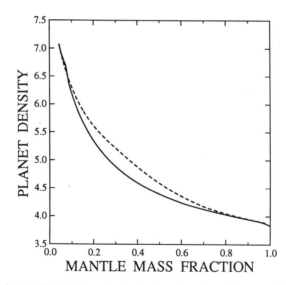

Fig. 4. The density of the Mercury protoplanet as a function of the remaining mantle mass fraction as vaporization proceeds. The dashed line corresponds to an ideal magma; the solid line is for a nonideal magma, both at 3000 K.

TABLE I
Predicted Mantle and Crust Compositions from
Different Vaporization Models[a]

Oxide	Mass % Oxide in Different Models[c]			
Model	1	2	3	4
SiO_2	1.2	32.1	25.9	19.9
CaO	9.9	14.6	13.7	11.7
MgO	67.7	34.5	42.3	50.0
Al_2O_3	12.4	18.1	17.0	14.4
FeO	8.3	0.0	0.4	3.4
TiO_2	0.5	0.7	0.8	0.6
K(ppb)	0.0	0.0	0.0	0.0
U(ppb)[b]	0.0	0.0	0.0	0.3
Th(ppb)[b]	278	401	377	322

[a]Mean uncompressed density = 5.3 g cm^{-2}.
[b]Primordial abundances 4.55 × 10^9 yr ago.
[c]Models: 1. Ideal magma, 3000 K; 2. Nonideal magma, 2500 K; 3. Nonideal magma, 3000 K; 4. Nonideal magma, 3500 K.

imately as volatile) or Th (less volatile). This is due to the formation of volatile U-oxide gases. In contrast, Pu and Th are essentially undepleted by vaporization by the time that the mean uncompressed planetary density of 5.3 g cm^{-3} is reached.

In each case in Table I, the calculated mean uncompressed planetary density is 5.3 g cm^{-3}, which matches the value deduced for Mercury. All four compositions in the table are depleted in the alkalis, FeO and SiO_2, and are enriched in CaO, MgO, Al_2O_3 and TiO_2 relative to chondritic material. There is a large difference, which brackets the end-member compositions expected from vaporization of a silicate magma, between the results of the ideal and nonideal models. Further discussion centers on the latter models because of the good agreement of these runs with the available data on silicate vaporization and activity coefficients.

The uncompressed planetary mean density was matched with a metal-to-silicate mass ratio of approximately 64 to 36, with the silicate phase considerably reduced in SiO_2 relative to some of the metal oxides. Inclusion of Ni in the metal phase will decrease slightly the calculated metal abundance because Ni is denser than Fe (8.90 vs 7.86 g cm^{-3}, respectively). However, this is a second-order effect; an alloy with the solar Ni abundance and Fe decreased by forming FeO = 0.1 MgO (as in our models) is only 0.8% denser than pure Fe.

The silicate phase is predicted to be depleted in the alkalis, FeO and SiO_2 and enriched in CaO, MgO, Al_2O_3 and TiO_2 relative to chondritic material. The vaporization models also predict that the silicate is depleted in SiO_2 and enriched in other oxides (CaO, MgO, Al_2O_3, TiO_2 and FeO) relative to silicate compositions calculated for equilibrium condensation models of Mercury (Basaltic Volcanism Study Project 1981). Another prediction of the vaporization models calculated by Fegley and Cameron (1987) is the production of trace-element abundance patterns which are depleted in easily oxidized elements relative to other refractory trace elements of similar volatilities. In particular, the depletion of U, Ce and other easily oxidized rare earth elements is expected. These abundance patterns are unique signatures due to vaporization and are not produced in other models (e.g., equilibrium condensation). These differences will influence several aspects of Mercury's composition and structure including: (a) the mineralogy of the silicate phase; (b) the nature of a crust (if a distinct, differentiated crust exists); (c) the chemical composition and physical properties, such as viscosity, of volcanic magmas; and (d) trace element partitioning between putative magmas and their source regions.

IV. PLANETARY COLLISIONS

The scenario in which a major planetary collision may have played a role in producing the strange density of Mercury is that of planetary accumulation from planetesimals. It is assumed that the primitive solar nebula has largely dissipated with the formation of the Sun, during which time it is highly turbulent and produces dissipation via a large turbulent viscosity (Cameron 1985b). After the turbulence dies away, small solid bodies, presumably in the form of clumps of interstellar grains, are able to settle to midplane, where they are likely to become gravitationally unstable to the formation of planetesimals of small asteroidal size (a few kilometers in radius) (Goldreich and Ward 1973). There follows an accumulation of these planetesimals into bodies of increasing size (Wetherill 1980a). This accumulation tends to produce a largest body in a given feeding zone, with the next largest body less in mass by a factor of a few, and so on down from there. The term feeding zone is inexact, since the source material for the accumulation of any one of the terrestrial planets is actually spread over the inner solar system. We use it here to refer to the volume of space dominated by the final accumulated planet.

Toward the end of the accumulation process the second largest body is likely to collide with the largest, with potentially surviving observable effects. In the case of the Earth this is the collision that may have initiated the formation of the Moon (Hartmann and Davis 1975; Cameron and Ward 1976; Benz, et al. 1986). In fact, more recent calculations (Benz et al. 1987b) have narrowed down the set of acceptable parameters to a collision between the proto-Earth and a body of 0.13 to 0.17 times its mass, taking place at rather low velocity. For the case of Mercury we have carried out similar collisional calculations. We have assumed that the target protoplanet Mercury was 2.25 times the mass of the present Mercury, in order to give it approximately a chondritic silicate to iron ratio, and we have hit it with a projectile of one-sixth that mass at somewhat higher velocities.

These calculations have required the use of a 3-dimensional hydrodynamic code run on Cray computers at Los Alamos. The method used is called "smoothed particle hydrodynamics," and has been described by Benz et al. (1986a). Briefly, the mass distribution in the target and the projectile is represented by mass points distributed in proportion to the density; these mass points are actually considered to have a finite distribution in space as defined by the shape of their "kernels". In condensed matter these kernels overlap, and one obtains bulk properties like pressure and density by a suitable averaging over the mass points. In Benz et al. (1986a) the kernels were given an exponential shape, but here they were given a polynomial form because the resulting sharp cutoff in the spatial distribution made the calculations easier.

Also described in Benz et al. (1986) was the Tillotson equation of state that was used in the earlier calculations of the proto-Mercury collisions. With this equation of state, rocky material was represented by granite, for which the relevant data were available at Los Alamos. Beyond our Mercury run 7, the ANEOS equation of state was used which was developed at the Sandia Laboratories (Thompson and Lauson 1984). This equation of state has a better representation of mixed phase conditions, when both a vapor phase and a condensed matter phase are present at the same time. For rocky material we used dunite with this equation of state; the appropriate equation of state parameters for this material were determined by J. Melosh, and we are grateful to him for providing them to us.

For our earliest proto-Mercury collision runs we used a relatively small number of particles; these results were for our preliminary orientation to the problem, and we do not report them here. These early runs used rock plus an iron core for the target but just rock for the projectile. After run 4, the projectile had an iron core as well. In these runs the proto-Mercury target was represented by 3000 particles, of which 959 were iron and 2041 were silicate rock. The projectile of one-sixth the mass was represented by 1000 particles, each half the mass of the particles used in the target. In the projectile, 319 of the particles were iron (beyond run 4) and 681 were rock.

The results of the collision calculations are shown in Table II. In runs 3 and 4 we were still getting our bearings, and the projectile was made entirely of rock. In run 3 the projectile had an impact velocity of 27 km s^{-1} at infinity; it hit the target centrally and knocked off most of the silicate rock and some of the iron. This would be a very good candidate for the kind of collision that would produce the observed Mercury in this scenario, except that the composition of the projectile is unrealistic. In run 4 the velocity of the impactor was increased to 38 km s^{-1}, resulting in the complete disruption of the target.

With run 5, we started a realistic series of cases by putting an iron core into the projectile. We chose an impact velocity of 25 km s^{-1} and found that the target was once again totally disintegrated. It was thus evident that the denser core in this impactor enabled the target to be disintegrated at a significantly lower velocity than was true of the softer case in which the impactor

TABLE II
The Outcome of the Mercury Collision Scenarios

Case	Vel. km/sec	Impact Param.[a]	Targ. Iron[b]	Proj. Iron[b]	Targ. Sil.[b]	Proj. Sil.[b]	Sil./Iron Ratio	Mass Ratio[c]
3	27	0.00	76.5%		5.1%	0.4%	0.14	0.67
4	38	0.00			Disintegrated			
		Below here the impactor had an iron core						
5	25	0.00			Disintegrated			
6	20	0.00	56.1%	37.7%	5.9%	7.5%	0.25	0.58
7	22	0.46	93.1%	7.2%	22.0%	2.8%	0.51	1.18
		Below here the ANEOS equation of state was used						
8	20	0.00	48.8%	42.0%	8.8%	11.3%	0.41	0.60
9	15	0.00	86.2%	69.3%	42.1%	31.3%	1.03	1.67
10	10	0.00	99.2%	97.8%	79.1%	58.6%	1.64	2.65
11	15	0.59	99.4%	0.6%	70.3%	12.3%	1.55	2.33
12	20	0.51	94.3%	3.8%	47.1%	6.8%	1.08	1.76
13	28	0.53	80.0%	0.0%	20.0%	1.2%	0.54	1.03
14	35	0.54	57.0%	0.3%	7.8%	0.3%	0.29	0.59

[a]The impact parameter is given in units of the target radius.
[b]The values given for target and projectile iron and silicate are the percentages of the numbers of particles in the residual body coming from these four sources.
[c]The mass ratio is the residual mass relative to the present mass of the planet Mercury (3.3 × 10^{26} gm).

was just rock. In run 6 the velocity was lowered to 20 km s^{-1} and the target survived the collision, after once again losing nearly all its silicates. It is interesting to note, however, that the iron core in the target lost 43.9% of its mass but picked up 37.7% of the mass of the iron core in the projectile. This is an excellent candidate to be the scenario for the formation of Mercury. However, in run 7, with an impact velocity raised slightly to 22 km s^{-1} and an impact parameter of nearly half the planetary radius, the collision leaves more than the present mass of Mercury, and hence this is not a good candidate for the Mercury scenario, especially since some of the lost mass will be reaccumulated, as discussed below.

Starting with run 8, we used the ANEOS equation of state. Run 8 was essentially a repeat of the conditions in run 6. The results of these two runs were very similar, showing that the results do not depend sensitively on the character of the equation of state and the choice of rocky material. The amount of iron left in the residual body is nearly the same; there is a little less iron from the target and a little more from the projectile. The total mass left behind after the collision is nearly the same in the two cases, and is a satisfactory value of only about 60% of the present mass of Mercury.

Runs 9 and 10 were head-on collisions like run 8, but at progressively lower collision velocities, 15 km s^{-1} for run 9 and 10 km s^{-1} for run 10. The amount of mass ejected in the collision decreases with decreasing collision velocity, so that in the 10 km s^{-1} case the planet is left with more mass than it had at the beginning. Neither of these cases is a candidate to produce the current Mercury.

Runs 11 and above were further cases in which the projectile hit the target off-center, with impact parameters just over half the radius of the target. The striking characteristic of these runs is that such off-center collisions are much less effective in ejecting mass from the target. Whereas a central collision at 20 km s^{-1} leaves only 0.60 Mercury masses, a similar collision which is noncentral leaves 1.76 Mercury masses, too high to make this a candidate collision for the present Mercury. Only in run 14, where the impact velocity was 35 km s^{-1}, does the planetary remnant become reduced to 0.59 Mercury masses, which makes it a good candidate for forming the observed planet.

As discussed in the introduction, the present silicate/iron ratio in Mercury is about in the range 0.4 to 0.5, and this should be regarded as equal to or more than the value to be expected in the remnant produced in these calculations for the result to be considered a candidate collision to lead to the present Mercury. The only cases (see Table II) which satisfy this criterion and which result in nearly complete silicate loss and relatively little net iron loss from proto-Mercury is a nearly central collision at about 20 km s^{-1} and a noncentral collision at about 35 km s^{-1}. The amount of mass lost is a sensitive function of the collision velocity and the impact parameter.

In the course of a typical collision, a great deal of material is thrown out from the site of the collision, some to escape and the rest to fall back onto the planet. Of the material thrown out, much is clumped into clusters of particles and the rest consists of single particles. These "single particles" are likely to have passed through the vapor phase and many of them will have recondensed with expansion of the material. At the end of the Cray runs this is the typical situation: a cloud of material surrounding a planetary core, some of it rising and the rest falling back. From our post-Cray analyses of the results, we find that nearly all of the larger clumps of particles (which tend to be iron-rich) fall back onto the planet and nearly all of the single particles escape. In the cases where the planet is disintegrated, this clumpy situation still exists, but the largest clumps are only of moderate size and very few of them collide with one another.

In a central collision at about 20 km s^{-1} a striking pattern develops in the debris. The planetary core is deformed into a relatively thin sheet, and the material which is thrown out is largely ejected in essentially the same thin sheet (the silicate is somewhat more dispersed than the iron). Figure 5 shows a plot of the debris at the end of the Cray run 8, viewed from a direction at right angles to the direction of collision, which shows the sheet. Figure 6 is a plot of the same case but viewed from the direction of the collision. In these plots the open circles are silicate particles and the filled circles are iron particles; the silicate particles do not hide other particles beyond them, so that the iron particles are fully visible.

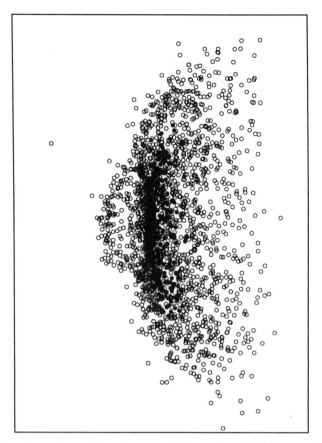

Fig. 5. A plot of the positions of the silicate (open circles) and iron (filled circles) particles at the end of Cray run 8, viewed along a right angle to the line of collision.

It is evident that a successful collision should be judged to be one in which the mass of the protoplanet is reduced well below that of the present planet Mercury and in which the remnant is composed predominantly of iron. All of the material which is ejected in the collision goes into independent orbit around the Sun; these orbits must necessarily cross the orbit of the remaining protoplanet. The present orbital elements of Mercury are subject to secular variation with time due to other planetary perturbations, so we do not know what the orbit of the remnant would have been. Nevertheless, we took the present orbit of Mercury as a suitable prototype of that orbit, and calculated whether any of the particles ejected from the collision would have been put into orbits that would cross the orbit of Venus. We found that, in general, only a few of the particles would do so if ejected at aphelion, and usually a few dozen and at most a few hundred would do so at perihelion. Thus, only a

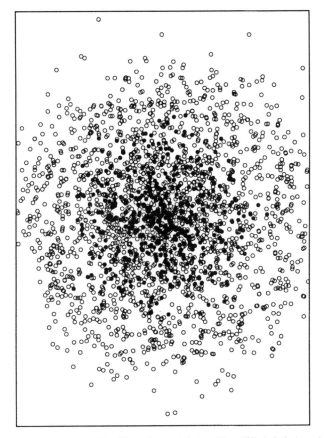

Fig. 6. A plot of the positions of the silicate (open circles) and iron (filled circles) particles at the end of Cray run 8, viewed along the direction of collision.

minority of the particles can be removed from Mercury-crossing orbits by other planetary perturbations.

The key to the question of whether the particles are reaccreted by Mercury probably lies in the fact that most of the ejected material has passed through the vapor phase. Upon expansion, this material will cool and condense into solid particles, but if the vapor is by that time at fairly low density, then the particle sizes will probably be very small, in the subcentimeter size range. Such small particles will be drawn into the Sun by the Poynting-Robertson effect in a time short compared to their expected collision time with Mercury, which is about 10^6 yr (G. Wetherill, personal communication). If a substantial part of the ejected material has sizes in the centimeter range, then the individual pieces would undergo a large number of collisions with one another, tending to grind themselves down to a size that can be removed by the Poynting-Robertson effect. Very much larger chunks ejected from the collision may not be ground down enough for Poynting-Robertson removal, and these are likely to be reaccumulated upon Mercury.

Thus, this scenario appears promising to account for the strange density of Mercury.

V. DISCUSSION

In this chapter we have discussed three scenarios which have been postulated to account for the high mean density of Mercury. We found little of merit in the equilibrium condensation scenario. We have found the planetary evaporation scenario to lead to some interesting predictions about the chemical composition of the present Mercury mantle. The efficiency of the postulated wind mechanism for the removal of the evaporated rock decomposition products from proto-Mercury is uncertain. A major planetary collision is a plausible occurrence in the late stages of planetary accumulation, and we have found that reasonable circumstances exist in which most of the silicates can be stripped from the protoplanet. We are uncertain about the amount of reaccumulation of these silicates that will take place from orbit. The chemical differences predicted between the second and third scenarios may provide a means for distinguishing between these scenarios observationally by means of spacecraft experiments.

Acknowledgments. This research has been supported in part by the National Aeronautics and Space Administration to Harvard University and to Massachusetts Institute of Technology. W. Benz also acknowledges support from the Swiss National Science Foundation. We are grateful to the Los Alamos National Laboratory for the provision of the computer time which has been used on this problem.

Color Section

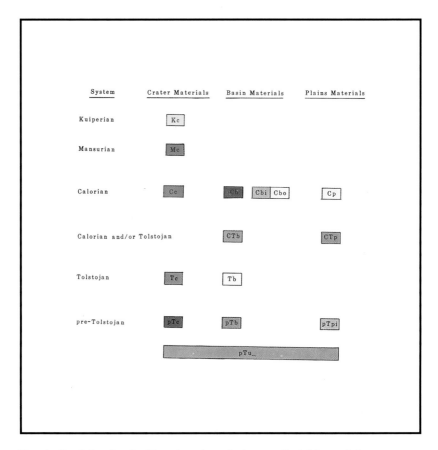

Plate 1. Correlation chart for Mercurian paleogeologic maps. Capital letters designate systems. Lowercase letters designate classes of rock-stratigraphic units: c — crater material; b — basin material; bi — basin continuous (inner) facies; bo — basin discontinuous (outer) facies; p — plains material; pi — intercrater plains material; u — undivided. Maps are Lambert Equal-Area projections centered on 0°, 100°; north at top.

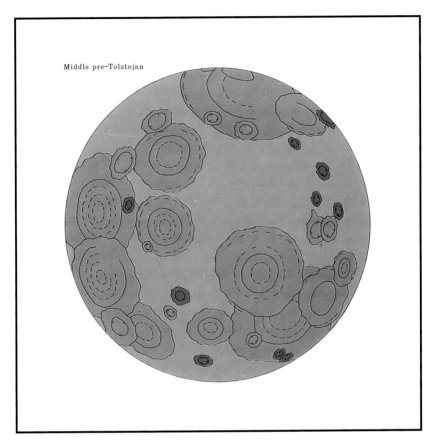

Middle pre-Tolstojan

Plate 2. Paleogeologic map of Mercury in middle pre-Tolstojan time. Undivided material probably consists of heavily cratered terrain.

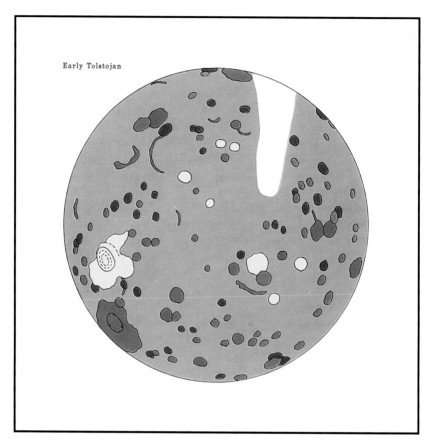

Early Tolstojan

Plate 3. Paleogeologic map of Mercury in early Tolstojan time. Tolstoj basin at lower left. Inter-crater plains have largely obliterated the ancient, pre-Tolstojan cratered terrain.

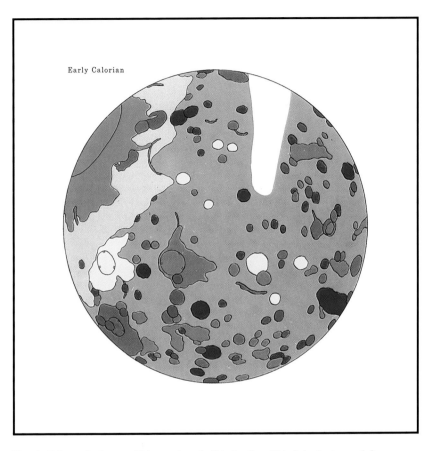

Early Calorian

Plate 4. Paleogeologic map of Mercury in early Calorian time. Caloris basin at upper left.

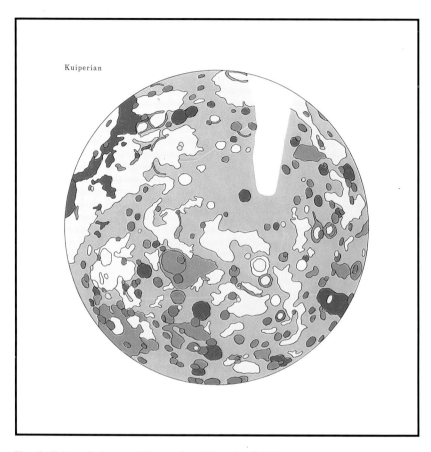

Kuiperian

Plate 5. Paleogeologic map of Mercury in middle to late Calorian time. Emplacement of smooth plains completed; cratering rate drastically declines.

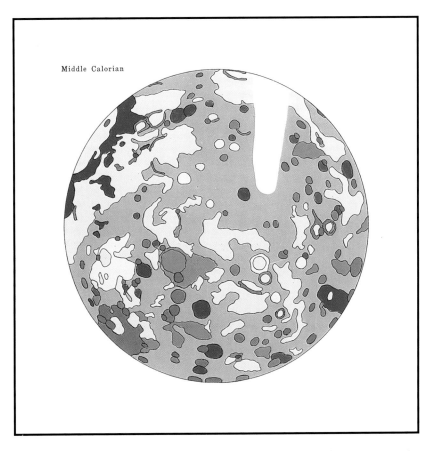

Middle Calorian

Plate 6. Geologic map of present-day surface of Mercury. Only craters have formed on Mercury's surface since middle Calorian time (Plate 5).

Bibliography

BIBLIOGRAPHY

(Compiled by Melanie Magisos)

Adams, J. B. 1974. Visible and near-infrared diffuse reflectance spectra of pyroxenes as applied to remote sensing of solid objects in the solar system. *J. Geophys. Res.* 79:4829–4836.

Adams, J. B., and Filice, A. L. 1967. Spectral reflectance 0.4 μm to 2.0 μm of silicate rock powders. *J. Geophys. Res.* 72:5705–5715.

Adams, J. B., and McCord, T. B. 1970. Remote sensing of lunar surface mineralogy: Implications from visible and near-infrared reflectivity of Apollo 11 samples. *Proc. Lunar Sci. Conf.* 1:1937–1945.

Adams, J. B., and McCord, T. B. 1973. Vitrification darkening in the lunar highlands and identification of Descartes material at the Apollo 16 site. *Proc. Lunar Sci. Conf.* 4:163–177.

Adams, W. S., and Dunham, T. J. 1932. Note on the spectrum of Mercury. *Publ. Astron. Soc. Pacific* 44:380.

Ahrens, T. J. 1979. Equations of state of iron sulfide and constraints on the sulfur content of the Earth. *J. Geophys. Res.* 84:985–999.

Ahrens, T. J., and O'Keefe, J. D. 1977. Equations of state and impact induced shock-wave attenuation on the moon. In *Impact and Explosion Cratering,* eds. D. J. Roddy, R. O. Pepin, and R. B. Merrill (New York: Pergamon Press), pp. 639–656.

Ahrens, T. J., and O'Keefe, J. D. 1978. Energy and mass distributions of impact ejecta blankets on the Moon and Mercury. *Proc. Lunar Planet. Sci. Conf.* 9:3787–3802.

Allen, C. C. 1977. Rayed craters on the Moon and Mercury. *Phys. Earth Planet. Int.* 15:179–188.

Allen, C. W. 1976. *Astrophysical Quantities* (London: Athlone Press).

Anders, E., and Ebihara, M. 1982. Solar-system abundances of the elements. *Geochim. Cosmochim. Acta* 46:2363–2380.

Anderson, E. M. 1951. *The Dynamics of Faulting* (Edinburgh: Oliver and Boyd).

Anderson, J. D., Colombo, G., Esposito, P. B., Lau, E. L., and Trager, G. B. 1987. The mass, gravity field and ephemeris of Mercury. *Icarus* 71:337–349.

Anglin, J. D., Dietrich, W. F., and Simpson, J. A. 1972. Solar flare accelerated isotopes of hydrogen and helium. In *High Energy Phenomena on the Sun,* eds. R. Ramaty and R. G. Stone, NASA SP-342, pp. 315–340.

Antoniadi, E. M. 1930. *La Planète Mars* (Paris).

Antoniadi, E. M. 1934. *La Planète Mercure et al Rotation des Satellites* (Paris). Trans. P. Moore, *The Planet Mercury* (London: K. K. Reid Ltd., 1974).

Arnold, J. R. 1965. The origin of meteorites as small bodies. II. The model. *Astrophys. J.* 141:1536–1547.

Ash, M. E., Shapiro, I. I., and Smith, W. B. 1971. The system of planetary masses. *Science* 174:551–556.

Baker, D. N. 1986. Jovian electron populations in the magnetosphere of Mercury. *Geophys. Res. Lett.* 13:789–792.

Baker, D. N., Higbie, P. R., Belian, R. D., and Hones, E. W., Jr. 1979. Do Jovian electrons influence the terrestrial outer radiation zone? *Geophys. Res. Lett.* 6:531–534.

Baker, D. N., Hones, E. W., Jr., Young, D. T., and Birn, J. 1982. The possible role of ionospheric oxygen in the initiation and development of plasma sheet instabilities. *Geophys. Res. Lett.* 9:1337–1340.

Baker, D. N., Akasofu, S. I., Baumjohann, W., Bieber, J. W., Fairfield, D. H., Hones, E. W., Jr., Mauk, B. H., McPherron, R. L., and Moore, T. E. 1984. Substorms in the magnetosphere. In *Solar Terrestrial Physics: Present and Future,* NASA Publ. 1120, pp. 8.1–8.55.

Baker, D. N., Blake, J. B., Klebesadel, R. W., and Higbie, P. R. 1986a. Highly relativistic electrons in the earth's outer magnetosphere. I. Lifetimes and temporal history 1979–1984. *J. Geophys. Res.* 91:4265–4276.

Baker, D. N., Borovasky, J. E., Burns, J. D., Gisler, G. R., and Zeilile, M. 1986b. Possible calorimetric effects at Mercury due to solar wind magnetosphere interactions. *J. Geophys. Res.* 92:4707–4712.

Baker, D. N., Simpson, J. A., and Eraker, J. H. 1986c. A model of impulsive acceleration and transport of energetic particles in Mercury's magnetosphere. *J. Geophys. Res.* 91:8742–8748.

Baldwin, R. B. 1949. *The Face of the Moon* (Chicago: Univ. of Chicago Press).

Baldwin, R. B. 1963. *The Measure of the Moon* (Chicago: Univ. of Chicago Press).

Baldwin, R. B. 1964. Lunar crater counts. *Astron. J.* 69:377–392.

Baldwin, R. B. 1971. On the history of lunar impact cratering: The absolute time scale and the origin of planetesimals. *Icarus* 14:36–52.

Baldwin, R. B. 1974. On the origin of the mare basins. *Proc. Lunar Sci. Conf.* 5:1–10.

Bame, S. J., Asbridge, J. R., Felthausen, H. E., Hones, E. W., and Strong, I. B. 1967. Characteristics of the plasma sheet in the earth's magnetotail. *J. Geophys. Res.* 72:113–129.

Banks, P. M., Johnson, H. E., and Axford, W. I. 1970. The atmosphere of Mercury. *Comments Astrophys. Space Phys.* 2:214–220.

Barlow, N. G. 1987. Relative ages and the geologic evolution of martian terrain units. Ph.D. Thesis, Univ. of Arizona.

Barrell, J. 1914. The strength of the earth's crust, part VI. *J. Geol.* 22:655–683.

Barshay, S. S. 1981. Combined Condensation-Accretion Model of the Terrestrial Planets. Ph.D. Thesis, Massachusetts Inst. of Technology.

Barshay, S. S., and Lewis, J. S. 1976. Chemistry of primitive solar material. *Ann. Rev. Astron. Astrophys.* 14:81–90.

Barshay, S. S., and Lewis, J. S. 1980. Accretion and the equilibrium condensation model. Unpublished.

Barth, C. A. 1969. Planetary ultraviolet spectroscopy. *Applied Optics* 8:1295–1304.

Bartholomay, H. A., and Larimer, J. W. 1982. Dust-gas fractionation in the early solar system. *Meteoritics* 17:180–181 (abstract).

Basaltic Volcanism Study Project. 1981. *Basaltic Volcanism on the Terrestrial Planets* (New York: Pergamon Press).

Basilevsky, A. T. 1981. On some peculiarities of structure of impact craters on planets and satellites of the solar system. *Dokl. USSR Acad. Sci.* 258(2):323–325. In Russian.

Basilevsky, A. T., and Ivanov, B. A. 1982. Impact cratering on stony and icy bodies: Different mechanisms of central peak formation? *Lunar Planet. Sci.* XIII:27–28 (abstract).

Baum, R. M. 1979. Historical sighting of the craters of Mercury. *The Strolling Astronomer.* 28:17–22.

Beatty, J. K., O'Leary, B., and Chaikin, A. 1987. *The New Solar System,* 3rd ed. (Cambridge: Sky Publ. Corp.).

Becker, R. H., and Pepin, R. O. 1984. The case for a Martian origin of the shergottites: Nitrogen and noble gases in EETA 79001. *Earth Planet. Sci. Lett.* 69:225–242.

Beebe, R. G., and Herzog, A. 1975. Surface properties and effective resolution from ground-based observations of Mercury. *Icarus* 25:555–560.

Beletskii, V. V. 1972. Resonance rotation of celestial bodies and Cassini's laws. *Celestial Mech.* 6:356–378.

Bell, J. F., and Hawke, B. R. 1981. The Reiner Gamma Formation: Composition and origin as derived from remote sensing observations. *Proc. Lunar Planet. Sci. Conf.* 12B:679–694.

Bell, J. F., and Hawke, B. R. 1987. Recent comet impacts on the Moon: The evidence from remote-sensing studies. *Publ. Astron. Soc. Pacific* 99:862–867.

Belton, M. J. S., Hunten, D. M., and McElroy, M. B. 1967. A search for an atmosphere on Mercury. *Astrophys. J.* 150:1111–1124.

Bender, P. L., Ashby, N., Vincent, M. A., and Wahr, J. M. 1986. Mercury relativity satellite. *Mercury: Program and Abstracts*, Tucson, Arizona, August.

Benz, W. Slattery, W. L., and Cameron, A. G. W. 1986*a*. Short note: Snapshots from a three-dimensional modeling of a giant impact. In *Origin of the Moon*, eds. W. K. Hartmann, R. J. Philips, and G. J. Taylor (Houston: Lunar and Planetary Inst.), pp. 617–620.

Benz, W., Slattery, W. L., and Cameron, A. G. W. 1986*b*. The origin of the Moon and the single-impact hypothesis I. *Icarus* 66:515–535.

Benz, W., Slattery, W. L., and Cameron, A. G. W. 1987*a*. Planetary collision calculations: Origin of the Moon. *Lunar Planet. Sci.* XVIII:60–61 (abstract).

Benz, W., Slattery, W. L., and Cameron, A. G. W. 1987*b*. The origin of the Moon and the single-impact hypothesis II. *Icarus* 71:30–45.

Berchem, J., and Russell, C. T. 1982. The thickness of the magnetopause current layer: ISEE-1 and -2 observations. *J. Geophys. Res.* 87:2108–2114.

Bergan, S., and Engle, I. M. 1981. Mercury's magnetic field revisited. *J. Geophys. Res.* 86:1617–1620.

Bergstralh, J. T., Gray, L. D., and Smith, H. J. 1967. An upper limit for atmospheric carbon dioxide on Mercury. *Astrophys. J.* 149:L137–L139.

Bernard, R. 1938. Observation d'un nouveau phenomene de fluorescence dans la haute atmosphere. Presence et variations d'intensité de la radiation 5893 Å dans la lumière du ciel au crepuscule. *Compt. Rend. Acad. Sci. Paris* 206:448–450.

Bernatowicz, T. J., Kennedy, B. M., and Podosek, F. A. 1985. Xe in glacial ice and the atmospheric inventory of noble gases. *Geochim. Cosmochim. Acta* 49:2561–2564.

Binder, A. B., and Cruikshank, D. P. 1967. Mercury: New observations of the infrared bands of carbon dioxide. *Science* 155:1135.

Birn, J. 1973. On the stability of the planetary system. *Astron. Astrophys.* 24:283–293.

Bischoff, A., and Keil, K. 1984. A1-rich objects in ordinary chondrites: Related origin of carbonaceous and ordinary chondrites and their constituents. *Geochim. Cosmochim. Acta* 48:693–709.

Bogard, D. D., Nyquist, L. E., and Johnson, P. 1984. Noble gas contents of shergottites and implications for the Martian origin of SNC meteorites. *Geochim. Cosmochim. Acta* 48:1723–1739.

Booker, J. R., and Stengel, K. C. 1978. Further thoughts on convective heat transport in a variable viscosity fluid. *J. Fluid Mech.* 86:289–291.

Boss, A. P. 1986. Protoearth mass shedding and the origin of the moon. *Icarus* 66:330–340.

Boyce, J. M. 1980. Basin peak-ring spacing on Ganymede and Callisto: Implications for the origin of central peaks and peak rings. *Repts. Planet. Geol. Program, 1979–1980*, NASA TM-81776, pp. 339–342.

Brandt, J. C., and Chamberlain, J. W. 1958. Resonance scattering by atmospheric sodium—VI. The analytic solution for the twilight intensity. *J. Atmos. Terr. Phys.* 13:99–106.

Bridge, H. S., Belcher, J. W., Lazarus, A. J., Sullivan, J. D., McNutt, R. L., Bagenal, F., Scudder, J. D., Sittler, E. C., Siscoe, G. L., Vasyliunas, V. M., Goertz, C. K., and Yeates, C. M. 1979. Plasma observations near Jupiter: Initial results from Voyager 1. *Science* 204:987–991.

Brinkmann, R. T. 1971. More comments on the validity of Jeans' escape rate. *Planet. Space Sci.* 19:791–794.

Broadfoot, A. L. 1976. Ultraviolet spectrometry of the inner solar system from Mariner 10. *Rev. Geophys. Space Phys.* 14:625–627.

Broadfoot, A. L., Kumar, S., Belton, M. J. S., and McElroy, M. B. 1974. Mercury's atmosphere from Mariner 10: Preliminary results. *Science* 185:166–169.

Broadfoot, A. L., Shemansky, D. E., and Kumar, S. 1976. Mariner 10: Mercury atmosphere. *Geophys. Res. Lett.* 3:577–580.

Broadfoot, A. L., Clapp, S. S., and Stuart, F. E. 1977*a*. Mariner 10 ultraviolet spectrometer: Airglow experiment. *Space Sci. Inst.* 3:199–208.

Broadfoot, A. L., Clapp, S. S., and Stuart, F. E. 1977*b*. Mariner 10 ultraviolet spectrometer: Occultation experiment. *Space Sci. Inst.* 3:209–218.

Broadfoot, A. L., Belton, M. J. S., Takacs, P. Z., Sandel, B. R., Shemansky, D. E., Holberg, J. R., Ajello, J. M., Atreya, S. K., Donahue, T. M., Moos, W. R., Bertaux, J. L., Blamont, J. L., Strobel, D. F., McConnell, J. C., Dalgarno, A., Goody, R. M., and McElroy, M. B. 1979. Extreme ultraviolet observations from Voyager 1 encounter with Jupiter. *Science* 204:979–982.

Brown, R. A., and Yung, Y. 1976. Io, its atmosphere and optical emissions. In *Jupiter,* ed. T. Gehrels (Tucson: Univ. of Arizona Press), pp. 1102–1145.

Brown, W. E., Adams, G. F., Eggleton, R. E., Jackson, P., Jordan, R., Kobrick, M., Peeples, W. S., Phillips, R. J., Procello, L. J., Schaber, G., Sill, W. R., Thompson, T. W., Ward, S. H., and Zelenka, J. S. 1974. Elevation profiles of the moon. *Proc. Lunar Planet. Sci. Conf.* 5:3037–3048.

Buckingham, R. A., and Dalgarno, A. 1952. Diffusion and excitation transfer of metastable helium in normal gaseous helium. *Proc. Roy. Soc. London* A213:506–519.

Buie, M. W. 1984. Lightcurve CCD Spectrophotometry of Pluto. Ph.D. Thesis, Univ of Arizona.

Bullen, K. E. 1952. Cores of terrestrial planets. *Nature* 170:363–364.

Buratti, B. J. 1985. Applications of a radiative transfer model to bright icy satellites. *Icarus* 61:208–217.

Burns, J. A. 1975. The angular momenta of solar system bodies: Implications for asteroid strengths. *Icarus* 25:545–554.

Burns, J. A. 1976. Consequences of the tidal slowing of Mercury. *Icarus* 28:453–458.

Burns, J. A., and Safronov, V. 1973. Asteroid nutation angles. *Mon. Not. Roy. Astron. Soc.* 165:403–411.

Burns, R. G. 1970. *Mineralogical Applications of Crystal Field Theory* (New York: Cambridge Univ. Press).

Bryan, J. B., Burton, D. E., Cunningham, M. E., and Lettis, L. A., Jr. 1978. A two-dimensional computer simulation of hypervelocity impact cratering: Some preliminary results for Meteor Crater, Arizona. *Proc. Lunar Planet. Sci. Conf.* 9:3931–3964.

Cadenhead, D. A., Wagner, N. J., Jones, B. R., and Stetter, J. R. 1972. Some surface characteristics and gas interactions of Apollo 14 fines and rock fragments. *Proc. Lunar Sci. Conf.* 3:2243–2257.

Cameron, A. G. W. 1978. Physics of the primitive solar accretion disk. *Moon and Planets* 18:5–40.

Cameron, A. G. W. 1985*a*. The partial volatilization of Mercury. *Icarus* 64:285–294.

Cameron, A. G. W. 1985*b*. Formation and evolution of the primitive solar nebula. In *Protostars and Planets II,* eds. D. C. Black and M. S. Matthews (Tucson: Univ. of Arizona Press), pp. 1073–1099.

Cameron, A. G. W., and Ward, W. R. 1976. The origin of the moon. *Lunar Sci.* VII:120–122 (abstract).

Cameron, A. G. W., DeCampli, W. M., and Bodenheimer, P. H. 1982. Evolution of giant gaseous protoplanets embedded in the primitive solar nebula. *Icarus* 49:298–312.

Camichel, H., and Dollfus, A. 1968. La rotation et la cartographie de la planète Mercure. *Icarus* 8:216–226.

Campbell, M. J., and Ulrichs, J. 1969. Electrical properties of rocks and their significance for lunar radar observations. *J. Geophys. Res.* 74:5867–5881.

Carey, W. C., and McDonnell, J. A. M. 1978. Monte Carlo sputter simulations and laboratory ion sputter measurements of lunar surfaces. *Proc. Lunar Sci. Conf.* 9:1725–1744.

Carpenter, R. L. 1966. Study of Venus by CW radar—1964 results. *Astron. J.* 71:142–152.

Carr, M. H., Saunders, R. S., Strom, R. G., and Wilhelms, D. E. 1984. *Geology of the Terrestrial Planets,* NASA SP-469.

Cassen, P., Young, R. E., Schubert, G., and Reynolds, R. T. 1976. Implications of an internal dynamo for the thermal history of Mercury. *Icarus* 28:501–508.

Castagnoli, G., and Lal, D. 1980. Solar modulation effects in terrestrial production of carbon-14. *Radiocarbon* 22:133–158.

Cathles, L. 1977. *The Viscosity of the Earth's Mantle* (New York: J. Wiley).

Chabai, A. J. 1977. Influence of gravitational fields and atmospheric pressures on scaling of explosion craters. In *Impact and Explosion Cratering*, eds. D. J. Roddy, R. O. Pepin, and R. B. Merrill (New York: Pergamon Press), pp. 1191–1214.

Chadderton, L. T., Krajenbrink, F. G., Katz, R., and Poveda, A. 1969. Standing waves on the Moon. *Nature* 223:259–263.

Chamberlain, J. W. 1961. *Physics of the Aurora and Airglow* (New York: Academic Press).

Chamberlain, J. W. 1963. Planetary coronae and atmospheric evaporation. *Planet. Space Sci.* 11:901–960.

Chamberlain, J. W., and Campbell, F. J. 1967. Rate of evaporation of a non-Maxwellian atmosphere. *Astrophys. J.* 149:687–705.

Chamberlain, J. W., and Hunten, D. M. 1987. *Theory of Planetary Atmospheres*, 2nd ed. (New York: Academic Press).

Chapman, C. R. 1968. Optical evidence on the rotation of Mercury. *Earth Planet. Sci. Lett.* 3:381–385.

Chapman, C. R. 1976. Chronology of terrestrial planet evolution: The evidence from Mercury. *Icarus* 28:523–536.

Chapman, C. R., and Gaffey, M. J. 1979. Reflectance spectra for 277 asteroids. In *Asteroids,* ed. T. Gehrels (Tucson: Univ. of Arizona Press), pp. 655–687.

Chapman, C. R., and McKinnon, W. B. 1986. Cratering on planetary satellites. In *Satellites,* eds. J. A. Burns and M. S. Matthews (Tucson: Univ. of Arizona Press), pp. 492–580.

Charters, A. C., and Summers, J. L. 1959. Comments on phenomena of high speed impact. *Rept. NOLR 1238,* Naval Ordnance Lab., White Oak, Silver Spring, Maryland, pp. 200–221.

Chase, M. W., Curnutt, J. L., Hu, A. T., Prophet, H., Syverud, A. N., Walker, L. C., Downey, J. R., Jr., McDonald, R. A., and Valenzuela, E. A., eds. 1971–1975. *JANAF Supplements,* (Midland, MI: Dow Chemical Co.).

Chase, S., Miner, E., Morrison, D., Münch, G., Neugebauer, G., and Schroeder, M. 1974. Preliminary infrared radiometry of the night side of Mercury from Mariner 10. *Science* 185:142–145.

Chase, S. C., Jr., Miner, E. D., Morrison, D., Münch, G., and Neugebauer, G. 1976. Mariner 10 infrared radiometer results: Temperatures and thermal properties of the surface of Mercury. *Icarus* 28:565–578.

Chenette, D. L., Conlon, T. F., Pyle, K. R., and Simpson, J. A. 1977. Observations of Jovian electrons at 1 AU throughout the 13-month Jovian synodic year. *Astrophys. J.* 215:L95–L99.

Cheng, A. F., Johnson, R. E., Krimigis, S. M., and Lanzerotti, L. J. 1987. Magnetosphere, exosphere and surface of Mercury. *Icarus* 71:430–440.

Chou, C.-L., Baedecker, P. A., and Wasson, J. T. 1976. Allende inclusions: Volatile-element distribution and evidence for incomplete volatilization of presolar solids. *Geochim. Cosmochim. Acta* 40:85–94.

Chou, C.-L., Sears, D. W., and Wasson, J. T. 1981. Composition and classification of clasts in the St. Mesmin LL chondrite breccia. *Earth Planet. Sci. Lett.* 54:367–378.

Christon, S. P., Daly, S. F., Eraker, J. H., Perkins, M. A., Simpson, J. A., and Tuzzolino, A. J. 1979. Electron calibration of instrumentation for low energy, high intensity particle measurements at Mercury. *J. Geophys. Res.* 84:4277–4288.

Christon, S. P., Feynman, J., and Slavin, J. A. 1986. Dynamic substorm injections: Similar magnetospheric phenomena at Earth and Mercury? In *Proc. Chapman Conf. on Magnetotail Physics,* in press.

Cintala, M. J. 1979. Mercurian crater rim heights and some interplanetary comparisons. *Proc. Lunar Planet. Sci. Conf.* 10:2635–2650.

Cintala, M. J. 1981. The Mercurian regolith: An evaluation of impact glass production by micrometeoroid impact. *Lunar Planet Sci.* XII:141–143 (abstract).

Cintala, M. J., and Grieve, R. A. F. 1984. Energy partitioning during terrestrial impact events: Melt production and scaling laws. *Lunar Planet. Sci.* XV:156–157 (abstract).

Cintala, M. J., and Hörz, F. 1984. Catastrophic rupture experiments: Fragment-size analysis and energy consideration. *Lunar Planet. Sci.* XV:158–159 (abstract).

Cintala, M. J., and Mouginis-Mark, P. J. 1980. Martian fresh crater depths: More evidence for substrate volatiles? *Lunar Planet Sci.* XI:143–145 (abstract).

Cintala, M. J., Head, J. W., and Mutch, T. A., 1976. Characteristics of fresh Martian craters as a function of diameter: Comparison with the Moon and Mercury. *Geophys. Res. Lett.* 3:117–120.

Cintala, M. J., Wood, C. A., and Head, J. W. 1977. The effects of target characteristics on fresh crater morphology: Preliminary results for the moon and Mercury. *Proc. Lunar Sci. Conf.* 8:3409–3425.

Clark, P. E. 1983. A look at the major terrains of Mercury through radar profiles. *Lunar Planet. Sci.* XIV:119 (abstract).

Clark, P. E., and Jurgens, R. F. 1984. Using new topographic profiles to interpret Mercury's surface structure. *Bull. Amer. Astron. Soc.* 16:668 (abstract).

Clark, P. E., and Jurgens, R. F. 1985. Radar derived morphology of linear features on Mercury. *Lunar Planet. Sci.* XVI:139 (abstract).

Clark, P. E., Strobell, M. E., Schaber, G. G., and Jurgens, R. F. 1984. Using new radar reflectivity maps to characterize features on Mercury. *Lunar Planet. Sci.* XV:166–167 (abstract).

Clark, P. E., Kobrick, M., and Jurgens, R. F. 1985. The use of radar and visual observations to characterize the surface structure of the planet Mercury. *IEEE Remote Sensing Trans.*, pp. 54–59.

Clark, P. E., Jurgens, R. F., Chan, F., Robinett, L., Brokl, S., Franck, C., and Stone, E. 1986. Mercury: New dual polarization measurements at X-band. *Lunar Planet. Sci.* XVII:135–136 (abstract).

Clark, R. N. 1979. Planetary reflectance measurements in the region of planetary thermal emission. *Icarus* 40:94–103.

Clayton, R. N., and Mayeda, T. K. 1984. The oxygen isotope record in Murchison and other carbonaceous chondrites. *Earth Planet. Sci. Lett.* 67:151–161.

Clayton, R. N., Grossman, L., and Mayeda, T. K. 1973. A component of primitive nuclear composition in carbonaceous meteorites. *Science* 182:485–488.

Clayton, R. N., Onuma, N., and Mayeda, T. K. 1976. A classification of meteorites based on oxygen isotopes. *Earth Planet. Sci. Lett.* 30:10–18.

Clayton, R. N., Onuma, N., Grossman, L., and Mayeda, T. K. 1977. Distribution of the presolar component in Allende and other carbonaceous chondrites. *Earth Planet. Sci. Lett.* 34:209–224.

Clayton, R. N., Mayeda, T. K., and Molini-Velsko, C. A. 1985. Isotopic variations in solar system material: Evaporation and condensation of silicates. In *Protostars and Planets II*, eds. D. C. Black and M. S. Matthews (Tucson: Univ. of Arizona Press), pp. 755–771.

Clow, G. D., and Pike, R. J. 1982. Statistical test of the $\sqrt{2}$ spacing rule for basin rings. *Lunar Planet Sci.* XII:123–124 (abstract).

Cohen, C. J., Hubbard, E. C., and Oesteruinter, C. 1973. Planetary elements for 10,000,000 years. *Celestial Mech.* 7:438–448.

Colombo, G. 1965. On the rotational period of the planet Mercury. *Nature* 208:575.

Colombo, G. 1966. Cassini's second and third laws. *Astron. J.* 71:891–896.

Colombo, G., and Shapiro, I. I. 1966. The rotation of the planet Mercury. *Astrophys. J.* 145:296–307.

Committee on Gravitational Physics. 1981. *Strategy for Space Research in Gravitational Physics on the 1980's.* Natl. Research Council, Natl. Academy of Sciences, Washington, DC.

Committee on Solar and Space Physics. 1980. *Solar System Space Physics in the 1980's: A Research Strategy.* Natl. Research Council, Natl. Academy of Sciences, Washington, DC.

COMPLEX. 1977. Strategy for Exploration of the Inner Planets: 1977–1987.

Conlon, T. F. 1978. The interplanetary modulation and transport of Jovian electrons. *J. Geophys. Res.* 83:541–552.

Connerney, J. E. P. 1981. The magnetic field of Jupiter: A generalized inverse approach. *J. Geophys. Res.* 86:7679–7693.

Cooper, J. F. 1986. Solar gamma ray and neutron observations from Mercury. *Mercury: Program and Abstracts*, Tucson, Arizona, August.

Cordell, B. M., and Strom, R. G. 1977. Global tectonics of Mercury and the Moon. *Phys. Earth Planet. Int.* 15:146–155.

Counselman, C. C. 1969. Spin-Orbit Resonance of Mercury. Ph.D. Thesis, Massachusetts Inst. of Technology.

Cox, L. P., and Lewis, J. P. 1980. Numerical simulation of the final stages of terrestrial planet accretion. *Icarus* 44:706–721.

Crabb, J., and Anders, E. 1981. Noble gases in E-chondrites. *Geochim. Cosmochim. Acta* 45:2443–2464.

Crawford, D. A., and Schultz, P. H. 1987. Electromagnetic emissions from low-angle hypervelocity impacts. *Lunar Planet. Sci.* XVIII:205–206 (abstract).

Croft, S. K. 1978. Lunar crater volumes: Interpretations of models of impact cratering and upper crustal structure. *Proc. Lunar Planet. Sci. Conf.* 9:3711–3733.

Croft, S. K. 1979. Impact Craters from Centimeters to Megameters. Ph.D. Thesis, Univ. of California Los Angeles.

Croft, S. K. 1981a. The excavation stage of basin formation: A qualitative model. In *Multi-Ring Basins: Proc. Lunar Planet. Sci.*, eds. P. H. Schultz and R. B. Merrill, 12A:207–225.

Croft, S. K. 1981b. The modification stage of basin formation: Conditions of ring formation. In *Multi-Ring Basins: Proc. Lunar Planet. Sci.*, eds. P. H. Schultz and R. B. Merrill, 12A:227–257.

Croft, S. K. 1981c. On the origin of pit craters. *Lunar Planet. Sci.* XII:196–198 (abstract).

Croft, S. K. 1983. A proposed origin for palimpsests and anomalous pit craters on Ganymede and Callisto. *Proc. Lunar Planet. Sci. Conf.* 14, *J. Geophys. Res.* 88:B71–B89.

Croft, S. K. 1985. The scaling of complex craters. *Proc. Lunar Planet. Sci. Conf.* 15, *J. Geophys. Res. Suppl.* 90:C828–C842.

Croxton, F. E., Cowden, D. J., and Klein, S. 1967. *Applied General Statistics* (Englewood Cliffs: Prentice-Hall).

Cruikshank, D. P., and Chapman, C. R. 1967. Mercury's rotation and visual observations. *Sky and Telescope* 34:24–26.

Curtis, S. A., and Hartle, R. E. 1978. Mercury's helium exosphere after Mariner 10's third encounter. *J. Geophys. Res.* 83:1551–1557.

Cuzzi, J. N. 1974. The nature of the subsurface of Mercury from microwave observations at several wavelengths. *Astrophys. J.* 189:577–586.

Danjon, A. 1933. *Ann. Obs. Strasbourg* 2:170.

Danjon, A. 1949. Photometrie et Colorimetrie des Planètes Mercure et Venus. *Bull. Astron. J.* 14:315.

Danjon, A. 1953. *Bull. Astron. J.* 17:363.

Davies, M. E. 1976. The control net of Mercury. RAND R-1914-NASA, pp. 1–20.

Davies, M. E., and Batson, R. M. 1975. Surface coordinates and cartography of Mercury. *J. Geophys. Res.* 80:2417–2430.

Davies, M. E., Dwornik, S. E., Gault, D. E., and Strom, R. G. 1978. *Atlas of Mercury* (Washington, DC: NASA). See also Photographic Ed., 1976.

Davis, J. C. 1986. *Statistics and Data Analysis in Geology,* 2nd ed. (New York: Wiley).

De Hon, R. A. 1978. In search of ancient astroblemes: Mercury. In *Reports of Planet. Geol. Program,* NASA TM-79729, pp. 150–152.

De Hon, R. A. 1979. Preliminary thickness study of plains-forming materials in the Caloris Basin region of Mercury. *Rept. Planet. Geol. Prog., 1978–1979,* NASA TM-80339, pp. 65–66.

De Hon, R. A., Scott, D. H., and Underwood, J. R. 1981. Geologic map of the Kuiper quadrangle of Mercury. U.S. Geol. Surv. Map I-1233.

Denardo, B. P. 1962. Measurements of Momentum Transfer from Plastic Projectiles to Massive Aluminum Targets at Speeds up to 25,600 Feet Per Second, NASA TN D-1210.

Denardo, B. P. 1966. Penetration of Polyethylene into Semi-Infinite 2024-T351 Aluminum up to Velocities of 37,000 Feet Per Second, NASA TN D-3369.

Denardo, B. P., and Nysmith, C. R. 1966. Momentum transfer and cratering phenomena associated with the impact of aluminum spheres into thick aluminum targets at velocities of 24,000 feet per second. In *The Fluid Dynamic Aspects of Space Flight* (New York: Gordon and Breach), pp. 389–402.

Dence, M. R. 1972. The nature and significance of terrestrial impact structures. *Intl. Geol. Congress, XXIV Session,* Section 15:77–89.

Dence, M. R., and Grieve, R. A. F. 1979. The formation of complex impact structures. *Lunar Planet. Sci.* X:292–294 (abstract).

Dence, M. R., Grieve, R. A. F., and Robertson, P. B. 1977. Terrestrial impact structures: Principle characteristics and energy considerations. In *Impact and Explosion Cratering,* eds. D. J. Roddy, R. O. Pepin, and R. B. Merrill (New York: Pergamon Press), pp. 247–275.

de Vaucouleurs, G. 1964. Geometric and photometric properties of the terrestrial planets. *Icarus* 3:187–235.

Devonshire, A. F. 1937. The interaction of atoms and molecules with solid surfaces VIII—The exchange of energy between a gas and a solid. *Proc. Roy. Soc.* A158:268–279.

Dial, A. L. 1978. The Viking I landing site crater diameter-frequency distribution. *Repts. Planet. Geol. Prog.*, NASA TM-79729, pp. 179–181.

Dienes, J. K., and Walsh, J. M. 1970. Theory of impact: Some general principles and the method of Eulerian codes. In *High-Velocity Impact Phenomena*, ed. R. Kinslow (New York: Academic Press), pp. 46–104.

Dobrovolskis, A. 1981. Ejecta patterns diagnostic of planetary rotations. *Icarus* 47:203–219.

Dollfus, A. 1950. Observation of an atmosphere around the planet Mercury. *Comptes Rend. Acad. Sci. Paris* 231.

Dollfus, A. 1961. Polarization studies of planets. In *Planets and Satellites*, eds. G. P. Kuiper and B. M. Middlehurst (Chicago: Univ. of Chicago Press), pp. 343–399.

Dollfus, A. 1985. Photopolarimetric sensing of planetary surfaces. *Adv. Space Res.* 5:47–58.

Dollfus, A., and Auriere, M. 1974. Optical polarimetry of the planet Mercury. *Icarus* 23:465–482.

Dollfus, A., and Titulaer, C. 1971. Polarimetric properties of the lunar surface and its interpretation. 3. Volcanic samples in several wavelengths. *Astron. Astrophys.* 12:199–209.

Dohnanyi, M. W. 1969. Collisional model of asteroids and their debris. *J. Geophys. Res.* 74:2531–2554.

Donahue, T. M., and Pollack, J. B. 1983. Origin and evolution of the atmosphere of Venus. In *Venus*, eds. D. M. Hunten, L. Colin, T. M. Donahue, and V. I. Moroz (Tucson: Univ. of Arizona Press), pp. 1003–1036.

Ducati, H., Kalbitzer, G., Kiko, J., Kirstein, T., and Muller, H. W. 1973. Rare gas diffusion studies in individual lunar soil particles and in artificially implanted glasses. *The Moon* 8:210–227.

Dunne, J. A., and Burgess, E. 1978. *The Voyage of Mariner 10: Mission to Venus and Mercury,* NASA SP-424.

Dyce, R. B. 1970. Radar studies of the planets. In *Surfaces and Interiors of Planets and Satellites,* ed. A. Dollfus (New York: Academic Press), pp. 140–168.

Dzurisin, D. 1976. Scarps, Ridges, Troughs, and other Lineaments on Mercury. Ph.D. Thesis, California Inst. of Technology.

Dzurisin, D. 1977. Mercurian bright patches: Evidence of physico-chemical alteration of surface material? *Geophys. Res. Lett.* 4:383–386.

Dzurisin, D. 1978. The tectonic and volcanic history of Mercury as inferred from studies of scarps, ridges, troughs, and other lineaments. *J. Geophys. Res.* 83:4883–4906.

Eichelberger, R. J., and Gehring, J. W. 1962. Effects of meteoroid impacts on space vehicles. *Amer. Rocket Soc. J.* 32:1583–1591.

Eichhorn, G. 1976. Analysis of the hypervelocity impact process from impact flash measurements. *Planet Space Sci.* 24:771–781.

Eichhorn, G. 1978. Heating and vaporization during hypervelocity particle impact. *Planet. Space Sci.* 26:463–467.

Elachi, C., Kobrick, M., Roth, L., Tiernan, M., and Brown, W. 1976. Local lunar topography from Apollo 17 Alse radar imaging and altimetry. *The Moon* 15:119–131.

Eppler, D. T., Erlich, R., Nummendal, D., and Schultz, P. H. 1983. Sources of shape variations in lunar impact craters. *Geol. Soc. Amer. Bull.* 94:274.

Eraker, J. H., and Simpson, J. A. 1979. Jovian electron propagation close to the sun (~0.5 AU). *Astrophys. J.* 232:L131–L134.

Eraker, J. H., and Simpson, J. A. 1986. Acceleration of charged particles in Mercury's magnetosphere. *J. Geophys. Res.* 91:9973–9993.

Evans, J. V. 1969. Radar studies of planetary surfaces. *Ann. Rev. Astron. Astrophys.* 7:201–248.

Evans, J. V., Ingalls, R. P., Rainville, L. P., and DeSilva, R. R. 1966. Radar observations of Venus at 3.8-cm wavelength. *Astron. J.* 71:902–915.

Fairfield, D. H., and Behannon, K. W. 1976. Bow shock and magnetosheath waves at Mercury. *J. Geophys. Res.* 81:3897–3906.

Fastie, W. C., Feldman, P. D., Henry, R. C., Moos, H. W., and Donahue, T. M. 1973. A search for far-ultraviolet emissions from the lunar atmosphere. *Science* 182:710–711.

Fedoretz, V. A. 1952. Photographic photometry of the lunar surface. *Publ. Karkov. Obs.*, vol. II, Vch. Zap. Karkov Univ. 42:49–179.

Fegley, B. 1982. A condensation-accretion model for volatile element retention. *Papers Conf. Planetary Volatiles* (Houston: Lunar and Planetary Inst.), pp. 37–38.

Fegley, B. 1983. Primordial retention of N_2 by terrestrial planets and meteorites. *Proc. Lunar Planet. Sci. Conf.* 13, *J. Geophys. Res. Suppl.* 88:A853–A868.

Fegley, B., Jr., and Cameron, A. G. W. 1987. A vaporization model for iron/silicate fractionation in the Mercury protoplanet. *Earth Planet. Sci. Lett.* 82:207–222.

Fegley, B., and Lewis, J. S. 1980. Volatile element chemistry in the solar nebula: Na, K, F, Cl, and P. *Icarus* 41:439–455.

Fegley, B., Jr., and Palme, H. 1985. Evidence for oxidizing conditions in the solar nebula from Mo and W depletions in refractory inclusions in carbonaceous meteorites. *Earth Planet. Sci. Lett.* 72:311–326.

Ferrari, A. J., Sinclair, W. S., Sjogren, W. L., Williams, J. G., and Yoder, C. F. 1980. Geophysical parameters of the Earth-Moon system. *J. Geophys. Res.* 85:3939–3951.

Feynman, R. P., Leighton, R. B., and Sands, M. 1963. *The Feynman Lectures on Physics* (Reading, MA: Addison-Wesley).

Field, G. 1964. The atmosphere of Mercury. In *The Origin and Evolution of Atmospheres and Oceans*, eds. P. J. Brancazio and A. G. W. Cameron (New York: Wiley), pp. 269–276.

Fielder, G. 1963a. Nature of the lunar maria. *Nature* 198:1256–1260.

Fielder, G. 1963b. Lunar tectonics. *J. Geol. Soc. London* 119:65–94.

Fielder, G. 1974. Lineament patterns on the Moon, Mars, and Mercury. In *Proc. Intl. Conference on New Basement Tectonics* 1:379–387.

Fink, U., Larson, H. P., and Poppen, R. F. 1974. A new upper limit for an atmosphere of CO_2, CO on Mercury. *Astrophys. J.* 287:407–415.

Fjeldbo, G., Kliore, A., Sweetnam, D., Esposito, P., Seidel, B., and Howard, T. 1976. The occultation of Mariner 10 by Mercury. *Icarus* 29:439–444.

Fleitout, L., and Thomas, P. G. 1982. Far-field tectonics associated with a large impact basin: Applications to Caloris on Mercury and Imbrium on the Moon. *Earth Planet. Sci. Lett.* 58:104–115.

Fortson, E. P., and Brown, F. R. 1958. Effect of soil-rock interface on crater morphology. *Tech. Rept. 20478*, U.S. Army Eng. Stn., Corp. of Eng., Vicksburg, Mississippi.

French, J. R., and Wright, J. 1986. Solar sail missions to Mercury. *Mercury: Program and Abstracts*, Tucson, Arizona, August.

Frey, H., and Lowry, B. L. 1979. Large impact basins on Mercury and relative crater production rates. *Proc. Lunar Planet. Sci. Conf.* 10:2669–2687.

Fricker, P. E., Reynolds, R. T., Summers, A. L., and Cassen, P. M. 1976. Does Mercury have a molten core? *Nature* 259:293–294.

Friedlander, A. L., and Davis, D. R. 1976. Penetrator mission concepts for Mercury and the Galilean satellites. Report No. SAI 1-720-399-M5, Science Applications, Inc., Schaumburg, Illinois.

Friedlander, A. L., and Feingold, H. 1977. Mercury orbiter transport study. Report No. SAI 1-120-580-T6, Science Applications, Inc., Schaumburg, Illinois.

Friedman, L., et al. 1978. Solar sailing: The concept made realistic. AIAA Paper No. 78-82, AIAA 16th Aerospace Sciences Meeting, Huntsville, Alabama, January 16–18.

Frost, H. J., and Ashby, M. F. 1982. *Deformation Mechanism Maps* (Oxford: Pergamon Press).

Gadsden, M. 1968. Sodium in the upper atmosphere: Meteoric origin. *J. Atmos. Terr. Phys.* 30:151–161.

Gaffey, M. J. 1984. Rotational spectral variations of asteroid (8) Flora: Implications for the nature of the S-type asteroids and for the parent bodies of the ordinary chondrites. *Icarus* 60:83–114.

Gaffey, M. J., and McCord, T. B. 1979. Mineralogical and petrological characterizations of asteroid surface materials. In *Asteroids*, ed. T. Gehrels (Tucson: Univ. of Arizona Press), pp. 688–723.

Gans, R. F. 1972. Viscosity of the Earth's core. *J. Geophys. Res.* 77:360–366.

Gardner, C. S., Voelz, D. G., Sechrist, C. F. J., and Segal, A. C. 1986. Lidar studies of the nighttime sodium layer over Urbana, Illinois. 1. Seasonal and nocturnal variations. *J. Geophys. Res.* 91:13,659–13,673.

Gault, D. E. 1973. Displaced mass, depth, diameter, and effects of oblique trajectories for impact craters formed in dense crystalline rocks. *The Moon* 6:32–44.

Gault, D. E. 1974. Impact cratering. In *A Primer in Lunar Geology*, eds. R. Greeley and P. Schultz, NASA TM-X-62359, pp. 137–176.

Gault, D. E., and Heitowit, E. D. 1963. The partitioning of energy for impact craters formed in rock. In *Proc. of the Sixth Hypervelocity Impact Symp.*, vol. 2 (Cleveland: Firestone Rubber Co.), pp. 419–456.

Gault, D. E., and Schultz, P. H. 1986. Oblique impact: Projectile ricochet, concomitant ejecta, and momentum transfer. *Meteoritics* 21:368–369 (abstract).

Gault, D. E., and Sonett, C. P. 1982. Laboratory simulation of pelagic asteroidal impact: Atmospheric injection, benthic topography and the surface wave radiation field. *Geol. Soc. Amer. Special Paper* 190:69–92.

Gault, D. E., and Wedekind, J. 1977. Experimental hypervelocity impact into quartz sand. II. Effects of gravitational acceleration. In *Impact and Explosion Cratering*, eds. D. J. Roddy, R. O. Pepin, and R. B. Merrill (New York: Pergamon Press), pp. 1231–1244.

Gault, D. E., and Wedekind, J. A. 1978. Experimental studies of oblique impact. *Proc. Lunar Planet. Sci. Conf.* 9:3843–3875.

Gault, D. E., Quaide, W. L., and Oberbeck, V. R. 1968. Impact cratering mechanics and structure. In *Shock Metamorphism of Natural Materials,* eds. B. M. French and N. M. Short (Baltimore: Mono Book Co.), pp. 87–99.

Gault, D. E., Guest, J. E., Murray, J. B., Dzurisin, D., and Malin, M. C. 1975. Some comparisons of impact craters on Mercury and the Moon. *J. Geophys. Res.* 80:2444–2460.

Gault, D. E., Burns, J. A., Cassen, P., and Strom, R. G. 1977. Merucry. *Ann. Rev. Astron. Astrophys.* 15:97–126.

Geake, J. E., and Dollfus, A. 1986. Planetary surface texture and albedo from parameter plots of optical polarization data. *Mon. Not. Roy. Astron. Soc.* 218:75–91.

Gehrels, T., Coffeen, T., and Owings, D. 1964. Wavelength dependence of polarization. III. The lunar surface. *Astron. J.* 69:826–852.

Gehrels, T., Landau, R., and Coyne, G. V. 1987. Mercury: Wavelength and longitude dependence of polarization. *Icarus* 71:386–396.

Geiss, J. 1987. First results of the Comet Halley exploration. *Terra Cognita* 7:39–41.

Gibson, E. K. J. 1977. Production of simple molecules on the surface of Mercury. *Phys. Earth Planet. Int.* 15:303–311.

Gilbert, G. K. 1893. The Moon's face: A study of the origin of its features. *Bull. Phil. Soc. Washington* 12:241–292.

Glasstone, S., Laidler, K. J., and Eyring, H. 1941. *The Theory of Rate Processes: The Kinetics of Chemical Reactions, Viscosity, Diffusion and Electrochemical Phenomena* (New York: McGraw-Hill).

Glushko, V. P., Gurvich, L. V., Bergman, G. A., Veitz, I. V., Medvedev, V. A., Khachkuruzov, G. A., and Yungman, V. S., eds. 1978–1982. *Thermodynamic Properties of Individual Substances,* 4 vols. (Moscow: High Temperature Inst.).

Goettel, K. A. 1981. Density of the mantle of Mars. *Geophys. Res. Lett.* 8:497–500.

Goettel, K. A. 1983. Present constraints on the composition of the mantle of Mars. *Carnegie Institution of Washington, Year Book 82*, pp. 363–366.

Goettel, K. A., and Barshay, S. S. 1978. The chemical equilibrium model for condensation in the solar nebula: Assumptions, implications and limitations. In *The Origin of the Solar System,* ed. S. Dermott (Chichester: John Wiley & Son), pp. 611–627.

Goldreich, P., and Peale, S. J. 1966. Spin-orbit coupling in the solar system. *Astron. J.* 71:425–438.

Goldreich, P., and Peale, S. J. 1967. Spin-orbit coupling in the solar system II: The resonant rotation of Venus. *Astron. J.* 72:662–668.

Goldreich, P., and Peale, S. J. 1968. The dynamics of planetary rotations. *Ann. Rev. Astron. Astrophys.* 6:287–320.

Goldreich, P., and Peale, S. J. 1970. The obliquity of Venus. *Astron. J.* 75:273–283.

Goldreich, P., and Soter, S. 1966. Q in the solar system. *Icarus* 5:375–389.

Goldreich, P., and Toomre, A. 1969. Some remarks on polar wandering. *J. Geophys. Res.* 74:2555–2567.

Goldreich, P., and Ward, W. 1973. The formation of planetesimals. *Astrophys. J.* 183:1051–1061.

Goldstein, B. E., Suess, S. T., and Walker, R. J. 1981. Mercury: Magnetospheric processes and atmospheric supply and loss rates. *J. Geophys. Res.* 86:5485–5499.

Goldstein, R. M. 1971. Radar observations of Mercury. *Astron. J.* 76:1152–1154.

Golitsyn, G. F. 1979. Simple theoretical and experimental studies of convection with some geophysical applications and analogies. *J. Fluid Mech.* 95:567–608.

Golombek, M. P. 1985. Fault type predictions from stress distributions on planetary surfaces: Importance of fault initiation depth. *J. Geophys. Res.* 90:3065–3074.

Gombosi, T. I., Nagy, A. F., and Cravens, T. E. 1986. Dust and neutral gas modeling of the inner atmospheres of comets. *Rev. Geophys.* 24:667–700.

Gooding, J. L. 1979. Petrogenetic Properties of Chondrules in Unequilibrated H-, L-, and LL-Group Chondritic Meteorites. Ph.D. Thesis, Univ. of New Mexico.

Gooding, J. L., Keil, K., Fukuoka, T., and Schmitt, R. A. 1980. Elemental abundances in chondrules from unequilibrated chondrites: Evidence for chondrule origin by melting of pre-existing materials. *Earth Planet. Sci. Lett.* 50:171–180.

Gooding, J. L., Mayeda, T. K., Clayton, R. N., and Fukuoka, T. 1983. Oxygen isotopic heterogeneities, their petrological correlations, and implications for melt origins of chondrules in unequilibrated ordinary chondrites. *Earth Planet. Sci. Lett.* 65:209–224.

Gordon, R. W. 1983. In *Yearbook of Astronomy 1983* (London: Sidgwick and Jackson), pp. 151–153.

Gosling, J. T., and Robson, A. E. 1985. Ion reflection, gyration and dissipation at supercritical shocks. In *Collisionless Shocks in the Heliosphere: Reviews of Current Research,* eds. B. T. Tsurutani and R. G. Stone (Washington, DC: American Geophysical Union), pp. 141–152.

Gradie, J., and Tedesco, E. 1982. Compositional structure of the asteroid belt. *Science* 216:1405–1407.

Greeley, R., Fink, J. H., Gault, D. E., and Guest, J. 1982. Experimental simulation of impact cratering on icy satellites. In *Satellites of Jupiter,* ed. D. Morrison (Tucson: Univ. of Arizona Press), pp. 340–378.

Greenberg, R., Wacker, J., Hartmann, W. K., and Chapman, C. R. 1978. Planetesimals to planets: Numerical simulation of collisional evolution. *Icarus* 35:1–26.

Greenspan, H. P., and Howard, L. N. 1963. On a time dependent motion of a rotating fluid. *J. Fluid Mech.* 17:385–404.

Greenstadt, E. W. 1985. Oblique, parallel and quasi-parallel morphology of collisionless shocks. In *Collisionless Shocks in the Heliosphere: Reviews of Current Research,* ed. B. T. Tsurutani and R. G. Stone (Washington, DC: American Geophysical Union), pp. 169–184.

Grieve, R. A. F., and Garvin, J. B. 1984. A geometric model for excavation and modification at terrestrial simple impact craters. *J. Geophys. Res.* 89:11,561–11,572.

Grieve, R. A. F., Dence, M. R., and Robertson, P. B. 1977. Cratering processes: As interpreted from the occurrence of impact melts. In *Impact and Explosion Cratering,* eds. D. J. Roddy, R. O. Pepin, and R. B. Merrill (New York: Pergamon Press), pp. 791–814.

Grieve, R. A. F., Robertson, P. B., and Dence, M. R. 1981. Constraints on the formation of ring impact structures, based on terrestrial data. In *Multi-Ring Basins: Proc. Lunar Planet. Sci.,* eds. P. H. Schultz and R. B. Merrill, 12A:37–57.

Grossman, J. N., and Wasson, J. T. 1983. The compositions of chondrules in unequilibrated chondrites: An evaluation of models for the formation of chondrules and their precursor materials. In *Chondrules and Their Origins,* ed. E. A. King (Houston: Lunar and Planetary Inst.), pp. 88–121.

Grossman, J. N., Clayton, R. N., and Mayeda, T. K. 1987. Oxygen isotopes in the matrix of the Semarkona (LL3.0) chondrite. *Meteoritics* 22:395–396 (abstract).

Grossman, L. 1972. Condensation in the primitive solar nebula. *Geochim. Cosmochim. Acta* 36:597–619.

Gubbins, D. 1974. Theories of the geomagnetic and solar dynamos. *Rev. Geophys. Space Phys.* 72:137–154.

Gubbins, D. 1977a. Energetics of the Earth's core. *J. Geophys.* 43:453–464.

Gubbins, D. 1977b. Speculations on the origin of the magnetic field of Mercury. *Icarus* 30:186–191.

Gudmundsson, O., Clayton, R. W., and Anderson, D. L. 1986. CMB topography inferred from ISC PcP travel times. *Eos Trans. AGU* 67:1100 (abstract).

Guest, J. E., and Gault, D. E. 1976. Crater populations in the early history of Mercury. *Geophys. Res. Lett.* 3:121–123.

Guest, J. E., and Greeley, R. 1983. Geologic map of the Shakespeare quadrangle of Mercury. U.S. Geol. Survey Map I-1408.

Guest, J. E., and O'Donnell, W. P. 1977. Surface history of Mercury: A review. *Vistas in Astron.* 20:273–300.

Gurnis, M. 1987. Spatial distributions of craters on Callisto and limits on the crater production population of the jovian system. *Icarus,* submitted.

Hagfors, T. 1967. A study of the depolarization of lunar radar echoes. *Radio Sci.* 2:445–465.

Hagfors, T. 1968. Relations between rough surfaces and their scattering properties as applied to radar astronomy. In *Radar Astronomy,* eds. J. Evans and T. Hagfors (New York: McGraw Hill), pp. 187–218.

Hagfors, T. 1970. Remote probing of the moon by infrared and microwave emissions and by radar. *Radio Sci.* 5:189–227.

Hale, W. S. 1983. Central structures in martian impact craters: Morphology, morphometry and implications for substrate volatile distributions. *Lunar Planet. Sci.* XIV:273–274 (abstract).

Hale, W. S., and Grieve, R. A. F. 1982. Volumetric analysis of complex lunar craters: Implications for basin ring formation. *Proc. Lunar Planet. Sci. Conf.* 13, *J. Geophys. Res. Suppl.* 87:A65–A76.

Hale, W., and Head, J. W. 1977. Impact melt on lunar crater rims. In *Impact and Explosion Cratering,* eds. D. J. Roddy, R. O. Pepin, and R. B. Merrill (New York: Pergamon Press), pp. 815–841.

Hale, W. S., and Head, J. W. 1979a. Lunar central peak basins: Morphology and morphometry in the crater to basin transition zone. *Repts. Planet. Geol. Prog., 1978–1979,* NASA TM-80339, pp. 160–162.

Hale, W., and Head, J. W. 1979b. Central peaks in lunar craters: Morphology and morphometry. *Proc. Lunar Planet. Sci. Conf.* 10:2623–2633.

Hale, W. S., and Head, J. W. 1980a. Central peaks in mercurian craters: Comparisons to the moon. *Proc. Lunar Planet. Sci. Conf.* 11:2191–2205.

Hale, W. S., and Head, J. W. 1980b. Crater central peaks on the Moon, Mercury, and Earth. *Repts. Planet. Geol. Prog., 1980,* NASA TM-82385, pp. 131–133.

Hale, W., and Head, J. W. 1980c. The origin of peak rings and the crater to basin transition. In *Papers Presented to the Conference on Multi-Ring Basins* (Houston: Lunar and Planetary Inst.), p. 27–29.

Hameen-Antilla, K. A. 1967. Surface photometry of the planet Mercury. *Ann. Acad. Sci. Fenn., Series VI: Physica* 252.

Hameen-Antilla, K. A., Laakso, P., and Lumme, K. 1965. The shadow effect in the phase curves of lunar type surfaces. *Ann. Acad. Sci. Fenn., Series VI: Physica* 172.

Hameen-Antilla, K., Pikkarainen, T., and Camichel, M. 1970. Photometric studies of the planet Mercury. *The Moon* 1:440–448.

Hanson, W. B., and Patterson, T. N. L. 1963. Diurnal variation of the hydrogen concentration in the exosphere. *Planet. Space Sci.* 11:1035–1052.

Hapke, B. 1977. Interpretation of optical observations of Mercury and the Moon. *Phys. Earth Planet. Int.* 15:264–274.

Hapke, B. W. 1981. Bidirectional reflectance spectroscopy I. Theory. *J. Geophys. Res.* 86:3039–3054.

Hapke, B. W. 1984. Bidirectional reflectance spectroscopy III. Correction for macroscopic roughness. *Icarus* 59:41–59.

Hapke, B. W. 1986. Bidirectional reflectance spectroscopy IV. The extinction coefficient and the opposition effect. *Icarus* 67:264–280.

Hapke, B., Danielson, G. E., Jr., Klaasen, K., and Wilson, L. 1975. Photometric observations of Mercury from Mariner 10. *J. Geophys. Res.* 80:2431–2443.

Hapke, B., Christman, C., Rava, B., and Mosher, J. 1980. A color-ratio map of Mercury. *Proc. Lunar Planet. Sci. Conf.* 11:817–821.

Hardorp, J. 1980. The Sun among the stars. III. Energy distributions of 16 Northern G-type stars and the solar flux calibration. *Astron. Astrophys.* 120:529–559.

Harlow, F. H., and Shannon, J. P. 1967. Distortion of a splashing liquid drop. *Science* 157:547–550.

Harmon, J. K., and Ostro, S. J. 1985. Mars: Dual polarization radar observations with extended coverage. *Icarus* 62:110–128.

Harmon, J. K., Campbell, D. B., Bindschadler, D. L., Head, J. W., and Shapiro, I. I. 1986. Radar altimetry of Mercury: A preliminary analysis. *J. Geophys. Res.* 91:385–401.

Harris, D. L. 1961. Photometry and colorimetry of planets and satellites. In *Planets and Satellites,* eds. G. P. Kuiper and B. M. Middlehurst (Chicago: Univ. of Chicago Press), pp. 272–342.

Harris, P. G., and Tozer, D. C. 1967. Fractionation of iron in the solar system. *Nature* 215:1449–1451.

Hartle, R. E. 1971. Model for rotating and nonuniform planetary exospheres. *Phys. Fluids* 14:2592–2598.

Hartle, R. E., Ogilvie, K. W., and Wu, C. S. 1973. Neutral and ion exospheres in the solar wind with applications to Mercury. *Planet. Space Sci.* 21:2181–2191.

Hartle, R. E., Ogilvie, K. W., Scudder, J. D., Bridge, H. S., Siscoe, G. L., Lazarus, A. J., Vasyliunas, V. M., and Yeates, C. M. 1975*a*. Preliminary interpretation of plasma electron observations at the third encounter of Mariner 10 with Mercury. *Nature* 255:206–208.

Hartle, R. E., Curtis, S. A., and Thomas, G. E. 1975*b*. Mercury's helium exosphere. *J. Geophys. Res.* 80:3689–3692.

Hartmann, W. K. 1965. Terrestrial and lunar flux of large meteorites in the last two billion years. *Icarus* 4:157–165.

Hartmann, W. K. 1966. Early lunar cratering. *Icarus* 5:406–418.

Hartmann, W. K. 1972. Interplanet variations in scale of crater morphology: Earth, Mars, Moon. *Icarus* 17:707–713.

Hartmann, W. K. 1976. Planet formation: Compositional mixing and lunar compositional anomalies. *Icarus* 27:553–559.

Hartmann, W. K. 1977. Relative crater production rates on planets. *Icarus* 31:260–276.

Hartmann, W. K. 1981. Discovery of multi-ring basins: Gestalt perception in planetary science. In *Multi-Ring Basins: Proc. Lunar Planet. Sci.,* eds. P. H. Schultz and R. B. Merrill, 12A:79–90.

Hartmann, W. K. 1984. Does crater "saturation equilibrium" occur in the solar system? *Icarus* 60:56–74.

Hartmann, W. K., and Davis, D. R. 1975. Satellite-sized planetesimals and lunar origin. *Icarus* 24:504–515.

Hartmann, W. K., and Kuiper, G. P. 1962. Concentric structures surrounding lunar basins. *Comm. Lunar Planet. Lab.* 1:51–66.

Hartmann, W. K., and Raper, O. 1974. *The New Mars: The Discoveries of Mariner 9,* NASA SP-337.

Hartmann, W. K., and Wood, C. A. 1971. Moon: Origin and evolution of multi-ring basins. *The Moon* 3:3–78.

Hartmann, W. K., Strom, R. G., Weidenschilling, S. J., Blasius, K. R., Woronow, A., Dence, M. R., Grieve, R. A. F., Diaz, J., Chapman, C. R., Shoemaker, E. M., and Jones, K. L. 1981. Chronology of planetary volcanism by comparative studies of planetary cratering. In *Basaltic Volcanism on the Terrestrial Planets,* ed. W. M. Kaula (New York: Pergamon Press), pp. 1048–1127.

Hashimoto, A. 1983. Evaporation metamorphism in the early solar nebula—Evaporation experiments on the melt $FeO-MgO-SiO_2-CaO-Al_2O_3$ and chemical fractionations of primitive materials. *Geochem. J.* 17:111–145.

Hastie, J. W., and Bonnell, D. W. 1985. A predictive phase equilibrium model for multicomponent oxide mixtures. Part II. Oxides of Na-K-Ca-Mg-Al-Si. *High Temp. Sci.* 19:275–306.

Hastie, J. W., Horton, W. S., Plante, E. R., and Bonnell, D. W. 1982*a*. Thermodynamic models of alkali-metal vapor transport in silicate systems. *High Temp. High Press.* 14:669–679.

Hastie, J. W., Plante, E. R., and Bonnell, D. W. 1982*b*. Alkali vapor transport in coal conversion and combustion systems. In *Metal Bonding and Interactions in High Temperature Systems,* eds. J. L. Gole and W. C. Stwalley (Washington, DC: American Chemical Society), pp. 543–600.

Hastie, J. W., Bonnell, D. W., Plante, E. R., and Horton, W. S. 1984. Thermodynamic activity and vapor pressure models for silicate systems including coal slags. In *Thermochemistry and Its Applications to Chemical and Biochemical Systems,* ed. M. A. V. Ribeiro da Silva (Dordrecht: D. Reidel), pp. 235–251.

Hawke, B. R., and Cintala, M. J. 1977. Impact melts on Mercury and the Moon. *Bull. Amer. Astron. Soc.* 9:531 (abstract).

Hawke, B. R., and Head, J. W. 1977. Impact melt on lunar crater rims. In *Impact and Explosion Cratering,* eds. D. J. Roddy, R. O. Pepin, and R. B. Merrill (New York: Pergamon Press), pp. 815–841.

Hawke, B. R., Coombs, C., and Cintala, M. J. 1986. Impact melt deposits on Mercury and the Moon. *Mercury: Program and Abstracts,* Tucson, Arizona, August.

Haxby, W. F., and Turcotte, D. L. 1976. Stresses induced by the addition or removal of overburden and associated thermal effects. *Geology* 4:181–184.

Head, J. W. 1976. The significance of substrate characteristics in determining morphology and morphometry of lunar craters. *Proc. Lunar Sci. Conf.* 7:2913–2929.

Head, J. W. 1977. Origin of outer rings in lunar multi-ring basins: Evidence from morphology and ring spacing. In *Impact and Explosion Cratering,* eds. D. J. Roddy, R. O. Pepin, and R. B. Merrill (New York: Pergamon Press), pp. 563–573.

Head, J. W. 1978. Origin of central peaks and peak rings: Evidence from peak-ring basins on Moon, Mars, and Mercury. *Lunar Planet. Sci.* IX:485–487 (abstract).

Heide, F. 1964. *Meteorites* (Chicago: Univ. of Chicago Press).

Heidenreich, J. E., III, and Thiemens, M. H. 1985. The non-mass-dependent oxygen isotope effect in the electrodissociation of carbon dioxide: A step toward understanding NoMaD chemistry. *Geochim. Cosmochim. Acta* 49:1303–1306.

Helfenstein, P. 1988. Geologic interpretation of photometric surface roughness. *Icarus* 73.

Helfenstein, P., and Veverka, J. 1987. Photometric properties of lunar terrains derived from Hapke's equation. *Icarus* 72:343–357.

Herbert, F., Wiskerchen, M., Sonett, C. P., and Chao, J. K. 1976. Solar wind induction in Mercury: Constraints on the formation of a magnetosphere. *Icarus* 28:489–500.

Heymann, D., and Mazor, E. 1968. Noble gases in unequilibrated ordinary chondrites. *Geochim. Cosmochim. Acta* 32:1–19.

Hill, T. W. 1974. Origin of the plasma sheet. *Rev. Geophys. Space Phys.* 12:379–388.

Hill, T. W., Dessler, A. J., and Wolf, R. A. 1976. Mercury and Mars: The role of ionospheric conductivity in the acceleration of magnetospheric particles. *Geophys. Res. Lett.* 3:429–432.

Hirschfelder, J. O., Curtiss, C. F., and Bird, R. B. 1964. *Molecular Theory of Gases and Liquids* (New York: Wiley).

Hodges, C. A. 1978. Central pit craters on Mars. *Lunar Planet. Sci.* XI:521–522 (abstract).

Hodges, C. A., and Wilhelm, D. E. 1978. Formation of lunar basin rings. *Icarus* 34:294–323.

Hodges, C. A., Shew, N. B., and Clow, G. 1980. Distribution of central pit craters on Mars. *Lunar Planet. Sci.* XI:450–451 (abstract).

Hodges, R. R., Jr. 1973a. Differential equation of exospheric lateral transport and its application to terrestrial hydrogen. *J. Geophys. Res.* 78:7340–7346.

Hodges, R. R., Jr. 1973b. Helium and hydrogen in the lunar atmosphere. *J. Geophys. Res.* 78:8055–8064.

Hodges, R. R., Jr. 1974. Model atmospheres for Mercury based on a lunar analogy. *J. Geophys. Res.* 79:2881–2885.

Hodges, R. R., Jr. 1980a. Methods for Monte Carlo simulation of the exospheres of the Moon and Mercury. *J. Geophys. Res.* 85:164–170.

Hodges, R. R., Jr. 1980b. Reply. *J. Geophys. Res.* 85:223.

Hodges, R. R., Jr., and Hoffman, J. H. 1974. Measurements of solar wind helium in the lunar atmosphere. *Geophys. Res. Lett.* 1:69–71.

Hodges, R. R., Jr., and Johnson, F. S. 1968. Lateral transport in planetary atmospheres. *J. Geophys. Res.* 73:7307–7317.

Hodges, R. R., Jr., Hoffman, J. H., Yeh, T. T. G., and Chang, G. K. 1972. Orbital search for lunar volcanism. *J. Geophys. Res.* 77:4079–4085.

Hodges, R. R., Jr., Hoffman, J. H., Johnson, F. S., and Evans, D. E. 1973. Composition and dynamics of lunar atmosphere. *Proc. Lunar Sci. Conf.* 4:2855–2864.

Hodges, R. R., Jr., Hoffman, J. H., and Johnson, R. S. 1974. The lunar atmosphere. *Icarus* 21:415–426.

Hoffman, J. H., and Hodges, R. R., Jr. 1975. Molecular gas species in the lunar atmosphere. *The Moon* 14:159–167.

Hoffman, J. H., Hodges, R. R., Johnson, F. S., and Evans, D. E. 1973. Lunar atmospheric composition results from Apollo 17. *Proc. Lunar Sci. Conf.* 4:2865–2875.

Holsapple, K. A., and Schmidt, R. M. 1979. A material strength model for apparent crater volume. *Proc. Lunar Planet. Sci. Conf.* 10:2741–2756.

Holsapple, K. A., and Schmidt, R. M. 1982. On the scaling of crater dimensions 2: Impact processes. *J. Geophys. Res.* 87:1849–1870.

Holt, H. E. 1978. Mercury geologic mapping. In *Repts. Planet. Geol. Prog.*, NASA TM-79729, p. 327 (abstract).

Hood, L. L., and Schubert, G. 1979. Inhibition of solar wind impingement on Mercury by planetary induction currents. *J. Geophys. Res.* 84:2641–2647.

Hood, L. L., and Sonett, C. P. 1982. Limits on the lunar temperature profile. *Geophys. Res. Lett.* 9:37–40.

Hood, L. L., and Vickery, A. 1984. Amplification and generation in hypervelocity impacts with applications to lunar paleomagnetism. *Proc. Lunar Planet. Sci. Conf.* 15, *J. Geophys. Res. Suppl.* 89:C211–C223.

Hood, L. L., Coleman, P. J., and Wilhelms, D. E. 1979. Lunar nearside magnetic anomalies. *Proc. Lunar Planet. Sci. Conf.* 10:2235–2257.

Hoppe, M. M., and Russell, C. T. 1982. Particle acceleration at planetary bow shock waves. *Nature* 295:41–42.

Horedt, G. P., and Neukum, G. 1984. Planetocentric versus heliocentric impacts in the jovian and saturnian satellite system. *J. Geophys. Res.* 89:10,405–10,410.

Hörz, F., Ostertag, R., and Rainey, D. A. 1983. Bunte breccia of the Ries: Continuous deposits of large impact craters. *Rev. Geophys. Space Phys.* 21:1667–1725.

Houseley, R. M., and Cirlin, E. H. 1983. On the alteration of Allende chondrules and the formation of matrix. In *Chondrules and Their Origins*, ed. E. A. King (Houston: Lunar and Planetary Inst.), pp. 145–161.

Housen, K. R., Schmidt, R. M., and Holsapple, K. A. 1983. Crater ejecta scaling laws: Fundamental forms based on dimensional analysis. *J. Geophys. Res.* 88:2485–2499.

Howard, H. T., Tyler, G. L., Esposito, P. B., Anderson, J. D., Reasenberg, R. D., Shapiro, I. I., Fjeldbo, G., Kliore, A. J., Levy, G. S., Brunn, D. L., Dickinson, R., Edelson, R. E., Martin, W. L., Postal, R. B., Seidel, B., Sesplaukis, T. T., Shirley, D. L., Stelzried, C. T., Sweetnam, D. N., Wood, G. E., and Zygielbaum, A. I. 1974. Mercury: Results on mass, radius, ionosphere, and atmosphere from Mariner 10 dual-frequency radio signals. *Science* 185:179–180.

Howard, K. A. 1974. Fresh lunar impact craters: Review of variation with size. *Proc. Lunar Sci. Conf.* 5:67–79.

Howard, K. A., Wilhelms, D. E., and Scott, D. H. 1974. Lunar basin formation and highland stratigraphy. *Rev. Geophys. Space Phys.* 12:309–327.

Hubbard, W. B. 1984. *Planetary Interiors* (New York: Van Nostrand-Reinhold).

Hughes, G. H., App, F. M., and McGetchin, T. R. 1977. Global seismic effects of basin-forming impacts. *Phys. Earth Planet. Int.* 15:251–263.

Hunten, D. M. 1967. Spectroscopic studies of the twilight airglow. *Space Sci. Rev.* 6:493–573.

Hunten, D. M. 1971. Airglow: Introduction and review. In *The Radiating Atmosphere*, ed. B. M. McCormac (Dordrecht: D. Reidel), pp. 3–16.

Hunten, D. M. 1984. A meteor-ablation model of the sodium and potassium layers. *Geophys. Res. Lett.* 8:369–372.

Hunten, D. M., and Wallace, L. 1967. Rocket measurements of the sodium dayglow. *J. Geophys. Res.* 72:69–79.

Hunten, D. M., Pepin, R. O., and Walker, J. C. G. 1987. Mass fractionation in hydrodynamic escape. *Icarus* 69:532–549.

Huss, G. R., Keil, K., and Taylor, G. J. 1981. The matrices of unequilibrated ordinary chondrites: Implications for the origin and history of chondrites. *Geochim. Cosmochim. Acta* 45:33–51.

Image Processing Laboratory. 1976. *1973 Mariner-10 Mission Balloon Suppl. Exper. Data Record* (Pasadena: Jet Propulsion Laboratory).

Ingalls, R. P., and Rainville, L. P. 1972. Radar measurements of Mercury: Topography and scattering characteristics at 3.8 cm. *Astron. J.* 77:185–190.

Ingersoll, A. P. 1971. Polarization measurements of Mars and Mercury: Rayleigh scattering in the Martian atmosphere. *Astrophys. J.* 163:121–129.

Ip, W.-H. 1986. The sodium exosphere and magnetosphere of Mercury. *Geophys. Res. Lett.* 13:423–426.

Irvine, W. M., Simon, T., Menzel, D. H., Pikoos, C., and Young, A. T. 1968. Multicolor photoelectric photometry of the brighter planets: III. Observations from Boyden Observatory. *Astron. J.* 73:807–828.

Istomin, V. G., Grechnev, K. V., and Kochnev, V. A. 1980. Mass spectrometer measurements of the lower atmosphere of Venus. *Space Res.* 20:215–218.

Ivanov, B. A. 1976. The effect of gravity on crater formation: Thickness of ejecta and concentric basins. *Proc. Lunar Sci. Conf.* 7:2947–2965.

Ivanov, B. A., Basilevsky, A. T., Kryuchkov, V. P., and Chernaya, I. M. 1986. Impact craters of Venus: Analysis of Venera 15 and 16 data. *Proc. Lunar Planet. Sci. Conf.* 16, *J. Geophys. Res.* 91:D413–D430.

Jackson, D. J., and Beard, D. B. 1977. The magnetic field of Mercury. *J. Geophys. Res.* 82:2828–2836.

James, O. B. 1980. Rocks of the early lunar crust. *Proc. Lunar Planet. Sci. Conf.* 11:364–393.

Jessberger, E. K., and Kissel, J. 1987. Bits and pieces from Halley's Comet. *Lunar Planet. Sci.* XVIII:466–467 (abstract).

Jet Propulsion Laboratory Publ. 77–51. 1977. *Report of the Terrestrial Bodies Science Working Group, Volume II, Mercury.*

Johnson, H. L. 1965. The absolute calibration of the Arizona photometry. *Comm. Lunar Planet. Lab.* 53:73–77.

Johnson, R. G. 1983. *Energetic Ion Composition in the Earth's Magnetosphere* (Tokyo: Terra-Reidel).

Jurgens, R. F. 1974. A survey of ground-based radar astronomical capability employing 64 and 128 meter diameter antenna systems at S- and X-band. *JPL Publication 890–44.*

Jurgens, R. F. 1982. Earth-based radar studies of planetary surfaces and atmospheres. *IEEE Trans.* GE-28:293.

Jurgens, R. F., Goldstein, R., Rumsey, H., and Green, R. 1980. Images of Venus by 3-station radar interferometry, 1977 results. *J. Geophys. Res.* 85:8282–8294.

Kallemeyn, G. W., and Wasson, J. T. 1981. The compositional classification of chondrites: I. The carbonaceous chondrite groups. *Geochim. Cosmochim. Acta* 45:1217–1230.

Kaula, W. M. 1964. Tidal dissipation by solid friction and the resulting orbital evolution. *Rev. Geophys.* 2:661–685.

Kaula, W. M. 1966. *Theory of Satellite Geodesy* (New York: Blaisdell).

Kaula, W. 1968. *Introduction to Planetary Physics: The Terrestrial Planets* (New York: J. Wiley).

Kaula, W. M. 1976. Comments on the origin of Mercury. *Icarus* 28:429–433.

Kaula, W. 1979. Thermal evolution of the Earth and Moon growing by planetesimal impacts. *J. Geophys. Res.* 84:999–1007.

Kaula, W. M. 1980. The beginning of the Earth's thermal evolution. In *The Continental Crust and Its Mineral Deposits*, ed. D. W. Strangway, Geol. Assoc. of Canada Special Paper 20.

Kaula, W. M., Schubert, G., Lingenfelter, R. E., Sjogren, W. L., and Wollenhaupt, W. R. 1974. Apollo laser altimetry and inferences as to lunar structure. *Proc. Lunar Planet. Sci. Conf.* 5:3049–3058.

Keays, R. R., Ganapathy, R., and Anders, E. 1971. Chemical fractionations in meteorites—IV. Abundances of 14 trace elements in L chondrites; implications for cosmothermometry. *Geochim. Cosmochim. Acta* 35:337–363.

Keihm, S. J., and Langseth, M. G. 1975. Lunar microwave brightness temperature observations reevaluated in the light of Apollo program findings. *Icarus* 24:211–230.

Kiefer, W. S., and Murray, B. C. 1987. The formation of Mercury's smooth plains. *Icarus* 72:477–491.

Kirby, S. H. 1983. Rheology of the lithosphere. *Rev. Geophys. Space Phys.* 21:1458–1487.

Kirsch, E. 1973. Estimation of an upper limit for the solar neutron emission during large flares. *Solar Phys.* 28:232–246.

Kirsch, E., and Richter, A. K. 1985. Possible detection of low energy ions and electrons from planet Mercury by the Helios spacecraft. *Ann. Geophys.* 3:13–18.

Klaasen, K. P. 1976. Mercury's rotation axis and period. *Icarus* 28:469–478.

Kopal, Z. 1974. *The Moon in the Post-Apollo Era* (Boston: D. Reidel).

Kornacki, A. S., and Wood, J. A. 1984. Petrography and classification of Ca,Al-rich and olivine-rich inclusions in the Allende CV3 chondrite. *Lunar Planet. Sci. Conf. 14, J. Geophys. Res. Suppl.* 89:B573–B587.

Kozlovskaya, S. V. 1969. On the internal constitution and chemical composition of Mercury. *Astrophys. Lett.* 4:1–3.

Krimigis, S. M., Sarris, E. T., and Armstrong, T. P. 1975. Observations of Jovian electron events in the vicinity of Earth. *Geophys. Res. Lett.* 2:561–564.

Kuiper, G. P. 1952. *The Atmospheres of the Earth and Planets* (Chicago: Univ. of Chicago Press).

Kumar, S. 1976. Mercury's atmosphere: A perspective after Mariner 10. *Icarus* 28:579–592.

Kunc, J. A., and Shemansky, D. E. 1981. The interaction of helium with alpha-quartz. *J. Chem. Phys.* 75:2406–2411.

Kunc, J. A., and Shemansky, D. E. 1985. The potential curve of the He-alpha-quartz surface interaction. *Surface Sci.* 163:237–248.

Kupo, I., Mekler, Y., and Eviatar, A. 1976. Detection of ionized sulfur in the Jovian magnetosphere. *Astrophys. J.* 205:L51–L53.

Kuznetsova, M. M., and Zeleny, M. 1986. Spontaneous reconnection at the boundaries of planetary magnetospheres. In *Proc. Workshop on Plasma Astrophysics* (Noordwijk, Netherlands: ESA Publ. NSP251), pp. 1–10.

Kyser, T. K., O'Neil, J. R., and Carmichael, I. S. E. 1982. Genetic relations among basic lavas and ultramafic nodules: Evidence from oxygen isotope compositions. *Contrib. Mineral. Petrol.* 81:88–102.

Lal, D. 1972. Hard rock cosmic ray archaeology. *Space Sci. Res.* 14:3–102.

Landau, R. 1975. The 3.5 micron polarization of Mercury. *Icarus* 26:243–249.

Lane, A. P., and Irvine, W. M. 1973. Monochromatic phase curves and albedos for the lunar disk. *Astron. J.* 78:267–277.

Langevin, Y., and Maurette, M. 1978. Plausible depositional histories for the Apollo 15, 16, and 17 drill core tubes. *Proc. Lunar Planet. Sci. Conf.* 9:1765–1786.

Larimer, J. W., and Bartholomay, M. 1979. The role of carbon and oxygen in cosmic gases: Some applications to the chemistry and mineralogy of enstatite chondrites. *Geochim. Cosmochim. Acta* 43:1455–1466.

Leake, M. A. 1982. The intercrater plains of Mercury and the Moon: Their nature, origin, and role in terrestrial planet evolution. In *Advances in Planetary Geology,* NASA TM-84894, pp. 3–535.

Leake, M. A., Chapman, C. R., Weidenschilling, S. J., Davis, D. R., and Greenberg, R. 1987. The chronology of Mercury's geological and geophysical evolution: The vulcanoid hypothesis. *Icarus* 71:350–375.

Lecar, M., and Aarseth, S. J. 1986. A numerical simulation of the formation of the terrestrial planets. *Astrophys. J.* 305:564–579.

Le Comte, C. L., and Schall, R. 1966. Etude d'impacts à grande vitesse à l'aide d'un canon à gaz léger. In *The Fluid Dynamic Aspects of Space Flight* (New York: Gordon and Breach), pp. 315–330.

Lee, T., Papanastassiou, D. A., and Wasserburg, G. J. 1976. Demonstration of ^{26}Mg excess in Allende and evidence for ^{26}Al. *Geophys. Res. Lett.* 3:109–112.

Leinert, C., Richter, I., Pitz, E., and Plank, B. 1981. The zodiacal light from 1.0 to 0.3 AU as observed by the Helios space probes. *Astron. Astrophys.* 103:177–188.

Lentner, M. 1972. *Elementary Applied Statistics* (Tarrytown-on-Hudson: Bogden and Quigley).

Lepping, R. P., Behannon, K. W., and Howell, D. R. 1975. A method of estimating zero level offsets for a dual magnetometer with flipper on a slowly rolling spacecraft: Application to Mariner 10. NASA-GSFC X-692-75-268, October.

Lepping, R. P., Ness, N. F., and Behannon, K. W. 1979. Summary of Mariner 10 magnetic field and trajectory data for Mercury I and III encounters. NASA TM-80060, November.

Lewis, J. S. 1972. Metal/silicate fractionation in the solar system. *Earth Planet. Sci. Lett.* 15:286–290.

Lewis, J. S. 1974. The temperature gradient in the solar nebula. *Science* 186:440–443.

Lewis, J. S., and Prinn, R. G. 1980. Kinetic inhibition of CO and N_2 reduction in the solar nebula. *Astrophys. J.* 238:357–364.

Lewis, J. S., and Prinn, R. G. 1984. *Planets and Their Atmospheres: Origin and Evolution* (New York: Academic Press).

Lewis, J. S., Barshay, S. S., and Noyes, B. 1979. Primordial retention of carbon by the terrestrial planets. *Icarus* 37:190–206.

Lingenfelter, R. E. 1979. Cosmic ray origin and propagation. *Proc. 16th Intl. Cosmic Ray Conf.* 14:135–145.

Liu, L., and Basset, W. A. 1975. Melting of iron up to 200 kbar. *J. Geophys. Res.* 80:3777–3782.

Lockheed Missiles and Space Company, Inc. 1984. Space telescope design reference mission, SE-01, LMSC/D613561B.

Lockwood, J. A., Efedili, S. D., and Jenkins, R. W. 1973. Upper limit to the 1-20 MeV solar neutron flux. *Solar Phys.* 30:183–191.

Loper, D. E. 1978. The gravitationally powered dynamo. *Geophys. J. Roy. Astron. Soc.* 54:389–404.

Lowes, F. J. 1974. Spatial power spectrum of the main geomagnetic field, and extrapolation to the core. *Geophys. J.* 36:717–730.

Lucchitta, B. K. 1976. Mare ridges and related highland scarps: Result of vertical tectonism? *Proc. Lunar Sci. Conf.* 7:2761–2782.

Lumme, K., and Bowell, E. 1981*a*. Radiative transfer in the surfaces of atmosphereless bodies. I. Theory. *Astron. J.* 86:1694–1704.

Lumme, K., and Bowell, E. 1981*b*. Radiative transfer in the surfaces of atmosphereless bodies. II. Interpretation of phase curves. *Astron. J.* 86:1705–1721.

Lumme, K., and Irvine, W. M. 1982. Radiative transfer in the surfaces of atmosphereless bodies. III. Interpretation of lunar photometry. *Astron. J.* 87:1076–1082.

Lyot, B. 1929. Studies of the polarization of the light of planets and some terrestrial substances. *Ann. Obs. Paris* 8(1):169. In French.

Lyot, B. 1930. La polarisation de Mercure comparée à celle de la Lune: Résultats obtenus au Pic-du-Midi en 1930. *Comptes Rend. Acad. Sci.* 29.

MacDonald, T. L. 1931. The distribution of lunar altitudes. *Brit. Astron. Assoc. J.* 41:172–183, 228–239.

Macek, W., and Grzedielski, S. 1986. Planetary magnetotails: Model based on ISEE-3 and Voyager 2 evidence. *Adv. Space Res.* 6(1):283–289.

Mackin, J. H. 1969. Origin of lunar maria. *Geol. Soc. Amer. Bull.* 80:735–748.

Majeva, S. V. 1969. The thermal history of the terrestrial planets. *Astrophys. Lett.* 4:11–16.

Malin, M. C. 1976*a*. Observations of intercrater plains on Mercury. *Geophys. Res. Lett.* 3:581–584.

Malin, M. C. 1976*b*. Comparison of a large crater and multiringed basin populations on Mars, Mercury, and the Moon. *Proc. Lunar Sci. Conf.* 7:3589–3602.

Malin, M. C. 1978. Surfaces of Mercury and the Moon: Effects of resolution and lighting conditions on the discrimination of volcanic features. *Proc. Lunar Planet. Sci. Conf.* 9:3395–3409.

Malin, M. C., and Dzurisin, D. 1977. Landform degradation on Mercury, the moon, and Mars: Evidence from crater depth/diameter relationships. *J. Geophys. Res.* 82:376–388.

Malin, M. C., and Dzurisin, D. 1978. Modification of fresh crater landforms: Evidence from the moon and Mercury. *J. Geophys. Res.* 83:233–243.

Marsch, E., Muhlhauser, K.-H., Schwenn, R., Rosenbauer, H., Pilipp, W., and Neubauer, F. M. 1982. Solar wind protons: Three dimensional velocity distributions and derived plasma parameters, 0.3 and 1.0 AU. *J. Geophys. Res.* 87:52–72.

Mason, B. 1966. The enstatite chondrites. *Geochim. Cosmochim. Acta* 30:23–39.

Mason, R., Guest, J. E., and Cook, G. N. 1976. An Imbrium pattern of graben on the Moon. *Proc. Geol. Assoc. London* 87:161–168.

Masson, P., and Thomas, P. G. 1977. Preliminary results of structural lineaments pattern analysis of Mercury. *Rept. Planet. Geol. Prog.*, NASA TM-3511, pp. 54–55.

Matsuhisa, Y., Goldsmith, J. R., and Clayton, R. N. 1978. Mechanisms of hydrothermal crystallization of quartz at 250°C and 15 kbar. *Geochim. Cosmochim. Acta* 42:173–182.

Maxwell, T. A., and Gifford, A. W. 1980. Ridge systems of Caloris: Comparison with lunar basins. *Proc. Lunar Planet. Sci. Conf.* 11:2447–2462.

Mayer, C. H. 1970. Thermal radio emission of the planets and Moon. In *Surfaces and Interiors of Planets and Satellites*, ed. A. Dollfus (New York: Academic Press).

Mazor, E., Heymann, D., and Anders, E. 1970. Noble gases in carbonaceous chondrites. *Geochim. Cosmochim. Acta* 34:781–824.

McAfee, J. R. 1967. Lateral flow in the exosphere. *Planet. Space Sci.* 15:599–609.

McCammon, C. A., Ringwood, A. E., and Jackson, I. 1983. Thermodynamics of the system Fe-FeO-MgO at high pressure and temperature and a model for the formation of the Earth's core. *Geophys. J. Roy. Astron. Soc.* 72:577–595.

McCauley, J. F. 1977. Orientale and Caloris. *Phys. Earth Planet. Int.* 15:220–250.

McCauley, J. F., and Wilhelms, D. E. 1971. Geological provinces of the near side of the Moon. *Icarus* 15:363–367.

McCauley, J. F., Guest, J. E., Schaber, G. G., Trask, N. J., and Greeley, R. 1981. Stratigraphy of the Caloris Basin, Mercury. *Icarus* 47:184–202.

McCord, T. B., and Adams, J. B. 1972a. Mercury: Surface composition from the reflection spectrum. *Science* 178:745–747.

McCord, T. B., and Adams, J. B. 1972b. Mercury: Interpretation of optical observations. *Icarus* 17:585–588.

McCord, T. B., and Clark, R. N. 1979. The Mercury soil: Presence of Fe^{2+}. *J. Geophys. Res.* 84:7664–7668.

McCord, T. B., Adams, J. B., and Johnson, T. V. 1970. Asteroid Vesta: Spectral reflectivity and compositional implications. *Science* 168:1445–1447.

McDonnell, J. A. M. 1977. Accretionary particle studies on Apollo 12054,58: In situ lunar surface microparticle flux rate and solar wind sputter rate defined. *Proc. Lunar Sci. Conf.* 8:3835–3857.

McFadden, L. A., Gaffey, M. J., and McCord, T. B. 1984. Mineralogical-petrological characterization of near-Earth asteroids. *Icarus* 59:25–40.

McGrath, M. A., Johnson, R. E., and Lanzerotti, L. J. 1986. Sputtering of sodium on the planet Mercury. *Nature* 323:694–696.

McKinnon, W. B. 1978. An investigation into the role of plastic failure in crater modification. *Proc. Lunar Planet. Sci. Conf.* 9:3965–3973.

McKinnon, W. B. 1980. Large impact craters and basins: Mechanics of syngenetic and postgenetic modification. Ph.D. Thesis, California Inst. of Technology.

McKinnon, W. B. 1981. Application of ring tectonic theory to Mercury and other solar system bodies. In *Multi-Ring Basins: Proc. Lunar Planet. Sci.*, eds. P. H. Schultz and R. B. Merrill, 12A:259–273.

McKinnon, W. B. 1986. Tectonics of Caloris Basin, Mercury. *Mercury: Program and Abstracts*, Tucson, Arizona, August.

McKinnon, W. B., and Melosh, H. J. 1980. Evolution of planetary lithospheres: Evidence from multiringed basins on Ganymede and Callisto. *Icarus* 44:454–471.

McSween, H. Y. 1985. Constraints on chondrule origin from petrology of isotopically characterized chondrules in the Allende meteorite. *Meteoritics* 20:523–540.

Meier, R. R., and Donahue, T. M. 1967. Distribution of sodium in the daytime upper atmosphere as measured by a rocket experiment. *J. Geophys. Res.* 72:2803–2829.

Melosh, H. J. 1977a. Crater modification by gravity: A mechanical analysis of slumping. In *Impact and Explosion Cratering*, eds. D. J. Roddy, R. O. Pepin, and R. B. Merrill (New York: Pergamon Press), pp. 1245–1260.

Melosh, H. J. 1977b. Global tectonics of a despun planet. *Icarus* 31:221–243.

Melosh, H. J. 1978. Tectonics of mascon loading. *Proc. Lunar Planet. Sci. Conf.* 9:3513–3525.

Melosh, H. J. 1980. Tectonic pattern on a reoriented planet: Mars. *Icarus* 44:745–755.

Melosh, H. J. 1981. Atmospheric breakup of terrestrial impactors. In *Multi-Ring Basins: Proc. Lunar Planet. Sci.*, eds. P. H. Schultz and R. B. Merrill, 12A:29–35.

Melosh, H. J. 1982a. A schematic model of crater modification by gravity. *J. Geophys. Res.* 87:371–380.

Melosh, H. J. 1982b. A simple mechanical model for Valhalla Basin, Callisto. *J. Geophys. Res.* 87:1880–1890.

Melosh, H. J. 1983. Reply. *J. Geophys. Res.* 88:2505–2507.

Melosh, H. J. 1984. Impact ejection, spallation, and the origin of meteorites. *Icarus* 59:234–260.

Melosh, H. J., and Dzurisin, D. 1978a. Mercurian global tectonics: A consequence of tidal despinning? *Icarus* 35:227–236.

Melosh, H. J., and Dzurisin, D. 1978b. Tectonic implications for the gravity structure of the Caloris Basin, Mercury. *Icarus* 33:141–144.

Melosh, H. J., and McKinnon, W. B. 1978. The mechanics of ringed basin formation. *Geophys. Res. Lett.* 5:985–988.

Mendell, W., and Low, F. 1975. Infrared orbital mapping of lunar features. *Proc. Lunar Sci. Conf.* 6:2711–2719.

Mercer-Smith, J. A., Cameron, A. G. W., and Epstein, R. I. 1984. On the formation of stars from disk accretion. *Astrophys. J.* 279:363–366.

Merrill, P. T., and McElhinny, M. W. 1985. *The Earth's Magnetic Field* (London: Academic Press).

Mewaldt, R. A., Stone, E. C., and Vogt, R. E. 1976. Observation of Jovian electrons at 1 AU. *J. Geophys. Res.* 81:2397–2400.

Milton, D. J. 1974. Geological map of the Lunae Palus quadrangle of Mars. U.S. Geol. Survey Map I-894.

Milton, D. J., and Roddy, D. J. 1972. Displacements within impact craters. *Proc. Intl. Geol. Congress 24th,* Section 15:119–124.

Milton, D. J., Barlow, B. C., Brett, R., Brown, A. R., Glikson, A. Y., Manwaring, E. A., Moss, F. J., Sedmik, E. C. E., Van Son, J., and Young, G. A. 1972. Gosses Bluff impact structure, Australia. *Science* 175:1199–1207.

Mink, D. J. 1987. Appulses and occultations of SAO stars by Mercury: 1987–1995. *Icarus* 71:478–481.

Mizuno, H., and Boss, A. T. 1985. Tidal disruption of dissipative planetesimals. *Icarus* 63:109–133.

Moffatt, H. K. 1978. *Magnetic Field Generation in Electrically Conducting Fluids* (Cambridge: Cambridge Univ. Press).

Moore, P. 1984. The mapping of Mars. *J. British Astron. Assn.* 94:45–54.

Moore, T. E., Arnoldy, R. L., Feynman, J., and Hardy, D. A. 1981. Propagating substorm injection fronts. *J. Geophys. Res.* 86:6713–6726.

Morgan, J. W., and Anders, E. 1980. Chemical composition of Earth, Venus, and Mercury. *Proc. Natl. Acad. Sci. USA* 77:6973–6977.

Morgan, T. H., Potter, A. E., and Zook, H. A. 1987. Impact driven supply of sodium and potassium to the atmosphere of Mercury. *Lunar Planet. Sci.* XVIII:663–664 (abstract).

Moroz, V. I. 1965. Infrared spectrum of Mercury (1.0–3.9 microns). *Soviet Astron. A. J.* 8:882–889.

Morris, R. V., and Mendell, W. W. 1984. Scattering coefficients and optical penetration depths for powders of San Carlos olivine. *Lunar Planet. Sci.* XV:573–574 (abstract).

Morrison, D. 1970. Thermophysics of the planet Mercury. *Space Sci. Rev.* 11:271–307.

Mouginis-Mark, P. J. 1979. Martian fluidized crater morphology: Variations with crater size, latitude, altitude, and target material. *J. Geophys. Res.* 84:8011–8022.

Mouginis-Mark, P. J., and Wilson, L. 1979. Photoclinometric measurements of Mercurian landforms. *Lunar Planet. Sci.* X:873–875 (abstract).

Mouginis-Mark, P. J., and Wilson, L. 1981. MERC: A FORTRAN IV program for the production of topographic data for the planet Mercury. *Computers & Geosci.* 7:35–45.

Muller, G. 1893. *Publ. Potsdam Obs.* 8(4):326.

Münch, G., Trauger, J., and Roesler, F. 1976. Interferometric studies of the emissions associated with Io. *Bull. Amer. Astron. Soc.* 8:468 (abstract).

Munk, W. H., and MacDonald, G. J. F. 1960. *The Rotation of the Earth* (Cambridge: Cambridge Univ. Press).

Murchie, S., and Head, J. W. 1986. Global reorientation and its effect on tectonic patterns on Ganymede. *Geophys. Res. Lett.* 13:345–348.

Murray, B. C., Belton, M. J. S., Danielson, G. E., Davies, M. E., Gault, D. E., Hapke, B.,

O'Leary, B., Strom, R. G., Suomi, V., and Trask, N. 1974. Mercury's surface: Preliminary description and interpretation from Mariner 10 pictures. *Science* 185:169–179.

Murray, B. C., Strom, R. G., Trask, N. J., and Gault, D. E. 1975. Surface history of Mercury: Implications for terrestrial planets. *J. Geophys. Res.* 80:2508–2514.

Murray, J. B. 1980. Oscillating peak model of basin and crater formation. *Moon and Planets* 22:269–291.

Murray, J. B., Dollfus, A., and Smith, B. 1972. Cartography of the surface markings of Mercury. *Icarus* 17:576–584.

Musmann, G., Neubauer, J. M., and Lamers, E. 1977. Radial variations of the interplanetary field between 0.3 and 1.0 AU: Observations by Helios 1. *J. Geophys. Res.* 82:551–562.

Mutch, T. A. 1970. *Geology of the Moon: A Stratigraphic View* (Princeton: Princeton Univ. Press).

Nash, D. B., Carr, M. H., Gradie, J., Hunten, D. M., and Yoder, C. F. 1986. Io. In *Satellites*, eds. J. A. Burns and M. S. Matthews (Tucson: Univ. of Arizona Press), pp. 629–688.

Natrella, M. G. 1963. *Experimental Statistics*, Natl. Bureau Standards Handbook, 91:4,1–4,14.

Ness, N. F. 1977. The magnetic field of Mercury. In *Highlights of Astronomy*, vol. 4, part I, ed. E. A. Muller (Dordrecht: D. Reidel), pp. 179–190.

Ness, N. F. 1978. Mercury: Magnetic field and interior. *Space Sci. Rev.* 21:527–554.

Ness, N. F. 1979a. The magnetic fields of Mercury, Mars, and Moon. *Ann. Rev. Earth Planet. Sci.* 7:249–288.

Ness, N. F. 1979b. The magnetic field of Mercury. *Phys. Earth Planet. Int.* 20:204–217.

Ness, N. F. 1979c. The magnetosphere of Mercury. In *Solar System Plasma Physics*, eds. C. F. Kennel, L. J. Lanzerotti, and E. N. Parker (Amsterdam: North-Holland Publ. Co.), pp. 183–205.

Ness, N. F., Behannon, K. W., Lepping, R. P., and Schatten, K. H. 1971. Use of two magnetometers for magnetic field measurements on a spacecraft. *J. Geophys. Res.* 76:3564–3573.

Ness, N. F., Behannon, K. W., Lepping, R. P., Whang, Y. C., and Schatten, K. H. 1974a. Magnetic field observations near Mercury: Preliminary results from Mariner 10. *Science* 185:151–160.

Ness, N. F., Behannon, K. W., Lepping, R. P., Whang, Y. C., and Schatten, K. H. 1974b. Magnetic field observations near Venus: Preliminary results from Mariner 10. *Science* 183:1301–1306.

Ness, N. F., Behannon, K. W., Lepping, R. P., and Whang, Y. C. 1975a. The magnetic field of Mercury: Part I. *J. Geophys. Res.* 80:2708–2716.

Ness, N. F., Behannon, K. W., Lepping, R. P., and Whang, Y. C. 1975b. Magnetic field of Mercury confirmed. *Nature* 255:204–205.

Ness, N. F., Behannon, K. W., Lepping, R. P., and Whang, Y. C. 1976. Observations of Mercury's magnetic field. *Icarus* 28:479–488.

Neugebauer, M., Fisk, L. A., Gold, R. E., Lin, R. P., Newkirk, G., Simpson, J. A., and Van Hollebeke, M. A. I. 1978. The energetic particle environment of the solar probe mission. JPL Publ. 78-64.

Neukum, G. 1971. Untersuchung ueber Einschlagskrater auf dem Mond. Dissertation, Univ. of Heidelberg.

Neukum, G. 1981. Surface history of the terrestrial-type planets. *Proc. Alpbach Summer School* 19.7-7.8.1981, ESA SP-164, pp. 129–137.

Neukum, G. 1982. Ancient cratering records of the terrestrial-type planets. *Lunar Planet. Sci.* XIII:588–589 (abstract).

Neukum, G. 1983. Meteoritenbombardement und Datierung Planetarer Oberflächen. Habilitation dissertation for faculty membership, Univ. of Munich.

Neukum, G. 1985. Cratering records of the satellites of Jupiter and Saturn. *Adv. Space Res.* 5(8):107–116.

Neukum, G., and Dietzel, H. 1971. On the development of the crater population on the Moon with time under meteoroid and solar wind bombardment. *Earth Planet. Sci. Lett.* 12:59–66.

Neukum, G., and Wise, D. U. 1976. Mars: A standard crater curve and possible new time scale. *Science* 194:1381–1387.

Neukum, G., and 41 others. 1985. Mercury Polar Orbiter: A proposal to the European Space Agency in response to a call for new mission proposals issued on 10 July 1985.

Ng, K. H., and Beard, D. B. 1979. Possible displacements of Mercury's dipole. *J. Geophys. Res.* 84:2115–2117.

Oberbeck, V. R. 1971. Laboratory simulation of impact cratering with high explosives. *J. Geophys. Res.* 76:5732–5749.

Oberbeck, V. R. 1975. The role of ballistic erosion and sedimentation in lunar stratigraphy. *Rev. Geophys. Space Phys.* 13:337–362.

Oberbeck, V. R., and Morrison, R. H. 1974. Laboratory simulation of the herringbone pattern associated with lunar secondary crater chains. *The Moon* 9:415–455.

Oberbeck, V. R., Hörz, F., Morrison, R. H., Quaide, W. L., and Gault, D. E. 1975. On the origin of the lunar smooth-plains. *The Moon* 12:19–54.

Oberbeck, V. R., Quaide, W. L., Arvidson, R. E., and Aggarwal, H. R. 1977. Comparative studies of lunar, martian, and mercurian craters and plains. *J. Geophys. Res.* 82:1681–1698.

O'Donnell, W. P. 1980. The Surface History of the Planet Mercury. Ph.D. Thesis, Univ. of London.

Offield, T. W., and Pohn, H. A. 1977. Deformation of the Decaturville impact structure, Missouri. In *Impact and Explosion Cratering*, eds. D. J. Roddy, R. O. Pepin, and R. B. Merrill (New York: Pergamon Press), pp. 321–341.

Ogilvie, K. W., Scudder, J. D., Hartle, R. E., Siscoe, G. L., Bridge, H. S., Lazarus, A. J., Asbridge, J. R. Bame, S. J., and Yeates, C. M. 1974. Observations at Mercury encounter by the plasma science experiment on Mariner 10. *Science* 185:146–152.

Ogilvie, K. W., Scudder, J. D., Vasyliunas, V. M., Hartle, R. E., and Siscoe, G. L. 1977. Observations at the planet Mercury by the plasma electron experiment: Mariner 10. *J. Geophys. Res.* 82:1807–1824.

O'Keefe, J. D., and Ahrens, T. J. 1976. Impact ejecta on the Moon. *Proc. Lunar Sci. Conf.* 7:3007–3025.

O'Keefe, J. D., and Ahrens, T. J. 1977. Impact-induced energy partitioning, melting, and vaporization on terrestrial planets. *Proc. Lunar Planet. Sci. Conf.* 8:3357–3374.

O'Keefe, J. D., and Ahrens, T. J. 1982. Cometary impact on planetary surfaces. *J. Geophys. Res.* 87:6668–6680.

O'Keefe, J. D., and Ahrens, T. J. 1987. The size distribution of fragments ejected at a given velocity from impact craters. *Intl. J. Impact Eng.* 5:493–499.

O'Leary, B. T., and Rea, D. G. 1967. On the polarimetric evidence for an atmosphere on Mercury. *Astrophys. J.* 148:249–253.

Onuma, N., Clayton, R. N., and Mayeda, T. K. 1972. Oxygen isotope cosmothermometer. *Geochim. Cosmochim. Acta* 36:169–188.

Onuma, N., Clayton, R. N., and Mayeda, T. K. 1974. Oxygen isotope cosmothermometer revisited. *Geochim. Cosmochim. Acta* 36:189–191.

Öpik, E. 1960. The lunar surface as an impact counter. *Mon. Not. Roy. Astron. Soc.* 120:404–411.

Orowan, E. 1969. Density of the Moon and nucleation of planets. *Nature* 222:867.

Orphal, D. L. 1977. Calculations of explosion cratering—I. The shallow-buried nuclear detonation JOHNIE BOY. In *Impact and Explosion Cratering*, eds. D. J. Roddy, R. O. Pepin, and R. B. Merrill (New York: Pergamon Press), pp. 897–906.

Orphal, D. L., Borden, W. F., Larson, S. A., and Schultz, P. H. 1980. Impact melt generation and transport. *Proc. Lunar Planet. Sci. Conf.* 11:2309–2323.

Orphal, D. L., Roddy, D. J., Schultz, P. H., Borden, W. F., and Larson, S. A. 1981. Energy coupling for meteoritic and cometary impacts. In *Papers Presented to the Conference on Large-Body Impacts and Terrestrial Evolution*, LPI Contrib. No. 449, p. 41 (abstract).

Osterbrock, D. E. 1974. *Astrophysics of Gaseous Nebulae* (San Francisco: Freeman).

Ozima, M., and Podosek, F. A. 1983. *Noble Gas Geochemistry* (Cambridge: Cambridge Univ. Press).

Palme, H., and Fegley, B., Jr. 1987. Formation of FeO-bearing olivines in carbonaceous chondrites by high temperature oxidation in the solar nebula. *Lunar Planet. Sci.* XVIII:754–755 (abstract).

Passey, Q. R., and Melosh, H. J. 1980. Effects of atmospheric breakup on crater field formation. *Icarus* 42:211–233.

Passey, Q. R., and Shoemaker, E. M. 1982. Craters and basins on Ganymede and Callisto:

Morphological indicators of crustal evolution. In *Satellites of Jupiter*, ed. D. Morrison (Tucson: Univ. of Arizona Press), pp. 379–434.

Peale, S. J. 1969. Generalized Cassini's laws. *Astron. J.* 74:483–489.

Peale, S. J. 1972. Determination of parameters related to the interior of Mercury. *Icarus* 17:168–173.

Peale, S. J. 1973. Rotation of solid bodies in the solar system. *Rev. Geophys. Space Phys.* 11:767–793.

Peale, S. J. 1974. Possible histories of the obliquity of Mercury. *Astron. J.* 79:722–744.

Peale, S. J. 1976. Does Mercury have a molten core? *Nature* 262:765–766.

Peale, S. J. 1981. Measurement accuracies required for the determination of a Mercurian liquid core. *Icarus* 48:143–145.

Peale, S. J., and Boss, A. P. 1977a. A spin-orbit constraint on the viscosity of a Mercurian liquid core. *J. Geophys. Res.* 82:743–749.

Peale, S. J., and Boss, A. P. 1977b. Mercury's core: The effect of obliquity on the spin-orbit constraints. *J. Geophys. Res.* 82:3423–3429.

Pechernikova, G. V., and Vityazev, A. V. 1979. Masses of the largest bodies and relative velocities during the accumulation of planets. *Pis'ma Astron. Zh.* 5, *Soviet Astron. Lett.* 5:31.

Pechernikova, G. V., and Vityazev, A. V. 1980. The evolution of orbital eccentricities of the planets in the process of their formation. *Soviet Astron.* 24:460–467.

Pechernikova, G. V., Majeva, S. V., and Vityazev, A. V. 1984. On the dynamics of circumplanetary swarms. *Pis'ma Astron. Zh.* 10(9):702–709.

Pechmann, J. B., and Melosh, H. J. 1979. Global fracture patterns of a despun planet: Application to Mercury. *Icarus* 38:243–250.

Pettengill, G. 1969. A review of radar studies of planetary surfaces. *Radio Sci.* 69D:1617–1623.

Pettengill, G. 1978. Physical properties of the planets and satellites from radar observations. *Ann. Rev. Astron. Astrophys.* 16:265–292.

Pettengill, G. H., and Dyce, R. B. 1965. A radar determination of the rotation of the planet Mercury. *Nature* 206:1240.

Pieters, C. M. 1978. Mare basalt types on the front side of the Moon: A summary of spectral reflectance data. *Proc. Lunar Planet. Sci. Conf.* 9:2825–2849.

Pieters, C. M. 1983. Strength of mineralogical absorption features in the transmitted component of near-infrared reflected light: First results from RELAB. *J. Geophys. Res.* 88:9534–9544.

Pieters, C. M., Adams, J. B., Mouginis-Mark, P. J., Zisk, S. H., Smith, M. O., Head, J. W., and McCord, T. B. 1985. The nature of crater rays: The Copernicus example. *J. Geophys. Res.* 90:12,393–12,413.

Pike, R. J. 1974. Depth/diameter relations of fresh lunar craters: Revision from spacecraft data. *Geophys. Res. Lett.* 1:291–294.

Pike, R. J. 1977a. Size-dependence in the shape of fresh impact craters on the moon. In *Impact and Explosion Cratering*, eds. D. J. Roddy, R. O. Pepin, and R. B. Merrill (New York: Pergamon Press), pp. 489–509.

Pike, R. J. 1977b. Apparent depth/apparent diameter relation for lunar craters. *Proc. Lunar Sci. Conf.* 8:3427–3436.

Pike, R. J. 1980a. Control of crater morphology by gravity and target type: Mars, Earth, Moon. *Proc. Lunar Planet. Sci. Conf.* 11:2159–2189.

Pike, R. J. 1980b. Formation of complex impact craters: Evidence from Mars and other planets. *Icarus* 43:1–19.

Pike, R. J. 1980c. Geometric interpretation of lunar craters. U.S. Geol. Survey Prof. Paper, 1046-C.

Pike, R. J. 1982a. Crater peaks to crater rings: The transition on Mercury and other bodies. *Repts. Planet. Geol. Program, 1982,* NASA TM-85127, pp. 117–119.

Pike, R. J. 1982b. Morphologic transitions for craters and basins on 13 solar system bodies. *Lunar Planet Sci.* XIII:627–628 (abstract).

Pike, R. J. 1983a. Large craters or small basins on the Moon. *Lunar Planet. Sci.* XIV:610–611 (abstract).

Pike, R. J. 1983b. Comment on "A schematic model of crater modification by gravity" by H. J. Melosh. *J. Geophys. Res.* 88:2500–2504.

Pike, R. J. 1985. Some morphologic systematics of complex impact structures. *Meteoritics* 20:49–68.

Pike, R. J., and Arthur, D. W. G. 1979. Simple to complex impact craters: The transition on Mars. *Repts. Planet. Geol. Prog., 1978–1979,* NASA TM-80339, pp. 132–134.

Pike, R. J., and Clow, G. D. 1982. Geomorphology of craters on Mercury: First results from a new sample. *Repts. Planet. Geol. Prog., 1982,* NASA TM-85127, pp. 120–122.

Pike, R. J., and Clow, G. D. 1984. Ode to gravity: Depth/diameter for fresh craters on Mercury. *Repts. Planet. Geol. Prog., 1983,* NASA TM-86246, pp. 104–106.

Pike, R. J., and Davis, P. A. 1984. Toward a topographic model of Martian craters from photoclinometry. *Lunar Planet. Sci.* XV:645–646 (abstract).

Pike, R. J., and Spudis, P. D. 1984. Ring spacing of Mercurian multi-ring basins and basin ring formation. *Repts. Planet. Geol. Prog., 1983,* NASA TM-86246, pp. 90–92.

Pike, R. J., and Spudis, P. D. 1985. Ring-diameter ratios for multi-ring basins average $2.0^{0.5}D$. *Repts. Planet. Geol. Prog., 1984,* NASA TM-87563, pp. 192–194.

Pike, R. J., and Spudis, P. D. 1987. Basin-ring spacing on the Moon, Mercury, and Mars. *Earth, Moon, Planets* 39:129–194.

Pike, R. J., Spudis, P. D., and Clow, G. D. 1985. Average spacing for rings of individual multi-ring basins is $2.0^{0.5}D$. *Repts. Planet. Geol. Prog., 1984,* NASA TM-87563, pp. 189–191.

Pikkarainen, T. 1969. The surface structure of the Moon and Mercury derived from integrated photometry. *Ann. Acad. Sci. Fenn., Series VI: Physica* 316.

Plagemann, S. 1965. A model of the internal constitution and temperature of the planet Mercury. *J. Geophys. Res.* 70:985–993.

Plescia, J. B., and Boyce, J. M. 1985. Impact cratering history of the saturnian satellites. *J. Geophys. Res.* 90:2029–2037.

Plescia, J. B., and Golombek, M. P. 1986. Origin of planetary wrinkle ridges based on the study of terrestrial analogs. *Geol. Soc. Amer. Bull.* 97:1289–1299.

Pohn, H. A., and Offield, T. W. 1970. Lunar crater morphology and the relative age determination of lunar geologic units. Part I. Classification. *U.S. Geol. Survey Prof. Paper* 700-C:C153–C162.

Post, R. L. 1977. High temperature creep of Mt. Burnet dunite. *Tectonophys.* 42:75–110.

Potter, A., and Morgan, T. H. 1985a. Discovery of sodium in the atmosphere of Mercury. *Science* 229:651–653.

Potter, A. E., and Morgan, T. H. 1985b. Observations of sodium on Mercury. *Bull. Amer. Astron. Soc.* 17:711–712 (abstract).

Potter, A., and Morgan, T. H. 1986a. Potassium in the atmosphere of Mercury. *Icarus* 67:336–340.

Potter, A. E., and Morgan, T. H. 1986b. Variations of sodium and potassium in the atmosphere of Mercury. *Bull. Amer. Astron. Soc.* 18:781–782 (abstract).

Potter, A. E., and Morgan, T. H. 1987. Variation of sodium on Mercury with solar radiation pressure. *Icarus* 71:472–477.

Preston, G. 1967. The spectrum of Comet Ikeya-Seki (1965f). *Astrophys. J.* 146:718–742.

Quaide, W. L., and Oberbeck, V. R. 1968. Thickness determinations of the lunar surface layer from lunar impact craters. *J. Geophys. Res.* 73:5247–5270.

Quaide, W. L., Gault, D. E., and Schmidt, R. A. 1965. Gravitative effects on lunar impact structures. *Ann. N.Y. Acad. Sci.* 123:563–572.

Quessette, J. A. 1972. Atomic hydrogen densities at the exobase. *J. Geophys. Res.* 77:2997–3000.

RAND Corporation. 1955. *A Million Random Digits with 100,000 Normal Deviates* (Glencoe: Free Press).

Rasool, S. I., Gross, S. H., and McGovern, W. E. 1965. The atmosphere of Mercury. *Space Sci. Rev.* 5:565–584.

Ratier, G. 1972. Observation de Mercure au voisinage de la phase nulle. *Icarus* 16:318–320.

Rava, B., and Hapke, B. 1987. An analysis of the Mariner 10 color ratio map of Mercury. *Icarus* 71:397–429.

Reasenberg, R. D., Goldstein, R. B., MacNeil, P. E., and Shapiro, I. I. 1977. The pole direction and precession of Mars (A). *Bull. Amer. Astron. Soc.* 9:520 (abstract).

Reedy, R. C. 1977. Solar proton fluxes since 1956. *Proc. Lunar Sci. Conf.* 8:825–839.

Reedy, R. C., Arnold, J. R., and Lal, D. 1983. Cosmic-ray record in solar system matter. *Ann. Rev. Nuc. Part. Sci.* 83:505.

Reynolds, R. T., and Summers, A. L. 1969. Calculations on the composition of the terrestrial planets. *J. Geophys. Res.* 74:2494–2511.

Richards, M. A., and Hager, B. H. 1984. Geoid anomalies in a dynamic earth. *J. Geophys. Res.* 89:5987–6002.

Ringwood, A. E. 1966. Chemical evolution of the terrestrial planets. *Geochim. Cosmochim. Acta* 30:41–104.

Ringwood, A. E. 1977. Composition of the core and implications for the origin of the Earth. *Geochem. J.* 11:111–135.

Ringwood, A. E. 1979. *The Origin of the Earth and Moon* (New York: Springer-Verlag).

Roddy, D. J. 1968. The Flynn Creek crater, Tennessee. In *Shock Metamorphism of Natural Materials,* eds. B. M. French and N. M. Short (Baltimore: Mono Books), pp. 291–322.

Roddy, D. J. 1977. Large-scale impact and explosion craters: Comparisons of morphological and structural analogs. In *Impact and Explosion Cratering,* eds. D. J. Roddy, R. O. Pepin, and R. B. Merrill (New York: Pergamon Press), pp. 185–246.

Roddy, D. J. 1979. Structural deformation at the Flynn Creek impact crater, Tennessee: A preliminary report on deep drilling. *Proc. Lunar Planet. Sci. Conf.* 10:2519–2534.

Ross, M. N., and Schubert, G. 1986. Tidal dissipation in a viscoelastic planet. *Proc. Lunar Planet. Sci. Conf. 16, J. Geophys. Res. Suppl.* 91:D447–D452.

Runcorn, S. K. 1962. Convection in the Moon. *Nature* 195:1150–1151.

Runcorn, S. K. 1975. On the interpretation of lunar magnetism. *Phys. Earth Planet. Int.* 10:327–334.

Runcorn, S. K. 1977. Convection in Mercury. *Phys. Earth Planet. Int.* 15:131–134.

Runcorn, S. K., Libby, L. M., and Libby, W. F. 1977. Primeval melting of the Moon. *Nature* 270:676–681.

Russell, C. T. 1977. On the relative locations of the bow shocks of the terrestrial planets. *Geophys. Res. Lett.* 4:387–390.

Russell, C. T. 1985. Planetary bow shocks. In *Collisionless Shocks in the Heliosphere: Reviews of Current Research,* eds. B. T. Tsurutani and R. G. Stone (Washington, DC: American Geophysical Union), pp. 109–130.

Russell, C. T., and Elphic, R. C. 1979. ISEE observations of flux transfer events at the dayside magnetopause. *Geophys. Res. Lett.* 6:33–34.

Russell, C. T., and Hoppe, M. M. 1983. Upstream waves and particles. *Space Sci. Rev.* 34:155–172.

Russell, C. T., and McPherron, R. L. 1973. The magnetotail and substorms. *Space Sci. Rev.* 15:205–266.

Russell, C. T., and Walker, R. J. 1985. Flux transfer events at Mercury. *J. Geophys. Res.* 90:11,067–11,074.

Russell, C. T., Luhmann, J. G., and Phillips, J. L. 1985. The location of the subsolar bow shock of Venus: Implications for the obstacle shape. *Geophys. Res. Lett.* 12:627–630.

Sabadini, R., Yuen, D. A., and Boschi, E. 1982. Polar wandering and the forced response of a rotating, multilayered, viscoelastic planet. *J. Geophys. Res.* 87:2885–2903.

Safronov, V. S. 1969. *Evolution of the Protoplanetary Cloud and Formation of the Earth and Planets* (Moscow: Nauka). In Russian. Trans. NASA TT-F-677, 1972.

Safronov, V. 1978. The heating of the Earth during formation. *Icarus* 33:1–12.

Safronov, V. S., and Vityazev, A. V. 1985. Origin of the solar system. *Astrophys. Space Phys. Rev.,* ed. R. A. Syunyaev, *Soviet Sci. Rev., Section E* 4:1–98.

Sagan, C. 1974. Mercury. In *Encyclopedia Britannica.*

Samson, J. A. R. 1982. Atomic photoionization. *Handbuch der Physik* 31:123–213.

Sandner, W. 1963. *The Planet Mercury* (London: Faber & Faber).

Schaber, G. G., and McCauley, J. F. 1980. Geological map of the Tolstoj quadrangle of Mercury (H-8). U.S. Geol. Surv. Map I-1199.

Schaber, G. G., Boyce, J. M., and Trask, N. J. 1977. Moon-Mercury: Large impact structures, isostasy and average crustal viscosity. *Phys. Earth Planet. Int.* 15:189–201.

Schield, M. A. 1969. Pressure balance between solar wind and magnetosphere. *J. Geophys. Res.* 74:1275–1286.

Schmidt, O. Yu. 1957. *Four Lectures on the Theory of the Earth's Origin,* 3rd ed. (Iz datel'stvo AN SSSR).

Schmidt, R. M., and Holsapple, K. A. 1980. Theory and experiments on centrifuge cratering. *J. Geophys. Res.* 85:235–252.

Schmidt, R. M., and Holsapple, K. A. 1981. An experimental investigation of transient crater size. *Lunar Planet. Sci.* XII:934–936 (abstract).

Schmidt, R. M., and Holsapple, K. A. 1982a. Bounds on crater size for large body impact: Gravity-scaling results. *Geol. Soc. Amer. Special Paper* 190:93–102.

Schmidt, R. M., and Holsapple, K. A. 1982b. Dynamic scaling relationships for impact crater formation. *Lunar Planet. Sci.* XIII:687–688 (abstract).

Schneider, N. M., Wells, W. K., and Hunten, D. M. 1985. Sunward displacement of Mercury's sodium emission. *Bull. Amer. Astron. Soc.* 17:712 (abstract).

Schoenberg, E. 1929. Theoretisch Photometrie. In *Handbuch der Astrophysik,* vol. 2 (Berlin: Springer-Verlag), pp. 1–280.

Schubert, G., Turcotte, D. L., and Oxburgh, E. R. 1969. Stability of planetary interiors. *Geophys. J. Roy. Astron. Soc.* 18:441–460.

Schubert, G., Cassen, P., and Young, R. E. 1979. Subsolidus convective cooling histories of the terrestrial planets. *Icarus* 38:192–211.

Schubert, G., Stevenson, D., and Cassen, P. 1980. Whole planet cooling and the radiogenic heat source content of the Earth and Moon. *J. Geophys. Res.* 85:2531–2538.

Schubert, G., Spohn, T., and Reynolds, R. 1986. Thermal histories, compositions, and internal structures of the moons of the solar system. In *Satellites,* eds. J. A. Burns and M. S. Matthews (Tucson: Univ. of Arizona Press), pp. 224–292.

Schultz, L., and Signer, P. 1977. Noble gases in the St. Mesmin chondrite: Implications to the irradiation history of a brecciated meteorite. *Earth Planet. Sci. Lett.* 36:363–371.

Schultz, P. H. 1976a. *Moon Morphology* (Austin: Univ. of Texas Press).

Schultz, P. H. 1976b. Floor-fractured lunar craters. *The Moon* 15:241–263.

Schultz, P. H. 1977. Endogenic modification of impact craters on Mercury. *Phys. Earth Planet. Int.* 15:202–219.

Schultz, P. H. 1981. Evidence and mechanisms for the non-local contribution to ejecta deposits. In *Workshop on Apollo 16,* eds. O. B. James and F. Hörz, LPI Tech. Rept. 81-01 (Houston: Lunar and Planetary Inst.), pp. 120–122.

Schultz, P. H., and Crawford, D. A. 1987. Impact vaporization by low-angle impacts. *Lunar Planet. Sci.* XVIII:888–889 (abstract).

Schultz, P. H., and Gault, D. E. 1975. Seismic effects from major basin formation on the Moon and Mercury. *The Moon* 12:159–177.

Schultz, P. H., and Gault, D. E. 1982. Impact ejecta dynamics in an atmosphere: Experimental results and extrapolations. *Geol. Soc. Amer. Special Paper* 190:153–174.

Schultz, P. H., and Gault, D. E. 1983. High-velocity clustered impacts: Experimental results. *Lunar Planet. Sci.* XIV:674–675 (abstract).

Schultz, P. H., and Gault, D. E. 1984. Effects of projectile deformation on cratering efficiency and morphology. *Lunar Planet. Sci.* XV:730–731 (abstract).

Schultz, P. H., and Gault, D. E. 1985a. Clustered impacts: Experiments and implications. *J. Geophys. Res.* 90:3701–3732.

Schultz, P. H., and Gault, D. E. 1985b. Impact-induced vaporization: Effects of impact angle and atmospheric pressure. *Lunar Planet. Sci.* XVI:740–741 (abstract).

Schultz, P. H., and Gault, D. E. 1986a. Impact vaporization: Late time phenomena from experiments. *Lunar Planet. Sci.* XVII:779–780 (abstract).

Schultz, P. H., and Gault, D. E. 1986b. Experimental evidence for non-proportional growth of large craters. *Lunar Planet. Sci.* XVII:777–778 (abstract).

Schultz, P. H., and Gault, D. E. 1987. Transition diameters for crater shape in laboratory experiments and on planets. *Lunar Planet. Sci.* XVIII:890–891 (abstract).

Schultz, P. H., and Mendell, W. 1978. Orbital infrared observations of lunar craters and possible implications for impact ejecta emplacement. *Proc. Lunar Planet. Sci. Conf.* 9:2857–2883.

Schultz, P. H., and Spudis, P. D. 1983. The beginning and end of lunar mare volcanism. *Nature* 302:233–236.

Schultz, P. H., and Spudis, P. D. 1985. Procellarum basin: A major impact or the effect of Imbrium? *Lunar Planet. Sci.* XVI:746–747 (abstract).

Schultz, P. H., and Srnka, L. J. 1980. Cometary collisions on the Moon and Mercury. *Nature* 284:22–26.

Schultz, P. H., Orphal, D. L., Miller, B., Borden, W. F., and Larson, S. A. 1981. Multi-ring basin formation: Possible clues from impact cratering calculations. In *Multi-Ring Basins: Proc. Lunar Planet. Sci.*, eds. P. H. Schultz and R. B. Merrill, 12A:181–195.

Schultz, P. H., Schultz, R. A., and Rogers, J. 1982. The structure and evolution of ancient impact basins on Mars. *J. Geophys. Res.* 87:9803–9820.

Schultz, P. H., Gault, D. E., and Crawford, D. A. 1986. Impacts of hemispherical granular targets: Implications for global impacts. *Lunar Planet. Sci.* XVII:783–784 (abstract).

Scott, D. H. 1977. Moon-Mercury: Relative preservation states of secondary craters. *Phys. Earth Planet. Int.* 15:173–178.

Scott, D. H., and Carr, M. H. 1978. Geologic map of Mars. U.S. Geol. Survey Map I-1083.

Scott, D. H., and King, J. S. 1987. Geologic map of the Beethoven quadrangle of Mercury. U.S. Geol. Survey Map, in press.

Scott, E. R. D., and Rajan, R. S. 1981. Metallic minerals, thermal histories and parent bodies of some xenolithic, ordinary chondrite meteorites. *Geochim. Cosmochim. Acta* 45:53–67.

Seek, J. B., Scheifele, J. L., and Ness, N. F. 1977. GSFC magnetic field experiments: Mariner 10. NASA-GSFC X-695-77-256, October.

Settle, M., and Head, J. W. 1976. Excavation of depths of large lunar impacts: Shallow or deep? *Interdisciplinary Studies by the Imbrium Consortium* 1:139–146.

Settle, M., and Head, J. W. 1979. The role of rim slumping in the modification of lunar impact craters. *J. Geophys. Res.* 84:3081–3096.

Shapiro, I. I., Zisk, S. H., Rogers, A. E. E., Slade, M. A., and Thompson, T. W. 1972. Lunar topography: Global determination by radar. *Science* 178:939–948.

Sharpe, H. N., and Strangway, D. W. 1976. The magnetic field of Mercury and models of thermal evolution. *Geophys. Res. Lett.* 3:285–288.

Shemansky, D. E. 1980. Comment on the article "Methods of Monte Carlo simulation of the exospheres of the moon and Mercury" by R. R. Hodges, Jr. *J. Geophys. Res.* 85:221–222.

Shemansky, D. E., and Broadfoot, A. L. 1977. Interaction of the surface of the moon and Mercury with their exospheric atmospheres. *Rev. Geophys. Space Phys.* 15:491–499.

Shirley, D. N. 1983. On the origin of the oxygen isotopic variations in ordinary chondrite chondrules. *Meteoritics* 18:396–398 (abstract).

Shizgal, B., and Blackmore, R. 1986. A collisional kinetic theory of a plane parallel evaporating planetary atmosphere. *Planet Space Sci.* 34:279–291.

Shoemaker, E. M. 1959. Address to earth sciences colloquium. *Proc. Lunar Planet. Exploration Coll.* 2(1):20–28.

Shoemaker, E. M. 1962. Interpretation of lunar craters. *Physics and Astronomy of the Moon* (New York: Academic Press), pp. 283–359.

Shoemaker, E. M. 1977. Why study impact craters? In *Impact and Explosion Cratering*, eds. D. J. Roddy, R. O. Pepin, and R. B. Merrill (New York: Pergamon Press), pp. 1–10.

Shoemaker, E. M. 1983. Asteroid and comet bombardment of the Earth. *Ann. Rev. Earth Planet. Sci.* 11:461–494.

Shoemaker, E. M., and Hackman, R. J. 1962. Stratigraphic basis for a lunar time scale. In *The Moon*, ed. Z. Kopal and Z. K. Mikhailov (New York: Academic Press), pp. 289–300.

Shoemaker, E. M., and Wolfe, R. F. 1982. Cratering time scales for the Galilean satellites. In *Satellites of Jupiter*, ed. D. Morrison (Tucson: Univ. of Arizona Press), pp. 277–339.

Shoemaker, E. M., Hackman, R., and Eggleton, R. 1962. Interplanetary correlation of geologic time. *Adv. Astron. Sci.* 8:70–89.

Shorthill, R. W., Saari, J. M., Baird, F. E., and LeCompte, J. R. 1969. *Photometric Properties of Selected Lunar Features*, NASA Contractor Report CR-1429.

Siegfried, R. W., and Solomon, S. C. 1974. Mercury: Internal structure and thermal evolution. *Icarus* 23:192–205.

Simonds, C. H., Phinney, W. C., Warner, J. L., McGee, P. E., Geeslin, J., Brown, R. W., and Rhodes, J. M. 1977. Apollo 14 revisited, or breccias aren't so bad after all. *Proc. Lunar Sci. Conf.* 8:1869–1893.

Simpson, J. A., Eraker, J. H., Lamport, J. E., and Walpole, P. H. 1974. Electrons and protons accelerated in Mercury's magnetic field. *Science* 185:160–166.

Singer, R. B., and Roush, T. L. 1985. Effects of temperature on remotely sensed mineral absorption features. *J. Geophys. Res.* 90:12,434–12,444.

Siscoe, G., and Christopher, L. 1975. Variations in the solar wind stand-off distance at Mercury. *Geophys. Res. Lett.* 2:158–160.

Siscoe, G. L., Ness, N. F., and Yeates, C. M. 1975. Substorms on Mercury? *J. Geophys. Res.* 80:4359–4363.

Slavin, J. A., and Holzer, R. E. 1979a. The effect of erosion on the solar wind stand-off distance at Mercury. *J. Geophys. Res.* 84:2076–2082.

Slavin, J. A., and Holzer, R. E. 1979b. On the determination of the Hermean magnetic moment: A critical review. *Phys. Earth Planet. Int.* 20:231–236.

Slavin, J. A., and Holzer, R. E. 1981. Solar wind flow about the terrestrial planets 1. Modeling bow shock position and shape. *J. Geophys. Res.* 86:11,401–11,418.

Smith, B. A., and Reese, E. J. 1968. Mercury's rotation period: Photographic confirmation. *Science* 162:1275–1277.

Smith, E. I. 1976. Comparison of the crater morphology-size relationship for Mars, Moon, and Mercury. *Icarus* 28:543–550.

Smith, E. I., and Hartnell, J. A. 1978. Crater size-shape profiles for the moon and Mercury: Terrain effects and interplanetary comparisons. *Moon and Planets* 19:479–511.

Smith, G. R., Shemansky, D. E., Broadfoot, A. L., and Wallace, L. 1978. Monte Carlo modeling of exospheric bodies: Mercury. *J. Geophys. Res.* 83:3783–3790.

Smith, J. V. 1979. Mineralogy of the planets: A voyage in space and time. *Mineral. Mag.* 43:1–89.

Smith, W. B., Ingalls, R. P., Shapiro, I. I., and Ash, M. E. 1970. Surface-height variations on Venus and Mercury. *Radio Sci.* 5:411–423.

Smyth, W. H. 1986. Nature and variability of Mercury's sodium atmosphere. *Nature* 323:696–699.

Snedecor, G. W., and Cochran, W. G. 1967. *Statistical Methods* (Ames: Iowa State Univ. Press).

Soderblom, L. A. 1970. A model for small impact erosion applied to the lunar surface. *J. Geophys. Res.* 75:2655–2661.

Solar System Exploration Committee. 1983. *Planetary Exploration through Year 2000: A Core Program* (Washington, DC: U.S. Govt. Printing Office).

Solomon, S. C. 1976. Some aspects of core formation in Mercury. *Icarus* 28:509–521.

Solomon, S. C. 1977. The relationship between crustal tectonics and internal evolution in the Moon and Mercury. *Phys. Earth Planet. Int.* 15:135–145.

Solomon, S. C. 1978. On volcanism and thermal histories on one-plate planets. *Geophys. Res. Lett.* 5:461–464.

Solomon, S. C. 1979. Formation, history and energetics of cores in the terrestrial planets. *Phys. Earth Planet. Int.* 19:168–182.

Solomon, S. C., and Chaiken, J. 1976. Thermal expansion and thermal stress in the Moon and terrestrial planets: Clues to early thermal history. *Proc. Lunar Sci. Conf.* 7:3229–3243.

Solomon, S. C., and Head, J. W. 1979. Vertical movement in mare basins: Relation to mare emplacement, basin tectonics and lunar thermal history. *J. Geophys. Res.* 84:1667–1682.

Solomon, S. C., and Head, J. W. 1980. Lunar mascon basins: Lava filling, tectonics, and evolution of the lithosphere. *Rev. Geophys. Space Phys.* 18:107–141.

Solomon, S. C., Comer, R. P., and Head, J. W. 1982. The evolution of impact basins: Viscous relaxation of topographic relief. *J. Geophys. Res.* 87:3975–3992.

Sonett, C. P., Colburn, D. S., and Schwartz, K. 1975. Formation of the lunar crust: An electrical source of heating. *Icarus* 24:231–255.

Sonnerup, B. U. Ö., and Cahill, L. J., Jr. 1967. Magnetopause structure and attitude from Explorer 12 observations. *J. Geophys. Res.* 72:171–183.

Sonnerup, B. U. Ö., Paschmann, G., Papamastorakis, I., Sckopke, N., Haerendel, G., Bame, S. J., Asbridge, J. R., Gosling, J. T., and Russell, C. T. 1981. Evidence for magnetic field reconnection at the Earth's magnetopause. *J. Geophys. Res.* 86:10,049–10,067.

Soter, S. L., and Ulrichs, J. 1967. Rotation and heating of the planet Mercury. *Nature* 214:1315–1316.

Spinrad, H., Field, G., and Hodge, P. 1965. Spectroscopic observations of Mercury. *Astrophys. J.* 141:1155–1160.

Spreiter, J. R., Summers, A. L., and Alksne, A. Y. 1966. Hydromagnetic flow around the magnetosphere. *Planet. Space Sci.* 14:223–253.

Spudis, P. D. 1984. Mercury: New identification of ancient multi-ring basins and implications for geologic evolution. In *Repts. Planet. Geol. Prog., 1983*, NASA TM-86246, pp. 87–89.

Spudis, P. D. 1985. A mercurian chronostratigraphic classification. In *Repts. Planet. Geol. Prog.*, NASA TM-87563, pp. 595–597.

Spudis, P. D. 1986. The materials and formation of the Imbrium basin. In *Workshop on the Geology and Petrology of the Apollo 15 Landing Site*, LPI Tech. Rept. 86-03 (Houston: Lunar and Planetary Inst.), pp. 100–104.

Spudis, P. D., and Hawke, B. R. 1986. The Apennine Bench Formation revisited. In *Workshop on the Geology and Petrology of the Apollo 15 Landing Site*, LPI Tech. Rept. 86-03 (Houston: Lunar and Planetary Inst.), pp. 105–107.

Spudis, P. D., and Prosser, J. G. 1984. Geologic map of the Michelangelo quadrangle (H-12) of Mercury. U.S. Geol. Survey Map I-1659.

Spudis, P. D., and Strobell, M. E. 1984. New identification of ancient multi-ring basins on Mercury and implications for geologic evolution. *Lunar Planet. Sci.* XV:814–815 (abstract).

Srnka, L. J. 1976. Magnetic dipole moment of a spherical shell with TRM acquired in a field of internal origin. *Phys. Earth Planet. Int.* 11:184–190.

Srnka, L. J. 1977. Spontaneous magnetic field generation in hypervelocity impacts. *Proc. Lunar Sci. Conf.* 8:785–792.

Srnka, L. J., and Mendenhall, M. H. 1979. Theory of global thermoremanent magnetization of planetary lithospheres in dipole fields of internal origin. *J. Geophys. Res.* 84:4667–4674.

Stacey, F. 1977a. A thermal model of the Earth. *Phys. Earth Planet. Int.* 15:341–348.

Stacey, F. 1977b. *Physics of the Earth* (New York: J. Wiley).

Standish, E. M., Keesey, M. S., and Newhall, X. X. 1976. JPL development ephemeris number 96. JPL Tech. Rept. 32-1603.

Stephenson, A. 1976. Crustal remanence and the magnetic moment of Mercury. *Earth Planet. Sci. Lett.* 28:454–458.

Stern, S. A., McClintock, W. E., and Skinner, T. E. 1986. Obtaining Mercury's UV spectrum from a sounding rocket mission. *Mercury: Program and Abstracts*, Tucson, Arizona, August.

Stevenson, D. J. 1974. Planetary magnetism. *Icarus* 22:403–415.

Stevenson, D. J. 1979. Turbulent thermal convection in the presence of rotation and a magnetic field: A heuristic theory. *Geophys. Astrophys. Fluid Dyn.* 12:139–169.

Stevenson, D. 1980. Applications of liquid state physics to the Earth's core. *Phys. Earth Planet. Int.* 22:42–52.

Stevenson, D. 1981. Models of the Earth's core. *Science* 214:611–619.

Stevenson, D. J. 1983. Planetary magnetic fields. *Rept. Prog. Phys.* 46:555–620.

Stevenson, D. J. 1984. The energy flux number and three types of planetary dynamo. *Astron. Nachr.* 305:257–264.

Stevenson, D. J. 1987. Mercury's magnetic field: A thermoelectric dynamo? *Earth Planet. Sci. Lett.* 82:114–120.

Stevenson, D. J., Spohn, T., and Schubert, G. 1983. Magnetism and thermal evolution of the terrestrial planets. *Icarus* 54:466–489.

Stöffler, D., Gault, D. E., Wedekind, J., and Polk, G. 1975. Experimental hypervelocity impact into quartz sand: Distribution and shock metamorphism of ejecta. *J. Geophys. Res.* 80:4062–4077.

Stolper, E. 1979. Trace elements in shergottite meteorites: Implications for the origins of planets. *Earth Planet. Sci. Lett.* 42:239–242.

Strom, R. G. 1964. Analysis of lunar lineaments I: Tectonic maps of the Moon. *Comm. Lunar Planet. Lab.* 2:205–216.

Strom, R. G. 1977. Origin and relative age of lunar and mercurian intercrater plains. *Phys. Earth Planet. Int.* 15:156–172.

Strom, R. G. 1979. Mercury: A post-Mariner 10 assessment. *Space Sci. Rev.* 24:3–70.

Strom, R. G. 1984. Mercury. In *The Geology of the Terrestrial Planets*, ed. M. H. Carr, NASA SP-469, pp. 13–55.

Strom, R. G. 1987a. The solar system cratering record: Voyager 2 Uranus results and implications for the origin of impacting objects. *Icarus* 70:517–535.

Strom, R. G. 1987b. *Mercury: The Elusive Planet* (Washington, DC: Smithsonian Inst. Press).

Strom, R. G., Murray, B. C., Belton, M. J. S., Danielson, G. E., Davies, M. E., Gault, D. E.,

Hapke, B., O'Leary, B., Trask, N. J., Guest, J. E., Anderson, J., and Klaason, K. 1975a. Preliminary imaging results from the second Mercury encounter. *J. Geophys. Res.* 80:2345–2356.

Strom, R. G., Trask, N. J., and Guest, J. E. 1975b. Tectonism and volcanism on Mercury. *J. Geophys. Res.* 80:2478–2507.

Strom, R. G., Woronow, A., and Gurnis, M. 1981. Crater populations on Ganymede and Callisto. *J. Geophys. Res.* 86:8659–8674.

Stuart-Alexander, D. E., and Howard, K. A. 1970. Lunar maria and circular basins: A review. *Icarus* 12:440–456.

Stull, D. R., and Prophet, H., eds. 1971. *JANAF Thermochemical Tables*, 2nd ed., NSRDS-NBS 37, Washington, DC.

Suess, S. T., and Goldstein, B. E. 1979. Compression of the Hermean magnetosphere by the solar wind. *J. Geophys. Res.* 84:3306–3312.

Summers, J. L. 1959. *Investigation of High-Speed Impact: Regions of Impact and Impact at Oblique Angles*, NASA TN D-94.

Sung, C. M., Singer, R. B., Parkin, K. M., and Burns, R. G. 1977. Temperature dependence of Fe^{2+} crystal field spectra: Implications to mineralogical mapping of planetary surfaces. *Proc. Lunar Sci. Conf.* 8:1063–1079.

Swindle, T. D., Caffee, M. W., and Hohenburg, C. M. 1986. Xenon and other noble gases in shergottites. *Geochim. Cosmochim. Acta* 50:1001–1015.

Tedesco, E. F., Tholen, D. J., and Zellner, B. 1982. The eight-color asteroid survey: Standard stars. *Astron. J.* 87:1585–1592.

Teegarden, B. F., McDonald, F. B., Trainor, J. H., Webber, W. R., and Roelof, E. C. 1974. Interplanetary MeV electrons of Jovian origin. *J. Geophys. Res.* 79:3615–3622.

Thomas, G. 1974. Mercury: Does its atmosphere contain water? *Science* 183:1197–1198.

Thomas, P. G. 1978. Structural lineament pattern analysis of the Caloris surroundings: A pre-Caloris pattern on Mercury. *Rept. Planet. Geol. Prog.*, NASA TMX-79-929, pp. 79–82.

Thomas, P. G. 1980. Etudes géologiques et structurales de la planète Mercure. Thèse de 3e cycle, Univ. Paris XI.

Thomas, P. G., and Masson, P. 1983. Tectonic evolution of Mercury: Comparison with the Moon. *Ann. Geophys.* 1:53–58.

Thomas, P. G., and Masson, P. 1984a. Geology of the Argyre area on Mars: Comparison with other basins in the solar system. *Earth, Moon, Planets* 31:25–42.

Thomas, P. G., and Masson, P. 1984b. Tectonics of the Caloris area on Mercury: An alternative view. *Icarus* 58:396–402.

Thomas, P. G., Masson, P., and Fleitout, L. 1982. Global volcanism and tectonism on Mercury: Comparison with the Moon. *Earth Planet. Sci. Lett.* 58:95–103.

Thompson, C., and Stevenson, D. J. 1988. Two-phase gravitational instabilities in thin disks with application to the origin of the moon. *Astrophys. J.* 333, October.

Thompson, D., Clark, P. E., and Leake, M. 1986. Characterization of linear structures on Mercury. *Mercury: Program and Abstracts*, Tucson, Arizona, August.

Thompson, S. L., and Lauson, H. S. 1984. Improvement in the chart D radiation-hydrodynamic code III: Revised analytic equations of state. Sandia Laboratories report SC-RR-71 0714.

Tidman, D. A., and Krall, N. A. 1971. *Shock Waves in Collisionless Plasmas* (New York: Wiley-Interscience).

Toksöz, M. N., and Johnston, D. H. 1977. The evolution of the Moon and the terrestrial planets. In *The Soviet-American Conference on Cosmochemistry of the Moon and Planets*, NASA SP-370, pp. 245–328.

Toksöz, M. N., Hsui, A. T., and Johnston, D. H. 1978. Thermal evolutions of the terrestrial planets. *Moon and Planets* 18:281–320.

Toomre, A. 1966. On the coupling of the Earth's core and mantle during the 26,000 year precession. In *The Earth-Moon System*, eds. B. G. Marsden and A. G. W. Cameron (New York: Plenum), pp. 33–45.

Tozer, D. C. 1967. Towards a theory of thermal convection in the mantle. In *The Earth's Mantle*, ed. T. F. Gaskell (London: Academic Press), pp. 325–353.

Trafton, L. M. 1977. Periodic variations in Io's sodium and potassium clouds. *Astrophys. J.* 215:960–970.

Trask, N. J. 1975. Cratering history of the heavily cratered terrain on Mercury. Proc. Intern. Coll. Planet. Geol. *Geol. Res.* 15:471.

Trask, N. J., and Guest, J. E. 1975. Preliminary geologic terrain map of Mercury. *J. Geophys. Res.* 80:2462–2477.

Trask, N. J., and Strom, R. G. 1976. Additional evidence for mercurian volcanism. *Icarus* 28:559–563.

Tsuchiyama, A., Nagahara, H., and Kushiro, I. 1981. Volatilization of sodium from silicate melt spheres and its application to the formation of chondrules. *Geochim. Cosmochim. Acta* 45:1357–1367.

Turcotte, D. L. 1983. Thermal stresses in planetary elastic lithospheres. *Proc. Lunar Planet. Sci. Conf.* 13, *J. Geophys. Res. Suppl.* 88:A585–A587.

Turcotte, D., and Oxburgh, E. R. 1967. Finite amplitude convective cells and continental drift. *J. Fluid Mech.* 28:29–42.

Turcotte, D. L., and Oxburgh, E. R. 1969. Convection in a mantle with variable physical properties. *J. Geophys. Res.* 74:1458–1474.

Turcotte, D. L., Willeman, R. J., Haxby, W. F., and Norberry, J. 1981. Role of membrane stresses in the support of planetary topography. *J. Geophys. Res.* 86:3951–3957.

Tyler, A. L., Schneider, N. M., Wells, W. K., Hunten, D. M., and Kozlowski, R. W. H. 1986. Mercury observations indicating spatial variations in atmospheric sodium and the presence of atmospheric potassium. *Bull. Amer. Astron. Soc.* 18:781 (abstract).

Ullrich, G. W. 1976. The mechanics of central peak formation in shock wave cratering events. *U.S. Air Force Report*, ARWL-TR-75-88.

United States Geological Survey. Shaded relief map of Mercury. Misc. Inves. Series, H-11,H 5M, 45/45R, H 12, H 5M, 45/135R, H 15, H5M, 90/OR.

United States Geological Survey. 1979. Mercury: Relief and albedo markings visible on Mariner-10 images. Misc. Inves. Map I-1171.

Urey, H. C. 1951. The origin and development of the Earth and other terrestrial planets. *Geochim. Cosmochim. Acta* 1:209–277.

Urey, H. C. 1952. *The Planets: Their Origin and Development* (New Haven: Yale Univ. Press).

Usselman, T. M. 1975. Experimental approach to the state of the Earth's core. *Amer. J. Sci.* 275:278–290.

van Diggelen, J. 1959. Photometric properties of lunar crater floors. *Rech. Obs. Utrect* 14(2).

Vening-Meinesz, F. A. 1947. Shear patterns of the earth's crust. *Trans. AGU* 28:1–61.

Veverka, J. 1970. Photometric and Polarimetric Studies of Minor Planets and Satellites. Ph.D. Thesis, Harvard Univ., Cambridge, Massachusetts.

Veverka, J., and Burns, J. A. 1980. The moons of Mars. *Ann. Rev. Earth Planet. Sci.* 8:527–558.

Veverka, J., Thomas, P., Johnson, T. V., Matson, D., and Housen, K. 1986. The physical characteristics of satellite surfaces. In *Satellites,* eds. J. A. Burns and M. S. Matthews (Tucson: Univ. of Arizona Press), pp. 342–402.

Vickery, A. 1986. Effect of an impact-generated gas cloud in the acceleration of solid ejecta. *J. Geophys. Res.* 91:14,139–14,160.

Vilas, F. 1985. Mercury: Absence of crystalline Fe^{2+} in the regolith. *Icarus* 64:133–138.

Vilas, F., and McCord, T. B. 1976. Mercury: Spectral reflectance measurements (0.33–1.06 μm) 1974/75. *Icarus* 28:593–599.

Vilas, F., and Smith, B. A. 1985. Reflectance spectrophotometry (~0.5–1.0 μm) of outer-belt asteroids: Implications for primitive, organic solar system material. *Icarus* 64:503–516.

Vilas, F., Leake, M. A., and Mendell, W. W. 1984. The dependence of reflectance spectra of Mercury on surface terrain. *Icarus* 59:60–68.

Vincent, M. A., and Bender, P. L. 1986. Orbit determination and gravitational field accuracy for a Mercury transponder satellite. *J. Geophys. Res.,* submitted.

Vityazev, A. V., and Pechernikova, G. V. 1981. Solution of the problem of planet's rotation in the statistical theory of accumulation. *Soviet Astron.* 25:494–499.

Vityazev, A. V., and Pechernikova, G. V. 1984. Shergottites, ALHA 81005 and variations of Fe/Si ratio in terrestrial planets. "Abstracts of the XIX Soviet Conf. on Meteorites and Cosmochemie," Moscow, p. 4 (abstract).

Vityazev, A. V., Pechernikova, G. V., and Safronov, V. S. 1984. Fractionation of matter in the protoplanetary cloud. "Abstracts of the 27th Intl. Geol. Congress," Moscow, pp. 429–430 (abstract).

Wagner, J. K., Hapke, B. W., and Wells, E. N. 1987. Atlas of reflectance spectra of terrestrial, lunar and meteoritic powders and frosts from 92 to 1800 nm. *Icarus* 69:14–28.

Wai, C. M., and Wasson, J. T. 1977. Nebular condensation of moderately volatile elements and their abundances in ordinary chondrites. *Earth Planet. Sci. Lett.* 36:1–13.

Wänke, H. 1987. Chemistry and accretion of Earth and Mars. *Bull. Soc. Geol. France* 1:13–19.

Wänke, H., and Dreibus, G. 1985. The degree of oxidation and the abundance of volatile elements on Mars. *Lunar Planet. Sci.* XVI:28–29 (abstract).

Wänke, H., and Dreibus, G. 1986. Mercury, a highly reduced planet? *Mercury: Program and Abstracts,* Tucson, Arizona, August.

Ward, W. R. 1975. Tidal friction and generalized Cassini's laws in the solar system. *Astron. J.* 80:64–69.

Warner, J. L. 1972. Metamorphism of Apollo 14 breccias. *Proc. Lunar Sci. Conf.* 3:623–643.

Warren, P. H. 1985. The magma ocean concept and lunar evolution. *Ann. Rev. Earth Planet. Sci.* 13:201–240.

Wasson, J. T. 1972. Formation of ordinary chondrites. *Rev. Geophys. Space Phys.* 10:711–759.

Wasson, J. T. 1977. Relationship between the composition of solid solar-system matter and distance from the Sun. In *Comets, Asteroids, Meteorites: Interrelations, Evolution and Origins,* ed. A. H. Delsemme (Toledo: Univ. of Toledo Press), pp. 551–559.

Wasson, J. T. 1985. *Meteorites* (San Francisco: W. H. Freeman).

Wasson, J. T., and Warren, P. H. 1979. Formation of the moon from differentiated planetesimals of chondritic composition. *Lunar Planet. Sci.* X:1310–1312 (abstract).

Wasson, J. T., and Wetherill, G. W. 1979. Dynamical, chemical and isotopic evidence regarding the formation locations of asteroids and meteorites. In *Asteroids,* ed. T. Gehrels (Tucson: Univ. of Arizona Press), pp. 926–974.

Watts, A. B., Bodine, J. H., and Steckler, M. S. 1980. Observations of flexure and the state of stress in the oceanic lithosphere. *J. Geophys. Res.* 85:6369–6376.

Wedekind, J. A., Gault, D. E., and Greeley, R. 1970. Model studies of isostatic adjustment of large impact craters. *Trans. AGU* 51:342 (abstract).

Weertman, J., and Weertman, J. R. 1975. High temperature creep of rock and mantle viscosity. *Ann. Rev. Earth Planet. Sci.* 3:293–315.

Weidenschilling, S. J. 1978. Iron/silicate fractionation and the origin of Mercury. *Icarus* 35:99–111.

Weidenschilling, S. J. 1980. Dust to planetesimals: Settling and coagulation in the solar nebula. *Icarus* 44:172–189.

Wetherill, G. W. 1975. Late heavy bombardment of the moon and terrestrial planets. *Proc. Lunar Sci. Conf.* 6:1539–1561.

Wetherill, G. W. 1976a. The role of large bodies in the formation of the Earth and Moon. *Proc. Lunar Sci. Conf.* 7:3245–3257.

Wetherill, G. W. 1976b. Comments on the paper by C. R. Chapman: Chronology of terrestrial planet evolution—The evidence from Mercury. *Icarus* 28:537–542.

Wetherill, G. W. 1977. Evolution of the Earth's planetesimal swarm subsequent to the formation of the Earth and Moon. *Proc. Lunar Sci. Conf.* 8:1–16.

Wetherill, G. W. 1978. Accumulation of the terrestrial planets. In *Protostars and Planets,* ed. T. Gehrels (Tucson: Univ. of Arizona Press), pp. 565–598.

Wetherill, G. W. 1979. Steady-state population of Apollo-Amor objects. *Icarus* 37:96–112.

Wetherill, G. W. 1980a. Formation of the terrestrial planets. *Ann. Rev. Astron. Astrophys.* 18:77–113.

Wetherill, G. W. 1980b. Numerical calculations relevant to the accumulation of the terrestrial planets. In *The Continental Crust and Its Mineral Deposits,* ed. D. W. Strangway, Geol. Assoc. of Canada Special Paper 20, pp. 3–24.

Wetherill, G. W. 1981. Solar wind origin of ^{36}Ar on Venus. *Icarus* 46:70–80.

Wetherill, G. W. 1985a. Occurrence of giant impacts during the growth of the terrestrial planets. *Science* 228:877–879.

Wetherill, G. W. 1985b. Asteroidal source of ordinary chondrites. *Meteoritics* 20:1–22.

Wetherill, G. W. 1985c. Remarks at the Lunar and Planetary Science Conference, Houston.

Wetherill, G. W. 1986. Accumulation of the terrestrial planets and implications concerning lunar origin. In *Origin of the Moon,* eds. W. K. Hartmann, R. J. Philips, and G. J. Taylor (Houston: Lunar and Planetary Inst.), pp. 519–550.

Wetherill, G. W., and Chapman, C. R. 1988. Asteroids and meteorites. In *Meteorites and the Early Solar System,* eds. J. F. Kerridge and M. S. Matthews (Tucson: Univ. of Arizona Press), pp. 35–67.

Wetherill, G. W., and Cox, L. P. 1984. The range of validity of the two-body approximation in models of terrestrial planet accumulation. I. Gravitational perturbations. *Icarus* 60:40–55.

Wetherill, G. W., and Cox, L. P. 1985. The range of validity of the two-body approximation in models of terrestrial planet accumulation. II. Gravitational cross-sections and runaway accretion. *Icarus* 63:290–303.

Wetherill, G. W., and Stewart, G. R. 1986. The early stages of planetesimal accumulation. *Lunar Planet. Sci.* XVII:939 (abstract).

Wetherill, G. W., and Stewart, G. R. 1987. Factors controlling early runaway growth of planetesimals. *Lunar Planet. Sci.* XVIII:1077 (abstract).

Whang, Y. C. 1977. Magnetospheric magnetic field of Mercury. *J. Geophys. Res.* 82:1024–1030.

Whang, Y. C., and Ness, N. F. 1975. Modeling the magnetosphere of Mercury. NASA-GSFC X-690-75-89.

Wiens, R. C., Becker, R. H., and Pepin, R. O. 1986. The case for a martian origin of the shergottites, II. Trapped and indigenous gas components in EETA 79001 glass. *Earth Planet Sci. Lett.* 77:149–158.

Wilhelms, D. E. 1970. Summary of lunar stratigraphy: Telescopic observations. U.S. Geol. Survey Prof. Paper 599-F.

Wilhelms, D. E. 1972. Geologic mapping of the second planet. U.S. Geol. Survey, Astrogeologic Studies 55.

Wilhelms, D. E. 1973. Comparison of martian and lunar multiringed circular basins. *J. Geophys. Res.* 78:4084–4095.

Wilhelms, D. E. 1976. Mercurian volcanism questioned. *Icarus* 28:551–558.

Wilhelms, D. E. 1984. Moon. In *The Geology of the Terrestrial Planets,* ed. M. H. Carr, NASA SP-469, pp. 106–205.

Wilhelms, D. E. 1987. The geologic history of the Moon. U.S. Geol. Survey Prof. Paper 1348.

Wilhelms, D. E. 1988. Geologic mapping. In *Planetary Mapping,* ed. R. Greeley and R. M. Batson (Cambridge: Cambridge Univ. Press), in press.

Wilhelms, D. E., and El-Baz, F. 1977. Geologic map of the east side of the Moon. U.S. Geol. Survey Map I-948.

Wilhelms, D. E., and McCauley, J. F. 1971. Geologic map of the near side of the Moon. U.S. Geol. Survey Map I-703.

Wilhelms, D. E., Hodges, C. A., and Pike, R. J. 1977. Nested crater model of lunar ringed basins. In *Impact and Explosion Cratering,* eds. D. J. Roddy, R. O. Pepin, and R. B. Merrill (New York: Pergamon Press), pp. 539–562.

Willemann, R. J. 1984. Reorientation of planets with elastic lithospheres. *Icarus* 60:701–709.

Willis, D. M. 1971. Structure of the magnetopause. *Rev. Geophys. Space Phys.* 9:953–985.

Wilshire, H. G., Offield, T. W., Howard, K. A., and Cummings, D. 1972. Geology of the Sierra Madera cryptoexplosion structure, Pecos County, Texas. U.S. Geol. Survey Prof. Paper 599-H.

Wisdom, J. 1985. Meteorites may follow a chaotic route to Earth. *Nature* 315:731–733.

Wollenhaupt, W., and Sjogren, W. 1977. Lunar laser altimetry map. *Proc. Lunar Planet. Sci. Conf.* 8:Plate 14.

Wood, C. A. 1980. Martian double-ring basins: New observations. *Proc. Lunar Planet. Sci. Conf.* 11:2221–2241.

Wood, C. A., and Andersson, L. 1978. New morphometric data for fresh lunar craters. *Proc. Lunar Planet. Sci. Conf.* 9:3669–3689.

Wood, C. A., and Head, J. W. 1976. Comparison of impact basins on Mercury, Mars, and the Moon. *Proc. Lunar Sci. Conf.* 7:3629–3651.

Wood, C. A., Head, J. W., and Cintala, M. J. 1978. Interior morphology of fresh martian craters: The effects of target characteristics. *Proc. Lunar Planet. Sci. Conf.* 9:3691–3709.

Wood, J. A. 1981. The interstellar dust as a precursor of Ca,Al-rich inclusions in carbonaceous chondrites. *Earth Planet. Sci. Lett.* 56:32–44.

Woronow, A. 1977. Crater saturation and equilibrium: A Monte Carlo simulation. *J. Geophys. Res.* 82:2447–2456.

Woronow, A. 1978. A general cratering history model and its implication for lunar highlands. *Icarus* 34:76–88.

Woronow, A., and Strom, R. G. 1981. Limits on large-crater production and obliteration on Callisto. *Geophys. Res. Lett.* 8:891–894.

Woronow, A., Strom, R. G., and Gurnis, M. 1982. Interpreting the cratering record: Mercury to Ganymede and Callisto. In *Satellites of Jupiter,* ed. D. Morrison (Tucson: Univ. of Arizona Press), pp. 237–276.

Wu, H. H., and Broadfoot, A. L. 1977. The extreme ultraviolet albedos of the planet Mercury and of the Moon. *J. Geophys. Res.* 82:759–761.

Yen, C.-W. 1985. Ballistic Mercury orbiter mission via Venus and Mercury gravity assists. In *AAS/AIAA Astrodynamics Specialist Conf. Proc.* (San Diego, CA: AAS Publ.), AIAA No. 85-346.

Yen, C.-W. 1986. Ballistic Mercury mission options. *Mercury: Program and Abstracts,* Tucson, Arizona, August.

Yoder, C. F. 1981. The free librations of a dissipative Moon. *Phil. Trans. Roy. Soc. London* A303:327–338.

Young, A. 1978. Mercury's craters from Earth. *Icarus* 34:208–209.

Zeller, E., Ronca, L., and Levy, P. W. 1966. Proton-induced hydroxyl formation on the lunar surface. *J. Geophys. Res.* 71:4855–4860.

Zohar, S., and Goldstein, R. M. 1974. Surface features on Mercury. *Astron. J.* 79:85–91.

Zohar, S., Goldstein, R. M., and Rumsey, H. C. 1980. A new radar determination of the spin vector of Venus. *Astron. J.* 85:1103–1111.

Zollner, J. C. F. 1815. *Photometrische Untersuchungen* (Leipsig: W. Engleman).

Zook, H. A. 1975. The state of meteoritic material on the Moon. *Proc. Lunar Sci. Conf.* 6:1653–1672.

Zschau, J. 1978. Tidal friction in the solid Earth: Loading tide vs. body tides. In *Tidal Friction and the Earth's Rotation,* eds. P. Brosche and J. Sündermann (Berlin: Springer-Verlag).

Zwickl, R. D., and Webber, W. R. 1977. Solar particle propagation from 1 to 5 AU. *Solar Phys.* 54:457–504.

Glossary

GLOSSARY*

Compiled by Melanie Magisos

aberration angle
: the direction that the solar wind is apparently rotated away from the solar direction because of the finite velocity of the planet perpendicular to the solar wind velocity.

absolute magnitude
: a measure of the intrinsic brightness of a star given by the apparent magnitude it would have if it were moved to a distance of 10 parsecs from the observer; in the solar system, the apparent magnitude of a comet or asteroid at 1 AU distance from both the Earth and the Sun.

accretion
: the agglomeration of matter together to form larger bodies such as stars, planets, and moons.

activation energy (E_a)
: the minimum energy required for chemical reactions of certain molecular-scale processes such as diffusion.

adiabat
: a curve or line plotted using coordinates selected to represent the pressure and volume or the temperature and entropy of a parcel of matter during a process in which no heat enters or leaves the parcel.

adsorption
: physisorption: physical-scale bonding of a gas atom to a surface with a duration of at least one vibrational period. Bonding energies are less than 0.5 cV. Chemisorption:

---

*We have used some definitions from *Glossary of Astronomy and Astrophysics* by J. Hopkins (by permission of the University of Chicago Press, copyright 1980 by the University of Chicago), from *Astrophysical Quantities* by C. W. Allen (London: Athlone Press, 1973), from *Glossary of Geology*, edited by M. Gary, R. McAfee, and C. L. Wolff (Washington, D.C.: American Geological Institute, 1972), and from *Mercury: The Elusive Planet* by R. G. Strom (Washington, D.C.: Smithsonian Institution Press, 1987). We also acknowledge definitions and helpful comments from various chapter authors.

chemical-scale bonding of a gas atom to a surface. Bonding energies are greater than 0.5 eV.

albedo, bond — fraction of the total incident light reflected by a spherical body.

albedo, geometric — ratio of planet brightness at zero phase angle to the brightness of a perfectly diffusing disk with the same position and apparent size as the planet.

albedo, hemispherical — fraction of incident light scattered by a surface as a function of angle of incidence.

albedo, normal — the brightness of a surface at zero phase angle relative to a perpendicularly illuminated, perfectly diffusing (Lambert) surface at the same distance as the surface.

albedo, physical — *see* geometric albedo.

albedo, single-particle scattering — the fraction of incident light scattered by a particle.

albedo, spherical — *see* albedo, bond.

Alfvén wave — an oscillation in a highly electrically conducting, magnetized fluid or plasma in which the magnetic field direction fluctuates but neither the magnetic field strength nor mass density changes.

alkali — an element in group IA in the periodic table (Li, Na, K, Rb, Cs, Fr).

angstrom (Å) — the unit in which the wavelength of visible light is often expressed. One angstrom unit is defined as 10^{-8} cm.

angular momentum — the angular momentum of a system about a specified origin is the sum over all the particles in the system (or an integral over the different elements of the system if it is continuous) of the vector products of the radius vector joining each particle to the origin and the momentum of each particle. For a closed system it is conserved by virtue of the isotropy of space.

antipodal point	point on the opposite side of a planet (i.e., 180° apart on a sphere).
aphelion	distance of greatest separation between a body orbiting in an eccentric orbit around the Sun and the Sun.
apparent magnitude	measure of the observed brightness of a celestial object as seen from the Earth. In the solar system, it is a function of the body's intrinsic brightness (due to its size, albedo and distance from the Sun) and its distance from the Earth. A sixth magnitude star is just visible to the naked eye.
arcuate scarp	a geological formation with a cliff-like structure with a curved trace as viewed from above and mapped.
asthenosphere	a weak spherical shell located below the lithosphere, in which isostatic adjustments take place, magmas may be generated, and seismic waves are strongly attenuated.
atmospheric atoms	atoms actually in the ballistic atmosphere, with near-thermal energies.
AU	Astronomical Unit, the mean distance between Earth and Sun (1.496×10^{13} cm $\simeq 500$ light seconds).
auroral zones	the regions mapped onto the surface of the planet in which most of the energy of charged particles in the magnetosphere is deposited.
backscatter	scattering of radiation (or particles) through angles greater than 90° with respect to the original direction of motion.
basins	very large impact crater structures, usually with concentric structural rings.
bolide	a term used to describe a meteor, the streak of light in the sky produced by the transit of a meteoroid through the Earth's atmosphere, that approaches the brightness of the full Moon.
bow shock	a thin current layer across which the solar wind velocity drops, and the plasma is heated, compressed and de-

flected around the planetary obstacle. Ahead of the bow shock, the component of velocity along the shock normal is supersonic and behind it, subsonic.

bulk composition   chemical composition of the entire planet.

Caloris basin   the most prominent multiring basin on Mercury.

central peaks   one or more mountain peaks located near the center of a crater.

chondritic composition   chemical composition like that of the chondritic meteorites (i.e., relatively unfractionated, except for the most volatile species).

commensurabilities   orbital frequencies that are in a ratio of small whole integers.

condensation   the physical process by which a vapor becomes a liquid or solid.

Coriolis effect   the acceleration which a body in motion experiences when observed in a rotating frame. This force acts at right angles to the direction of the angular velocity.

cosmochemical abundances   the general pattern of elemental abundances in the Sun, planets and meteorites.

crater classes   the UA classification system designates Class 1 craters as the freshest and Class 5 craters as the most degraded. The USGS classification system is just the reverse, Class 5 being the freshest and Class 1 being the most degraded.

crater size-frequency distribution plot   a graph showing the number of craters, per unit diameter interval, over a range of diameter.

crust   the outermost, highly differentiated, solid layer of a planet or satellite, mostly consisting of crystalline rock or ice.

Curie temperature   the temperature marking the transition between ferromagnetism and paramagnetism, or between the ferroelectric phase and the paraelectric phase.

CW radar	radar system employing unmodulated (continuous wave) transmission.
degraded crater	a relatively old crater, significantly eroded, filled in, or superimposed by other craters so that its morphological appearance is changed from its original form.
delay-Doppler radar	radar technique employing modulated transmission in which both the time delay and Doppler shift of the echo are measured. It is used for reflectivity mapping and altimetry.
depolarization	randomization of radar wave polarization by multiple scattering or single scattering off wavelength-scale structure.
deviatoric stress	the full stress tensor minus hydrostatic pressure. This portion of the stress induces distortion, as opposed to volume compression or extension.
differentiation (in a planet)	a process whereby the primordial substances are separated. Generally metal sinks to the center to form a core, displacing the lighter silicates which form the crust plus mantle.
diffuse echo	partially depolarized component of a planetary radar echo associated with diffuse, high-angle backscatter off wavelength-scale surface structure.
Doppler spectrum	spectrum of radar echo Doppler shifts due primarily to planet rotation.
eccentricity	a number which defines the shape of an ellipse. It is the ratio of the distance from center to focus to the semimajor axis.
ecliptic	plane of the Earth's orbit.
ejecta blanket	the deposit surrounding an impact crater composed of material ejected from the crater during its formation.
emissivity	ratio of the radiation emitted by a body to that emitted by a blackbody at the same temperature.

energy
accommodation
coefficient (α)

coefficient of fractional energy exchange of a gas atom with a surface, including both adsorption collisions and free-free collisions. The equation for α at the macroscopic level is defined by

$$\alpha = (E_2 - E_o)/(E_1 - E_o)$$

where E_o is the mean energy per atom of the impacting particle, E_2 is the mean energy per atom leaving the surface, and E_1 is the mean energy per atom in the limiting case of thermal equilibrium with the surface. Similar coefficients can be constructed to specify the fractional momentum exchange of gas atoms with the surface.

equilibrium
condensation
model

a model for the chemical composition of the planets in which solids are hypothesized to have condensed from an initially hot nebula of solar composition, which cools slowly enough so that chemical equilibrium is maintained, and in which accretion takes place rapidly enough so that the solids may be characterized as being due to condensates at a particular temperature, which decreases with increasing distance of the planet from the Sun.

eutectic

an alloy or solution having its components in such proportions that the melting point is the lowest possible temperature with those components.

exosphere,
exobase

the outermost part of an atmosphere, in which collisions can be neglected to first order and the atoms can be regarded as executing ballistic orbits. Its bottom, the exobase, is taken as the level where the local mean free path is equal to the scale height.

facies

the aspect, appearance and characteristics of a rock unit, usually reflecting the conditions of its origin; especially as differentiating the unit from adjacent or associated units.

fault

a fracture or zone of fractures along which the sides are displaced relative to one another, parallel to the fracture.

felsic

used to describe light-colored silicate minerals such as quartz, feldspar and feldspathoids.

Fe/Si fractionation processes which separate Fe from Si, either in the solar nebula prior to incorporation of materials into a planet or during subsequent processes that strip mantle silicates from an assembled planet.

flux transfer event an apparent linkage of the interplanetary magnetic field and a planetary magnetic field that is finite in extent and duration. These events appear to be initiated on the dayside magnetopause near the subsolar point and proceed to higher latitudes.

fractionation the physical separation of one phase, element or isotope from another. Mass fractionation: a process which causes the proportions of the isotopes to change in a manner that is dependent on the differences in their mass. Most chemical and physical processes are of this type. Thus the $^{18}O/^{16}O$ ratio changes twice as much, for a given physical or chemical process, as the $^{17}O/^{16}O$ ratio. Consequently, samples which were originally similar in their proportion of oxygen isotopes spread along a line with slope 0.5 (the "mass fractionation line") on a plot of $^{17}O/^{16}O$ against $^{18}O/^{16}O$ when the samples have experienced the usual chemical and physical processes.

free air anomaly gravity acceleration anomalies projected onto a sphere or spheroid without corrections for density of topography or underlying terrain.

fresh crater a relatively young crater, with morphological structure similar to its form immediately after formation; there has been insufficient erosion, deposition or subsequent cratering to degrade its morphology very much.

FWHM full width at half maximum. The full width of a spectral line at half-maximum intensity.

galactic cosmic rays (GCR) energetic charged particles that diffuse into the solar system from outside. GCR ions generally have energies above 100 million electron volts.

Gauss the cgs unit of magnetic flux density. 1 Gauss = 1 Maxwell per square centimeter = 10^{-4} Tesla.

geoid	the equipotential surface ("mean sea level") of the Earth's gravitational field.
geomagnetic storm	a sudden change in the magnetospheric field resulting from the field's interaction with the solar wind. Such storms generally involve change in the field's topology, as well as the release of large amounts of energy and the accelerations of hot plasma and particles.
graben	an elongate crustal depression bounded by normal faults on its long sides.
Grüneisen parameter	parameter appearing in the Debye theory which is given by $\gamma = d(\ln \theta)/dd(\ln V)$ where θ is the Debye temperature and V is the volume.
heat of adsorption (D_o)	the dissociation energy measured from the zero-point energy level.
heavy bombardment	the period of time, beginning during planetary formation and apparently lasting from about 4.2 to 3.8 Gyr ago, when the cratering rate was high throughout at least the inner solar system and most of the basins and other craters were formed on the Moon and terrestrial planets.
heliocentric	centered on the Sun.
Hermean	of or related to Mercury (synonymous with Mercurian).
Hill sphere	a spherical region (extending out roughly 100 radii) within which a body's gravity dominates the orbital motion of satellites or other slowly moving objects in its vicinity.
Holsapple-Schmidt crater scaling law	a class of power-law scaling relations for various features of a crater (size, ejecta distribution, etc.) that have been shown by K. A. Holsapple and R. M. Schmidt to arise from a point-source approximation of the deposition of the energy and momentum of the impactor into the impacted body. As an example, the crater volume V in the large-scale, gravity-dominated regime scales with the impactor radius and velocity as $V \sim [ga/U^2]^\alpha$ where the coefficient α is from the appropriate measure of the

point-source approximation, and is about 0.65 for non-porous and about 0.5 for porous materials.

horst — an elongate, uplifted crustal block bounded by normal faults on its long sides.

hot poles — the subsolar points on Mercury at 0° and 180° that face the Sun at perihelion.

impact melt — target material that was melted by the heat generated by an impact.

inferior conjunction — the geometry when Mercury and the Sun are at the same ecliptic longitude, but Mercury is between the Sun and the Earth.

infrared (IR) — that part of the electromagnetic spectrum that lies beyond the red, having wavelengths from about 7500 Å to a few millimeters (about 10^{11} to 10^{14} Hz). Infrared radiation can be produced by atomic transitions, or by vibrational (near-IR) and rotational (far-IR) transitions in molecules. Planetary thermal emissions peak in the infrared.

intercrater plains — ancient level to gently rolling plains interspersed between craters in the heavily cratered regions of Mercury.

interplanetary magnetic field (IMF) — the magnetic field of the solar wind.

interstellar grains — small solid particles (including silicates) that exist in interstellar space; some may have become incorporated into comets and meteorite parent bodies and preserved (i.e., not melted nor vaporized through the formative and later periods of solar system history).

ionosphere — the region of ionized atmosphere close to a planetary surface. Formed from the neutral atmosphere through photoionization by ultraviolet radiation and through impact ionization by energetic particles.

isotopic ratios — ratios of two atomic masses which have the same number of protons, hence are the same chemical element, but differ in the number of neutrons.

Jovian electrons	electrons with a hard spectrum, i.e., with large fluxes of energetic electrons, presumed to be coming from Jupiter because of their resemblance to electrons seen just outside the Jovian magnetosphere.
kinematic viscosity	the ratio of the viscosity coefficient to the density of a material.
Kolmogorov spectrum	a spectrum in a homogeneous and isotropic turbulent medium with energy continually transferred between turbulent eddies of different sizes. The distribution of energy cascading down the different scales is referred to as the Kolmogorov spectrum.
Lambert's law	a simple scattering law according to which the intensity of scattered light is independent of the emission angle. An ideal Lambert surface scatters light uniformly in all directions (i.e., a diffuse scatterer).
laminar flow	smooth, nonturbulent flow.
lava	molten rock erupted onto the surface of a planet.
libration	a small oscillation around an equilibrium configuration, such as the angular change in the face that a synchronously rotating satellite presents towards the focus of its orbit.
limb	the edge of the apparent disk of a celestial body, as of the Sun, the Moon, a planet or a satellite.
lithosphere	the stiff upper layer of a planetary body, including the crust and part of the upper mantle, lying above the weaker asthenosphere.
lobate scarps	*see* arcuate scarps.
lobe of magnetotail	region of unidirectional magnetic field behind a planet, pointing either toward or away from the Sun.
Lommel-Seeliger surface	a surface with large-scale roughness where shadowing effects are important.
lunar fines	lunar soil.

lunar grid	a global system of preferential alignments of quasi-linear geologic features on the Moon, such as straight sections of crater walls, scarps or graben.
Mach number	the ratio of the flow velocity to a characteristic wave velocity of a fluid or plasma. Mach numbers often used are fast and slow magnetosonic and Alfvénic Mach numbers.
mafic	of or relating to a group of minerals characterized by high magnesium and iron content.
magma	a mobile or fluid rock material, lava, generalized to refer to any material that behaves like silicate magma in the Earth.
magnetic substorm	a sudden release of energy into the night hemisphere, presumed to be caused by the process of reconnection between the two oppositely directed magnetic lobes of a planetary magnetotail.
magnetism, remanent	that component of a rock's magnetization whose direction is fixed relative to the rock and which is independent of moderate applied magnetic fields.
magnetopause	the outer boundary of a planetary magnetosphere. Generally a strong thin current flows on this boundary.
magnetosheath	the region between a planetary bow shock and magnetopause in which the shocked solar wind plasma flows around the magnetosphere.
magnetosonic wave	an oscillation in a highly electrically conducting, magnetized fluid or plasma in which the magnetic field strength and the mass density change, and the field direction may change. In the fast magnetosonic wave, the field strength and mass density change are in phase. In the slow magnetosonic wave they are out of phase.
magnetosphere	the magnetic cavity created by the planet. For a planet with an intrinsic magnetic field, magnetospheric field lines have at least one end intersecting the planetary surface. For an unmagnetized planet or a comet, the magnetosphere may be defined as the region of draped magnet-

ic field in which planetary or cometary ions dominate over solar-wind ions.

magnetotail — the antisolar part of a planetary magnetosphere in which the ends of the magnetic field lines have been pulled away from the planet by the solar wind. A magnetotail consists of two oppositely directed magnetic lobes separated by a region of denser hot plasma called a plasma sheet.

magnitude — an arbitrary number, measured on a logarithmic scale, used to indicate the brightness of an object. If l_i is the brightness of star i, and m_i its magnitude, then $m_2 - m_1$ = 2.5 log (l_2/l_1). Two stars differing by 5 magnitudes differ in luminosity by a factor of 100. One magnitude is the fifth root of 100, or about 2.512. The brighter the star, the lower the numerical value of the magnitude.

mantle — the interior zone of a planet or satellite below the crust and above the core.

Mariner 10 — a U.S. spacecraft, launched in November 1973, which encountered Venus once and Mercury three times.

mascons (mass concentrations) — the dozen or so large-scale gravity anomalies that are found on the Moon. Most are gravity enhancements and are associated with maria.

Maxwell-Boltzmann distribution — the distribution function that any species of particle has if it is in thermodynamic equilibrium. This distribution function describes both the equilibrium in velocity space or kinetic energy, and the equilibrium in potential energy (Maxwell distribution).

mean density — the mass of a planet divided by its volume.

Mercury I, II, III — the three encounters of Mariner 10 with Mercury. These occurred on 29 March 1974, 21 September 1974 and 16 March 1975.

MHD — magnetohydrodynamics, dynamics of a conducting medium (i.e., plasma) coupled to a magnetic field.

micrometer — micron, $\mu m = 10^{-6}$ m = 10,000 Å.

microwave	an electromagnetic wave (in the radio region just beyond the infrared) with a wavelength of from about 1 mm to 30 cm (about 10^9 to 10^{11} Hz).
moment of inertia	the product of the mass of a body and the square of its radius of gyration.
nm	nanometer = 10^{-9} m = 10Å.
nodes (of a solar system body)	the two points where the body's orbit intersects the ecliptic.
normal fault	a fault resulting from tension that pulls the crust apart causing crustal lengthening.
nT	nanotesla = 10^{-9} Tesla, a unit of magnetic field strength, also called gamma.
obliquity	the angle between an object's axis of rotation and the pole of its orbit.
opposition	the position of a planet or asteroid when its ecliptic longitude is 180° from that of the Sun. At opposition, the Earth lies approximately between the body and the Sun.
opposition surge	an enhancement in the brightness of an object when observed with phase angle $< 1°$, in excess of that predicted by extrapolation of the brightness versus phase relation from larger phase angles.
oxidation	*see* oxidation/reduction processes.
oxidation/reduction processes	the process by which elements gain or lose electrons, thus changing their affinities for metallic (reduced) versus silicate (oxidized) phases.
perihelion	distance of least separation between a body orbiting in an eccentric orbit around the Sun and the Sun.
phase angle	the angle between observer, object and source of illumination, subtended at the object.
phase function	amount of light scattered into unit solid angle by a body as a function of phase angle, normalized to unity at zero phase angle.

phonon the quantum associated with lattice vibrations in a solid. Phonons are sound quanta.

photometric function the brightness of a surface as a function of angles of illumination, viewing and phase, normalized to unity at zero phase.

photometry the measurement of light intensities.

pi theorem a theorem that forms the foundation of dimensional analysis used in fluid dynamics and engineering. It is used to transform relations between physical variables in dimensional form into a set of dimensionless independent variables thereby permitting applications of physical processes over broad ranges in scale.

planetesimal small rocky or icy body formed from the primordial solar nebula, perhaps 1 to 10 km in diameter. Planets form, in part, from aggregation of planetesimals.

planetocentric centered on a planet. A satellite is a planetocentric (as opposed to heliocentric) body. A planetocentric coordinate system is subtended at the planet's center.

plasma sheet the region of plasma between the two lobes of a planetary magnetotail. The plasma here is generally denser and hotter than in the lobes and the magnetic field strength is lower than in the lobes.

plasmasphere the high-altitude magnetic field-aligned extension of the ionosphere. On the Earth this region extends out to about 6 planetary radii.

plasmoids aggregates of magnetized plasma and relativistic particles.

P_{max} the maximum positive amount when the polarization is plotted as a function of solar phase angle (Sun-Mercury-observer). P_{max} occurs near 100° phase.

P_{min} the minimum in the polarization-phase relation (see P_{max}). This negative polarization (having the electric vector maximum in the plane through Sun, Mercury and the observer) occurs near 10° phase.

polarimetry	the study of polarization of reflected sunlight or emitted thermal radiation.
poloidal field	vector fields, such as magnetic fields, can be decomposed into two families of components: "toroidal" and "poloidal." In simple terms, the poloidal field is part of the field that extends outside of a planet.
positive mass anomaly	localized excess mass on or beneath a planetary surface. Manifested as a positive free air anomaly. *See* mascons for the lunar case.
precession	a slow, periodic conical motion of the rotation axis of a spinning body.
prograde motion	motion in the same direction as the prevailing direction of motion.
protoplanet	a precursor body from which a planet develops.
quasi-parallel shock	a collisionless shock in a plasma for which the magnetic field forms an angle of less than 45° with the shock normal.
quasi-perpendicular shock	a collisionless shock in a plasma for which the magnetic field forms an angle of greater than 45° with the shock normal.
quasi-specular scatter	strong, completely polarized component of a planetary radar echo associated with specular reflections from the near-subradar region.
radar cross section	effective projected area of a radar target calculated on the assumption of a perfect, isotropic reflector. It is often expressed as a dimensionless quantity normalized by the true projected area of the target (planet).
radar footprint	surface regions sensed by radar during radar remote sensing, either by the width of the radar beam itself or by virtue of the delay/Doppler resolution of the echoes.
radar imaging	delay-Doppler maps produced from radar echoes.
radiation	energy propagating as electromagnetic waves, i.e., (in

order of increasing wavelengths) gamma rays, x-rays, ultraviolet light, visible light, infrared light, radio waves.

rare-earth elements (REE)
one of the elements with atomic numbers from 57 to 71, inclusive, in the lanthanide series of the periodic table.

ray
any of a system of bright elongated streaks associated with a crater.

Rayleigh number
a nondimensional combination of parameters involving the temperature gradient, and the coefficients of thermal conductivity and kinematic viscosity, which determines when a fluid, under specified geometrical conditions, will become convectively unstable.

reconnection
the process by which independent bundles of oppositely directed magnetic fields become joined resulting in the conversion of energy stored in the magnetic field to heating and acceleration of plasma; this can occur at the magnetopause and in the magnetotail.

reflectance
the fraction of the total radiant flux incident upon something (like a planetary atmosphere or surface) that is reflected; in general, it varies according to the wavelength distribution of the incident radiation. Reflectance is usually measured for a narrow range of wavelengths through a filter or other dispersive device.

refractory
element or phase stable at high temperatures. The opposite of volatile.

regolith
a layer of fragmentary debris produced by meteorite impact on the surface of the Moon, planet or asteroid.

Restrahlen effect
reflection of light from solid surfaces has two basic components: a surface component which follows the laws of Fresnel reflectance, and a volume component in which photons pass through particles of the solid and are either reflected by multiple scattering or absorbed. If the solid material has a very large absorption coefficient, such as in the center of an intense absorption band, Fresnel reflectance approaches unity, so that a large fraction of the incident light is reflected. Very little light actually enters

the solid particle to be lost to absorption. This high reflectance in regions of large absorption coefficient is known as the Restrahlen effect.

retrograde
motion

motion opposite to the prevailing direction of motion.

Reynolds number

a dimensionless number ($R = Lv/\nu$ where L is a typical dimension of the system, v is a measure of the velocities that prevail, and ν is the kinematic viscosity) that governs the conditions for the occurrence of turbulence in fluids.

ring current

electrical current that flows around the equator of a planetary magnetosphere when energetic charged particles are trapped inside the magnetosphere and drift around it.

saturation
equilibrium

the case where a surface has accumulated so many craters (of a particular size) that subsequent craters tend to destroy (by overlapping and other processes) roughly equal numbers of pre-existing craters.

S-band radar

radar observations made at 12.5-cm wavelength.

scarp

a line of cliffs produced by faulting or erosion, or a cliff-like face or slope of considerable linear extent.

secondary impact
crater

craters formed by the impact of material thrown out during the excavation of a larger crater.

single-particle
scattering
function

amount of light scattered into unit solid angle by a particle as a function of phase angle, normalized so that its integral over solid angle is π.

smooth plains

Mercurian plains with a smooth appearance due to a small number of superposed craters.

SNC meteorites

a group of uncommon, but apparently genetically related meteorite types which are highly differentiated (the shergottites, nahklites, and chassigny). They may originate from Mars.

solar composition material
the composition of the Sun, presumed to be the same as that of the solar nebula from which the planets were formed.

solar cosmic rays (SCR)
energetic charged particles originating on the Sun with energies below about 100 million electron volts.

solar flare
a sudden release of energy in or near the Sun's photosphere. Solar flares can accelerate energetic charged particles to energies of millions of electron volts.

solar nebula
the gas-dust disk that surrounded the proto-Sun. Mass of the solar nebula is usually assumed to be in the range from 0.02 to 0.05 $M_\odot$. The term protoplanetary cloud is also sometimes used as a synonym for the solar nebula.

solar neutrons
energetic neutral particles believed given off by the Sun during solar flares but not observed at 1 AU because of their short half-lives.

solar wind
the supersonically expanding outer plasma envelope of the Sun. It has a median velocity of about 440 km s^{-1}. The density at 1 AU is about 7 particles cm^{-3}.

source atoms
atoms freshly ejected from the surface or freshly generated in some other way, such as neutralization of an ion. In practice, they are distinguished from atmospheric atoms by their higher energies.

spin-orbit resonance
when the spin and orbital angular velocities are in the ratio of small integers.

sticking coefficient (S)
refers to chemical-scale bonding to the surface. The quantity S is the rate of adsorption per incident particle. The adsorption lifetime is undefined in the usage of this parameter, and it is therefore a useful quantity only for chemical-scale bonding in which residence time is long compared to typical experimental time scales.

stratigraphy
in planetary usage, the study of the order in which rock units were deposited on a planetary surface.

strike-slip faults
faults where the crust on either side of the fault plane has moved laterally.

subradar point   the center of a planet's disk as seen by the observer (radar) at any given time. The locus of such points is called the "subradar track."

subsurface   the region within $\sim$ 500 μm of the surface.

superior conjunction   the geometry when Mercury and the Sun are at the same ecliptic longitude, but the Sun is located between Mercury and the Earth.

superposition   the position of one feature on top of another feature or surface.

tectonics   the subject of large-scale movements and deformation of a planet's crust.

terminator   the line of sunrise or sunset on a planet or satellite.

Tesla (T)   unit of magnetic field, 10^4 Gauss.

thermal conductivity   the proportionality constant that gives the amount of heat conducted through a unit cross section in unit time under the influence of unit heat gradient. In cgs units the thermal conductivity is expressed in calories per centimeter squared per second per degree Celsius.

thermal diffusivity   the process of heat distribution and diffusion.

thermal inertia   a material parameter which indicates the rate at which a body's temperature responds to changing heat input. It is proportional to the square root of the product of thermal conductivity and volume heat capacity.

thrust fault   a fault resulting from compression that pushes one part of the crust over another and causes crustal shortening.

UBV   a system of stellar magnitudes devised by Johnson and Morgan at the Yerkes observatory which consists of measuring an object's apparent magnitude through three color filters: the ultraviolet (U) at 3600 Å; the blue (B) at 4200 Å; and the "visual" (V) in the green-yellow spectral region at 5400 Å. It is defined so that, for AO stars,

$B-V = U-B = 0$; larger values indicate "redder" colors.

uncompressed density
: *see* zero-pressure density.

upstream wave
: oscillations in the magnetic field and plasma observed near a planet in the solar wind on magnetic field lines which connect to the bow shock. They are caused by particles that are either reflected by the bowshock or leak out of the magnetosheath and flow back up along these field lines.

van der Waals forces
: the relatively weak attractive forces operative between neutral atoms and molecules.

viscosity
: the measure of a liquid's resistance to flowing.

VLA
: Very Large Array, an array of radio telescopes sited in New Mexico.

volatile
: an element that condenses from a gas or evaporates from a solid at a relatively low temperature.

vulcanoids
: a population of potential Mercury-specific impactors.

X-band radar
: radar observations made at 3.5-cm wavelength.

zero-pressure density
: density the materials constituting a planet would have if they were uncompressed, i.e., at zero pressure.

ACKNOWLEDGMENTS TO FUNDING AGENCIES

The editors acknowledge the support of National Aeronautics and Space Administration Grant NAGW-1019, the University of Arizona, and the University of Arizona Press for the preparation of this book. The following authors wish to acknowledge specific funds involved in supporting the preparation of their chapters.

Cameron, A.G.W.: NASA Grants NSG 22-007-269 and NAG 9-108.
Fegley, B., Jr.: NASA Grant NAG 9-108.
Hapke, B.: NASA Grant NSG-7147.
Hunten, D.M.: Nasa Grant NSG-7558.
Lewis, J.S.: NASA Grant NAGW-340.
Peale, S.J.: NASA Grant NGR 05-010-062.
Russell, C.T.: NASA Contract NAS 2-12383 and NASA Grant NAGW-717.
Schubert, G.: NASA Grant NSG-7315.
Strom, R.G.: NASA Grant NSG-7146.
Veverka, J.: NASA Grant NSG 7156.
Wasson, J.T.: NSF Grant EAR 84-08167.
Wetherill, G.W.: NASA Grants NSG-7437 and NAGW-398 (Departmental Program).

Index

INDEX